WILLIAM COLLINS' dream of knowledge for all began with the publication of his first book in 1819. A self-educated mill worker, he not only enriched millions of lives, but also founded a flourishing publishing house. Today, staying true to his spirit, Collins books are packed with inspiration, innovation, and practical expertise. They place you at the centre of a world of possibility and give you exactly what you need to explore it.

Collins. Do more.

COLLINS

Internet-linked

Dictionary of Physics

ERIC DEESON

Collins

HarperCollins*Publishers* Ltd.
77–85 Fulham Palace Road
London w6 8jb

The Collins website address is:
www.collins.co.uk

Collins is a registered trademark of
HarperCollins*Publishers* Ltd.

First published in 2007

Text © 2007 Eric Deeson
Illustrations by Cara Wilson

12   11   10   09   08   07
10   9   8   7   6   5   4   3   2   1

A catalogue record for this book is available from the British Library

ISBN 13 978-0-00-724226-9
ISBN 10 0-00-24226-3

Typeset by Rowland Phototypesetting Ltd, Bury St Edmunds, Suffolk
Printed and bound in Great Britain by Clays Ltd, St Ives plc

Mixed Sources
Product group from well-managed
forests and other controlled sources
www.fsc.org Cert no. SW-COC-1806
© 1996 Forest Stewardship Council

Introduction

The *Collins Internet-linked Dictionary of Physics* is aimed primarily at students and teachers. It will also be valuable to people wishing to read about aspects of the subject in some depth.

Most entries are more than just a definition. Explanations are clear, straightforward and up to date, and illustrations are provided to make topics easier to understand. Entries often give some ideas about uses, or some notes about history or work in the laboratory; all are designed to help understand the concepts. There are many cross-references (given in italic) to related entries. If there are a number of related terms, special linked entries show each of the terms in bold so they can be found quickly.

There are also a number of longer entries on a single theme, such as **atom** and **electric**, set in single column. Inside these entries, the main words and phrases given in bold are often in logical order rather than in alphabetical order. Also, italic terms show by such things as '(above)' and '(see below)' that they refer to other places in the theme entry rather than elsewhere in the book.

Appendix 1 lists the international system of units for measurement; appendix 2 shows the Greek alphabet. Appendix 3 shows the circuit symbols and appendix 4 lists all the Nobel Laureates in physics and what they were awarded the prize for, from 1901 to the present day. Appendix 5 lists some of the best websites that contain information about physics at about the level of this dictionary.

Dedication

I was very fortunate to have worked with two very good physics teachers at school – Mr Jardine and Mr Tucker loved the subject, and worked hard to share what they felt. But I dedicate this dictionary to two other physics teachers: my father-in-law, Bill Sing, one of those many unsung heroes who worked long and hard for his physics students in Britain and in Nigeria; and to Bill Jarvis, who is now retired, but still a lovable eccentric in the best sense of the word – I worked with the latter in many contexts, including the journals *School Science Review* and *Physics Education* and the project in the early 1990s that led to this dictionary. I also dedicate the dictionary to someone close to the other end of the educational ladder: Yani is a sort of adopted granddaughter just starting Year 10; she encouraged me very much during the early stages of writing this book.

Thanks also to Christopher Riches, my editor at Collins over many years, to various people in the Institute of Physics, and to the owners of and contributors to the websites listed throughout the book.

Eric Deeson
March 2007

a **a)** unit prefix symbol for atto- (*Appendix 1*), 10^{-18}, a quintillionth. The prefix comes from atten, eighteen in Danish and in Norwegian.
b) symbol for the year, unit of time – this symbol is correct but not much used

A **a)** symbol for absolute, old name for the kelvin temperature scale and its degree
b) symbol for amagat, unit of amagat (a measure of relative density for a gas) though that is a pure ratio
c) symbol for ampere, the unit of electric current (charge flow)
d) incorrect symbol for the ångström, non-SI unit for wavelength and atomic size
e) symbol for atomic, as in A-bomb
f) symbol for a hot spectral type of star

a **a)** symbol for acceleration, unit: metre per second per second, $m\ s^{-2}$
b) rare symbol for radius (*r* preferred), as with a_0 – the Bohr radius, the radius of the ground state of the hydrogen atom
c) out-dated symbol for (linear) absorptivity, the modern form being α, unit: per metre, m^{-1}

A **a)** symbol for the absorbance of a sample of a substance; this is a number, so has no unit
b) symbol for the (radio)activity of a sample, unit: becquerel, Bq (= 1 event per second)
c) symbol for the amplitude of a wave or other vibration; the unit depends on the context
d) symbol for area, unit: square metre, m^2
e) symbol for Helmholtz free energy, unit: joule, J
f) symbol for the interstellar extinction value for a given wavelength, unit – stellar magnitude
g) symbol for a nuclide's nucleon number (once mass number, before that atomic mass), the number of protons and neutrons in its nucleus; this too is a number, so has no unit

α symbol for alpha, in particular in physics for alpha-particle (the normal helium nucleus) and alpha radiation

α **a)** symbol for (linear) absorptivity; unit: per metre, m^{-1}
b) symbol for angular acceleration, unit: radian per second per second, $rad\ s^{-2}$
c) symbol for attenuation constant
d) symbol for the fine structure constant 1/137

Å symbol for the ångström, out-dated unit for wavelength, atomic size and chemical bond length – 10^{-10} m, 0.1 nm or 100 pm

abampere out-dated (cgs) unit of electric current, equal to 10 A

abaxial for rays in an optical system, away from the axis, so likely to suffer geometrical aberration

Abbe condenser an apochromatic compound lens designed by Ernst Abbe for use as a microscope condenser (lens to concentrate light on the object). It consists of two thick lenses and is still in use, despite its chromatic aberration problems.

Abbe criterion the condition for successful resolution by an optical instrument, an outcome of Abbe theory. The angular separation of two points to be resolved should be no less than the ratio of the wavelength used to the diameter of the object lens. This is widely used in electron microscopy, photography, spectroscopy, and so on – in all cases where diffraction limits resolving power.
See also *Rayleigh criterion*.

Abbe, Ernst (1840–1905) Involved from 1866 in the work of the world famous Carl Zeiss optical works at Jena, Germany, Ernst Abbe became director of research then a partner, and took the firm over as owner in 1888. During his life, he developed a number of optical and astronomical instruments (such as those below), and became professor of physics in Jena in 1870 and director of its observatories a few years later. **Abbe**, the large crater on the far side of the Moon, bears his name.

Abbe number (or **Abbe constringence**) V, no unit the reciprocal of the *dispersive power* of a transparent substance to white light

Abbe prisms the many prisms designed by Ernst Abbe to process input light in various ways, some of which are still in wide use

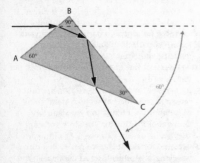

The path of select monochromatic light is as shown. Turning the Abbe prism has the output light still leaving along the same path, but its wavelength changes.

Perhaps the most simple **Abbe prism** is the one shown, a dispersive prism made of a glass giving it an angle of minimum deviation of 60°. That applies to a single wavelength, light of all others being deviated more than 60°. As you turn the prism around point O, in the plane of the sketch, the output light still leaves along the same line, but its wavelength changes. Thus, the system has a monochromatic output – and you can choose the wavelength (colour) of the output.

Two other Abbe prisms in use today are the non-dispersive systems for telescopes and view finders that invert the image (flip it both horizontally and vertically, in effect turning it through 180°):

• the **Abbe–König prism** – in which the output light follows the same line as the input. This makes the device of most use in telescopes, but it is also of growing importance in 'streamlined' prism binoculars.

• the **Abbe–Porro prism** – a simpler design, made up (in effect) of four 45°, 45°, 90°

prisms. It shifts the light sideways from the input path, so can reduce the total length of the system without loss of magnifying power (and also invert the image). The design is most common in 'traditional' prism telescopes (monoculars) and binoculars.

You can find further details of these two prisms, including ray sketches and photos of the two types of prism binocular, by following these links in the *Wikipedia* (http://wikipedia.org/): Abbe > Ernst Abbe > Abbe prism >.

Abbe refractometer a device, first developed by Ernst Abbe, to measure the refractive index of transparent liquid and solid samples. Various designs have appeared since that time. The sample is a thin section (solid) or in a thin cell (liquid) between two identical prisms, one upside down, and you measure the total refraction at the output. Other versions allow accurate reading of the critical angle; most use the sodium D line as a bright source of the necessary monochromatic light.

Abbe theory description of how an optical system forms an image of an object by diffraction – because of diffraction, the system cannot capture all the light from the object, so the image cannot be fully true to the object. However, Abbe started from the assumption that any system can produce a point image from a point object (given paraxial rays and other techniques to avoid aberration) – thus the system will produce an image space that maps exactly to the object space.

aberration a) the small cyclic change over a year of the apparent direction of light from a close star. This parallax-like effect of the Earth's annual motion (discovered by James Bradley in 1729) shows that the speed of light in empty space is finite.

the tangent of the angular displacement θ = Earth's orbital speed v/speed of light c → 20.5 arc seconds, the constant of aberration

There is a far smaller form of this, diurnal aberration, caused by the Earth's rotation round its axis.

b) a defect in an optical or electron-optical system that makes the image not a true copy of the object, as in c) below – the image of a point object is not a point

c) a defect in the image produced by the optics of an optical system (instrument). It is possible (though often complex and costly) to correct most aberrations. There are these six types, mostly having analogues in images produced other than with light. All apply to monochromatic light except for chromatic aberration.

1 **Astigmatism** gives two separate line images at a right angle of an object in the form of a cross. This occurs with single lenses, and in an eye with abnormal curvature of the cornea and/or tilt of the lens. We can correct astigmatism using a toric lens, one or both of whose surfaces are cylindrical (as if, but not, taken from large torus). The shape of a toric contact lens lets it correct for other eye defects as well as astigmatism and keeps it the right way up on the cornea.

2 **Chromatic aberration** may occur with axial and non-axial objects and leads to colour fringing of the image. The cause is dispersion – the difference in refractivity of the lens material at different wavelengths, so the image of a point source of white light is a tiny spectrum. The measure of this aberration is the distance between the red and blue images. Correction involves a second lens, but there is no chromatic aberration even with such a doublet at only two wavelengths.

3 **Coma**, which often relates to spherical aberration, can appear in the image of an off-axis point object formed by a lens or mirror: the image appears like a comet. The paraxial rays from the object form a point image; rays from larger and larger circles round the axis produce larger and larger circles next to the point image.

4 **Distortion** causes the image geometry to differ from that of the object. The sketches show the barrel and pin-cushion

distortion of a square object, either being quite common with projectors.

5 (Petzval) **field curvature** is when the image of a plane object at a right angle to the axis is not plane; instead, it tends to be parabolic. People use compound lenses and/or stops to correct, or reduce, this defect.

6 In the case of **spherical aberration**, the image of a point source formed by non-paraxial rays is not a point but a small disc. Using a parabolic mirror, rather than a spherical one, avoids the problem; in the case of a lens, stopping it down helps, as can the choice of a suitable radius of curvature for each lens surface.

Note that both chromatic and spherical aberrations can appear on the axis; the others are off-axis only.

Other types of optical aberration are

• **de-focus** – image blurring due to small scale imperfections of optical surfaces

• **tip**, or **tilt** – in which part of the image of a plane object is not parallel with other parts

• **piston** – in which part of the image is parallel with other parts, but shifted in or out.

Dynamic optical systems can reduce such problems.

ablation the loss of surface layers of an object, for instance, of a spacecraft as it enters an atmosphere at speed (where the loss process is vaporisation and heat shields – made of an **ablative polymer** – try to prevent it). Ablation causes the complete vaporisation of meteors – though meteorites, much reduced, still reach the Earth's surface.

abnormal reflection the reflection by the ionosphere of a radio wave whose frequency is greater than the critical value (normal limit)

abohm the 'absolute', or cgs, unit of resistance. An abohm is 10^{-9} ohms.

Abrikosov, Alexei A (1928–) joint winner of the 2003 Nobel Prize 'for pioneering contributions to the theory of superconductors and superfluids'

absolute independent of anything else. An absolute measure – such as absolute temperature – is not relative to anything else (in this case, the melting or boiling temperature of any substance).

absolute electrometer a precision electrometer of the attracted disc type which

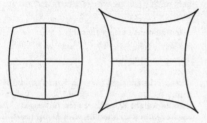

Barrel and pin-cushion distortion are common aberrations in projection.

sets, as in a balance, the attraction between two charged discs against the weight of the disc that is free to move

absolute expansion the expansion of a sample of a fluid with temperature, taking into account the expansion of its container

absolute humidity no standard symbol, unit: kilogram per cubic metre, $kg\,m^{-3}$ the water vapour content of a gas (mainly of air) given as the mass of vapour per cubic metre of gas. Compare with *relative humidity*.

absolute magnitude M, no unit The *magnitude* – brightness – an object in the sky would appear to have at a distance of ten parsecs (32.6 light years) from Earth. It depends on the object's absolute magnitude (which relates to its luminosity), the distance from the Earth, and how much the space dust between the two absorbs and scatters the radiation used in the measurement.

absolute motion the motion of an object, in particular the Earth, through absolute, fixed space – i.e. through the so-called ether. The Michelson–Morley experiment was set up to measure this – but failed, so proved there is no ether.

absolute permeability μ, unit: henry per metre, $H\,m^{-1}$ for a medium, the ratio of magnetic flux density B in it to the applied magfield strength H. In other words, it measures how much the medium modifies – focuses or repels – the external magfield. The unit, the henry per metre, is the equivalent of the volt-second per ampere-metre, $V\,s\,A^{-1}m^{-1}$. The absolute permeability of empty space, μ_0, $4\pi \times 10^{-7}\,H\,m^{-1}$, is a fundamental quantity in electromagnetic theory. See also *relative permeability*.

absolute permittivity ε, unit: farad per metre, $F\,m^{-1}$ for a medium, the ratio of the electric displacement D (or electric flux density) in it to the applied electric field strength E. A second relationship is the charge at a point in the medium to the product of the electric field strength at a second point and the square of the distance between them, with a constant as in

$$\varepsilon = D/E$$

$$\varepsilon = Q/4\,\pi\,E\,r^2 \text{ (from } E = Q/4\,\pi\,\varepsilon\,r^2)$$

In either case, the value tells us how well the medium can transfer ('permit') electric field effects. The unit is the henry per metre, the name of the ampere-second per volt-metre, $A\,s\,V^{-1}m^{-1}$ (or $C^2\,N^{-1}\,m^{-2}$). The absolute

permittivity of free space is ε_0, 8.854×10^{-12} $F\,m^{-1}$, a fundamental quantity of electromagnetic theory. See also *relative permittivity*.

absolute refractive constant μ (or n), no unit an alternative name for *absolute refractive index* – a better name, but not strictly correct as the values may vary to an extent with context (with wavelength and in liquids and gases, with temperature)

absolute refractive index μ (or n), no unit of a transparent substance, the refractive index for radiation of a given wavelength that enters it from vacuum. A better term for radiation of a given wavelength is 'constant' rather than 'index'. Its value is the ratio of the speed of light in empty space, c_0, to the speed, c, in the substance.

$$\mu = c_0/c$$

As the absolute refractive constant of air is close to 1 (it is in fact 1.000 292 for the sodium D line at standard temperature and pressure), it is normal to measure μ for a liquid or solid substance in air rather than in empty space.

absolute temperature T, unit: kelvin, K temperature on the absolute (or Kelvin) scale, often called thermodynamic temperature as it relates closely to thermodynamic models. The absolute temperature of a sample of a substance is the Celsius temperature +273.15. Thus 100 °C = 373 K.

absolute unit **a**) a measure – such as absolute temperature – that is not relative to anything else (in this case, the melting or boiling temperature of some substance)
b) a unit in the cgs (absolute) unit system, such as abohm and abvolt
c) a unit of an SI base measure (*Appendix* 1) – such as the second for time – and of a measure which is a pure function of those – like the metre per second for speed. In general, if $a \ldots n$ are base measures and x is an unknown measure, we have

for the measures: $x = f(a \ldots n)$
so for the units: $U_x = k\,f(U_a \ldots U_n)$ where k is a constant

Then, if the constant $k = 1$, U_x is an absolute unit.

absolute weight W, unit: newton N weight measured in a vacuum, i.e. with no upthrust to reduce the value

absolute zero 0 K, zero on the thermodynamic (absolute, Kelvin) scale of temperature, the lowest temperature normal matter can reach, outcome of the first and second laws of thermodynamics (for at 0 K matter would have zero thermal energy). However (as with the speed of light in empty space), a matter sample can never actually reach 0 K, as the third law of thermodynamics predicts; the record is a few tens of picokelvin above that limit (a picokelvin is a million-millionth of a kelvin). The value on the Celsius scale is −273.15 ... °C. See also *zero point energy* (point is an old word for temperature), the quantum thermal energy of atoms at 0 K.

absolutism theory, first proposed in the seventeenth century by Thomas Hobbes and then (for physics) by Isaac Newton, that space and time are fixed, i.e. absolute

absorbance *A*, no unit the log of the absorptance, i.e. the log of the ratio of the radiant energy reaching a sample's surface and that transmitted by the sample. Older terms are optical density and optical extinction.

absorbate substance absorbed by an absorber (strictly an absorbent) – see *absorption* a).

absorbed dose unit: gray, Gy, = joule per kilogram See *radiation absorption*.

absorbent an absorber of particles – see *absorption a)*.

absorber something which absorbs, as a sponge is an absorber of water and graphite is an absorber of neutrons, e.g. in a pile. See also *shock absorber*.

absorptance no symbol, no unit the fraction of the radiant energy passing through a sample's surface that the sample does not transmit. There are other more precise uses of the word, but those in fact define absorptivity.

absorption a) the penetration of particles of one substance (the absorbate) through an interface into a second (the absorbent), e.g. by diffusion
b) taking radiation into the body, where it may cause harm by ionisation, either as a result of exposure to direct radiation or by such means listed in *absorption of radioactive materials*
c) the transfer into a different form of energy when radiation meets and enters a sample of an absorber (see, for instance, *incident*); absorbance and absorptivity measure how much this happens for a given substance and a specific radiation. In the case of a

homogeneous absorber, the radiation falls off exponentially. No substance can transmit 100% of any radiation that enters it – many processes lead to radiation absorption. It is common to include scattering as such a process, though this doesn't fit the above statement about energy transfer. Absorption that excludes scattering sometimes has the name 'true' absorption; 'total' absorption includes scattering.

For instance, there is **sound absorption** as sound passes through matter (although sound cannot pass without matter): the sound energy falls off as viscous forces oppose the motion of the particles; there is also some transfer to thermal energy between compressions and rarefactions by conduction (and, at low frequencies, re-radiation), and there is scattering.

In the case of particle and electromagnetic radiation, the absorbing processes depend on the radiation energy; they involve scattering and can include nuclear reactions and the excitation, even ionisation, of atoms and molecules. They tend to lead to re-radiation of the absorbed energy – but rarely in the same form and in the same direction as the incident radiation. **Absorption bands** and **absorption lines** mark an **absorption spectrum**, discovered by Wollaston in 1802 and explored by Fraunhofer in the 1810s and 1820s: these will be low intensity (dark) bands and lines on the normal high intensity continuous spectrum of the original incident radiation. Absorbing molecules of a given type cause each dark band by absorbing radiation in a specific frequency range; absorbing atoms cause loss at a specific frequency – a dark line.

Note that this description can be over-simple in that, in practice, most absorption lines in the spectrum of a star come from its own atmosphere. See *stellar spectral classes*.

absorption cell a box with two parallel, thin, transparent walls designed to hold a gas or liquid to measure its absorptivity

absorption constant or **absorption coefficient** *a* or *α* strictly linear absorption constant (coefficient), a measure of the absorbing power of an absorber for a specific radiation type and energy – all, like **absorption factor**, being old and sometimes loose terms for the *molar absorptivity* of the absorber for that radiation

absorption discontinuity or **absorption edge** in a graph of the absorptivity of a substance against incident x-ray wavelength, a sharp change. For instance, one such change occurs at the 'K edge', after which the x-ray photons no longer have enough energy to raise electrons in the substance from the K level.

absorption lines See *absorption* c).

absorption of light See *light absorption*.

absorption of neutrons See *neutron absorption*.

absorption of radiation cause of the effect in the body of ionising radiation – a biological hazard as ionisation can damage living cells, in particular their growth and reproduction. See *absorption* b).

absorption of radioactive materials taking these into the body (where they may do more damage than if outside), by ingestion (eating, drinking), inhalation (breathing) or entry through the skin. Once in the body, atoms of active nuclides may stay for a long time adding greatly to an exposed person's dose and causing damage by ionisation.

absorption of sound See *sound absorption*.

absorption of x-rays See *x-ray absorption*.

absorption spectrum See *absorption* c).

absorptivity (or absorption constant) a or ε, unit: per metre, m^{-1} of a transparent substance, the absorptance of a pure sample per unit sample thickness – a measure of the absorbing power of the substance for the radiation concerned. A better – and now much more common – term is *molar absorptivity*.

abundance or **abundance ratio**, C, no unit the relative population of one species within a set, such as of an element within a rock or a nuclide in an element. See also *natural abundance*.

ac (still sometimes AC) alternating current, current that changes direction at a regular rate. Some people use the word as an adjective to mean 'alternating', as in 'ac current'; this is not correct. See also *electric circuits*.

acceleration a, unit: metre per second per second, $m s^{-2}$ the rate of change of an object's speed (as a scalar) or the rate of change of its velocity (as a vector). In other words, it is the change of the object's speed or velocity, in metres per second per second. (It is neither correct nor wise to say the unit as 'metre per second squared'.) See also *force and motion*.

acceleration, angular ω, unit: radian per second per second, $rad s^{-2}$ the rate of change of an object's angular velocity. See *angular acceleration* and *force and motion*.

acceleration, centripetal 'centre-seeking acceleration', the acceleration towards the centre of a curve of an object moving along that curve. See *angular acceleration* and *force and motion*.

acceleration in simple harmonic motion See *simple harmonic motion* and *force and motion*.

acceleration of free fall g, unit: metre per second per second, $m s^{-2}$, or newton per kilogram, $N kg^{-1}$ the 'acceleration due to gravity', 9.81 (about 10) ms^{-2} at the surface of the Earth, strictly in a vacuum (see *terminal speed*). It is the rate of change of an object's velocity as it falls freely in a gravfield.

acceleration, radial, unit: metre per second per second, $m s^{-2}$ the acceleration along the radius of a curve in which an object is moving. See *angular acceleration*.

acceleration, tangential, unit: metre per second per second, $m s^{-2}$ the component of an object's acceleration along the tangent to the curve in which it is moving. See *angular acceleration*.

accelerator a) type of catalyst

b) less common term for *activator*

c) device designed to cause acceleration, as the accelerator ('gas pedal') of a car controls the rate of fuel ('gas') transfer to the engine

d) short for particle accelerator – see *nuclear physics* and see also *photon accelerator*.

accelerator, particle device designed to accelerate charged particles to a high speed relative to the speed of light in empty space. See *nuclear physics* and see also *photon accelerator*.

accelerometer device to measure acceleration. Typical designs measure the force acting on a known mass – they are types of forcemeter – and use $a = F/m$. In an **integrating accelerometer**, a chip integrates the acceleration readings to give speed and integrates those values to give distance moved.

acceptor type of impurity which makes an intrinsic semiconductor p-type – able to accept, i.e. attract and absorb, electrons (compare with *donor*). Common acceptors used with silicon and germanium are trivalent aluminium, gallium and indium. See also *acceptor level*.

acceptor circuit a resonant circuit (often an inductor and a capacitor in series) whose impedance is a minimum at a certain frequency. It thus accepts (i.e. passes) signals only at and close to that frequency. As such it is a detector circuit (to use an old radio term);

if the capacitor is variable, it becomes a tuning circuit.

acceptor level the energy level of an acceptor (p-type impurity in a semiconductor). For the doping to be effective, the acceptor level needs to be close to the top of the semiconductor's valence band (see *band theory*) – the level can then accept, and fill with, electrons from the valence band to leave plenty of positive 'holes'.

ac circuits alternating current circuits, those with an alternating supply (input), as with most electronic and microelectronic circuits and sub-circuits (networks). See *electric circuits*.

accommodation the power of the eye to handle light from objects at different distances – in particular between its near point and its far point, its extremes of clear vision. An eye accommodates by changing the thickness (and thus the focal distance) of the lens so it can focus on objects that differ in distance. As the eye ages, the lens becomes less flexible – accommodation takes place more slowly and also the distance between the near point and the far point becomes less. Accommodation is the same as the eye's depth of field, except that the latter has subjective as well as objective aspects.

accumulator a secondary electric battery (or cell), one you can re-charge again and again, in particular a lead–acid battery as used in cars

accuracy how close a measured or calculated value is to the true value, the deviation of the former from the latter being a common measure. Accuracy is not the same as precision – 'my height is 2.345 m' is precise but not accurate; 'my height is 1.8 m' is accurate but not precise.

achromat short name for achromatic lens, a doublet of two lenses of different refractive constants glued together to reduce chromatic aberration

for lenses A, B, ... N of powers P_A, P_B ... P_N and dispersive powers ω_A, ω_B ... ω_N:

there is *achromatism* at the '*achromatic condition*'
$$\omega_A P_A + \omega_B P_B + ... + \omega_N P_N = 0$$

achromatic = lacking colour, for instance not showing chromatic aberration as applied to lenses and prisms, i.e. to optical elements that depend on refraction. Also, **achromatic colours** contain no colour, so are shades of grey.

An **achromatic condenser** is a condenser (lens that gathers and concentrates light, e.g. in a microscope) that doesn't add artificial colour to the image. The same applies to any **achromatic lens** (or *achromat*) and **achromatic prism**.

for prisms A and B of angles of deviation d_A and d_B and dispersive powers ω_A and ω_B:

there is *achromatism* at the '**achromatic condition**'
$$\omega_A d_A + \omega_B d_B = 0$$

achromatism absence of chromatic aberration in a system. This is possible at only two input wavelengths in the case of an achromatic lens – this is, in fact, a double lens (see *achromat*); it is possible at three wavelengths in the case of a triplet . . . and so on. Achromatic lenses of as much as a ten or a dozen elements are common for top cameras, but they are very costly; also each extra lens in the set absorbs a few per cent of the light reaching it, so such a lens may pass much less than two-thirds of the light from the object.

aclinic line line on the Earth's surface joining points at which the angle of dip ('inclination') is zero – the magnetic equator

acoustic concerned with sound, sound energy, sound waves

acoustic absorption coefficient or **acoustic absorption constant** measure of how much a given substance absorbs sound that travels through it. It is usual to include scattering, though this is not true absorption; the value depends on the substance and on the wavelength of the sound in question. See *absorptivity*.

acoustic array set of nearby sound sources or receivers arranged to give a directional effect – to send out a beam of sound or to detect sound coming only from a certain direction. Arrays like this depend on interference effects.

acoustic capacitance or **acoustic stiffness** one imaginary part of the acoustic impedance of a sample of a substance. It relates to the elasticity (inverse of stiffness) k of the substance, being given by A^2/k, where A is the section area of the sample at 90° to the sound waves.

acoustic delay line See *delay line*.

acoustic filter sound channel that contains a number of elements with the right impedance to transfer only a certain range of wavelengths; as in electronics, there are low-pass, band-pass, and high-pass filters. The elements used are often resonant skins (diaphragms) or tubes which vary in cross-section and may have branches.

acoustic grating a row of rods which affects sound waves in the same way as a diffraction grating affects light waves

acoustic impedance, Z_A, unit: acoustic ohm the opposition of some barrier to sound waves – the ratio of the mean sound pressure to the velocity of an infinitesimal sample of the medium (strictly a particle) at that point in a channel. In the case of a plane sound wave, the impedance of a channel is the product of its section area, the speed of sound in the substance, and its density.

As with ac impedance, this is a mathematically complex measure with two imaginary components: the acoustic capacitance (which relates to the stiffness of the substance) and the acoustic inertance (or acoustic mass, equivalent to inductance, which relates to its inertia). (The inertance L of a mass m of gas in a channel of area A is m/A^2.)

Knowledge of acoustic impedance is important in ultrasound imaging (echo sounding), where we also need to know how boundaries (e.g. tissue boundaries) reflect and transmit sound waves.

acoustic inertance, L unit: kg m^{-4} See *acoustic impedance*.

acoustic instrument musical instrument that does not include its own artificial amplification, like an acoustic guitar as opposed to an electric guitar. The design of an acoustic instrument includes sound boards and volumes of air that resonate throughout the instrument's frequency range.

acoustic mass, L unit: kg m^{-4} same as acoustic inertance. See *acoustic impedance*.

acoustics the science and applied science of sound sources, the properties and transfer of sound waves, how a surface affects all these, and how to sense them. Acoustics is important in the design of public spaces like meeting rooms and concert halls.

acoustic stiffness same as *acoustic capacitance*

acoustic surface any surface designed to affect input sound in a given way. For instance, **acoustic tiles** absorb sound.

acoustoelectronics the science and applied science of how sound waves interact with the free electrons and holes of a semiconductor. The field includes the use of acoustic delay lines and of transducers to convert electric signals to sound signals.

actinic able to absorb ultraviolet, light and/or infrared to cause chemical change, as in film cameras. **Actinic radiation** is a radiation which causes a chemical change.

actinide an element in the second 'rare earth' series, with proton numbers from 89 (actinium, Ac, discovered in 1899 by André Debierne, a colleague of Marie Curie) to 103 (lawrencium, Lr, discovered 1961). The **actinide series** is one of the three natural radioactive series: it starts with uranium 235 (^{235}U, half-life 7.1×10^8 years), passes through actinium 227 (^{227}Ac, 21.6 years) and ends with stable lead 207 (^{207}Pb). The nuclides in the chain have nucleon numbers given by $4n+3$, where n ranges from 58 to 51.

actinium Ac highly active element found in radium ores, proton number 89, a transitional rare-earth from the actinide series, period 7. Its relative density is 10.07, and its melting and boiling temperatures are 1100 °C and 3200 °C. The half-life of the most stable isotope, the alpha-emitter ^{227}Ac, is 21.8 years. André Debierne (1874–1949) discovered actinium in 1899; the name comes from the Greek word 'aktis', meaning ray.

actinometer a meter that measures radiation intensity, used mainly in meteorology, for instance pyrheliometer and pyranometer. Most actinometers used in the past as atomic and nuclear radiation detectors worked by measuring the fluorescence or decomposition produced by the absorbed energy. **Actinometry** is the measurement of light absorption by photochemical effects, or, more generally, radiant energy absorption.

actinon old name for radon, in particular radon-219

action a) old name for force, with reaction being opposing force. **Action at a distance** used to describe a force that involves no contact, i.e. a force with a field effect. The action at a distance between quantum particles is much harder to explain: Einstein called it 'spooky'. John Bell was the first in the 1960s to prove that photons leaving an event in opposite directions act in such a way as to imply they exert an influence on one another. The proof was mathematical and Einstein did not accept it; however, twenty

years later, in the early 1980s, a team measured the effects in practice and showed that these agreed with Bell's theory.
b) the absorption of radiation by living cells to produce a specific effect
c) symbol I, the product of energy transfer and time during an interaction in a conservative dynamic system (unit: joulesecond; the name planck was once proposed for this), more formally as

$$\text{action} = 2 \cdot \int_{t_1}^{t_2} W \, dt$$

Here W is the system's total energy and t_1 and t_2 are the starting and ending times of the interaction.

See also *least action* and *Planck constant*.

action of points the power of sharp points and edges to discharge a charged object. When an object is charged, the charge is all at the surface – and the more curved the surface, the greater the charge density there.

action potential wave of chemical depolarisation that travels through a living cell or series of cells when a nerve passes a signal or a muscle contracts. The cause is the entry into the cell of sodium ions, and the outcome is a change of potential from about –70 mV to about +40 mV; the whole process can take place several hundred times a second (e.g. in order to keep a muscle in tension). An important example is the action of the heart muscle, with a wave starting at the sino-atrial node.

action spectrum much the same shape as the absorption spectrum of a substance to a radiation, a graph of the ability of the radiation to cause a specific effect. An example is the action spectrum for photosynthesis in given plant tissue.

activation a) the irradiation of a sample to make it radioactive. It is the basis of various kinds of **activation analysis**, for the spectra of the activity produced (e.g. by neutrons or gamma rays) relates closely to the nuclides in the sample.
b) adding enough energy to a system to allow a given process to take place, the **activation energy** being the minimum needed

activator (or **optical activator** or **accelerator**) substance which, when added in tiny amounts to a phosphor, causes the phosphor

to give out the desired colour during *phosphorescence*

active a) short form of radioactive, as in 'active salts'
b) the opposite of passive, as in (for instance) an **active circuit element** (one which can provide switching or add gain, so may need its own power supply, like a transistor as opposed to a resistor) and **active noise control**. In the latter system, a microphone senses sound in the environment, an electronic circuit passes only the unwanted sound (the noise), and a speaker outputs the exact same signal – except for a 180° phase change. Destructive interference around the device thus cancels out the noise. Uses for this include making meeting rooms and lorry engines quieter.
c) in phase with the supply in an ac circuit. An **active circuit** has resistance and/or zero net reactance, while an **active circuit element** has only resistance. People also use the terms **active output** to mean **active current** or **active voltage**. The product of input current and active voltage, or of input voltage and active current, is the circuit's **active power**, unit: voltamp, VA.

activity a) short form of radioactivity, as in 'the activity in the soil of Cornwall'
b) symbol A, unit: becquerel, Bq (= per second, s^{-1}), the degree of radioactivity of an active sample, give by the mean number of decays per second (now called the Becquerel, though the curie Ci is still sometimes used). **Specific activity** is the activity of unit mass of the substance, unit $kg^{-1} s^{-1}$.
c) the output of a reactor or accelerator in wanted particles per second; the unit is protons per second, for instance.
d) (or optical activity) the ability of a substance to rotate the plane of polarisation of light passing through a sample
e) the thermodynamic (= ideal) concentration of a chemical in a reaction for use with the law of mass action. The **activity ratio** is the actual concentration divided by that ideal.

activity, optical See *optical activity*.

actuator device with electrical input and corresponding mechanical output (normally force or torque), the opposite of a sensor. Actuators put the power, speed or motion of a system under electrical control, as in computer controlled machines and robots.

acuity the eye's resolving power, which depends on the details of its optical system and the density of sensor cells in the retina. A 'perfect' eye can resolve points 0.5 minutes of arc apart.

adaptation the eye's ability to change sensitivity when moving between light and dark. As a measure, the unit is the second.

adaptive control automatic control of which a human operator can set some aspects (such as flight height, speed and bearing in the case of an autopilot). Sensors provide the necessary feedback loops.

adc, or ADC analog-to-digital converter, device that changes an analog signal to its digital equivalent

addition of vectors finding the single vector (the resultant) with the same effect as those being added

additive process the mixing of lights of different colours by addition, to produce new colours. For addition, the three primary colours are usually considered to be red, green and blue; the colour white consists of a roughly equal mix of these. Colour television as it produces light uses this principle.

adhesion attractive force between two substances across their interface, an **adhesive** being something that aids this. The adhesion energy is the energy needed to separate a square metre.

adiabatic change or **adiabatic process** a change within a system with no net energy loss to or gain from outside the system, which is therefore 'isolated'. Such systems are usually gas samples; compare with isothermal change. During adiabatic change, the pressure p and volume V of an ideal gas relate as

$$P V^{\gamma} = \text{constant}$$

where γ is the ratio of the principal specific thermal capacities of the gas, c_p/c_V. The preferred term for **adiabatic** is isentropic (meaning no change of entropy).

adiabatic chart a graph showing how a pair of thermodynamic variables (such as the pressure and volume of a gas sample) relate to each other during an adiabatic change

adiabatic cooling or **adiabatic demagnetisation** an important process for bringing a system close to absolute zero. The most common method uses a paramagnetic salt. When cooled as far as possible by other means, the salt is strongly magnetised isothermally and then adiabatically.

adiabatic lapse rate the lapse rate – rate of change of temperature with height – for a

small volume of air rising adiabatically through the atmosphere

adiabatic process See *adiabatic change.*

admittance Y, unit: reciprocal ohm, Ω^{-1} the reciprocal of the impedance of a circuit or circuit element. Like impedance, admittance has a complex value; the real part is the conductance Q and the imaginary part is the susceptance B:

$$Y = Q + B$$

adsorption rather like *absorption* a), but a process in which the 'absorbed' particles (the **adsorbate**) do not enter the host substance (the **adsorbent**) but attach as a thin layer only to its surface. The bonding may be chemical (chemisorption) or involve van der Waals forces.

advanced gas-cooled reactor (agr, or AGR) type of fission reactor, the second generation built in Britain. The moderator is graphite and the coolant is carbon dioxide (with the heat exchanger inside the pressure vessel). See *nuclear power.*

aerial or **antenna** the part of a radio system which accepts radio waves as the input (a receiving aerial) or outputs them (a transmitting aerial); there is exchange between an alternating electric signal and a radio wave. Most aerials consist of a number of individual **aerial elements** (wires, parabolic reflectors or horns, and a link to the actual circuit) forming an **aerial system**; this can be an array, which has directional properties, as does the parabola ('dish'). The simplest design is the **half-wave dipole aerial** – this is a stiff metal wire clamped at the centre (which is therefore both a node and the connector to the radio circuit); the two free ends are antinodes, so the aerial is most effective for sending or receiving signals whose wavelength is double the rod's length. The dish has the actual sensor or radiator at its focus, so in the former case it is just like a radio telescope – very good at collecting the radio waves reaching the whole surface from a given direction. In the latter case, a dish used to transmit radio waves sends out a parallel beam; this can travel much further than a diverging beam, which follows the inverse square law of attenuation.

aerial gain, unit: decibel, dB **a)** the ratio of the power absorbed by an aerial to that absorbed by a dipole to give the same signal strength in the same direction
b) a measure of the efficiency of a given

aerial design compared with a half-wave dipole

aerial impedance, unit: ohm, Ω a measure of a transmitter's losses to the signal passed – the rms power supplied divided by the mean square of the current supplied

aerodynamic heating the transfer of energy to an object from friction with the fluid through which it is moving (often air in the case of a vehicle)

aerodynamics the study and use of the motion of air (or any gas) in different contexts, and of the motion of, and consequent forces acting on, solid shapes passing through air. **Internal aerodynamics** is the study of the flow of gases through holes, tubes, valves, fans, and so on.

aerofoil See *airfoil*.

aeronautics the science and applied science of aircraft design, sometimes spacecraft design as well

aerophysics the study of the properties of moving gases, and of the behaviour of objects moving in gases, where the speed of movement is greater than the speed of sound

aerosol a fairly stable suspension of fine solid particles or liquid droplets in a gas

aether old spelling of ether, the medium once thought to fill empty space

afc automatic frequency control, system by which a radio set stays tuned to a given station, even though various effects cause the input wavelength to change slightly from moment to moment

afterglow a) radiation from a phosphor after excitation ceases
b) (which is really the same thing) radiation from a fluorescent tube after switching off the current

afterheat the radiation and energy output by a reactor core after shut-down, due mainly to active fission products

after image an image perceived by a tired retina after removal of the object. After images may appear dark on bright or the reverse, and often change from one to the other.

agate a hard, crystalline form of silica once widely used for making 'knife-edges' for instruments such as balances and pendulums

agc automatic gain control, system giving an output of much more uniform amplitude than the input

age equation equation for the age of the Earth based on the relative abundance of lead isotopes in the rocks of the crust

ageing change in the properties of a substance with time, for instance the magnetic nature of iron and the strength of the pressure vessel of a nuclear power station

age of the Earth estimated at just over 4.5 thousand million years, the oldest rocks so far found being almost four thousand million years old (these are rare but some include minerals several hundred million years older). See *dating*. Life appeared less than one thousand million years ago.

age of the Universe See *big bang*.

age theory See *Fermi*.

agonic line line on the Earth's surface joining all points of zero magnetic declination, i.e. points in which the directions to true north and magnetic north are the same

agr advanced gas-cooled reactor – see *nuclear power*.

air the mixture of gases in the atmosphere, the percentage mix by volume when dry being:

nitrogen	78.1
oxygen	20.9
argon	0.933
carbon dioxide	0.03 (and rising)

There are also traces of the other 'inert' gases than argon: in descending order of abundance they are neon, helium, krypton, xenon and radon. Radon, being radioactive (the half-life of its most long-lived isotope is only a few days), is present in negligible amounts (less than $1:10^{19}$). On the other hand, being active, it is a hazard, especially when trapped in buildings after seeping from the ground.

air capacitor capacitor with only air between the plates – low capacitance but hardly any dielectric loss (an inefficiency)

air cell See *air wedge*.

air coil inductor with only air within and around the coil – low inductance but hardly any hysteresis loss (an inefficiency)

air equivalent the thickness of air at STP which would absorb a given radiation to the same extent as some other absorber

airfoil object shaped to have a large lift force when moving through the air, as well as a small drag (due to air friction). The cause is the Bernoulli effect: the shape is such that the air speed over the top of the foil is greater than that below; the Bernoulli effect is that the pressure in a fluid is less the faster it flows. Also relevant is Newton's third law of force: the foil shape deflects passing air downwards, so the air deflects the foil upwards. An aircraft wing is the most well known airfoil, but all of a plane's control surfaces have the same shape (for instance the rudder and a tailplane). Hydrofoils act the same way in water (e.g. to lift the boats called hydrofoils).

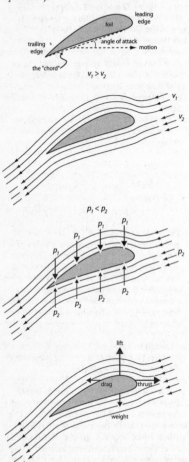

The shape of a foil is such as to produce large lift and only small drag

air pressure, or **atmospheric pressure** the pressure of the atmosphere, at the Earth's surface or at some depth or height within it. The standard value is defined as 101.325 kPa. *Barometers* measure air pressure for weather purposes or to give a measure of altitude above mean sea level.

air wedge a sample of air trapped in a triangular space with a very small apex angle. It produces interference effects, by reflection at the two edges of the wedge. These are much like those of Newton's rings, for these arise from a wedge of varying angle. Monochromatic light of wavelength λ input to a wedge produces a series of light and dark bands (interference fringes):

for dark fringes: $2t/\lambda = 0, 1, 2, ...$
for light: $2t/\lambda = 0.5, 1.5, 2.5, ...$

Interference occurs between rays reflected from the top and bottom of the air wedge, the two sides being t apart at that point. The factor of 0.5 follows the 180° phase change of rays reflected from an optically denser surface. To obtain, observe and measure air wedge fringes, it is common to form the wedge between two microscope slides kept apart at one end with a slip of thin paper.

Airy disc the circular image of a point source produced by a perfect (aberration-free) system, due to Fraunhofer diffraction. The name is that of Sir George Airy (1801–1892, astronomer – and Astronomer Royal – and engineer).

albedo the reflecting power of a diffuse surface at a given wavelength – the intensity of light reflected as a fraction of the incident light – mainly used in astronomy in the context of cold objects like planets. There is a measure that's much the same for the reflecting power of a substance to neutrons.

alchemy from the Arabic *al khimia*, meaning much the same as the English word chemistry, and the origin of that word. Alchemy was the pseudo-science of Europe in the Middle Ages, having degraded from the science of Arabia. Prior to that, **alchemists** of the Egyptian tradition hoped to discover *transmutation* using the 'philosopher's stone' (changing, for instance, lead into gold); those of the Chinese looked for the 'elixir of [eternal] life'; those based in India hoped to find the 'panacea', the cure for all disease; and the Greeks brought astrology and mathematics.

alcohol thermometer thermometer in which the thermometric liquid is coloured ethanol rather than water (as in the first instruments).

Using ethanol makes it possible to measure temperatures below 0 °C, indeed down to about − 117 °C (156 K). On the other hand, the upper limit of 79 °C is too low for most uses in science – a mercury thermometer (invented by Gabriel Fahrenheit) is better: its useful range is − 39 °C to 357 °C.

Alferov, Zhores (1930–) shared the 2000 Nobel Prize with Herbert Kroemer 'for developing semiconductor heterostructures used in high-speed electronics and opto-electronics'

Alfvén, Hannes (1908–1995) a physicist whose work on plasma physics gained him the 1970 Nobel Prize: 'for fundamental work and discoveries in magneto-hydrodynamics with fruitful applications in different parts of plasma physics'. He is often called the founder of magnetohydrodynamics (mhd). An **Alfvén wave** is an mhd wave with energy transfer in the direction of the magfield, but mass displacement at 90° to that.

al Kindi, Abu Yusuf (800–873) 'the philosopher of the Arabs', so called as al Kindi, born near Basra and working mainly in Baghdad, produced original work in a great range of fields – including philosophy – at the height of Muslim science. His main work in physics was to write a book on optics (translated as *De aspectibus*) which had great influence for 500 years.

allobar a form of an element in which the isotopes are in proportions different from the natural form. Enriched uranium is an allobar.

allochromy process in which a substance outputs electromagnetic radiation with a different wavelength from that input. See, for instance, *fluorescence* and *Raman effect*.

allotropy the existence of two or more physically different forms of a given substance under the same conditions. In the case of carbon, the room temperature allotropes are diamond and graphite. Some people call the fullerenes (*bucky balls*) C_{60} and C_{70} and/or the bucky tubes allotropes too, but not all agree (see also *graphene*).

allowed band a band of energies for a substance in which an electron can exist, according to the band model of the solid state. An allowed band relates to a single atomic energy level of the substance.

alloy any metal that is not an element, such as
 • a mixture of metals
 • a mixture of metals produced by dissolving one or more in a second when

in the liquid state – an alloy is often called a metal solution
 • an intermetallic compound
It is common to describe alloys of a particular type in terms of an **alloy phase diagram**, showing the structure over a range of temperatures. **Alloy theory** tries to explain such diagrams in terms of the atomic properties of the alloy components – but not with full success. See also *amalgam*.

alloyed junction a germanium transistor junction between the semiconductor and metal contacts; on heating, metal atoms diffuse into the germanium to form the collector and emitter.

alpha α a common name for the alpha particle, the helium nucleus $^4_2$H. Alphas make up alpha radiation, one of the three common products of radioactivity, first being recognised by Becquerel in 1896 (and confirmed by Rutherford in 1989). The cause is **alpha decay**, a form of nuclear fission in which a nucleus of X splits into a nucleus of Y and an alpha; those fission products carry away excess energy (as kinetic energy), called **alpha decay energy**. See *nuclear physics*.

alpha radiation (α radiation) one of the three common forms of radioactive radiation, being made up of alphas (alpha particles). It carries away a great deal of kinetic energy but, as the alpha's mass is large, the speed is not high. This means that, as the rate of ionisation of atoms along the path is very large, the radiation is short range and easily absorbed by matter. Alpha radiation may have a range of a few centimetres in air, but even a thin sheet of paper will absorb it. See *nuclear physics*.

alphatron a vacuum gauge which contains an alpha source and an ionisation detector. The number of ions detected in unit time gives a measure of the gas density.

Alpher–Bethe–Gamow theory how the big bang must have produced hydrogen, helium and more massive elements in the ratios observed. While this theory of nucleosynthesis is still in the main valid, the 1948 paper that describes it is much more famous for its name. In fact, it was written by Ralph Alpher, a PhD student of George Gamow, and the latter decided to include the name of Hans Bethe; Bethe was not upset by this.

alternating current (ac) current which changes direction regularly, at a constant (or fairly constant) frequency. For **alternating current**

13

circuits, see *electric circuits: ac circuits*. See also *generator* and *transformer*.

alternative sources of power large-scale sources of electricity (in most cases) that aren't too costly, cause less CO_2 pollution than fossil and other organic fuels, and/or use less precious reserves of those fuels. The move to these is driven in the UK by the Government's determination to reduce CO_2 output by 20% by 2010; however, although onshore and offshore wind farms are predicted (2006) to produce 5% of UK electricity by then, most people expect the reduction to be no greater than 18%. As well as wind, other such sources used more in other countries include solar power, hydroelectric and hydrothermal energy, coastal tides, and waves. In most developed countries, the main alternative source is nuclear (fission) power; research continues into nuclear fusion as a large-scale source of energy.

alternator an electric generator with an alternating output. A common design is the **synchronous alternator**, in which the shaft carries one or more coils linked through a pair of slip rings with carbon brushes to the output; a turbine or some other mechanical link spins the shaft at a given frequency. The coil(s) turn in the magnetic field of one or more pairs of poles (often of electromagnets supplied by a separate dc source); induction causes a voltage to appear at the output.

altimeter device designed to show height above mean sea level (aneroid barometer type) or above ground (radar type). The traditional design (an aneroid barometer marked in metres) is inaccurate, insensitive and irrelevant (in that pilots like to know how far they are above ground level rather than sea level). Other systems, such as the radar altimeter, are now common. This works much like an echo sounder, but uses radio waves rather than sound waves.

aluminium (US: aluminum) Al stable metallic element, proton number 13, found as aluminium oxide in bauxite (its main ore) and many clays. Indeed, it is the most common metal in the Earth's crust. Group III, period 3; relative density 2.58, melting and boiling temperatures 658 °C and 1800 °C. The most common isotope is ^{27}Al. Hans Oersted (1777–1851) in 1824 and then Friedrich Wöhler (1800–1882) in 1827 discovered aluminium; the name comes from alum, a major salt of the metal.

Alvarez, Luis (1911–1988) winner of the 1968 Nobel Prize for work in cosmic rays and nuclear physics: 'for his decisive contributions to elementary particle physics, in particular the discovery of a large number of resonance states, made possible through his development of the technique of using hydrogen bubble chambers and data analysis'

am, or AM amplitude *modulation*, changing a radio carrier wave's amplitude to mirror the required signal

amagat, A the density of a gas sample compared to that at standard temperature and pressure. This non-standard unit is a ratio, so has no dimension.

Amagat's experiments experiments on real (as opposed to ideal) gas behaviour, designed much like those of Andrews and deemed as important in France, where Émile Amagat (1841–1915) worked

amalgam a solid or liquid alloy one of whose components is mercury

ambient surrounding, in the area – as **ambient air** means the air around some experiment (or, sometimes, moving air), and **ambient temperature** is the room temperature at the time

americium Am trans-uranic (and, therefore, radioactive) metallic element of the actinide series, proton number 95. Period 7, relative density 13.67, melting and boiling temperatures 995 °C and 2607 °C. The most stable isotope is the alpha-emitter ^{243}Am, half-life 7370 years. Glenn Seaborg (1912–1995) and his team first made it in 1944 (though it was later found to be a small part of uranium ore); it is named after its country of first manufacture. A major use is in smoke detectors.

ammeter, symbol: ———(A)——— meter used to measure the current in *electric circuits*, the unit of current being the amp. An analog meter – one with a pointer on a scale – may work magnetically (a moving iron meter, with a non-uniform scale) or electromagnetically by the motor effect (a moving coil meter, with a linear scale). A digital meter, with a digital display, measures the current electronically. In any case, the meter should have a very low resistance so it affects what it measures as little as possible. See also *current balance*.

It is normal to use a single basic analog meter, able to read very small currents, with a chosen 'multiplier' or 'shunt' for the range required at a given moment. (This is cheaper than having many different meters.) The shunt carries the excess current around the meter. When the meter resistance is R_m and its full scale deflection current is I_m, the largest pd it can take is $V_m = I_m R_m$. To allow the meter to read up to I, a higher value than I_m, use a shunt in parallel with the meter to carry the excess, $I - I_m$. The pd between the ends of the shunt equals V_m (as meter and shunt are in parallel). Therefore, the shunt resistance must be $R_s = V_m/(I - I_m)$.

Note that digital meters sometimes, and multi-meters always, have a range of shunts built in and a switch to let the user choose the correct range of readings.

Point contact electrometers are the meters able to measure the smallest currents – as low as 10^{-18} A, which means they can detect individual electrons. A problem was that these meters could not tell which way a passing electron was going, but a new design in mid-2006 was able to overcome that.

amorphous not having a crystal structure (i.e. no long range order) in the solid state. Note that some apparently amorphous solids consist of very small crystals distributed at random.

amount of substance unit: mole, mol the number of basic entities in a sample, strictly as such or as compared to the number of ^{12}C atoms in 0.012 kg of ^{12}C. The entities may be (for instance) photons, elementary particles, ions, atoms or molecules. Amount of substance (not the same as amount of matter) is a pure number, so has no unit; however, the mole is the standard, 6.023×10^{23}. See also *Avogadro number*.

amp or **ampere** A the unit of electric current, a flow of one coulomb per second, $C\,s^{-1}$. It is the current which, in two parallel infinite straight conductors a metre apart in empty space, gives a force of 2×10^{-7} N m^{-1}. An amp is also the common short form for amplifier.

Ampère, André (1775–1836) mathematician and physicist whose major work was on the links between electricity and magnetism. Most notably, he was the first person to provide, in 1820, an explanation of Oersted's discovery of the magnetic effect of current.

ampere balance name for the *current balance*

ampere-hour, Ah old unit of electric charge, the charge transferred by a current of one amp in an hour, 3600 C

Ampère–Laplace law, Ampère–Laplace theorem or **Ampère's law** the law which relates the force between two nearby current elements. A current element is a small part of

a circuit carrying a current; the two elements need not be in the same circuit. Say the elements are parallel (as in the sketch) and distance d apart (offset by angle ϕ) and of lengths $\mathrm{d}l_1$ and $\mathrm{d}l_2$, and they carry constant currents I_1 and I_2.

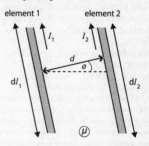

The law gives the element of force between two currents

Then the force element $\mathrm{d}F$ between them in a medium of permeability μ is:

$$\mathrm{d}F = (\mu/4\pi d^2)I_1\,\mathrm{d}l_1\,I_2\,\mathrm{d}l_2\,\sin\Theta$$

It follows that the field strength $\mathrm{d}B$ at a point near a single current element $\mathrm{d}l$ of length l is:

$$\mathrm{d}B = (\mu/4\pi d^2)I\,\mathrm{d}l\,\sin\Theta$$

Last, for an infinitely long straight conductor with current I, this gives:

$$B = \mu I/2\pi d$$

Ampère's law/theorem or **Ampère's circuital law/theorem** that the line integral $\int B\,\mathrm{d}l$ over all the line of force elements $\mathrm{d}l$ on a particular *complete* line of force round a current is the product of the current I and the permeability:

$$\int B\,\mathrm{d}l = \mu\,I$$

This is still sometimes called **Ampère's work law** as it describes the work done (energy needed) in moving a monopole around a closed path.

Ampère's rule a mnemonic (or rule of thumb) for the field direction outside a current: imagine swimming in the (conventional) current direction towards a compass: the N-pole will then be on your left. The corkscrew rule is better for non-swimmers who know how to use a corkscrew! Otherwise use the right-hand grip rule.

ampere-turn, At unit of magnetomotive force (mmf), that produced by 1 A in one turn of a coil

amplification process providing gain (increase of signal amplitude) in a system, circuit or circuit element. **Amplification factor** (no unit) was a particular definition in the case of an electron (vacuum) tube used as a triode amplifier (somewhat equivalent in effect to a transistor), but is now widely used in most contexts: it is the ratio of a small change of output pd to the (smaller) change of input pd that causes it.

amplifier ('amp' is a common short form) a device whose output has a higher amplitude than its input, with output and input signal forms being the same unless there is significant distortion; this is therefore an active device which needs a power supply. Formerly the triode electron (vacuum) tube, or valve, was the standard system; now it is the transistor. Transistors and field effect transistors (often supplied as complete chips) are the main types of linear amplifier for low powers at frequencies up to about a gigahertz. At higher frequencies the most common types of device are the parametric amplifier and special valves, while for high power the best basis is the thyristor.

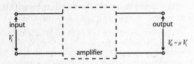

The output of an amplifier matches the input but has a greater amplitude

The **amplification**, or gain – the ratio of the output pd (or current, or amplitude) to the input signal value – is frequency-dependent; amplifiers are therefore designed for particular frequency ranges as well as for particular purposes. On the other hand, a suitable chain of amplifiers can handle a very wide band width. The **amplification factor** μ is the gain when the source has no internal impedance z and the output is open (infinite impedance). Another important measure for such a system is the efficiency – the ratio of the output power to that input. See also *dc amplifier*.

amplitude or peak value, A the maximum displacement in a cyclic measure (like a wave or harmonic motion); the unit depends on the context. If the measure varies like a simple harmonic motion (i.e. as a sine wave), the amplitude is half the peak-to-peak value.

amplitude distortion See *distortion*.

amplitude fading type of signal fading in telecommunications – all frequencies in a communication band are cut by a similar factor which varies unpredictably over periods that can be as long as some hours. The main cause is the movement of large-scale irregularities in the ionosphere; this causes a focusing or defocusing effect on the received signal.

amplitude modulation (am, or AM) type of *modulation* in which changes of the carrier wave's amplitude carry the required signal, common with medium- and long-wave radio systems

amplitude of accommodation unit: dioptre, $D (= m^{-1})$ the range of accommodation of an eye. It is the difference between the reciprocals of the distances (in metres) from the eye to its near point and to its far point.

amu atomic mass unit, replaced by mass number and then by nucleon number

analog modern spelling in English of **analogue** in the context of a quantity with a continuous range of possible values (such as the height of an adult female). This is as opposed to digital, which describes a measure with a number of quite distinct possible values.

The output of an **analog circuit** is a continuously varying linear function of its input: it is a linear circuit. An **analog computer** works entirely with analog (rather than digital) values – for instance, input from a sensor and output to an analog device; it can carry out almost any mathematical operation. It is not possible to design an analog computer to handle a wide range of other tasks, however: they are, therefore, special-purpose rather than general-purpose systems and, overall, more costly than digital computers (which can be mass-produced). Analog computers run flight training simulators. For a digital computer to work with analog inputs, there must be an analog-to-digital converter.

analogous pole the end of a pyroelectric crystal or device that gains positive charge when the temperature rises. Compare with *antilogous pole*.

analog-to-digital converter (adc, or ADC) device that changes an analog input (e.g. from a sensor) to its digital equivalent output (e.g. to a computer) – see *pulse-code modulation* (pcm)

analyser one of the two 'crossed' polarising sheets (or crystals) in a *polarimeter*, for finding the plane of polarisation of a (plane polarised) wave. The analyser passes most radiation if its plane is parallel to the plane of polarisation, and least if the two are at 90°. In structure, an analyser is just the same as a polariser for the radiation concerned: in the case of light, for instance, it would be a sheet of Polaroid.

anamorphic lens optical system in which cylindrical lenses (or mirrors or prisms) give an image scaled differently in the vertical and horizontal planes. Such lenses are common in cinematography, and some artists use them to produce unusual effects.

anaphoresis See *electrophoresis*.

anastigmat or **anastigmatic lens** a triplet or quadruplet lens used as a camera's object lens. The design is to reduce astigmatism and all other aberrations except distortion.

Anaxagoras (*c.*500–430 BCE) philosopher and physician who thought much about astronomy and physics. He produced an early atomic theory and believed that all heavenly bodies are much like the Earth in composition.

Anderson bridge a bridge circuit used to compare a capacitance C and an inductance L. The values of resistors R_1 and R_2 are fixed; the user adjusts those of R_3 and R_4 to give a zero reading on the meter. Then

$$L = C\,[(R_1\,R_2) + (R_2 + R_3)\,R_4]$$

Anderson, Carl (1905–1991) received a share of the 1936 Nobel Prize for his discovery of the positron in 1932, the first antiparticle found. He went on to discover the first meson, in 1947.

Anderson, Philip (1923–) shared the 1977 Nobel Prize for 'fundamental theoretical investigations of the electronic structure of magnetic and disordered systems'

AND gate, symbol: ⟤ gate whose output 0 is 1 only if its two or more inputs $I_1 \dots I_n$ are all 1. Like the symbol, this truth table is for a two-input AND gate:

I_1	I_2	O
0	0	0
0	1	0
1	0	0
1	1	1

Andrews' experiments experiments carried out in the 1860s by Thomas Andrews (1813–1885), on the behaviour of a real gas (carbon dioxide) in order to compare this with Boyle's ideal gas law. A dry CO_2 sample is trapped in a thick glass capillary by a plug of mercury; a similar tube carries a sample of dry nitrogen. The open ends of both tubes are inside a case of water whose pressure can be raised as high as 10 MPa by a large screw. A constant temperature bath surrounds the whole system. The volume of nitrogen (assumed an ideal gas) gives the pressure throughout the system, and the volume of CO_2 can be both read directly and observed.

The pressure–volume, p-V, graphs are a series of isothermals (curves at constant temperature); at high temperatures they show near-ideal behaviour. Below the critical temperature T_c there is interchange between unsaturated vapour, saturated vapour and liquid phases.

See also *Amagat*, who carried out much the same experiments independently a few years later.

anechoic chamber sometimes called dead room, a box or room whose walls absorb all sound waves that reach them. Thus there are no reflections (echoes) and no reverberation. Doing this is often of value for audio recording, so audio studios tend to be dead rooms.

anelasticity property of a solid with no elastic region (see *elasticity*): under stress it moves straight to plastic behaviour. By extension, people also use the term for any imperfectly elastic behaviour.

aneroid barometer type of barometer which gives readings based on the behaviour of low pressure air in a small box rather than on that of a large volume of toxic, costly liquid in a large delicate tube. As the air pressure changes, the box expands and contracts; levers and pulleys translate the small movements to a lever on a scale. See also *altimeter*. The aneroid barometer is much more compact, cheaper and safer than mercury-in-glass systems.

angle, no standard symbol, unit: radian, rad, or degree, ° how much a line or plane inclines to a second. The unit is the radian (rad), although the degree system (degree (°), minute ('), second (")) is far more common in practice. 1 rad = 57.30° (as 2π rad = 360°). See also *solid angle*.

angle of deviation See *deviation* b) and d).

angle of dip δ, unit: degree, ° also called

(angle of) inclination, the angle at a place between the Earth's magfield and the horizontal, as measured by a dip circle (a magnetic compass that pivots in the vertical plane)

angle of friction μ, unit: degree, ° the largest angle from the horizontal that two surfaces in contact can remain without sliding – see *friction*.

angle of incidence i, unit: degree, ° the angle between an *incident* light ray and the normal at the point of incidence – see *incident*.

angle of polarisation i_B, unit: degree, ° the angle of incidence of unpolarised light on a reflecting surface at which the reflected light is fully polarised. See *Brewster angle*.

angle of reflection symbol r, unit: degree, ° the angle between a reflected light ray and the normal at the point of incidence – see *incident*.

angle of refraction symbol r, unit: degree, ° the angle between a refracted light ray and the normal at the point of incidence – see *incident*.

Ångström, Å out-dated unit of length used mainly for light wavelength, atomic size and chemical bond length: 10^{-10} m, 0.1 nm

Ångström, Anders (1814–1874) physicist who helped found spectroscopy and was the first to identify hydrogen in the solar spectrum

Angstrom pyrheliometer a pyrheliometer in which thermocouples measure the temperature difference between two identical platinum strips, one in the radiation whose strength is to be measured, and the other shielded

angular acceleration ω, unit: radian per second per second, rad s^{-2} the rate of change with time of an object's angular velocity as it moves in an arc. See *force and motion: force and motion through an angle*. See there too for radial and tangential acceleration.

angular dispersion the same as dispersive power

angular displacement, θ, unit: radian, rad the angle (from the centre) between two points on the path of an object moving in a curve. See *force and motion: force and motion through an angle*.

angular frequency or **angular velocity** ω, unit: radian per second, rad s^{-1} the angular displacement in unit time of an object moving in an arc of a circle, $d\theta/dt$. It is the

product of 2π (the number of radians in one revolution) and the cyclic frequency ν.

angular impulse the time integral of a torque, the product of the torque (if constant) and the time applied. It equals the change in angular momentum it would cause to a free mass.

angular magnification, no unit the magnifying power of an optical system measured as the ratio of the angle to an off-axis point on the image to the angle to the same point on an object, both being measured from the optical centre of the system

angular momentum (still sometimes called moment of momentum) L, unit: newton metre per second, N m s^{-1} the product of an object's moment of inertia I and its angular velocity/frequency ω: $L = I \omega$. See *force and motion: force and motion through an angle*. Angular momentum is a conserved pseudo-vector measure. Two types (orbital and spin) are of great importance in atomic and nuclear systems; both being quantised with quanta units of $\hbar$ ('h bar': Planck's constant divided by 2π).

angular velocity ω, unit: radian per second, rad s^{-1} the same as angular frequency – the rate of change with time of angular displacement, $d\theta/dt$, the angle from the centre swept out in unit time by an object moving in a curve. See *force and motion: force and motion through an angle*.

anharmonic motion a) the oscillation of an object whose restoring force is not directly proportional to its distance from a central point. An **anharmonic oscillator** is an object with such an oscillation. An example is the vibration of atoms in molecules rather than in samples of a pure element.
b) chaotic, completely unpredictable motion – unlike simple, or even compound, harmonic motion (where there is some regularity and, therefore, predictability). An example is the Brownian motion of a speck of dust, for instance, in a fluid.

anhysteretic magnetisation the remanence of a ferromagnetic sample after being subject, while placed in a constant field, to a second outside alternating field which is reduced to zero from a strength greater than that of the constant field. This remanence differs from that found in the hysteresis cycle, which is what the name means.

anion a negative ion, i.e. one attracted by an anode

anisotropy the property of having some feature whose value varies with the direction in which it is measured (e.g. the strength of a wood sample). Compare with *isotropy*.

annealing the heating of a solid sample to a little below its melting temperature followed by slow cooling. This relieves strain and removes crystal imperfections, making the sample stronger.

annihilation reducing matter to 'nothing', by letting it interact with antimatter. When any particle meets its antiparticle, they annihilate each other; the process releases energy equal to the two rest masses plus the input kinetic energy. To conserve momentum in the case of **electron–positron annihilation**, two photons carry off the energy in opposite directions: this is **annihilation radiation** in the gamma region (first observed in 1933). Rarer cases involve the production of three rather than two photons, or result from the interaction of a positron with two electrons. See also *nullification* and *nucleon annihilation*.

annual variation any cyclic change in a geomagnetic or astronomical measure with a period of a year, caused by the orbit of the Earth around the Sun

annular eclipse solar eclipse in which the eclipsing object (Moon) and eclipsed object (Sun) are in a straight line from the observer (Earth), but the former is too small to cover the latter. There is little loss of perceived brightness, therefore, and eclipse observation hardware shows the Sun as a bright ring round the dark Moon. It occurs when the Moon is unusually far from the Earth. This means its angular size is less than that of the Sun, so it covers less than the Sun's full disc.

anode of a pair of electrodes, the one that is more positive than the other (the cathode). If there is a current between two electrodes, the anode is the one at which the electrons enter the rest of the circuit. In a discharge in a *discharge tube*, the **anode dark space** is a gap between the anode glow and the anode surface. The cause is the slowing of the electrons (cathode rays) as they crowd together on the approach to the cathode.

anode drop or **anode fall** the pd of perhaps 20–40 V between an anode and the point in a low pressure gas a few millimetres from it where the anode glow starts. The **anode glow** is a bright region in the gas discharge close to the surface of an anode – see *discharge tube*.

anode rays ultraviolet radiation from an anode in a process of secondary emission caused by high energy electrons hitting it when the discharge is through a fairly high vacuum. At even higher levels of vacuum and electron energy, most of the radiation is of x-rays.

anode saturation a condition which occurs in certain discharge and electron tubes (valves). A cloud of electrons builds up round the anode, to form a space charge. This may be to such an extent that further electrons cannot reach the anode. The circuit current then falls to zero.

anomalous dispersion a very rapid change in the refractive constant of a substance, at wavelengths near values associated with absorption bands

anomalous expansion of water the expansion (fall in density) of liquid water as it cools through the range 4 °C to the freezing temperature, so that water is densest at just under 4 °C. There are many theories as to why this happens – most to do with H_2O's hydrogen bonds – but none is widely agreed. This means that a layer of water at 4 °C can remain at the bottom of ponds and lakes in even extremely cold weather, thus giving water-dwelling plants and animals a far better chance of survival in severe winter conditions.

anomalous viscosity viscosity (fluid friction) that falls with flow rate or velocity gradient. Also called thixotropy, this is a characteristic of non-Newtonian liquids.

anorthic system See *crystal system*.

antenna alternative term for aerial

anthropic principle paradox of modern astronomy which comes in several versions. One is that the Universe is the way it is only because we – carbon-based living creatures – are here to observe it, rather than that we are here because the Universe is the way it is. All versions depend on the belief that if the Universe isn't the way it is, we couldn't live in it, so we couldn't observe it. 'The way it is' may include the values of the cosmological and fine structure constants, the number of dimensions, the structure of strings, and so on. Some people link the principle to that of intelligent design (that the Universe has, or had, a creator).

antibody a large powerful protein that works within the immune system. Some modern vaccines involve these, including to take radioactive atoms into cancers.

anticathode large, cooled metal block in an x-ray tube which outputs x-rays when it absorbs very high energy electrons. In some designs this is the anode itself; in others it is linked to the anode.

anticoincidence circuit radiation detector circuit giving an output pulse only if just one of the two or more inputs receives a pulse in a given time. Compare with *coincidence circuit* and *XOR* gate.

anti-crack line or plane of higher density in a solid (rather than one of lower density – see *crack*). In an anti-crack, the stress is compressive, leading perhaps to slippage and spreading. Anti-cracks can form and grow at pressures greater than a few gigapascals. The formation of anti-cracks in the mineral olivine at around 400 km below the Earth's surface, followed by the superplastic growth and joining of the anti-cracks, may explain earthquakes at such depths.

anticyclone a weather system, sometimes over a million square kilometres in extent, with a high pressure at the centre and closed isobars around that. Winds pass clockwise around the system in the northern hemisphere and anticlockwise in the southern.

antiferroelectricity the dielectric behaviour of some solids with high permittivity but no net natural polarisation. The cause is considered to be antiparallel polarisations within the substance. This behaviour disappears above the antiferroelectric Curie temperature.

antiferromagnetism a property of some materials (for instance, FeO, MnO and MnS) with a low positive susceptibility up to their Néel temperature. Above that, they are paramagnetic – the susceptibility falls with temperature, as described by the Curie–Weiss law. The atoms of antiferromagnetic materials interact to give an antiparallel pairing of magnetic dipole moments. Compare with *ferromagnetism* and see also *ferrimagnetism*.

antilogous pole the end of a pyroelectric crystal or device that gains negative charge when the temperature rises. Compare with *analogous pole*.

antimatter matter whose atoms consist of anti-nuclei (anti-protons and anti-neutrons) with a cloud of anti-electrons (positrons) around, and therefore having all physical properties reversed other than mass and those that depend on mass (compare with negative matter, which – if it exists – has negative mass and repulsive gravitation). No one has yet found proof that antimatter exists; however, there is no theoretical reason against the existence of whole anti-galaxies, made of anti-atoms around whose anti-nuclei positrons orbit.

Some theories claim there must be as much antimatter in the Universe as there is 'normal' matter. However, when an antiparticle meets its twin, they annihilate each other with the release of the appropriate packet of energy – and at the time of the big bang, the matter and antimatter then created would have annihilated itself *if* the quantities were the same *and* matter and antimatter are equal and opposite. Current thinking is that the two are not fully equal and opposite – a small asymmetry between them favours matter over antimatter. Such an asymmetry may exist between the masses of B-mesons and B-anti-mesons, as shown by the frequency at which they oscillate between each other (or mix).

In 2006, the general belief was that matter and antimatter make up 4% of the Universe in total, but, to repeat, antimatter has not yet been found to exist in bulk. The nearest reached was in a CERN experiment in 2002 – this produced very large numbers of anti-hydrogen atoms; these were cold – low energy – which allows time to study them.

antimony Sb metalloid element (period 5) with proton number 51, with two allotropes and whose most common isotope is ^{121}Sb. Stibnite, a sulphide ore, has been in use since ancient times as a pigment, though the element was first isolated in the seventeenth century; the name is from the Greek 'antimonos', meaning not alone, as it was hard to make in a pure form. The relative density is 6.7 and the melting and boiling temperatures are 630.5 °C (and international fixed temperature) and 1440 °C.

antinode a point of maximum displacement, halfway between two nodes, in a normal standing wave. It is a point of maximum amplitude in an interference pattern: maximum brightness in the case of light, maximum loudness in the case of sound (for instance, at the open end of a sounding air pipe), strongest signal in a resonant circuit, maximum particle probability in an electron cloud. An **antinodal line** joins the antinodes in a two- or three-dimensional interference pattern. See *sound sources*.

antiparallel parallel, but paired in opposite

directions (like two pins next to each other with the heads at opposite ends)

antiparticle elementary particle with opposite but equal charge, baryon/lepton number, magnetic moment, strangeness, and isospin number to those of the corresponding 'normal' particle. Mass and spin are the same (but see negative matter), as is the half-life if unstable. Only photons and neutral pions have no antiparticles; in fact they are their own antiparticles. See also *antimatter*.

antireflection coating a bloomed surface of a lens or prism, designed to reduce unwanted reflection

antiresonant frequency that frequency, when driving a vibrating system, at which its amplitude of vibration is least. Compare with *resonance*.

aperiodic damped so that any oscillation has no simple harmonic nature. In tele-communications, an **aperiodic circuit** is one used only within a frequency band far away from any of its resonant frequencies.

aperture the effective diameter (i.e. the diameter of the used light-collecting part) of a lens, mirror or other optical system – the diameter of the exit pupil when the system is ready to work with an object at infinity. It is normal to quote aperture as a fraction of the focal distance f. This then is the focal ratio, which has no unit; its numerical value is the f-number of the system. For example, $f4$ means aperture = $f/4$ (the focal ratio), the f-number being 4 . The larger the f-number of a system, the smaller the aperture. See also *numerical aperture*. A system's **aperture angle** is the angle between the axis and the edge of the entrance pupil at the object, the specimen surface in the case of a microscope.

aperture distortion the loss of definition in a television screen image because the flying spot that scans the screen (the 'aperture') is not a point: the picture cannot show any detail smaller than the spot

aperture ratio in the case of a lens used with a close object, the value $n \sin \alpha$; n is the refractive constant of the medium in which the lens lies and α is the angle between the axis and the line from the foot of the object to the edge of the lens. Its importance relates to the fact that for such a close object the light-passing ability of the lens depends on that ratio.

aperture synthesis a system in radio astronomy that gives the effect of a large aperture telescope by linking the outputs of a

number ('array') of fixed or movable small ones. The same technique is possible at light wavelengths, but it is then much harder to synchronise the signals suitably.

aphelion that point in the orbit of the Earth or other planet that is furthest from the Sun. Compare with *perihelion*.

apochromatic lens a multiple lens system able to correct most chromatic aberration and some spherical aberration and coma

apogee that point in the orbit of the Moon (or artificial satellite) round the Earth at which it is farthest from the Earth (perigee is the closest point).

apostilb an old unit of luminance, equal to $1/\pi$ candela per square metre, now rarely used

apparent depth the perceived depth of a sample of transparent liquid, which differs from the real depth because of refraction. In other words, it is the distance of an image beyond a plane refracting interface (such as the surface of water or of a block of glass) when viewed from within a second medium (such as air). In the situation shown in the sketch, the refractive constant n of the second medium at the temperature concerned is object distance/image distance, i.e. real depth/apparent depth.

$$n = \text{real depth/apparent depth}$$

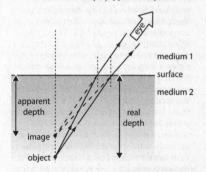

A transparent substance appears less thick/deep than it really is because of refraction.

apparent expansion less than the actual (absolute) expansion of a liquid or gas sample because its container expands as well

apparent magnitude the perceived *magnitude* (brightness) of an object, such as a star, in the sky. This is likely to differ from the absolute magnitude as the object may be further away from the observer, or closer to, than the 'standard' distance of ten parsecs (32.6 light years).

appearance energy (or appearance potential) See *ionisation energy*.

Appleton, Edward (1892–1965) gained the 1947 Nobel Prize 'for his investigations of the physics of the upper atmosphere especially for the discovery of the so-called Appleton layer[s]'

Arago, Dominique (1786–1853) astronomer and physicist who helped confirm Young's transverse wave theory of light (though he didn't agree with it), discovered the polarisation of light by quartz crystals, and also worked on electromagnetism and the ether

arc or **electric arc** gas discharge between two electrodes (of carbon in most cases) with a high current density and high luminous energy output but a fairly low potential gradient, still sometimes used as a source for spectrographic analysis. The various kinds of **arc furnace** also use this energy source, as does the **arc lamp**. However, a major problem with arc lamps is that, as much of the light comes from the bright glowing electrodes (for example, carbon rods at up to 5000 °C), these erode quickly – particularly at the anode, in which a crater forms.

Archimedes (*c.*287–*c.*212 BCE) a 'natural philosopher' who did experiments (a rare thing to do in those days) – in particular on *levers* and other machines and in the field of hydrostatics. **Archimedes' law** is in that latter field and relates to the Eureka story: the upward force on an object in a fluid equals the weight of fluid displaced by the object.

arcing the process that occurs in an (electric) arc discharge. Opening a circuit in which there is a heavy current can cause momentary arcing between the switch contacts (which therefore erode); indeed, the way to start an arc lamp is to make the two electrodes touch, then to pull them a few millimetres apart. **Arcing contacts** are temporary contacts that bypass those of the main switch in a heavy current circuit and suffer the damage which can be caused by this momentary arcing, thus preventing damage to the switch. An **arcing ring** is a metal ring placed around an insulator that may experience an arc: the ring carries the arc so that the insulator does not suffer damage.

area, *A*, unit: square metre, m² the extent in space of a real or imaginary two-dimensional (or equivalent) surface. Dimensionally it is the product of two lengths at right angles; surprisingly, it is a vector – whose direction is at 90° to the surface. See, for instance, *magnetic moment*, sense c).

Argand diagram a two-dimensional extension of the number line concept. At 0 (the origin) on the (real) number line an axis at 90° represents the imaginary number line. (An imaginary number is a multiple of *i*, $\sqrt{-1}$.) Points on either of these axes represent real and imaginary numbers; points in the **Argand space** between them represent complex numbers (numbers of the form *a* + *bi*). The mathematician Jean Argand (1768–1822) devised the concept. See *electric circuits: ac circuits*; one approach to the analysis of these involves Argand diagrams. See also *phase*, sense c).

argon Ar inert (noble) gas in period 3 whose name comes from the Greek 'argos', meaning idle – it has very few compounds. The proton number is 18, and the main isotope is ⁴⁰Ar. Its melting and boiling temperatures are –189 °C and –186 °C. Lord Rayleigh (1842–1919) discovered argon, the most common inert gas in the atmosphere, in 1894.

Aristarchus of Samos (310–230 BCE) the astronomer believed to be the first to propose a Universe with the Sun, rather than the Earth, at the centre

Aristotle (384–322 BCE) like his master, Plato, a great philosopher; unlike him, a philosopher who also worked in 'natural philosophy', i.e. what we now call science; unlike the scientists who followed almost 2000 years later, an armchair philosopher, many of whose ideas were later shown to be wrong. Wrong as it was, the **Aristotelian** view of the Universe and the physics of the Universe retained great power through those many centuries.

armature a) the rotating part of an electric generator or motor. By extension, it is any moving part of an electric machine which either carries an induced voltage or closes a magnetic circuit.
b) (rare) the keeper of a magnet
c) (rare) the iron core of an electromagnet

Armstrong, Edwin (1890–1954) engineer who first devised the concept of feedback for use in amplifiers and devised important new techniques of radio broadcasting

array set of sources or sensors arranged to give a directional effect – to send out a tight beam or to detect radiation coming only from a certain direction. Arrays like this depend on interference effects, so the units need to be about as close to each other as a few wavelengths. See *acoustic array*; the technique is also very common in radio telescopy. In radio communications, the structure sometimes has the name beam aerial.

arsenic As metalloid element of group V period 4, with three allotropes, proton number 33. Relative density 5.73, sublimes at a temperature of 613 °C.

artificial atom common name for the quantum dot. This is much less hard to work with than a normal atom, as it is much bigger, but still has its own set of discrete energy levels.

artificial line model of an electric power transmission line made from a circuit with resistance, capacitance and inductance. The model should behave exactly like the real thing, and so is of value for research, design, development and maintenance.

artificial radioactivity radioactivity not observed in nature, i.e. a result of human activity. In most cases, this is because the nuclides concerned have decayed in nature since their creation because of their short half-lives.

asdic (or **ASDIC**) 'allied submarine detection investigation committee' – devised to detect shipping under water, now called echo sounding or sonar

aspheric or **aspherical** in the case of the surface of a lens or mirror, having the form of a section not of a sphere or cylinder but of a paraboloid, or other nonspherical solid. Although harder to make, such a lens (sometimes known as a bispherical lens) or mirror suffers less or not at all from such aberrations as spherical aberration. Paraboloid mirrors are near-universal as reflecting telescope objectives.

astable circuit or **astable multivibrator** circuit with two states between which it jumps at regular intervals. It is a pulse generating circuit that needs no outside trigger to produce each pulse. The clock circuit of a computer is of this type.

astatic coils pair of identical coils linked in series and hanging as a unit with the same axis. The interaction between the current in each and any outside magnetic field is equal and opposite: the unit will not turn whatever the current or outside field.

astatic pair or **astatic system** pair of

permanent magnets hanging antiparallel so that a current in a coil around one will provide the only torque

astatine At element with proton number 85, but extremely rare as it is radioactive – even the most stable of the twenty isotopes, ^{210}At, has a half-life of only 8.3 h. Discovered in 1940, it was given its name from the Greek 'astatos', meaning unstable. Its melting and boiling temperatures are 250 °C and 350 °C.

astigmatic lens lens with the appropriate degree of astigmatism to correct the astigmatism of the wearer's eye

astigmatism type of image *aberration* found in optical systems (including the eye); it follows the system having a different refractive power in two directions at 90°.

Aston, Francis (1877–1945) nuclear physicist who invented the mass spectrograph and used it to separate isotopes. This led to his being awarded the 1922 Nobel Prize for Chemistry. The **Aston dark space** is the very narrow dark space in a discharge tube between the cathode surface and the cathode glow, the distance the electrons need to gain enough energy to ionise the gas.

astronomical telescope telescope designed for astronomical use, so needing no image-inverting lens or system. This makes it more efficient, cheaper and less bulky than one that does (a terrestrial telescope). Astronomical telescopes may be refractors or reflectors in the case of light astronomy; there are other types of telescope for the study of various other radiations that reach the Earth.

astronomical unit au the mean distance between the centres of the Sun and the Earth (just under 150 million km), sometimes used as the unit of distance in the solar system

astronomy the study of the physical Universe beyond the Earth's atmosphere. This involves ground-based and space-borne telescopes, to observe radiation of all wavelengths and types. There have been close links between physics and astronomy for centuries: technical developments lead to improvements in instrumentation, and increase our knowledge of the Universe, while astronomical observations provide evidence for new theories of physics.

astrophysics the study of the structure of, and processes in and between, planets, stars, and other astronomical systems. An important field is the physics of stars and how they obtain the energy they give out.

asynchronous motor motor with an

atmosphere

alternating supply whose speed depends on load rather than on supply frequency. The most important type is the induction motor.

atmosphere a) See *standard atmosphere*.
b) the gassy layers of the Earth (and of any other astronomical object) between the surface and 'outer space'. Pollution from human activity leads to various possible problems, from asthma to global warming through the greenhouse effect.

atmospheric absorption how the various gases in the various layers of the atmosphere take in the various kinds of radiation (this includes the science of the greenhouse effect): the absorption by the atmosphere of radiation passing from outside the Earth towards the surface, or the other way about. Much radiation from outside cannot reach the surface because the windows are fairly small. The main windows for electromagnetic radiation are:

- optical: 200 nm (ultraviolet, below which the ozone in the stratosphere is the absorber) up to about 800 nm (just past the red end of visible)
- infrared: 8–11 μm (water vapour absorbing most other infrareds)
- radio: 8 mm – 20 m (short radio waves)

atmospheric electricity the electrical properties of the air, studied both in normal weather and during electrical storms. Normally there is a downward field of around 130 V m^{-1}, with a current from air to Earth of 2×10^{-14}A m^2. The free charges in the air result from ionisation by radioactivity in rocks, the tiny but significant level of radon gas (see also *air*) and cosmic radiation; the field is between the positively charged ionosphere (charged by absorption of ultraviolet radiation from the Sun) and the negative solid Earth. See also *lightning*.

atmospheric layers the different gas layers between the Earth's surface and the hard vacuum of ('outer') interplanetary space. The different layers have significantly different properties as shown:

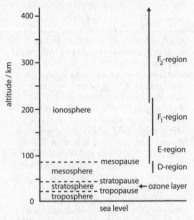

The various layers of the Earth's atmosphere differ greatly.

atmospheric pressure the pressure of the atmosphere, at the Earth's surface or at some depth or height within it. The standard is 101 325 Pa. See *air pressure*.

atmospherics a) old name for static, the interference with radio signals, heard as crackling or hissing in receivers, caused by electrical disturbances, such as lightning, and the solar wind. This background noise of numerous short bursts of radio waves can interfere greatly with radio signals.
b) background hiss (white radio noise) that comes from radio waves from the Sun and galaxy: see *radio astronomy*.

Atom

atom with the charged atom (ion), the smallest particle of an element that shows the properties of that element. Throughout history (from the times of Democritus and Lucretius to the late nineteenth century) people believed that atoms have no structure. Since the discovery of radioactivity at the end of the nineteenth century, various models have developed – to the positive, massive structured nucleus in the centre of a quantised negative electron cloud. Atomic properties involve the electron cloud of the ion and the neutral atom, and are the basis of chemistry; nuclear properties ignore the electron cloud and are the basis of nuclear physics. See also *atomic physics* and *nuclear physics*.

atom, artificial common name for the quantum dot. This is much less hard to work with than a normal atom, as it is much bigger, but still has its own set of discrete energy levels.

atom bomb weapon of mass destruction designed for the very rapid release of large quantities of energy from the fission of a critical mass or more of a fissile substance

atomic clock timing system based on the (very high) regular frequency of atomic or molecular events giving a sharp spectral line. Early atomic clocks were masers that used the radiation from certain transitions in ammonia (23.9 kHz) and caesium (9.19 GHz), and the principle is still the same. Modern systems, such as the time standards used in most advanced countries, maintain time within a second in several thousands of years. The most reliable type of *caesium clock*, however, has a much better self-correcting feedback system so is reliable within a second in sixty million years. A different type, announced in mid-2006, which combines an ultraviolet laser with a single mercury atom, takes us to a second in 400 million years.

atomic core a particularly stable positive ion: an atom with no conduction electrons (i.e. it has only closed electron shells), or one which has lost all its conduction electrons (as in the case of metallic conductor). See also *atomic physics*.

atomic energy old term for nuclear energy/nuclear power

atomic energy levels the few separate values of energy allowed in an atom's electron cloud by *quantum physics*

atomic heat dated term for molar thermal capacity (see also *Dulong and Petit's law*)

atomic mass, *A*, unit: atomic mass unit, amu the mass of an atom (in practice of its nucleus) as a multiple of the mass of a hydrogen (^{1}H) atom. The **atomic mass unit** (amu, sometimes called the dalton) is the value of that multiple. The mass of a ^{1}H atom is in fact 1.008 on the scale on which the hydrogen nucleus – the proton – has a mass of one unit. The more recent definition of atomic mass is one-twelfth of the mass of an atom of carbon-12 (^{12}C, 1.6605×10^{-27}kg, 930 MeV). However, instead of atomic mass, we now use nucleon number, which has the same value but no unit.

atomic number, Z, no unit now the proton number of a species, the number of protons in the nucleus. This defines what the element is, and from that comes its chemical nature. That's because the number of electrons in the neutral atom's electron cloud equals the proton number.

atomic orbital wave function that defines the (quantised) energy of an electron in an atom. The main 'shells', or energy levels, are K, L, M ..., while the sub-shells are s, p, d and f. See *quantum physics*.

atomic pile out-dated term for the central part of a nuclear fission plant, the core that contains the nuclear fuel. See *nuclear power*.

atomic physics or **atomic theory** See, in particular, *atom* and *nuclear physics*.

atomic units non-standard units of atomic measurement still in occasional use. For atomic unit of energy, see *hartree, rydberg*. The atomic unit of length is the radius of the first hydrogen atom orbital (5.3×10^{-9} m). The atomic unit of mass is the rest mass of the electron, 9.1×10^{-31} kg; see *nucleon number*. Its **atomic volume** is the volume of one mole of a substance's solid phase. **Atomic weight** an old term for nucleon number.

atomicity the number of atoms in the molecule of a gas. If the atomicity is 1, the gas is monatomic; if the atomicity is 2, it is diatomic; and if the atomicity is greater than 2, the gas is polyatomic.

attenuation, unit: decibel, dB negative gain, or negative *amplification*, for instance, the way a signal weakens as it passes through a channel

attenuation coefficient or **attenuation constant** a often the same as absorption constant, the measure of attenuation by absorption as a given radiation, current or signal type transfers through a distance through a medium or system; its value depends on the medium or system and the radiation energy. There are linear, mass, atomic and molar attenuation constants that depend on the measure used for distance (length) in the medium. In some cases, people count attenuation as a result of scattering as well as of absorption. In this case, the attenuation constant is greater than the absorption constant. See also *geometric attenuation*.

attenuation distortion the distortion of a complex wave that follows the way the channel attenuates different frequencies differently. See *distortion*. An **attenuation equaliser** is a circuit designed to correct for attenuation distortion.

attenuator electronic system designed to attenuate (reverse of amplify) an input signal with no distortion. It is normal to measure its effect in decibels.

atto-, a unit prefix for 10^{-18}, a quintillionth. It comes from atten, eighteen in Danish and in Norwegian.

attracted disc electrometer an absolute

electrometer (type of voltage meter) that consists of a large fixed disc and a smaller disc (the 'attracted disc') parallel to it hanging on a spring over the centre. The system measures the force on the attracted disc and relates it to the pd between the discs.

au symbol for *astronomical unit*, still often used to give distances in the solar system

audibility the ability of the human ear to detect sound (pressure, audio) waves. The standard limits of audibility are 20 Hz (below which is infrasound) and 20 kHz (above which is ultrasound).

audio frequency af frequency in the audio range, roughly 20 Hz to 20 kHz

Auger effect or **Auger ionisation** the escape from an excited atom or positive ion of an electron, called an **Auger electron**, to leave a less energetic positive or double positive ion. The original excitation may follow the absorption by the electron cloud of a gamma photon from the nucleus, or of a photon of gamma, x-, or other radiation from outside. The effect was discovered in 1925. See also *internal conversion*. **Auger emission spectroscope** is the study of the energies of Auger electrons from a substance in order to obtain information about its composition.

Auger, Pierre (1899–1989) physicist whose main work concerned x-rays, cosmic radiation, and neutrons

Auger shower See *cosmic radiation*.

aurora an extensive electric discharge in the upper atmosphere sometimes observed from high latitudes. It results from the movement between the Earth's poles of charged particles from the solar wind trapped in the Earth's magfield. The discharge takes many forms (such as arcs, 'curtains', rays, streamers) and various colours. The effect in the Antarctic is called the **aurora australis** (or southern lights; 'aurora' means dawn, from the appearance). That around the North Pole is called the **aurora borealis** (or northern lights), which a few times a year is visible even from southern Britain.

autodyne type of radio receiver – see *beat receiver*.

autoelectronic emission or **autoemission** the radiation of electrons from a cold surface. See *field emission*.

autoionisation the process of atomic de-excitation with the escape of an orbital electron. Unlike the similar Auger effect, the original excitation is the transfer of an electron to an energy level above the atom's ionisation energy.

automatic frequency control (afc) **a)** use of an electronic circuit in a radio transmitter or receiver (for instance) to lock onto a given frequency so it stays tuned to a given station, even though various effects try to change the wavelength slightly from moment to moment **b)** system for keeping any alternating pd within set frequency limits

automatic gain control (agc) or **automatic volume control** (avc) system which uses an electronic circuit to maintain a radio receiver's output volume within limits when the input signal strength varies

autoradiograph image formed on a photographic plate by direct contact of an object that contains natural or artificial radioisotopes

autotransductor transductor (type of current transformer used in a magnetic amplifier) with the same coil for both the main and the control signals

autotransformer a voltage transformer with a single coil rather than two or more. Tappings (contacts) at points within the coil give the same effect as having several separate output coils (the more traditional design).

avalanche runaway effect in which electrical output increases exponentially (at least up to some limit). It is most important in ionisation processes, where each ion ionises several atoms, each of which ionises several more atoms (but see also chain reaction). In the case of ionisation (which needs a field strength above a certain value which depends on the substance concerned), the result can be **avalanche breakdown**: a process which disrupts the structure of the substance, at least for a time. This is the case with a spark discharge, and is used in Geiger-Müller and spark detectors. In the **avalanche detector**, the substance concerned is a semiconductor in a strong field: input radiation causes a certain degree of ionisation which triggers the avalanche to produce a very strong output pulse.

An **avalanche diode** is a photomultiplier in which a single photon absorbed in the semiconductor when just at the point of avalanche breakdown triggers a heavy current. Support circuitry detects this within a few nanoseconds, and quenches the current by cutting back the applied electric field. The system is then ready for the next event. The avalanche diode is potentially the most effective photomultiplier. Compare with *Zener diode*.

avc See *automatic volume control*.

average life or **average lifetime** See *mean life*.

Avogadro, Amadeo (1776–1856) physics professor at Turin from 1834 to 1850 who proposed what we now call **Avogadro's hypothesis** or **Avogadro's law** in 1811. This is that equal volumes of (ideal) gases contain equal numbers of particles at the same temperature and pressure.

Avogadro constant or **Avogadro number** N_A or L the number of particles in one mole of gas – indeed, by extension and definition, the number in one mole of any substance. It is about 6.023×10^{23}. See also *Loschmidt number*, this being the correct name as it was Loschmidt who worked it out.

axial length a) the length of one of the three axes (a, b, or c) of a crystal's unit cell. The crystal's **axial ratio** is the ratio of its axial lengths, with that of the b-axis taken as 1, i.e. in the form $a:1:c$.

b) a major measure used in eye tests

axino the fermionic superpartner of the axion according to supersymmetric theory

axion a so-called pseudo-Goldstone boson, particle proposed to explain the lack of violation of CP-symmetry in certain string interactions predicted by quantum chromodynamics and also required by string theory. The axion would have no charge and a tiny mass (little more than $1\ \mu eV/c^2$). At the time of writing, no test has shown an axion. However, a very slight rotation of the plane of polarisation of light in vacuum in a strong magfield (observed in 2006) could be evidence for it; other tests of this nature are now being devised.

axis a) line about which an object rotates, or (**axis of symmetry**) about which it is symmetrical

b) one of two or three lines (in most cases at 90°) used as reference lines to give the coordinates of a point in 2D or 3D space, called x, y, and z

c) one of the three edges of a crystal's unit cell, called a, b, and c

d) principal axis of an optical system

azeotropic mixture a mixture of two liquids that boils with no change of composition. This means you cannot separate the liquids by distillation.

azimuth an angle measured in the horizontal plane

azimuthal quantum number See orbital angular momentum *quantum number*.

b symbol for bit, binary digit (0 or 1)

B a) symbol for bar, old unit for pressure, 100 000 Pa

b) symbol for barn, now quite rarely used unit for atomic and nuclear cross-section, $10^{-28}m^2$

c) symbol for bel, unit of loudness (though the decibel is much more common) and, by extension, unit of (for instance) amplification and comparative power

d) chemical symbol for boron

e) symbol for byte, set of eight binary digits (0s or 1s)

f) unit prefix symbol for (US) billion, as in 10 BV, which we now give as 10 GV

b a) symbol for breadth or width

b) symbol for the impact parameter in Rutherford scattering

B a) symbol for binding energy

b) symbol for bulk modulus

c) symbol for magfield strength (magnetic field strength or magnetic flux density)

d) symbol for (degree of) magnetisation

e) symbol for susceptance

β symbol for beta particle (the electron as radiated in radioactivity) and beta radiation

β symbol for speed as a fraction of the speed of light, or for the square of this

Babinet compensator device able to produce plane or elliptically polarised light output with a path length the user can change. It consists of two identical quartz prisms that have no net refractive effect, but slide over each other to provide a 'parallel block' of variable thickness. The main use is in polarising microscopes (for instance, those used to examine thin sections of rock).

back emf a) induced voltage in an electromagnetic system that, in line with Lenz's law, tends to oppose the cause – the applied voltage and thus the normal current.

Back emf occurs only when the current changes, so is most significant in ac systems and during switch on or off of a dc system. See also *eddy current*.

b) voltage in an electrolytic cell that occurs as a result of polarisation and opposes the voltage applied from outside

back focal distance the distance in a complex optical system between the final reflecting or refracting surface and the nearest focal point

background noise a) perhaps low level of unwanted sound that makes it harder to detect wanted sound

b) by extension, the same in the case of any type of signal. For instance, a stray current and/or voltage in an electric circuit or channel may interfere with the desired output. There are many causes. Noise in digital systems is easier to combat than that in analog systems. See also *signal-to-noise ratio*.

background radiation the low level of ionising radiation that always exists; indeed, it may well be the cause of the random mutations in the DNA of living cells that leads to evolution. The natural sources of background radiation are the environment (rocks and soils, air and water) and space (cosmic radiation); radiation from x-ray systems and fallout from nuclear weapons tests and nuclear power has added slightly to these in recent decades.

back scattering scattering of radiation by more than 90°, i.e. back towards the direction from which it came

Bacon, Roger (*c.*1214–1294) believed to be one of the first European practical scientists, doing major work in optics such as on reflection and refraction (and influenced strongly by al Kindi who wrote 300 years earlier)

baffle absorbing device: a board, plate or more complex system designed to control (for instance, reduce or improve) the transfer of a fluid, or of energy in the form of light or sound, between two points

Baird, John Logie (1888–1946) engineer who started research into television and radar in the 1920s, both demonstrating the former and patenting the latter in 1926. Baird produced the first three-dimensional and colour television systems in 1944.

balance weighing system used to compare an unknown mass with a known one (but see *current balance*). Strictly, the word applies only to a two-armed balance (the arms being of equal or unequal length according to the design); it is, however, common to use the term for any weight or mass meter, or indeed any sort of meter at all. See also *spring balance*.

balanced amplifier a push–pull amplifier

ballistic galvanometer a moving-coil meter that measures the total charge of a system by passing it all as a brief current pulse through a lightly damped coil. Assume the length of the pulse is much less than the period of the coil's oscillation; then the maximum deflection ('throw') of the coil and attached needle is proportional to the charge passed.

ballistic pendulum massive suspended 'bob' used to measure the speed of an object that collides with it. The calculation depends on the two masses and the distance the bob moves, and follows the law of constant momentum.

ballistics the study of the motion of projectiles, objects in free fall near the Earth's surface

ballast resistor resistor made of a substance with a high temperature coefficient of resistance; as a result, it keeps the circuit current constant over a wide range of applied voltage. See also *thermistor*.

Balmer, Johann (1825–1898) physicist whose study of the wavelengths of lines in the hydrogen spectrum led to the spectral series relation, later extended to other spectra by Bohr and Sommerfeld using the quantum theory. For the **Balmer band**, **Balmer continuum** and **Balmer limit** see *hydrogen spectrum*.

Balmer series a set of lines in the *hydrogen spectrum* caused by electron transitions to or from the second lowest energy level (principal quantum number $n = 2$). The series lies mainly in the visible region, whereas the Lyman and Paschen series (for the lowest and third lowest energy levels) lie in the ultraviolet and near infrared.

band set of consecutive very close or continuous values of wavelength (in a spectrum) or of energy (in the electronic system of a sample of matter)

band head the sharply defined end of a band spectrum

band model or band theory

the core of solid state physics, in which the orbital electrons of an atom in the solid state can exist only at certain 'allowed' *energy levels* (below) or *energy bands* (bands of close energy levels – see below); between these are bands of 'forbidden' energies. In addition, the number of nearby electrons that can exist at a given energy level at the same time is restricted to two (one with each spin value). These two statements arise from early *quantum physics*. That also explains atomic spectral lines as being the result of electrons jumping between levels, with the absorption or emission of specific quanta of energy W.

The band theory is central to much of modern electronics and related fields; it depends on the model of a single atom as a nucleus surrounded by a number of electron energy levels (rather than a diffuse cloud). However, the energy levels are not in real space – so we do not see precise orbits; they are in *momentum space* (below).

The first sketch below shows the energy level pattern (in momentum space) for a single atom. It should be clear from the above that if two atoms of the same species are very close, they cannot have the same energy levels. Therefore, the levels split (as in the second sketch); the degree of splitting depends on how close the two atoms are. In the solid state atoms are very close, which is why the model applies mainly to the solid state. In a solid, the split levels therefore become energy bands. Atomic electrons can rest only in these bands; they need specific ranges of energy to jump between bands. As in all physical systems, the total energy is the sum of the component energies; in this case, this means that electrons fill the bands starting at the bottom. The band pattern of any solid at 0 K is therefore a set of full bands with a set of empty bands above.

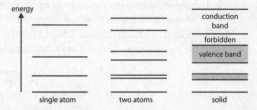

At absolute zero temperature, all the electrons in a solid would lie in the lowest possible energy states. When the temperature rises, in a process a bit like evaporation, a number of electrons jump to higher states, leaving holes (empty levels) behind. The Fermi level is the energy level at which the probability of finding an electron is 0.5. That is to say, it is the energy of the highest energy electron in the system at absolute zero – in this case, since all the electrons are in the lowest possible states, none is above the Fermi level. As the temperature rises, so does the Fermi level.

We use that concept to show how conducting metals, semiconductors and insulators differ; we also name the bands, shown above, as follows:

The **valence band** is the highest set of electron energy levels where there are electrons at absolute zero.

The **forbidden band** (or gap) is the electron energy range above the valence band. Here there can be no electrons at absolute zero – or at any higher temperature.

The **conduction band** is the electron energy range above the forbidden band, where there can be no electrons at absolute zero, but can be at higher temperatures – and if there *are* electrons there, they can move around and form an electric current.

In metals, there is no forbidden band (or even a negative one as in the first sketch below). That means it needs very little energy to move electrons up into the conduction band: metals conduct well at low temperatures. In this case, the Fermi level is in the conduction band in all contexts above zero. (Note that superconductivity is not part of the band model.)

Intrinsic semiconductors have a forbidden energy gap, but one small enough that a few electrons can 'jump' it at normal temperatures. An electron that enters the conduction band leaves a mobile 'hole' in the valence band, as in the second sketch. At any temperature above zero, the Fermi level is half-way up the forbidden band. (Note that we come to extrinsic semiconductors shortly.)

Last, an insulator has such a large forbidden gap that electrons can't jump up over it at any temperature. There are no conduction electrons ever, therefore – and the Fermi level is in the valence band.

The sketches below show how conductors, intrinsic semiconductors and insulators differ in the band picture. They differ in the closeness of the valence and conduction bands.

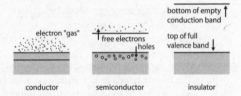

conductor   semiconductor   insulator

It is clear that insulators (with Fermi levels in the valence band) have negligible conductivity except at *very* high temperatures (when electrons can gain enough energy to jump the large gap). Conductors (Fermi level in the conduction band) can pass current well at any temperature, while (intrinsic) semiconductors (Fermi level imaginary, i.e. within a forbidden band) have a small conductivity that rises quite fast with temperature. As the next sketches show, extrinsic semiconductors contain impurity atoms that insert extra narrow energy bands into the forbidden gap. These extra bands are just above the valence band in the case of a p-type semiconductor (acceptor impurity), and just below the conduction band for an n-type semiconductor (donor impurity). In either case, the Fermi level is at the impurity level. In a p-n junction, the Fermi level must have the same energy throughout – as a result, the edges of the valence and conduction bands bend as shown. (Note that, unlike the previous band sketches – which are all one-dimensional energy graphs – this is two-dimensional: energy against distance x either side of the junction.)

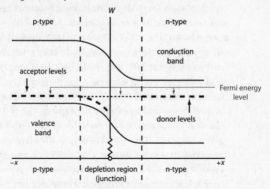

Now follow notes on other aspects of the band model of matter.

Brillouin zone one of a number of energy bands for electrons in a solid when plotted in a certain way. It is a region in momentum space that relates to the band model but is more important to x-ray crystallography.

density of states curve graph showing how, in a solid, the number of energy levels in a band varies with energy. The figure shows a typical case, that of a substance like a metal with two overlapping bands. The shading shows possible occupied levels while the dashed lines show the effective edges of the bands. Note that in this case energy rises along the x-axis rather than the y-axis as above.

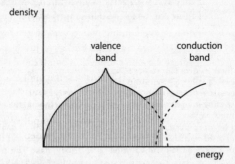

The density of states curve is very useful in solid state physics

energy band a range of allowed electron energy values, with forbidden energy ranges on each side; each is really a set of very close *energy levels* (below). In a single atom, according to the quantum theory, the electrons can have only certain values of energy; these are the atom's energy levels. The Pauli exclusion principle is that there can be only two electrons (with opposite spin) at each level. When two or more atoms are close, therefore, the levels split into energy bands.

energy levels the set of discrete values allowed by the quantum theory for the electrons of a single lone atom. A similar set applies within each nucleus, and molecules have energy levels for vibration and rotation too. In each case, each energy level has a quantum number associated with it. An **energy level diagram**, a stretched one-dimensional graph with energy on the (y-) axis, plots the allowed levels and may show where the atom's electrons are at a given moment. **Energy level degeneracy** (symbol g) is the number of possible states of a system with the same energy level. No energy level is perfectly sharp, however. The uncertainty principle means that it has a range of energies whose size is inversely proportional to its life time. **Energy level breadth** (or **energy level width**) is the size of that energy range in a given case.

forbidden band range of energies in which no electron of the system in question may exist. Electrons can pass between allowed energy levels and bands unless the change involves a **forbidden transition**. This is a jump that requires a change of quantum number that may not occur as it breaks a conservation rule.

momentum space the mathematical space in which we plot the graphs of the band model of solids. It is complementary to the usual position-space but now the coordinates at any point are electron momentum coordinates (and the origin stands for an electron with zero momentum, i.e. at rest). The concept is of value for x-ray crystallography as much as for the band model.

band-pass filter system that passes all input frequencies that fall in a given band of frequencies and rejects all others. Such filters are of great value in optics and electronics and have uses in other contexts too; the filters may be broad band or narrow band on the basis of band width. As with all filters, there is not a sudden cut-off at the band edges.

band spectrum or **molecular spectrum** like a line spectrum but with bands rather than lines, with each band made up of many very close lines. This is because the source (emission spectrum) or the absorber (absorption spectrum) consists of molecules rather than elements, and each line relates to a molecular electron orbit transition between different vibrational states whose energies are very close.

band theory the same as the *band model* of matter

band width or **bandwidth** the range of frequencies transferred by a channel, device or system. The cut-offs at the lower and upper limits are never sharp, so it is normal to define the band edges as the frequencies at which the transfer effectiveness is half the maximum. See also *band-pass filter* and *base band*.

bar B old unit of pressure (in particular of air pressure), equal in value to 100 000 Pa – rather less than mean air pressure at sea level, now defined as 101 325 Pa. The millibar (mB), 100 Pa, still sometimes appears in meteorology as a unit of air pressure.

Bardeen, John (1908–1991) physicist who developed the transistor in 1947 with Walter Brattain and shared with him and lab director William Shockley the 1956 Nobel Prize that followed. He was the first (and, in 2005, still the only) person to receive the physics prize twice. The second time was in 1972, when he shared it with Leon Cooper and John Schrieffer for the 'BCS' theory of superconductivity.

barium, Ba alkaline earth metal, element 56, discovered by Humphry Davy in 1808; the name comes from the Greek word for 'heavy'. Group II, period 6; relative density 3.5, melting and boiling temperatures 725 °C and 1640 °C; most common isotope ^{138}Ba. As barium sulphate does not transmit x-rays, it is the basis of the '**barium meal**' doctors use to study a patient's gut.

Barkhausen effect important early evidence for the domain theory of ferromagnetism. First carried out by Heinrich Barkhausen in 1919, the test involves smoothly but quickly magnetising and demagnetising an iron bar using a coil, with a second ('search') coil feeding an amplifier and speaker. Each time the magnetising field changes, the output is a 'rushing noise': a very large number of tiny clicks and crackles, each being the induced effect of a single domain switching direction. In other words, the magnetisation of a ferromagnetic sample is not smooth, but in steps, because of the domain structure.

Barkla, Charles (1877–1944) physicist who won the 1917 Nobel Prize for work that led to the crucial field of x-ray spectroscopy

barn b (or B) outdated, but still quite often used, unit of atomic and nuclear cross-section, equal to 10^{-28} m^2. The significance of nuclear cross-section is in the field of firing particle beams at targets to explore nuclear reactions; the geometric cross-section of a neutron is around 40 mb, so the barn is large on a nuclear scale. First used in the days of firing alphas at atoms to explore nuclei, the name comes from 'it's as easy as hitting a barn door'.

Barnett effect the slight magnetisation of a ferromagnetic cylinder spun at high speed. The magnetisation is parallel to the spin axis, varies with the speed, results from deformation of the atomic electron shells, and also appears (to a much lesser extent) in

paramagnetics. Discovered by Samuel Barnett in 1915, this was the first gyromagnetic effect; quantum theory explains it, though some people later saw it as evidence for antigravity.

barograph aneroid barometer giving a plotted chart on paper of the air pressure reading over a period of time

barometer device designed to measure the pressure of the air. As air pressure depends on altitude (among other factors), barometers have long been in use as altimeters. In the first design, *Torricelli*'s mercury barometer, the length of a mercury column – the **barometric height** h – gives the air pressure. The unit is inches or centimetres or millimetres of mercury, the standard value is 760 mm Hg (millimetres of mercury). That last unit is still quite widely used, while mercury-in-glass instruments are not – though giving absolute readings, they are large, costly, hard to use, and hazardous. The pascal is the correct unit (given by the product of the length in metres and the density of the mercury). See also the modern instrument, the *aneroid barometer*.

barostat device designed to allow the user to control the gas pressure inside a space, in particular to correct for changes in atmospheric pressure

barrel distortion a form of *aberration* c) sometimes observed in projection

barrier or **potential barrier** local high energy peak between two lower energy regions. If the charge carriers on one or both sides do not have enough energy to cross the barrier it is in effect an insulator: the system can then store charge like a capacitor (with a **barrier capacitance**) or allow one-way flow only – **barrier rectification**. There is also a barrier round the potential well of a nucleus as far as concern alpha decay and bombardment by positive particles. See also *barrier height* and *barrier width*.

barrier height measure of the size of a barrier in energy terms on an energy/distance graph

barrier layer photo-cell type of photo-voltaic cell. See *photo-diode*.

barrier potential an electric charge energy barrier (i.e. a Coulomb barrier) expressed as electric potential energy:

$$V = (1/4\,\pi\,\varepsilon_0)\,Q_1\,Q_2/d$$

Here the charges are Q_1 and Q_2, distance d apart in empty space (ε_0 is its permittivity).

barrier width a measure of the size of a barrier

in terms of distance on an energy/distance graph

Bartlett force the exchange force between nucleons that follows the transfer between them of spin direction

barycentre old term for centre of mass or centre of gravity

baryon massive fundamental particle (*barys* is Greek for 'heavy') – a nucleon or any other fermion that decays into a nucleon with the emission of a meson. The proton is the only stable baryon – current theory has it, however, that all consist of three tight-bound quarks (anti-quarks in the case of anti-baryons). As we now see far more baryons than anti-baryons, the very early Universe would have needed a phase of baryogenesis. *See particle physics.*

baryonic matter any kind of matter whose mass is mainly from baryons: normal matter and antimatter, therefore, but not electrons, photons, dark matter or black holes (all of which are non-baryonic matter)

baryon number B a quantum number used to describe nuclear reactions and interactions. Its value for baryons is 1, for anti-baryons –1, and for other particles 0 (see also lepton). The baryon number, constant (conserved) in any system, is a third of the number of quarks minus a third of the number of anti-quarks. The actual conserved quantum number is quark number – but baryon number was invented long before quarks were dreamed of.

base a) the control terminal of a bi-polar junction transistor, the region between collector and emitter. A small amplitude signal applied between the base and the emitter (the input) causes a large output signal between emitter and collector.
b) a substance whose molecules tend to gain protons in chemical reactions, in particular an alkali (which dissolves in water to produce OH^- ions and reacts with an acid to produce a salt)
c) an adjective meaning fundamental. A **base band** is a communications channel able to support signals in only one frequency band. **Base units** in a unit system are those in the system on which all others depend. There are seven in SI (see *Appendix 1*).

Basov, Nikolai (1922–2001) physicist who won the 1964 Nobel Prize for the work on quantum electronics that led to the invention of the maser and the laser

Bateman equations set of equations that give the abundances of the various nuclides in a radioactive series at various times after starting with a pure parent

battery symbol: ⊣⊢–⊣⊢ a linked set of one or more electric (electrochemical) cells, to increase the output voltage if in series (first symbol) or the output power or usage time (capacity) if in parallel (second symbol). We can always use the third symbol, which stands for any dc source (see *Appendix 3*).

Like a single cell, a battery can be non-rechargeable (primary) or rechargeable (secondary). Secondary cells and batteries are far more cost-effective than primary ones, and much research goes into developing ones light enough and with enough output capacity for use in portable electronic hardware (such as laptop micros) and vehicles. The traditional lead–acid battery (invented in 1859, by Gaston Planté) is still able to compete against the nickel–cadmium type, whose lifetime is fairly short and whose components are more costly and less easy to dispose of. Reducing the thickness of the plates and leads, for instance, further raises the power/weight ratio and the performance at high discharge rates. Indeed, various flat lead–acid batteries have been produced, able to work with high efficiency and reliability.

Bayer pattern grid of red-green and green-blue filters that covers the single ccd of a cheap digital camera in order to produce a colour image

beam wide, perhaps parallel, ray of radiation (though in theory a ray has zero cross-section). The **beam factor** (for a light source such as a searchlight) or **beam width** (for a directional aerial) is the angular (or solid angular) extent of the main beam lobe.

beam aerial directional aerial: a horn, dish or array

beam coupling effect of an interaction between a beam of charged particles and an alternating electric field (see, for instance, *bunching*)

beam current the current in a cathode ray tube carried between gun and screen by the electron beam that forms the spot

beat(ing) the oscillating volume observed with two sound waves of very close frequency, such as the two engines of a twin-engined plane, a result of interference. Similar effects occur with other wave radiations, as well as with alternating currents. If the frequencies of the two waves are v_1 and v_2, the *beat frequency* v_b is their difference:

$$v_b = | (v_1 - v_2) |$$

beat frequency oscillator generator of audio or radio signals by the **beat frequency process**: the system's inputs are a fixed radio frequency signal and one of variable frequency, and the output is a beat signal. The system design ensures this is of constant strength over a wide range of frequency.

beat oscillator radio frequency generator in a type of radio receiver called a **beat receiver**. This produces audio beats between the input radio frequency signal from the aerial and the oscillator signal. In a heterodyne system, the output then goes to rectifier, amplifier and speaker. In the far more common super-heterodyne set, there is an extra amplification stage before rectification. Autodyne systems have the rectification and amplification carried out by the beat mixing process. All are forms of **beat reception**.

beats the result of any beating process

Beckmann thermometer very accurate design of mercury-in-glass thermometer. It has a very large bulb and can read to 0.01 K over a range of a few kelvin. To change the range, one must either use a different thermometer or adjust the volume of mercury in the bulb using a reservoir.

becquerel, Bq the unit of *activity* b), the number of decays in a sample per second. Therefore, $1 \text{ Bq} = 1 \text{ s}^{-1}$.

Becquerel, Antoine Henri (1852–1908) physics professor in Paris who followed his father **Edmond Becquerel** (1820–1891), whose main work was in the fields of magnetism and solar radiation, to the post, and *he* followed *his* father **Antoine César Becquerel** (1788–1878), the first person to use electrolysis to separate metals from their ores. Becquerel did research in a number of fields, but shared the 1903 Nobel Prize with the Curies for his accidental discovery, in 1896, of natural radioactivity. The radiation produced, which he went on to explore, carried the name **Becquerel rays** for some years.

Bednorz, Georg (1950–) materials scientist awarded, with Karl Müller, the 1987 Nobel Prize following their development of the first high-temperature superconductors

Beer–Lambert law or **Beer's law** or the **Beer–Lambert–Bouguer law** that the light absorptance of a dilute solution is proportional to the product of path length, concentration, and molar absorptivity

Békésy, Georg (1899–1972) physicist given the 1961 Nobel Prize for physiology for his new theory of hearing, which was based on the study of waves formed in the fluid in the cochlea. It replaced that of Helmholtz.

bel B named after Alexander Graham Bell, unit of relative power (energy content) used with radiations and signals, circuit inputs/outputs. Two signals of power P_1 and P_2 differ by n bels, where $n = \log_{10} (P_2/P_1)$. In the input/output case, n is the circuit's gain (loss if negative). See also *decibel* and *neper*.

Bell, Alexander Graham (1847–1922) applied scientist who was the major inventor and developer of the telephone, inventor of the phonograph (an early audio player), and developer of the hydrofoil craft. He later worked in aeronautics.

Bell, John Stewart (1928–1990) physicist who, in 1964, designed a crucial test for the uncertainty aspects of the quantum theory. The test was carried out successfully in the 1980s, and showed the reality of quantum action at a distance.

Benz, Karl (1844–1929) engineer who invented the first successful internal combustion engine, in 1885, for use in a car – his 'Tri-Car' ran first on gas and later on petrol (gasoline). He also invented all the main parts of the engine as well as support items such as the water-cooled radiator, the clutch and the accelerator (gas pedal).

berkelium Bk element 97, trans-uranic metal in the rare earth actinide series, discovered in 1949 at the Berkeley campus of the University of California (hence the name) by Glenn Seaborg. Relative density calculated to be around 14, melting temperature 986 °C, boiling temperature unknown, most stable isotope ^{247}Bk (alpha emitter with half-life of 1400 years).

Bernoulli, Daniel (1700–1782) mathematician, botanist and anatomist, most famous for his application of calculus to fluid flow. He described the Bernouilli effect in 1738. He was unusual in being professor at Basle first of anatomy, then of botany, and then of physics; he also gained the French Academy prize ten times. His uncle, **Jacques Bernoulli** (1654–1705) was a mathematician who calculated the **Bernoulli numbers** (a sequence of coefficients in a certain power series), and is also known for the **Bernoulli theorem** in probability theory. **Jean Bernoulli** (1667–1748), brother of Jacques and father of Daniel, was also a mathematician; he extended Leibnitz's calculus.

Bernoulli effect named after Daniel Bernoulli (though earlier stated by others), the relationship between the pressure in a moving fluid and how fast it is moving: the pressure is inversely proportional to the square of the speed. This is the basis of, for instance, the force of lift on a moving air- or hydrofoil, and follows from the **Bernoulli equation** (or **Bernoulli principle**) – in any volume through which there is a steady fluid flow, the total energy is constant. In this case, energy contains terms for pressure and kinetic and potential energies. All this theory follows Newton's laws of force.

Berthelot, Marcellin (1827–1907) chemist who was one of the founders of thermochemistry. He devised the standard method to measure the specific latent thermal capacity of steam and carried out much important early work in organic chemistry.

Berthelot's equation of state an equation of state for a gas that better fits practice than does that of van der Waal – but still fails at the critical temperature of the substance. The equation is:

$$(p + a/TV^2)\,(V - b) = RT$$

Here p, V and T are the pressure, volume and absolute temperature of the gas sample, R is the universal gas constant, and a and b are constants.

beryllium, Be alkaline earth metal, element 4, group II, period 2, discovered in 1797 by Louis Vauquelin in the form of BeO and isolated in 1828 independently by Friedrich Wöhler and Antoine Bussy. Relative density 1.85, melting and boiling temperatures 1285 °C and 2469 °C; most common isotope ^{9}Be.

Bessel, Friedrich (1784–1864) mathematician and astronomer who developed the study of what came to be called the **Bessel functions**. These are a set of functions with many uses in advanced physics, in particular in wave motion, thermal conduction, sound output from a drum skin. Each function is a power series which solves the **Bessel differential equation**; this has the form $x^2 d^2y/dx^2 + xdy/dx + (x^2 - n^2)\,y = 0$. Here n is a real number, an integer in special cases.

beta current gain factor β the current

amplification factor of a transistor in common-emitter mode when short circuited. (Its value can be as high as 500.)

with I_C and I_B being the collector and base currents, and V_{CE} kept constant:

$$\beta = \delta I_C / \delta I_B$$

beta decay a form of radioactivity, a weak nuclear reaction, with half-life of the order of several minutes. During the process, a nucleus emits a beta particle β; its nucleon number remains the same, but its proton number increases by 1. The process involves the decay of a nuclear neutron (n) into a proton (p), electron (e$^-$, the **beta particle**) and an anti-neutrino $\bar{v}$, with the release of (kinetic) energy W. See also *nuclear physics*.

beta-ray spectroscopy the capture of beta radiation, and its analysis in terms of the range of beta-particle energy. It is the greatest range that helps identify the source. See also *nuclear physics*.

betatron a cyclic accelerator for electrons in which the energy transfer to the particles involves a process of induction rather than the more usual electric field acceleration. In other words, the electrons gain energy as they move through an increasing alternating magnetic field. The system can produce beams of hundreds of megavolts. The **betatron synchrotron** was a proton accelerator that also uses a high frequency magfield. The term now appears only in **betatron synchrotron coupling** or **betatron synchrotron resonance**. This is a mode of saturation in free electron lasers.

Bethe, Hans (1906–2005) physicist who directed the Los Alamos research into the atomic bomb. He was awarded the Nobel Prize in 1967 for his work on nuclear reactions, which included the theory of cosmic radiation cascades and the **Bethe cycle** (the thermonuclear carbon cycle).

bevatron type of proton synchrotron. The name comes from **BeV** (billion electron volts), an old term for GeV.

B/H loop the thick S-shaped graph obtained by plotting the induced magnetism (B) in a sample as the applied magnetic field (H) changes through a complete cycle. See *hysteresis*.

bi- prefix meaning two-fold, or double, as in a **biaxial crystal** – one with two axes, parallel to which the two components of polarised light travel at the same speed (rather than at different speeds as is usual). Also, **bi-concave** applies to a lens with both faces concave, and **bi-convex**, to a lens with both faces convex. See also other examples below.

bias a) outside influence on a system that alters its normal state and thus affects how it behaves. For instance, in electronics, a **bias voltage** is a pd applied to an active device to fix where on its input/output characteristic curve it is to operate.
b) in magnetic recording, a high frequency alternating magfield added to the signal being recorded in order to reduce distortion

bi-filar made with two threads or wires, as in the case of the **bi-filar electrometer**; by means of the separation of two taut conducting quartz fibres, this gives a measure of the voltage between two nearby plates. The **bi-filar pendulum** shown is not much use as a pendulum – but the study of how it vibrates (around an axis parallel to the threads) is of value in civil engineering.

That pendulum is a device based on **bi-filar suspension** – hanging from two threads, or from one thread doubled. The latter meaning appears in the case of the **bi-filar winding** of some resistance coils – the wire is doubled (so the two ends are together) before forming the coil; then the coil has no magfield and no inductance when it carries a current.

big bang theory

theory, first proposed by George Gamow in the 1940s, that our Universe started as the explosion (the big bang) of an infinitely dense, infinitely hot point object, nearly 14 000 million years ago. Apart from matter, the other remnant we can detect is the cosmic background radiation – however, the matter we can detect is less than 5% of what there must be for the Universe to expand as slowly as it does (*Hubble law*). There are various ideas as to what made up the Universe before the big bang. For instance, the study of worm holes leads to the suggestion that the event took place when our Universe branched out from an overall infinite 'multiverse', while general relativity predicts a gravitational singularity. The problem is that no big bang theory includes quantum effects.

A variation of the theory is that the Universe is cyclic with a big bang every so often, after the **big crunch** of the previous Universe. The 2006 version of this has a parallel Universe of dark matter; such a theory can explain why the cosmological constant is slightly positive (why the expansion of the Universe seems to be getting greater). Another version (most lately coming from Stephen Hawking, also in 2006) is that ours is only one of the Universes that started around the same time, but by now there is little nearby evidence of the others (whatever happened to them). (This is not the same as the 'parallel Universe' theories.) See http://physicsweb.org/article/news/10/6/16.

The big bang theory depends on three assumptions; people are testing each one with great care:
● that physical laws are absolute, unchanging and universal (and therefore that physical constants had the same values at the start of the Universe)

- that the large scale Universe is homogeneous and isotropic (the same in all directions and in all parts – the cosmological principle)
- that neither we nor any other observers of the Universe are in any way special (which comes from the Copernican principle)

According to the theory, the Universe evolved through a number of phases. In the first two, the particles of the Universe were far hotter than those we can now explore in accelerators, and we do not have the power to apply quantum physics – as we have neither the theory nor the practice to describe these states, there are many different theories. Here are the stages of the Universe since the big bang.

1 In the first picosecond or so, the very early Universe was exponentially expanding ('inflation') from the superdense stage just after the initial singularity. It would pass from a phase in which there was just one fundamental force, through the separation of gravity from the gauge interactions (electromagnetism and the strong and weak nuclear forces), to the current picture of four different fundamental forces. It would also need a phase of baryogenesis, as we now see far more baryons than anti-baryons.

2 Up to about a second after the big bang, the Universe changed from a very hot, superdense plasma of quarks and gluons to a 2 GK (2×10^9 K) ball of neutrinos and photons (plus a few, but more and more, protons, neutrons and electron–positron pairs); its energy density was now some 2 GJ m^{-3}, with matter at 10 kg m^{-3}. Now we can use known high energy physics to discuss the growth of the Universe. In the next few seconds, the temperature became low enough to allow deuterons to form, leading to a nuclear chain reaction that produced alpha particles (helium nuclei), plus very low levels of a few other light nuclei. About a year after the bang, the Universe's temperature had fallen to some 700 MK, too low to create or destroy electromagnetic photons. Inhomogeneities begin to grow.

3 Some 400 000 years later came the next stage, when the temperature was low enough for electrons and nuclei to combine to form neutral atoms of hydrogen (76–77%) and helium (23–24%; supernovae are believed to be the main source of heavier elements). Now electromagnetic radiation would interact much less with particles than before. It is believed that the microwave background pattern (with a red-shift of around 1500) was set at that time. How this differs from 'pure' 3 K black body radiation is a major field of research: differences could depend on black holes, cosmic strings, gravitational waves, and the decay of particles as

yet unknown. This phase would have been followed by one in which the only radiation would have been the 21 cm hydrogen line, close study of which would have great value – but as yet no-one has found this very faint cosmic background.

4 The next major event may have occurred about 200 million years later: the formation of population III stars, quasars, galaxies, clusters and superclusters. We can observe most of these, so can gain a picture of the structure of the lumpy Universe at any desired time from that date. The date corresponds to a red-shift of around 10. Some astronomers are concerned that objects with greater shifts may be found (the current highest values are well over 8), in which case the theory would need major revision. Revision would also be required if further research were to lead to the discovery of much younger galaxies, as is possible, and/or if the microwave background is uniform to one part in several hundred thousand (currently it is so to 1 in 10 000): a uniform background implies a uniform early Universe from which galaxies and stars could not condense.

The current supporting evidence for the big bang theory is the 23–24% helium content of stars; Gamow's calculation of this has needed little refining, and the theoretical value is extremely close to that observed. The exact fraction depends on the lifetime of the neutron, the expansion speed of the early Universe, and the number of neutrinos.

See also http://www.umich.edu/~gs265/bigbang.htm

big science the funding and use of very large facilities, such as space programmes, major accelerators and large telescopes, in order (initially) to explore pure science. The capital and running costs of such projects are so high, and the benefits so uncertain, that individual governments find their funding very hard to bear. International cooperation, rather than competition as a few decades ago, is now common in big science.

Billet split lens system that gives two close images of a single slit, and thus shows interference fringes. It consists of a bi-convex lens split at 90° to the axis; the two halves are shifted slightly along the common surface, so that neither of their centres is quite axial.

billion a thousand million, 10^9. Strictly, this is the US billion – called in Europe milliard – the traditional British billion (and that in most other languages) being a million million, 10^{12}.

bi-metal made of two metals. A **bi-metal strip** consists of two strips the same size fixed side by side – the device bends with temperature change as a demonstration that metals vary in their expansivity. It is used for switching in thermostats, while some old clocks have a **bi-metal pendulum** – in this case, there are two rods of different metals mounted side by side in a U-shape; their relative lengths cancel out expansion effects so the clock keeps better time. Much the same applies to the balance wheel of a small analog clock or watch.

bimorph the piezoelectric cell at the heart of vinyl record player pickups, 'crystal' microphones and speakers, and pressure-activated gas cooker lighters. The cell consists of two piezoelectric crystals fixed next to each other. A voltage across the cell causes one to expand and the other to contract so the cell bends by a very precise amount that depends on the voltage. Conversely, bending the cell causes a voltage to appear across the crystal.

binary consisting of two roughly equal states; for instance, **binary fission** is the fission of an active nucleus into two roughly equal nuclei.

binary notation numbering system with base 2, and only two digits (bits), 0 and 1. As almost all digital electronic systems involve binary pulses of current, binary values of pd, and two-state methods of storage (for instance, magnetic storage), binary notation is of great importance.

binary optics See *diffractive optics*.

binary star or **binary system** a common two-star system in which each star is in orbit around the other. From the study of such systems, astronomers gain much information about the two stars concerned.

binding energy W_B **a)** for a nucleus, the energy associated with nuclear mass defect . Binding energy becomes available when nucleons combine to form a nucleus. On the other hand, to cause nuclear breakdown (for instance, fission), the nucleus must obtain from somewhere the corresponding quantity of binding energy.

b) for a nucleus, the **binding energy per nucleon**, W_B/A (where A is the nucleon number). A graph of this ratio against A rises quickly to a peak at around $A = 55$ (in the region of iron and nickel, the most stable nuclei); it then falls off slowly. The curve is not smooth, however: there tend to be peaks for species with even numbers of both protons and neutrons (taken as evidence for the alpha particle model of nuclei). The shape of the graph also shows that there can be a net energy release if light nuclei fuse to

produce more massive ones (the process of nuclear fusion), and if massive nuclei split (suffer fission) to lighter products. Fusion outputs more energy per nucleon than fission as that part of the curve is steeper.

c) for an orbital electron, the energy needed to remove it from its site in an atom or molecule. See also *ionisation energy*.

Binnig, Gerd (1947–) physicist who worked with Rohrer on the first scanning tunnelling electron microscope (1980), for which they shared half the 1986 Nobel Prize

binoculars optical instrument that consists of a pair of terrestrial telescopes (monoculars) mounted side by side, for use with both eyes. In the most common design, a pair of prisms (e.g. Abbe–Porro prisms) in each half folds the path to reduce the distance between the object and eye lenses; the prisms also invert the upside down image, making it the right way up. The design also allows the object lenses to be further apart than the eye lenses (and therefore the user's eyes): this improves the stereoscopic effect of binocular vision, i.e. the ability to obtain a three-dimensional view by combining the images from each eye (compare with *parallax*).

biological shield the thick concrete (and/or perhaps steel) box around the core of a nuclear reactor. The shield should absorb all radiation from the core so that people working outside receive no dose from it at all.

bioluminescence the production of visible light by living organisms such as fireflies and many marine creatures, caused by a chemical change in the tissues

biomagnetism the study of the tiny magnetic signals (rarely more than a millionth of the Earth's field strength) that result from electric currents in living cells. Since the development of the highly sensitive superconducting quantum interference device (squid) in 1964, there have been various important developments of biomagnetism in medical physics; examples include monitoring the activity of the heart and other muscles and the study of neural activity in the brain (magnetoencephalography).

bionics the design and use of systems based on those found in living organisms, the applied science of integrating electronic and mechanical systems into living organs to overcome problems of disease or disability or to enhance normal functions. Bionic systems

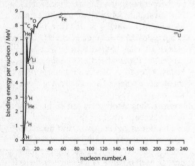

The binding energy per nucleon curve peaks at iron-56, so this is the most stable nuclide of all.

being developed in 2006 range from a bionic eye made of a video camera and a light-sensing chip, a mimic insect eye with nearly 9000 lenses in a dome the size of a pin-head, and 'exo-skeleton' legs whose wearer can carry heavy loads for a great distance.

biophysics the application of aspects of physics to living systems. Examples are the mechanics of movement and the electronics of nervous networks. Compare bionics and medical physics.

Biot–Fourier equation equation for thermal conduction, i.e. thermal energy transfer:

the power, rate of energy transfer, P is given by

$$P = (\lambda / c \, \rho) \, dT/dl$$

Here λ is the thermal conductivity of the medium, c its specific thermal capacity and ρ its density, while dT/dl is the temperature gradient.

Biot, Jean Baptiste (1774–1862) astronomer and physicist who was the first to recognise the extraterrestrial nature of meteorites. With Gay-Lussac, he went up in balloons to perform physics experiments at altitude (in his case on magnetism). He carried out major early investigations into polarisation, and invented the polariscope.

Biot–Savart law law which states, in particular, that the magfield near a long straight current-carrying wire varies with the current and is inversely proportional to the square of the distance to the wire. The following equation gives the field element dB due to a current element IdI a perpendicular distance d away in air. Integrating will give the overall field strength at the point due to the current in the whole wire.

$$dB = (\mu_0/4\pi) \, IdI/d^2$$

Here dB is the output magfield element, IdI is the current element, and d is the perpendicular distance between them in air.

Biot's law The angle of rotation of a plane polarised ray passing through an active medium is proportional to the path length and to the concentration (if the active substance is in solution), and inversely proportional to the square of the wavelength.

bi-polar having two poles or other areas of opposite properties. A **bi-polar electrode** sits in an electrolytic cell between anode and cathode, so that, by induction, one face

becomes a secondary anode and the other a secondary cathode. **Bi-polar integrated circuits** work with bi-polar transistor logic (field effect transistor logic is more common now). See *transistor*.

bi-polar transistor type of transistor with two electrodes, the emitter and the collector

bi-prism unnecessary name for a triangular prism with one extremely obtuse angle (i.e. one very close to 180°). See *Fresnel bi-prism.*

birefringence the appearance of two images of an object viewed through one of a small class of crystals (the **birefringent crystals**). The incident light splits into 'ordinary' and 'extraordinary' rays – these have different speeds and follow different paths (unless they enter parallel to an optic axis). The 'ordinary' ray follows Snell's law; the other does not. We can say the substance has an 'ordinary' refractive constant and an 'extraordinary' refractive constant. See *polarisation.*

bismuth Bi metal of group V, period 6, with proton number 83, first isolated by Claude Geoffroy the Younger in 1753 (though described as a metal called *Wismut* several times in the Middle Ages, including by the engineer Agrocola in 1546, these forms may have been compounds; the name may come from old German *weisse Masse* = white stuff). Relative density 9.7, melting and boiling temperatures 270 °C and 1560 °C; only natural isotope ^{209}Bi, an alpha emitter with half-life 10^{19} years.

bispherical lens same as *aspheric lens*

bi-stable having two stable states, between which the system concerned can flip. For **bi-stable circuit** or

bi-stable multivibrator , see *flip-flop*, the common name.

bit b binary digit, i.e. 0 or 1

Bitter pattern pattern on the polished surface of a ferromagnetic substance dusted with very fine magnetic powder. The powder gathers at the domain boundaries (and thus is good evidence for domains). Similar techniques help detect cracks and faults.

black body or **cavity** object whose surface absorbs all incident radiation: ie, it has 100% absorptance. A black body also has 100% emissivity – the radiation given off depends only on the temperature (Kirchhoff's law). No ideal black body exists, but a cavity – a small hole in the wall of a hollow container kept at a uniform temperature inside – is a good approximation. Gustav Kirchhoff first used the name, in 1862, as such a surface appears jet black below about 650 K. See also *Planck's formula.*

black body radiation or **cavity radiation** the radiation emitted by a black body, which has a completely smooth spectrum. This is because the radiation depends only on the temperature. The family of graphs of energy/wavelength versus temperature in the figure shows that as the temperature rises, so does the radiant power, but the peak wavelength falls towards the visible region. Classical physics – thermodynamics and electro-magnetism – cannot explain these curves, though the Wien displacement law combined with the observation that the peak value of W_λ varies with T^5 does better. Instead nineteenth-century physics predicts the heavy dashed curve at the right of the family: whatever the surface temperature, the surface gives off more and more radiation at shorter and shorter wavelengths. In other words, any object at any temperature would appear white hot. This is the 'ultraviolet catastrophe', solved only in 1900 by Max Planck's quantisation of the vibration of the atoms (the radiation sources) – the start of quantum physics.

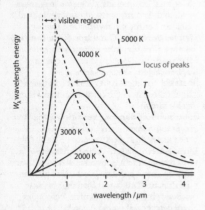

Black-body radiation curves avoid the ultraviolet catastrophe.

black body temperature the temperature of a surface as measured by a radiation pyrometer; in most cases, this is significantly less than the true value. The cosmic background radiation of the Universe

has a black body temperature of just under 3 K.

Blackett, Patrick, Lord (1897–1974) physicist who won the Nobel Prize in 1948 for various aspects of his work in nuclear physics (such as cosmic radiation). He also confirmed the existence of the positron and obtained the first cloud chamber photographs of the decay of a gamma photon into an electron–positron pair.

black hole object whose gravitational field is so high that neither particles nor electromagnetic radiation can escape its 'event horizon' – except perhaps by tunnelling (Hawking radiation) or maybe through a wormhole. The relativistic curvature of space–time leads to self-closure. The mass of a black hole can vary from that of a few suns (a stellar black hole) to that of hundreds of millions of suns (a galactic black hole, such as may exist at the core of a quasar). Some theories predict black holes on an atomic scale. Such **mini black holes** (smaller than an atom but with masses of thousands of tonnes) could explain all manner of puzzles, from geological mascons to the tests tending to support the so-called 'fifth force' of nature.

Geologist John Michell first proposed the concept and worked out what it would mean, in 1783, while in 1915 Einstein proved black holes could exist, and in 1967 Stephen Hawking proved they must exist. John Wheeler first used the name in 1967. While we still have no direct proof that any type of black hole exists, the x-ray spectra of certain binaries and galactic nuclei supports the belief that they do. Black holes found this way come in three mass ranges – stellar (about ten times the solar mass), intermediate (about a thousand times), and supermassive (10^5–10^{10} times, as thought to be at the cores of all galaxies). The x-rays produced from around an astronomical black hole suggest that matter does not fall straight into it, but passes first through a surrounding 'accretion disc' (rather like matter falling into a whirlpool). The size of the disc depends on the mass of the black hole. Note that the main force pulling matter into a black hole seems not to be gravity but a magnetic process.

The physics of merging black holes – being so extreme – is one of the frontiers of our subject; study of the gravitational waves produced may be a way to explore it. In 2006, there were high hopes about detecting these waves with enough detail to be able to recognise the signatures of different types of event (already modelled with the use of supercomputers).

Black, Joseph (1728–1799) chemist who carried out much important early work in thermodynamics

blanket layer of fissile material around the core of a breeder nuclear reactor. Its main purpose is to breed a new generation of fuel, but it also has design advantages, for instance in shielding.

blazed grating diffraction grating ruled to encourage the output to concentrate in only one or two directions (orders of spectrum)

blind spot the part of the retina without photo-cells (rods or cones), where the optic nerve and blood vessels enter

blink comparator or **blink microscope** device used, mainly in astronomy, to compare two almost identical pictures. The system jumps to and fro at a high rate between the two images, so that a flicker is produced at points where they differ – e.g. to show the motion of a single minor planet.

Bloch, Felix (1905–1983) professor of physics at Stanford University, USA, from 1934 to 1971. His work in nuclear magnetic resonance led to his sharing of the 1952 Nobel Prize with Edward Purcell. He discovered **Bloch bands** which are the close energy levels produced when a non-degenerate gas becomes solid. **Bloch's functions** also play a role in the band theory of solids: they are solutions to the Schrödinger wave equation for an electron which moves in a periodic potential field (one that cycles with time).

Bloch wall region which separates two adjacent magnetic domains; perhaps a hundred atoms thick, it is where the particle spin directions change gently from that of one domain to that of the next.

block and tackle pulley system with one or more fixed large pulleys and one or more moving large pulleys (the blocks), plus two sets of small pulleys (the tackle) between them. See also *pulley*.

blocking oscillator oscillator which produces single cycles separated by periods of inaction whose length depends on the circuit constants. A form of *squegging oscillator*, it has uses for time-base and pulse generation.

Blömbergen, Nicolaas (1920–) physicist whose 1981 Nobel Prize followed his research

into the non-linear optics of laser spectroscopy

Blondel–Rey law the relation between the apparent intensity I and the real intensity I_0 of a slowly flashing source (frequency below about 5 Hz):

$$I = I_0 t/(t + k)$$

Here t is the actual flash duration and k is a constant (about 0.2 s).

blooming process that aims to reduce to nothing the energy loss by reflection of light at a transparent surface. (Such losses can reach 8%.) The surface (of a medium with refractive constant n) is coated with a layer of a substance, such as magnesium fluoride, with constant n_b (where n_b is less than n). The thickness of the layer is $\lambda/4n_b$, where λ is the wavelength at the centre of the range whose reflection is unwanted. In the case of visible light, this wavelength is in the green region; thus bloomed lenses (for instance, those of cameras) appear purple because of the deficiency of green. Blooming cuts down reflection by a process of destructive interference. In some cases, there are several such layers, to reduce reflection in the red and blue regions too.

blue shift Doppler shift of electromagnetic radiation to a higher frequency (i.e. to the blue end of the visible spectrum), when viewed from a point towards which the source is moving. This phenomenon is relevant mainly in astronomy. Compare with *red shift*.

blue sky the effect of scattering of sunlight as it passes through the atmosphere – the air scatters blue light more than red. As a result, the sun appears reddish (especially if near the horizon where the air path is longer) and the rest of the sky appears blue. A fairly crude survey in 2006, using a portable spectrometer, claimed to show that the sky of Rio de Janeiro in Brazil is the bluest in the world. For a full report, perhaps showing the survey wasn't *too* crude, try http://www.npl.co.uk/blueskies

Blumlein, Alan (1902–1942) electronic engineer and inventor, with 128 patents. He invented stereo sound recording, achieved the first fully electronic television system, and helped develop Britain's radar techniques.

B mesons group of three mesons (quark/anti-quark pairs) and their three anti-mesons –

each consists of a b-anti-quark (or a b-quark) and either an u- or a d- quark (or an u- or a d- anti-quark). B mesons are massive – over five times the proton mass – because of the great mass of the b-quarks they contain. B mesons decay by the decay of their b-quarks (and, perhaps, also that of the u- or d- quarks) through the weak interaction. A beam can also oscillate (or mix) by turning into their antiparticles and back again through a quantum-mechanical process.

Board of Trade Unit (BoTU or BOTU) unit of energy once used with UK house power supplies, equivalent to the kilowatt-hour, 3.6×10^6 J

body force force, such as gravity, which acts throughout an object (once called 'body') rather than only at the surface

Bohr, Aage (1922–) son of Niels Bohr, nuclear physicist who shared the 1975 Nobel Prize with Rainwater and Mettleson for work on the liquid drop model of the nucleus

Bohr–Breit–Wigner theory theory that nuclear reactions consist of two stages: in the first, the nucleus captures the inward particle to produce a new, excited nucleus; in the second, this breaks down to give the reaction products and energy.

bohrium Bh transitional radioactive earth metal – actinide series, period 7 – with proton number 107, created by a Russian team in 1976 and confirmed by a German team in 1981; the name comes from Niels Bohr. Highly radioactive – longest lived isotope being ^{262}Bh, half-life 17 s – so no properties known.

Bohr magneton μ_B fundamental constant, the magnetic moment due to the current associated with an electron, whose spin is $h/4\pi$.

$$\mu_B = Q_e\, h/4\pi\, m_e$$

Here Q_e and m_e are the electron charge and mass, and h is the Planck constant.

It is common to relate the magnetic moments of other elementary particles and nuclei, and those of electrons in orbit, to this 'unit'; its value is nearly 10×10^{-24} J T^{-1}. (There is, however, a much smaller 'nuclear magneton' for the proton, given as above but using the proton mass and charge values.)

Bohr, Niels (1885–1962) theoretical physicist who based his work on quantum concepts to devise an effective model of the atom. The **Bohr atom** is his 1913 model of atomic structure, with electrons in orbit (the **Bohr orbit**) around a positive nucleus; he developed this theory (known as the **Bohr theory**) in Manchester from the Rutherford model, by quantising orbital electron energies. The work followed study of spectra and earned him the 1922 Nobel Prize.

Bohr orbit See *Bohr, Niels.*

Bohr radius a_0 the radius of the ground state orbit in the hydrogen atom, a little over 50×10^{-12} m

Bohr theory See *Bohr, Niels.*

boiling the change of state (phase) to vapour throughout the body of a liquid; unlike evaporation, a surface phenomenon, this takes place at a fixed **boiling temperature** (or boiling point) which depends only on the substance, its purity and the pressure. Boiling occurs when the vapour pressure of the substance equals the pressure outside. As well as being of great value in analysis, boiling temperatures were long a major source of fixed temperatures for thermometric scales, though triple points are now mainly used.

boiling water reactor (bwr, or BWR) type of thermal fission reactor in which water acts as both moderator and coolant; the water boils when in contact with the fuel rods in the core. See *nuclear power.*

bolometer highly sensitive detector of radiant thermal energy, mainly in the microwave and infrared regions. Bolometers depend on the change of resistance with temperature of a thermistor or a platinum sample. A **linear bolometer** consists of a single thermistor or a very narrow strip of platinum; its use is to determine the input powers at different wavelengths of a black body distribution. In the **surface bolometer**, the sensor is in one arm of a Wheatstone bridge circuit, with an identical one in a second arm acting as a control. Such a system is used to measure total radiation input, and can detect temperature changes much smaller than 10^{-6} K. For **bolometric**, see *magnitude.*

Boltzmann constant k ratio of the universal gas constant to the Avogadro constant, in value just under 1.4×10^{-23} J K^{-1}

Boltzmann distribution See *Boltzmann's formula.*

Boltzmann entropy hypothesis concept which relates a system's entropy S in a given state to the probability p of its being in that state:

$$S = k \ln p + \text{constant}$$

Boltzmann, Ludwig (1844–1906) physicist whose main contributions to knowledge were in various areas of thermal physics

Boltzmann's formula expression that relates the number n of particles in a system of each energy value W at temperature T to the number n_0 with the lowest energy:

$$n = n_0 \, e^{-W/kT}$$

This relates to the equipartition of energy, in the development of which law Boltzmann played a major role, and to the energies of the different levels in, for instance, an atom. The distribution of energies in the system as given by the formula is the **Boltzmann distribution**, while **Boltzmann's law** is the law of equipartition of energy.

Boltzmann's law the law of equipartition of energy

Boltzmann statistics approach to the description and analysis of large groups of particles, which includes much of Boltzmann's theory (see *Boltzmann constant, Boltzmann entropy hypothesis, Boltzmann's formula*) and led to great advances in thermodynamics. The theory is classical but relates to the form of quantum statistics where quantum effects are very small.

bomb device containing an explosive (chemical fuel with very rapid reaction or combustion rates) which is designed to explode, i.e. to release a large quantity of energy in a small volume in a short time. See also *nuclear weapons*.

bomb calorimeter thick-walled, totally enclosed container in which a small sample of a fuel is burned quickly, in order to find the fuel's specific thermal capacity

bond link between atoms in a molecule, **bond energy** being the energy needed to break the link. Various techniques, including x-ray diffraction and spectroscopy, can explore the details (energies, lengths, angles) of bonding in particular cases. The main types of bond (here described for a diatomic molecule) are:
covalent – one electron from each atom orbits the pair of cores;
electrovalent, or ionic – bonding between oppositely charged ions formed by the transfer of an electron from one atom to the other; and
coordinate – like covalent, but with both shared electrons coming from a single atom.

Bondi, Hermann (1919–2005) mathematician and astrophysicist who, in 1948, devised (with Fred Hoyle and Thomas Gold) the steady state theory of the Universe. This was more accepted than the big bang until the discovery of the cosmic background radiation.

bonding a) formation of a *bond* between atoms to form a molecule

b) adhesion, a **bonding pad** being a strip along the edge of an integrated circuit (ic) on to which fine gold wires are fixed (bonded) to link the ic to the pins of the microchip

boojum a) transient surface pattern on superfluid helium-3 (^{3}He), perhaps concerned with the decay of a supercurrent. The name comes from Lewis Carroll's *Hunting of the Snark.*

b) similar feature sometimes observed with liquid crystals at room temperature

booster a) amplifier, especially one working with radio frequency signals

b) first stage of a multi-stage rocket (i.e. the launch vehicle)

c) radio signal relay station which amplifies the input before sending it on

d) generator in a circuit designed to output a higher voltage than the input (**positive booster**) or, much less often, a lower one (**negative booster**)

bootstrap independent. A **bootstrap circuit** or **bootstrap device** needs no input to control it. A **bootstrap theory** is one that requires that a wider theory should be consistent with it.

borated of H_2O used for safety in some designs of nuclear reactor, with a high level of dissolved borium ions (such as from the salt borax). Borated water is a very good neutron absorber.

Born, Max (1882–1970) physicist who made many advances in quantum physics and other areas of modern theory. He received the Nobel Prize in 1954.

boron B metalloid of group 3, period 2, number 5, with two allotropes, isolated by Louis Gay-Lussac and by Humphry Davy in 1808 and named after its common compound borax. Relative density 2.35, melting and boiling temperatures 2080 °C and 3927 °C; most common isotope ^{11}B.

Bose–Einstein distribution distribution of bosons between energy states when modelled by **Bose–Einstein statistics** (see *statistics*). The Bose–Einstein distribution law gives the number of bosons n in a state of energy W as

$$n = 1/(e^{W/kT} - 1)$$

Here k is the Boltzmann constant and T is the temperature.

Bose developed the statistical theory for photons, and Einstein developed it for particles with mass. The latter saw that close to absolute zero a sample of bosons would become a superfluid (like ^{4}He below 2.17 K) – a **Bose–Einstein condensate** (see also *Bose gas*). This has quantised energy and such properties, not yet well explained, as flowing against gravity out of a container and slowing light to much less than a millionth of c_0. Such an exotic form of matter appears when the wavelengths of nearby atoms overlap: the whole set of atoms then acts like a single atom. Common Bose–Einstein condensates come from working with ^{87}Rb vapour and may extend to 150 or 200 atoms in a line; however, two- and three-dimensional equivalents have been made. Common or not, Bose–Einstein condensates have a physics that is still far from well understood.

Bose gas An ideal Bose gas is a boson system – one that consists of bosons – and is the quantum mechanical equivalent of an ideal gas in the classical sense; the sample follows a Bose–Einstein distribution. In 2006, a one-dimensional Bose gas (a *Bose–Einstein condensate* of around 150 rubidium atoms close to o K constrained by laser beams) was found never to come to thermal equilibrium because of its quantisation. Instead, the atoms behave rather like the balls of a Newton's cradle – with no change after even many thousand impacts, even though quantum effects means some of the atoms tunnel through others. This is the first non-chaotic system found – see http://plus.maths.org/latestnews/may-aug06/newtonscradle/index.html for more details.

Bose glass a 'fifth state of matter' observed in 2007 in a group of rubidium atoms held in a crystal lattice very close to o K – the group did not behave like the usual Bose–Einstein condensate or as the also usual Mott insulator.

Bose, Jagdish (1858–1937) physicist whose early work, developed by Einstein, led to the **Bose–Einstein statistics** system (see *statistics*) that describes the behaviour of bosons on a quantum physics basis

boson elementary particle, such as a meson or a photon, with unit spin (compare fermion); in bulk bosons have a Bose–Einstein distribution

boson system system of particles, such as the carbon nucleus, that obey Bose–Einstein statistics though they contain no elementary bosons

Bothe, Walter (1891–1957) physicist whose work in the 1920s on detectors and cosmic rays led to his being awarded the 1954 Nobel Prize

BoTU or **BOTU** (British) Board of Trade Unit, old name for the kilowatt-hour, kWh, 3.6 MJ

Bouguer's law See *Lambert's law* b).

bound bonded, in other words fixed, constrained or contained in such a way that only input energy can effect a release. See also *binding energy* and *bond*.

boundary layer the layer of a moving low-friction fluid next to the walls of a container, like a pipe, in which friction effects are not negligible. The thickness of the layer varies with the square root of the kinematic viscosity – see *fluid friction*.

Bourdon tube curved or coiled tube with an oval section that tends to straighten as the pressure inside rises (for instance, with temperature) compared with that outside. The system is much like the design of a coiled party squeaker. A common use is in a thermograph, the tube being filled with a liquid of high expansivity. The **Bourdon gauge** is a type of aneroid pressure gauge that works in the same kind of way. **Eugène Bourdon** (1808–1884) was an instrument- and model-maker.

Boyle, Robert (1627–1691) one of the group of scientists and thinkers who was one of the prime movers of the seventeenth century 'golden age' of British science, being, for instance, one of the founders of the Royal Society in 1660. He is most well known in physics for his work on gas pressure around the same time, with Robert Hooke as his assistant. This work led to *Boyle's law*.

Boyle's law or **Boyle–Mariotte law** law (Boyle, or maybe his assistant, Robert Hooke, 1662; Edme Mariotte 1676) that the product of the pressure and volume of a constant mass of (ideal) gas at constant temperature is constant, i.e. $pV = $ constant.

In the case of a real gas, the product is pressure-dependent, being of the form

$$pV = RT (A + Bp + Cp^2 + \ldots$$

Here A, B, C ... are the virial constants, A being the most significant and in this case taking the value 1.

At low temperatures, B has a large negative value; as the gas temperature rises, the value falls, passes through zero, and becomes positive. The **Boyle temperature** of a given gas is that at which B is zero, making the Boyle relation, $pV = RT$, then strictly correct. See also *Andrews' experiment*.

Boys, Charles Vernon (1855–1944) physicist and prolific inventor, knighted in 1935. **Boys' experiment** to measure the gravitational constant G replaced that of Cavendish because of its higher accuracy. It involves measuring the oscillation of a short beam hanging on a quartz fibre when the gold spheres at each end are near different large masses. **Boys' radiometer**, still sometimes used, is a very effective sensor of radiant energy. A small blackened platinum disc fixed to the junction of a thermocouple absorbs the radiation.

Bq the becquerel, the unit of *activity* b), the number of decays in a sample per second. Therefore, $1 \text{ Bq} = 1 \text{ s}^{-1}$.

Brackett series a series of lines in the *hydrogen spectrum*

Bradley, James (1693–1762) astronomer who followed Halley as Astronomer Royal and discovered *aberration* a) while trying to measure the speed of light in empty space

Bragg, William (1862–1942) physicist who shared the 1915 Nobel Prize with his son Lawrence (1890–1971) for their work on x-ray crystallography. (So far they're the only parent and child to share a Nobel Prize.) In the diffraction of x-rays of wavelength λ by a crystal, the **Bragg angle** is the angle of incidence that leads to constructive interference – given by $90 - \sin^{-1}(2d/n\lambda)$. This expression follows from **Bragg's law**, which states that there is constructive interference when n is an integer; d is the distance between the atomic planes in the crystal.

braking radiation English for the German term 'Bremsstrahlung', which we tend to use

branch a) the path between two **branch points** in a circuit or network. A branch point is any junction of more than two conductors (i.e. a point where the circuit or network splits)
b) one of the two or more possible decay modes of an active nuclide. Then, **branch fraction** or **branch ratio** is the fraction of the nuclide's number of decays by a given branch to the total number of decays by all branches.

branching the existence of more than one decay process for a nuclide

brane or **p-brane** object that appears in string, superstring, and M-theories, where p is its number of dimensions. A 0-brane is a particle with no dimensions, a 1-brane is a string, a 2-brane is a membrane, etc. There are many other types.

brasses alloys, mainly of copper and zinc, with high golden sheen (and therefore widely used for decoration), and fairly low melting temperature and strength. Brasses were once widely used for scientific equipment and in machines.

Brattain, Walter (1902–1987) physicist who shared the 1956 Nobel Prize with Bardeen and Shockley for the development of the transistor

Braun, Karl (1850–1918) physicist who shared the 1909 Nobel Prize with Marconi. This followed his discovery of the rectifying properties of certain crystal junctions, and the idea that this feature could be of value in radio reception.

Braun, Wernher von (1912–1977) rocket engineer who designed the V2 system used near the end of the Second World War. In 1945 he went to the US to lead the country's artificial satellite programme.

Bravais lattice repetitive pattern of points such that the environment of each point is the same. Physicist **Auguste Bravais** (1811–1863) was the first to show that there can be only fourteen unique basic patterns like this in three dimensions, and that these relate to the seven crystal systems.

breakdown a) sudden, and disruptive, electric discharge through an insulator (for instance, a spark in air)
b) still sudden (but less disruptive) decrease of resistance, when the applied pd rises to a certain level, called the **breakdown potential**. See also *Zener breakdown*.

breeder reactor or **converter reactor** fission reactor with a conversion factor greater than 1, i.e. one that produces more fissile material than it consumes. There are various types, all much more cost-effective in use than conventional thermal reactors. A common scheme converts non-fissile uranium-238 (^{238}U) to fissile plutonium-239 (^{239}Pu). See also *fast breeder reactor*.

breeding any process of nuclear transformation with a conversion factor greater than 1

breeding ratio (or **conversion factor**) the ratio

of the number of fissile nuclei produced in a nuclear reaction to the number consumed.

Breeding gain is breeding ratio – 1.

Breit–Wigner formula expression for the absorption cross-section of a given nuclear reaction used mainly in cases where the intermediate excited nucleus has more than one possible mode of decay

Bremsstrahlung 'braking radiation' (German) – the electromagnetic radiation produced when an electron decelerates at a high rate near a nucleus. It is the major cause of energy loss from high energy electrons passing through matter, with each close encounter leading to the output of a single photon. 'Bremsstrahlung' is the source of the continuous background of an x-ray spectrum, and is an important part of cosmic radiation at the Earth's surface.

Brewster angle i_8 or angle of polarisation, the angle of incidence of unpolarised light on a reflecting surface at which the reflected light is fully polarised. **Brewster's law** states that at this value the incident and refracted rays are at 90°. The **Brewster windows** of some gas lasers are windows set at the Brewster angle to minimise reflection losses at the output.

Brewster, David (1781–1868) physicist responsible for many advances in optics, and a founder of the British Association for the Advancement of Science. He invented the kaleidoscope in 1816, at about the same time as his major work on polarisation.

bridge electric circuit section whose elements form the sides of a square; the input comes to two opposite corners, with the output taken from the other two. Most **bridge circuits** have a power supply at the input and a meter at the output. Adjustment of the resistance or reactance of the element(s) in one or two arms of the bridge gives zero output, at which point the user can work out an unknown resistance or reactance. See, for instance, *Wheatstone bridge*.

bridge rectifier bridge circuit with a diode in each arm and an alternating input; the output is full-wave rectified (see top right)

Bridgman effect the absorption or release of thermal energy that results from an electric current in an anisotropic crystal. The cause is the non-uniform current density.

Bridgman, Percy (1882–1961) physicist whose 1946 Nobel Prize followed his research into the effects on matter of very high pressures; this led to the production of artificial diamonds in 1955.

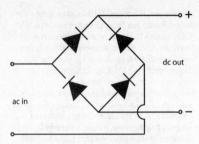

This bridge circuit is a full-wave rectifier.

brightness everyday term for the power of light, i.e. the luminous intensity of a source, the energy content of the light produced, or the illumination of a surface. The term brightness is, however, subjective, while the other terms are objective.

Brillouin functions series of complex expressions for the magnetisation of a paramagnetic substance at a given thermodynamic temperature, B_j, where j is the angular momentum quantum number

Brillouin, Leon (1899–1948) physicist whose main work lay in quantum and wave statistics

Brillouin zone energy band for electrons in a crystal when plotted in a certain way. It is a region in momentum space that can hold two electrons.

Brinell test test for the hardness of a surface that involves pressing a standard steel ball into the surface with a set force. The user can deduce the hardness from the size of the indentation produced.

British thermal unit BTu old unit of energy: the energy required to raise the temperature of one pound (0.454 kg) of water by 1 °F (0.56 °C), therefore equal to 1055.06 J. Please do not confuse this with the Board of Trade unit, BoTU, which is 3.6 MJ.

broadening the increase in width of a sharp spectral line as a result of any of various effects:

a) natural broadening – the result of the uncertainty principle

b) Doppler broadening – the result of the Doppler effect as the line's source particles are moving

c) resonance broadening – when there is energy exchange between the particles during photon emission

d) opacity broadening – when there is significant absorption and re-emission in the source sample

Brockhouse, Bertram (1918–2003) Nobel Prize winner for 1994, sharing this with Clifford Shull, for his work on the development of neutron spectroscopy

Broglie See *de Broglie.*

bromine Br liquid halogen, in group VII and period 4, with proton number 35, found by Antoine Balard in 1826, named from Greek 'bromos', meaning 'stink' (which well describes the vapour). Density 3.1 kg m^{-3}, melting and boiling temperatures –7 °C and 59 °C; only two, almost equal, isotopes: ^{79}Br and ^{81}Br. Bromine has many minor uses in the chemical industries.

Bronson resistor or **Bronson resistance** high resistance circuit element – an ionisation chamber that contains a constant source of ionisation. The applied voltage is very low compared to that giving saturation, so the output current is close to being ohmic.

bronzes alloys of copper, mainly with tin, used for strength and toughness since the Bronze age (which started in Mesopotamia about six thousand years ago). Bronzes are still in wide use for springs, bearings, bushings and similar, while fine bronze wires are still common for suspension.

brown dwarf a star so small (not much more massive than Jupiter, m_J) that nuclear reactions in its core provide very little observable radiation output. Objects larger than around 13 m_J can produce deuterium by fusion. Because it is hard to observe brown dwarfs, no one is sure how common they are. The first two confirmed were both in 1995; since then, people have observed brown dwarf x-ray, flare and radio sources and gained some evidence of brown dwarf stellar system formation and very violent weather patterns.

Brownian motion a) the random motion of microscopic solid specks suspended in a fluid – which provides evidence for the random thermal motion of the fluid's molecules. Working with pollen grains, the botanist **Robert Brown** (1773–1858) first studied the effect in 1827 (though first observed in 1765); it gives a theoretical limit to the sensitivity of instruments such as balances and needles. While the motion is random, and therefore tends to reduce any temperature differences, a nano-motor (one using a single 'chiral' molecule powered by Brownian motion) was designed in 2006 to act as a frig.

b) the mathematics of this motion, since applied to many other such supposedly stochastic ('random walk') systems, like stock market movements and fossil evolution. Einstein first explained the effect and studied the maths, in 1905, as an example of anharmonic motion. For **Brownian noise**, see *red noise.*

brownout or **sag** power cutback that involves supply of reduced power rather than no power (blackout)

Brunel, Isambard Kingdom (1806–1859) engineer who carried out many important works, building bridges, tunnels, railways and steamships, and set up a number of mass production factories. His father Marc Brunel (1769–1849) built the first underwater tunnel (beneath the River Thames, now used by the underground railway).

brush electric contact between a fixed and a moving conductor. The first brushes were bundles of wires looking like besoms. Nowadays spring-loaded blocks of graphite are usually used.

brush discharge discharge which appears at the surface of a conductor when the non-uniform field outside is strong (for instance, during an electric storm) but below that needed for sparking. The luminous discharge consists of many glowing branching threads, and can reduce the chance of a lightning strike. The traditional name is St Elmo's fire (from St Erasmus, patron of sailors: the masts and rigging of sailing ships often show the effect).

BTU or **BTu** British thermal unit, old unit for energy

bubble a) thin film of liquid enclosing a mass of gas, with surface tension holding the excess pressure inside. The excess pressure of a bubble of radius r in a film of surface tension σ is $4\sigma/r$.

b) small drop of air or vapour within a mass of liquid or solid. In the liquid, the bubble tends to collapse unless the pressure inside (for instance, vapour pressure) equals, or is greater than, the pressure outside. See also *boiling* and *cavitation.*

c) small, stable magnetic domain of one polarity within a 'sea' of opposite polarity in the surface of a magnetic material, once seen as the basis for a robust and non-volatile (but costly) magnetic data storage system called **bubble storage**. See also *mram,* the non-volatile magnetoresistive read and write chips first sold in 2006.

bubble chamber radiation detector that encourages the growth of tiny bubbles in a superheated liquid (for instance, liquid hydrogen) along the path of an ionising particle. The charges deposited as the particle loses energy act as nuclei for the initial formation of the bubbles; the bubbles do not collapse at once as the liquid is just above the boiling temperature. Invented by Donald Glaser in 1952, the bubble chamber was a major advance on the cloud chamber (the basis of which is much the same), as the much higher density of working substance greatly raises the likelihood of seeing significant events. Also, liquid hydrogen is 'pure protons'. Modern chambers may contain several cubic metres of liquid.

bubble fusion a recent type of *cold fusion* based on cavitation

bucket-brigade device bbd delay line formed from a series of small capacitors (for instance, in the surface of a microchip) with transistor switches between each pair. As each switch closes, the stored charge transfers to the next capacitor. There is a bucket-brigade process in a number of modern chips (such as the charge coupled device); the output is analog so is of value in analog as well as digital circuits.

buckminster-fullerenes the spherical structures C_{60} and C_{70}, most often called **bucky-balls**. These are a form of *graphite* – some say they are distinct allotropes of carbon – with zero-dimensional crystals. **Bucky-tubes** are much the same, but the crystals have one dimension (length).

buffer method of isolating two systems to prevent or reduce unwanted interaction. A **buffer circuit** (for instance, an emitter follower) has a high input impedance and a low output impedance, which minimises the interaction between two circuits.

bulk matter in sufficient quantity that lets one ignore particle effects, including as compared with, for instance, a thin film. Thus, **bulk lifetime** is the mean time between the generation and re-combination of electron-hole pairs in a bulk semiconductor sample.

bulk modulus K the (volume) elastic constant of a substance: the applied force per unit section area divided by the **bulk strain**, the change of volume per unit volume that results

bumping violent form of boiling that occurs when bubbles form at pressures significantly

above the outside pressure and then grow very quickly

bunching also called beam coupling, the clumping of particles in a beam of charged particles as a result of velocity modulation. The particles accelerate or slow to form bunches.

Bunsen cell a primary electric cell with carbon plates in nitric acid for the cathode and a zinc amalgam rod in dilute sulphuric acid for the anode

Bunsen ice calorimeter calorimeter which allows the user to measure the specific thermal capacity of small solid and liquid samples

Bunsen, Robert (1811–1899) chemist who invented the field of spectroscopy (by this means discovering, with Kirchhoff, caesium and rubidium). His **Bunsen burner** was the first burner to have an air supply the user could control, to provide a wide range of output power; in fact Bunsen had nothing to do with it – Michael Faraday was the inventor and Bunsen's lab assistant brought it to the form we now know, in 1855. The design has since changed little.

buoyancy the upward force (old name 'upthrust') on an object fully or partly immersed in a fluid that makes it float, or at least appear less heavy. The upward force equals the weight of fluid displaced (Archimedes' law); it equals the object's weight for it to float (law of flotation); it acts through the object's **centre of buoyancy**.

bus a) (also called **busbar**) a conductor of high current-carrying capacity compared with other conductors in the system, used, for instance, to carry the power supply **b)** (sometimes called highway) a major path for data transfers in an information technology system, carrying the bits of the data units in parallel

buzzer electromagnetic device that works much like an electric bell (i.e. with a current make-and-break system); its output is a buzzing sound as a result of the vibration of a metal strip or plate.

bwr boiling water reactor

bypass capacitor capacitor used to provide a low impedance path around a circuit for alternating current (ac). It allows the supply of direct current to a given point, or extracts from a signal a wanted ac component.

c unit symbol for centi-, 10^{-2}

C a) chemical symbol for carbon
 b) unit symbol for coulomb, unit of electric charge

c **a)** symbol for molar concentration, unit: mol dm^{-3}
 b) symbol for specific thermal capacity, unit: joule per kilogram per kelvin, J kg^{-1} K^{-1}
 c) symbol for speed (*u* or *v* sometimes used), unit: metre per second, m s^{-1}
 d) symbol for speed of light in empty space (sometimes c_0), unit: metre per second, m s^{-1} (the reason is that *c* was a symbol in an equation by Weber in 1856 – Weber's constant – that later turned out to be $\sqrt{2}$ × the speed of light; by the end of the century many physicists used it as now, with the justification that Latin *celeritas* means (great) speed)

C **a)** symbol for capacitance, unit: farad, F
 b) symbol for molecular concentration, unit: kg dm^{-3}
 c) symbol for thermal capacity, unit: joule per kelvin, J K^{-1}

cable a wire or bundle, often thick, that carries electric current, often for power supply. A **coaxial cable** has a central conductor for a high frequency signal (such as a video or broadband IT signal), surrounded by an earthed conducting tube that acts as a faraday cage to reduce interference. An insulator separates core and tube, the latter being foil, woven mesh or braid.

cadmium Cd toxic transition metal, in period 5, proton number 48, first isolated independently by Friedrich Strohmeyer and Karl Hermann, in 1817, from the zinc ores once called calamine (from the Latin and Greek words for which cadmium derives); relative density 8.7, melting and boiling temperatures 321 °C and 765 °C; most common isotope ^{114}Cd. Used to coat iron, in nickel–cadmium dry cells (below), in photocells, in tv tube phosphors, and as a neutron absorber.

cadmium cell original design for the standard *Weston cell*, with cadmium sulphate solution as electrolyte and output a little below 1.02 V

cadmium sulphide cell photoconductive photo-cell with much higher sensitivity and much wider frequency range than the selenium type. Used in light exposure meters, infrared security sensors and street lamps, for instance, it consists of a sample of cadmium sulphide between two electrodes, one being transparent.

caesium (US: cesium) Cs metal, proton number 55, the first element to be discovered using spectroscopy, in 1860 by Robert Bunsen and Gustav Kirchhoff – its two bright blue lines give it its name (from the Latin *caesius*, = heavenly blue). Group I, period 6; relative density 1.9, melting and boiling temperatures 28 °C and 680 °C; only natural isotope ^{133}Cs. Used in the caesium clock (below), as a tracer (with ^{134}Cs, half-life two years), in photo-cells, and in ion propulsion systems.

caesium clock type of *atomic clock*, used to define the second as the unit of time, and accurate to 1 part in 10^{13}. In a magfield, atoms of caesium-133 have a certain two hyperfine ground state energy levels; excitation between these involves absorbing energy at the radio frequency of exactly 9.192 631 770 GHz. The caesium clock was the first type of atomic clock to work, in the 1950s (though the concept dates back to the 1930s). See http://www.sciencemuseum.org.uk/on-line/atomclocks/page3.asp for history and working details.

calcite natural form of calcium carbonate,

whose transparent variety is Iceland spar. Calcite shows birefringence, and its crystals were once widely used as sources of polarised light.

calcium Ca alkaline earth metal, proton number 20, discovered in 1808 by Humphry Davy, who named it after *calx*, the Latin word for lime: group II, period 4; relative density 1.55, melting and boiling temperatures 840 °C and 1484 °C; most common isotope ^{40}Ca.

calculus branch of mathematics (developed independently by Newton and Leibnitz) concerned with the effect on a function $y = f(x)$ of very small changes in the independent variable x. **Differential calculus** concerns the resulting rate of change of the dependent variable y; it gives the slope dy/dx of the graph of y against x. **Integral calculus** develops a continuous function from the sum of a finite number of discrete values, and gives the area between the curve and the x-axis. Both are essential tools in all branches of physics. The **calculus of variations** is the calculus of definite integrals and their stationary (minimum and maximum) values.

Calder Hall Britain's – and the world's – first nuclear power station, a Magnox type, opened in 1956 and closed in 2003

calibration process that assigns absolute values to the readings of a measuring device

californium Cf transitional rare earth element, period 7, in the trans-uranic actinide series, proton number 98, first made in 1950 by Glenn Seaborg and Albert Ghiorso. Relative density 15.1, melting temperature 900 °C; most stable isotope is the alpha-emitter ^{251}Cf, half-life 898 years. As ^{252}Cf is a very intense neutron emitter (half-life 2.6 years), it is a start-up source for some fission reactors, a moisture gauge for deep wells and a source for portable activation analysis systems (e.g. for airport security and precious metal prospecting).

Callendar and Barnes apparatus a continuous flow calorimeter that measures the specific thermal capacity c of a fluid. The fluid passes at a constant rate m/t through an insulated tube that contains an electrical heater along the axis. Steady state occurs when the input and output temperatures T_1 and T_2 are constant. In this case, the input electric energy $V I t$ is equal to $m c (T_2 - T_1) + W_1$; here W_1 is the energy leakage, found by conducting a second experiment with the

same temperature values while changing the others.

Callendar's air thermometer a constant pressure gas thermometer which compensates for the fact that the gas in the manometer is not at the same temperature as that in the bulb. It is of use up to 450 °C. Physicist Hugh Callendar (1863–1930), also devised an accurate platinum resistance thermometer.

calomel electrode a half-cell using mercury (I) chloride ('calomel') whose potential against a hydrogen electrode is very well known and so can act as a standard

caloric theory widely held theory of the eighteenth century, published by Antoine Lavoisier in 1783 after he disproved the phlogiston theory. It is that 'heat' is a weightless ('subtle') fluid that flows from hotter to cooler objects and is conserved throughout the Universe; the theory competed with the less modern one based on atoms – the kinetic theory. The nineteenth century work of Count Rumford, Heinrich Helmholtz and James Joule led to the theories being joined, with energy the main linking factor, into what we now call thermodynamics.

calorie cal old unit of thermal energy, equivalent to close to 4.2 J, the energy involved in changing 1g of pure water at 15 °C by 1 °C. A **kilocalorie** (kcal, or Cal), 1000 cal, is what many people mean by 'calorie'.

calorific value old term for energy value, often measured in a bomb calorimeter

calorimeter device used to contain a sample of a substance for **calorimetry** (the measurement of specific thermal capacities and other energy values). There are many designs of calorimeter, such as the *bomb calorimeter* and the *Callendar and Barnes apparatus*.

camera optical instrument giving an image of an object or scene from the input light (or other radiation), the simplest being a pin-hole camera, with no lens. A **lens camera** consists of a light-tight box and a lens system designed for minimum aberration and which may be adjustable for object distance; it also commonly contains a shutter, which opens for a variable time to allow light through, and a photosensitive surface to record the image. In a photographic camera, that surface may be a film (or 'emulsion') on which an image may be developed and fixed by chemical means, or it may be a photo-ccd read electronically (as in a digital camera). Both types have 'movie' versions – the movie camera and the video camera. (The orthicon and the vidicon were the two types of – very costly – video camera in use until ccds took over in the early 1990s.)

A **camera lucida** fits on to the eye lens of a microscope to combine the image with one of a sheet of paper for tracing. It is much the same in principle as a **camera obscura**, a system that projects a pin-hole, lens or telescope image on to a large screen in a dark room.

Campbell's bridge ac bridge circuit used to compare a mutual inductance with a capacitance

Canada balsam transparent cement of refractive constant 1.55, used for joining optical components such as lens elements

canal rays the radiation of positive ions produced through holes (channels) in the cathode of a discharge tube. This is a common process for forming beams of positive ions for mass spectroscopy.

candela cd unit of luminous intensity, one of the seven base units of SI (*Appendix* 1). The normal intensity of a black body surface element of area 0.167 mm^2 at 1769 °C is 1 cd. The formal definition, however, is the luminous intensity in a given direction of a monochromatic source of 540 THz with a radiant intensity in that direction of 1/683 watts per steradian W sr^{-1}.

candle power old term for luminous intensity and for its unit (more often called **candle**, 1.02 candela). Britain first set such a measure, in the 1860 Gas Act; this defined the standard candle's composition and rate of burning.

canonical distribution function in statistical mechanics giving the distribution of particles in a **canonical ensemble** (a set of particles that can be described in this way) – the fractions of the particles that have each sub-range of momentum and position. (Maxwell's molecular distribution law is a special case.) The **canonical equations** are somewhat similar expressions of classical mechanics – but they involve Hamiltonians, so lead to quantum physics.

capacitance C unit: farad F **a)** the ability of a system to store charge; the change of charge Q with unit change of applied potential V: $C = Q/V$
b) the (negative) imaginary part of acoustic impedance, exactly analogous to the electronic case

capacitive coupling coupling – i.e. linking sub-circuits to allow power transfer between them – through a common capacitor

capacitive reactance X_C unit: ohm, Ω the reactance of a pure capacitor of capacitance C at frequency v:

$$X_C = 1/(2 \pi v C)$$

Strictly this is imaginary – there is a factor of i ($\sqrt{-1}$) in the denominator, as appears in an Argand diagram.

capacitive tuning the use of a variable capacitor in a system's tuning circuit

capacitor (once condenser) system designed to have a given capacitance, for instance for the storage of charge. A **parallel plate capacitor** consists of a pair of conducting surfaces ('plates', though these may be rolled-up foil) with an insulator ('dielectric') between, whose capacitance depends on the plates' area of overlap A, their separation d, and the relative permittivity (η) of the insulator:

$$C = \eta \ A/d$$

However all conductors have some capacitance, and this may be significant in certain cases. For instance, the capacitance of a lone sphere of radius r in a medium of relative permittivity η is $C = 4 \pi \eta r$.

In an ac circuit, a capacitor allows a current, but resists it with a capacitive reactance Xc

$$X_C = 1/(2 \pi i v C)$$

Here C is the capacitance of the capacitor, v is the frequency of the circuit supply, and i is the square root of –1.

In a dc circuit, the capacitor offers an infinite reactance (as the frequency is 0). It charges up to the supply voltage, so there is a temporary current.

capacitor microphone microphone which uses a flexible plate for one electrode of a capacitor, the other being fixed. Input sound waves vibrate the plate and change the capacitance; the output signal is weak but noiseless and has a good frequency response.

capacity a) old name for capacitance
b) the maximum energy of a primary or secondary cell available in set conditions (often given, as charge, in ampere-hours)
c) the volume of a container available for holding a fluid, unit: cubic metre, m^3

capillarity the range of visible effects of surface tension, of which the most notable is

capillary action. This is the rise of a liquid (or fall in some cases, such as mercury in glass) inside a narrow tube (**capillary tube**) or wedge.

capillary constant a measure of the surface tension of a liquid, the product of the height it rises in a tube and the radius of the tube. The value is $2\,\gamma/g\,d$; γ is the liquid's surface tension and d the difference between its liquid and vapour densities, while g is the acceleration of free fall.

capillary electrometer an electrolytic cell, one electrode being the surface of a sample of mercury and the other the end of a capillary thread of mercury. The end of the thread moves in response to an applied voltage up to around 1 V – a very sensitive and yet very accurate system. Capillary electrometers were common in the past and still have uses, e.g. in medicine.

capillary waves liquid surface waves of such small length that gravity has no effect on their speed. The speed then depends only on the capillary constant and density of the liquid, and on the wavelength.

capture the absorption of one particle by a second; for instance, of a nucleon by a nucleus, or of an electron by a positive ion to form an atom. In most cases, capture results in excess energy (excitation), therefore being followed by the release of a photon; compare with *binding energy*. In **electron capture** (a form of beta-decay – see *nuclear physics*), **capture radiation** is the x-radiation produced from the excited nucleus.

carbon C non-metal with a number of allotropes (diamond, graphite and the fullerenes), proton number 6. Group IV, period 2; relative density 2.2 (graphite) or 3.5 (diamond), sublimes at about 3500 °C; most common isotope ^{12}C. The name comes from the Latin 'carbo' = charcoal. Apart from being the basis of all life on Earth ('organic'), carbon has many special features –

a) one allotrope is one of the softest solids known, while a second is one of the hardest;

b) it stays solid to a higher temperature than any other element but has no liquid form (at standard pressure);

c) its atom is very small so can form multiple bonds and has great affinity for other small atoms (so carbon forms over ten million compounds, much more than all the other elements together);

d) as well as having a number of allotropes,

carbon has about a dozen distinct forms, from soot to carbon nano-foam.

carbon black fine soot used to coat a surface in order to maximise its radiation-absorbing power

carbon cycle or **carbon-nitrogen cycle** sequence of six nuclear reactions widely believed to be the major source of energy in many stars. The net outcome is:

$$4p^+ \rightarrow \alpha^{++} + 2e^+ + 2\nu + W$$

Here W is the binding energy involved when the four protons combine into the alpha particle. That energy appears as gamma photons (γ) and in the escape of the positrons e^+ and neutrinos ν. The cycle involves carbon-12 as a catalyst. Here are the six steps:

$$
\begin{aligned}
^{12}\text{C} + {}^1\text{p} &\rightarrow {}^{13}\text{N} + \gamma \\
^{13}\text{N} &\rightarrow {}^{13}\text{C} + e^+ + \nu \\
^{13}\text{C} + {}^1\text{p} &\rightarrow {}^{14}\text{N} + \gamma \\
^{14}\text{N} + {}^1\text{p} &\rightarrow {}^{15}\text{O} + \gamma \\
^{15}\text{O} &\rightarrow {}^{15}\text{N} + e^+ + \nu \\
^{15}\text{N} + {}^1\text{p} &\rightarrow {}^{12}\text{C} + \alpha
\end{aligned}
$$

See also *beta decay*, *fusion* and *proton-proton chain*.

carbon dating or **radiocarbon dating** See *carbon-14* and *dating*.

carbon fibre generic term for fibrous forms of carbon ('whiskers') with high strength and stiffness but low density; they are of value when bound in composite materials. The type of fibre depends on the temperature reached in the manufacturing process.

carbon-14 or **radio-carbon** ^{14}C carbon's only radioactive isotope, a beta-emitter with a half-life of 5570 years: $^{14}\text{C} \rightarrow {}^{14}\text{N} + e^- + \bar{\nu}$

If we assume that ^{14}C appears at a constant rate (not quite the case – see *dating*), its ratio in the environment to the stable ^{12}C will be constant too. Thus, all living creatures take it in in the same ratio and have that ratio in their cells. On death, however, the ^{14}C decays without being replaced: the ratio of ^{14}C to ^{12}C in dead material becomes a measure of time since death. Carbon dating is of great value for checking the age of archeological remains: the ashes of wood fires, the timbers of buildings, the bones of food animals, the remains of people and much of what they owned, and so on.

Carbon dating has a current limit of 60 000 years for two reasons. First, it is hard to measure such a small ratio as exists after (say) ten half-lives; second, as the rate of ^{14}C production is not quite constant, fairly effective calibration of the technique goes back over only that period.

carbon microphone microphone in which inward sound waves vibrate a plate on the surface of a sample of carbon grains; this changes the resistance of the system, and thus the current obtained in the circuit. It was one of the first microphones developed, and used for a hundred years in phone handsets and repeaters in phone lines.

carbon–nitrogen cycle See *carbon cycle*.

carbon resistor normal resistor for use in electric circuits; it consists of a rod of carbon grains in some type of cement. Few resistors now are wire wound as all once were. Indeed, for microelectronics, high accuracy and low cost metal film resistors are now the norm.

Cardew voltmeter hot-wire ammeter with resistance in series so it acts as a voltmeter

cardinal points the six focal, nodal and principal points on the axis of an optical system

Carey–Foster bridge modification of the Wheatstone bridge able to give the difference in resistance of two almost equal circuit elements. The two elements serve as the two arms of the bridge to give a balance point. Swapping the elements results in a new balance point. The difference between their resistances is proportional to the distance between those two balance points.

Carnot cycle cycle of action of a perfect heat engine (**Carnot engine**), working between temperatures T_1 and T_2 – a reversible sequence of adiabatic and isothermal changes in the pressure and volume of the gas used which returns the system to its starting point

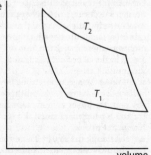

The Carnot cycle is how a perfect heat engine works.

with a net transfer of input to output energy ('work'). See also *Otto cycle*.

Carnot, Nicolas (1796–1832) mathematician and engineer who helped found thermodynamics by writing the first theory of heat engines, which includes the Carnot cycle

Carnot's principle (or **Carnot's theorem**) uses the second law of thermodynamics to derive the maximum efficiency of the Carnot engine: $T_2 – (T_1/T_2)$. No heat engine working over a given temperature range can have a greater efficiency than the perfect reversible engine. It follows that all perfect engines have the same efficiency no matter what their working substance.

carrier a) (or **current carrier**) charged particle (such as an electron, ion or hole) that carries current in a system. There is a current in any net transfer of charges. **Carrier concentration** is the number of current carriers per unit volume of conductor.

b) (or **carrier wave**) uniform electromagnetic wave (or current) that transfers a signal (information) from place to place. Some process of modulation adds the input signal (analog or digital) to the carrier; demodulation extracts it at the far end. Radio and microwave carriers are involved in radio and television communications; infrared and light waves carry signals through fibres. **Carrier frequency** is the frequency of the carrier in a given case. It is much higher than the carried signal's frequency, but see *power line transfer*.

c) non-active substance, such as a cement, used in bulk to hold small amounts of a radioisotope for use as a tracer

Cartesian coordinate system the *x*, *y*, *z* system of naming points in 3D space – see *coordinate*.

Cartesian sign convention standard approach to drawing and using ray sketches and object/image equations in optics: inputs pass from left to right; distances are positive when in that direction or above the axis and negative otherwise.

cascade a) set of capacitors in series, this giving the total capacitance C:

$$1/C = 1/C_1 + 1/C_2 + 1/C_3 + ...$$

b) series of circuits, sub-circuits or elements, the output of one being the input of the next

c) set of similar processes in series, as in the case of uranium enrichment by diffusion or centrifugation, and in the **cascade liquifier**. Here the input is a mix of gases which differ in critical temperature T_c. The gas with the highest T_c liquefies by compression; its evaporation at a lower pressure cools the next gas below it and so on. Any such process is a **cascade process**.

d) the *cascade particle* is the xi (hyperon), Ξ^0 and Ξ^-.

e) in a discharge, the same as avalanche

cascade shower branching cascade of particles through the air, produced in cosmic radiation, a sequence of alternating electrons slowing by 'Bremsstrahlung' and photons slowing by pair production. The **cascade unit** is the distance in which an electron's energy in such a shower falls to $1/e$ of its initial value. This is almost the same as the mean life distance of the photon it produces by 'Bremsstrahlung' before it decays by pair-production to produce further electrons.

Casimir effect attractive force – the **Casimir force** – between two *very* close uncharged metal plates. The cause is an imbalance in the radiation pressure outside and between the plates due to zero point energy: the plates are so close that some zero point particle wavelengths cannot fit between them. Thus, there are more possible states outside, and as each state has energy, an inward pressure results. The effect can cause problems with very small systems, but may also be able to help decide between unified field theories for spaces with ten or eleven dimensions. We can control the force by changing the density of charge carriers in the plates. This may be of use in microelectromechanical systems.

In 1948, physicist **Hendrik Casimir** (1909–2000) deduced the effect and worked out the size, to explain very small forces observed between particles in a colloid. Casimir was also known for his important early theory of superconductivity.

Cassegrainian telescope common type of reflector with a central hole in the object (main) mirror. A secondary mirror in front of the main mirror reflects the converging rays through the hole to focus at the eye lens system behind it.

cataphoresis the motion of positively charged colloid particles towards the cathode in an electric field, used to purify proteins. Compare this form of electrophoresis with the *Dorn effect*.

catastrophe sudden discontinuity (for instance a shock wave) in the otherwise smooth behaviour of a system. The **ultra-violet catastrophe** is an outcome of black body theory which led to the quantum theory. **Catastrophe theory** classifies, for instance, the form and folding of surfaces in mathematics, where too a sudden change in one aspect follows a very small change in another. People also apply it to analyse – even predict – events such as earthquakes on a fault line.

cathetometer device for measuring vertical distances using a microscope or telescope able to move up and down along a marked scale

cathode electrode in any device at which electrons enter it, i.e. the one(s) with the most negative potential with respect to the anode(s). The symbol K is sometimes used.

cathode fall the potential drop between the cathode surface of a *discharge tube* and the end of the Crookes space

cathode follower circuit used in the output of a recorder to allow the use of longer cables – originally a radio triode (electron tube) circuit in which the load (output) is in the cathode section rather than in the anode section; now any circuit used this way in this context (such as emitter follower and source follower). The circuit gives 100% negative feedback, very high input impedance and very low output impedance; the output voltage 'follows' the grid voltage closely.

cathode ray oscilloscope (or **oscillograph**) cro, or CRO device based on a cathode ray tube and used to show on a fluorescent screen an image of the signal in a circuit. The associated electronics lets the user study signals, even of very high frequency and low amplitude. The deflection system shown is electrostatic (voltage-driven) with deflection of the beam in the vertical and horizontal directions being the result of voltage between the corresponding pairs of plates. Most cro deflection systems work this way. Alternatively, coils around the neck of the tube may carry currents which produce electromagnetic deflection; this is the system found in the crts of old tv sets and IT monitors.

cathode rays stream of electrons leaving the surface of the (often heated) cathode in a very low pressure gas or vacuum. Discovered in the late nineteenth century, and studied by a number of people, cathode rays' nature not clear until, in 1897, J J Thomson showed that the 'rays' are in fact streams of charged particles, of mass about 1/1840 that of the hydrogen atom. This was the first indication that atoms must have parts rather than being indivisible, so was a major step towards modern physics.

Cathode rays carry current in various systems, including the great range of vacuum (electron) tubes ('valves') in wide use before solid state physics, and in high power electronics, for instance radio and television transmitters, radar sets, and microwave power devices.

cathode ray tube crt, or CRT mushroom-shaped thick-walled glass vacuum tube with, at the neck, an electron gun that focuses a narrow beam of electrons ('cathode rays') onto a screen. It is the only electron tube now in common use, though flat screens are fast taking over. The crt screen has a fluorescent layer that emits a spot of bright light where the beam meets it. A deflection system points the beam at any part of the screen and so can build up an image. There are various ways of scanning the screen with this 'flying spot' to produce an overall picture; see, for instance, television.

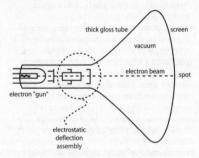

Signals on the pairs of plates move the beam, and thus the spot, anywhere on the screen.

cathode space narrow, low brightness gap in a low pressure gas discharge, between the **cathode glow** and the negative glow. In the early days of research into the low pressure gas discharge, this was the name given to the main 'dark' space, that now called after its discoverer, William Crookes. In those early days, people thought the cathode glow ran from the cathode surface to the Crookes space; in fact, it consists of two glowing regions, the cathode space, and the Aston space. See *discharge tube* for illustration.

cation positive ion, one that the cathode of a system attracts. Positive ions are atoms or molecules with one or more missing

electrons. In some contexts, we can view the proton (positive hydrogen ion) as a cation.

catoptric power the focusing power of a reflector, i.e. the reciprocal of its focal distance. The unit is the dioptre, D.

cat scan computer-assisted tomography scanning process, using x-rays to image sections (slices) of a medical or industrial object and a computer to form a 3D image for analysis of internal structures. See also *x-ray microtomography*.

cat's whisker early crystal rectifier (see *crystal radio set*) that consists of a fine copper wire pressing on the surface of a semiconducting crystal (often of galena). Users had to move the point of contact to find the best effect.

Cauchy, Baron Augustin (1789–1857) mathematician and mathematical physicist who did much work on calculus and the theory of functions (which he helped to found) and on wave theory and optics. The **Cauchy dispersion formula** relates the refractive constant n of a substance and wavelength λ; it holds in most cases for a limited region of the spectrum.

With constants for the substance a, b and c

$$n = a + b/\lambda^2 + c/\lambda^4$$

causal dynamical triangulation (cdt) a new theory of the universe that uses a novel approach to unifying quantum mechanics and gravity. Some people believe it may replace string theory, which has tried to do that since the 1980s; it has already had more success than the string theories in modelling our four-dimensional view. A feature is that the number of dimensions depends on the scale of viewing the model.

causality the concept of classical physics that all events have precise causes, though in a given case the causes may be so many or so complex that one cannot predict their actual outcomes. Aspects of the quantum theory mean that we can *never* describe in full the causes of a small scale event. Thus, we must now replace causality with probability.

caustic or **caustic curve** the curve of points at which rays cross after passing through an optical system – only paraxial rays pass through a single focus if the system's elements have spherical surfaces. The caustic curve is rather heart-shaped. A **caustic surface** is a caustic curve in three dimensions.

Cavendish experiment the crucial early work that allowed **Henry Cavendish** (1731–1810), chemist and physicist, to 'weigh' (i.e. find the mass of) the Earth. In 1798, he used a torsion balance (designed and part built ten years before by John Michell) to measure the gravitational force between two small hanging lead spheres and two large fixed ones; the data allow a calculation for the gravitational constant G – and this lets us find the mass of the Earth and thus that of the other objects in the solar system. A century later (1895) Charles Boys improved the experiment by much reducing the scale (and therefore the errors) with no loss of sensitivity. A century after that, we still use much the same approach to measure G.

cavitation the formation of small, short-lived, low-pressure regions (cavities) in a liquid as a result of local reductions in pressure during turbulent flow or the passage of ultrasonic waves. As the cavities collapse, they produce shock waves which can cause much damage to nearby solid surfaces. A **cavitation tunnel** is a water tunnel which uses cavitation for testing. See also *cold fusion*.

cavity a local short-lived low-pressure region in a liquid as a result of cavitation. *Cavity flows* are the liquid flows around such cavities.

cavity radiation the same as black body radiation

cavity resonator closed box in a radio frequency resonant circuit, an electric form of Helmholtz resonator. The circuit's resonant frequency is that of one of the lower modes of oscillation of the box. Such systems are very important when used with microwaves – for instance, in klystrons and magnetrons.

Cayley, George (1773–1857) engineer who was a pioneer of the science of aerodynamics. He produced the first glider able to carry a person and developed many of the concepts later used in powered flight.

ccd charge coupled device

cdt causal dynamical triangulation, a new theory of the universe

celestial mechanics study of the motion of objects in space under the forces of gravity of others

celestial sphere infinite sphere centred on the Earth that rotates once every 24 hours – our view of the sky and of the celestial objects used for mapping in astronomy. The sphere has celestial poles, prime meridian and equator; relative to these, we quote the **celestial coordinates** of each object in terms of right ascension (between 0 h and 24 h, in hours, minutes and seconds) and declination (between +90° and –90°). See also *ecliptic*.

cell a) the base unit which repeats in some system such as on a screen (**picture cell** or pixel) and in crystallography (**unit cell**: see crystal structure);
b) unit source of electricity that involves chemical energy (for instance, electrochemical cell and fuel cell) or light energy (e.g. photo-cell). We may join such cells to form a battery.
See also *electrolysis*, the reverse of electrochemical cell action.

cell constant the area of the electrodes of a chemical cell divided by the distance between them. The pd produced, V, is
$V = E\ R/(R + r)$
Here E is the cell's electromotive force (the pd between the terminals when it supplies no current), r is its own (source) resistance, and R is the circuit resistance. Both E and r are base constants for a given cell in given conditions.

Celsius scale the most widely used everyday temperature scale, in the past known as the centigrade scale; it has the temperature of melting ice as 0 °C and the temperature of condensing steam as 100 °C. The Celsius degree is the equivalent of the kelvin, the unit of absolute temperature. On the original Celsius scale, devised in 1742 by astronomer **Anders Celsius** (1701–1744), the ice and steam temperatures were 100° and 0° respectively.

cent a) interval on a musical scale equal to 1/1200 of the octave
b) one hundredth of a dollar, i.e. 1/100 of the reactivity needed to make a nuclear reactor critical

centi- unit prefix for 10^{-2} (which strictly should rarely be used in science), as in centimetre, 10^{-2}m

centigrade scale old term for Celsius scale

centimetric waves electromagnetic waves in the radar region, with wavelengths from 1 to 10 cm

centrad one hundredth of a radian, i.e. about 0.57°

central force force whose size is a function of distance from a fixed centre and whose direction is towards or away from that point. Coulomb (electrostatic) and gravitational forces are of this type; nuclear forces are central only when the system has spherical symmetry.

centred optical system set of one or more optic elements with spherical surfaces whose centres lie on a single line (the system's axis). Most simple standard systems are centred, but see *binoculars*.

centre the point in or near a system to which some property of the system relates

centre of area the geometric centre of a system (in particular, a surface). Compare with *centroid*.

centre of buoyancy the point through which acts the upward force of buoyancy ('upthrust') of an object partly or fully immersed in a fluid. For stability, the centre of buoyancy must lie below the object's centre of mass.

centre of curvature the centre of the sphere of which a surface is part. The term most often refers to lenses and mirrors with spherical surfaces. See also *curvature*.

centre of gravity the point of an object or system in a gravitational field through which we can consider the weight to act; it is the same as the centre of mass if the gravitational field in the region is uniform

centre of mass the point in the object or system at which one can consider the whole mass to be concentrated when dealing with the effect of any outside force. It is the same as the centroid if the object or system is uniform.

centre of pressure of a surface partly or fully immersed in a fluid, the point through which we can consider the sum of all the pressures to be acting. If the surface is horizontal, the centre of pressure is at the centroid.

centrifugal force fictitious force (and therefore the term is best not used) deemed to act radially outwards on an object moving in a curved path; the word means 'centre-fleeing'. In fact, for the object to move in a curve, there must be a force *towards* the centre, otherwise the object would move in a straight line (Newton's first law of force). This real centre-seeking force is the *centripetal force* on the object. It is an invalid use of Newton's third law to claim that centrifugal force is the reaction to centripetal force as both would act on the same object. See also *centrifuge*.

centrifuge container spinning at a high rate in order to separate the mixture contained within. The mixture may be a suspension of solid particles in a liquid, or a mixture of gases that differ slightly in density. Other types of centrifuge, such as those used in aerospace research, investigate the effect of high centripetal force on the person or object under test.

A centrifuge separates by simulating the effect of a high gravitational field in the container (up to 10^9 g in some cases). Denser matter moves 'down' the field, i.e. away from the central axis; less dense matter 'rises' towards the centre.

centripetal force the force needed to keep an object moving in a curve; it acts towards the centre: the word means 'centre-seeking'. Newton's first law of force tells us that a moving object will travel in a straight line unless a net outside force acts on it to change its direction. If the path is part of a circle of radius r and the object moves at constant speed v (constant angular speed ω) the change of direction is a change of velocity; this implies that the object accelerates. The acceleration a is v^2/r, or ω^2 r. Using Newton's second law of force), this gives the centripetal force on the object as m v^2/r or m $\omega^2 r$.

There is a centripetal force in a centrifuge, in the orbit of a charged particle in a Coulomb (electric) field, on a mass freely moving in a gravitational field, and whenever a car travels around a bend.

centrobaric of a uniform object, having the centre of mass at the geometric centre (centroid)

centroid a) the geometric centre of a surface, object, or system

b) the same as a system's centre of mass if the system is uniform

centrosymmetry the symmetry of an object or system with respect to a point (rather than, say, a line)

cerium Ce the most common transitional rare earth element in the lanthanide series, number 58, discovered independently in 1803 by Jöns Berzelius and Martin Klaproth, and named after the newly found minor planet Ceres. Period 6; relative density 6.8, melting and boiling temperatures 798 °C and 3443 °C; most common isotope ^{140}Ce. The main uses are in making alloys and with carbon in arc lamps.

CERN 'Organisation Européenne pour la Recherche Nucléaire' – the European Organisation for Nuclear Research, a fully international facility, based on the border between Switzerland and France. It is the site of various accelerators – including the large hadron collider – and the labs they need, and was the birthplace of the world-wide web.

cgs system inconsistent 'metric' system of measures developed from that produced in Napoleonic France a couple of centuries ago, and that eventually led to the units we use in science now (see *Appendix 1*). Widely used in European science until the 1960s, the system's name comes from its three base measures: centimetre, gram and second (for length, mass and time). Later, the calorie, an inconsistent unit of thermal energy, was added.

The mks (metre, kilogram, second) system took over from cgs in the 1960s.

Chadwick, James (1891–1974) discoverer of the neutron (1932), which led to his Nobel Prize in 1935

chain reaction cascade of neutron-induced nuclear fissions, each of which gives further neutrons. The neutron-induced fission of uranium-235 can output up to two or three new neutrons (the number depends on the input energy and the fission mode). Those neutrons may escape from the uranium sample, or non-fissile nuclei may absorb them; however, if one or more (on average) causes a new fission, we have a chain reaction.

The design of a fission reactor (see *nuclear power*) aims to have exactly one new net fission produced per fission. If the branching ratio is much greater than one, the chain becomes an explosive avalanche as in the case of a fission bomb.

Chamberlain, Owen (1920–2006) nuclear physicist who carried out research into the anti-proton (which he discovered in 1955) and was awarded (with Emilio Segrè) the 1959 Nobel Prize

Chandrasekhar, Subrabmanyan (1910–1995) astrophysicist who gained the Nobel Prize in 1983 for his work on nuclear processes in, and the structure and evolution of, stars

change of state change of a sample of substance from one state (phase), such as solid, to a second, such as liquid. In each case there is a transfer of latent energy. The common names of the various changes are:

change	name
solid → liquid	melting (once called fusion)
solid → gas	sublimation, evaporation
liquid → gas	vaporisation, evaporation, boiling
gas → liquid	condensation, liquefaction
gas → solid	condensation, freezing
liquid → solid	freezing

channel a) region between source and drain in a field effect transistor. The **channel conductivity** depends on the voltage at the one or more gates (control electrodes).
b) path or waveband used to transfer a signal between two points, the **channel width** being the band width involved
c) name for a mode of decay (de-excitation) of an atomic or nuclear system. **Channel width** relates to the initial energy level.

chaos the way certain non-linear dynamic systems behave, the result of feedback at values that produce reproducible, non-random, but quite unpredictable effects. Chaos theory describes this in cases such as the atmosphere, solar system and plate tectonics. The system $X(t)$ at time t is a function of that at the previous moment of measurement – it depends entirely on the past system.

$$X(t+1) = f(X(t))$$
$$= fk(X(t))$$
$$= k\ X(t)\ (1 - X(t))$$

Here k is a parameter for each given system that relates to the feedback strength. Thus, given values for it and for $X(0)$ (the initial value of X, i.e. when t is 0), we can find $X(t)$ in any case.

Despite this deterministic feature, we cannot always predict the outcome. In many cases, X will settle down as a constant, while in many others it will oscillate between two constant values. In yet others – where 'chaos reigns' – the values of X as time passes show no pattern at all. Moreover, even very small changes in the starting values of X and k give totally different results. (Putting k to 4 and $X(0)$ to, for instance, 3.999 and 4.000 shows both aspects of chaotic behaviour.)

The first example found was a century ago, when Henri Poincaré worked on the gravitational three-body problem. The advent of computers has allowed us to explore such systems at speed and in detail. Plots of X in two and three dimensions produce the many beautiful pictures (such as Mandelbrot sets) that are now recognised as the product of chaos analysis. Most deterministic physical systems are chaotic in this sense – which means their future can never be predicted.

Quantum chaos theory deals with non-deterministic systems that follow the laws of quantum mechanics; workers in the field still try to apply the correspondence principle.

characteristic curve a) plot of how the voltage between a pair of points varies with the voltage between a second pair or with the current between the same pair
b) graph that relates the density of photochemical product on a photographic film to the logarithm of the exposure

characteristic equation an *equation of state* that link the pressure, volume and temperature of a gas sample

characteristic functions solutions to the equations for certain boundary conditions of an atomic or nuclear system. In wave mechanics, they are the feasible solutions of the Schrödinger equation; in quantum mechanics, the corresponding functions are eigenfunctions. See also *characteristic value*.

characteristic impedance a) the square root of the ratio of the series impedance of a power transfer system to its shunt (parallel) admittance
b) for an infinitely long uniform cable, the input impedance

characteristic radiation or **characteristic x-radiation** the x-region line spectrum produced by irradiating a sample of a given substance with x- or gamma-rays

characteristic temperature for a solid, the ratio of h ν_m to the Boltzmann constant. Here h is the Planck constant and ν_m is the solid's maximum frequency in the Debye theory.

characteristic value the value of some parameter that corresponds to the characteristic function of a system. It is the value for particle energy in the case of wave mechanics, and the system's eigenvalue in quantum mechanics.

characteristic x-radiation See *characteristic radiation*.

charcoal non-crystalline form of carbon produced by heating organic matter (most often wood) in the absence of air. Further heating to drive off absorbed gases gives activated charcoal – samples of this have a huge surface area, so are very good gas adsorbers.

charge Q unit: coulomb, C

property of some particles, and of some samples of matter, that leads to a force between them and others. The force is called Coulomb force, electric force or electrostatic force. Tests show that there are two types of charge, given the names positive (as on protons) and negative (as on electrons). Charges of the same type repel each other; opposite charges attract each other. The force in either case varies with the size of each charge and is inversely proportional to the square of the distance between them.

charge, weak See *weak charge*.

charge conjugation mathematical process of changing the signs of all particles and all fields in a law of physics, a system or a reaction to produce the anti-version; the original and its conjugate are the same under **C-symmetry**. There C is **charge conjugation parity** – a quantum number with values –1 and +1 associated with particles with zero baryon number, charge and strangeness. (The eta and neutral pions and their antis are C-particles.) C is constant (conserved) in all interactions that involve gravity, the strong nuclear and the electromagnetic forces. Cronin and Fitch in 1964 showed that **C-parity** does not apply to the weak interaction.

Under charge conjugation, the signs of electric charge, magnetic moment, baryon/lepton number, flavour, and the z component of iso-spin all change. Mass, linear momentum and spin do not change – they are invariant.

charge coupled device ccd, or CCD important (highly sensitive and cheap) pixel-based integrated circuit detector of light and other radiations. Invented in 1969 (for data storage in IT) and based on silicon, the ccd is common in digital video and still cameras, scanners, astronomical instruments, and particle detectors.

The image area on the chip is a grid of up to several million separate cells (each, in essence, a capacitor); during a set time period, the radiation input to each cell builds up free electrons. An applied voltage allows the charges, once freed, to remain for a long time, to give a permanent image of the input. At any stage, the system can scan the cells (taking perhaps a few seconds by a bucket-brigade process) to produce a corresponding signal; at the same time, the cells lose the charge so can start the process again.

To produce a colour image, the device either has three ccds (costly), or a single ccd covered with such a layer as a Bayer pattern, a special grid of red-green and green-blue filters (less efficient and less accurate).

A ccd cell has an efficiency of up to 70%, so can often react to a single photon; this gives the chip its extreme sensitivity even to exceedingly low radiation levels (though noise can be a problem). CCDs can cope also with very high radiation levels as a cell becomes saturated only when it holds several hundred thousand electrons. (The image then 'blooms' as excess electrons spill into other cells.)

charge density a) the electric charge per unit length of a line – **linear charge density** λ, in C m^{-1}

b) the charge per unit area of a surface – **surface charge density** σ, unit C m^{-2}

c) the charge per unit volume of a substance – **volume charge density** ϱ, C m^{-3}

Note – the net charge of a charged conductor concentrates on the surface (because of its mutual repulsion); the charge density varies with the curvature.

charge independence or iso-spin symmetry, the principle that the strong nuclear force between two nucleons depends only on their angular momentum and spin. Related to this is the principle of

charge invariance: the nuclear force between two nucleons is invariant during rotation in iso-spin space. Both imply that the forces do not relate to charge as such.

charge-transfer device semiconductor device in which there is transfer of packets of charge from one cell to the next. There are several designs, such as the bucket-brigade device, the charge coupled device and the surface charge transistor (a variant of the ccd).

Charles' law or **Gay-Lussac law** the volume of a fixed mass of ideal gas at constant pressure is proportional to the absolute temperature: $V \propto T$, or $V = V_0 (1 + \alpha_v \Theta)$. Here V_0 is the volume at 0 °C, Θ is the Celsius temperature, and α_v is the volume expansivity of the gas (around $1/273$ °C^{-1} in all cases). All these relations also apply closely to a real gas above its critical temperature.

To test gases, physicist Jacques Charles (1746–1823) made the first ascent in a hydrogen balloon, a few years before he stated the law in 1787. More accurate results were published in 1802 by Joseph Gay-Lussac (1778–1850), which is why some people name the law after him.

charm one of the *flavours* (internal quantum numbers) of the quark family, paired with strangeness. The discoveries in 1974 of the **charm quark** (a massive quark with charge $+(2/3) Q_e$) and of the J/ψ particle (the first **charmed particle**) confirmed that this property exists. In fact, the J/ψ is a charm/anti-charm quark pair – see *psi-particle* – its strange name being the result of its having been discovered by two teams at the same time.

charmonium a form of quarkonium – a flavourless meson made up of a quark and its anti-quark; in this case the quarks are charm quarks. The J/ψ is an example – see *psi-particle*.

Charpak, Georges (1924–) nuclear physicist whose invention in 1968 of the multi-wire proportional chamber gained him the 1992 Nobel Prize

chemical concerned with the bonds (forces) between atoms in molecules, i.e. concerned with the science of chemistry. For **chemical energy** see *energy*.

chemical hygrometer a device which measures the humidity (moisture content) of

air by drying a sample of a 'chemical' substance (such as silica gel), absorption of the moisture on the surface, and finding the increase in mass of the drying agent

chemical shift the small shift of some spectral lines as the result of small changes in energy levels after chemical change

chemiluminescence form of luminescence caused by chemical change, which occurs in, for instance, the tissues of living creatures such as fireflies. It is the fluorescence (or de-excitation) of excited atoms or molecules.

chemisorption form of adsorption in which chemical bonds hold the adsorbed layer

chemosphere old name for thermosphere, the layer of the atmosphere in which inbound uv radiation ionises atoms

Cherenkov, Pavel (1904–1990) physicist who discovered the radiation named after him in 1934. For this, he shared the 1958 Nobel Prize with Ilya Frank and Igor Tamm (who explained its cause).

Cherenkov radiation the electromagnetic analog of the sonic boom shock wave of an object moving faster than the speed of sound in that medium: the light radiation produced by a particle moving through a substance at a speed v greater than the speed of light in that substance c. The light is bluish, and forms a cone-shaped wave front in advance of the particle. The angle of the cone is $\cos^{-1} (c/v)$; here $c = c_0/n$, n being the refractive constant of the substance and c_0 the speed of light in free space. Measuring the cone angle gives the particle speed and thus its energy.

Chernobyl nuclear power station in the Ukraine where, in 1986, the world's worst nuclear accident took place – when one of the four badly designed reactors exploded during an experiment, sending 14×10^{18} Bq of active fallout over much of Europe and elsewhere. Estimates of the resulting deaths range from below 10 000 (IAEA/WHO 2005) to perhaps 50 000 (EU 2006) or 100 000 (Greenpeace 2006). The reactor has never been cleaned up, nor has the surrounding evacuated area (about the size of Greater London), which has returned to nature and now shows levels of background radiation that are little higher than the norm. The event's release of ^{131}I caused a large local increase in cancer of the thyroid – since 1992, 3000 extra cases in Ukraine among people aged 0–18 at the time, about forty times the normal rate; the increase has not yet peaked (mid-2000s), though in theory it should have done. Other health problems, such as blood defects and leukemias, seem to be extremely rare, and that applies too to wild life (even to the birds nesting inside the holed pressure vessel).

The Chernobyl information site http://www.davistownmuseum.org/cbm/Rad7.html includes much useful information from the physics viewpoint. Perhaps the best site looking at Chernobyl after twenty years is the BBC's http://news.bbc.co.uk/1/hi/in_depth/europe/2006/chernobyl/default.stm

chief ray the ray from the tip of the object to the centre of the entrance pupil of an optical system

child the product of some process or reaction, also called daughter or son. Thus, the product nucleus of a decay is sometimes called a child (or daughter). The preferred term is decay product.

chi-meson χ^0 old name for the eta particle, a boson resonance of mass 958 MeV

chip or **microchip a)** integrated circuit – a small piece of semiconductor (mainly silicon) into whose surface have been built one or more layers of circuits (the main elements being diodes and transistors)
b) the container which mounts and protects this chip and links its terminals to pins (legs)

chirality a) that property of **chiral** objects in which they differ from their mirror image, so allow left-handed and right-handed forms. This applies to some molecules, where it is significant in the context of their optical activity.
b) highly abstract symmetry observed within elementary particle interactions, where the property in question is spin. The **massive chiral fermion** is a virtual particle found in *graphene* – it has mass, but should not have, as particle physics theory predicts that chiral particles have no mass.

chi-squared test test for statistical significance used where the single test statistic has a χ^2 distribution

Chladni figures the shapes formed by the nodal lines of a surface when it vibrates. **Chladni's plates** are a set of metal sheets of various shapes which vibrate when stroked with a bow to produce sound. Dust sprinkled on the surface migrates to the nodal lines (lines of minimum vibration), thus showing the Chladni figures. The figure observed depends on the shape and size of the plate and on where the plate is bowed relative to the support.

chlorine Cl toxic yellow-green gas, a halogen with proton number 17, group VII, period 3, discovered in 1774 by Carl Scheele and named from *chloros*, Greek for pale green. Density at stp 3.2 kg m^{-3}; melting and boiling temperatures −102 °C and −34 °C; most common isotope ^{35}Cl.

choke common term for inductor, in particular one used to reduce ('choke off') unwanted high frequencies of current in a circuit

chopper fast turning shutter in a beam of radiation that passes the radiation in short pulses

chp combined heat and power

chroma the attribute of an observed colour that allows the observer to judge how much pure colour it contains (as opposed to white and grey), i.e. to judge the saturation. **Chroma key** is a video technique in which taking away a colour from a scene (such as a newsreader's blue background) reveals a new image (such as film of the event in question).

chromatic concerned with colour aspects of light radiation. **Chromatic aberration** is a type of aberration caused by an optical system – giving the image falsely coloured fringes.

chromaticity the colour quality of a viewed light sample, independent of its brightness (see *colour*). It is normal to give the chromaticity of a light in terms of its three **chromaticity coordinates**: r (for red), g (for green), and b (for blue), on a **chromaticity diagram**. This diagram is a plot of all possible values of any two of the three coordinates (though there are three dimensional versions). The maximum value for any coordinate is close to 1, and the three always sum to 1; if each is around 1/3, the light is white. Thus the higher a coordinate is above 1/3, the more saturated is the light in that colour.

chromatic resolving power the resolving power of a spectroscope, i.e. how close two wavelengths can be while still being separated

chromatic scale the 13-tone musical scale given by the white and black keys of a piano; the frequencies of adjacent notes – a semi-tone apart – are in the ratio $1:^{12}\sqrt{2}$. Compare with *diatonic scale.*

chromatism old term for chromatic *aberration*

chromatography method of separating the components of a mixture to allow their identification (i.e. **chromatographic analysis**). The components of the mixture, suitably dissolved, can migrate along a column in a stream of liquid or gas solvent; they form different bands along the column. Chromatography is important, and there are many techniques.

chrominance signal the part of a colour video signal that carries the colour information (see *chromaticity*). Compare with *luminance.*

chromium Cr transition metal in period 4, proton number 24, discovered in 1797 by Louis Vauquelin and named from the Greek *chroma*, meaning colour, as the salts have many strong colours. Relative density 7.2, melting and boiling temperatures 1900 °C and 2700 °C; most stable isotope ^{52}Cr.

chromodynamics quantum theory of the fundamental interaction, with the gluon as mediator, that tries to explain the strong nuclear force – see *quantum chromodynamics*

chrono- prefix denoting time. Thus, a **chronograph** is a pen-recorder, in which the lengthwise axis shows the time of the events it records (e.g. the temperature in a room in the case of a chrono-thermograph). A **chronometer** is a very accurate clock (i.e. a good time-keeper); a **chronoscope** is a mechanical or electronic device to measure short time intervals (down to hundredths of a second); and a **chronotron** is software that works at much the same precision to allow users to manipulate audio files or synchronise audio with video streams.

chronon the proposed quantum of time, which, if it exists, would be about 10^{-24} s, the time taken by a photon to travel the diameter of an electron

Chu, Steven (1948–) one of the three joint Nobel Prize winners in 1997, for developing ways to use laser light to cool and trap atoms

Ci symbol for the curie, former name of the

becquerel, Bq, the unit of *activity* b) – equal to one decay in a radioactive sample per second

circle of least confusion the image with least distortion in cases of spherical and chromatic aberration. Both processes lead to a range of images between extremes, for a given point object. The circle of least confusion lies between these, but although it suffers least distortion it is still not a perfect image (hence the name). In the case of chromatic aberration of white light from the object, the extreme images are those formed by extreme red and extreme blue light. All other images show blue and red fringes, except the circle of least confusion which is near white.

circuit electric circuit, complete conducting path between the two terminals of a source of electric energy which is therefore able to carry a current when the circuit is switched on (**closed circuit**). As well as the source (which may supply direct or alternating power), switch (or **circuit breaker**), and conductors, a circuit contains some combination of active and passive **circuit elements** (components), each of which affects the current in some way. (If it did not, it would have no function.) Energy from the source appears in some other form, such as thermal and light radiation, or a field – this depends on the purpose of the circuit.

We describe each element in terms of how much it will oppose current – its resistance (or reactance); the circuit current depends on the rate of energy supply (the source emf) and the overall resistance (or impedance) of the circuit. A number of **circuit rules** exist to help analyse a given circuit into its simplest form (its **equivalent circuit**) and then to find the current with a given source. See, for instance, *Ohm's law* and *Kirchhoff's rules.* Such rules, linking some kind of flow in some kind of system to the force or pressure applied against some kind of resistance, are widely valid. See, for instance, *magnetic circuit.*

circuital of a field, having a non-zero curl at some points

circuit symbols world-wide agreed set of simple icons for all the circuit elements, so that users can build any circuit from a sketch with no knowledge of any special language. See *Appendix 3.*

circular involving, resembling or shaped like circles or arcs (part circles). The **circular functions** are the trigonometric functions (sin, cos, tan, etc.) if their arguments are real, and the hyperbolic functions (sinh, cosh, tanh, and so on) if they are complex. **Circular measure** is the measure of angles in radians.

circular motion the motion of an object in a circular arc or path under the action of a centripetal force (i.e. one acting towards the centre of the circle). All the physics to do with such motion is fully comparable with that for linear motion (motion in a straight line). See *force and motion*.

circular polarisation the form of polarisation shown by a transverse wave if the plane of polarisation rotates uniformly round the axis as the wave travels. We can view a circularly polarised ray as the sum of two linearly polarised rays at 90° and out of phase by 90°. The direction of rotation ('left', i.e. clockwise when viewed from in front, or 'right') depends on which linearly polarised ray leads the other. In the case of electromagnetic radiation viewed as particles (photons), left and right circular polarisation relates to the spin direction of the photons.

circular wave number k See *wave number*. Alas, many people call k the 'wave number', but they are not the same

CKM matrix Cabibbo–Kobayashi–Maskawa matrix, a 3×3 matrix that shows the nine possible weak interaction quark decays (between u or c or t – those with charge $+\frac{2}{3} Q_e$ – and d or s or b – the quarks whose charge is $-\frac{1}{3} Q_e$). The V-values in the matrix are the coupling constants concerned, each being a complex number associated with the weak charge of that decay.

quarks	b	d	s
c	V_{bc}	V_{dc}	V_{sc}
t	V_{bt}	V_{dt}	V_{st}
u	V_{bu}	V_{du}	V_{su}

cladding the bonding of one metal surface to another to prevent the latter's corrosion. The process helps, for instance, to protect the surfaces of the metal cylinders that contain pellets of fuel in a nuclear reactor.

clamping diode or **catching diode** diode used to keep a voltage above or below a set value

Clark cell electrochemical cell used as a standard until superseded by the Weston cell

classical physics the physics in vogue until the 1890s. In that decade, came discoveries such as cathode radiation and radioactivity. These, added to concerns about spectra and

the photoelectric effect, led to the modern physics of atomic and nuclear structure, relativity, and quantum effects and statistics. Classical physics is highly deterministic: the philosophy implicit in it is that it is possible in theory to know everything about any system, and thus to predict its future completely (but see *chaos*). Modern physics denies such perfect knowledge, and therefore forbids the ability to predict the future. In any case, classical physics cannot explain all we know; for instance, it predicts that an electron in orbit round a nucleus would radiate its energy in a very short time (as it has constant acceleration). Quantum physics allows this not to happen.

class of amplifier or **amplifier class** mode of amplifier circuit action. The output of a **class A amplifier** closely follows the input, i.e. it carries power throughout the input cycle: there is low distortion but low efficiency. The output of **class B** is half-wave rectified, i.e. like class A for alternate half-cycles, with a cut-off at zero; in this case, distortion can be quite high, but so is efficiency. **Class AB** is part way between these two – there is an output for more than half the time, but not all the time (in other words, the cut-off is below zero); such amplifiers tend to act as class A at low input levels and as class B at high input levels.

A **class C amplifier** gives an output for less than half the cycle, giving it higher efficiency than the others, but significant distortion. A **class D amplifier** is one which acts by means of pulse-width modulation. Efficiency is theoretically high, but as yet such systems are not fast enough to avoid significant distortion.

There are other classes, and this is a big topic anyway – please see http://users.telenet.be/educypedia/electronics/amplifierclasses.htm for more detail.

classon name sometimes given to the massless bosons (which are elementary particles) – the graviton and the photon

clathrate substance with a crystal structure, some of the atoms of which form a cage. Each cage can trap unrelated molecules without normal bonding in the process of **clathration**.

Clausius–Clapeyron equation equation that relates the melting or boiling temperature T of a substance to the outside pressure p, in terms of the associated change in volume ΔV and the specific latent thermal capacity L:

$$dp/dT = L/T \, \Delta V$$

Clausius, Rudolf (1822–1888) physicist who put thermodynamics on a sound scientific footing and also worked in optics and electricity

Clausius's equation equation that relates the specific thermal capacities c_1 and c_v of the liquid and vapour of a substance to the specific latent thermal capacity of vaporisation L at temperature T:

$$c_v - c_1 = T\, d(L/T)/dT$$

Clausius's virial law a system's mean kinetic energy equals its virial equation. From it follows the Clausius equation of state, which is related to that of Debye.

cleavage the splitting of a crystal *along* a plane of atoms (easier than doing *across* a plane). A crystal's **cleavage planes** depend on its crystal structure.

clinical thermometer thermometer designed to work with high accuracy over a restricted range of temperatures (often 35–43 °C), near that of the normal human body. The traditional design is a mercury-in-glass thermometer with a kink or bulge in the tube near the bulb (reservoir) to retain the liquid at the highest level. In other words, it is a type of maximum thermometer. Cheaper and safer systems – electronic or photo-chromic – are now the norm.

cloaking some technique to channel radar or light (for instance) around an object to make it invisible. Techniques in research include using superlenses, light fibres, and meta materials. A team had good success with the technique in 2006, using a meta-material to cloak a small copper cylinder from micro-waves. In the first such device to be made, late in 2006, a meta material object only very weakly reflected a beam of microwaves and thus cast very little shadow.

clock system for displaying or recording the passage of time. Most modern systems depend on some automatic cyclic process, unlike, for instance, earlier candle, sand, and water clocks.

The pendulum clock follows Galileo's discovery that the period of swing of a pendulum does not depend on amplitude. (By repute, he tested this theory on a swinging chandelier, using his pulse rate as a clock.) Some modern pendulum clocks can keep time to within 0.01 s per day.

Resonant electrostriction in a quartz crystal in an alternating field is the basis of many modern crystal clocks and 'quartz' watches.

These can be accurate to within 0.001 s per day (0.000 01%).

Atomic clocks use the frequency of a line in the radio or microwave spectrum of a gas such as ammonia; their accuracy is even higher. Most accurate are caesium clocks working with the 9.1 ... GHz radiation associated with a transition between two specific energy levels of solid caesium nuclei at 0 °C. That spectral line defines the second in SI, with caesium clocks able to work within 10^{-15}%.

closed circuit electric (or other) circuit with a complete path between the two terminals (sides) of the energy source. **Closed circuit television** is non-broadcast television: the signal sources (for instance, receiver, camera, recorder) link directly in a network, usually cable, to the displays. Such systems may allow two-way communication, as is the case with cable television, a comparatively wide-area video network (over a small town, for instance).

closed cycle cycle of a heat engine that circulates the same thermal energy transfer fluid repeatedly

closed shell shell in an atomic or nuclear system with no empty quantum states

closed system system with, in general, no interaction with the outside – there is no energy transfer to or from it

close-packed structure crystal system which fits the (identical) atoms together as tightly as is possible. There are two such structures: the face-centred cubic (fcc) and hexagonal close-packed (hcp). In each case, each atom is symmetrically placed within twelve others.

cloud chamber early, and still important, system able to display the tracks of ionising particles. Such particles leave a trail of ions and electrons on which the supersaturated vapour condenses (much like an aircraft vapour trail).

There are two types of cloud chamber, which differ in the method of obtaining the supersaturated vapour. In the **diffusion cloud chamber** there is a gradient of vapour density away from the source (for instance, a layer of solid CO_2), and only in a thin layer within the chamber are conditions right for track formation. On the other hand, the whole of an **expansion cloud chamber** (invented by C P R *Wilson* in 1912) is available to show tracks; however, this approach cannot work continuously. Instead, the system suddenly expands the working volume to cool it all below its dew temperature. This makes the vapour inside supersaturated, and droplets form on the tracks of ionising radiation. After photographing any tracks, the system returns to the non-saturated state. A variant is the **cloud-ion chamber**; this combines features of the expansion system with a means of collecting the ions produced.

cmos complementary metal oxide semiconductor, basis of many types of integrated circuit (chip), such as micro-processors, static read-and-write stores, and the sensors used in many digital cameras. Such a chip uses very little power and gives out little waste energy – this is because it needs power only when any of the transistors in the circuit change state. The basis of cmos circuits is the field effect transistor or mosfet – in this design, the metal oxide layer insulates the metal of the gate from the semiconductor substrate. The names remain, though neither metal nor metal oxide are now in use.

coaxial having the same axis. So the two conductors of **coaxial cable** have the same axis – one is the core; the other is a tube round the outside of the central insulator.

cobalt Co ferromagnetic metal, transitional element in period 4, number 27, first isolated in 1737 by Georg Brandt and named from the German 'Kobold' = goblin full of mischief (the miners used to blame goblins for all that went wrong in the mines). Relative density 8.9, melting and boiling temperatures 1495 °C and 2900 °C; only stable isotope ^{59}Co. ^{60}Co (a beta-emitter with half-life 5.3 years) has been very widely used in radio-therapy, though linacs are taking over for safety reasons.
Cobalt blue is a salt used since ancient times to give a deep blue colour to glass and artists' paints.

Cockcroft, John (1897–1967) nuclear physicist, for whose work on particle acceleration he shared the 1951 Nobel Prize with Walton. He also worked on the Cavendish Laboratory cyclotron, and became the first director of Britain's Atomic Energy Establishment.

Cockcroft–Walton accelerator development of a standard voltage-doubler system to produce a very high pd for accelerating a beam of protons or other ions. Designed in 1931, it was the first particle accelerator. The system worked at 800 kV and produced the first artificial nuclear disintegration ('splitting the atom') – they bombarded lithium with protons, each disintegration

producing two alpha particles. Systems of this type can now produce beam currents of several milliamperes over several million volts.

Coddington lens powerful magnifying lens (simple microscope), in essence a sphere with a central stop

coefficient a) constant factor in a term of an expression. For instance, in 2 *a s*, the coefficient is 2.
b) value (parameter) in a physical system which is constant under given circumstances, but has a different value for other circumstances
 In all such cases, the modern way is to use the word 'constant'; the context will make it clear whether the constant is a fundamental constant of the Universe or not.

coefficient of coupling the mutual inductance of two electromagnetically linked coils as a fraction of their maximum possible mutual inductance

coefficient of expansion old name for expansivity

coefficient of friction *n* or μ, no unit measure of how much *friction* there is to oppose motion between two given surfaces in contact – the ratio of the force of friction between them to the normal force between them. Values range from very close to 0 in the case of a lump of smooth ice on a smooth ice surface to around 1 for the rubber of a car tyre on the concrete of a road.

coefficient of restitution a measure of the elasticity of a substance, particularly in cases of collision

coefficient of viscosity in the case of a given fluid, the measure of its fluid friction

coercive force the magfield strength needed to reduce a sample's magnetisation to zero. When the sample's magnetisation is at saturation, the coercive force is a constant for the substance, i.e. it is its *coercivity*.

cog a) (small) gear wheel, sometimes called a **cog wheel**
b) tooth of a gear (wheel). See also *machine*.

Cohen-Tannoudji, Claude (1933–) one of the three joint winners of the 1997 Nobel Prize, for developing ways to work with laser light to trap and cool atoms

coherent a) of two or more waves, having a constant phase relation. The waves of a **coherent beam** (as output by a laser, for instance) remain in step, whereas the light from an incandescent lamp is incoherent.

b) of units in a system, mutually consistent. SI (*Appendix 1*) is a system of **coherent units** – all derived units come from the base units by multiplication or division. Earlier systems, such as the cgs system, have been incoherent.

cohesion the force between particles of the same kind that resists separation. Compare with *adhesion* and see *surface tension*.

coincidence the occurrence of two or more events at or close to the same time. A **coincidence circuit** is a circuit with two or more inputs that gives an output only when there is coincidence between input pulses. The output may feed a **coincidence counter** in the case of particle detection, where the inputs are from separate detectors. This technique, **coincidence counting**, much reduces the problems of background radiation.

cold cathode cathode with a low electron emission threshold, therefore able to emit electrons at room temperature. Compare with *thermionic emission*, which is much more common. Such **cold emission** will, however, require a higher voltage than when the cathode temperature is high; see also *field emission*. In the case of light sources, cold emission is the output of light at room temperature; see, for instance, *chemiluminescence*.

cold fusion a) the apparent fusion of nuclei in a small electrochemical cell at room temperature, a process claimed to have been discovered early in 1989 and which was explored in depth for most of that year. The original experiment, performed by Martin Fleischmann and Stanley Pons, involved the electrolysis of heavy water using palladium electrodes. On the basis of the apparent excess of thermal energy and neutrons produced, they concluded that deuterons were fusing. Some other people found similar effects, but by 1990 most workers had decided that these outcomes were meaningless experimental accidents. The second claim came from Rusi Taleyarkhan in 2002 and involves cavitation, the energetic collapse of bubbles generated by ultra-sound in acetone; strong doubts about this had surfaced by 2006. Even so, while donors remain cautious, research in this field remains possible.

b) the fusion of deuterium and tritium nuclei with muons as a catalyst (see also *exotic atoms*). Though this is certainly a real effect, it is still far from any practical application in the 2000s – all systems so far (since the 1940s) produce less power than they consume.
c) For a newly found process (2006) that involves mixing the nuclei concerned with metal and cooling to a few kelvin, see http://physicsweb.org/article/news/10/7/13

cold trap container of low-temperature substance (such as liquid nitrogen or a solution of solid CO_2 in acetone). It is used to condense, and thus to trap, passing vapour.

collapsar old term for black hole

collecting power old term for the (converging) power of an optical system

collective magnetism the magnetism of very fine powder. See *supermagnetism*.

collector the electrode of a transistor which 'collects' minority carriers from the base; they become majority carriers in the collector.

collector ring old term for slip ring

collimator device able to produce a parallel beam from the diverging radiation from a nearby source (from an early misprint for collineator), as in a spectroscope

collision interaction (by impact or close passage) between two particles or objects. During the interaction, there is some exchange of energy, momentum and/or charge. This depends on the collision type – elastic, inelastic, superelastic. **Collision broadening** is the increase in width of a spectral line or energy level as a result of collisions between the particles of its source. In the material of a nuclear reactor, **collision density** is the number of neutron collisions per unit volume per second.

colloid substance with two fully inter-mixed continuous phases. A colloid consists of a mixture of very small particles (taken to be in the range 50 nm – 50 μm) and a continuous medium (fluid or transparent solid). Examples are sols, gels and emulsions. In a **stable colloid**, the particles do not interact, clump or settle.

colorimeter See below, after *colour*.

a) in full, **colour charge**, a classifying property of quarks (with colour labels red, green and blue) and anti-quarks (with complementary or anti-colours), of the hadrons they form, and of gluons. See *quantum chromodynamics*.

b) subjective sensation produced in the eye-brain system by light (as well as by disease, drugs, electric fields, pressure, and so on). It is not possible to measure colour as such. Colorimetry measures colour in terms of colour values, while spectroscopy measures the light wavelengths to which viewed hues relate. A **colour analyser** is a colorimeter designed to work with, or even control, processes that involve colour change.

We describe a viewed colour in terms of brightness (or luminance, which relates to colour shade), hue (which relates to light wavelength), and saturation. Compare with *chromaticity*. An unsaturated colour is a hue mixed with white or grey – this is a tint. The table gives the main sets of hues in the visible spectrum, with their wavelength ranges. However, in reality, the spectrum contains hundreds of hues that the normal human eye can distinguish. In each case, it can also distinguish some twenty tints. Note that

- the size of a wavelength range in the table relates to the eye's sensitivity to those hues;
- eyes differ in their ability to detect extreme reds and blues: the values given are defined as such for convenience;
- purples lie on the line that joins deepest red to deepest violet on the chromaticity diagram.

hue range	wavelength range/nm
violets	390–435
blues	435–500
greens	500–575
yellows	575–585
oranges	585–620
reds	620–740

colorimeter device designed to measure the colour content of a light; the device may form comparison colours by addition or by subtraction. See also *chromaticity*. **Colorimetry** involves the use of a colorimeter and the study of the results.

colour blindness defective colour vision. There are various types, which are thought to be the result of the absence in the retina of one or more sets of cones (colour-sensitive cells). With the most common type of colour blindness, it is hard to distinguish between reds and greens; some 8% of white males and 0.1% of white females suffer from this.

colour confinement the same as confinement, the theory that we can never see most quarks as they are confined in pairs or triplets held by a force that rises with separation

colour content colour saturation

colour coupler substance used in a developer to produce the dyes needed for colour photography

colour equation expression that gives a colour in terms of the sum (addition) of three primary components and white. This involves a process of **colour mixture** by addition which applies to the effects of combined coloured lights. Colour mixture by subtraction obtains coloured pigments (dyes) by mixing substances that absorb lights of different colours – see, for instance, *filter* b). See also *colour triangles*.

colour mixing See *additive process* and *subtractive process*.

colour photography the process of making colour slides and colour prints from suitable negatives. The film contains several layers of dye, which react differently to the primary colours. Full **colour printing** requires four passes of the paper over four plates; these carry images (separations) in black and the three primary colours cyan, magenta and yellow.

colour picture tube the cathode ray tube of a non-flat colour tv set. The screen pixels consist of close triplets of tiny dots of phosphors that emit primary colours when they absorb the cathode rays (high energy electrons). There are various ways to produce the effect of three separate beams of electrons; most use three electron guns and a metal mask of fine holes between guns and screen.

colour quality See *chromaticity*.

colour space or **colour system** chromaticity diagram in two or three dimensions. In any of those contexts, a **colour specification** gives the values of the coordinates of a given colour in the diagram.

colour television video signal transfer process that carries signals with chrominance as well as luminance data; the receiver must react to both aspects of the signal.

colour temperature the temperature of a light source, found by measuring the temperature of a black body that has roughly the same colour, i.e. the same spectral distribution of visible light

colour triangles diagrams used in colour mixing to give the results of mixing primary colours. There is a triangle for the addition of coloured lights, and one for the subtractive effects of pigments, as shown. Note that the old name for cyan is peacock blue.

colour vision the process that provides the sensations of colour in the human eye/brain system. The retina (light-sensing surface) of the eye contains two types of cell. The output of the rods depends only on the luminance (brightness) of the light reaching each point; cones are much less sensitive (so do not react to low light levels) – but they *do* output colour information. The details of how colour vision works are not yet clear, but most people assume that there are three types of cone, which react to the three primary colours of light. This is the three-colour (trichromatic) theory of vision.

columbium Cb original name for *niobium*, named after Columbus (it being first found in an ore from America) by its discoverer Charles Hatchett (1801)

column of gas the gas (commonly air) within a tube which, when suitably excited, acts as a *sound source* as a result of standing pressure waves in the column. This applies to the production of speech and to wind and brass instruments (amongst others).

coma a type of aberration caused by an optical system in which the image of an off-axis point source looks like a comet

combination tones extra tones in the subjective impression of notes with two or more different frequencies. The two clearest combination tones have frequencies which are respectively the difference between and the sum of the frequencies of any pair of sources, but many people observe at least a few others. The effect is not the same as beats.

combined heat and power chp system using some of the 'waste heat' from power generation to heat the power station and nearby buildings

comet small non-magnetic lump of ice, rocks and dust in the solar system (and, doubtless, in the systems of other stars) that moves in a highly elliptical orbit spending most of its period far from the Sun. When a comet is close to the Sun (less than about 3 au distant), the warmth causes jets of neutral gas and dust which move away from the Sun in the solar wind in the form of the comet's plasma 'tail'; the comet also leaves a trail of debris.

Halley showed that comets' orbits are elliptical rather than parabolic (Newton's theory), part of his evidence being the periodic return of **Halley's comet** (most recently in 1986). Much study now goes into the nature and structure of their nuclei, as this seems likely to reflect the make-up of the very early solar system (some 4500 million years ago).

common shared between two or more otherwise separate entities (circuits, for instance). In the **common base connection** (CB) of a transistor, the base is common to both input and output circuits. This means that in this case, the emitter is the input and the collector the output. We can describe the **common collector** (CC) and **common emitter** (CE) modes of transistor usage in much the same way, as in the table. The CE mode is perhaps in widest use; in all three cases, the common lead often goes to the earth, as in the CE circuit shown.

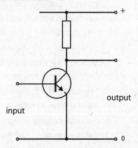

It is normal to earth the common lead of a transistor.

mode	common	input	output
CB	base	emitter	collector
CC	collector	base	emitter
CE	emitter	base	collector

common impedance coupling the linkage of two ac circuits by a common capacitor or inductor, to give capacitative coupling and inductive coupling

communications satellite artificial Earth satellite used to aid communications between points on the Earth's surface (or the same for another planet or the Moon). There are two main types, depending on orbit. Geosynchronous satellites have equatorial orbits of radius some 36 000 km. Their period is then the same as the period of the Earth's rotation; each one therefore stays very close to a fixed point in the southern sky (northern hemisphere) or northern sky (southern hemisphere). Aerials used in the ground stations to send signals to the satellites, and aerials used to collect those sent from the craft, can therefore be fixed. However, because of the great distances involved, the signals need a lot of amplification.

Satellites in near-Earth orbit have periods of some ninety minutes. Thus each one passes across the sky very quickly. While such a satellite can handle much weaker signals, users need tracking (non-directional) aerials. To provide its global portable phone coverage, the Iridium system uses 66 near-Earth satellites; this is very close to the minimum for the purpose. (There are only 24 GPS satellites – which includes three spare – but these are in much more distant orbits.)

commutative relation a) equation of the form A [operation] B = B [operation] A that applies where [operation] is commutative. A **commutative operation** is one for which the relation is true, as in the case of A × B = B × A (but not A – B = B – A) when A and B are numbers. A × B is not equal to B × A if A and B are matrices that differ in dimension. For a pair of **commuting variables** (non-conjugate variables), the operation in question is commutative.

b) (also called **commutator**) expression, in cases of non-commutative operations, for the difference between A [operation] B and B [operation] A

commutator a) the two half-rings on the shaft of a direct current generator or motor that link the coils (which carry ac) to the outside circuit

b) same as *commutative relation* b)

compact disc cd small optical disc (up to about 130 mm across) able to store analog or digital audio and/or video signals and/or data for future use. In the last case, it is therefore a form of backing storage, and current standards allow for some 580 megabytes on a 130 mm cd or around 4.7 GB in the dvd version. These discs consist of a surface of plastic on a rigid base. The recording process involves pitting the surface to match the signal or data in question, while a semiconductor laser and a photo-cell handle the reading process.

compander electronic circuit which combines the features of a *compressor* b) and an expander (to reverse the action of the compressor)

comparator a) device that involves some kind of microscope for the accurate measurement of differences in length or changes such as take place during expansion

b) circuit (based, for instance, on an operational or differential amplifier) whose output is the result of comparing its two input signals in some way

compass device which shows direction relative to the Earth's surface as required for navigation. Traditionally, i.e. for at least a thousand years, the compass was a freely suspended magnetised rod or needle which aligned (magnetically) north/south in the Earth's field. Inertial techniques, such as the gyrocompass, are now very common.

compensated pendulum pendulum design, based on a bi-metal approach, that makes the effective length (that between support and centre of mass of bob) independent of temperature

compensating eyepiece eyepiece of an optical instrument that attempts to correct the chromatic aberration caused by the object lens

compensation temperature of some ferrimagnetic materials, the temperature below the Néel temperature at which there is no moment. The effect happens when the two types of moment are equal and opposite.

complementarity principle statement made by Bohr that in quantum systems there are pairs of **complementary measures** whose values we cannot measure or know exactly. (This is not the case with classical systems.) Such complementary measures are for instance a particle's position and momentum, or its energy and the time of measurement. In practice, the complementarity principle leads to **complementary views**: valid but mutually exclusive views of a system, such as the wave and particle views of electromagnetic radiation. See also *Heisenberg's uncertainty principle*.

complementary colours pair of pure spectral colours (hues) that mix to give white; there are infinitely many such pairs, though greens have no complements. This is because the members of the pairs lie on opposite sides of the chromaticity diagram with white between them; however, opposite the greens is the purple boundary.

complementary metal oxide semiconductor design of integrated circuits (in which metal oxide is no longer in use) – see *cmos*.

complex composed of simpler components. A **complex ion** is a charged radical (stable group of atoms, such as the sulphate ion $(SO_4)^{2-}$), an **ionic complex** being an ionic compound with complex ions.

complex number number that has real and imaginary parts: one of the form $a + ib$, i being $\sqrt{-1}$

complex plastics substance whose rate of flow is proportional to the difference between the applied stress and the yield stress

compliance the reciprocal of the stiffness of a sample, the elastic extension produced by unit load

components two or more vectors with the same effect as a single vector. In other words, a vector is the resultant of its components. Any vector has an infinite number of sets of components; of most use are the pair parallel to and at 90° to a given direction, e.g. horizontal and vertical.

compound made up stably of more simple parts, as a chemical compound consists of elements

compound lens lens system that consists of two (a doublet) or more simple lenses on the same axis, close or in contact (even glued together) so that one lens may reduce aberrations caused by another. (An insect's **compound eye** and bionic copies, while sometimes called compound lenses, are not of this type – the simple lenses are not on the same axis.)

compound microscope microscope built with two or more lenses on the same axis and thus able to produce large accurate images of small close objects by processing the light rays in the lens system. There are two main lenses, both normally also compound: the object lens is closer to the object and the eye lens is closer to the eye.

compound nucleus excited nucleus (one with an excess of energy over the ground state) formed as a result of a collision in which the original nucleus absorbs the inbound particle

compound pendulum pendulum with a rigid link between suspension point and bob; unlike a simple pendulum, not all its mass is in the bob.

compound-wound machine electric generator or motor which has field coils both in series and in parallel (shunt) with the outside circuit; this gives it a better characteristic than either series-wound or shunt-wound machines.

compressibility k the reciprocal of the bulk modulus (bulk elastic constant) of a sample; the change in volume produced by unit applied pressure. The **linear compressibility** of a solid sample is its change in length in a given direction per unit applied stress; normally the direction chosen is that of the stress.

compression the reduction in volume of a sample of matter that follows from the particles being moved closer together as a result of an outside force, or pressure. A sound wave travels by means of a series of compressions and rarefactions (lower pressure regions) in matter; indeed, a sound wave often appears as a plot of pressure in a substance from point to point in time or in space.

compressor a) device designed to raise the pressure of gas in a container
b) amplifier circuit which amplifies high amplitude inputs less than it does low amplitude ones

Compton, Arthur (1892–1962) nuclear physicist awarded the Nobel Prize in 1927 for his work on the **Compton effect**. Final proof that electromagnetic waves can act as particles, this is the scattering of a photon by an electron in matter, particularly significant in the case of x- and gamma radiations. The electron, called the **Compton electron**, gains some energy and momentum from the photon, and so suffers **Compton recoil**; in some cases, it may even leave its atom. At the same time, this **Compton scattering** results in modification of the photon, which now has a longer wavelength than before the interaction (the **Compton wavelength** λ_0): $\lambda = \lambda_0(1 - \cos\phi)$, where ϕ is the photon's angle of deflection (scattering).

In the case of gamma photons, 1 in about 137 Compton interactions shows the **double Compton effect**. A second, lower energy, photon appears out of the scattered one; its spectrum has a distribution similar to that of 'Bremsstrahlung'.

computerised tomography cat (or ct) technique for scanning an object in three dimensions. The source (for instance, of x-rays) and detector revolve around the target on opposite sides and a computer collates the output to produce a three-dimensional image on screen or printout. Major application areas are medicine and metallurgy. A similar method (though with no source) scans the radiation output from lamps, for instance, to build up a detailed picture of their action.

concave curving inward, i.e. away from the viewpoint. A **concave lens** is an imprecise term for a lens which is thinner in the centre than at the edge. A lens which is thinner in the centre than at the edge may have both faces concave (a **bi-concave lens**) or just one (a **concavo-convex lens**); all are diverging lenses for glass in air. A **concave mirror**, on the other hand, is a converging mirror.

concentration any measure of the amount of solute in unit quantity of solvent. Most common of the very many such measures are **molar concentration** (symbol c, or molarity) and **molecular concentration** (symbol C), respectively the number of moles and the mass of solute per cubic decimetre (litre) of solution – units mole and kilogram per cubic decimetre, mol and kg dm^{-3}.

concentration cell cell in which the two electrodes sit in solutions of the same electrolyte which are at different concentrations; a porous partition keeps the solutions apart.

condensation the change from vapour to liquid (or, sometimes, to solid), the reverse of evaporation

condensation pump former name for diffusion pump, often found as the second stage of a vacuum pump system

condensed matter bulk matter in the solid or liquid phase – where there are many particles close to each other. **Condensed matter physics** (one of the largest fields of physics) is the study of such systems – but also of more exotic samples, such as the Bose–Einstein

condensate in 1, 2 or 3D and superfluids. Solid-state physics is its main branch.

condenser a) rare term for the main lens, compound lens or mirror of an optical or electron-optical instrument. It collects and allows control of the intensity (and aperture) of the radiation coming into the system. **b)** lens or mirror used to concentrate the light from a source for use in a microscope (in particular). See also *Abbe condenser*. **c)** old term for capacitor, a **condenser microphone** being a capacitor microphone **d)** the part of a system such as refrigerator or a still in which vapour condenses to liquid

conductance G unit: siemens S ($= \Omega^{-1}$) the ability of a matter sample to carry electric current. (There are analogs for other types of conduction.) It is the reciprocal of the sample's resistance in dc contexts. In ac circuits, it is the real part of admittance Y, B being the susceptance: $Y = G + iB$.

conduction the net transfer of energy between two points, without net motion of the medium and, in most cases, in a restricted path. The power (rate of energy transfer) depends on the length and section area of the path, the 'pressure' (some form of potential energy difference, for instance, temperature difference), and the nature of the medium. The principles apply in a number of contexts, such as acoustic (sound), electrical, and thermal conduction.

See *band theory* for the process of electrical conduction by a solid – the net motion of electrons and/or of holes. In liquids other than liquid metals, ions carry current during conduction in electrolytes (see *electrolysis*). Gases tend to be very good insulators as samples contain very few free charges. Those few charges are ions of the gas atoms (and a corresponding number of free electrons), created for a short time by ionisation by cosmic and ultraviolet rays and lightning. An applied electric field shows the gas to be ohmic at first, until the field is strong enough to collect charges as quickly as they form. There is then a plateau in the voltage/current curve. At a much higher voltage, each ion gains enough energy from the field to cause further ionisations in an avalanche (cascade) process: the current rises with voltage again, but faster now. At a critical field strength (the breakdown value), discharge occurs.

The breakdown voltage depends on the nature of the gas(es) and the pressure. At pressures above a small fraction of atmospheric pressure, discharge is in the form of an intermittent spark. (It is intermittent because the spark current is high enough to reduce the field strength far below the critical value, sometimes even to discharge the source.) The discharge in gases at low pressure produces a continuous gentle glow throughout the space between the electrodes. (Because it is continuous, it does not discharge the voltage that causes it.) The colour of the discharge depends on the gas(es) present; depending on pressure, there may also be a complex structure of glowing and darker regions. Taking the pressure to still lower values (below about 1 Pa) causes the glow to disappear – there are no longer enough gas particles to produce ions to carry the current. In such gases, and in vacuum, given a high enough field strength, current passes in the form of *cathode rays*. The process involves the direct transfer from the cathode of fast moving electrons. See *discharge tube*.

conduction current current through matter, carried by electrons, holes (in effect, missing electrons) and/or ions, down a potential gradient (i.e. caused by an electric field rather than being in free fall)

conduction electrons electrons that move down a potential gradient, and so contribute to the net current (flow of charge); electrons in atomic and molecular shells are not free to move.

conductivity a) (symbol σ, unit: siemens per metre S m^{-1}, = per ohm per metre, Ω^{-1} m^{-1}) the ability of a substance to carry electric current (**electrical conductivity**). It is the reciprocal of its resistivity, ϱ.
b) (symbol λ, unit: watt per metre kelvin, W m^{-1}K^{-1}) the ability of a substance to transfer thermal energy (**thermal conductivity**). It is the rate of energy transfer (i.e. power p) per unit section area A of the conductor per unit temperature gradient dT/dl along the conductor. The Searle's bar and Lee's disc are traditional methods for measuring λ in the cases respectively of good and poor thermal conductors. The highest thermal conductivities are those of ultra-pure diamond.

conductor substance with a high conductivity, or a specific sample in use. Most metals are good electric and thermal conductors. See *electric circuit*.

cone a) cell in the retina of the human eye receptive to a limited waveband of light

b) solid object with a plane base in the form of a smooth curve (often a circle), every point on which joins a single point (the apex) outside the base. The volume of a cone is A h/3; A is the base area (π r^2 in the case of a circle) and h is the perpendicular distance between base and apex.

cone of friction the cone above the central point of contact between two surfaces, with axis the normal (perpendicular) to the surfaces and with angle equal to the angle of friction α. The resultant of the normal force and the force of friction between the surfaces must lie within this cone.

cone of silence cone-shaped region in which there is no signal from an omnidirectional transmitting aerial. It is the result of destructive interference between the output signal and that reflected from the ground.

configuration arrangement of the parts of a whole, for instance the arrangement of atoms in a molecule and the distribution of electrons between the states available

confinement or **colour confinement** the theory that we can never see most quarks as they are confined in pairs or triplets held by a force that rises with separation. This is a somewhat dubious theory in which what we observe 'explains' what we observe: no theory yet devised *requires* that there be confinement.

conical refraction the effect observed in a solid made of biaxial crystals for inbound rays in a particular direction: the refracted rays lie on the surface of a cone

conjugate linked in pairs with inter-changeable properties. Thus **conjugate complex numbers** are pairs of complex numbers of the form $a \pm ib$. Their product is real: $a^2 + b^2$.

 canonical conjugate variables a pair of variables of any type (often, in quantum theory, matrices) which are non-commuting. The Heisenberg uncertainty principle applies to the pair.

conjugate impedances impedances with the same resistive component but equal and opposite reactive components

conjugate planes planes in optical systems such that the image of any point object in one plane lies in the other; **conjugate points** are points on the conjugate planes with that property, while **conjugate focal points** are the two points on the axis at which paraxial rays come to a focus when inbound from the two sides of the system.

consequent poles one or more poles in a magnet additional to the main pair

conservation law law which states that in any closed system (i.e. one with no interchange with the outside), some property or measure stays constant during any interaction. The laws of conservation of (constant) angular momentum, constant (conservation of) charge, constant energy, constant (linear) momentum, constant mass, and constant mass-energy are all of this type in classical and quantum systems. In the latter field, there are many others too.

conservative field force field in which the energy transfer in moving an object between any pair of points is constant, no matter what the path. It follows that there is no net energy transfer when an object moves all the way round a closed path. A **conservative force** is the force in such a field, and a **conservative system** is a system in which there are only conservative forces acting.

consonance combination of two or more musical notes which are generally agreed to have a pleasant effect. The frequencies of the notes are usually simple numerical ratios (e.g. $1 : 1.25 : 1.5 : 2$, for the major chord). Compare with *dissonance*.

constant a) not changing during an interaction, i.e. staying the same while other values change. This applies to the law of constant angular momentum, the law of constant energy, and so on.
b) an unchanging value, for instance a fundamental constant, as opposed to a variable

constantan alloy widely used in coil windings and thermocouples. It consists of copper and nickel in roughly equal amounts and has high resistivity and low temperature coefficient of resistance.

constant pressure gas thermometer standard thermometer whose active substance is a fixed mass of gas; the pressure is constant, and the gas volume varies with absolute temperature. The **constant volume gas thermometer** is also a standard whose active substance is a fixed mass of gas; in this case, the volume is constant, and the pressure varies with temperature.

constrain to restrict the position and/or motion of parts of a system. A **constraint** is any condition (such as 'reaction' force) that reduces the number of degrees of freedom in the system.

constringence or **Abbe constringence** V the

Abbe number of an optical system, the reciprocal of its dispersive power to white light. The constringence of a transparent substance is the ratio $(n - 1)/ \Delta n$, where n is the refractive constant at one wavelength, and Δn is the difference between the refractive constants at set wavelengths on either side.

constructive interference interference in which the interfering waves reinforce each other, to give constant high intensity (energy content) at the point concerned

contact a) state of touching
b) (also called **contactor**) any type of switch or electric link in a circuit

contact angle θ the angle between the edge of a liquid drop and a clean, horizontal, solid surface. The value depends on how much the liquid wets the surface; it therefore sometimes bears the name wetting angle. The limits are for the contact angle to be $0°$ (perfect wetting) and $180°$ (zero wetting); neither happens in practice. Mercury does not wet glass well so has a contact angle above $90°$; water does: its angle is less than $90°$.

contactor See *contact* b).

contact potential (or **contact potential difference**) the voltage that appears between two solid surfaces in contact, or between a solid surface and a liquid. The value (perhaps a few tenths of a volt) depends on the substances concerned, and on temperature: see *thermocouple* and *work function*. The resistance between the two touching surfaces, the **contact resistance**, affects the current between them, so is relevant in circuit design.

containment a) the prevention of the escape and release of radioactivity from the core of a nuclear reactor, by means of the **containment system** (for instance, shielding)
b) the process of preventing a plasma from touching the walls of its container during a fusion reaction. **Containment time** is how long the plasma ions remain trapped within the plasma.

continuity principle the change of mass of fluid within a space during continuous flow is the difference between the net mass flows into and out of the space. The **continuity equation** expresses the principle during a small interval of time in terms of the fluid density and the three components of the fluid flow velocity. A more general version of this equation describes meteorological cases and the flow of neutrons from a reactor core.

continuous flow calorimeter device (such as

that of Callendar and Barnes) that measures the specific thermal capacity of a fluid flowing through a tube rather than trapped in a container

continuous radiation radiation that shows a **continuous spectrum**, i.e. it includes all wavelengths in the range of interest rather than a number of sharp peaks. It may be called 'white', as in white light and white noise.

continuous wave wave that is uniform in time – neither pulsed nor of only short duration

control the use of one system or process (programs included) to affect the behaviour of another. A **control electrode** (such as a **control grid**) in an electron tube is one whose potential determines the current. The **control rods** of a fission reactor control the reaction rate by absorbing a greater or lesser fraction of the neutrons involved in the process.

convection a) process of thermal energy transfer by the net flow of fluid particles. In **natural convection**, the flow is the result only of temperature differences – and therefore density differences – within the fluid ('hot air rises'); it requires a gravitational field or equivalent to set up the closed **convection currents**. **Forced convection**, on the other hand, requires an energy supply from outside (for instance, from a fan or pump) to cause the fluid to flow (or to increase the natural rate). In forced convection the rate of transfer of energy to or from a surface in the fluid varies with ΔT – the temperature difference between the surface and the fluid. In natural convection, the rate varies with $\Delta T^{5/4}$. See also *Newton's law of cooling*.
b) in electricity, the net transfer of charges in free fall (for instance, outside the electron gun of a cathode ray tube), this forming an electric **convection current**

conventional current the electric current 'down' a conventional potential gradient. The convention, from the early days of work on currents at the start of the nineteenth century, is that electric current is from positive to negative. Electron current (as in metals and vacuum, for instance) is in the other direction.

convergent old term for converging, i.e. coming together, as do the rays of a **convergent beam** of radiation. Once the rays meet, they diverge unless absorbed: rays do not halt at a focus or image!

converging a) coming together, the same as convergent

b) bringing rays together, i.e. changing the directions of rays in a beam so they tend to come together, towards a focal point F, or to diverge less quickly. A **converging lens** is one that is thicker at the centre than at the edge (in the case of glass in air for instance); a **converging mirror** is a concave reflector. Either may be called positive, as its power is positive.

conversion nuclear process of changing a fertile substance into a fissile one, as in a breeder reactor (converter)

conversion electron See *internal conversion*.

conversion factor the same as breeding ratio

converter reactor the same as breeder reactor

convex curving outward, toward the viewpoint. A **convex lens** is one with two convex surfaces, and is thus thicker in the centre than at the edge; in other words it is a converging lens (if glass in air or similar). A **convexo-concave** lens has one convex and one concave surface, with the former more curved, so is converging. A **convex mirror** is a reflector with a convex surface – it is a diverging mirror.

coolant fluid used to transfer energy from a place at high temperature, such as the core of a nuclear reactor (where the system uses the energy gained by the coolant) or the engine block of a car (where it usually does not)

Coolidge tube early x-ray generator. See *x-ray tube*.

cooling fall in temperature as a result of energy loss by conduction, convection, evaporation or radiation. A **cooling correction** is a correction made to allow for such energy losses in experiments that involve temperature change. For instance, in calorimetry, the final temperature reached is lower than it should be in theory because of cooling; plotting a **cooling curve** (graph of temperature against time) allows one to assess the size of the correction needed. The **cooling method** is a way to measure the specific thermal capacity of a liquid by comparing the rate of cooling of identical volumes of the liquid and of water in identical situations.

Cooper, Leon (1930–) physicist who shared the 1972 Nobel Prize (with John Bardeen and John Schrieffer) for work on superconductivity

Cooper pair bound pair of particles involved in, for instance, the superfluidity of liquid helium-3. In this case, the particles are helium-3 atoms, and there are two phases, called A and B.

coordinates set of numbers representing the position of a point within a **coordinate system** with reference to the origin, the zero point of the system; each number describes one dimension. In the **Cartesian coordinate system**, each coordinate is the distance from a line ('axis') at 90° through the origin. We traditionally call the axes x, y and z in the case of three dimensions. **Polar coordinates** include one distance and an angle, in two dimensions – in 3D, the system becomes that of **cylindrical coordinates**. These are a version of spherical coordinates that place a point in space in terms of the polar coordinates of the foot of the perpendicular from the point to the x,y plane and the length z of that perpendicular; however, there are at least seven systems of cylindrical coordinates in common use.

In the system of **laboratory coordinates**, the observer is the point of reference. However, it is often more convenient to use the centre of mass of the interacting system being studied.

The coordinates of a point on the Earth's surface are its latitude and longitude; those of a point on the sky are declination and right ascension (see *celestial sphere*). In each case, both coordinates are angles and both systems are forms of **spherical coordinates** that require no third, distance, dimension.

coordination in physics, the same as linkage

Copenhagen interpretation The attempt by Bohr and Heisenberg in around 1927 to show that quantum mechanics can make sense to human minds. The basis is the interpretation by Born of the wave function as an expression of probability. The wider view of the Copenhagen model looks for instance at what would happen in the Young two-slit experiment if the input beam consists of only one photon and concludes that the next step in every aspect of our Universe depends on the probabilities of a finite number of possibilities.

Copernican principle that neither we nor any other observers in the Universe are in any way special – the Earth is *not* the centre of the Universe.

Copernican system the system of astronomy that envisages the Sun as the centre of the solar system. First proposed in India some ten thousand years ago (and repeated by Aristotle in the fourth century bce), the concept was revived by Copernicus; it came to replace the Earth-centred view that had developed in Europe since the time of the ancient Greeks.

Copernicus, Nicolas (1473–1543) scientist and mathematician (sometimes regarded as the founder of modern astronomy) who developed the theory of the sun-centred – rather than Earth-centred – Universe. This is the *Copernican system*: the sun at the centre and all the planets in orbits that are perfect circles. He wrote the crucial work *De revolutionibus* ('Of the revolutions [of the heavenly spheres]') in 1530, but it was not published until just before he died.

coplanar forces forces in the same plane (as must be the case with two parallel forces or two forces that act through a single point)

copper Cu malleable and ductile transitional metal of period 4, proton number 29, one of the nine elements known since ancient times (first uses some 10 000 years ago, the name coming from the Latin *Cyprium*, for Cyprus, its main source). Relative density 8.96, melting and boiling temperatures 1084 °C and 2565 °C; most common isotope ^{63}Cu. Its softness and good electrical conductivity make the metal of great value for electrical wiring and similar (including in micro-circuits where it is starting to replace aluminium); it is also part of many widely used alloys, such as brass and sterling silver.

copper loss the resistive loss of energy in a circuit, particularly in a motor, transformer or generator

Corbino effect the current around the edge of a conducting disc in an axial magfield that results from a current between centre and edge

core central, main part, such as
a) the nucleus and filled shells of an ion, a **core electron** being an electron in a filled shell
b) the block of a nuclear reactor that contains the fuel, etc.
c) the iron part of an electrical machine (such as generator, motor, transformer). In the last case hysteresis effects and eddy currents while the machine is working result in the **core loss** (or iron loss) of energy and reduce efficiency.

core-type transformer type of transformer in which the coils surround most of the core, rather than only a small fraction of it

Coriolis effect the effect that results in the scouring of river banks and the circular motion of air (winds) around cyclones and anticyclones. First described by Gaspard de Coriolis (1792–1843), the fictitious **Coriolis force** causes the path of objects moving at constant velocity to appear to curve when viewed from a rotating reference point. The apparent **Coriolis acceleration** is at 90° to the true motion, to the right in the northern hemisphere and to the left in the southern; it varies with the true velocity and the rate of rotation.

corkscrew rule turning a corkscrew in the direction of conventional current in a wire makes the handle turn in the direction of the field round the wire

cornea the transparent outer layer of the eye, through which light passes to the iris that surrounds the pupil. Its curvature provides the main focusing effect.

Cornell, Eric (1961–) one of the three joint winners of the 2001 Nobel Prize, for being able to obtain Bose–Einstein condensates in 'dilute gases of alkali atoms', and for research into the properties of these condensates

corner cube device like one corner of a cube that reflects inward radiation directly back towards its source. The reflectors used on cycles and cars work on the same principle, and so, in effect, does the type of road marking reflector called a cat's eye.

Cornu prism prism of, for instance, quartz, used at minimum deviation to overcome the effect of double refraction. The **Cornu spiral** is a double spiral curve used to determine the intensities of radiation at different points in a Fresnel diffraction pattern.

corona faint glow around objects that follows

ionisation of the nearby air as a result of a high potential gradient (e.g. in an electrical storm). St Elmo's fire (see *brush discharge*) is an example.

corpuscular theory the view, held from the time of Newton until the early nineteenth century, that light travels as fast 'corpuscles' (matter particles). Light (and some other forms of electromagnetic radiation) is now thought of in terms of particles called photons which also have wave properties.

correcting plate large, thin lens, often of complex shape, used to correct the aberrations of mirrors. The Schmidt lens used in a type of astronomical telescope is a well known example.

correlation the closeness of association between two variable values. The **correlation coefficient** (symbol r) is the measure of this – a value of 1 indicates complete correlation, 0 none at all, and –1 a complete inverse correlation. **Correlation energy** is the energy associated with the repulsion between electrons in atoms and molecules.

correspondence principle principle, first stated by Niels Bohr, that when the values of quantum numbers are high, quantum systems reduce to their classical form – in other words, that quantum mechanics applied to relevant systems will extrapolate to a classical description of the corresponding non-quantum systems as we cross the boundary between them. So, are the classical laws of physics limiting cases of the laws of quantum physics? At the moment, this is of special interest at the limits of quantum chaos theory.

cosmic

involving or associated with the whole Universe (the 'cosmos'), rather than just our small part of it. Thus a **cosmic law** is one that has universal validity.

cosmic inflation stage in the very early Universe – see *inflation*.

cosmic microwave background a highly uniform microwave radiation from the Universe that provides evidence for the big bang theory. Its effective temperature is around 3 K, and its red shift is about 1500. See *microwave background radiation*.

cosmic radiation radiation from space observed at the Earth's surface, called **cosmic rays** because the source is outside the Earth (but it's not the same as cosmic background radiation). The **primary**

cosmic rays (i.e. those that enter the atmosphere) are nuclei of high energy (a few million eV to perhaps 10^{20} eV, about 1 J). Most of these nuclei are protons (hydrogen nuclei) but people have come across nuclei of elements up to nickel. In the atmosphere, the particles quickly lose energy; each one leads to a shower of **secondary cosmic rays**, a brief but intense beam of electrons, nucleons, neutrinos, mesons, x-photons and gamma photons, and so on, formed when one high-energy cosmic ray particle interacts with a molecule high in the atmosphere. Such a secondary shower (or Auger shower) is a cone that may cover tens of square kilometres at the Earth's surface.

The Pierre Auger **cosmic ray telescope** in Argentina, which started work in 2006, has water tanks and detectors spread over 3000 km^2. The staff hope to detect some thirty showers a year – and to be able to pinpoint the direction from which each one's primary particle came. See also *nuclear physics: accelerators*.

cosmic strings See *strings*.

cosmogony the science of the origin and development of the whole Universe (cosmos) and of such parts as galaxies the solar system

cosmological coincidence the fact, first noted by Paul Dirac that the ratio of the strengths of the Coulomb (electric) force to the gravitational force between two charged particles is very close to the ratio of the size of the Universe to the size of an electron. Whether this is a coincidence or not, a number of later theories 'prove' it; see, for instance, *zero point energy*.

cosmological constant Λ constant that appears in a term in Einstein's gravitational field equations that relates to the expansion of the Universe; interpreted as the energy density of vacuum, it is constant throughout the age of the Universe. Einstein's reasons for using it no longer apply (he called it 'the greatest blunder of my career'), but modern theorists still cannot get rid of it; it therefore provides major problems of interpretation. For instance, it is not clear why the value is as small as it is, though current theories of the cyclic Universe (see **big crunch**) are better at explaining that than most others.

cosmological principle that the large scale Universe is homogeneous and isotropic (the same in all directions and in all parts)

cosmology the branch of astronomy concerned with the structure and history of the whole Universe (cosmos)

Cosmotron the 3 GeV proton synchrotron built at the Brookhaven Laboratory (USA) in the 1960s, in its time the world's most powerful accelerator. The name follows its use to reveal particles found before only in cosmic radiation.

Cotton–Mouton effect the double refraction observed in some solids and pure liquids in a transverse magnetic field. See also *Kerr effect*.

Coudé system or **Coudé telescope** reflecting telescope in which light passes from the main mirror (objective) to a diverging mirror on its axis, and from there to a plane mirror (also on the axis) which reflects it through the side of the tube to the eye lens

coulomb C the unit of electric charge; the charge transferred in one second by one ampere (i.e. it is one ampere-second, As). It is equivalent to the charge on 6.24×10^{18} electrons.

Coulomb barrier the potential barrier around a charge, in particular, around a nucleus. A charged particle that enters or leaves the nucleus must pass through this.

Coulomb, Charles Augustin de (1736–1806) engineer who started to do research during the French Revolution. His main work was in static electricity and magnetism, which is why the unit of charge bears his name. He invented the torsion balance to measure electric, electrostatic and magnetic forces.

Coulomb energy the binding energy in a system (such as a nucleus or a solid) associated with the electric forces between its particles

Coulomb excitation the energy gained by a nucleus from a passing charged particle that does not pass through the Coulomb barrier

Coulomb field common name for an electric field

Coulomb force or **Coulomb interaction** the electric force between two charges, thought to involve the exchange of a photon. The **Coulomb law** defines that force F in terms of the charges Q_1 and Q_2 and the distance d between them:

$$F = Q_1 \, Q_2 / 4 \, \pi \, \varepsilon \, d^2$$

Here ε is the permittivity of the medium around the charges. This force leads to, among other things, **Coulomb collisions** or **Coulomb scattering**, the deflection of positive particles (such as protons and alpha particles) by the Coulomb barriers of the nuclei they approach or of negative particles by electron clouds.

Coulomb's theorem relationship that gives the strength E of an electric field near a surface of charge density σ:

$$E = \sigma / \varepsilon$$

coulometer or **coulombmeter** device able to measure electric charge (or current) by the electrolysis produced in a cell ('voltameter')

counter device for counting the number of current pulses in a circuit, or the number of ionising particles or photons passing through it. In the latter case, the name applies to the detector only, or to the detector and the electronics (with output). **Counter efficiency** is the fraction of valid events that are counted; **counter recovery time** is the time after it detects an event before the system can detect again.

counter meter or **frequency meter** instrument that contains a frequency standard (such as a quartz oscillator) that allows it to be used for timing; for counting pulses or events in a given time; or for frequency measurement

couple a) pair of equal antiparallel forces **b)** their resultant net torque (turning effect) T. With d the perpendicular distance between the lines of actions of the forces each of size F, $T = F \, d$. The unit is the newton-metre, N m.

coupled systems separate systems of the same type – for instance, pendulums or ac circuits – which interact so that energy passes between them. This coupling affects the two amplitudes and frequencies of oscillation. Most musical instruments involve coupled systems.

coupling the interaction between two coupled systems. It is clear that various properties of nuclear, atomic and molecular systems couple: they affect each other through energy transfer. Chemical bonds, for instance, are the result of coupling.

The design of the coupling between two **coupled circuits** in electronics is to allow the correct type and size of energy transfer. Often the link is a reactance; then the **coupling constant** K is:

$$K = X_C / X_1 \, X_2$$

Here X_C is the reactance common to both circuits while X_1 and X_2 are the individual reactances of the same type.

The value of the coupling constant gives the degree of coupling between the two circuits. This approach is **capacitative coupling** if the reactors are capacitors and **inductive coupling** if they are inductors.

C-parity or **CP** charge parity, in full *charge conjugation*/parity, a quantum number with values –1 and +1, whose symmetry under the weak nuclear force was shown to be broken by Cronin and Fitch in 1964

CPT charge conjugation/parity/time reversal, trio of measures individually thought to be conserved under reversal in nuclear events – but one by one found not to be. Now, it is believed (or hoped?) that the three together must be preserved if current quantum physics is correct – **CPT symmetry** is a fundamental basis of all laws of physics.

crack line or plane of lower density, and thus of weakness, in a solid sample. As the greatest stress is at the ends or edges of the crack, once formed it will tend to propagate. The stress is tensile, so cracking tends to cause the sample to break; compare with *anti-crack* where the stress is compressive.

CR circuit circuit that contains only capacitance and resistance (or the equivalent of these)

creep plastic flow: the slow, permanent deformation of a plastic sample when under constant stress. There are three types. In **primary creep**, creep rate falls with time; in **secondary creep** the creep rate is constant; while **tertiary creep** involves a creep rate that rises until the sample fractures. **Thermal creep** is the increase of creep rate with temperature.

critical concerned with a turning point in some aspect of some feature of a substance or system

critical angle the angle of incidence on the surface, for light passing into a medium which is optically less dense, at which the emerging light grazes the surface (i.e. leaves it at just below 90°). The sine of the critical angle is the reciprocal of the refractive constant of the first medium relative to the second. The **critical angle refractometer** is a device in which grazing occurs at the interface between a substance of known refractive constant and one whose refractive constant is unknown, thus allowing the user to find the unknown value. See also *total internal reflection*.

critical constants the critical temperature, critical pressure and critical volume of a substance

critical damping See *damping*.

critical density the mass of a sample of substance divided by its critical volume

critical field for a superconductor at a given temperature, the magnetic field strength at which superconductivity will vanish

critical flicker frequency the frequency above

which the eye/brain system can no longer detect the flicker of an image

critical frequency or **cut-off frequency** the signal frequency in an electric or acoustic circuit or transfer system at which there is a sudden change between low and high attenuation

critical isothermal the isothermal on a substance's pressure/volume graph at the critical temperature. The curve shows a single point of inflection there. See *Andrews' experiments*.

critical mass the smallest mass of a fissile element able to maintain a chain reaction or critical reaction

critical point same as critical temperature

critical potential the excitation energy of a nucleus, atom or molecule, i.e. the energy involved in a jump between specific energy levels (in most cases, between the ground state and a higher level)

critical pressure the saturated vapour pressure of a substance at its critical temperature

critical reaction reaction in, for instance, a nuclear reactor, in which each transformation leads to exactly one more such transformation, i.e. the multiplication factor is 1. If the nuclear reaction depends only on prompt neutrons, it is **prompt critical**; if, on the other hand, it needs delayed neutrons as well, it is **delayed critical**.

critical specific volume the reciprocal of the critical density of a substance

critical speed (sometimes called **critical velocity**) the speed of a flowing fluid at which smooth (laminar) flow changes to turbulent flow. See also *Reynold's number*.

critical stage the stage in a process at which there is a sudden change of property or output (see, for instance, critical angle)

critical state the state of a fluid at which its liquid and gas phases have the same density; this occurs at the critical temperature

critical temperature the temperature above which it is not possible to convert a substance from gas to liquid form at a given pressure

critical velocity See *critical speed*.

critical volume the volume of a sample (or often of a mole) of substance at the critical temperature and critical pressure

cro or **CRO** cathode ray oscilloscope

Cronin, James (1931–) particle physicist who shared the Nobel Prize in 1980 with Val Fitch. Their crucial experiment of 1964 showed that the decay of neutral kaons is not symmetrical with respect to time reversal – the half-lives of the K^0 and its antiparticle are not quite the same.

Crookes' glass glass which is highly opaque to ultraviolet; it has a high content of cerium and other rare earth elements, and is used in spectacle lenses to reduce damage

Crookes space (sometimes incorrectly called **cathode space** or **Faraday space**) the comparatively dark region between the negative glow and the positive column in the discharge of a low-pressure gas, discovered by William Crookes in the late 1870s. It is the main region in a discharge tube where there is little or no glow – it almost fills the tube at very low pressures. In the past, the name applied to the whole region between the cathode and the positive column, but we now know this has a fine structure – see *discharge tube*.

Crookes' tubes low-pressure gas tubes used in the study or discharge

Crookes, William (1832–1919) chemist and physicist (also deeply interested in the occult) who invented the radiometer and spinthariscope; improved the design of electron tubes; discovered helium, thallium, cathode rays, and scintillation; and advanced our knowledge of spectra, gas discharge, and canal and cathode rays.

crossed cylinder weak lens with the effect of equal converging and diverging cylindrical lenses crossed at the common axis at 90°

crossed lens spherical lens with minimum spherical aberration to inbound parallel light

crossed Polaroids pair of Polaroid sheets crossed at 90° – light cannot pass through

cross linking the bonding between the sides of the chain molecules of a polymer

crossover network filter circuit with two output paths, one for frequencies above a certain value (the **crossover frequency**), and the other for those below. At the crossover frequency, the two outputs are equal. A major use is in audio systems, to feed the bass and treble in a signal to the correct speakers.

cross product symbol × the vector product of two vectors, i.e. the product that is also a vector, one that is at 90° to both the original two. For instance, torque is force times distance from the turning point – this is a cross product: $\mathbf{T} = \mathbf{F} \times \mathbf{d} = F\,d \sin \theta$. ($\theta$ is the angle between F and d.)

cross-section σ unit: square metre m² measure of the probability of a given interaction, for instance between a target nucleus and a particle in a beam, expressed as an area. There are various expressions and definitions. One is that if d is the greatest distance for interaction between given target and particle at a given particle energy, $\sigma = \pi\, d^2$. See also *barn*, a common non-SI unit of section area in this context.

cross-talk interference between the signals in nearby channels of communication, such as close radio frequency bands or phone cables

crown glasses range of optical glasses with high refractive constants and densities. Flint glasses are another such range.

crt or **CRT** cathode ray tube

cryogen refrigerant for work at temperatures below 123 K

cryogenics the science and applied science of temperatures below 123 K and their effects on and in matter

cryometer thermometer designed to work at temperatures below 123 K

cryoscopy measuring low freezing temperatures and the effect on these of impurities (which gives the relative molecular masses of the impurities)

cryostat device to maintain a very low temperature within a set range

cryotron switch that works by superconductivity: a current in a superconducting coil (the control) switches the current in an axial wire (the gate). Cryotrons have much potential in superconducting computers, for they can carry out binary logical operations and act as storage cells.

crystal a) the form of a substance with long-range order in the three-dimensional arrangement of its atoms

b) sample of such a solid substance with visible crystal faces at characteristic angles. See also *liquid crystal*.

crystal analysis See *x-ray crystallography*.

crystal base the entire contents of one unit cell of a crystal – see *crystal structure*.

crystal clock highly accurate clock, able to work within 1 ms per day. Feedback between electrostriction and the piezoelectric effect causes a carefully cut crystal (often of quartz) to resonate at a precise frequency around 100kHz. The circuit, then, is a **crystal-controlled oscillator** or **crystal oscillator** which offers a very high frequency stability.

crystal counter early semiconductor counter of ionising particles and photons – its working substance was a crystal.

crystal diffraction the reflection by crystal planes of a wave passing through a crystal, leading to constructive and destructive interference effects. It appears only if the wavelength is of the same order of magnitude as the inter-plane spacing. See also *x-ray crystallography*.

crystal dimension all crystals have three dimensions, and most can grow in all three directions. However, some crystals cannot grow at all, so we call them zero-dimensional; others – one- and two-dimensional crystals – can grow in only one or two directions. See *graphite* for examples.

crystal filter electronic circuit that contains one or more piezoelectric crystals, the resonance or antiresonance of which filters the input signal

crystal grating crystal used as a diffraction grating; see *crystal diffraction*.

crystal habit the visible appearance of a solid crystal, in terms of the faces and the angles between them. It reflects the structure in terms of the unit cell in the **crystal lattice**, this being the arrangement in space of points that represent the centres of the particles of the crystal. There are fourteen forms of lattice.

crystalline of a solid substance, having a regular crystal structure and, perhaps, matching external appearance. See also *liquid crystal*.

crystallography the study of the structure, forms and properties of crystals, in the last century using x-rays

crystal microphone microphone that uses the piezoelectric effect. Input sound waves stress the crystal to produce a corresponding voltage (signal) at the output.

crystal oscillator See *crystal clock*.

crystal parameters the distances from the origin of the axes of a solid crystal to where each axis cuts a face; the unit is arbitrary (as there is no fixed size for a crystal of a given shape). See also *axial length* a).

crystal radio set first design of radio set (c.1906), which used a galena crystal rectifier as a semiconductor diode. The **crystal rectifier** (a name no longer in use) was a crystal mounted on one electrode in a circuit made up of a long wire aerial, a simple tuning capacitor, and a passive headset (one that needs no power); indeed the whole circuit is passive – the only power comes from the input radio signal. The other electrode was the 'cat's whisker', a fine, sharp pointed wire that one jiggled over the crystal surface while trying to find a signal. The design is still in use to an extent today, but with a modern solid-state diode rather than a crystal.

crystal structure the regular, repeated, three-dimensional pattern of particles (atoms, ions, or molecules) and planes of particles in a crystalline sample. The unit cell is the smallest three-dimensional pattern that can repeat to build up the macroscopic structure; it consists of a set of points in the *crystal lattice*, and the angles between its lines and planes describe the **crystal system**. In any given case, this is one of seven groupings of crystals based on the fourteen forms of crystal lattice and the 32 possible relevant symmetry operations. Each has its own internal structure (that unit cell that relates to the three crystal axes) and perhaps a corresponding visible form (crystal habit). The seven crystal types are

a) *triclinic* (or *anorthic*) – the axes need be neither equal nor perpendicular

b) *cubic* – the axes are equal and perpendicular

c) *hexagonal* – there are two equal axes at 120°

d) *monoclinic* – one axis is perpendicular to the plane of the other two

e) *orthorhombic* – the axes are perpendicular but need not be equal

f) *tetragonal* – the axes are perpendicular and two are equal

g) *trigonal* (or *rhombohedral*) – the axes are equal and at equal angles less than 120°

C-symmetry the symmetry of physical laws, systems and reactions under *charge conjugation* – the initial state and the conjugate (anti-state) are the same. Weak interactions are the only ones known to break C-symmetry. See **CPT**.

ct (or **CT**) computerised tomography

cubic expansivity γ unit: per kelvin, K^{-1} the volume expansivity of a substance: the change of volume of unit volume per degree temperature change

cubic system see *crystal system*

curie old unit of radioactivity equal to 3.7×10^{10} becquerel (disintegrations per second). The curie is the activity of about 1 g of pure radium.

Curie balance device to measure the susceptibility of dia- and paramagnetic substances. It compares the force on a sample in a field with that on a similar sample of distilled water.

Curie constant

Curie constant C the product of the volume susceptibility χ_m of unit mass or mole of a paramagnetic substance and its temperature T. The susceptibility is the ratio of the sample's degree of magnetisation, M (in A m^{-1}), to the applied field strength, B (also in A m^{-1}). It is approximately constant in many cases, up to the Curie temperature T_C, as stated by the **Curie law**:

$$C = \chi_m\, T = M\, T/B$$

Curie, Irène Joliot- (1897–1956) physicist, elder daughter of Pierre and Marie Curie, who took over her mother's work in 1932 (in which year they jointly discovered the neutron, although they did not recognise it). In 1934, Irène and her husband Jean-Frédéric Joliot (later known as **Joliot-Curie**) (1900–1958), produced the first artificial isotopes; for this the couple received the Nobel Prize for chemistry the next year. Both died from radiation-induced cancer.

Curie, Marie (born Marja Sklodowska, 1867–1934) wife of **Pierre Curie** (1859–1906), with whom she discovered radium in 1895 and then polonium in 1898. They went on to discover artificial radioactivity and to share (with Becquerel) the 1903 Nobel Prize. After her husband died in a road accident, she took over his work as professor of physics – but in 1911 became the first and only person to receive a second Nobel Prize in a different field – this time in chemistry (but still for her work in radioactivity). She died from radiation sickness, as did her daughter Irène Joliot-Curie and *her* husband Frédéric Joliot.

Curie range of a ferroelectric substance, the temperature band in which it shows ferroelectricity most strongly; there is also a large relative permittivity in this region.

Curie temperature T_C **a)** for a ferromagnetic substance, the temperature above which it has no ferromagnetism. Symbol T_C, it is the constant temperature that appears in the Curie law for the susceptibility χ of a given substance at temperature T:

$$\chi = C/(T - T_C)$$

In this case, C is the Curie constant for the substance concerned.

b) in the case of paramagnetism, the temperature T_C in the Curie–Weiss law, a modification of the Curie law
c) by extension to piezoelectricity, the

temperature above which a piezoelectric has no polarisation
Curie–Weiss law for a paramagnetic substance, the version of the Curie law that applies above the Curie temperature T_C, as in

$$C = \chi_m\,(T - T_C)$$

This has a singularity at the Curie temperature, at and below which the substance shows a degree of spontaneous magnetisation.

curium Cm trans-uranic actinide rare earth of period 7, proton number 96, first made in 1944 by Glenn Seaborg and Albert Ghiorso and named after Marie and Pierre Curie. Relative density 13.51, melting and boiling temperatures 1340 °C and 3110 °C; most stable isotope (of 19) ^{250}Cm, alpha emitter with half-life 15.6 million years. The metal has a major use as a fuel for nuclear thermoelectric generators.

curl or **rotation** vector operator that shows the rotation of a vector field **F**. It is the vector product of the vector differential operator del, Δ, and **F**.

$$\text{curl } \mathbf{F} = \Delta \times \mathbf{F}$$

If the value is zero at all points the field is irrotational, otherwise it is a rotational field. In all this, rotation is a field property that depends on position – it does not change with time.

current a) rate of flow of any fluid (liquid or gas), no special symbol. The unit is the cubic metre per second, m^3 s^{-1}.
b) symbol I unit: amp or ampere, A, the net flow of charge Q between two points, caused by a voltage (or electric field). Its value is the rate of flow of charge dQ/dt; the unit is the coulomb per second, called the ampere, equivalent to the transfer of some 6.2×10^{18} electrons per second.

The charge carriers are not always electrons, though they are in metals, n-type semiconductors and vacuum; ions are the charge carriers in electrolytes and gases, while holes carry current in intrinsic and p-type semiconductors. See *conduction* and *conduction current*, which involves the net flow of free charges through a voltage (in a field); in this context, a *convection current b)*, on the other hand, is the net transfer of charges in free fall.

In the case of ac, or the transitory charge/discharge current in dc, a **displacement current** inside the dielectric of a capacitor

equals the conduction current in the outside circuit; it follows the changing electric field in the substance. Part of this *displacement* current is the **polarisation current**, the effect of the movement of bound charges.
c) See also *neutral current*.

current balance device for 'weighing' current; it measures the force between the current in two coils, one fixed, the other an arm of a balance. It is thus an absolute ammeter.

current carrier See *carrier* a).

current density j unit: ampere per square metre, A m^{-2} the current in a conductor or beam of charged particles per unit area (overall or at a point)

current transformer electrical transformer used with a meter to isolate the meter from the supply, and/or to step the input current up or down into the meter's range

curvature R unit (same as that of power): the dioptre, D the reciprocal of the radius of curvature, r, of an arc or spherical surface. The radius is the distance between the arc or surface and the system's centre of curvature; the smaller the radius, the greater is the curvature. A plane wave has zero curvature ($r = \infty$); a spherical surface gives it a curvature equal to its reflecting or refracting power, i.e. to $1/f$, where f is the focal distance.

curvature of field See *aberration*.

cut-off bias the overall grid voltage in a given cathode ray tube that will just reduce the electron beam to zero. The value depends not only on the design of the electron gun but also on the pds between other electrodes.

cut-off frequency same as *critical frequency*

cybernetics the science and applied science of control and information transfer in living and non-living systems

cycle single complete oscillation in any oscillating system, or the complete set of changes in the periodic function that shows it. The unit of frequency is the hertz (Hz), one cycle per second.

cyclic accelerator accelerator in which the charged particles (in most cases, positive ions) in the beam pass again and again through the same (or much the same) path in a magfield in a vacuum. During each cycle, they gain energy. See *nuclear physics: accelerators*.

cyclotron the classic example of a cyclic accelerator, first devised by Lawrence and others in 1930. It works on the principle of **cyclotron resonance:** each particle gains energy from the alternating electric field through which it passes as it crosses the gap between the dees (hollow D-shaped sections). See *nuclear physics: accelerators*. **Cyclotron resonance frequency** is the orbit frequency of a given charged particle in a given magnetic field; it depends on the field strength and the mass and charge of the particle. See also *betatron*.

cyclotron radiation is electromagnetic radiation output from any charged particles moving for any reason in a curve. The cause is the acceleration of the particles when moving this way even at constant speed. As well as accelerators, there are many astronomical sources, the closest being the solar wind focussed by Jupiter's magfield. See also *synchrotron radiation*, which is much the same except that now the particles are moving at relativistic speeds.

cylinder solid object with two identical parallel plane ends (often circles), joined by parallel straight lines. The volume is the product of the area of one end (or any parallel, or principal, cross-section) and the perpendicular distance between the ends. A cylindrical object has the shape of a cylinder or with one or more surfaces being circles. **Cylindrical coordinates** are a version of spherical coordinates; there are a number of types – one places a point in space in terms of the polar coordinates (r, θ) of the foot of the perpendicular from that point to the x/y plane and the length z of that perpendicular.

cylindrical lens 'toric' lens, one whose two refracting surfaces are sections of a cylinder. As with a spherical lens (where the surface sections are of a sphere), the lens can be bi-concave, concavo-convex or bi-convex – converging or diverging. The lenses used to correct for astigmatism have one cylindrical and one spherical surface. A **cylindrical mirror** has a cylindrical surface.

cylindrical wave wave whose wave fronts form a set of coaxial cylinders

cylindrical winding method of winding a coil of a transformer – the coil is wound in a helix, and may be single-layer or multilayer

d unit symbol for deci-, 10^{-1}

D a) symbol for dioptre, unit of power in optics and unit of curvature
b) symbol for dimension, as in 3D and 11D
d **a)** symbol for diameter
b) symbol for distance (*s* more usual)
c) symbol for relative density, no unit
D **a)** symbol for electric displacement
b) symbol for electric field strength
c) symbol for diffusion constant (or coefficient)
δ symbol for a very small change in some measure
Δ symbol for a finite change in some measure
da unit symbol for deca, or (in mainland Europe) deka, 10^1

Daguerre, Louis (1789–1851) scenery painter who, in 1837, developed the first practical approach to photography, the pictures being **daguerrotypes**. A daguerreotype, common in the 1840s and 1850s, was a non-replicable photograph on a copper plate coated with silver and sensitised in iodine vapour. It was exposed for several minutes and then developed by fuming with mercury vapour. The fixing process uses common salt solution or sodium hyposulphite.

d'Alembert's principle for a system of linked moving particles, the difference between the forces from outside and the internal forces forms a system in kinetic equilibrium. The concept (1742), for which there are many statements and formulas, follows from Newton's third law of force. **Jean d'Alembert** (1717–1783), physicist and mathematician, also helped Diderot edit a major encyclopedia.

Dalén, Gustaf (1869–1937) engineer who received the Nobel Prize in 1912 for inventing the 'sun valve'. This used the level of natural light to control the output of gas lamps in buoys and unstaffed light ships. The award

was controversial for a number of reasons, but is closer to the spirit of Nobel Prizes than many, for it is applied physics rather than pure. (Nobel's will spoke of 'the benefit to mankind'.)

dalton Da modern name for the unified atomic mass unit, one-twelfth of the mass of a carbon-12 atom

Dalton, John (1766–1844) schoolteacher and active natural historian (biologist), who first described colour blindness. The most common red-green type is still sometimes called **daltonism**. His research later moved to what we now call physics and chemistry, and his ideas on the atomic structure of matter were the first useful ones in modern times. He believed that the atoms of different elements differed in their weights (masses), and in 1803 he published a table of relative atomic mass values; the data were the result of his work on the laws of chemical composition.

Dalton's law the total pressure of a mixture of gases is the sum of the partial pressures of the gases under the same conditions, where the partial pressure of a gas is the pressure it would have if alone in the same space. Strictly the law (Dalton 1801) applies only to ideal gases.

damped of a vibration, falling in amplitude with time as the result of some friction or resistance effect – this is a **damped oscillation**

damper shock absorber, device or system designed to increase damping, in most cases to just below critical

damping causing the fall in amplitude of a *damped oscillation* (a) as a result of some kind of opposition. If the resistance is just enough to prevent the oscillation, the system shows *critical damping* (b); lower resistance than that

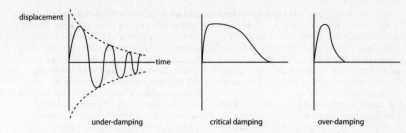

displacement

time

under-damping   critical damping   over-damping

gives under-damping (as in a), while a higher value leads to over-damping (c).

For small amplitudes, the force of resistance F varies with velocity v or, more generally, rate of change of state x):

$$F = m\,a = -k\,x - \mu\,v$$

Here m, k and μ relate respectively to inertial, elastic and resistive factors.

The system's **damping factor** α is $\mu/2\,m$. Some people use this term for the ratio of the amplitude of one cycle to that of the next cycle.

Daniell cell a primary cell once used as a standard voltage source. The positive electrode is a copper rod in copper (II) sulphate solution; the other is zinc–mercury amalgam in dilute sulphuric acid (to give 1.08 V) or in zinc sulphate solution (1.11 V). A porous wall keeps the liquids apart but lets the ions through.

dark adaptation See *adaption*.

dark current the (very) small current through a photo-cell when in the dark

dark energy the main component of the Universe, about 74%, after 'normal matter' (4%) and '*dark matter*' (22%). While the effects of dark energy are clear, its nature is not. Dark energy is likely to have been the 'fuel' for the expansion of the universe for at least nine billion years. It may well relate to Einstein's *cosmological constant*, the concept that even 'empty' space has density and pressure.

dark field illumination or **dark ground illumination** approach to microscopical viewing of small transparent objects that are hard to see against a bright background. An opaque disc covers the centre of the condenser lens so that only light from its edge reaches the object; non-scattered light does not then enter the object lens.

dark matter the fraction of the matter in the whole Universe that is very hard to observe. It is shown by how gravity calculations differ between using the calculated mass of the Universe and using the mass of all observed objects – gravity is the only force that affects it. Dark matter was first proposed in the 1930s; the missing mass could consist of dust, objects hidden by dust, unobserved particles (such as *wimps*) and/or energy, very faint stars (such as brown dwarfs), neutron stars and strange stars (like pulsars), as well as forms of matter unknown at present. One way to explain part of the gap is to accept that *neutrinos* have mass; while that mass is sure to be very small, neutrinos are very common. No particles have yet been found with that property of being subject only to gravity, but see *axion* and 'sterile *neutrino*'.

'The calculated mass of the Universe' comes from the rate at which the Universe expands. The missing mass figures come now from the study of the harmonics of the sound waves that caused tiny variations in the *microwave background* radiation. Current theories of the proportion and importance of this cold dark matter are under potential threat from various types of observation of, for instance, quasars and the cosmic microwave radiation background. The only clear evidence for dark matter in the skies is in the distribution of mass in the 'Bullet cluster', two colliding clusters of galaxies some four billion light years away. The study of gravitational lensing shows the distribution of the mass of the two clusters through space; in 2006, that there must be dark matter there became very clear (though this source had been suspected for some years). The only way to explore it on Earth is using very powerful accelerators – there are hopes that the large hadron collider (due to start working early in 2008) will produce particles with enough energy for this.

There are many good introductions to the subject on the web – this one is a good start: astro.berkeley.edu/~mwhite/darkmatter/dm. For an early report on dark matter and the 'Bullet cluster' see http://physicsweb.org/article/news/10/8/17; there's also a very good *Scientific American* article on this – February 2007, pp. 24–31.

dark space any part of the continuous electric discharge through a gas at low pressure where any glow is faint. See *discharge tube*.

Darmstadt German town, site of the world's most powerful heavy ion accelerators. Set up in 1969, the research campus has used its main linear accelerator and 200 m synchrotron, for instance, to produce the six elements with the highest proton numbers known. Plans are to complete by 2014 a new pair of 1100 m synchrotrons, fed from the older machines. Called FAIR (Facility for anti-proton and ion research), this international centre should produce beams of up to 10^{11} ions per second at energies up to around 35 GeV/c^2.

darmstadtium, Ds transitional trans-actinide metal, period 7, with proton number 110, prepared in 1994 by Sigurd Hofmann and team, who named it after the German town Darmstadt where they were working. No properties are known or uses – even the most stable isotope, 281 Ds, has a half-life of only 11 s.

data a) information inside an IT system (where it can have no meaning). There the system may transfer, process and/or store it. **Data capture** is the automatic input of data from one or more sensors into an IT system; **data logging** is the same on a large scale over time.

b) the results of some research, whether in an IT system or not

dating the use of the knowledge of some process of radioactive decay in a substance to estimate the time since it was formed. Different decay processes help us estimate the age of organic archeological remains, fossils and rocks.

a) Carbon dating (or radiocarbon dating) is the most common method for archeology, as it can give results for organic remains (for instance, cooked foods, charcoal, bone tools, wood in building works). Carbon exists in the environment as stable ^{12}C with a fairly constant ratio of radioactive ^{14}C. All living things therefore contain carbon in the same ratio. On death, however, they no longer take

in carbon so, as the ^{14}C decays, the ratio falls. Radiocarbon dating involves finding the ratio of ^{14}C in the carbon of the sample of organic matter; knowing its half-life – 5730 years – one can then obtain a value for time since death. 'Finding the ratio' used to involve finding the radioactivity; now, on the other hand, accelerator-based mass spectroscopy lets one count all the ^{14}C atoms left in the sample – this is far more sensitive, so can work with milligram carbon samples.

Until the late 1980s, radiocarbon dating was accurate to only about 11 000 years bp (before present, in fact before 1950), as accuracy could be confirmed only by correlation with tree ring dates. Now research into such decays as that of uranium into thorium in ancient coral beds has increased the limit to around 60 000 years bp. The limit exists because, without calibration, radiocarbon dating must assume that the ratio of ^{14}C to ^{12}C in the atmosphere remains constant with time, and this is far from the case, even before 1950. In addition, it is hard to measure the tiny fractions of ^{14}C that remain after five or six half-lives. For these reasons, it is believed that radiocarbon ages of 30 000 years may be as much as 5000 years in error.

b) Potassium–argon dating, used in geology to date rocks, depends on the radioactive decay of potassium-40, ^{40}K; this is present as a tiny fraction of potassium (which is widespread in rocks and soil). With a half-life of 1.28×10^9 years, this decays to ^{40}Ar. This dating method, believed to offer fair reliability to about 10^7 years, involves measuring the ratio of ^{40}K :^{40}Ar. As the half-life is so long, the method is of value up to a few billion years ago.

c) Rubidium–strontium dating works in the same way. The ratio of rubidium-87 to strontium-87 is used to date rocks to several thousand million years. The half-life of ^{87}Rb (which exists as just over a quarter of natural rubidium) to ^{87}Sr is 5×10^{11} years.

d) Uranium–lead dating depends on the decay of uranium-238 to lead-206. Because of the very long half-life (4.5×10^9 years) of ^{238}U, the technique is of value for dating even the oldest rocks. A variant depends on the decay of ^{235}U to ^{207}Pb.

daughter out-dated term for the product of some process or reaction, also called child or son. Thus, the nucleus left by a radioactive decay is sometimes called a daughter (or child); the preferred term is decay product.

Davis, Ray (1914–2006) physical chemist who shared the 2002 Nobel Prize with Masatoshi Koshiba 'for pioneering contributions to astrophysics, in particular for the detection of cosmic neutrinos'

Davisson-Germer experiment the first experiment that showed that moving particles have wave features. Performed in 1927 by Clinton Davisson (1881–1958) and Lester Germer (1896–1971), it involved the diffraction of electrons in a beam by a nickel crystal. Davisson received the Nobel Prize for this work in 1937.

Davy, Humphry (1778–1829) popular teacher, chemist and physicist who founded the science of electrochemistry. In the first decade of the nineteenth century, he used electrolysis to discover barium, calcium, magnesium, potassium, sodium and strontium. In 1816, he invented the **Davy lamp**, a safety oil lamp for use underground (where the level of flammable methane may be dangerously high); a fine copper mesh around the flame conducts energy so well that the flame cannot heat nearby gases and cause an explosion. From 1813, Davy worked with Michael Faraday.

dB decibel, common unit of sound intensity (loudness), with **dB(A)** the unit of perceived loudness (though most people in the field now seem to think this is an unnecessary distinction)

dc (or **DC**) direct current

dc amplifier direct-coupled amplifier

deadbeat of a meter, almost critically damped: it reaches the final reading very quickly

dead room Common term for *anechoic chamber*

dead time time needed after an electronic circuit (such as a gate or sensor) reacts to an input, before it can react to a second input. **Dead time correction** is a factor added to the output of a detector to allow for this.

de Broglie equation equation which is the basis of wave mechanics. It relates the wavelength λ of a particle of mass m to its velocity v:

$$\lambda = h/mv$$

Here h is the *Planck constant*, and mv is the momentum p of the particle. **Louis Victor, Duc de Broglie** (1892–1987), physicist, received the Nobel Prize in 1929 for this work in wave mechanics, which he carried out for his PhD in 1924. His father, *Louis César de Broglie* (1875–1960), was also a physicist and did useful research, particularly into x-ray spectra.

de Broglie wave matter wave, i.e. one that in most cases we can view as a beam of one or more particles, but which also shows wave features and has a **de Broglie wavelength** (see *de Broglie equation*).

debye D old unit of dipole moment, somewhat over 3×10^{-30} coulomb metre

Debye length in a plasma, the maximum distance over which an electron can react to the electric field of an ion

Debye, Peter (1884–1966) physicist who gained the Nobel Prize for chemistry in 1936 for his work on molecular structure and x-ray powder photography

Debye scattering the incoherent background scattering of x-rays from a crystal in directions other than at the Bragg angles

Debye–Scherrer ring the uniform diffraction ring pattern obtained by passing a narrow beam of monoenergetic x-rays through a crystal powder. See also *x-ray diffraction*.

Debye–Sears effect effect in which a vibrating piezoelectric crystal in a liquid sets up supersonic (high frequency acoustic) waves in the liquid. If the liquid is in a suitable cell, standing waves result, to produce a diffraction grating for inward light; the spacing of the grating is the wavelength of the sound waves.

Debye theory of specific thermal capacity theory which gives a solid's c_v, its specific thermal capacity at constant volume, in terms of a characteristic frequency v, the temperature T, and the Boltzmann and Planck constants k and h. Here the **Debye temperature** (characteristic temperature) θ is $h\,v/k$. As a result, the theory gives c_v as a function of θ/T, which broadly agrees with experimental results. The frequency v is the maximum frequency of thermal vibration of the atoms of the solid.

The theory leads to the **Debye T^3 law** which applies at low temperatures: the specific thermal capacity of a substance varies with the cube of the temperature. See also *Dulong and Petit's law*.

deca- or **deka-** da unit prefix for 10^1, as in decagram (dag), 10 g. Strictly speaking, this should rarely be used in pure science as it is not part of SI.

decay a) the exponential fall-off of any type of activity or effect, as in the light output of a phosphor after illumination ceases, or the current in a capacitative dc circuit after switch-on. Similar terms and relations to those used in sense b) apply in each case.

b) the spontaneous disintegration of an unstable (radioactive) nucleus to give a second nucleus (*decay product*), and one or more other particles or photons. Energy leaves in the movement of the decay products.

The activity (number of decays per second) varies with the number of parent nuclei. In other words, decay is exponential:

$$n = n_0 \exp(\lambda\ t)$$

Here n is the number of parent nuclei that remain after time t, n_0 being the number at the (arbitrary) start; λ is the **decay constant**, the reciprocal of mean life.

The decay constant (reciprocal of the mean life) relates to the half-life:

$$t\tfrac{1}{2} = 0.693\ 15\ \lambda$$

It also relates to the activity dn/dt, being the probability of the decay of a parent nucleus in unit time:

$$\lambda n = dn/dt$$

See also *radioactivity* and *quantum Zeno effect*.

If the main decay product itself is radioactive, we have a **decay chain**. There are four major decay chains, three being still found in nature and one (the neptunium series) being extinct. The mass numbers of the main steps in each chain reduce by four (because of alpha decay); the values of the mass numbers of the chain isotopes then are as in the table at the bottom of this page:

For full details of a major decay chain, see *actinium series*.

deceleration negative acceleration. The **deceleration parameter**, q, is the rate at which the expansion of the Universe is slowing, taken as negative (unless the rate is rising); there is no unit.

deci- d unit prefix for 10^{-1}, as in decimetre (dm), 100 mm. Strictly speaking, this should rarely be used as it is not part of SI.

decibel dB the normal unit of power or sound level, one tenth of a *bel*. It is a ratio of the level in question to some standard. (However, people often use it as if it were absolute.) For instance, we use the decibel to measure the gain of an amplifier, by treating the input and output power levels P_i and P_o as follows:

$$\text{gain in decibels} = 10\ \log(P_o/P_i)$$

It follows that a 3 dB rise or fall occurs when the power doubles or halves. This rather strange system follows a classical law of physiology: if one sound appears twice as loud as a second, it is 3 dB greater. Similar laws apply to all the physiological senses. See also *neper*.

The **decibel (A-weighted)**, or dB(A), is the unit of perceived loudness as shown by many sound level meters, though there is no evidence that the meters and the system give valid results for all types of sound at all ranges of frequency.

decile one of the ten equal parts into which we divide a population, listed in order, for statistical purposes; also one of the nine points that separate these parts.

decimal balance balance the lengths of whose arms are in the ratio 10:1. It avoids the need for large counter weights.

declination a) the angle between the magnetic meridian and the geographical meridian at a given point on or near the Earth. Its value depends on where the point is, and changes slowly with time (see *variation*).

b) the angle of an object in the sky above the celestial equator, one of the two coordinates on the *celestial sphere*. If the object lies below the equator, the declination is negative.

declinometer device for measuring magnetic declination

decomposition voltage the greatest pd one can apply to the terminals of an electrolytic cell without causing a steady current

chain	mass numbers	start	end	
thorium	$A = 4\ n$	thorium 232	lead 208	natural
neptunium	$A = 4\ (n + 1)$	plutonium 241	thallium 205	extinct
radium	$A = 4\ (n + 2)$	uranium 238	lead 206	natural
actinium	$A = 4\ (n + 3)$	uranium 235	lead 207	natural

decoupling a) the use of an inductor in series or a capacitor in parallel to remove unwanted ac signals from part of a circuit
b) the breaking of chemical bonds. See also *coupling.*

decrement gauge quartz-fibre manometer, used to measure gas pressures in the range 1–0.01 Pa

de-emphasis at a signal's receiver, the reduction or enhancement of higher frequencies introduced into the signal by pre-emphasis

defect departure from regular pattern in a crystal. All crystals at a temperature above absolute zero have some defects. **Line defects** are defects extending for some distance in one direction; see *dislocation.* **Point defects** involve a single atom or molecule, and may be a vacancy (missing particle), an impurity atom, or a crystal atom not in its normal place. A **Frenkel defect** is a vacant site, with the missing ion or atom in a nearby interstice. A **Schottkv defect** consists of a vacancy alone.

defect conduction conduction in a semiconductor by the transfer of holes (electron vacancies)

definition measure of the detail (e.g. sharpness of focus) of an image, high definition systems being able to offer more detail. See also *resolution.*

deflection change of angle or or direction of motion

deflection coils coils in cathode ray tube systems that change the direction of the electron beam from the electron gun. In **magnetic deflection** (used in television, IT monitors, and radar sets), the coils are around the neck of the tube. Compare with *deflection plates,* which are electrodes for the same purpose.

deflection magnetometer magnetometer with a mounted needle for giving readings of magfield strength

deflection plates electrodes inside the neck of, for instance, oscilloscopes, which provide **electrostatic deflection.** Compare with *deflection coils,* and see *cathode ray tube* for a sketch.

deformation change of shape of a solid object. The deformation is elastic if the object returns to the original shape when the deforming force is removed; otherwise, it is plastic. The main deformations are compression (squashing), tension (stretching), shear (slanting), and torsion (twisting);

bending is not a fifth form as it is the result of a mix of compression and tension.

deformation potential voltage that appears between the faces of a solid sample when deformed, as in the piezoelectric effect

degaussing neutralising the magfield of, or in, an object by using coils to produce an equal and opposite field. Degaussing systems used on ships prevent them from detonating magnetic mines, and with cathode ray tubes prevent image distortion.

degeneracy a) condition in which matter has such a high energy (temperature) that all atoms have lost all electrons. This 'fourth state of matter' (plasma) occurs, for instance, in the centres of stars.
b) condition of a gas below about 20 K when the molecular thermal energy is less than $3R/2$, R being the gas constant
c) state of a system in which two or more of the states have the same energy level
d) in such a degenerate system (sense c), the number g of *different* states with the same, given energy level
e) the classical equivalent of sense c): the number of *different* states of a system with the same vibration frequency

degenerate a) showing *degeneracy, senses a), b), c), or d)*
b) of an energy level in a degenerate system, shared by more than one *different* state of the system. For instance, the three p states of an atom have the same energy, though they differ in magnetic quantum number m. In much the same way, this applies when a wave function (or an eigenfunction) has more than one solution (or eigenvalue) with the same value.

degenerate semiconductor semiconductor whose Fermi level is inside the valence or conduction band; this gives it near metallic features over a wide range of temperatures.

degradation a) loss of quality, as in the loss of energy of a particle passing through matter
b) the loss of energy available for use when a system's entropy rises
c) the loss of quality of a signal as a result of noise

degree a) value on a scale
b) unit of angle °. The degree of angle is 1/360 of a complete revolution (circle).
c) unit of temperature °. The Celsius degree has the same value as the kelvin (though the scale zeros are not the same).

degrees of freedom the independent ways in which a system and/or its parts can differ;

each degree needs a variable to describe it. In the case of a monatomic gas, there are three degrees of freedom given by the three coordinates of the space that contains it, as it can move in those directions. A diatomic gas on the other hand has six: of the additional three, two are for rotation round an axis at 90° to the link between each pair of atoms, and one is for vibration along that axis. The concept of equipartition of energy implies that the energy in each degree of freedom is the same.

Dehmelt, Hans (1922–) physicist who developed the ion trap technique for working with mass spectroscopy and shared the 1989 Nobel Prize for that with Wolfgang Paul

deionisation the process of recombination of ions and electrons formed by ionisation. The **deionisation time** of a detector is the time required for sufficient deionisation to occur before the detector can work again.

deka- dk the same as deca-

dekatron in valve days, a display tube for a counter, with ten cathodes and transfer electrodes in a circle around an anode. Each pulse transfers the current to the next cathode, which then glows. After a complete circle, a pulse goes to the next tube in the set.

del or **nabla** ∇ the differential vector operator $i(\partial/\partial x) + j(\partial/\partial y) + k(\partial/\partial z)$. Here [i], [j] and [k] are the unit vectors in the x, y and z directions See also *Laplace operator*.

delayed critical See *critical reaction*.

delayed neutrons neutrons produced during fission, not as a direct result of the fission but from the decay of excited fission products

delay lens a lens used with low frequency microwaves – see *dielectric lens*

delay line line down which a signal (or data) passes and returns to delay its transfer, for instance, to put it in step with some other signal (or to provide data storage as in the very first radar sets and computers). An **acoustic delay line** converts the signal to sound (which travels much more slowly than electric effects). An **electromagnetic delay line** is a set of capacitors and inductors which slows the signal but keeping it in electric form; a **magnetic delay line** uses magnetic waves for the same purpose. **Delay time** is the time taken by a pulse to pass through any circuit or element.

Delbruck scattering the scattering of light (and sometimes higher energy) photons by the Coulomb fields of nuclei

delta circuit common way to link the three phases of an ac supply. Three coils form a triangle (like the Greek letter delta, Δ); one corner goes to each phase. Compare with *star*.

delta E effect a form of magnetostriction, with E for elasticity

delta function a Dirac function used to describe quantum scattering

delta radiation once thought to be a new form of radioactivity (after alpha, beta and gamma). It appeared as many small side tracks leading from the heavy cloud chamber track of an alpha. These **delta particles** are in fact electrons knocked from the atoms by the ionising radiation with enough energy to cause further ionisations.

demagnetisation process of removing the permanent ferromagnetism of a sample. This involves making the effects of the domains random. Standard methods include rough treatment – such as raising the magnet to above its Curie temperature and letting it cool while lying east-west – or leaving it in a coil while dropping the ac in the coil to zero.

demagnetising field the field produced by the induced poles s and n when a ferromagnetic material is magnetised. As this tends to oppose the outside field, that applied field weakens as a result.

Democritus (*c.*460–*c.*370 BCE) ancient 'natural philosopher' who proposed that matter consists of tiny, indivisible 'atoms' (= things that can't be cut up)

demodulation inverse of modulation, i.e. the removal of the carrier to leave the original signal. A **demodulator** (or 'detector'), as for instance in a modem or radio set, is a device that has this function.

demultiplexing opposite of multiplexing, i.e. separating previously multiplexed signals. A **demultiplexer** has this function.

denaturant isotope added to a fissile sample to prevent its use in nuclear weapons

densitometer device that measures the reflection or transmission density of photographic film. It provides one way to digitise a photographic image.

density ϱ unit: kilogram per cubic metre $kg\ m^{-3}$ of a substance, the mass of unit volume of a sample under standard conditions

density of states curve for a solid, a graph showing how the number of energy levels in a band varies with energy. For more detail and a typical sketch, see *band model*.

density parameter Ω for the Universe, the

ratio of its density to the critical density; there is no unit. If the value is greater than 1, the Universe is closed (it has enough mass to stop expanding and then shrink again); if it is less than 1, it is open. Because of dark matter, whose nature (and therefore mass) is unknown, the value of the density of the Universe is not yet clear. The critical density is:

$$\rho_c = 8 \pi \, G/3 \, H^2$$

In this case, G is the gravitational constant and H is the Hubble constant.

depletion a) reduction of the fraction of uranium-235 in a uranium sample. This can be by enrichment (where a second sample gains ^{235}U) or by use in a reactor. Depleted uranium, being very dense and inert, has many uses, for instance as shielding and for alloys, though there are concerns that it may be more toxic than thought.
b) reduction of the density of any type of particle compared with the normal value. Thus a so-called 'ozone hole' is a region of the upper atmosphere in which the ozone density is less than (1975) normal; by the mid-2000s it had dropped over 65% in the main hole in winter – the figure in the early 1990s was at most about 10%.
c) the reduction of the density of free charges in a sample. The **depletion layer** in a semiconductor diode, for instance, is a region that has too few free charges to balance the fixed charge density of donors or acceptors. The **depletion mode** of a field effect transistor is the mode in which the transistor conducts with zero gate voltage ('bias').
depolariser substance used in an electrochemical cell to reduce polarisation. Here, polarisation results from a layer of gas bubbles on the surface of an electrode, cutting the output emf; the depolariser would reduce or remove that layer.
depth of field or **depth of focus** for a given setting of a camera, that range of object distance that gives adequately sharp images. Field depth becomes larger as the aperture becomes smaller.
derivative or **differential** the rate of change of a function of an independent variable with respect to that variable, the process involved being differentiation. If the function is $y = f(x)$, the derivative is dy/dx. If the variable is time t, the derivative dy/dt is often written as $\dot{y}$.

Higher order derivatives follow from the differentiation of a derivative. The nth-order derivative is $d^n y/dx^n$, the derivative of the $(n-1)$th derivative.

A **partial derivative** may be of value when the function is of two or more independent variables. For instance, in the case of $y = f(x, z)$, there are two partial derivatives. We obtain each by differentiating with respect to one variable while keeping the other(s) constant; in this case, this gives $\partial y/\partial x$ and $\partial y/\partial z$.
derived unit unit in a system that depends on a base unit, rather than being found from a standard (as base units are). For instance, the pascal is a derived unit: it is equivalent to the newton per square metre, the newton in turn being a kilogram metre per second per second; kilogram, metre and second are base units.
Descartes, René (1596–1650) so-called father of modern philosophy ('I think, therefore I am.') and father of modern mathematics (inventing our Cartesian system of 3D coordinates and analytical geometry, the basis of later calculus)
Destriau effect See *electroluminescence.*
destructive interference process that occurs at a set of points in an interference pattern that results in zero radiation intensity. At each such point (node) the two equal input waves are exactly out of phase so have no net effect.
detector a) demodulator device – one that removes the carrier from a modulated signal to leave the original signal. See also *acceptor circuit.*
b) sensor, any device able to respond to a physical effect, and thus to mark the presence of, or even measure, a signal of some kind. Thus, in this sense, a radio aerial is a detector.
c) device or system used to detect and perhaps to count and/or measure aspects of high energy radiations (including those of subatomic particles). A range of such systems appeared in the early decades of the last century. Modern detectors also have many other uses in science and applied science outside atomic and nuclear physics: in astronomy (special types of telescope); biomedicine (see, for instance, positron emission tomography); engineering; materials science, and so on. See *nuclear physics.*
deuterated of a compound, having deuterium atoms in place of the normal hydrogen atoms. **Deuterated water** is heavy water.

deuterium D the hydrogen-2 isotope ^{2}H, a stable isotope, found in natural hydrogen at a level of around 0.015%. It has one neutron in the nucleus as well as the one proton held by all hydrogen nuclei. This gives a deuterium atom about twice the mass of one of hydrogen: hence the common name of heavy hydrogen. Therefore, while its chemical features are the same as those of hydrogen, reactions take place more slowly. Its physical properties also differ somewhat.

deuteron the nucleus of deuterium, one proton bound to one neutron. Deuteron beams are widely used in nuclear physics.

deviation *a* **a)** in statistics, the difference between a measured value and the true value. In practice, the 'true value' is usually the mean of all the results. In that case, the deviation of each individual value is its *residual*. See also *mean deviation* and *standard deviation*.

b) the change of frequency of the carrier in frequency modulation

c) in the context of the Earth's magnetism, **angle of deviation** is the angle at a place between the directions of magnetic north and true North. (Strictly, this is variation, but the error is common – it doesn't help that there is even a third term for this: declination.)

d) the turning of a ray during reflection or refraction, the **angle of deviation** *d* being the angle through which it turns.

The angle of deviation produced by a refracting prism depends on the angle of incidence. It has a minimum value *D* (when the ray passes symmetrically through the prism). *D* is a function of the prism's refractive constant *n* and its angle *A*:

$$n = \sin\left[(A + D)/2\right]/\sin(A/2)$$

If the prism angle *A* is small, this leads us (approximately) to :

$$D = (n-1)\,A$$

dew film of small liquid droplets on a surface, formed as vapour in cooled moist gas (in most cases, air) condenses. This happens if the gas temperature falls below its **dew temperature**; often called the dew point, this is the temperature at which the moisture content of the sample becomes maximum (it becomes saturated: ie, relative humidity becomes 100%). Below the dew temperature, the air cannot hold all the vapour it contains and the excess must condense, as dew. Dew is therefore a form of precipitation; it relates to

mist and fog, where the droplets form throughout the gas volume, and to frost where the surface is below the ice temperature.

Dewar flask or **Dewar vessel** the lab equivalent of the homely vacuum flask. It consists of a double-walled container of thin silvered glass with vacuum between the walls, and a small insulating stopper. The design much cuts down transfer of energy between the flask's contents and the outside (by conduction, convection, evaporation and radiation). **James Dewar** (1842–1923), physicist and chemist, invented the device in 1892. He also pioneered various techniques of low-temperature physics and discovered the magnetic properties of liquefied gases.

dextrorotatory (or **dextrogyric**) of an optically active substance, able to rotate the plane of polarisation of passing light 'to the right' (i.e. clockwise when viewed from the output side). The reverse is laevorotatory.

dialysis the diffusion of ions through a porous barrier down a concentration gradient. A voltage on each side speeds up this process, in that case being electrodialysis.

diamagnetism magnetic property of those substances which have a negative susceptibility and relative permeability just less than 1. This means that in a magnetising field, the magnetisation that results is such as to oppose the applied field. The effect is the result of the forces on the magnetic fields associated with the orbiting electrons.

A **diamagnetic material** is a material that **a)** in bar form turns to lie across an external field;

b) in any form tends to move towards where the outside field is weaker;

c) has a weaker field inside than outside (the 'lines of force' are further apart). Diamagnetism is far from rare but is never strong – so paramagnetism and ferromagnetism often mask it. Bismuth, copper and hydrogen are common pure diamagnetics. The Langevin theory of diamagnetism, which attempts to explain the property, arose from classical thinking; Pauli corrected it to allow for quantum effects. The explanation depends on the Larmor precession of the electron orbits.

diameter *d* the greatest distance across a circle, i.e. twice the radius

diamond form (allotrope) of carbon with colourless cubic crystals. It is the hardest natural substance known and artificial versions have the highest known thermal conductivity.

Diamond Light Source UK's biggest investment in pure science since the 1970s, a research facility designed to explore synchrotron radiation. The radiation comes from passing a beam of 3 GeV (3×10^{-11} J) electrons from a linear accelerator and a synchrotron into a storage ring where they lose energy by synchrotron radiation (mainly at x-ray wavelengths) in the strong fields of large magnets. Seven 'beamlines' channel the radiation into research areas (though the plan is to raise the number to about forty over the years). Planned research projects are in such fields as biotechnology, materials science, atomic structure and medicine. Diamond's web site is http://www.diamond.ac.uk/default.htm

diaphragm a) skin, a 2D source of sound waves
b) (also called stop) device (e.g. iris type) used to reduce the aperture of an optical system or element; the effect of this is to increase the depth of field and to cut down aberration.

diatonic scale eight-note musical scale that consists of five tone and two semitone intervals. A scale played on the white keys of a keyboard is diatonic.

dichroic having the property of dichroism, as in the case of a sheet of Polaroid or crystal of tourmaline. A **dichroic mirror** is a transparent sheet coated with layers of substances of alternating high and low refractive constant; for inward white light the colours of the reflected and transmitted beams are complementary.

dichroism a) property of a substance that lets pass light which is polarised in one plane but blocks light polarised at 90° to this plane. Tourmaline is a naturally dichroic substance, Polaroid a synthetic one.
a) the property of showing one colour in reflected light and a second in transmitted light

dichromaticism property of some substances whereby the colour of the light passed depends on the thickness of the sample

dielectric substance able to maintain an electric field, once set up, without loss of energy. This is an ideal, for it has to be a perfect insulator, one with no free charges.

dielectric constant old term for relative permittivity

dielectric heating the increase of temperature, and its use, in a dielectric in an alternating electric field. The main uses are in moulding plastics and wood, and in cooking.

dielectric hysteresis 'memory effect' that appears as a dielectric reacts to an alternating electric field. The effect of the field in the dielectric depends not only on the field but on the past history; see also *electret* and compare with magnetic *hysteresis*. It entails a loss of energy: **dielectric loss** is all the energy loss in such a substance in an alternating electric field, including the result of dielectric hysteresis.

dielectric lens lens for work with radio frequency radiation, and made of a dielectric material. At longer wavelengths, the same design would work but would be too large to be useful; instead, the lens (now called a delay lens) consists of an array of conductors – which has the same effect.

dielectric polarisation the effect of an electric field on the electron orbits in a substance. See *polarisation*.

dielectric strength or **disruptive strength** the highest electric field an insulator can sustain without breakdown. The usual unit is the volt per millimetre, V mm^{-1}

Diesel cycle the cycle of action of an engine using air as the working substance and a heavy oil (such as 'diesel oil') as the fuel. The usual four stages of the cycle involve four strokes of the piston:
a) *induction*: air at normal pressure enters the cylinder
b) *compression* of the air, which adiabatically raises it to a high temperature

Here are the usual four stages of an engine cycle for the Diesel engine.

c) *injection* of fuel at high pressure into the cylinder (* in the sketch), leading to
power: expansion of gases at constant pressure as the fuel burns, followed by adiabatic expansion to the end of the piston stroke
d) *exhaust*: a valve opens so the pressure falls to normal air pressure, after which the waste products are emitted

Engineer **Rudolf Diesel** (1858–1913) first demonstrated the system in 1897. Compare it with the Otto cycle.

Dieterici equation of state version of the van der Waals equation of state for a real gas. The basic van der Waals equation is $(p + k)$ $(V - b) = R\,T$. Here p, V and T are the pressure, volume and temperature of a gas sample, R is the gas constant, and k and b are other constants that concern the attractive forces (van der Waals forces) between the particles and their actual volume respectively. The (first) Dieterici equation puts k as another constant a, divided by $V^{5/3}$.

difference tone subjective combination tone whose frequency is the difference between the frequencies of two real tones. A difference tone is not the same as a beat.

differential a) operating on the basis of the difference between two values, as in, for instance differential air thermometer, differential amplifier, differential galvanometer
b) same as derivative, the result of differentiating a function. See also *differential equation*.

differential air thermometer simple detector for radiant energy. A tube which contains a thread of liquid links two equal glass bulbs, one clear and the other black. Radiation causes the air in the blackened bulb to expand more than that in the other, and this causes the index thread to move.

differential amplifier two-input amplifier whose output is a function of the difference between the inputs

differential analyser electronic circuit or analog computer able to integrate input data. It can thus handle integral and differential equations.

differential calculus study of the function $y = f(x)$ – in particular, the rate of change of the dependent variable y that results from very small changes in the independent variable x. This gives the slope dy/dx of the graph of y against x.

differential equation equation that involves derivatives of an unknown function. Where the function $f(x, y)$ has a constant solution, an exact' differential equation has the form $(\partial f/\partial x)\,dx + (\partial f/\partial y)\,dy = 0$

differential galvanometer magnetic needle meter inside two coils. The deflection of the needle depends on the difference between the currents in the coils.

differentially compound-wound of an electric motor or generator, having compound field winding – i.e. coils both in series and in parallel – with the series coil field opposing that of the parallel coil

differential vector operator ∇ See *del*.

differentiation the process of finding the derivative of a function

differentiator circuit whose output is the time differential (derivative) of the input, i.e. one that measures the input's rate of change

diffraction the effect of an obstacleon passing radiation, which tends to cause interference fringes that make the edge of the shadow less sharp than expected. Some radiation bends round into the shadow region; in some parts of the non-shadow region the intensity is reduced. The result is a diffraction pattern. Note, however, that diffraction is not the same as interference. Diffraction appears with all waves, particularly if the obstacle is much the same in size as the wavelength and the waves are coherent. The diffraction of beams of particles gave strong evidence for the wave–particle duality that lies behind quantum physics.

diffraction analysis study of the diffraction of x-rays or particle beams to explore the structure of matter

diffraction grating device able to produce diffraction patterns from a series of parallel lines or reflecting surfaces with separation of the same order as the wavelength λ of the radiation in question. For that radiation, it then gives a series of maxima and minima of intensity at different angles; if the radiation contains a number of wavelengths the result is a set of spectra. The maxima appear at angles θ given by

$$d\,(\sin i + \sin \theta) = n\,\lambda$$

Here i is the angle of incidence of the radiation on the grating of line or plane spacing d, while n is an integer.

In fact, n there is the 'order' of the maximum or spectrum in question; the zeroth order ($n = 0$) is for the maximum obtained when the waves pass straight through the grating to make a bright central spot.

In the case of x-rays it is normal to use crystals to provide the grating effect. With light the grating lines are the plateaus between fine grooves drawn on the surface of glass or metal by a diamond point; it is common to make replicas in plastic of surfaces prepared in this way to keep grating costs down. Gratings exist for many other types of wave as well.

diffraction pattern pattern of high and low intensity produced over a plane by any diffraction process. Some people find it useful to discuss two main classes of diffraction pattern. Plane wave fronts produce Fraunhofer diffraction; spherical ones, from a point source, lead to Fresnel diffraction, whose patterns are concentric circles. The figures (*right*) show typical diffraction patterns of plane wave fronts; in each case, λ is the wavelength.

diffraction spectrum spectrum of radiation produced by a diffraction grating. A **normal diffraction spectrum** is the output of a grating with waves incident at 90° (i.e. along the normal). In any case, the angle between spectral lines varies with the difference in their wavelengths.

diffractive optics also called binary optics, the use of a combination of lenses which work by refractive and diffractive effects (such as a Fresnel lens combined with a standard one). Applications include implanted eye lenses which give cataract sufferers both near and far vision, infrared night vision systems, optical disc heads, and scanners for copiers and fax machines. A diffractive surface can also compensate for refractive aberrations and can split or combine beams in, for instance, optoelectronics. On the other hand, diffraction is highly wavelength dependent – this becomes no problem when refraction provides 80% or more of the total effect.

diffractometer instrument used for diffraction analysis of crystals. It is normal to have a monochromatic incident beam and to rely on an ionisation detector to measure the intensity at different angles. The input beam is of x-rays for standard use, electrons to study near surface, or neutrons.

diffuse (as adjective or verb) spread out in random or apparently random directions

diffused junction semiconductor junction (for instance, pn) formed by a diffusion process.

diffuse refraction

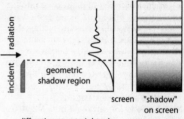
diffraction at a straight edge

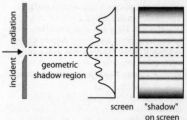
diffraction at a slit (width 2Ωλ)

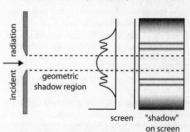
diffraction at a wide slit (width 5λ)

The Fraunhofer diffraction of plane waves by straight edges gives patterns like these.

The substrate is raised to a high temperature in an atmosphere that contains the desired impurities in gas form.

diffuse reflection also called scattering, the effect that occurs when rays are reflected from a surface with irregularities of about the same size as the wavelength: the rays of an incident parallel beam can leave at (almost) any angle. Any single case, of course, follows the laws of reflection.

diffuse refraction sometimes wrongly called diffusion, process similar to diffuse reflection except that the rays, rather than bouncing back from the surface, pass through a block of 'translucent' material. Frosted glass is a well-known **diffuser**, and lamp-shades, especially those around fluorescent tubes, have a similar effect (and, to many, the same name). However, the name wrongly implies that the process is diffusion.

diffusion a) incorrect name for the apparently random spreading of a beam of radiation by scattering during diffuse reflection or diffuse refraction or a similar process in some other context
b) the passage of particles through matter in cases where the probability of scattering is high compared with that of capture. This particularly applies to the diffusion of neutrons in a nuclear pile.
c) much the same as evaporation, the net escape of solid, liquid or gas from a surface, as a result of the random thermal motion of its particles

diffusion cloud chamber one of the two main types of cloud chamber

diffusion length the mean free path of diffusing particles between when they appear and when they vanish. Typically this applies to minority carriers in a semiconductor (the loss being by recombination) and to neutrons in a pile (the loss being by capture).

diffusion pump pump similar to a filter pump, but using a jet of silicone oil; it is common as the second stage of a vacuum pump system. It has very low efficiency, but is robust and can reach pressures down to about 10^{-8} Pa.

diffusion tensor imaging a medical scanning technique that depends on tensors (higher order vectors) to give the target organ's permeability to water in different directions in tensor form.

diffusivity, diffusion constant or **diffusion coefficient** D the constant in *Fick's law*, which applies in any diffusion process. The value m/t or $m/A\,t$ sometimes has the name **diffusion current**. This name also applies to the maximum current in an electrolytic cell; above it the ions do not diffuse towards the electrodes quickly enough to maintain the process.

digital able to take only a restricted number of values between the bottom and top of the range allowed, as compared with analog. Most IT systems, including digital circuits and the digital computer, are *binary digital* – their states have only two allowed values (called 0 and 1).

digital camera electronic camera, for still or moving pictures. It has a digital output from each pixel in the sensing device (with perhaps

32 bits to store the luminance value) and stores an image in digital form. Note, however, that the actual output from the sensor (e.g. a ccd chip) is analog – there is an analog-to-digital converter between that and the data store.

digital display display consisting of numbers on a screen (for instance, a liquid crystal display) rather than a needle on a dial (which is analog)

digital inverter logic circuit whose output has the opposite value to its input it is a NOT gate

digital meter meter which shows the measured value on a digital display

digital speech synthesised speech, each phoneme (speech sound unit) going to the loudspeaker as a digital pattern. There needs to be a digital-to-analog converter in the interface between synthesiser and output, as speakers and headsets are analog output units.

digital television video system which uses digital techniques in at least some parts, resulting in a range of advances. In particular, an analog-to-digital converter (modulator) and compressor prepare the video signals.

digitron cold cathode display unit whose cathodes, lying over each other, have the shapes 0 to 9. An electronic switch links the correct cathode to the supply at any moment so that it alone glows.

dihedral formed by or having two intersecting planes or sides. The angle between two intersecting faces of a crystal is the **dihedral angle**.

dilatation small increase in volume, as measured (for instance) by a **dilatometer**. This is any device for the study of cubic thermal expansion.

dimension a) one of a set of independent mutually perpendicular directions in Euclidean space, called x, y and z
b) the same in any model of a space with more than three dimensions
c) the size of an object in terms of two or three measures at 90° to each other
d) one of a set of base measures – i.e. ones used to define any other measure – see *dimensional analysis*.

We tend to view our Universe as having three mutually perpendicular physical dimensions (called x, y and z) plus the 'fourth dimension', that of time t. However, many modern theories propose or demand more than three physical dimensions – perhaps claiming that we cannot 'see' the others as we had no need to do so during evolution.

Extra dimensions often make the mathematical expressions of physical structures simpler, in particular those outside the Earth. For instance, the view of gravity as a distortion of four-dimensional space–time is widely known (with a 3D well for each object, e.g. planet). Perhaps the other forces of nature involve bending of other dimensions? Thus, even in the 1920s, Theodor Kaluza (1885–1954) showed that if we express Einstein's theory of gravity in five dimensions, Maxwell's electromagnetic equations appear very simple.

The grand aim of modern physics is to produce a single (unified) theory of all the four (known) forces of nature. There have been various attempts to do this, almost all of which require a fairly large number of dimensions. For instance, the string theories (strong contenders for twenty years) propose at least ten dimensions. It is possible to assess (if not to prove or disprove) each new theory by comparing how its multi-dimensional space describes such objects as neutron stars and black holes against observational data.

dimensional analysis technique for checking the validity of an equation, solution to a problem, or unit. It follows the fact that in a coherent units system, such as the one we use (SI – *Appendix 1*), any measure can appear in only one way in terms of the base measures in the system. In this context, those base measures are dimensions.

The table (*above right*) shows the three dimensions used in mechanics (where this form of analysis is most common) followed by some others expressed in terms of these.

Thus we can use the technique to verify that, apart from any dimensionless constants, pressure = force/area (for instance). The dimensions of the right-hand side are those of force divided by those of area: $[M L T^{-2}/L^2] = [M L^{-1}T^{-2}]$ – ie, the same as those of pressure.

It is no problem to extend the system to cover the other four base measures, for instance by having [Q] for charge.

di-neutron unstable particle that consists of two bound neutrons. Di-neutrons often

measure	dimension	unit	normal unit
length	[L]	m	m
mass	[M]	kg	kg
time	[T]	s	s
area	[L²]	m²	m²
volume	[L³]	m³	m³
density	[M L⁻³]	kg m³	kg m⁻³
velocity	[L T⁻¹]	m s⁻¹m	s⁻¹
acceleration	[L T⁻²]	m s⁻²m	s⁻²
force	[M L T⁻²]	kg m s⁻²	N
pressure	[M L⁻¹ T⁻²]	kg m⁻¹ s⁻²	Pa
momentum	[M L T⁻¹]	kg m s⁻¹	N s

appear when a triton collides with a target nucleus to produce a proton: the remaining two neutrons remain bound; they also appear in helion reactions. Di-neutrons exist for only a short time as the force between the two neutrons is very small.

diode device that passes current one way only, and so acts as a *rectifier* (something which converts input ac to dc). The solid-state diode is a basic element of transistors and of electronic and microelectronic circuits, and there are many types. The figures show (a) typical structure, (b) circuit symbol, (c) full-wave rectification effect, and (d) characteristic curve.

a)
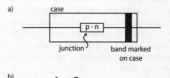

b)

c)
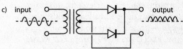

Some aspects of the solid-state diode.

diode transistor logic dtl approach to the design of a large class of electronic and microelectronic circuits, better than the resistor–transistor logic that came before because of the increased fan-in, but still slower than the more modern transistor–transistor logic. Even so, dtl designs are still common; the input goes through a diode with the output taken from the collector of an inverting transistor.

dioptre D the unit for the power of an optical element or system, the inverse of focal distance in metres. Thus a diverging lens of focal distance 0.2 m has a power of –5 D. As power measures the ability of the system to converge a parallel beam, the dioptre is the same as the radian per metre, a name which many people prefer.

dioptric system optical system built of lenses designed to produce an image, distinct from one which uses mirrors, both mirrors and lenses, or optical fibres

dip δ magnetic inclination, the angle on or near the Earth's surface between the Earth's magfield and the horizontal. A **dip circle** measures this; it consists of a magnetic needle free to rotate in a vertical plane at the centre of a 360° angle scale.

dipole Note that there is much confusion between senses a) and b); obtain the correct sense from the context where the context does not include the full name.
a) (in full, magnetic dipole) pair of equal and opposite poles a small distance apart, such as in a simple bar magnet
b) (in full, electric dipole) pair of small, close, equal but opposite charges
c) (in full, dipole aerial) simple non-directional *aerial* that consists of a straight metal rod half as long as the wavelength of the radio signals it receives or transmits. The signal feed is at the centre (or at one end for wavelengths four times the rod length).

dipole moment a) the same as magnetic moment (all senses)
b) (of an electric dipole) the product of one of the two charges and the distance between them. The unit is the coulomb metre C m^{-1}; see also *debye*.

di-proton the ^{2}He nucleus, an unstable, bound pair of protons briefly observed in proton–proton scattering

Dirac, Paul (1902–1984) physicist whose main early work was in the field of general relativity. He was awarded the Nobel Prize in 1933 for linking Heisenberg's matrix mechanics and Schrödinger's wave mechanics into the new field of quantum mechanics.

The **Dirac function** or **Dirac constant** ℏ is the Planck constant h divided by 2π. It comes from the **Dirac equation** in quantum mechanics which derives, in turn, from the **Dirac field theory** of the electron. This theory describes how the electron behaves and led Dirac to predict the positron in 1930 (discovered by Anderson in 1932); he had viewed the positron as a 'hole' in a universal sea of electrons. At the same time, symmetry between electricity and magnetism led Dirac to suggest the magnetic equivalent of charge particles – the **Dirac monopole**; see *magnetic monopole*. While the theory does not fit fully into modern quantum electrodynamics, it covers all particles with spin ½ (fermions).

direct-coupled amplifier or **direct current (dc) amplifier** amplifier whose working frequency range (band width) goes down to 0 Hz (dc). A good design can output dc levels several hundred thousand times the input values.

direct current dc continuous current whose charge flow is in one direction only and which (according to some views) has a steady value. A good direct current source, like a chemical battery, can provide a constant current; dynamos and ac power supply units tend to give less smooth output. It is the task of a rectifier to produce dc from alternating current.

direct current restorer device which restores the dc or low frequency part of a signal which has been lost in a circuit with high impedance to dc or low frequencies

directional aerial or **director** aerial able to receive from, or transmit in, one or more direction(s) better than others

direct stroke lightning stroke that hits any part of a communications or power line. It therefore deposits in it more charge and energy than does an **indirect stroke** (one that comes only near enough to induce a flow of charge in the conductor).

disappearing-filament pyrometer device to measure high temperature – the design lets the user compare the glow of a filament with the glow of the hot source. This is one type of *optical pyrometer*, and the names are sometimes confused.

discharge a) reduction of the net charge of an object (such as a capacitor), perhaps to zero
b) transfer of chemical to electrical energy in a cell, causing it to 'run down'
c) the rate of flow of fluid at a point in a channel

d) the (effect of the) passage of charge (current) through a gas or other insulator, in particular the light output. The discharge in a low pressure gas involved much fascinating physics – see, for instance, http://science-education.pppl.gov/SummerInst/SGershman/Structure_of_Glow_Discharge.pdf for a good introduction. It is also the cause of the aurora, but instead of cathode rays, the particles involved are the solar wind. Various types of **discharge tube** allow the study of gas discharge – conduction through gases at reduced pressure – and cathode rays. See also *spark*.

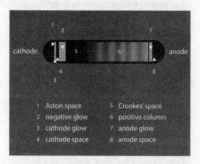

1 Aston space	5 Crookes' space
2 negative glow	6 positive column
3 cathode glow	7 anode glow
4 cathode space	8 anode space

The glow, or discharge, in a low pressure gas consists of a number of regions.

discomposition effect same as *Wigner* effect

discriminator or **discriminator circuit a)** circuit that selects certain input signals and not others on the basis of, for instance, amplitude or phase. A pulse height analyser, for instance, selects within an amplitude range, while a frequency discriminator takes in signals only if they are in a set frequency range.
b) circuit with a frequency-modulated or phase-modulated input and an amplitude-modulated output

dish reflecting, parabolic aerial element with directional properties – used to collect radio waves from, or to radiate them into, a certain direction

disintegration the breaking up of a nucleus into two or more fragments, either spontaneously (as in radioactivity) or as the result of an interaction (fission in a reactor). The first artificial disintegration ('splitting the atom') was in the Cockroft–Walton accelerator. The **disintegration constant** of an isotope is a name for its decay constant.

dislocation extensive fault in a crystal lattice, a line defect. Deformation causes dislocation and dislocations are the sites at which further plastic deformation takes place under stress. There are two types. An **edge dislocation** is a line of weakness that appears at the end of an extra 'plane' of atoms. In a **screw dislocation** there is a helix effect: a given plane in the lattice rotates and advances so that it looks like the thread of a screw. In practice, dislocations often have features of both types.

dispersion a) the spreading of waves of different wavelengths in a beam as a result of refraction: in refraction the magnitude of the deviation depends on wavelength. The result is a spectrum, of colour in the case of light, provided the input beam is narrow and parallel; if it is not, chromatic aberration results.
b) the variation of the refractive constant n with wavelength λ to cause dispersion. In general for a given substance, n falls as λ rises. In the case of visible light, the **mean dispersion** is the difference in n for the F and C lines of the hydrogen spectrum, $n_F - n_C$.
c) loose synonym for dispersive power

dispersion forces the attractive van der Waals forces between electron–nucleus dipoles

dispersive power measure of how much samples of a transparent substance – such as a given type of glass – can disperse white light. It is the ratio shown below, and the reciprocal for the substance of the Abbe constringence or Abbe number, V.

$$1/V = (n_F - n_C)/(n_D - 1)$$

Here n is the refractive constant of the substance for the F, C and D lines of the hydrogen spectrum.

displacement a) (symbol s) the distance and direction between two points, the unit being the metre, m. Displacement is the vector form of the scalar distance (it has direction as well as size).
b) (symbol D) the electric field strength at a point in a dielectric medium as a result of an outside field.

$$D = \varepsilon E$$

Here ε is the absolute permittivity of the medium and E is the outside field strength in free space.

The medium increases the field strength by way of the distortion of its atoms, which polarise – they gain positive and negative ends ('poles'). The name displacement comes from this redistribution of electric charge.

If, for instance, the medium is the dielectric between the plates of a capacitor, and the applied field changes, so too does the polarisation displacement. The effect of this is a changing current in the dielectric; this **displacement current** is exactly the same as that in the outside circuit from moment to moment. Indeed, Clerk Maxwell predicted that displacement currents must exist when he developed the Maxwell equations in 1865. He also used the concept of **displacement current density**, J (given below in vector form).

$$J_D = \frac{\partial D}{\partial t} = \varepsilon \frac{E\partial}{\partial t} \text{ (as } D = \varepsilon E)$$

displacement law a) See *Wien displacement law*.

b) rule linking the type of radioactive decay of a nucleus with the displacement on the graph of nucleon number against proton number. For instance, in negative beta decay the product ('child') nucleus is one place to the right on the graph (see *actinium series*).

b) The arc spectrum of an element is similar to the spark spectrum of the element one place higher in the periodic table (i.e. of that with proton number 1 greater). It is also similar to the second spark spectrum of the element two places higher.

disruptive causing breakdown. For instance, a **disruptive discharge** is a current forced through an insulator when the voltage is so high that the material breaks down in some way – the current is the result of an applied field greater than the **disruptive strength** (or dielectric strength) of the substance.

dissipation release of energy from a system in which some resistive force has been overcome. There is no dissipation from a current in a superconductor (as there is no resistance) or from an object moving through empty space (as there is no friction).

dissociation a) the breakdown of molecules in solution into ions (negative anions and positive cations

b) the chemical breakdown of molecules into smaller molecules (or atoms). The **dissociation constant** is the equilibrium constant of such a reaction if it is reversible (as most are).

dissonance the sound of two or more musical notes that combine to produce an unpleasant effect. Compare with *consonance*.

distance *l* or *d* unit: metre, m the length of the path between two points. This is the scalar that relates to the vector displacement *s*.

distance ratio (a better term than velocity ratio) base measure for the action of a machine (e.g. lever or pulley). The ratio of the distance moved by the input force (effort) to the distance moved by the output force (load), it has no symbol or unit. See also *force ratio*.

distortion any effect which prevents the output of a process from being a faithful copy of the input. There are many types depending on the process, as follows.

a) optical distortion – type of aberration caused by an optical system in which straight lines in the object are not straight in the image. Barrel and pin-cushion distortion are quite common with projectors.

b) acoustic distortion – and this in turn has a number of types. **Amplitude distortion** occurs when the gain of a system with a fixed frequency input varies with amplitude; in most cases an amplifier cannot properly handle high amplitudes. In much the same way, **frequency distortion** (or **attenuation distortion**) occurs when a system's gain depends on frequency for an input of fixed amplitude. **Phase distortion** occurs when there is a phase shift in the system between inputs that differ in frequency, and **transient distortion** when there is a change in the duration of some components of a multi-signal input.

c) electronic distortion – which has amplitude distortion, frequency distortion and phase distortion as above, as well as others. **Harmonic distortion** involves the appearance of harmonics between input and output, while **non-linear distortion** is the result of the dependence of gain on instantaneous input value; it is a mix of amplitude and harmonic distortions.

distortional wave old term for shear wave (S-wave)

distribution or **frequency distribution** the frequency with which each of the possible values of a variable measure occurs in a given sample, the most well known being the *normal distribution*. A **distribution function** is a mathematical expression for a frequency distribution. Maxwell's **distribution law** describes the frequency distribution of the speeds of the particles of a gas in equilibrium; it follows classical statistical theory. See *Maxwell distribution*.

diurnal motion the apparent cycling during each 24 hour period of objects across the sky as a result of the Earth's rotation

divergence div the scalar product ∇F of a vector field of strength, F, ∇ being the differential operator (del or nabla)

divergent a) of a beam of radiation, spreading apart, the reverse of convergent **b)** of a neutron chain reaction, supercritical (i.e. 'explosive')

diverging causing *divergence* a): spreading apart a beam of radiation. This is the effect of a **diverging lens** (such as a concave glass lens in air, which acts by refraction) and of a **diverging mirror** or **diverging reflector** (such as a convex mirror, which works by reflection).

D-layer or **D-region** the lowest layer of the ionosphere in the Earth's atmosphere. It has a low concentration of free electrons, and reflects (in effect) radio waves up to about 50 kHz.

D lines two close, intense lines in the yellow region of the sodium spectrum. The wavelengths are 589.6 nm (D_1) and 589.0 nm (D_2). The bright yellow D lines are widely used as a reference in spectroscopy as well as a test for resolving power.

dobson unit for the concentration of ozone in the ozone layer. The normal value is some 300 dobson, but it may fall below 100 units at times of greatest depletion.

dollar the reactivity that a reactor needs to become critical to prompt neutrons

domain region in a ferromagnetic sample in which the fields of all the atoms or molecules point the same way. If the sample is unmagnetised (has no net magnetic effect), the domain directions are random or in closed loops. Magnetisation involves lining up the domain fields to produce poles and an outside magfield. During this process, some domains grow at the expense of others, with the Bloch walls (**domain boundaries**) moving through the sample. If all the domain fields are in the same direction the sample is magnetically saturated. See *hysteresis*.

Techniques such as the Barkhausen effect and Bitter patterns provide evidence that domains exist and behave in this way.

The domain theory also applies to antiferromagnetic and ferroelectric matter.

dominant wavelength the wavelength of any pure colour (one at the edge of the chromaticity diagram) which, mixed with white, will produce that colour

donor type of impurity which, when added to an intrinsic (pure) semiconductor, gives it an excess of free conduction electrons. It is an element of valency 5 such as phosphorus. In the band theory, a **donor level** is the energy level of the donor atoms. The result is n-type semiconduction. The energy value is, in most cases, measured from the bottom of the conduction band: the impurity's ionisation energy must be at least as large as this. Compare donor with *acceptor*.

dopant impurity (such as a donor or an acceptor) which, when added to an intrinsic semiconductor, gives it an excess of one type of charge carrier

doping the process of adding a dopant to an intrinsic semiconductor to give it an excess of one type of charge carrier. If the semiconductor already contains an impurity that affects its conductivity, then adjustment of the doping is made by **doping compensation** to produce the desired effect. This effect is a function of **doping level** – the concentration of impurity atoms added to the semiconductor to produce the correct type and degree of conductivity; the lower the level the lower the conductivity.

Doppler broadening one type of broadening of spectral lines, an example of the Doppler effect. The width of the lines is larger than it should be because of the random motion of the particles that emit or absorb the radiation. At any moment, some particles are moving towards the observer and some away.

Doppler, Christian (1803–1853) physicist best known for the *Doppler effect* – that the perceived length of waves from a source depends on the relative motion between source and observer. It is interesting that Doppler first explored this with light rather than with sound.

Doppler effect the change of observed length of waves as a result of relative motion between source and observer; it applies to any wave motion. If source and observer are moving closer the observed frequency rises; if moving apart the frequency falls. In the case of sound and other waves in a fixed medium, the observed frequency v as opposed to the true frequency v_0 is

$$v = v_0(c + v_0)/(c - v_s)$$

Here c is the speed of the waves through the medium while v_0 and v_s are the velocities (or their components) of observer and source along the line that joins them.

Electromagnetic radiation in free space (and in effect in other contexts) has no fixed medium. The above relation becomes

$$v = v_0 \sqrt{(1 - v/c)/(1 + v/c)}$$

In this case c is the speed of light and v is the velocity of separation of source and observer.

The main importance of the effect with electromagnetic radiation is the **Doppler shift** or red-shift (the Doppler effect as observed in the radiation from stars and galaxies).

Doppler radar radar system that uses microwave beams that reflect from distant objects as in normal radar. If the target is moving the return signal differs in frequency from that sent; the system measures the change and gives a reading of the speed of motion of the target along the line from the observer. Police use the technique to measure the speed of moving road vehicles.

Dorn effect the appearance of a voltage between the ends of a liquid column when colloid particles pass through it (under gravity, for instance). The effect is the reverse of *electrophoresis*. Compare with *cataphoresis*.

dose measure of energy absorbed from ionising radiation. There are various types.

An **absorbed dose** is the energy absorbed from the radiation by unit mass of the sample; the unit is the gray ($1 \text{ Gy} = 1 \text{ J kg}^{-1}$) though the rad (**radiation absorbed dose**, 0.01 gray) is still quite common. Of more concern to those involved with radiation protection is **equivalent dose**. This is the product of absorbed dose and a 'quality factor', q, that depends on the radiation type and uniformity. For instance, the quality factor of medium energy x-rays is taken as 1; that of neutrons is 10. The unit is the rem (roentgen equivalent man): $1 \text{ rem} = 1 \text{ rad} \times q$.

The roentgen is the unit of **exposure dose** widely used with x- and gamma rays. This is the total charge of one sign produced in unit mass of dry air at the point in question. The SI unit of exposure dose is the coulomb per kilogram ($C \text{ kg}^{-1}$).

In any given case, the **dose rate** is the dose (defined as appropriate from above) per unit time. Also, the **volume dose** is the total dose, i.e. the product of dose and sample volume.

dosemeter or **dosimeter** device used to measure absorbed or exposed dose. In the case of absorbed dose, most units consist of some kind of ionisation detector; in the case of exposed dose, photographic film (in the form of a 'film badge') is common. **Dosimetry** is the field of dosemeter design and usage.

dot product the scalar product of two vectors. If the vectors are V_1 and V_2, the scalar product is written with a dot: $V_1 . V_2$, hence the name; it is a scalar of value $V_1 V_2 \cos \theta$. (Here θ is the angle between the vectors.) An example is energy transfer, given by force F times displacement s: $F . s = F s \cos \theta$.

double refraction common name for *birefringence*

double-sideband transfer the normal method of communication by an amplitude-modulated carrier. The effect of adding signal to carrier is to create two sidebands that flank the carrier in the frequency spectrum. See also *single sideband transfer*.

double-slit experiment *Young's double-slit experiment* of 1801, which gave major proof of the wave nature of light.

doublet a) two thin lenses, in contact (or very close), used together to reduce aberration. The power of the doublet is the sum of the powers of the lens elements.
b) pair of very close lines in a spectrum, such as the sodium D lines. The **doublet interval** is the difference between the wave numbers of the lines.
c) very small magnetic or electric dipole, often made up to help with the maths of some system

drag the force of resistance F to the movement of an object through a fluid. Here the **drag constant** C_d (or drag coefficient) is a number used in finding the drag in a given case; it is the value $2F/A \rho v^2$; ρ is the fluid's density, A is the section area of the object at 90° to the flow (the frontal area), and v is the relative velocity. It is a function of the Reynolds number; the drag on two objects with the same frontal area moving at the same speed through the same fluid is proportional to C_d.

drain a) the outlet for a flowing fluid
b) the outlet for energy lost from a system
c) the electrode of a field effect transistor through which charge carriers leave

drift a) how objects travel when there is no net force acting from outside – at constant speed in a constant direction. For instance, charged particles pass at constant speed through a Faraday cage, such as a **drift tube**, a

sort of delay line (see below and *Faraday tube*). For **drift chamber**, see *nuclear physics: detector*.
b) how electrons travel through a conductor to carry a current – at a constant mean net speed in a constant mean direction (along the electric field) but interacting with the atomic cores they pass close to. In metals, drift speed is a few millimetres per second; in graphene, the electrons reach relativistic speeds.

drift mobility the mean drift speed of diffusing charge carriers per unit applied field. The mobilities of electrons and holes differ in most media.

drift transistor type of transistor in which the impurity concentration (dopant level) varies across the base with a high level near the emitter. This gives the device a good gain at very high frequencies.

drift tube tube in a linear accelerator through which the particles move at constant speed: in effect, they are in free fall, as the tube shields them from the outside field. While the particles are in the tube, therefore, the outside field can change ready for the next stage of acceleration.

driver circuit whose output is the input to a more important circuit. For instance, a driver may be the amplifier before the output stage of a power amplifier.

driving point impedance the ratio of the value of the alternating voltage applied to a circuit to the current that results between the terminals

drum musical instrument whose source is a skin – a 2D diaphragm – with a large air resonator below. An **ear drum** is an area of skin which transfers sound waves from the outer ear (the tube) to the middle ear.

dry cell primary or secondary (rechargeable) voltaic cell whose electrolyte is in the form of a jelly or paste rather than being liquid; the container is leak-proof. This design makes the cell safe and easy to carry so is the norm for small scale portable equipment.

dry ice solid carbon dioxide, a very convenient refrigerant where there is no power to drive a refrigerator pump

dual-in-line package the most common form of chip mounting, having two rows of pins

duality a) wave–particle duality, the property of radiation of behaving like a wave in some contexts and like a particle or stream in others. Light and shorter wavelength electromagnetic radiations, for instance, shows wave features in interference and particle features in the photoelectric effect.

b) the way a description of one object in one of the five superstring theories matches the description of another object in one of the other four

Duane–Hunt relation the highest frequency (v_0) in the output of an x-ray tube varies with the voltage V applied:

$$h_v = V Q_e$$

Q_e is the electron charge and h is the Planck constant.

dubnium Db transitional rare earth metal in the actinide series, period 7, with proton number 105, first created in 1967 by a team led by Georgii Flerov at Dubna in Russia (hence the name, though it had the name hahnium for a while). No one has made enough of the metal to be able to find any of the physical properties (or to develop any uses) – its most stable isotope is ^{268}Db, half-life 32 hours.

ductile soft enough to be able to be drawn into wires (of a metal like copper). This **ductility** results from a combination of properties. In particular, high ductility follows low yield stress and a high rate of work hardening over a large range of plastic deformation. All true metals are ductile.

Dulong and Petit's law the molar thermal capacity of a solid element approximates to $3R$, i.e. is about 25 J K^{-1} mol^{-1}; R is the gas constant, and molar thermal capacity is the product of mass per mole and specific thermal capacity. The concept follows the law of equipartition of energy. **Pierre Louis Dulong** (1785–1838), a French chemist and physicist, announced the law in 1819 with his co-researcher **Alexis Petit** (1791–1820).

In practice, the law holds only for regularly crystalline elements at fairly high temperatures. At low temperatures, the molar thermal capacity falls below $3R$ and varies with the cube of the temperature. Einstein devised a theory of specific thermal capacities that explains this.

Dunning, John (1907–1975) leader of the team set up to develop the first cyclotron, the first person to measure the energy transfer of nuclear fission, and influential in the development of US nuclear weapons

duplexer two-channel multiplexer – especially as used in radar and radio systems – designed to let the output and input pulses pass through the same aerial, or for two signals that differ in frequency to share one channel

Duralumin hard, strong and light alloy of aluminium with a trace of silicon, 4% copper, and 5% each of magnesium and manganese. It was the first age hardened aluminium alloy, Alfred Wilm 1903, and much used first in airship and then aircraft frames.

Dushman's equation relation between the thermionic output of an object – j, the current density produced – and the object's temperature T.

$$j = a\,T^2\exp(-b/T)$$

Both a and b are constants.

dust core the powdered core of an electromagnetic device (such as an inductor) or machine designed to work at high frequencies. The ferrite dust is sintered into a block in which eddy currents are tiny.

dwarf star small, faint, high-density star in the main sequence of the Herzsprung–Russell diagram (classes G, K and M)

dynamic moving or changing – not static

dynamic characteristics for an active electronic element or device, the relationships between current and voltage in a set of conditions, expressed as a set of V/I curves. The currents are less than those in the corresponding static characteristics in most cases

dynamic equilibrium the state of a system which is in overall balance despite constant internal motion or change. For instance, a liquid in a closed space with its vapour may be in equilibrium – i.e. with no net change of state – but there is a constant interchange of particles between liquid and vapour and back. In this case, the rates of evaporation and condensation are the same.

dynamic friction the force needed between two surfaces in contact to keep them sliding over each other at constant speed

dynamic impedance the maximum impedance of a parallel capacitor-inductor circuit; it is found at one specific frequency.

dynamic optical systems optical systems which include movable parts as well as static lenses, mirrors and/or prisms. Those parts can be micro-mirrors on the surface of a silicon chip, controlled by electronic signals, as in some projectors.

dynamic pressure the pressure associated with a moving fluid

dynamic range the input amplitude range of an electronic element, device or circuit (such as a speaker) over which there is useful output. It is, in other words, the input amplitude range, given in decibels, between the minimum at which the signal-to-noise ratio is acceptable and the overload level.

dynamics the study of how objects in motion behave. Based on Newton's laws of force (motion), the main concern of dynamics is the effect of outside forces on how the objects move.

dynamo machine that converts input mechanical energy to electricity, i.e. a generator. A coil spins in a magnetic field (produced, except in very small machines, by one or more electromagnets). The motion induces an alternating voltage between the ends of the coil; the output passes to the outside circuit through sliding contacts, a pair of complete slip rings to give ac or a single split ring (into two halves) to give dc. People often use the term dynamo for the dc version only, the ac generator being called an alternator.

dynamometer torquemeter; a device that measures the torque output by a rotating system such as an electric motor. Some people wrongly use the term for a forcemeter, such as spring balance.

dynamotor type of motor-generator in which the same field coils and core are used for both actions; it acts as a rotating transformer.

dyne dyn old unit of force, 10^{-5} newton. The dyne was the standard unit of force in the old cgs system – the force able to accelerate one gram by one centimetre per second per second.

dynode electrode in a thermionic tube designed to produce a secondary emission of electrons. See, for instance, *electron multiplier* and *photomultiplier*.

dysprosium Dy transitional rare earth element (number 66) in the lanthanide series, suspected by several chemists in the 1870s and found as oxide in 1886 by Louis de Boisbaudran (who named it from the Greek *dysprositos* = hard to find); it was not isolated until over sixty years later. Relative density 8.54, melting and boiling temperatures 1408 °C and 2565 °C; most common isotope ^{164}Dy. Having a high absorption cross-section for thermal neutrons, the metal has various uses in the nuclear power industry.

e symbol for electron

E unit prefix for exa- 10^{-18}, as in e.g. exagram 10^{-18} g

e **a)** of an orbit, symbol for eccentricity, no unit

b) common symbol for the elementary charge – the electron charge Q_e, $1.602 \ 10 \times 10^{-19}$ C

c) symbol for the Euler number, the base of natural logarithms, 2.718 ...

E **a)** symbol for electric field strength

b) symbol for energy (W preferred)

c) symbol for electromotive force

d) symbol for illuminance

e) symbol for illumination

f) symbol for irradiance

g) symbol for (Young's) modulus of elasticity

ε **a)** symbol for electric constant, permittivity

b) symbol for energy density

c) symbol for a very small value

ear the sense organ of hearing in many large animals, that converts input sound waves to signals in the nerve to the brain, and has a function in balance as well. Sound waves pass through the outer ear and make the drum vibrate. The vibrations pass along the bones of the middle ear, the muscles having some control over intensity. The sensors themselves are on the inside of the cochlea, which gives frequency discrimination; the nerve links this structure to the brain. Sensors linked to small floating chalky grains in the labyrinth control balance.

earth **a)** the agreed zero of electric potential, that of the Earth

b) (also called ground in the USA and some other countries) a large conducting object such as the Earth, to which an object is linked to keep it at zero potential for reasons such as safety. In a multicore cable (such as a power lead), the low resistance earth line should link to an earth in use. The normal symbol is as shown.

$$\perp$$

c) old term for rock, or ore, or metal oxide, or metal, for instance, alkaline earth or rare earth – the sources of some metals are their oxides as found in an ore or rock

Earth the third major planet out from the Sun, at a distance of some 150 million kilometres. The mass is around 6×10^{24} kg and its diameter is nearly 13 000 km; this gives a mean relative density of just over 5.5. The rocky crust, about 35 km thick at the continents and 10 km thick under the oceans, lies over the mantle; the boundary is the Mohorovičić discontinuity. The outer mantle is around 375 km thick; the thickness of the lower mantle is about 1900 km, and there is a transition region some 590 km thick.

The magnetic nickel–iron core has two regions as well, though we know much less about these. It is believed that a transition layer about 140 km thick separates a solid inner core with a radius of some 1200 km from a 2000 km thick liquid core.

Earth inductor device to measure large uniform magfields. It consists of a large coil of known area and number of turns, linked to a meter. When the user flips the coil quickly through 180°, the reading of the meter gives the induced emf; from this we can find the field strength in question. The name arises from its common use in the past to measure the Earth's field strength at different places.

earthing the process of linking an object to an *earth* b). If the object and its charge is small, touching with the hand is enough: the human body is a large enough conductor for the purpose. The earthing of lightning rods, and of power supplies to buildings, involves connection to a large metal plate buried in the ground outside.

earthquake disturbance in the Earth's crust great enough to cause damage at the surface. In contrast, an earth tremor, although felt by people at the surface or picked up by sensors (seismographs), causes no damage. During a quake, up to around 10^{18} J may be released in an area of a few cubic kilometres. This is the focal region centred on a vague focus up to a few tens of kilometres deep.

The point on the Earth's surface above the focus of the quake is the epicentre. The intensity of earthquakes is measured on the Richter scale (or more useful versions).

Earthquakes produce various kinds of wave, in particular P-waves (pressure or push waves, which are longitudinal waves of the same type as sound) and S-waves (shear or shake waves, which are transverse). These pass through the different parts of the Earth (but not the liquid core in the case of S-waves), moving at different speeds as the medium changes, and showing reflection, refraction, interference and diffraction effects. The patterns provide much information about the Earth's structure.

Earth's magnetism See *geomagnetism*.

east–west effect the appearance of more cosmic showers from the east than from the west. The cause is the Earth's magnetic field; the effect shows that most of the particles that produce showers have a positive charge.

échelette or **échelette grating** type of grating used particularly with infrared waves. Its design, with up to 100 000 grooves per centimetre, concentrates most of the output radiation into a single order of spectrum. This gives maximum intensity.

echelon or **échelon grating** very high resolution grating that allows study of the fine and hyperfine detail of a spectrum. It consists of a few tens of thin, accurately parallel and equal glass plates, set in contact to give a stepped edge about two centimetres long. The stepped edge acts as the grating, being most effective by reflection.

echo weaker sound signal detected after an original sound, as a result of reflection of sound waves – i.e. there are two received signals, one coming direct from the source and the other by reflection. Their separation in time depends on the distance of the reflector. The term is also used, by extension,

of other waves. The effect is a source of interference in many situations, causing, for instance, ghosting in the case of television signals.

echo chamber reverberation chamber, with high values of reverberation. Used for special effects or for research, it is the opposite of the anechoic chamber.

echo location the use of the reflection of sound waves to learn about the environment

echo sounding use of radar-like systems which use ultrasonic waves to (for instance) detect fish or measure the depth of water below a boat or ship. A piezoelectric transducer outputs wave pulses; it then acts in reverse as a sensor to detect the return echoes and form an image. See also *sonar*, which (though still using sound waves) is closer to radar in being used in all directions. **Echo prospecting** is much the same process, used on land to explore the strata of the rocks under the surface, for instance in the search for oil. Ultrasound scanning works by forming an image of structures inside the body.

All these techniques depend on the reflection and transmission characteristics of sound waves at boundaries between different media (water/sea-bed, skin/organ tissue, and so on). In each case, the user obtains diagnostic information about a volume of space and structures inside it. See also *acoustic impedance*.

eclipse the shadow in sunlight of one planet or moon on a second. The most important examples are the **lunar eclipse** (when the shadow of the Earth falls on the Moon's surface) and the **solar eclipse** (when the shadow of the Moon falls on the Earth's surface).

There are three types of solar eclipse:
a) annular eclipse: the Moon is over the Sun in the sky but does not cover it entirely, being too far from the Earth – viewers see the edge of the Sun as a ring (Latin *annulus*) and all the shadow is penumbra.
b) partial eclipse: the Moon is not quite on the straight line between Sun and Earth so the viewer sees part of the Sun – the viewer is in the partial shadow region of a full shadow, the penumbra.
c) total eclipse: the Moon is directly between the Sun and the viewer and the Moon is close enough to cover the Sun's whole disc – now the viewer is in total shadow (umbra region). These terms also apply in theory to lunar

eclipses; however as the Earth is so much larger than the Moon – and also has a thick atmosphere which refracts the Sun's rays into the shadow region, the effects are far less clear. Eclipses also occur elsewhere in the solar system, but are not visible to the naked eye. Examples occur when the planet Mercury passes across the Sun's disc (an occultation) and when the shadows of Jupiter's moons fall on the planet's surface.

ecliptic the circle on the sky along which the Sun appears to move during the year. The ecliptic crosses the celestial equator at the points corresponding to the spring and autumn equinoxes (respectively the so-called first point of Aries and first point of Libra).

Eddington, Arthur (1882–1944) regarded by many as the greatest astronomer of the last century, a pioneer of the astrophysics of stars – and his popular books on astronomy and physics remain of great interest. The **Eddington theory** is that physical measures must take account of the observer, the techniques and their nature: in other words, they are not absolute.

eddy current circulating current in a metal sample as a result of electromagnetic induction by a changing magfield. The current takes energy from the system so is usually undesirable. It is therefore common to divide the sample (for instance, the core of an electrical machine such as a motor or transformer) into slices (called laminations). As the currents cannot pass between the slices, they are much smaller.

The effect is used to advantage in **eddy current damping** (or electromagnetic damping). Lenz's law states that an induced current tends to oppose its cause; eddy currents will therefore tend to slow down a metal object moving relative to a field. A common way to show this is by dropping a small strong magnet; it takes more time to fall through a metal tube. (It is not certain if this would happen if the tube were a superconductor.) In **Eddy current heating**, high frequency alternating fields cause a rapid rise in the temperature of a metal object while the surroundings remain cool. This is of use in certain kinds of oven and furnace.

eddy viscosity an extra component of viscosity (fluid friction) as a result of eddies when flow is turbulent

edge dislocation See *dislocation*.

edge tone sound produced when a thin sheet of fast moving gas from a slit or hole meets the edge of a barrier. Standing waves appear between the slit or hole and the edge.

Edison, Thomas Alva (1847–1931) inventor who became involved in telegraphy when he moved into newspaper publishing. He invented various telegraph units including several multiplexers. The profits allowed him to set up the first large-scale research lab/workshop. His inventions, of which there are more than a thousand (a record for any person), include the phonograph (an early sound recorder/player), the transmitter and receiver that made Bell's telephone viable, the first useful light bulb (lamp), and the **Edison cell**. This last is a nickel–iron ('NiFe') secondary cell (an accumulator, one that can be recharged); it is lighter than the lead–acid cell but outputs a somewhat lower voltage (of the order of 1.4 V).

His only important finding in physics, however, is the **Edison effect**. This is thermionic emission: the flow of charge from a hot electrode to a cold one in a vacuum but not the other way. This discovery, which was to be the basis of electronics for half a century, was not, however, developed by Edison, as he saw no use for it. Despite all those inventions, therefore, Edison is not renowned for insight. His famous saying sums him up: 'genius is 1% inspiration and 99% perspiration'.

effective energy the energy of a polyenergetic beam, which is the energy of a mono-energetic beam scattered to the same extent

effective force the equilibrant of a set of forces acting on an object. It is not, surprisingly, the resultant, which is equal to this but in the opposite direction.

effective mass function of the true mass of a charge carrier and its position in the energy band structure; it depends on the particle's mobility.

effective resistance the ratio of the power dissipation to the square of the current in a resistive element in an ac circuit. It may differ from the dc value, as a result of eddy currents and skin conduction.

effective value the root mean square of a measure in ac circuits

effective wavelength wavelength of a polyenergetic beam which is the wavelength of a monoenergetic beam scattered to the same extent

efficiency ξ the ratio, in any process of energy transfer, of the useful output to the total input. People often use power for energy here, power being the energy transfer in unit time. There is no unit, though the ratio often appears as a percentage. Here are some special cases:

a) in an electric motor: the ratio of output power to input electric power; the reverse in the case of a generator

b) in other machines (such as levers and pulleys): the force ratio divided by the distance ratio

c) in engines: the value $(T_i - T_0)/T_i$, T_i being the input (source) temperature and T_0 that of the output (sink)

d) in dc circuits: the value $R(R + r)$, where R is the circuit resistance and r the source resistance.

effusion the passage of gas particles through a hole or a porous barrier. Graham's law of diffusion applies at normal pressures if the particles' mean free path is much less than the size of the hole(s). In other circumstances, effusion is much more complex.

ehf extremely high frequency

Ehrenfest rule if there is an adiabatic change in a system described by quantised measures, the quantum numbers must either change suddenly or not at all. If the change is slow, they will not change; in this case, the measures are adiabatically invariant.

Ehrenfest theorem the motion of a wave packet is the same as that of the classical particle if none of the potentials acting changes significantly across the wave packet's length. Named after Paul Ehrenfest (1880–1933).

eigenspace a set of eigenvectors with the same eigenvalue! German 'eigen' means characteristic, first used in this sense by Hilbert in 1904. We can picture any transformation in space (e.g. stretching or reflecting) through its effect on vectors in the space concerned. For a transformation, **eigenvectors** are the vectors which do not change or change only in length. If a vector changes in length, its **eigenvalue** is the scale factor (e.g. 0.5 if it is halved in length). Then, an eigenspace is a set of eigenvectors with the same eigenvalue. For instance, squashing a brick to half its length will give the 0.5 eigenvalue to all vectors parallel to the stress.

eightfold way old term for the use of SU(3) group theory to describe the relationships between sets of baryons and mesons. The sets are of eight (sometimes ten) particles and the

theory closely relates to the later quark models. Named after the path to stopping suffering, in Buddhism, by Murray Gell-Mann, the picture had great predictive power – he proposed the A⁻ particle in 1962, and it was discovered, just as described, two years later. Yuval Ne'eman (1925–2006) came up with the same system at the same time.

The three patterns here plot the values of strangeness S and of charge Q, for, respectively, the octet of mesons, the octet of baryons with spin ½ and the decuplet of baryons with spin 3/2.

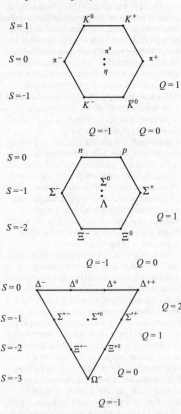

This method of classing mesons and baryons leads to the quark model.

Einstein, Albert (1879–1955) one of the greatest mathematical physicists of all time, who devised and developed the relativity theories and whose research into photoelectricity led to the quantum theory and his Nobel Prize in 1921. From around that time, his main work was an attempt to devise a unified field theory. He became more and more involved in politics and after the second world war was asked to become the president of Israel.

Einstein ranks with Galileo and Newton for his development of our view of the physical nature of the Universe. He started to publish his important work in quantum theory, Brownian motion and special relativity in 1905, while an examiner in the Swiss patent office. The special theory unified classical and more recent work, and was followed by the general theory in 1915. After the second world war, Einstein was still working hard to unify physics, in particular quantum mechanics and gravity – but without success. Einstein's first wife, Mileva Marić, was also a physicist; they married in 1904 and she may have had a significant role in his early work leading to relativity.

Einstein A constant the rate of de-excitation of a sample of excited atoms. It is not clear if this is constant under all circumstances for any given transition; the point is important in the development of efficient lasers.

Einstein and de Haas effect the slight rotation of an iron bar when suddenly magnetised; its reverse is the Barnett effect.

Einstein law the law of mass energy equivalence $W = m\,c^2$

Einstein ring the image obtained of a distant object in space – such as a distant galaxy or quasar – when there is a gravity lens from a massive object on our direct line of sight

Einstein's equation or **Einstein's photoelectric equation** relation that gives the highest energy of electrons released in the photoelectric effect: $h\,v - \chi$. Here $h\,v$ is the input photon energy and χ is the escape energy value (work function) of the surface.

Einstein's general theory of relativity See *relativity*.

Einstein shift the red-shift of lines in the spectrum of a star or galaxy that results from the object's gravitational field. It follows from Einstein's general theory of relativity.

Einstein's special theory of relativity See *relativity*.

Einstein's theory of specific thermal capacities theory that gave an expression for the specific thermal capacity of a

solid at a given temperature. It was quite successful, but Debye's theory took over in due course.

einsteinium Es transitional rare earth, proton number 99, in the actinide series, period 7; it was first found in 1952, in the radioactive fallout of an H-bomb test, by Albert Ghiorso and others and named after Einstein when he died a few years later. Relative density 8.84, melting temperature 860 °C; most stable isotope ^{252}Es, which decays by electron capture or alpha or beta emission, mean half-life 470 days.

Einthoven galvanometer a string meter that can measure current as low as 10^{-11} A. The current concerned passes through a very tight thread in a magfield, and a microscope measures the deflection of the thread. **Willem Einthoven**, a physiologist, invented the device to help research electrocardiography; he obtained the Nobel Prize for medicine in 1924 because it is so effective.

elastance the reciprocal of capacitance, the unit being the farad^{-1} (sometimes called daraf)

elastic able to return to its original state after an applied stress goes away. However, the sample may not return at once to its initial state: there may be an **elastic after-effect** (a form of *hysteresis*).

elastic collision collision in which there is no change in the total kinetic energy of the objects. See *restitution constant.*

elastic constant the strain per unit stress (compliance) or stress per unit strain (stiffness) of samples in the elastic region. There is a number of these – in fact, an anisotropic substance has 21 elastic constants.

elastic deformation deformation which will vanish when the stress that causes it goes; there is, in other words, no net displacement of the sample's particles.

elasticity the property of a sample of matter of being able to return to its original state after an applied stress goes away. Hooke's law states that, in this case, the strain (deformation) is proportional to the stress (load). This is true up to the elastic limit or the proportional limit of the sample, and does not depend on the type of stress or the strain that results.

In practice, the sample reaches its elastic limit at a slightly higher stress than the proportional limit: it remains elastic but strain increases faster than stress between the two. Past the elastic limit, deformation includes a plastic component as well as the elastic one; when the stress goes away, a net deformation remains. This results from net displacements of some of the particles, such as in crystal defects. Above the yield point, flow begins; now the sample continues to deform with time with no increase in stress. Flow continues, with necking in the case of tension, until the sample fractures.

elastic limit the upper end of the range of applied stress over which a sample shows elastic behaviour. See *elasticity*

elastic liquid liquid that is viscoelastic: this means that it has a 'memory' of past deformation so can store energy when sheared.

elastic scattering process in nuclear and atomic physics that results from elastic collisions between the particles of a beam and those of a target

elastic wave wave form that transmits in matter as a result of the elastic nature of the substance. There are two main types: compressional (which is longitudinal) and shear (transverse). These are the same as the P- and S-waves of an earthquake.

elastodynamics the study of elastic waves

elastoresistance the change of a sample's resistance while stressed within its elastic limit

E-layer or **E-region** the Heaviside layer, the middle layer of the Earth's ionosphere. Lying above the D-layer, about 90–150 km above the surface, it reflects medium-wave radio signals.

electret permanently charged sample of substance with net positive charge at one end or side and negative charge at the other. Named in 1885 by Oliver Heaviside, but known long before – for instance, see electrophorus – it is the electrical analog of the magnet and shows much the same properties. Some good electret materials are ferroelectric – they are stable in this regard, whereas others are only meta-stable.

In an **electret transducer**, the electret's permanent field allows energy transfer in various contexts, for instance microphone, pickup, speaker.

a system of electrical elements such as resistors, and one or more sources of electricity (electrical energy), set up as a closed loop to allow a complete path for the current (flow of electric charge). There are various types of source, all of which offer a voltage (or electromotive force) to force free charges to move and form a current. The main types are

- sensors or transducers (such as photo-cells and thermopiles),
- chemical sources (primary and secondary cells and batteries), and
- machines (generators and transformers).

It is normal to divide circuits into two types:

- dc circuits (dc being direct current), in which the current is only one way, and
- ac circuits (ac for alternating current), in which the current changes direction in a regular cycle.

We shall look at those in turn, after first taking care of more general terms. See also the major entry for *electricity* (and *electric*) that follows this one. See *transient* for dc circuits during switch-on and switch-off.

conductor anything which, given a suitable voltage, can carry a current (with, in brackets, what it is that carries the current):

- in the case of solids, metals (in which case the free charges involved are electrons and the system follows Ohm's law), and semiconductors – intrinsic (electrons and holes), n-type (electrons) and p-type (holes)
- liquid metals (electrons) and electrolytes (liquids with free ions, such as molten salts and solutions of salts)
- gases (electrons and ions – see *spark* and *discharge tube*)
- vacuum (electrons, in the form of cathode rays)

current except in the case of superconductivity, a continuous flow of electric charge caused by a *voltage* (below), with some kind of energy transfer from the supply. Symbol I (unit: amp or ampere, A), a current is the net flow of charge Q between two points. Its value is the rate of flow of charge dQ/dt; the ampere is the coulomb per second, the transfer of some 6.2×10^{18} electrons per second. Traditionally we take the current direction to be down the potential gradient, i.e. from what we call positive to less positive or negative.

voltage a difference in electrical potential, i.e. a potential difference (pd), between two points one of which is more positive than the other. If the two points are at the ends of a complete conducting path – a circuit or sub-circuit – there will be a current (net flow of charge). In the case of a source of electricity (energy transfer from some other form), the proper name for the voltage produced is electromotive force E. In the case of a circuit element (energy transfer to some other form), there is a pd between its ends as a result of the current; we tend to call this a voltage V. In all cases, the unit is the volt, V.

dc circuit an electric network with

- a source of electrical energy whose output makes free charge move in one direction only and a fairly constant rate – we tend to call the output the source's voltage (though strictly its name is 'electromotive force')
- one or more electric elements that absorb energy from the current and transfer it to some other 'form of energy' – such as lamps and motors
- a complete conducting path from one side of the source to the other
- a switch to allow simple control of the circuit (not required, but always needed in practice)
- maybe one or more meters (whose effect on how the circuit behaves must be very close to zero).

There are many kinds of electric circuit element (old name: component) – see the circuit symbols list in *Appendix 3*. However, in the case of dc circuits we need to know only the resistance of each element – how much it tries to oppose current. For this reason, we often call the elements resistors. (Strictly, a switch is an element – it should have zero resistance when closed and infinite resistance when open.) In circuit design, we need to work out the current taken from the supply – this depends on the source voltage and the resistance of the whole circuit. Circuit resistance depends in turn on the elements' resistances and whether they are linked in series (and carry the same current) or in parallel (in which case the current splits between them). See also *network analysis* and *Kirchhoff's rules*.

The table shows the measures used with dc circuits. Here 'element' can be a single element or any group of them (network or sub-circuit).

measure	symbol	unit	unit symbol
element resistance	R or R_i	ohm	Ω
circuit resistance (outside the source)	R	ohm	Ω
source resistance	r	ohm	Ω
voltage (between ends of an element or sub-circuit	V	volt	V
source voltage ('electromotive force')	E	volt	V
charge	Q	coulomb	C
current in an element	I or I_i	amp	A
circuit current	I	amp	A
power transfer in an element	p	watt	W
energy transfer in an element	W	joule	J

The following table shows how these measures relate to each other.

the resistance of resistors in series [1]	$R = R_1 + R_2 + \ldots$
the resistance of resistors in parallel[1]	$1/R = 1/R_1 + 1/R_2 + \ldots$
the output voltage of cells in series[2]	$E = E_1 + E_2 + \ldots$
any constant current (by definition)	$I = Q/t$
any voltage (by definition)	$V = W/Q$

the current in a resistor (from Ohm's law)	$I = V/R$
the circuit current (V and R are the circuit voltage and resistance)	$I = V/R$
the circuit current (E and r are the source voltage and resistance)	$I = E/(R + r)$
power transfer in an element	$p = V I = I^2 R = V^2/R$
energy transfer in an element	$W = V I t = I^2 R t = V^2 t/R$

Note 1 – the same applies to source resistances, r_1 ....

Note 2 – the output voltage of cells in parallel is the same as each one if they are all the same; if they are not the same, we use Kirchoff's rules.

$$R = \rho \, A/l$$

Here R is the resistance of a sample of the substance (for instance, a wire) whose resistivity is ρ, when the sample length is l and its cross-sectional area is A.

The inverse of resistivity is conductivity. See also the *temperature coefficient (constant) of resistivity*.

ac (still sometimes AC) alternating current, current that changes direction in a regular cycle. Some people use the word as an adjective to mean 'alternating', as in 'ac voltage'; this is not correct.

ac circuits alternating current circuits, those with an alternating supply (input), as with most electronic and microelectronic circuits and sub-circuits

alternating circuit a circuit with a source alternating in voltage at a certain frequency. In the simplest case, the alternating value is like a sine wave (it is simple harmonic), as in

$$V = V_0 \sin 2 \pi v \, t$$

Here V is the voltage at time t, V_0 the voltage amplitude (peak or maximum value), and v the frequency of alternation.

The effective value of an alternating quantity is often the root-mean-square value: in the case of simple harmonic alternation $V_{rms} = V_0/\sqrt{2}$. Thus the common UK mains supply of about 250 V (rms) at 50 Hz has a peak value of about 350 V. (A second common standard is 220 V rms, i.e. 310 V peak, at 60 Hz.)

alternating current (ac) a current which changes direction with a frequency that doesn't depend on the characteristics of the circuit which carries it. The size of the current depends on the alternating voltage supplied and on the circuit's impedance. As impedance is a complex value (with real and imaginary parts, respectively in phase and 90° out of phase with the current), the circuit current and supply voltage are not usually in phase.

alternating current circuit a circuit with an alternating supply, containing some combination of resistance R, capacitance C and inductance L. (A name sometimes used is an LCR circuit.) Each of those has a reactance X (opposition to AC) given by the values

resistive reactance	$X_R = R$
capacitative reactance	$X_C = 1/(2\pi\nu C)$
inductive reactance	$X_L = 2\pi\nu L$

The opposition of the whole circuit to AC is the impedance Z. In the simplest case in which L, C and R are in series:

$$Z = \sqrt{((X_L - X_C)^2 + X_R^2)}$$

The value of the current I comes from $I = V/Z$ (rms or peak).

However, in an ac circuit, the current and voltage sine waves are not in phase with each other. In fact, the current cycle is behind (it lags) the voltage cycle by a phase angle σ where:

$$\tan \sigma = (X_L - X_C)/R$$

The ohm is the unit of both impedance and reactance, as well as that of resistance.

Argand diagram a two-dimensional extension of the number line concept. At 0 (the origin) on the (real) number line an axis at 90° represents the imaginary number line. (An imaginary number is a multiple of i, $\sqrt{-1}$.) As inductive and capacitive reactances, X_C and X_C, respectively lead and lag X_R by 90°, we can use an Argand diagram to show how they relate. The resultant of the three values of X is the circuit impedance Z, which leads or lags X_R by the phase angle σ.

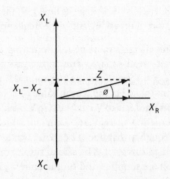

impedance Z unit: ohm, Ω the opposition of a complete circuit to alternating current: the ratio of the applied alternating voltage to the alternating current that results. In all cases, as noted above, impedance includes resistance and reactance terms and phase changes.

Where one circuit transfers a signal (or power) to a second, **impedance matching** is the process of ensuring maximum power transfer. In the case of a power transmission line, for instance, this requires the line impedance to equal both the output impedance of the source and the input impedance of the load; otherwise, reflections occur, with a loss of transfer efficiency.

loss In much the same way, an ac supply to an inductor does not produce the 90° phase shift of a perfect inductor. The **loss angle** is the difference between the 90° phase angle of theory and the phase angle met in practice with a given inductor. A similar term applies to capacitative circuits. The **loss factor** of an insulator across which there is an ac supply is the product of relative permittivity and power factor. (Power factor is the ratio of the power used to that supplied.)

electricity

the science and applied science of electric charge, current and energy. Indeed, it is the common name for charge, current or power. It links very closely with magnetism – the link being electromagnetism.

electric concerned with positive and negative charge and the flow of charge (current)

electric arc See *arc*.

electric bell device that produces a ringing or buzzing sound when it carries a current. It provides a standard example of the electromagnetic make and break circuit in use; the moving 'armature' vibrates the bell's hammer.

electric braking the use of the motor of an electric vehicle as a generator to slow the motor and therefore the vehicle. The current produced either passes through a high resistance (wasteful 'dissipative braking') or returns to charge the supply ('regenerative braking').

electric cars 'zero emission' (while driven) vehicles driven by current from on-board batteries feeding one or more motors. The first automobiles were electric, and successful, holding all records for

decades (e.g. an electric car broke the 100 km h⁻¹ barrier in 1899).
'Zero emission' applies only to driving, note – there are major
emissions when generating the electricity used to 're-fuel'. Even so,
this system is more efficient than internal combustion, with around
0.25 kWh km⁻¹ as opposed to about 0.8 – and there are other
advantages (such as quietness).

A **hybrid electric car** has a mix of electric and petrol drives, the
former using a battery; the petrol engine keeps the battery fully
charged, but most of that energy comes from braking and from
regeneration as the car slows. Thus, there is overall less waste
energy during a journey. Also, local emissions are less as the
electric motor provides power during idling and also helps the car
accelerate. See also *flywheel bus*.

electric charge Q Unit: coulomb, C fundamental measure, a property
of some elementary particles that causes electric forces between
them. The electron carries the elementary unit of charge, Q_e; so too
does the proton, but its charge is opposite in type. Traditionally the
two types of charge (none other has been found) are 'negative' and
'positive' respectively. Charges of the same kind repel each other,
while charges of opposite kind attract each other. The forces result
from the exchange of photons between the charges.

Charged objects have an excess of electrons over protons, or a
deficit.

electric circuit source of electric energy which needs a complete
conducting path between its terminals for there to be a current. For
the major circuit symbols, see *Appendix 3*.

electric conductance See *conductance*.

electric constant name sometimes used for the absolute permittivity
of empty space, ε_0 – about 8.85 F m⁻¹

electric current See *current*.

electric current density See *current density*.

electric degree 1/360 of a single full cycle in alternating current
circuits. Strictly, we should measure phase angle in electric degrees
rather than degrees.

electric dipole See *dipole*.

electric dipole moment See *dipole moment*.

electric discharge See *discharge* and *spark*.

electric displacement See *displacement* and *electric flux density*.

electric double layer layer believed to exist between two substances
in contact that differ in permittivity. Proposed by Helmholtz
and extended by Lenard, it explains the voltage that appears
across the interface. See *Helmholtz double layer* and *electrokinetic
potential*.

electric energy W unit: joule, J the ability to do something in some electrical context, better called **electric potential energy**. This does not mean that there is more than one form of energy – energy is energy, though it appears in different relationships in different contexts. In electricity, the main contexts with their relationships are as follows.

a) for a charge Q in a field with potential V at that point, energy is QV

b) when a current I passes through a potential difference (voltage) V for time t, the path resistance being R, energy is $V\,I\,t = V^2\,t/R = I^2\,R\,t$

c) with charge Q in a capacitor of capacitance C, forming a voltage V between its ends, energy is $Q^2/2C = \frac{1}{2}\,C\,V^2 = \frac{1}{2}\,Q\,V$

d) with an inductance L passing an ac I_0, the energy transfer in the inductor is $\frac{1}{2}\,L\,I_0^2$

electric eye old term for *photo-cell*

electric field region of space in which charges experience electric forces. The **electric field strength** (symbol E) is the force on unit charge at that point in a field; from $E = F/Q$, the unit is the newton per coulomb (N C^{-1}), equivalent to volt per metre (V m^{-1}).

electric flux Ψ_E the imaginary lines of electric force which people sometimes use to describe an electric field; though it does not exist as such, it can be useful. For a small area ΔA at angle θ to a field of strength E, we define the flux as $E\cos\theta\,\Delta A$. For a closed surface, the total flux leaving (positive) or passing in (negative) is as in the box.

$$\Psi_E = \int E\cos\theta\,dA$$

See *Gauss's theorem*.

electric flux density the electric flux per unit area (Φ/A), a measure still sometimes used for field strength

electric image concept sometimes of use in solving electrostatic problems. We can imagine a charge $+Q$ at a certain distance from a plane conducting surface to have an image $-Q$ as far behind the surface as the actual charge is in front. The laws of reflection give the positions of images in other situations.

electric intensity old term for electric field strength

electric motor transducer with an electric input and a force as output, an important kind of electric machine. There are various types. The simplest rotary motor has just the same structure as a generator but works in reverse (indeed the generator feature gives it a back emf, i.e. reverse voltage, which tends to oppose the input).

See also *induction motor, linear motor* and *synchronous motor*.

electric network a set of electric elements linked by conducting paths. The *network* becomes a circuit only if there is a source of electrical energy and a complete path for current between the two sides of the source.

electric polarisation P unit: coulomb per square metre, $C\,m^{-2}$ feature of a dielectric (insulating) substance in an electric field. Its particles distort to become dipoles with positive and negative ends. Electric polarisation links the displacement D observed to the original field strength E (i.e. the strength of the field if the substance were absent):

$$P = D - \varepsilon_0 E$$

electric potential V unit: volt, V, the joule per coulomb $J\,Q^{-1}$ of a point in an electric field, the energy involved in bringing unit positive charge to the point from infinity (i.e. from a region with zero field). See also *potential difference*.

electric power
a) symbol P, the rate of transfer of electric energy, i.e. the product $V\,I$, the unit being the watt W
b) current produced by generators on a commercial scale to help meet the energy needs of a community. **Power stations** contain very large generators (hundreds of megawatts being common); these are driven by steam turbines (the steam being produced either by burning a fuel or from the output of a nuclear reactor), or, in the case of hydroelectric plants, by water turbines. On a small scale, electric power comes from chemical batteries, small generators, fuel cells, solar cells, or thermoelectric units.

electric spark See *spark*.

electrical engineering the study and practice of applied electricity, electromagnetism and large current electronics. A major concern is the efficient and green generation and transfer of electric power.

electro- prefix for electric or electrical

electrocardiograph ecg device that detects and amplifies the electric potentials of the parts of the heart as it beats, and produces time graphs of these; detection is by means of electrodes attached to the chest.

electrochemical cell chemical source of electric power, in which some reaction takes place to make a negative and a positive electrode (terminal). Electrochemical cells are either primary or secondary. While a primary cell is available for use as soon as made, a secondary cell is not – but the user can charge and recharge it hundreds of times. In other words, the action of a primary cell is not reversible, while that of a secondary cell is.

electrochemical equivalent z unit: kilogram per coulomb kg C^{-1} the mass of an element freed from an electrolyte when a coulomb of current passes during *electrolysis* (below)

electrochemical series or **electromotive series** the set of elements arranged in increasing order of electrode potential; it is normal to use a scale on which the electrode potential of hydrogen is zero. We say that elements that tend to lose electrons in forming ions are electropositive; those forming anions are electronegative. Elements higher in the set, i.e. ones more electropositive, tend to displace lower ones from solution. For instance, putting zinc metal (high) in a copper (less high) sulphate solution will displace the copper: copper metal appears and zinc dissolves.

electrochemistry the science and applied science of electrode potentials, electrolysis, and the design and use of chemical ('voltaic') cells. Its main concern is how ions in solution and in molten matter behave.

electrochromism the change of colour that follows a chemical change in a substance under the influence of an applied voltage. The change is reversed by a reverse voltage. Uses include the control of window tint, variable decorations, flat displays for full colour (but fairly static) data, variable camera filters, and anti-glare driving mirrors.

electrode in an electric system, a device designed to produce (input), collect (output), or deflect charge carriers. It may be a metal plate, grid, mesh or wire or (in some old electrochemical cells) a tube of mercury. An anode is an electrode with a net positive charge; cathodes have negative charge. In an electron tube there may be quite a number (up to as many as ten) of different electrodes, as in the *electron gun* of the cathode ray tube in a cro.

electrode dissipation the transfer to thermal energy at an electrode in an electron tube, as a result of electron or ion impacts

electrode efficiency the ratio, in electrolysis, of the actual mass of element freed in unit time to the theoretical value

electrode potential the equilibrium voltage between a metal and a solution of its ions, in electrolysis. We cannot measure this directly – it is normal to compare the value with that of a hydrogen electrode; this is of value in any event as hydrogen's potential is defined as zero (see *electrochemical series*).

electrodeless discharge in a discharge tube (for instance, a fluorescent tube), the discharge obtained by placing it in a strong, changing electric field (such as near a spark source) or in an ion beam. The process is of value for giving the spectra of the gases or vapours in the tube.

electrodeposition the electrolytic process of freeing a metal, which tends to form a layer on the cathode. This is the basis of electroplating, electrotyping, the electrolytic recovery of a metal from an ore, and so on.

electrodialysis process of dialysis (the diffusion of ions through a porous barrier down a concentration gradient) accelerated by a voltage on each side

electrodisintegration rare term for the use of an electron beam to cause nuclei to break up

electrodynamic instrument or **electrodynamic meter** one of the many types of electric meter. It has a coil, with needle attached, that is free to rotate in the field of one or more fixed coils that carry the same current. There are various arrangements for different purposes.

electrodynamics

a) the study of the magnetic forces between currents in circuits near each other and the attempt to explain these by classical physics

b) quantum theory that uses the photon as mediator to try to explain the electromagnetic exchange force

electrodynamometer type of electrodynamic meter

electroencephalograph eeg system rather like an electrocardiograph (ecg) but used to measure and display currents and pds in the brain; the electrodes are attached to the scalp.

electroendosmosis old term for electrosmosis

electroforming forming metal parts by electrolytic deposition of a metal layer on a shaped cathode made of conducting wax; at the end of the process the wax is melted and removed.

electrokinetic potential the gradual change of potential in a liquid at an interface with a solid. (There is also a very sharp change in the potential of a layer about one ion thick right at the surface itself, see *Helmholtz double layer*.) The potential plays a role in such processes as electrosmosis. It also explains the mechanism of the action of the **electrokinetic transducer**; this is a transducer with an acoustic (sound) input and an electric output that depends on the electrokinetic potential generated by the sound wave as it causes liquid to slosh to and fro through a porous barrier. Electrodes each side of the barrier pick up a signal that matches the input sound wave.

electroluminescence the Destriau effect – the output of light from certain phosphors in a changing electric field. When used as a light source, putting an alternating voltage across a thin dielectric layer that contains such phosphors will produce a gentle glow.

electrolysis a chemical change that results from a current in an **electrolyte**. This is a liquid that contains ions and thus can conduct current. It can be a molten substance or solution; both types are in wide use. The solutions are of acids or salts (ionic compounds). (Liquid metals are not electrolytes: they do not contain ions and they conduct by the passage of free electrons.)

Electrolytic conductivity (symbol $\varkappa$) is the conductivity of an electrolyte. **Electrolytic dissociation**, or breakdown, is the appearance of ions when an ionic compound becomes an electrolyte. The process is reversible and not all such compounds convert fully to ions when molten or dissolved. Indeed, **electrolytic polarisation** is the tendency of the products of electrolysis to recombine. Because of this process, to achieve a steady electrolytic current, there must be a minimum voltage across the given cell.

In an **electrolytic cell** (container of electrolyte with two electrodes), drifting ions carry the current. When they reach the electrode of opposite sign, they lose their charge and thus change their nature. Thus water (with some acid added to increase its conductivity) contains H^+ and OH^- (hydroxyl) ions, these being positive (cations) and negative (anions) respectively; in the electric field between two electrodes, these move towards the negative electrode (cathode) and positive electrode (anode) respectively. The reactions cause the appearance of bubbles of hydrogen and oxygen:

At the cathode: $\quad H^+ + e^- \rightarrow H \quad$ ('nascent' = newborn, hydrogen)

$\qquad\qquad\qquad 4H \rightarrow 2H_2 \qquad$ (hydrogen gas bubbles)

i.e. four electrons leave the cathode in the 'breakdown' of $2H_2O$.

At the anode: $\quad OH^- \rightarrow e^- + OH \quad 2OH \rightarrow H_2O + O$

$\qquad\qquad\qquad 2O \rightarrow O_2 \qquad$ (oxygen gas bubbles)

i.e. four electrons enter the anode in the 'breakdown' of $2H_2O$.

Thus for every four electrons passing through the cell (and the outside circuit), two hydrogen molecules and one oxygen molecule appear at the electrodes.

This process is common and of great value in many contexts. It was a very early use of electricity in the nineteenth century and led to the science and technology of electrochemistry; as a result, the quantitative statements, **Faraday's laws of electrolysis**, appeared early on. These are that, during electrolysis:

a) The mass of an element freed is proportional to the charge passed.

b) The mass of element freed by a given charge depends on the ion concerned.

In fact, in the latter case, the mass depends on the ion's charge and its relative mass; these lead to the **electrochemical equivalent** z of the element. This is the mass freed by a coulomb of charge; see also faraday. The unit of z is the kilogram/coulomb (kg C^{-1}):

$$z = m/Q = m/I\,t$$

There are many uses of electrolysis; some of the main ones are as follows.

● **electrolytic polishing** – electroplating used to give a metal surface a very fine sheen

● **electrolytic rectification** (rare, though once common) – use of a special electrolytic cell which, because of the particular combination of electrolyte and electrode, conducts much better one way than the other; to provide rectification

● **electrolytic reduction** – to obtain metals from molten or dissolved salts (and that includes ores)

● **electrolytic refining** – to remove impurities from metal samples

● **electrolytic separation** (not very efficient) – to separate the isotopes of an element, possible because of the slightly different physical properties of isotopes

● **electrolytic tank** – electrolytic cell of value in the study of electrostatics. A conducting model of the object under test (an electrode assembly for instance) goes into the tank as cathode; the way it plates makes it easier to understand the variation in electric field over its surface.

● **electroplating** – process that coats the surface of the conducting cathode of an electrolytic cell with metal from solution. Thus, to coat a metal object with silver, the user should make it the cathode in a cell with a silver salt as electrolyte.

electrolytic to do with electrolysis, or based on an electrolyte

electrolytic capacitor common type of capacitor, with anode and cathode (often of aluminium foil) in an electrolyte. The dielectric (insulator needed between the electrodes of any capacitor) is a thin oxide layer formed on the anode in advance. In most electrolytic capacitors, the electrolyte is gauze or paper impregnated with a paste, the whole being rolled up in a leak-proof tube. This design offers safety and portability as well as large capacitance in a small space. An **electrolytic cell** – sometimes called voltameter – is any system in which electrolysis causes chemical change.

electrolytic polarisation the tendency of the products of electrolysis to recombine. In a chemical (voltaic) cell, whose action is the reverse of electrolysis, polarisation results in a fall in the output voltage. The cause is, for instance, a layer of gas on an electrode – and this needs a depolariser to reduce the effect.

electromagnet conducting coil, almost always containing a 'soft' iron core. When there is a current in the coil, a magnetic field appears. This magnetises the core (if any) to strengthen the field greatly; switching off the current removes the magnetism. In other words, an electromagnet (unlike a permanent magnet) gives a field that is under outside control. Because of this, electromagnets have many uses in industry, electronics, audio and so on. It follows that there are very many designs and sizes.

electromagnetic relating to electromagnetism, the interface between electricity and magnetism, where currents have magnetic effects and magnetic fields affect charges

electromagnetic damping See *eddy current*.

electromagnetic deflection the use of the fields of coils to control the electron beam of a cathode ray tube (as in a tv set). See *deflection coils*.

electromagnetic delay line See *delay line*.

electromagnetic field region of space in which there are electric and magnetic effects, in other words, there can be electric and magnetic forces. Maxwell's equations relate the field's properties (electric and magnetic field strengths and induction and electric charge and current densities). The Poynting vector gives the size and direction of energy flow, this being in the form of electromagnetic radiation.

Standard theory tells us that an electric field can arise only from electric charge, while a magfield is an effect of charge in motion. However, what an observer observes depends on the relative motion between observer and field: a magnetic field to one observer may be an electric field to a second.

electromagnetic focusing technique for focusing a beam of electrons in a tube by means of the magfield of the coil that surrounds the electron tube. It is similar to electromagnetic deflection (see *deflection coils*). Such an **electromagnetic lens** is important in, for instance, the electron microscope.

electromagnetic gun See *linear motor*.

electromagnetic induction the flow of charge that leads to the appearance of a voltage between the ends of a conductor in a changing magfield. It is as important as the reverse effect, that on which the electromagnet is based. The change may result from an increase or decrease in an electromagnetic field, or relative motion between conductor and field.

There are three **laws of electromagnetic induction:**

a) Faraday's first law: a voltage appears between the ends of a conductor when there is a change in the magfield around it.

b) Faraday's second law: an induced voltage is proportional to the rate of change that causes it.

c) Lenz's law: an induced voltage tends to oppose the change that causes it.

Faraday's second law gives the size of the effect and Lenz's law gives its direction; the latter follows from Le Chatelier's principle and other such statements of energy conservation. The relation that in turn follows from the laws is

$$E = - d\Phi/dt.$$

Here E is the induced pd (in fact an emf) and $d\Phi/dt$ is the rate of change of field ('flux', Φ) cutting the conductor.

Many devices depend on this effect; the most important are electric generators and transformers.

electromagnetic interaction any interaction between elementary particles that depends on the particles' electric and magnetic fields. An example is the Coulomb (electrostatic) force between two charged particles. In any such interaction, the following quantum measures remain constant (conserved): angular momentum (spin), charge, baryon number, iso-spin, strangeness, and charge parity.

electromagnetic lens same as *electron lens*

electromagnetic mass that part of the mass of a moving charged particle (or object) that depends on its charge; it results from the magfield of the moving charge – and such a field needs energy. For a particle of radius r and charge Q moving at slow speed in a medium of permeability μ, the value is $2/3\mu Q^2/r$. There is a relativistic increase in this component of mass at relativistic speeds.

electromagnetic moment of a magnet or current in a coil, the strength of its magnetism. It is the torque on it when set at 90° to a field.

electromagnetic pump pump used to accelerate the flow of a conducting fluid. The fluid passes through a tube between the poles of a strong magnet, and there is a current in the fluid along the axis of the pipe. The motor effect causes the fluid to accelerate as a result.

electromagnetic radiation range of transverse waves that travel in free space at 300 000 km s^{-1}. (The exact figure, c or c_0, is 2.997 927 × 10^8 m s^{-1}.) See *wave* for the basic properties; however the photoelectric effect shows they have particle aspects too. See also *quantum theory*.

The energy transfer process of electromagnetic radiation involves alternating electric (E) and magnetic (B) fields at 90° as shown in the figure and described by Maxwell's equations. See also *Poynting vector*.

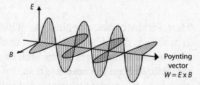

Electromagnetic radiation passes by way of alternating electric and magnetic fields.

electromagnetic spectrum method of classing for convenience the many electromagnetic radiations into broad bands, in which they appear in order of increasing frequency (inverse order of wavelength). The main bands are radio, microwave (which includes radar), infrared, visible light, ultraviolet, x-rays and gamma rays. See *spectrum*. For details of the bands, and the sub-bands of some of them, see their main entries.

electromagnetic system extension of the old cgs system, once widely used, based on **electromagnetic units**. In this system, the permeability of free space takes the value of unity (1). Compare with *electrostatic system* – the two systems are not compatible.

electromagnetic waves See *electromagnetic radiation*.

electromagnetism the science and applied science of all effects that concern the interaction of electric and magnetic fields. See also *electromagnetic field*.

electrometer device that detects or measures voltage or charge while taking a negligible current. It contains a very high impedance amplifier built with field effect transistors. Various non-electronic systems were common in the past.

electromotive force emf (or EMF) E (or W) unit: volt, V the sum of the voltage around a complete circuit, a measure of the energy supplied. It is in fact the energy needed to take unit charge around the whole circuit; the unit, the volt, is the joule per coulomb ($J\,C^{-1}$). An alternative definition is the power supplied per unit current; this gives the same unit, of course.

$$E = W/Q = P/I$$

The name is old and clumsy; more common now is potential difference (pd). On the other hand, the actual pd between the terminals of a source supplying current I to a circuit is less than the emf. This is because of the pd needed to pass the current through the source's own resistance r (such as the resistance of the coil of a generator). The relationship is:

$$E = I(R + r)$$

Here I is the current supplied and R is the resistance of the circuit. E is, in fact, the pd between the terminals when the source supplies no current. To avoid problems between using emf and using pd, the word 'voltage' is common (as in this book) – it should always be clear from the context which type of voltage is involved in a given case.

Common sources of emf in practice are voltaic (chemical) and fuel cells, electromagnetic induction, and the thermoelectric effect.

electromotive series See *electrochemical series*.

electron

stable elementary particle with a rest mass (m or m_e) of 9.1095×10^{31} kg (rest mass energy –0.511 MeV, about 8.2×10^{-24} J) and charge (Q_e, or e) 1.6022×10^{-19} C. The radius is given by $Q_e^2/m\,c^2$ (with certain assumptions as to shape and charge distribution); this has a value of around 3×10^{-15} m.

Unless otherwise stated, the name applies to the common negative electron (also called negatron); the antiparticle has a positive charge so is widely known by the name positron. Electrons are bound into the shells (K, L, etc.) of all atoms, molecules and ions, where they determine chemical behaviour. Free electrons appear at high density in metals (giving them their well-known sheen and electrical and thermal properties).

electron affinity
a) of an electron in an atom or ion, the energy needed to move the electron to infinity from the negative ion left behind
b) of an element, the extent to which an atom tends to capture electrons (i.e. its electronegativity, see *electrochemical series*)

electron beam narrow, closely parallel beam of electrons, found in cathode ray tubes (and other such sources). Electron beams have useful applications in industry (in, for instance, **electron beam machining** and **electron beam melting**) and in medicine (**electron beam therapy**).

electron capture form of *beta+ decay*

electron charge Q_e, or e one of the fundamental constants of physics, approximately $1\,6 \times 10^{-19}$ C

electron cloud model of an atomic electron that follows from the Schrödinger wave equation: each electron in an atom smears throughout the space available. Cloud density at any point is proportional to ω^2 (the probability density function). The cloud is a sphere only for a ground state system.

electron density the number of free electrons per unit mass of a sample (per unit volume in astronomy)

electron diffraction process of diffraction by, for instance, crystals, which shows that electron beams have a wave nature. In fact, study of electron diffraction patterns (which are exactly comparable to those of x-rays) gives an electron of momentum p a wavelength h/p (where h is the Planck constant).

electron drift speed the mean speed at which electrons *drift* through a conductor to form an electric current

electronegativity See *electron affinity.*

electron gas the set of free electrons in a sample of metal that behave like real gas particles in a liquid or solid sample. The concept is of value in many contexts, such as thermal and electrical conductivity.

electron gun set of electrodes able to produce an intense, tight electron beam that is easy to control. It is found in devices, such as the cathode-ray tube, that require a beam of electrons.

electronic concerned with electrons (as in electronic energy levels) or electronics (as in electronic circuit)

electronic engineering the application of the concepts of electronics and microelectronics; in its concern for communications and computing it overlaps with IT

electronic music the use of electric or electronic circuits to synthesise and amplify sounds treated as music. Early electric organs used various systems of discs turning in electric or magnetic fields to produce their effects. The Moog synthesiser of the 1960s was the basis of the modern device. In these, electronic oscillators are able to produce a wide range of frequencies and wave shapes (far more, in fact, than traditional musical instruments).

electronic energy levels the discrete (quantised) energies in an atom associated with the different electron orbitals. See *energy level, band theory* and *quantum physics.*

electronic lens same as *electron lens*

electronic switching high speed switching of a signal circuit using no moving parts. It is now available in the modern digital phone exchange and offers many services in addition to fast, reliable, noise-free call connection.

electronics the science and applied science of small currents (less than about 1 ampere) and of circuits working with them. In particular, the subject covers conduction through gases, vacuum and semiconductors. The last is the special field of microelectronics.

electron lens region of space that contains a shaped magnetic field (as in an electron microscope), or a shaped electric field or both – or the system of electric and/or magnetic fields involved. used to converge a beam of charged particles (see *electromagnetic focusing*). Just the same design of system works with ion beams – such a lens is not just for electrons. However, a given electron lens can tightly focus a beam of electrons or ions of only one energy; chromatic aberration, therefore, has its analog here. Note that there was a proposal in 2007 (*Science* **315** (5816)) to use an analogue of a p-n junction formed from grapheme to make an electron lens with negative refractivity.

electron mass $0.511 \text{ MeV}/c^2$ – see *electronvolt*.

electron microscope microscope which uses an electron beam rather than a light beam to let us study the fine structure of a surface (in a reflection system) or of a very thin sample (in a transmission system).

The standard form, the transmission microscope, uses a beam of energy around 100 kV to give image resolution down to less than half a nanometre (a magnification of up to a million diameters). Its main disadvantages are that it can work only with very thin samples and that it gives a two-dimensional image with no feel of depth. See *scanning electron microscope* and *scanning tunnelling microscope* for more useful – but more costly – systems; there are many other designs in use in the field of **electron microscopy** (the use of electron beams for magnification). Holographic techniques introduced from the early 1990s, for instance, allow three-dimensional 'snap shots' lasting a tenth of a microsecond each.

electron mirror an analogue of the electron lens, that reflects rather than passes an input beam of electrons, and may cause it to converge or diverge. A simple mirror consists of two tubes in a line, one positive and the other negative with respect to the cathode.

electron multiplier device used to obtain a very large amplification of a tiny input electron current. The design involves a series of secondary electrodes (called dynodes) between source and final anode. These have increasingly high positive potential – enough for each electron to gain such energy that on reaching a dynode it causes a high level of secondary emission. Compare with *photomultiplier*.

electron optics the study and design of electric and magnetic fields systems to control electron beams

electron paramagnetic resonance See *electron spin resonance*.

electron phase focusing technique that causes the clumping of the electrons in a uniform beam. In other words, the electron concentration along the beam becomes cyclic. The effect results from an outside high-frequency field which modulates the particle velocity – as in the klystron, for instance.

electron probe microanalysis highly effective method of analysing very small samples of matter, with masses as low as 10^{-15} kg. A high energy electron beam is focused to a spot as little as a micrometre across; the user can then study the spectrum of the x-rays produced.

electron radius See *electron*.

electron shell the one or more electrons in an atom with the same quantum number n. This is a concept that emerges from the Bohr planetary model of the atom; although that model is very simple, the shell concept remains of use. The values of n are 1 (for the K shell), 2 (L shell) ... The K shell can contain only two electrons (spins +/- $\frac{1}{2}$). The L shell can hold eight, so needs a second quantum number to describe the electrons; this is the orbital angular momentum quantum number l – the values (0, 1, 2 ...) of this relate to the names s, p, d, f, g. See the *Pauli exclusion principle*. This tells us that no more than two electrons can exist in a given orbital (i.e. can have the same set of the three main quantum numbers), and those two differ in spin quantum number.

electron spectroscopy the study of the energies of electrons from some source; it gives information about the source such as its energy level structure. There are many sources of electrons (see for instance *beta decay*), and electron spectroscopy can be applied to all.

electron spin resonance esr (or ESR) or **electron paramagnetic resonance** effect that appears in paramagnetic samples with unpaired electrons in the atoms when held in a strong magfield and irradiated with microwaves. The unpaired electrons have net magnetic moment so each aligns parallel or antiparallel to the field; these two states have an energy difference. An absorbed microwave photon can flip an electron to the higher state.

The frequency absorbed for this resonance depends on various factors; its value gives information about the chemical bonds in the molecules in question.

electron synchrotron type of particle accelerator – see *nuclear physics*.

electron telescope telescope that uses a photomultiplier – which is, in fact, an electron multiplier – to amplify the weak infrared, visible and ultraviolet signals from stars and galaxies. It gives a much enhanced image and more information than other techniques working at these wavelengths.

electron temperature of the plasma of electrons and ions formed by a discharge in a gas, the temperature at which the gas particles have the same kinetic energy as the electrons. As such electrons are not in equilibrium, the electron temperature may be far higher than that of the bulk matter; for instance, the electron temperature in a normal fluorescent tube is around 15 000 K.

electron trap crystal defect in a solid, such as a semiconductor, that can trap free electrons and thereby reduce conductivity

electron tube formerly known as thermionic valve, a device able to control the electron current (flow of charge) in the vacuum between its heated cathode and an anode. The cathode ray tube is the only electron tube now in common use. The others have been superseded by semiconductor diodes and transistors. However, integrated circuit scale vacuum devices may become widespread.

electron tunnelling See *Josephson effect* and *scanning tunnelling electron microscope*.

electronvolt (or, being phased out, **electron volt**) eV non-standard unit of energy still widely used in atomic, solid state, and nuclear physics. It is the energy transfer when an electron (or other particle with unit elementary electric charge) passes through 1 V: 1.6022×10^{-19} J. Because the energies of particles are so small when given in joules, use of the electronvolt is still very common – even 1 GeV is tiny: 16 nJ. See, for instance *nuclear physics: accelerator*.

As we often interchange mass and energy at this scale of matter, the electronvolt also often appears as a unit of energy. In fact, from $W = m c^2$, this non-standard mass unit should be the electronvolt divided by the square of the speed of light – eV/c^2. For instance, on this scale the electron mass is 0.511 MeV/c^2; in speech, it is quite common – though not correct – to quote this as 0.511 MeV.

electro-optics the science and applied science of light (and similar electromagnetic) waves as a means of carrying signals. The term is now rare – see *optoelectronics*.

electrophonic instrument old term for an electronic music instrument

electrophoresis the drift of fine solid particles through a fluid (such as a colloid) across which there is a voltage. If the particles move towards the anode, the process is anaphoresis; if to the cathode, it is cataphoresis. Electrophoresis is the converse of the Dorn effect.

electrophorus ancient device for supplying

charge by electrostatic induction, consisting of a flat insulating base plate and a separate metal plate with an insulating handle. The user charges the base by friction (rubbing), places the metal plate on it for a moment and then earths it, then removes the charged plate. The base does not lose its charge, so the process can be repeated as often as required.

electroscope device able to detect voltage and also to measure it crudely; some models can also detect and give the sign of the charge on an object. A common traditional type has a pair of very thin leaves of a metal (e.g. gold) at the end of a rod in an earthed container. After linking the rod to the object under test, the separation of the leaves shows the voltage compared to earth, or the charge of the object. Modern systems are electronic.

electrosmosis or **electro-osmosis** the passage of an electrolyte through a porous barrier with a voltage across it

electrostatic concerned with static (meaning stationary) charge, as opposed to moving charge (which is electric current). A static charge gives rise to an electric or **electrostatic field** (while a moving one has a magnetic effect). That electric field is of use in many situations (and a nuisance in many others).

electrostatic deflection technique used in many designs of cathode ray tube to control the direction of the passing electron beam: see *deflection*. In much the same way, **electrostatic focusing** involves a shaped field to converge the beam: see *electron lens*.

electrostatic generator device that uses static electricity – produced by friction or induction – to produce a high voltage (and, sometimes, a continuous current. The Wimshurst and van de Graaff machines are traditional systems of the two types.

electrostatic induction the separation of charge on a neutral object as a result of an electric field. The object gains negative charge on one side and an equal positive charge on the other; the process involves either the motion of free charges (in a conductor) or the polarisation of the particles (in an insulator or dielectric substance).

electrostatic lens electron lens that uses electric rather than magnetic fields to produce its focusing effect

electrostatic machines See *frictional electricity*.

electrostatic precipitation the use of induced charge to remove small, unwanted solid or liquid particles from a fluid. This technique

is important in, for instance, pollution control. In a strong field, the particles gain induced charge and collect on the nearest electrode. (In a second system, the particles first gain charge by passing the fluid through a charging field.)

electrostatics the science and applied science of charges at rest, and of the forces between them. The fundamental principle is that charges of the same type repel each other, and unlike charges attract each other. Any charge therefore has a force field around it.

$$F = Q_1 \, Q_2/4 \, \pi \, \varepsilon \, d^e$$
$$E = Q/4 \, \pi \, \varepsilon \, d^e$$

The force F between two charges Q_i distant d apart is as above, which also gives the field strength E due to one charge. Here ε is the permittivity of the medium (the product of its relative permittivity and the permittivity of empty space).

The subject also includes the concept of electrical, or electrostatic, potential, Φ (sometimes, wrongly, called voltage (= potential difference), V); the confusion is that for both the unit is the volt V, the joule per coulomb J C^{-1}. At a given point in a field, the potential is the energy needed to bring unit charge to that point from infinity (hence the unit). The electric field at a point is the gradient of the potential at that point.

electrostatic shield conducting container for an object that is to be protected from outside electric fields. An example is the Faraday cage used to protect sensitive electronic hardware.

electrostatic system old unit system, extension of cgs, based on **electrostatic units** (esu) and widely used in the past. The permittivity of free space has the value unity; compare this with the *electromagnetic system* – where it is the permeability that takes this value: the two systems are not compatible.

electrostriction the change in the size of an insulating object when placed in an electric field, the reverse of the piezoelectric effect. If the outside field is not uniform, the object will tend to move towards the region of greatest field strength.

electroviscous effect the decrease in a colloid's viscosity (fluid friction) when a small amount of an electrolyte is added

electroviscous fluid fluid that shows an increase in viscosity when in an electric field. The increase can be so large that the liquid appears to become almost solid. This gives many uses in the transfer and control of forces in, for instance, hydraulic systems and couplings.

electroweak theory theory that, with some success, tries to combine electromagnetism and the weak force – it has these interactions becoming one (the **electroweak force**) at energies above about 100 GeV/c^2. Abdus Salam and colleagues shared the 1979 Nobel Prize for their work on this.

element a) species of matter whose atoms all have the same proton number Z. This is a chemical element; chemical elements join to make compounds but have no simpler parts (though their atoms do). Of the 92 elements up to uranium, 88 occur in nature. The other four and all the trans-uranic elements are artificial, having been produced by accelerators and other such systems – this is because none of their isotopes is stable enough to last thousands of millions of years.
b) the simplest part of a system: thus electric elements are the components of a circuit, each of which we can treat as a unit. The elements of a compound lens are the single lenses of which it consists. See also fuel element, a component of a nuclear pile.
c) a gate or other logic circuit, sometimes called a **logic element**
d) See *magnetic element*
e) (*colloquial*) the part of a 'heating' device in which the energy transfer takes place – for instance, the coil of a kettle or the bar of a fire

elementary basic or fundamental, relating to element (in any sense). Thus the **elementary charge** Q_e (or e) is the charge on an electron or proton, about 1.6×10^{-19} C. No smaller charges are known for certain, though quarks have fractional charge.

elementary particles (or **subatomic particles**) particles that make up atoms or are similar to atoms in mass. The term is not completely accurate; strictly an elementary particle has no components, but as physics makes progress it seems that we can always come up with a structure for what was once thought indivisible.

The first elementary particle found was the electron, the cathode ray particle (discovered by J J Thomson in 1897). The proton was identified a few years later by Rutherford in 1918, and, in 1932, Chadwick discovered the third subatomic particle, the neutron. However, that same year, Anderson discovered the first antiparticle, the positron,

and so on, and now we know well over a hundred elementary particles. There are many properties and quantum numbers that keep all these particles distinct: mass, charge, spin, strangeness, iso-spin and so on.

There are also many ways to classify these particles, but none is fully satisfactory yet. For instance, particles with integral spin are bosons; fermions have half-integral spin. Alternatively, we can distinguish classons (photons and, if they exist, gravitons), leptons (the electron family) and hadrons on the basis of interaction types. In turn, the hadrons fall into the two classes of baryons and mesons, while baryons may be either nucleons or hyperons. Then there are many so-called resonance particles – excited mesons and baryons. Last, each one has (or is) its own antiparticle. See all the named objects, the *eightfold way* and *quark*.

ellipsometer device used to measure the thickness of a thin film. Plane polarised light meeting the surface reflects with a component of elliptical polarisation. The size of this component depends on the film thickness.

elliptical polarisation form of polarisation of electromagnetic radiation in which the ray consists, in effect, of two plane polarised components out of phase by 90°. If the two amplitudes are the same, we have circular polarisation; if not, the ray is elliptically polarised.

em short for electromagnetic, applied particularly to radiation

e/m the specific charge of an elementary particle, strictly Q_e/m or Q_x/m. Knowing the value in a given case is helpful for identification. This was so in several contexts early in the history of atomic and nuclear physics: various new particles were found to be electrons by this measure. The value for the electron is about 1.759×10^{11} C/kg; it falls with speed as a result of relativistic mass increase.

emanation early name for any species of radon, a radioactive gas that diffuses from solid samples as a result of the alpha decay of the parent nuclei. This happens in several radioactive decay chains. **Actinium emanation** (or action) is radon-219; **radium emanation** (radon itself) is ^{222}Rn; **thorium emanation** (or *thoron*) is ^{220}Rn. The **emanating power** (also no longer in use) of the solid parent sample is the ratio of the rate of release of radon to the rate of production.

emission the output of radiation of any kind from any kind of surface, in particular output of electrons or of electromagnetic radiation with more energy than microwaves. In the case of electron emission from the 'gas' below a metal surface, there are several significant processes. They differ in how the electrons near the surface of a conductor gain the energy to leave ('evaporate'), and are useful in different contexts.

a) field emission – result of a strong electric field between the (negative) surface and outside, as in the field ion microscope
b) photoelectric emission – where the energy comes from electromagnetic radiation absorbed by the surface (in particular infrared, visible and ultraviolet, this being the photoelectric effect)
c) secondary emission – the energy source being matter radiation absorbed by the surface, in particular an electron beam (as in the electron multiplier)
d) thermionic emission – where the surface, often also made negative, is at a high temperature (for most surfaces, over 1500 K, though coating with certain oxides will reduce this value somewhat)

emission spectrum the spectrum of electromagnetic radiation output by the surface of a substance as a result of atoms or molecules within it falling from a higher energy state towards the ground state

emissivity (or **emissive power**) ε the power of thermal electromagnetic radiation from a surface (i.e. the rate of energy transfer in this form), compared with that from a black body under exactly the same conditions. Rather than power, one may use exitance.

According to Kirchhoff's radiation law, the emissivity of a given surface is the same as its absorptivity – in other words a perfect radiator ($\varepsilon = 1$) is a perfect absorber.

Strictly, all this should apply at a single defined wavelength; the term is then *spectral emissivity*.

emittance old term for exitance

emitter the region of a bi-polar transistor from which majority carriers pass into the base. There they become minority carriers.

emitter-coupled logic the mode of action of a family of fast logic circuits. The input is to an emitter-coupled pair of transistors (called a long-tailed pair) while the output is through an **emitter follower**. This common collector, bi-polar transistor circuit giving 100% negative feedback feeds its output to the emitter as shown – it is much the same as a cathode follower (an audio sub-circuit), but based on transistors rather than vacuum tubes.

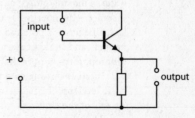

The emitter follower gives 100% negative feedback.

emphasis pre-emphasis and de-emphasis, selectively amplifying different bands of a signal to reduce the signal-to-noise ratio

emulsion dispersion of droplets of one liquid in a second. A **photographic emulsion** is a thin-layered film that contains photosensitive chemicals. It is also sensitive to ionising radiations (as Becquerel found). An **emulsion detector** is a device that may consist of a very thick layer of emulsion able to give a three-dimensional view of particle tracks (in, for instance, cosmic ray studies).

end correction c (or x) the distance between the open end of a column of air and the effective end of the column, as far as sound wave production is concerned.

In theory, the air in a pipe that is open at both ends vibrates in fundamental mode to produce sound with wavelength twice the length l of the tube. In practice, the wavelength is greater than this: it is $2 (l + c)$. For a simple tube c is about 0.6 times the radius for almost any length. See *sound*.

endo-ergic able to absorb energy during an interaction, the converse of exo-ergic. If the result tends to be a fall in temperature, people sometimes use endo-thermic instead.

energy W unit: joule, J

that property of some radiation, particle or system that can make things happen. Energy can cause an object to change its speed, for instance; for that to happen there is energy transfer from the source to the object. While in that 'source' – a tank of fuel or of very hot water, a moving bullet, or an atom bomb inside a missile, for instance – the energy has the potential to cause such a change. In fact, all energy is potential energy: until it causes something to happen, and in that case, after the transfer, we tend to have a different form of potential energy.

Interconvertible with mass m – and entwined with it at the most basic level (see the list below) – energy is one of the fundamental aspects of the Universe.

Though energy is a fundamental, it is convenient to speak of energy types (or forms) that appear to differ with context. Such contexts are as follows.

a) chemical energy – the energy involved in chemical change (as bonds form, change or break)

b) electric energy – the energy associated with electric charges and their fields and with moving charges (currents) and their magnetic fields: $W = V Q = V / t$ (but for the energy associated with capacitance and inductance, please see *electric energy*)

c) internal energy (symbol W_U, or U) – the energy of a system due to the motion of its particles and their potential in each other's force fields

d) kinetic energy (symbol W_k, or T) – the energy involved in a moving mass $W_k = \frac{1}{2} m c^2$ (m and c are the mass and the speed of the moving object.)

e) mass energy, the energy involved in a mass at rest – Einstein's special theory of relativity gives the relation for that conversion: '$E = m c^2$'; c is the speed of light in empty space. A stricter statement, $W = \Delta m\, c_0^2$, uses mass and energy changes (Δ is the symbol for change) and the more usual symbols.

f) mechanical energy – the energy of a working machine (like a pulley system), either 'kinetic' or 'potential' or both in any given case

g) nuclear energy – the energy associated with nuclear force fields and changes of nuclear structure

h) potential energy (symbol W_p, or U or V) – the energy of an object or system that results from its state (such as its position in a gravitational field, in which case $W = m g h$)

i) radiant energy – the energy of radiation (which is energy in transfer). $W_p = m\,g\,h$ (g is the acceleration of free fall and h is the object's height above ground)

j) thermal energy – the energy of thermal motion of the particles of matter, which depends on temperature

Study of even these simple statements may make it clear that the so-called forms of energy are not distinct. In fact, the names exist only for convenience: all energy by definition is potential. There is even less need to use the word work – energy transfer is fine.

energy band in the *band model* of matter, a range of allowed electron energy values, with forbidden energy ranges above and below. Each band is really a set of very close energy levels.

energy density the energy in unit volume of a substance in a given context

energy levels the set of discrete values allowed by the quantum theory for the electrons of an atom. A similar set applies within each nucleus, and molecules have energy levels for vibration and rotation too. See *band model* and *quantum physics*.

energy management working with a real time processor and software to control the use of fuel in, for instance, a building or process. (The term derives from the fact that, to the economist, energy means fuel.) There is a set of temperature and/or other sensors for input, and a feedback loop to the control switches, pumps and motors.

energy value unit: joule per kilogram, J/kg^{-1} the energy that burning (combustion) of unit mass of a fuel can provide. Energy value was formerly known as calorific value; sometimes it still is in the case of food stuffs (biological fuels).

engine machine whose action involves the burning of a fuel to obtain useful 'mechanical' energy. There are two main types. In the **external combustion engine** the burning takes place outside the machine itself. The main examples are the reciprocating steam engine (as with old, steamy train 'engines'), and the rotary steam turbine (the norm in thermal power stations). In both cases, the energy from the fuel converts water to steam; pipes then lead this into the cylinder to drive a piston (reciprocating) or to push on the blades of the turbine (rotary). The efficiency of these engines is low (not more than 10 or 15%), but they can work with almost any fuel.

In the **internal combustion engine**, the fuel burns where the mechanical energy appears. There are three types in this case: reciprocating (in which a piston moves up and down in a cylinder), rotary (with a *turbine* fan or a rotating piston, as in the *Wankel engine*), and reaction (in which the forward force comes from the emission of a jet of hot gas, as in the rocket and jet engines).

The first internal combustion engine was a design by Huygens in 1680; the fuel was gunpowder, and the system was not usable in practice. Otto's four-stroke engine of 1876 was the first effective internal combustion system; it soon started to replace the steam (external combustion) systems and is the basis of most modern engines.

Whatever its type, the action of an internal combustion engine has four stages:
a) induction (= drawing in): input of air and perhaps fuel to the combustion chamber or cylinder
b) compression: squashing the air or air/fuel mixture in order to raise its temperature
c) power: combustion (burning) of the fuel in the air, leading to expansion of the hot gases
d) exhaust: output of the waste products
In the reciprocating (petrol and diesel) engines these four stages are shared over two or four strokes of the piston in the cylinder. In reaction engines, there is a continuous release of hot gases in one direction which drives the engine the other way (Newton's third law of force); some jet engines also include a turbine.

In all these cases, the efficiency of transfer of energy is 20 or 30%, but, compared with external combustion engines, the range of useful fuels is much lower. There have been various attempts over the years to design an efficient internal combustion engine able to use a wide range of fuels.

engineering the structured (technological) use of the concepts of science in practice to design and build products that are as effective and efficient as possible. The main fields are as follows, with (in each case) a cycle of activities: needs assessment, design, implementation (building), running and testing (evaluation):
a) aeronautical engineering: working with flying vehicles (and that includes airships and spacecraft as well as aircraft)
b) chemical engineering: working with large-scale chemical reactions to produce 'chemicals' (e.g. fuels and some food stuffs) in bulk
c) civil engineering: concerned with large works such as buildings, bridges and roads
d) electrical engineering: in particular power generation and transfer (currents above about an amp)
e) electronic engineering: involved with light currents (less than about an amp)
f) human engineering: ergonomics and other ways of working at interfaces between people and their modern environment
g) marine engineering: working with vehicles floating on water
h) mechanical engineering: concerned with machines (mechanisms)

i) motor engineering: the field of work being land vehicles
j) nuclear engineering: the production of nuclear power on any scale
enrich to increase the ratio of a given component in a mixture, e.g. a given isotope in an element refined from nature
enrichment increasing the fraction of fissile nuclei in a nuclear fuel. For instance, natural uranium (from its ores) contains 0.7% of the fissile species uranium-235. There needs to be a ratio of some 3% to produce useful chain reactions. (In the past, the natural ratio was higher; this is why there is evidence of various natural reactors in ore bodies millions of years old.) Isotopes differ slightly in physical properties – enrichment processes depend on this; the most common involve repeated centrifugation or diffusion.
enthalpy H unit: joule, J the energy content of a matter sample (in most cases, a gas). It is the sum of the sample's internal energy W_U (or U) and the product of its pressure p and volume V:

$$H = W_U + p V$$

There is no specific meaning other than as defined above – the use of the concept follows the frequent appearance of that term $(W_U + p V)$ in thermodynamics. The **specific enthalpy** (symbol h) of a substance is the enthalpy of a sample per unit mass; the unit is the joule per kilogram.
entrance pupil or (rarely, but in some ways better) **entrance port** the image of the aperture of an optical system, viewed from the object. It is the largest bundle of diverging rays from the object that can pass through the system. Compare with *exit pupil*.
entropy S a measure of the lack of useful energy available in a system. In fact, it is entropy change, ΔS, that has real meaning. In particular, for a system that absorbs energy ΔW reversibly:

$$\Delta S = \Delta W/T$$

Here T is the system's temperature at the time.

If (as is usual) the energy transfer is not reversible, the system's entropy will increase with time. This is so although the total energy of the system remains constant: the available energy falls. This concept follows from the second law of thermodynamics, and we can expand it to equate entropy with disorder: only a system with increased order shows a rise in entropy. As a result, we can also define the entropy of a system in a given state by way of the probability p of finding it in that state:

$$S = k \ln p$$

where k is the Boltzmann constant.

environment surroundings in general

environmental control the use of processors, software and various hardware units to sense aspects of the environment and take the desired action. It includes energy management, air-conditioning control, and surveillance.

EOR the exclusive OR gate, more often called XOR, whose output is 1 only if just one of the two or more inputs is 1

epicentre See *earthquake*.

epitaxy means of growing a thin layer of crystal on a single crystal base (substrate) so that the structures of both are identical. The technique is important in semiconductor engineering.

epithermal neutron neutron with slightly more energy than a thermal neutron, approximately the same as that of chemical bonds. The energy range concerned often appears as 1.6×10^{-21} to 1.6×10^{-17} J (10^{-2} to 10^2 eV).

EQ symbol the equivalence gate, whose output is 1 only if both inputs are the same.

inputs		output
I_1	I_2	O
o	o	1
o	1	o
1	o	o
1	1	1

equalisation the reduction or removal of the frequency distortion of an electronic signal by using an **equaliser** circuit to compensate

equation a) old term for a reaction in chemistry or nuclear physics. The word reaction is better. Such expressions are now rarely written with an equals sign (=); $\rightarrow$ is

the norm with $\leftarrow \rightarrow$ showing a reversible reaction.
b) mathematical statement that two values or expressions are equal. If the two are always equal, the correct word is identity; in other cases, it is a conditional equation. Thus $x^2 - x = x(x-1)$ is an identity; $x^2 - x = 2$ is conditional: it is correct only if $x = -1$ or $x = 2$.

equation of continuity spatial equation that follows from the law of constant mass. For a mass of fluid of density ϱ with particles moving in it at speed c, the equation of continuity is

$$\delta\rho/\delta t + \rho \text{ div} v = 0$$

equation of state characteristic equation that relates the pressure p, volume V, and temperature T of a sample of matter of given mass. The best known is that of an ideal gas:

$$p V = R T$$

Here R is the gas constant.

One feature of an ideal gas is that the particles are mathematical points (they have no volume); a better equation allows for the finite volume b of all the particles of the gas. A second feature is that in practice the particles attract each other; the pressure inside the sample is therefore higher by a factor k than that measured. The equation becomes

$$(p + k)(V - b) = R T$$

Various attempts have been made to refine this further, to make it match real gases more closely; see, for instance, *van der Waals*. (There are also equations for solid samples.)

equations of motion equations that describe the motion of a particle in a straight line whose acceleration is constant during a certain time period. A similar set applies to motion in a circle. See *force and motion*.

equator, magnetic the line on the Earth's surface joining points at which the angle of dip is zero – see *magnetic equator*.

equilibrant (or, rarely, **equilibriant**) the single force able to keep an object in equilibrium under the action of a given set of two or more co-planar forces. It is equal to, but in the opposite direction from, the resultant of the set.

equilibrium state of balance. In mechanics, an object or system (hereafter system) is in equilibrium if its linear and angular momenta do not change. This means that there are no net linear or turning forces acting on it from outside. The concept applies much more widely than for a system acted on by forces, however. In general, a system is in equilibrium if its properties or behaviour do not change. Then, its state of equilibrium may be stable, unstable, or neutral; this depends on the effect of a small interaction that tends to move it away from its equilibrium state:

a) neutral equilibrium: the system is in equilibrium in its new state, so it stays there after the interaction

b) stable equilibrium: the system tends to return to its original state after the interaction

c) unstable equilibrium: the system tends to move further from its initial state after the interaction

In the last case, therefore, the system moves into a state where it is no longer in equilibrium.

All this applies to chemical as well as to a wide range of physical systems (and to biological populations). In the case of a chemical interaction, a reaction can come to a **chemical equilibrium**. The **equilibrium constant** [K] depends on the concentrations [...] at equilibrium of the reactants and products; it also depends on temperature.

For the reaction $xX + yY \longleftrightarrow wW + zZ$

equilibrium constant is $K = [W]W [Z]Z / [X]X [Y]Y$

Here, as with balance between a sample of liquid and its vapour in a closed space, there is **dynamic equilibrium**. The overall effect is balance, but all the time, particles within the system are changing state.

Two systems are at **thermal equilibrium** or **thermodynamic equilibrium** if they are at the same temperature; there is then no net transfer of energy between them. Also, some people use the latter term in cases where the internal energy of the system is a minimum.

equipartition of energy the equal sharing of the energy of a system among its various degrees of freedom. The energy of each degree of freedom is (on average) $\frac{1}{2} k T$, where k is the Boltzmann constant and T is the system's temperature.

equipotential a) at the same potential (energy)

b) line or surface joining points with the same potential. An equipotential is at 90° to the lines that show the force field in question.

equivalent having the same value or effect. The **equivalent circuit** or **equivalent network** of a given circuit has the same effect as the circuit but is easier to analyse. We can always reduce a real circuit to an equivalent with this in mind – by, for instance, swapping all series resistors with a single resistor whose resistance is the sum of theirs.

equivalent dose See *dose*.

equivalent length the distance between the poles of a bar magnet, approximately 5/6 the length of the bar, since the poles are not at the ends. Its value is half the ratio of its magnetic moment to its pole strength.

equivalent network the same as an *equivalent circuit*.

equivalent resistance in a circuit or network, the resistance of a single resistor that would have the same effect (would transfer the same power) as all the resistive elements in the circuit

equivalent sine wave for a given complex signal, the sine wave that would have the same fundamental frequency and root mean square value (energy content) as that signal

erbium Er transitional rare earth in the lanthanide series, proton number 68 and period 6, discovered in 1843 by Carl Mosander, and one of the four elements named after Ytterby, a small village near Stockholm (in a quarry in which a number of other new elements were found). Relative density 9.07, melting and boiling temperatures 1530 °C and 2860 °C; most common isotope ^{166}Er.

erect of an image, upright, ie, the same way up as the object. Often an optical system inverts the image; adding an **erecting lens** (as in a terrestrial telescope) or an **erecting prism** (as in binoculars) overcomes this.

E-region See *E-layer*.

erg the unit of energy in the old cgs system, 10^{-7} J – the energy transfer of one dyne moving one centimetre

ergon a quantum of energy of an oscillator, given by $h \nu$, where h, is the Planck constant and ν the frequency

ergonomics the study of how people relate to a particular environment. It includes the biology and engineering aspects of humans working with machines and equipment, and thus the design of systems, furniture, lighting, vehicles, homes and heating/air conditioning; it also extends into the area of health and safety.

error any deviation from a correct or expected result. The **probable error** of a set of values is the deviation between the set's mean and its mode. A **random error** is a small unpredictable error in each reading; if the set is large enough one can expect these to become insignificant. The **standard error** is a measure of the accuracy of the set – if the distribution of the n readings is normal, the standard error of the mean is σ/n; σ is the standard deviation. **Systematic error** is an inbuilt error of method or technique, such as poor zero calibration, or bad reading technique (for instance, always viewing a meter needle from the left rather than the front).

error equation or **error function** equation that gives the probability that a given value will lie between given limits, provided that a set of readings of the given value has a normal distribution

Esaki, Leo (1925–) physicist who shared the 1973 Nobel Prize with Giaever and Josephson for his work on electron tunnelling. He designed the **Esaki diode**, old name for tunnel diode, on that basis.

escape speed (or **escape velocity**) c_e unit: metre per second, m s^{-1} the lowest speed an object needs to escape the gravitational field of a planet (or star or moon). The escape speed of the Earth is just over 11 km s^{-1}. In any given case, the value is

$$c_e = 2G\,m/r$$

Here m and r are the planet's mass and radius while G is the gravitational constant.

esr (or **ESR**) *electron spin resonance*

esu electrostatic units. See *electrostatic system*.

etalon device whose output is a set of interference fringes. The light passes at a non-zero angle through a parallel pair of half-silvered glass plates a few millimetres apart. Because of its high resolving power and sharp fringes, the device is widely used in light spectroscopy.

The **Fabry–Pérot etalon** is a version with variable air gap devised by the French physicists Charles Fabry (1867–1945, discoverer of the ozone layer) and Alfred Pérot (1863–1925) early last century. It is now the only version used, with the air gap

set as required to cause resonance at the wavelength used; the device then becomes a Fabry–Pérot interferometer. In turn, this is the basis of the ruby (and other types of) laser design.

eta particle once called the chi-meson ψ°, a neutral boson resonance of mass 958 MeV/c^2. Its mean lifetime is about 35 nanoseconds.

ether (former spelling: aether) massless medium once thought to fill all space – people felt a need for something other than vacuum to carry electromagnetic radiation as well as a basis for absolute motion in the Universe. The 1887 Michelson–Morley experiment showed no evidence for it and the concept disappeared within a decade. See also *Lorentz transformations*.

Ettingshausen effect the small temperature difference between the sides of a current-carrying semiconductor sample at 90° to an applied magfield

Euclid (c.330–c.225 bce) mathematician whose geometry remains the basis of modern non-relativistic **Euclidean geometry**, that of angles and straight lines. **Euclidean space** is traditional three-dimensional space – also called Cartesian, from René Descartes, who devised the usual coordinate system based on x, y and z.

Euler, Leonhard (1707–1783) the major mathematician of the eighteenth century, who studied under Jean Bernouilli and moved with that family to St Petersburg. There he became in turn professor of physics and of mathematics, and published many hundreds of papers and books. **Euler's angle** helps describe how a rigid body rotates in Cartesian space. **Euler's equations** are expressions for this motion. **Euler's formula** is $e i\pi = -1$. Here e is the **Euler number** – a transcendental number of value 2.718 28 ... (the limit of $(1 + x)n$ as n tends to infinity). It is the base of natural logarithms (ln) and of exponential expressions (exp), both of which are common in physics.

eureka according to legend, Archimedes's shout ('I have found it!') when he devised his law of immersed objects, and still used for a great leap of inspiration

europium Eu transitional rare earth metal in the lanthanide series, number 63, period 6, first identified by Paul de Boisbaudran in 1890 and William Crookes in 1895 (though it is normal for Eugène Demarçay, who made a purer version and named it after the continent in 1901, to receive the credit).

Relative density 5.25, melting and boiling temperatures 822 °C and 1597 °C; two common isotopes – ^{151}Eu and ^{153}Eu. Some salts have uses in phosphors.

eutectic mainly used by metallurgists, a way to show how a property of an alloy (most often, its melting temperature) depends on the mix from 100% metal A (left-hand y axis) to 100% metal B (right-hand y axis). We can view the uppermost line on the graph (the solidification curve) as having solid below and liquid above – so often call this a phase diagram in physics, where it may refer to only a single pure substance and have pressure/volume axes.

evaporation the change of state from solid or liquid to vapour that takes place to some extent at any temperature below the boiling temperature of the substance. The process transfers energy from the bulk to the vapour – it is mostly the higher energy particles that tend to escape.

evaporator the part of a refrigeration system in which the cycled fluid evaporates, thus cooling the surroundings

even–even nucleus nucleus with an even number of protons and an even number of neutrons, and thus with a particularly high degree of stability. This point is evidence for the alpha-particle model of the nucleus. An **even–odd nucleus** (compare with *odd–even nucleus*) – which has an even number of protons and an odd number of neutrons – is less stable, but more so than an odd–odd nucleus.

event horizon the closed surface round about a mass like a black hole at which the escape speed reaches (and then exceeds) the speed of light. No radiation, signals or matter can therefore escape (but see *Hawking radiation*), so we can gain no information about events within. The radius of the boundary is the Schwarzschild radius, r_S; the name comes from Karl Schwarzschild, who, in 1916, devised this solution to one of Einstein's gravitational field equations. (It is a coincidence that the name is German for 'black shield'.)

$$r_S = 2 \ G \ m/c^2$$

G is the gravitational constant and c the speed of light in empty space; m is the mass of the central object.

If a mass is of larger size than this, it is not a black hole. If a star contracts to that size at the end of its life, it becomes a black hole – a zero-radius singularity in space-time.

exa E unit prefix for 10^{18} – as in exametre (Em), which is 10^{18} m (just over 100 light years)

exchange the theory that holds that an interaction 'at a distance' between particles involves virtual particles passing between them all the time. In the case of nucleons, the intermediary particles of the nuclear force are pions. Electron exchange explains covalent bonding and ferromagnetism. Each such interaction involves **exchange energy** and the force is an **exchange force**. See *force*.

excitation **a)** feeding current to the coil(s) of an electrical machine
b) raising a system to a higher energy state or level. If the system is quantised, there's a specific energy change involved in each such transfer. Thus, each energy level of an atom has its **excitation energy** (or **excitation potential**); this is its energy above the ground state.
c) feeding enough energy to a sensor to produce an output. This relates to the **excitation purity** of a colour, a measure of how close it is to the edge of the chromaticity diagram.

excited state state of an atom or other system with energy higher than that of the ground state

exciton the modern name for an electron-hole pair, which is a low energy excitation in a solid. There are two types – with parallel and anti-parallel spin.

exclusive OR gate See *EOR gate*.

exitance M the rate of luminous (or, more generally, radiant) energy transfer from unit area of a surface. See also *luminous exitance* and *radiant exitance*.

exit pupil (or **exit port**) the image of the aperture of an optical system viewed from the image. It is the largest bundle of converging rays from the system that can reach the image. Compare with *entrance pupil*.

exo-ergic having a net energy output during an interaction, the converse of endoergic. If the result tends to be a rise in temperature, we may use the term **exo-thermic** instead, so an **exo-thermic process** is a process that can provide energy.

exo-planet a *planet* in orbit round a star other than our Sun (but see *planemo*), some two hundred being known at the time of writing (mid-2006)

exotic strange or bizarre in nature, unusual

exotic atoms atom with a more massive negative particle than an electron around the nucleus. Such atoms, and molecules formed from them, have been made with muons, pions, anti-protons, kaons, sigmas and so on.

None of these particles is stable (the muon atom – with a half-life of two microseconds – is the most long-lived); therefore exotic atoms do not last long. They decay by the fall of the negative particle through the atom's energy levels, and its final collision with the nucleus. As they fall, they release x-ray photons.

Despite the short lifetime, the study of the x-ray spectra of exotic atoms gives useful information about, in particular, the electromagnetic and strong nuclear interactions. Thus, muon atoms help map the nuclear shape and charge distribution; also, muon molecules can catalyse certain nuclear fusions at low temperatures (real 'cold fusion'); pionic atoms and molecules in living tissue are an aid to radiotherapy. All exotic particles add to our knowledge and understanding of quantum physics.

A second class of exotic atoms has electrons around a nucleus which consists of some other positive particle than one made up of nucleons. For instance, so-called muonium and positronium atoms have muons and positrons (respectively) for their nuclei.

It is also possible to create a mix of the two types.

exotic chemistry the study of reactions that involve exotic atoms or molecules

exotic matter a) matter formed from exotic atoms and molecules Do not infer, however, that such matter may exist stably anywhere in the Universe.
b) matter whose tension (negative pressure) is greater than its energy density. (The breaking tension of steel is some 10^{12} times smaller than its energy density.) A wormhole through which matter could safely pass would have to include some such exotic matter according to current theories, but no one knows if such matter can in fact exist.
c) matter which breaks one or more 'rules' of classical physics (such as having negative mass)
d) matter which is unusual, even if now quite easy to make – as is the case with nanotubes and Bose–Einstein condensates

exotic particles particles that behave in unusual ways. This rather vague term can be applied, for instance, to the 'heavy electrons' that can be regarded as the current carriers in a superconductor.

expander circuit that increases the range of the amplitude variations in a signal transfer system. A useful increase in signal-to-noise ratio follows the use of a compressor (the inverse circuit) at the sending end and an expander at the receiving end.

expanding Universe theory any theory that holds that the Universe must be expanding because the red-shifts of distant galaxies are greater than those of closer ones. The cause is thought to be the big bang. The alternative view of a steady-state Universe is rare now – though it is not certain that red-shift can be equated with speed of separation.

expansion increase of volume, as a result of, for instance, a force from outside or a temperature increase. See also *thermal expansion.*

expansion cloud chamber See *cloud chamber.*

expansivity or **coefficient of expansion** unit: per degree, deg^{-1} or K^{-1} the ability of a substance to show thermal expansion, i.e. the fractional change in size of a sample with unit temperature change. There are various important cases.

In solid samples, change of length, area or volume may be of interest; the expansivities in question are α, β and γ respectively. To a good degree of approximation, $\beta = 2\alpha$ and $\gamma = 3\alpha$. The relationship that defines α is

$$\alpha = \Delta l / l \Delta T$$

Here l is the sample's initial length and Δl the change of length observed with a temperature change ΔT.

Exactly similar relations apply to β, the change of area per unit area per unit temperature change, and to γ for volume.

In the case of liquids, only the change of sample volume has any meaning. Here, however, there are two values, which depend on whether or not one takes into account the change of volume of the container. Thus, a liquid's **absolute expansivity** is the change in volume per unit volume per unit temperature change, taking the container into account; on the other hand, its **apparent expansivity** does not take the container into account.

Charles' law gives the expansivity of gases.

experimental physics the use of practical tests for theories and concepts rather than thought experiments. It involves saying 'Let's see what happens when ...' rather than 'I think this is what happens when ...'. By tradition, Aristotle and students of natural science (what we now call physics) until the late middle ages preferred to be armchair thinkers rather than getting their hands dirty on real tests. There is a little truth in this, but it is not likely that Galileo ever really changed the approach by dropping equal sized balls that differ in mass from the top of the tower of Pisa.

expert system software which stores 'human wisdom' (expertise) in statement form, applies that to new situations, and learns better and better how to make relevant decisions. See, for instance, *robot scientist*.

exploring coil old term for search coil

explosive chemical fuel with a very rapid combustion rate

exponent the part of a floating point number that gives the power of 10 by which the mantissa is multiplied. Thus, the number of particles in a mole is 6.025×10^{23} where 23 is the exponent.

exponential of an expression, involving the exponential function exp, where exp x = ex. A figure would show two common forms of exponential change, the growth of pd V across a capacitor during charging (after switch-on at time t_0), and the fall of current I in it during the same time. Both measures approach their final levels (V_0 and 0 respectively) asymptotically.

With R being the circuit resistance and C the capacitance the two expressions concerned are

$$V_t = V_0 (1 - \exp(t/R\,C))$$

$$I_t = I_0 \exp(t/R\,C)$$

The graph of the current is the curve of **exponential decay**, also observed in the breakdown of a radioactive sample. For the exponential growth curve, see the saturated vapour graph under *vapour*. There are many examples of exponential change in physics.

exposure a) the total energy received from radiation by a surface. There are a number of ways to define it – sometimes per unit area and/or per unit time. The corresponding units are joule (J) or J m^{-2}, and watt (W) or W m^{-2}.
b) the time for which a surface receives radiation. In this sense, the term **exposure time** is better.
c) in the context of ionising radiation, see *dose*.

exposure meter device that measures the intensity of input light for photography (though there are other types for other contexts). It then displays the correct aperture for a given camera speed, or the time for which the shutter should open given the aperture. This is sometimes called **exposure time**.

extensive shower cosmic ray shower with so much input energy that, by the time it reaches the ground, it can be detected over an area up to several square kilometres

extensometer device that measures small changes in sample length (for instance, when under strain)

external work old term for energy transfer, most often by a gas sample as it expands against an outside pressure p (as in a steam engine). It is the integral of p with respect to V between the limits of initial and final volume V_1 and V_2.

extinction constant old name for *molar absorptivity*

extraordinary ray ray that does not appear to follow the normal laws of refraction as it passes through a *birefringent* crystal

extrapolation the process of estimating values outside a known range. We can do this, for instance, by using a graph of the known range and extending it.

extremely high frequency ehf in the range 3–30 GHz. See *radio*.

extremely low frequency elf in the range 3–30 Hz. See *radio*.

extrinsic semiconductor type of semiconductor which works because of the presence of an impurity

f a) unit prefix symbol for femto-, 10^{-15}

F a) symbol for farad, the unit of capacitance

b) symbol for the Faraday constant, just under 10^5 C mol^{-1}

c) symbol for the fermi, old name for the femtometre, fm, 10^{-15} m

d) symbol for a vector field

e) chemical symbol for fluorine

_f_ a) non-standard symbol for force

b) symbol for force of friction

c) symbol for focal distance (focal length), unit: metre, m

d) non-standard symbol for frequency v

e) symbol for function, as in $f(x)$, meaning a function of x

F a) symbol for flow rate, unit: cubic metre per second ('cusec'), m^3 s^{-1}

b) symbol for force, unit: newton, N

φ symbol for angle and angular displacement (in effect the same thing), symbol: radian, rad

Φ symbol for field strength (flux)

Fabry–Pérot etalon etalon with an air gap the user can vary, basis of the ruby laser design. See *etalon* also for **Fabry–Pérot interferometer**.

factor of a given integer (whole number), another integer which divides into the given one with no remainder; a **prime factor** or prime number has no factors (other than 1 and itself). Thus, the factors of 12 (apart from 1 and 12) are 2, 3, 4 and 6; of these 2 and 3 are prime: we can therefore write 12 in terms of prime factors as $2^2 \times 3$.

factorial of a number n, is the product $2 \times 3 \times 4 \times \ldots \times (n-1) \times n$. It is written $n!$ and spoken of as 'factorial en' or 'en shriek'.

factor of merit or **figure of merit** measure of the sensitivity of a mirror meter. It is the (corrected) value of the deflection in millimetres at 1 m distance for a current of 10^{-6} A.

fading the fall-off in the strength of a received signal, usually short lived. The normal cause is destructive interference between signals taking slightly different paths See also *amplitude fading* and *selective fading*.

Fahrenheit, Gabriel (1686–1736) physicist who was the first to use mercury rather than alcohol in a liquid-in-glass thermometer, and the first to explore how boiling temperatures depend on pressure. His temperature scale is still in wide use in the USA. On it, the ice temperature is 32 °F and the steam temperature is 212 °F (though the two values were the other way round in the original version). To convert to Celsius, use either the fact that –40° is the same temperature on both scales, or the relation $T_c = (T_f - 32) \times 5/9$.

Fahrenheit hydrometer balance method of measuring the density of a liquid, now rarely, if ever, used

fail-safe device device that switches off a system if something goes wrong, such as a trip switch in a power circuit or a 'dead man's handle' in a train cabin

FAIR planned new double synchrotron for work with heavy ions at *Darmstadt*, due to start working in 2014

fallout the deposition of particles on the ground at some distance from the source. There are several main types. **Local fallout** consists of large particles that return to the ground less than 500 km downwind and do so within a few hours. **Tropospheric fallout** involves particles fine enough to reach the troposphere; there winds parallel to lines of latitude carry them round the Earth. These particles fall back to Earth a month or more later at any place with about the same latitude as the source. The finest particles reach the stratosphere; they may return anywhere on the Earth as **stratospheric**

fallout and may do so even many years later.

Fallout particles come from volcanic eruptions and from explosions. Fallout from the explosion of nuclear weapons, and from accidents at nuclear power stations, consists of hazardous radioactive particles. Such diverse sources as conventional power stations, very large fires and engine exhausts also produce fallout. In the case of radioactive fallout, the longer the particles are in the air the less their activity when they reach the ground.

The *Chernobyl* information site http://www.davistownmuseum.org/cbm/Rad7.html includes much about the event's very well studied world-wide nuclear fallout effects.

fall time measure of rate of decay, the time for an amplitude or quantity to fall from 90% to 10% of its peak or starting value

fan-in the number of inputs to a logic circuit, **fan-out** being the number of outputs. In either case, the term applies only where the number is greater than about four.

far infrared that part of the infrared spectrum the energy of whose radiations relate to changes in the rotational energy of molecules

far point, unit: metre, m the greatest distance at which an eye can clearly see objects without aid, i.e. with the image in focus on the retina. The far point of a perfect eye is infinity – the perfect eye sees stars as sharp points, for instance. See also *accommodation*.

far sight common name for the vision defect hypermetropia

far ultraviolet that region (about one third) of the ultraviolet spectrum furthest in frequency from visible light, centred around 10^{12} Hz

---

Faraday, Michael (1791–1867)

'the father of electrical engineering', a great chemist whose work in physics was even more crucial – in electromagnetism leading to the voltage transformer, the dynamo and the electromagnetic theory of light; in electrochemistry (in particular electrolysis); in electrostatics; in magnetism (discovering diamagnetism); and in light (working on polarisation). Famous quote (when asked the value of his early work on electromagnetism): 'What use is a new-born baby?' In fact, Benjamin Franklin said this first – Faraday more likely said 'One day the Government will tax it.'

Faraday started as assistant to Humphry Davy at London's Royal Institution in 1813. As such, he devised and carried out many important experiments, and as a result took over Davy's chair in 1827. His work ranged far and wide through science and applied science: for instance diving, lighthouse lighting, education, liquefaction of gases, optical illusions, acoustics, the theory of force, combustion, and gravity theory.

farad F the unit of electric capacitance, the capacitance of a system that stores one coulomb with a pd of one volt. Such a large capacitance is rare (that of the Earth is only a few tens of farads, for instance); more common units are mF, μF, nF and pF.

faraday or **Faraday constant** F non-standard unit of charge, the charge able to release one mole of a singly charged ion in electrolysis. The value is almost 10^5 coulomb.

Faraday balance type of balance whose design allows it to measure the magnetic susceptibilities of diamagnetic and paramagnetic materials, including gases

Faraday cage shield against electric fields, an earthed conducting container made of foil or netting and used for such things as drift tubes in accelerators and secret meeting rooms

Faraday constant old name for *faraday*

Faraday cylinder type of Faraday cage used to collect and measure charged particles

Faraday dark space name sometimes given to the Crookes' dark space (or, incorrectly, to the cathode dark space), the main region in a *discharge tube* where there is no glow; it almost fills the tube at very low pressures. William Crookes discovered this in his work on discharge tubes in the late 1870s, over twenty years after Faraday's research stopped.

Faraday disc old term for homopolar generator

Faraday effect effect that appears with plane polarised light passing through a transparent insulator in a magfield: the plane rotates by an angle that varies with field strength and path length.

Faraday–Neumann law another name, quite common in mainland Europe, for Faraday's second law of electromagnetic induction

Faraday's laws of electrolysis See *electrolysis*.

Faraday tube same as drift tube, or field tube – a hollow, cylindrical Faraday cage through which charged particles can pass with no Coulumb force

faradmeter device able to measure capacitance directly.

fast axis the optic axis of a birefringent crystal parallel to which travels the (faster) extraordinary ray

fast breeder reactor high-temperature breeder reactor, such as one using liquid sodium as coolant

faster than light travel (ftl travel) hypothetical concept concerning speed. A major prediction of relativity is that it is not possible to accelerate any object or particle with non-zero rest mass up to the speed of light in free space (some 300 000 km s^{-1}); the particles of light itself (photons) have zero rest mass. Thus, it is not deemed possible for spacecraft to accelerate above that speed. Such concepts as tachyons and wormholes may help the science fiction fans, but after a hundred years there has been little change in this position. See http://math.ucr.edu/home/baez/physics/Relativity/SpeedOfLight/FTL.html for much on this.

fast neutron neutron with energy above some given limit; the limit depends on context. In a nuclear reactor a fast neutron has too much energy to cause a normal fission.

fatigue the weakening (by an increase in density of defects) in a crystal that results from frequent deformations (from, for instance, expansion, vibration or flexion) over a period of time. **Metal fatigue** is the most well known type – when this happens in a metal sample so that it loses strength.

fdm frequency division multiplexing

feed or **feeder** any line into or out of a circuit or system, including those which supply power (*power feed*) as well as those which carry a signal

feedback the return of part of an output

signal to the input of the circuit concerned. The effect is to change the input signal. **Negative feedback** occurs in many kinds of amplifier and control system. In this case, part of the output returns to the input in anti-phase to stabilise the true output (this reduces gain and distortion but improves the frequency response). In **positive feedback**, a crucial aspect of control, the portion of the system's output which the circuit returns to the input adds to the input in phase, thus raising gain or causing oscillation.

fel *free electron laser*, a system in which the laser action takes place in a very high energy electron beam

Felici balance device used to measure the mutual inductance between two coils. It involves an ac bridge circuit.

femto- f unit prefix for 10^{-15}, as in femtometre, fm, which is rather over the radius of a nucleon. The term comes from the Danish *femten*, meaning fifteen.

Fermat's principle the path of a ray between two points in an optical system is the one that takes the least time. Modern versions specify the least or greatest time, as there are a few cases where the latter occurs. **Pierre de Fermat** (1601–1665) was a lawyer and secretive mathematician who did much useful work in this field, and in the development of differential calculus and probability theory (in this last case, with Pascal).

fermi old name for the femtometre, fm, still sometimes used in nuclear physics

Fermi age theory theory that provides a method – the **Fermi age equation** – for working out how neutrons slow with time since their creation. It is important in the applied science of nuclear reactors.

Fermi–Dirac distribution function or **Fermi–Dirac distribution law** expression that gives the probability that an electron will be in a certain energy state. It is part of **Fermi statistics** (or **Fermi–Dirac statistics**) – see *quantum statistics*.

Fermi energy the same as Fermi level

Fermi, Enrico (1901–1954) nuclear physicist, the first to cause artificial nuclear disintegration. He achieved this in 1934 by firing neutrons at uranium. Straight after his 1938 Nobel Prize ceremony he moved to the USA; there, in 1942, he built the first nuclear reactor, for research into chain reactions.

Fermi gas or **Fermi sea a)** the set of free electrons in a metal (the electron gas) **b)** the set of free nucleons in a large nucleus In either case, one can analyse how they behave as if the particles are those of a gas in a closed space.

Fermi level or **Fermi energy** the energy level in the band model at which the chance of finding an electron is 50%. It is the maximum energy of a free electron in a metal at 0 K.

fermion particle with half-integral spin, i.e. $\pm \frac{1}{2} h/2\pi$, $\pm \frac{3}{2} h/2\pi$, and so on. Fermions obey Fermi–Dirac statistics (hence the name), as the Pauli exclusion principle applies to them. The class includes all leptons and baryons. A **fermionic condensate** is a superfluid phase of fermions that can form at a low temperature (and, often, very high pressure); it was first seen in the case of electrons, still the most important case of this type of quantum fluid.

Compare with *boson*, a particle that follows Bose–Einstein statistics and can partake in a Bose–Einstein condensate. See also *quantum statistics*.

fermium Fm trans-uranic, transitional rare earth metal in the actinide series, period 7, with proton number 100, first found in the fallout of a 1952 hydrogen bomb and named after Enrico Fermi. Melting temperature 1520 °C, but little else known: the most stable isotope, alpha emitter ^{257}Fm, has a half-life of only 100 days.

Ferranti effect the sudden increase of voltage along a power line when the load falls quickly: a result of electromagnetic induction. **Sebastian de Ferranti** (1864–1930) was an engineer who campaigned for the use of ac in British national electricity generation and distribution, and invented many electromagnetic machines and systems.

ferrimagnetism type of magnetic behaviour shown by certain solids (such as the ceramic ferrites). It is similar to that of ferromagnetic materials, but in ferrimagnetism, the magnetisation is weaker – the susceptibility is small and positive and increases with temperature). The effect is due to there being more than one type of magnetic atom or ion present in the crystal lattice; the unequal magnetic moments align antiparallel. Such a solid shows no magnetic moment above its Néel temperature. In some cases (e.g. the garnets), there is also a lower temperature at which there is no moment, when the two types of moment are equal and opposite; this is the compensation temperature of the substance. Compare with *antiferromagnetism*.

ferrites solids which have chemical composition of the form $Fe_2O_3.XO$, where X is a divalent metal (such as cobalt, manganese, or nickel; XO is Fe_3O_4 in one case – that of magnetite). Ferrites may be net ferro- or ferri-magnetic. Their importance is that they have magnetic properties but are electrical insulators with low density; thus, though there is a degree of permanent magnetisation in any sample ('hardness'), in use there are no eddy current losses. The main use of ferrites is as cores in systems such as inductors, aerials and microwave circuits.

ferroelectricity property of certain crystals, in which, over a certain temperature range (the Curie range), there is hiqh permittivity and high piezoelectricity, often mainly in one direction. It is somewhat analogous to ferromagnetism. See also *electret*. A **ferroelectric capacitor** uses a ferroelectric solid as its insulator ('dielectric').

ferrofluids liquids which respond to a magfield. They consist of suspensions of magnetic nanoparticles in a carrier and can show magfield structures on a microscopic scale (e.g. on the surface of steels and magnetic tape).

ferrohydrodynamics the science and applied science of the behaviour of fluids with high susceptance in strong magfields

ferromagnetism set of magnetic properties of certain solids that have a high positive susceptibility; this allows even quite weak magnetic fields to magnetise them to saturation. Above the Curie temperature of the substance, the behaviour is paramagnetic – its susceptibility follows the Curie–Weiss law.

The main examples are iron, cobalt and nickel, as well as a number of alloys of these. A (magnetically) 'hard' material retains a high fraction of its ferromagnetism when the magnetising field becomes zero; a 'soft' material retains only a little – see *hysteresis*. Also, see *domain* for the model of the behaviour (though there are various theories to explain the details of the cause of domains).

fertile of a nuclide, able to capture neutrons to become fissile. A **fertile material** is a material that contains one or more fertile nuclides.

Féry spectrograph spectrograph (spectrometer with output to paper or a device like a ccd) which has just one optical element in the system. This is a prism with cylindrical faces one of which is a reflector.

Féry total radiation pyrometer instrument that can measure temperatures up to about 1400 °C by sampling all wavelengths from the surface and measuring their total energy

fet, or **FET** field effect transistor. See also *metal oxide semiconductor*.

Feynman, Richard (1918–1988) physicist who received the 1965 Nobel Prize for work on quantum electrodynamics. With Gell-Mann he proposed the quark and also produced a valuable theory of the weak interaction. His work on 'Bremsstrahlung' and similar radiations is also important. However, Feynman became, and remains, most well known as a populariser of science.

Feynman diagrams diagrams that give a graphic view of interactions between elementary particles and fields; they have no physical significance, note, as they are more like graphs. They offer so much power that they are the only way used to represent interactions in many branches of physics. Each line in the diagram, called a **Feynman propagator**, shows the average (i.e. highest probability) behaviour of an interacting particle. In practice, each aspect of each particle's nature (such as momentum and spin) appears on a Feynman diagram in a simple, clear way. (This can be so because, for instance, the lines – called edges – are tensors.)

The simple example shown is the diagram for the interaction (wiggly line, for the

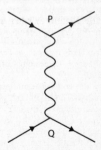

Feynman diagrams can show every detail of how two particles interact. This one is a sketch of two electrons meeting.

exchange photon) between two electrons at points P and Q; time passes to the right.

fibre optic gyroscope type of gyroscope in which one rotating light fibre coil lies in each of the three planes; for each one, a sensor monitors the light signal in the loop, reading changes as changes in the craft's motion in that plane. Light from a source splits to pass in both directions around the coil; the system compares the wavelengths of the two beams.

Fibre optic gyroscopes are more and more common in aircraft and spacecraft; they are much cheaper, less bulky, and more reliable than conventional gyroscopes as used for inertial navigation (a method of dead reckoning). See also *laser gyroscope*, one design of which also uses optical fibres.

fibre optics the science and applied science of transfer of information or data by light waves through a highly transparent thread of, in most cases, glass. (See *light guide*.) A single light wave travels through the fibre by a series of internal reflections as shown; this is so even if the fibre bends through a sharp curve. The process makes it easy to transfer a signal over a long distance. By 1990 the longest fibre optics system planned is to link Japan and Europe via Russia, with a capacity to transmit 560 million bits per second over 17 000 km. The capacity of systems is increasing, and one special technique has achieved sending 2.5 gigabits per second over 2000 km without amplification (regeneration). Also data sent through a fibre is very secure as it is so hard to 'tap'.

The fact that a light wave has a very large band width compared with, say, a radio wave is just as significant as the transfer method – it can carry a vast amount of data in a short time. See also modulation and multiplexing; by these means a single fine fibre can convey

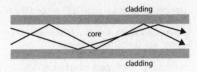

Light can travel tens of kilometres through a fibre light guide without much loss.

dozens of video signals and hundreds, even thousands, of phone, fax and data transfers. It is crucial that the source of the light should produce a tightly parallel beam of single frequency waves. The laser (including the light emitting diode, a semiconductor laser) can do this: such sources are simple, small, reliable and cheap.

Other uses or optic fibres include displays – large as in airports and on motorways; and small as in flat television screens; and the distribution of light to all the 'lamps' of a vehicle from a central, efficient, high-power source. Other uses exploit the effect on the output signal of bending or pressing the fibre. When suitably set up, the output falls as the fibre bends or squashes. One application of this is in an optic cable set in a road surface that can distinguish between different types of traffic and measure vehicle speeds and weights.

Fick's law the mass of particles that will diffuse in unit time across unit area of a plane of constant concentration varies with the concentration gradient (if this does not change with time). The constant of proportionality is the diffusivity.

$$m = D \, A \, t \, dC/dx$$

Here m is the mass of diffused particles crossing a surface of area A at 90° to the concentration gradient dC/dx (i.e. in the x direction) in time t.

field

a) the distribution of a scalar or vector measure through space, described in terms of the size, and direction if a vector, at each point

b) region of space in general. Thus, in an optical system, a **field lens** is the one furthest from the eye in a compound eyepiece. Its function is to increase the **field of view**, the overall region of object space that the system can image. Field of view also depends on the aperture of a stop called the **field stop**.

c) in the case of force, a region of space in which there is a force on a suitable object. For instance, a mass affects the space round it so that other masses experience forces; the gravitational (force) field of the mass is the region in which those forces are significant. In fact, gravitational fields (and electric and magnetic fields which we can describe in just the same way) extend to infinity. Nuclear fields do not. As force is a vector, force fields are vector fields: we can – and should – describe every point in terms of direction as well as strength. (Magnetic force field patterns made with iron dust, whether in two or three dimensions, happen to model this very well – though lines of force are not real.)

field coil coil in a generator or motor to produce the magfield needed

field curvature type of aberration caused by an optical system in which the image of a plane object at 90° to the axis is not plane

field effect transistor fet junction transistor that has (in effect) three terminals – called source, gate and drain – with the gate field in control of the current in the channel between source and drain. The field between gate and source controls how well the channel conducts, in effect by changing its resistance. (In practice, there are four terminals in most types of fet – but the fourth, the body, is almost always kept at the same voltage as the source. As well as that, some versions have more than one gate.) The most common type by far is the **mosfet**, metal oxide semiconductor fet: the gate insulator is a metal oxide layer; almost the same is the **igfet**, insulated gate fet, though in some cases the insulator may not be metal oxide.

Unlike the junction transistor, the fet has only one current carrier (the carrier in the channel); therefore, the two types are n-channel and p-channel, as in the symbols shown for the basic junction fets.

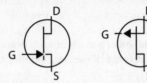

In the symbols for the two types of basic (junction) semiconductor fet – the n-channel and the p-channel – the electrode names do not usually appear.

field emission the release of electrons from a surface (in most cases, a metal, where the supply is better) as a result of a strong electric field near the surface. The field strength needs to be of the order of $10^{10}\,\mathrm{V\,m^{-1}}$; the normal technique is to use a very sharply pointed surface, as shown in the **field emission microscope**. This system (designed by Erwin Müller in 1936) allows detailed study of the structure at the tip of the point (which is cooled by, for instance, liquid helium to reduce the effect of thermal vibration). The magnification is the ratio of the radius of the screen to that of the point, and may be well over a million.

field ionisation process in which electrons 'tunnel' into a surface (usually a very sharp point) from the gas outside the tip of the point into the positively charged point itself, under the influence of a strong electric field. Fields of the order of $10^7\,\mathrm{V\,m^{-1}}$ cause this effect, which also leads to the high acceleration of the ions produced away from the point. The ions produce an image of the surface structure in the **field ion microscope** – to produce a sketch of the design, use that for field emission microscope and reverse the field; the rays passing to the screen are then ions rather than electrons. Field ionisation has many similarities to field emission in theory and practice.

field lens see *field* b).

field magnet permanent magnet in a small electrical machine, the function of which is just the same as that of the field coil in larger systems

field of force for a force that acts at a distance (exchange force), the same as force field

field of view see *field* b).

field stop stop (aperture) in an optical instrument that affects its field of view

field strength the size and direction of any force field at a given point, or, less helpfully, its intensity; the definitions vary with field type

field theory in the context of a force that acts at a distance, one of a number of theories that attempt to relate the potentials of that field and the transfer within it of that force and energy. A **quantum field theory** describes the features of a field in terms of mathematical operators. The view of any quantised field is of a set of particles with given properties: the field properties are the sum of these particle properties. On the other hand, we can view any particle as the quantum of some field.

The **unified field theory** was originally an attempt to relate Maxwell's electromagnetic field theory with relativity; it is now any theory which tries to relate any two or more separate theories. The three broad types of field are gravitational, electromagnetic, and quantum. In each case, analysis involves tensor maths.

field tube modern name for the *Faraday tube*.

fifth force a fundamental force additional to the four known (electromagnetism, the two nuclear forces, and gravity). With a range of about 100 m, it was proposed in the late 1980s to explain some results of tests (on gravity), but only two teams have since been able to repeat these results.

figure of merit same as factor of merit

filament fine drawn thread of metal or glass. A major use of glass filaments is for use as optic fibres. A metal filament, being so fine, has high resistance per unit length; it therefore becomes very hot, even with a fairly small current. It glows in the case of light bulbs (globes or lamps), and provides the energy for thermionic emission in most electron tubes.

film thin layer of solid or liquid on a surface (for instance, photographic film: see *emulsion*) or in air (for instance, a soap film)

film dosimetry method of monitoring people's exposure to ionising radiation over a period. Each person wears a **film badge** (a piece of photographic film – this is sensitive to ionising radiation). The badges are developed at the end of each monitoring period (such as a week) and the amount of radiation is measured from how fogged they are.

film resistor resistor that consists of a conducting layer on the surface of an insulating core

filter a) electronic system which passes signals only if their frequencies are in a certain pass band (range). A **band-pass filter** allows through signals with frequencies between a low and a high cut off value; a **high-pass filter** transfers signals above a certain (threshold) frequency; a **low-pass filter** transmits only those below a given frequency. **b)** similar arrangement in the case of light (and other electromagnetic) waves. For instance, a **colour filter** is a film transparent to only a certain range of wavelengths; it absorbs the others. The best colour filters work by interference rather than by simple absorption: their outputs can be very narrow bands. A **neutral filter** has a smaller output than its input at all frequencies. In much the same way, a **Thoraeus filter** works with ionising radiations: it has specific thin layers of aluminium, copper and tin. **c)** mesh able to remove solid particles from a fluid

filter pump fast acting vacuum pump using a jet of water to remove air from an enclosed space. It works down to a pressure about the same as the water vapour pressure at the temperature concerned.

fine structure the appearance of very close lines in a spectrum viewed at high resolution. The cause is either electron spin, electron relativistic motion, or molecular vibration – see also *hyperfine structure*.

The **fine structure constant** (symbol α) appears in expressions for the fine structure of the states of a single electron. Its value, about $1/137$ (and known to better than $1 : 10^9$ by mid-2006), is a fundamental constant, a dimensionless number that is a measure of the strength of the electromagnetic interaction.

$$a = 2\pi\, Q_e^2 /\, h\, c_0$$

Here Q_e is the elementary (electron) charge, h is the Planck constant, and c_0 is the speed of light in empty space.

first sound sound waves that appear in liquid helium II when the normal and superfluid components oscillate in phase with each other (giving a cyclic density variation)

fissile of a nuclide, able to undergo fission – spontaneously or induced by absorption of a slow (thermal) neutron. (Some people use the term only for the second case; others go further, and apply it only where there can be a chain reaction.) A **fissile material** contains one or more fissile nuclides.

fission the splitting of a fissile nucleus into two (or sometimes more) large nuclei (*fission fragments* or **fission products**) and a small number of neutrons; the nuclear binding energy falls, the excess being carried away by the fragments and the neutrons.

Fission is a form of radioactivity: a fissile nucleus splits this way on its own or after it absorbs a slow ('thermal') neutron – in this latter case, we call the nucleus **fissionable**. The best known fissionable nuclide is uranium-235. It is fissile too, slightly so, with 0.25 spontaneous fissions per second per kilogram. See *nuclear physics*. Where there can be a chain reaction, the energy made available by fission can be used in nuclear power piles and atom bombs. The latter are **fission reactors**. There are many types; see *nuclear power*.

Spontaneous fission, truly a form of radioactivity, is very common among the trans-uranic elements. The process is just the same, except it needs no input neutron. Uranium-238 is fissile in this sense with 7 fissions $s^{-1}\, kg^{-1}$ – see *fission-track dating*.

fission-track dating technique that gives a value for the age of a glass or other mineral that contains uranium. Each fission makes one or two fission fragment tracks appear inside the sample. Etching a polished surface shows the tracks, ready for counting; the density of tracks relates to the time since the substance became solid.

Fitch, Val (1923–) physicist who shared the Nobel Prize in 1980 with James Cronin. In 1964, their crucial experiment showed that the decay of neutral kaons is not symmetrical with respect to time reversal. This was the discovery of CP violation.

Fitzgerald–Lorentz contraction common name for the Lorentz(–Fitzgerald) contraction. **George Fitzgerald** (1851–1901), the actual proposer, made various important discoveries in electrolysis and electromagnetic radiation.

five-fourths power law the rate of loss of energy in the free cooling of an object, by convection and radiation, is proportional to the temperature excess of object over environment to the power 5/4

fixed point number number in standard form, with the decimal (denary) point always in a set position. Such numbers have two parts: the mantissa (which holds the actual value within the range 1 to 9.999 ...) and the exponent (which gives the size). Thus in the distance between Earth and Moon quoted as 3.844×10^5 km, 3.844 is the mantissa and 5 is the exponent.

fixed temperature (or **fixed point**) one of the number of constant and easy to produce temperatures used to define temperature scales. Many are the melting or boiling temperatures of elements at standard air pressure. On the 1968 International Practical Temperature Scale, there are eleven such values; they range from 13.81 K (the triple temperature of hydrogen) to 1337.58 K (the melting temperature of gold). See also *ice temperature* and *steam temperature*.

Fizeau, Armand (1819–1896) the first person to use the Doppler shift to measure the speed of stars, and the first to devise a successful terrestrial method to measure the speed of light (**Fizeau's wheel**, 1849). He also set up an (unsuccessful) experiment that led to the Michelson–Morley experiment.

flame region of high rate of chemical interaction between gases, in which one or more burn to emit light

flash barrier fireproof structure used to prevent arcing at certain points in electric circuits and machines

flashover unexpected spark or arc between two conductors. In a given case, **flashover voltage** is the pd at which flashover occurs.

flash temperature or **flash point** in the case of a fuel, ignition temperature – the lowest temperature at which there is enough flammable vapour for ignition by a spark or small flame

flavo(u)rdynamics (rare term for the) model that uses the W and Z bosons as mediators to try to explain the weak nuclear exchange force

flavour a quantum number of elementary

particles that relates to their weak inter-actions. In the electroweak theory, there are flavour-changing processes; this is not the case in quantum chromodynamics.

The concept was first named in 1968, by Murray Gell-Mann and a colleague (says the story) as they passed an ice-cream shop. They wanted a term to use for a set of quantum numbers that relates to hyper-charge, iso-spin, and strangeness in their quark model of hadrons. (Flavour applies to leptons too, but the main use is with quarks.) For the details of quarks' flavours – beauty (more often now, the less tasty name bottom), charm, iso-spin, strangeness and truth (more often: top) – see *quark*.

F-layer or **F-region** the highest (Appleton) layer of the Earth's ionosphere, with the highest concentration of free electrons. The height range is around 150–1000 km. The region is of value for radio communications around the clock (in the other layers, there is night-time recombination of electrons and ions).

Fleming, John Ambrose (1849–1945) electrical engineer whose most important advance was to invent the thermionic valve, in 1904. **Fleming's rules** concern electromagnetism in relation to conventional currents (+ to –):

- the left-hand (motor) rule – shows with the thumb the direction of motion of (i.e. force on) a current in the direction of the centre finger in a magfield in the direction of the first finger
- the right-hand (induction) rule – centre finger for induced current direction, first finger for the field, and thumb for the motion of the wire in which the current appears

Fleming–Kennelly law the reluctivity of a ferromagnetic substance near saturation varies linearly with applied field strength. **John Fleming** (1849–1945) pioneered the large scale use of electric heating and lighting and invented the thermionic valve (electron tube).

Flettner motor small motor used (for instance) on a boat, that depends on the Magnus effect

flicker photometer light meter system, now rarely used, in which the eye sees a rapid alternation between the illuminated surface under test, and a standard. When there is no flicker effect, the two surfaces have the same illumination.

F line reference line in the green-blue region

of the hydrogen spectrum. The wavelength is 486.133 nm.

flint glass range of hard, yet easy to polish, optical glasses, often used for lenses and prisms

flip-flop or **bi-stable** device or system with two stable states. The system flips from one state to the other on a suitable input, and stays there until another suitable input makes it flop back.

Such a circuit – the bi-stable circuit or bi-stable multivibrator – forms the basis of semiconductor storage, with a flip-flop for each bit. A typical flip-flop circuit has each output fed back to the other input. Note, however, that there are other designs, and also that circuits with somewhat different behaviour may have the same name.

floating a) of an object, in equilibrium in a fluid. The **floating objects law** (or 'law of flotation') states that a floating object displaces its own weight of fluid. This follows from Archimedes' principle (that the upward force on an object partly or fully immersed in a fluid equals the weight of fluid displaced); both apply in any non-zero gravitational field.

b) not fixed in value, as of a **floating link** in a circuit: one not tied to earth or a fixed supply voltage. A **floating point number** is one held or shown in true value ('normal' form), with the decimal (denary) point in the correct place. Compare 1500.56 (floating point form) with $1.500\ 56 \times 10^3$ ('standard' form, i.e. fixed point number).

flotation the state of floating in equilibrium in a fluid. The law of flotation is the floating objects law given in *floating a)*.

flow the transfer of bulk matter by the net independent motion of the particles. It occurs in fluids under shear stress. A **flow curve** shows either the rate of flow at points across the stream, or the total shear against time.

fluence a) the rate of flow of energy through an element

b) the number of particles passing through unit area per unit time

In other words, fluence is a measure of the intensity of radiation in a space (still sometimes called flux).

fluid substance that flows when under shear stress, i.e. a liquid or gas (but see *fluidise*). The major differences between a liquid and a gas are that a liquid tends to have a surface and to be hardly compressible while a gas has no surface and tends to be highly compressible. An **ideal fluid** is a liquid with zero fluid friction and no compressibility or is an ideal gas; a **perfect fluid** has zero fluid friction. See also *superfluidity*.

fluid coefficient or **fluid constant** the reciprocal of the value of fluid friction (see *viscosity*)

fluid dynamics the main sector of fluid mechanics, also called hydrodynamics (though strictly this applies only to liquids, as aerodynamics strictly applies only to gases). It is the science and applied science of the motion of fluids (liquids and gases, ideal, perfect, or real). The field includes the forces and torques on objects and systems in fluids, transport of a fluid through a pipeline, making weather forecasts, and so on, and people often apply the concepts of the field to areas like traffic and powder flow.

fluid friction a) viscosity, the friction within a fluid or across a surface with some other substance. This tends to oppose the flow of a sample and to oppose the motion of an object through it.

b) ξ the measure of fluid friction, also called (coefficient of) viscosity, unit: pascal second, Pa s

fluidics the science and applied science of fluid flow through pipes; it includes both hydraulics (liquids) and pneumatics (gases). The analogy with electric current is close. Indeed fluidic circuits are valuable where electric circuits face problems (for instance, at high temperatures or in high intensity ionising radiation). The disadvantage is that they react perhaps a million times more slowly. Fluidic systems are particularly good at switching, sensing and amplifying.
Microfluidics – a new field – is much the same except the channels concerned are no more than a few tens of micrometres across.

fluidise to make a mass of solid particles act as a fluid, often by high frequency vibration or by blowing a fluid through it. A **fluidised bed**, as sometimes used to make fluid models for research, for instance, is a container or conveyor for fluidised particles. In the latter case, such research is of great value as the power used to fluidise the solid is 5–10% less than that needed to transport the solid conventionally. For instance, one design of coal-burning power station fluidises the coal using supercritical steam, saving 5% of coal and cutting emissions significantly.

fluid mechanics the study of static and moving fluids and the forces acting. It includes mainly fluid dynamics, but aeronautics (the engineering of flight) and fluidics are important sectors as well.

fluorescence the absorption of energy by matter, with its immediate release in the form of electromagnetic radiation; see *luminescence* and compare with *phosphorescence*. The radiation follows the decay of short-lived excited atoms and molecules towards their ground states.

The input energy can come from various sources. If it is also electromagnetic, the input frequency is (in most cases) higher than that of the output. A particular example is the absorption of ultraviolet by certain compounds, with the release of visible light. Such compounds form fluorescent dyes and inks and some detergents. They also coat the inside of **fluorescent lamps** (or **fluorescent tubes**), in which discharge in the low pressure gas outputs a great deal of ultraviolet.
Fluorescent screens show images produced by electron beams and/or x-rays (in, for instance, tv sets and **fluoroscopes**).

fluorine F toxic, pale yellow gas, a halogen whose atomic form is the most reactive element (though the normal molecule is diatomic, F_2). Georgius Agricola, in 1529, first described a fluorine compound, the mineral fluorspar CaF_2, as used (in fact since ancient times) by miners as a flux (to help other solids melt), and a number of workers studied hydrofluoric acid HF after Carl Scheele first did so in 1771. (HF is highly corrosive, though a weak acid.) People tried to obtain pure fluorine for decades in the nineteenth century (some dying as a result); Henri Moissan succeeded in 1886, and received the 1906 Nobel Prize for chemistry. Density 1.7 kg m^{-3} at stp; melting and boiling temperatures –220 °C and –188 °C; only natural isotope ^{19}F.

Fluorine's main uses that relate to physics are in

a) halons (CFCs used as working fluids in refrigerators)

b) fluxes in steel making and working with ceramics

c) low friction (e.g. non-stick) plastics

d) in the case of ^{18}F, a positron emitter with half-life 110 minutes, in positron emission tomography (pet)

flutter variation at a fairly high frequency (i.e. above a few hertz) in the running speed of a tape, often caused by uneven wear in the drive. This causes annoying rapid changes of pitch in the output, especially if the signal on the tape is analog.

flux a) old term for flow (or apparent flow), used, for instance, for the rate of flow of particles, the power transferred by a wave, or the strength of a field. A term still sometimes used in the last case is **flux density** (flux per unit area) measured by a **fluxmeter** if the field is magnetic.

  Flux refraction is the sudden change of direction of an electric field across the interface between two dielectrics, and of a magnetic field across the interface between media that differ in permeability. See also *electric flux density*.

b) solid used to help some other solid melt, as common salt helps ice melt. Fluxes are of use in mining and in solders.

c) fluence, a measure of the intensity of radiation in a space

flyback the rapid return of the **flying spot** in a cathode ray tube and on the screen at the end of the electron beam. This is either to the start of the next line (**line flyback**) or to the start of the next frame (**frame flyback**).

flywheel massive wheel (i.e. one with a large moment of inertia) used to store energy in some mechanical systems. Once up to speed, inertia keeps the system working at constant speed even if the input energy rate varies. There are various designs of **flywheel bus** – an electric or hybrid bus with a flywheel whose inertia can keep the electrics working and start the bus moving again during a stop of several minutes.

fm short for frequency modulation (see *modulation*) or frequency modulated

f-number the numerical value of the focal ratio of a system, i.e. that of the ratio of the focal distance to the diameter of the aperture (or entrance pupil) of an optical system, in particular of a camera. Thus, if the focal ratio is $f/4$, the f-number is 4. There are various other ways to write focal ratio: for instance we can write an f-number of 8 as $f8$, $f.8$, or $f/8$ (but the spoken value is always '$f8$').

focal relating to a real or virtual focus of an optical system or element. Sometimes, the word applies to acoustics and to electron optics as well as to the way such systems affect visible light.

focal distance (often called **focal length**) f unit: metre, m the distance between the focus (or one of the foci) and the centre (or pole) of an optical system or element. The greater the

focal distance, the lower the (focusing or focal) power, unit: dioptre, D. It is normal to give the value as positive for converging lenses and mirrors and negative for diverging elements (compare that with the new Cartesian convention.) There are various expressions for focal distance:

a) for a reflecting surface: $f = r/2$ (r is the radius of curvature)

b) for a refracting surf ace: $f = n_1 \, r/(n_2 - n_1)$ (n is refractive constant)

c) for a thin lens in air: $1/f = (n - 1)\,(1/r_1 - 1/r_2)$

d) for two touching thin lenses: $1/f = 1/f_1 + 1/f_2$

focal plane in any optical system, the plane at 90° to the axis that contains the corresponding focus

focal point normal term for principal focus (or focus)

focal power P unit: dioptre, D how well an optical system can converge or diverge inward rays, the reciprocal of its focal distance

focal ratio the aperture of an optical system given as a fraction of the focal distance f. See also f-*number*.

focus a) also called principal focus, or principal focal point, the point through which rays input to a system or element close and parallel to the axis pass or appear to pass after reflection or refraction

b) as a verb, to converge or bring to a focus

c) of an image, 'in focus' if it is sharp enough for clear viewing. The **depth of focus** of an optical system set up in a certain way is the range of object distance over which the image is in adequate focus.

focusing the process of converging or bringing to a focus. In the case of the optics of charged particle beams, this may involve electrostatic focusing or electromagnetic focusing.

foil a wing on which there is a force of lift (which is downwards in some cases) when it moves through a fluid. See *airfoil* and *hydrofoil*.

food irradiation the process of bathing certain food stuffs in radiation (in most cases, from cobalt-60) to clean and/or preserve them. Irradiation, by x-rays, electrons or gamma rays, causes slight chemical changes which are claimed to kill bacteria and halt spoilage with no effect on taste. Opponents believe that irradiation may reduce vitamin and enzyme content, and in any event fear that some may abuse the technique to disguise bad foods. It is hard to detect whether a product has had such treatment (though electron spin resonance may provide the answer, as it can identify changes in the chemical bonding).

foot ft old (imperial) unit of length and distance equal to 12 inches, a third of a yard; it is equal to 0.304 8 m. Other old units relate to it: the **foot candle** is a unit of illumination (10.764 lux); the **foot-lambert** is a unit of luminance (that of a source with output 3.426 cd m^{-2}); the **foot-pound** is a unit of energy transfer (about 1.36 J). All are units of the **foot-pound-second system**, a system of units still widely used in the US.

footprint a) the surface area taken up by an object such as a building on land or a computer on a desk
b) the area of the Earth's surface to which an Earth satellite can send signals, or from which it can receive them

forbidden band range of energies in the band model in which no electron of the system in question may exist. Electrons can pass between allowed energy levels and bands unless the change involves a **forbidden transition**. This is a jump that requires a change of quantum number that may not occur as it breaks a conservation rule.

force F unit: newton, N

vector that tends to change an object's speed, direction, and/or shape. In a given case, its value is given by $F = dp/dt$ (the rate of change of momentum); this comes from Newton's second law of force (motion). The unit is the newton (N), the force able to give a mass of one kilogram an acceleration of 1 metre per second per second ($1\ N = 1\ kg\ m\ s^{-2}$).

Forces appear in various contexts, see for instance *body force*, *centripetal force*, *contact force* and *coriolis force*. However, the only four types of fundamental force known in nature, with their relative strengths, are gravity (10^{-38}), weak nuclear (10^{-13}), electromagnetic ($1/137$) and strong nuclear (1). See also *fundamental force*. In recent years, a number of people have claimed evidence for a fifth force, strong enough to distort the predicted effects of gravity over up to a few hundred metres. Most physicists now believe, however, that this effect is a result of variations in the density of rocks in the Earth's surface.

In this entry, we look at the effects of force on an object's linear motion and its angular motion (and then its motion in a circle). For acceleration in simple harmonic motion, see *simple harmonic motion*.
force and motion in a straight line the simplest case of dynamics (the effect of a force on the motion of an object) – the force and its effect are in the same direction. See also *components*. Newton's laws of force relate a net force that acts on an object from outside and the object's change of momentum, in most cases its acceleration. Given an acceleration, the equations of motion describe how the object's motion changes.

Newton's laws of force (or **laws of motion**, 1687) in modern terms are as follows:

a) An object's momentum is constant unless a net force acts on it from outside.

b) The net outside force is proportional to the rate of change of momentum produced.

c) If object A applies a force on object B, B applies the same force on A.

In fact, Galileo was the source of the first law which, like the third, is a statement of the law of constant momentum. In a rational system of units such as SI, the constant in the second law takes the value 1. Then we can extend the second law as in the box:

$$F = dp/dt \text{ (from the second law, in SI)}$$

As an object's momentum is the product of its mass and its velocity:

$$F = d(m\ v)/dt = m\ dv/dt = m\ a$$

In other words, the net force equals the rate of change of momentum produced. This leads to the unit of force, the newton N, given by 1 kg m s^{-2}. Note that when the object's speed is relativistic we can no longer take its mass to be constant. Then $F = d(mv)/dt = dm/dt + dv/dt$ – meaning we need more and more force to accelerate the object closer to the speed of light. (In fact, we need an infinite force to bring the object to the speed of light.)

acceleration a, unit: metre per second per second, m s^{-2} (it is best not to pronounce the unit as metre per second squared – that has two meanings and doesn't make sense anyway) the rate of change of an object's speed (as a scalar) or the rate of change of its velocity (as a vector). In other words, it is the change of the object's speed or velocity, in metres per second, per second.

The **acceleration of free fall** g is the vertical downward acceleration of any object free to move under gravity. Near the Earth's surface, the value is around 9.81 m s^{-2}. As g is constant for all objects at a given place in a given gravitational field, it is a good measure of field strength. Indeed, the unit newton per kilogram N kg^{-1} (field strength) is just the same as m s^{-2} (acceleration).

measures and equations of motion five measures and five equations that describe the motion of a object in a straight line. The measures are

● its acceleration a – taken to be constant during a certain time period

● the time period in question, t

● its velocity v_1 at the start of that time

- its velocity v_2 at the end
- the distance s it moves.

Note that some authors and teachers use u and v for v_1 and v_2 respectively.

Note too that of the five measures only t is a scalar: the other four are vectors (as are momentum and force).

Each equation uses four of those five measures of motion; we show the missing one in each case in square brackets at the right.

$$v_2 = v_1 + a\,t \qquad\qquad [s]$$
$$s = \tfrac{1}{2}\,(v_1 + v_2)\,t \qquad\qquad [a]$$
$$v_2{}^2 = v_1{}^2 + 2\,a\,s \qquad\qquad [t]$$
$$s = v_1\,t + \tfrac{1}{2}\,a\,t^2 \qquad\qquad [v_2]$$
$$s = v_2\,t - \tfrac{1}{2}\,a\,t^2 \qquad\qquad [v_1]$$

The first equation follows from the definition of acceleration and the second from the definition of distance (displacement); the others arise from the geometry of *velocity/time graphs*. See also *projectile*.

force and motion through an angle a second important case of dynamics. The object:

- moves to and fro in the arc of a circle (as in the case of pendulums), or
- cycles round and round a circle (which we call circular motion), or
- moves in some other way (e.g. to and fro in a straight line) that we can relate to circular motion (the name now being simple harmonic motion).

The force involved may be constant in size but always changing direction (as in the case of motion in a circular orbit) or it too may vary in a simple harmonic way (as when sliding a piston up and down a cylinder). Often, indeed, in such cases, we talk of torque rather than of force: torque is the turning effect of a force, sometimes still called its moment. In each case, we still have the five measures that describe the motion. Now, however (still except for time t), the measures describe angular motion rather than motion in a straight line.

measures and equations of angular motion five measures and five equations now describe the motion of a object through an angle. The measures are

- its angular acceleration (rate of change of angular velocity) α – taken to be constant during a certain time period; unit: radian per second per second, rad s^{-2}

- the time period in question, t unit: second, s
- its angular velocity (rate of change of angle) ω_1 at the start of that time
- its angular velocity ω_2 at the end
- the angle θ through which it moves; unit: radian, rad

Each equation uses four of those five measures of angular motion; we show the missing one in each case in [square brackets] at the right.

$$\omega_2 = \omega_1 + \alpha\, t \qquad\qquad [\theta]$$
$$\theta = \tfrac{1}{2}\,(\omega_1 + \omega_2)\, t \qquad\qquad [\alpha]$$
$$\omega_2{}^2 = \omega_1{}^2 + 2\,\alpha\,\theta \qquad\qquad [t]$$
$$\theta = \omega_1\, t + \tfrac{1}{2}\,\alpha\, t^2 \qquad\qquad [\omega_2]$$
$$\theta = \omega_2\, t - \tfrac{1}{2}\,\alpha\, t^2 \qquad\qquad [\omega_1]$$

angular acceleration α, unit: radian per second per second, rad s^{-2} the rate of change of an object's angular velocity. From the angular equivalent of the measures and laws of force, we can relate this to the torque T applied (rather than the force) and to the object's 'moment of inertia' I (a horrid term that gives the equivalent of mass). So, for $F = m\,a$, we now have $T = I\,\alpha$.

Linear acceleration is a vector, so it can help to split it into components. In this case, the two components concerned are along the tangent and along the radius. We call these **tangential and radial acceleration** respectively; their directions are the direction of the motion (that of the velocity from moment to moment) and 90° to that. The former is constant, at zero, if the object's speed stays the same. The radial acceleration is the centripetal acceleration, v^2/r.

angular displacement, θ, unit: radian, rad the angle (from the centre) of two points on the path of an object moving in a curve

angular frequency ω, unit: radian per second, rad s^{-1} the equivalent to angular velocity in the case of simple harmonic motion

angular momentum (still sometimes called moment of momentum) L, unit: newton metre per second, N m s^{-1} the angular analogue of linear momentum, the product of an object's moment of inertia I and its angular velocity/frequency ω: $L = I\,\omega$

angular velocity ς, unit: radian per second, rad s^{-1} the rate of change of angular displacement, $d\theta/dt$, the angle from the centre swept out in unit time by an object moving in a curve

circular motion the motion of an object in a curve under the action of a centripetal force (i.e. one pulling the object towards the centre of the curve). All the physics to do with such motion is fully comparable to that for linear motion (motion in a straight line). The centripetal force is given by either expression below.

$$F = m \, v^2 / \, r = m \, \omega^2 \, r$$

See also *rotation* (sense 2).

forced made to happen in a certain way rather than being allowed to happen in a natural way

forced convection See *convection*.

forced oscillation or **forced vibration** See *oscillation*.

force polygon scale drawing of the forces on an object in one plane. From it, the user can find the forces' resultant and equilibrant. See *parallelogram rule* and *triangle rule*.

force ratio no unit the ratio of a machine's output force (load) to that input (effort)

formula a general rule given in symbol form in order to express a relationship. Thus, the formula for pressure is $p = F/A$. Units follow the same formulae as the measures in question.

Fortin barometer accurate mercury barometer, now not often used, the user can adjust for changes in temperature, etc. **Jean Fortin** (1750–1831) was a physicist.

forward bias or **forward voltage** voltage applied across a circuit in such a way as to decrease distortion

fossil reactor the remains of a fission reaction found in ancient rocks. See *natural reactor*.

Foucault, Jean (1819–1868) mechanic turned physicist, the second person (after Fizeau) to devise a successful Earth-based method to measure the speed of light, using a rotating mirror. He also used the technique to show that light travels more slowly through water than through air. For his work on how the Earth rotates, he is remembered for the Foucault pendulum – but also invented the gyroscope, in 1852.

The **Foucault pendulum** is a simple pendulum with a massive bob on a long wire, first used in Foucault's cellar, in 1851. The swing takes so long to die away that it is easy to see the effect of the Earth's turning beneath it over a period of 24 hours: the plane of swing remains fixed in space. In fact, modern versions keep going using an electro-magnet switched on and off by a photo-cell.

four-dimensional continuum name for Einstein's space, i.e. having three spatial dimensions and one of time

Fourier, Jean Joseph de (1768–1830) mathematician who went with Napoleon to Egypt and was governor there for four years; he continued to publish in maths. **Fourier analysis** (and many associated mathematical concepts) appeared as part of his major work on the conduction of thermal energy (1822, an analysis based on Newton's law of cooling). **Fourier analysis** is a technique aimed at giving many types of periodic function, however complex, as the sum of a set (**Fourier series**) of sine waves; the frequencies of these are multiples of the frequency of the original function.

$$y = f(x)$$
$$= \Sigma a_n \sin (2\pi \, n \, x/T) + \Sigma b_n \cos (2\pi \, n \, x/T) + c$$

In each case, T is the period of the function $y = f(x)$, summation is from $n = 1$ to $n = \infty$; a, b and c are constants.

Fourier also worked out how the action of a greenhouse, and applied the thinking to the atmosphere – what we now call the greenhouse effect.

fourth dimension the time continuum in Einstein's four-dimensional space, deemed to be perpendicular to the three spacial dimensions

fovea the yellow spot, the most sensitive central region of the retina (though works very poorly in dim light). It subtends about 2°.

Fowler, William (1911–1995) physicist who received the 1983 Nobel Prize for work in theoretical nuclear physics as applied to forming the chemical elements in the early Universe

fps unit system the foot-pound-second unit system. Though the 'imperial' name refers to the British Empire, which spread round the world the system defined by Act of Parliament in 1824, both the foot and the pound go back to ancient times – to Sumeria in 1500 bce and to the Roman empire respectively. The US was no longer in the British Empire by 1824, which is why some US day-to-day measures differ to an extent from those in the fps system.

fracture the breaking of a solid sample into two parts, in most cases by the spread of cracks. **Brittle fracture** takes place with no prior plastic flow; **ductile fracture** follows a stage of plastic flow. See *elasticity*.

frame of reference viewpoint, a rigid framework of axes from which one can observe and measure the position, orientation and motion of all points in a system.

An inertial frame of reference moves at constant velocity (e.g. for all small scale intents and purposes, the Earth's surface); in such a frame, observers agree that Newton's first law of force applies.

A non-inertial frame accelerates (e.g. a lift in free fall, a funfair round-about, the Earth's surface on a large enough scale); in such a frame, observers agree that Newton's first law of force does not apply. For instance, if you could drop a French bowls ball from the top of the Burj Dubai tower on a totally windless day, it would reach the ground about a quarter of a metre East of the vertical line from the dropping point. On such a scale (this skyscraper is over 800 m high), we cannot say the Earth's surface is inertial: as the Earth turns from west to east at a significant speed.

francium Fr radioactive alkali metal, group 1, period 7, number 87, predicted in 1871 by Dmitry Mendeleev and discovered in 1939 by Marguérite Perey (who named it after her country). Relative density and boiling temperature not known, melting temperature 27 °C; most stable isotope ^{223}Fr, an alpha and beta emitter with half-life 22 minutes.

Franck–Condon principle Nuclei have no time to move during electronic interactions. This means that nuclear fields remain constant in their effect during electronic changes.

Franck–Hertz experiment crucial early experiment (1914) on a low pressure gas in a discharge tube that proved the concept of atomic energy levels. The current shows a series of peaks (resonances) as the applied voltage rises; each peak relates to an energy level in the gas atoms (see above right).

Franck, James (1882–1964) physicist who shared the 1925 Nobel Prize with Gustav

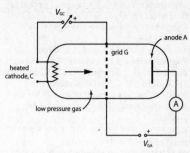

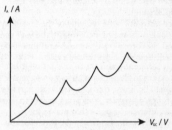

Using a set-up like that on top gives a characteristic curve like that on the bottom.

Hertz for their work on atomic energy levels – the *Franck–Hertz experiment*

Frank, Ilya (1908–1990) physicist who, in 1937, with Igor Tamm, explained Cherenkov radiation. Frank, Tamm and Cherenkov shared the 1958 Nobel Prize for this.

Franklin, Benjamin (1706–1790) statesman and scientist who invented bifocal lenses and (by way of a famous kite-flying experiment) showed that lightning is electrical. Franklin's theory of the 'electrical fluid' was influential for some years.

Fraunhofer, Joseph von (1787–1826) optical engineer who did much important work in applied optics and astronomy. His work to improve telescopes, prisms and lenses led to his re-discovery (compare with *Wollaston*) of the dark *Fraunhofer lines* on the solar spectrum caused by absorption. **Fraunhofer diffraction** is the type of diffraction observed with parallel light (i.e. with plane wavefronts). See *diffraction pattern* and compare with *Fresnel diffraction* which is more complex.

Fraunhofer lines the many fine dark lines against the bright background of the solar spectrum. These are due to absorption of radiations of specific energies in the Sun's atmosphere (and to a lesser extent in that of the Earth). There are thousands of these lines; Fraunhofer named the eight most obvious A–H.

free a) unforced (as opposed to forced). Thus, **free convection** is the cooling of a warm object in cooler air when there is no draught. See also *free fall* and *free oscillation*.

b) unbound (as opposed to bound). See *free electron*, *free energy*, *free magnetism* and *free radicals*.

c) empty, with no effect from force fields, as in the case of *free space*

free electron electron that is not attached to any atom or molecule in a medium, so able to move around alone under the influence of outside forces. In a **free electron laser** (fel) the laser action takes place in a very high energy electron beam; the output is high energy too, and coherent – and the user can tune the wavelength from the radar region to visible light. The beam, from an accelerator, passes through a 'wiggler'; this is a series of magnets with alternating poles that makes the electrons in the beam wiggle. **Free electron paramagnetism** appears in certain substances – they have a very small positive susceptibility as a result of the magnetic moments of these electrons.

free energy the energy available from a given change in a system. See *Gibbs free energy* and *Helmholtz free energy*; 'free energy' by itself means the latter in most cases.

free fall the state of an object that moves under the influence of gravity alone (i.e. its only acceleration is g); see also *microgravity* and *weightlessness*.

free magnetism an old model of the behaviour and properties of a permanent magnet, which regards a magnetic fluid as being concentrated at the poles

free oscillation or **free vibration** the motion of an object when it vibrates with no outside force acting; it then vibrates in its natural mode (fundamental and harmonics)

free radicals neutral atoms and molecules with single unpaired electrons. This gives them such high reactivity that their lifetimes are in most cases very short.

free space (imaginary) medium that contains no matter, gravitational fields or electromagnetic fields. No such region exists in the known Universe, but interstellar and intergalactic space approximate closely enough for most purposes. See also *zero point energy*.

free surface energy See *surface tension*.

freezing the change of state from the liquid to the solid phase (also from vapour to solid). The freezing of pure liquid water at 0 °C is a common example. **Freezing mixtures** absorb energy when they mix (i.e. their mixing is endoergic); the process therefore reduces the temperature of the environment, enough, for instance, to freeze water.

freezing temperature (often called, with less meaning, **freezing point**) the temperature at which a pure substance freezes at standard pressure. The value falls with increased impurity concentration, and rises with pressure (except in the important case of water). See also *fixed temperature*, *ice temperature* and *supercooling*. A maybe useless fact (if you collect these): 'Water has been found to form ice at room temperature if it is placed between a tiny tungsten tip and a graphite surface.' (2006: http://physicsweb.org/article/news/10/5/2)

F-region See *F-layer*.

Frenkel defect See *defect*.

---

frequency

a) the number of times a given event occurs per unit time ; the statistical measure of rate of occurrence. This is **absolute frequency**, unit: per second, s^{-1}, etc. It may be more useful to give **relative frequency**, the ratio of absolute frequency to the total number of relevant events during the period.

b) ν (or f) unit: hertz, Hz, the number of cycles of a vibration, oscillation, or other periodic change per unit time. The hertz is the per second, s^{-1}. See also *angular frequency* (pulsatance), given by 2π v. The **natural frequency** of a system (its fundamental) is the main one it shows when no outside force causes it to change. In the case of resonant electric circuits, the term for that is **resonant frequency**. given by $1/(2\pi \sqrt{(L\ C)})$; C is the circuit capacitance and L its inductance.

frequency changer system which absorbs either electromagnetic radiation or an alternating signal at one frequency and outputs the energy or signal at a new frequency. Such techniques have a number of major uses, such as in fibre optic communications (where it is electromagnetic radiation that is absorbed). See also *frequency doubler.*

frequency discriminator electronic circuit whose output frequency is proportional to the difference between the input frequency and a set reference frequency. It is used in automatic frequency control and in frequency modulation.

frequency distortion See *distortion.*

frequency distribution the set of relative frequencies of all events concerned. It is given as a table or as a graph (*frequency distribution curve* or *frequency distribution graph*) for discrete data the graph is a histogram (a *frequency polygon*). If the graph is 'bell-shaped' the distribution is 'normal'.

frequency divider form of electronic frequency changer whose output frequency is the input frequency divided by a whole number

frequency division multiplexing fdm See *multiplexing.*

frequency doubling the re-radiation of input electromagnetic radiation at twice the original frequency. It is a property of a number of substances as well as certain radio systems. Thus, potassium niobate fed with infrared outputs in the blue region. (This was the first way to obtain blue light from a semiconductor laser.) Frequency doubling results from the process of harmonic generation: the absorbed radiation deposits energy which, as a result of the dipole moment of the molecules of the substance, re-radiates in the same direction at twice the frequency. (The general process is second harmonic generation.)

In practice, the output is mainly (if not all) of the same frequency as the input, and when the second harmonic appears, so too do other frequencies. Only a few materials have the right properties to produce efficient frequency doubling; fewer still are cheap and robust in crystal form.

frequency function a probability density function: it gives the relative frequency in a distribution of any specific value where the possible values are continuous rather than discrete.

frequency meter circuit whose output is the frequency of the input signal

frequency modulation form of modulation which adds a signal to a carrier by changing the carrier frequency to match

frequency multiplier device which is, in effect, the opposite of the frequency divider: the frequency of its output is an exact multiple of that of the input.

frequency polygon a name for frequency distribution

frequency pulling the process of changing the output frequency of an oscillator to a slightly different outside frequency; it involves a type of resonance

fresnel now rarely used unit of frequency, equal to 10^{12} Hz

Fresnel, Augustin (1788–1827) physicist who devised many concepts and structures in optics. (Pronounce his name as fre-NEL.)

Fresnel bi-prism device which provides a neat way to produce interference fringes. The name is inaccurate as the 'bi-prism' is a single glass prism with angle very close to 180°. Light, from a slit S, incident on that apex gives two images S_1 and S_2 and therefore a field in which two coherent wave sets overlap (in the shaded region). This is the condition for a stable pattern of constructive and destructive interference.

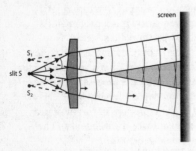

The 'bi-prism' has the effect of two close coherent sources, so makes a set of interference fringes.

Fresnel diffraction the process and effects of diffraction in cases where either source or sensor, or both, are close to the diffractor. The wave fronts are not plane, unlike the case of Fraunhofer diffraction, and this makes analysis less simple.

Fresnel lens (or simply **Fresnel**) large flat lens (often now made of a plastics substance), with a stepped surface to cause refraction as if by a much thicker (and therefore more massive and costly) standard lens. Each step has the same curvature as the corresponding part of the thick lens. First designed for lighthouse lenses (which must have a large area), smaller Fresnel lenses now also appear in, for instance, spot lamps, overhead projectors, headlamps, reading aids, and view-finders.

Fresnel mirror device that gives interference fringes by reflection from two touching mirrors inclined at almost 180°. This is much the same, in effect, as the Fresnel bi-prism.

Fresnel's rhomb glass rhombohedron set up to have two internal reflections (like a raindrop in a rainbow storm). It takes in plane polarised light and outputs circularly polarised light.

Fresnel zones or 'half-period zones', separately treated circular bands half a wavelength apart on a spherical wave front, used to help analyse Fresnel diffraction

friction force that tends to oppose the relative motion of two surfaces in contact: each surface applies a force on the other surface against the relative motion. In the case of two solids, the main cause seems to be the tiny hills and valleys in even the smoothest surface. The function of lubricants is to keep such surfaces slightly apart.

Sliding friction (or **kinetic friction**) is the friction that appears when there is actual relative motion; the force here is less than that of **static friction** (or **limiting friction**), the opposition to slip between two touching surfaces that are at rest. There is also **rolling friction** when one surface rolls rather than slides over the other (as in ball or roller bearings and wheels). The force here is even smaller, but never zero.

The *laws of friction* are as follows.

a) The friction force does not depend on the area of contact (though this is not always true).

b) The friction force varies with the 'reaction' force between the surfaces.

c) The sliding friction force does not depend on relative speed.

From the second of these comes the measure of friction (μ or n, no unit), sometimes called the **friction constant** (or **friction coefficient**). This is the ratio of the friction force F in question to the 'reaction' force R.

$$\mu = F/R$$

Its value is also the tangent of the **angle of (limiting) friction**, the angle of slope at which one surface will just slide over the other. Values range from very close to 0 in the case of a lump of smooth ice on a smooth ice surface to around 1 for the rubber of a car tyre on the concrete of a road.

See also *fluid friction*. For friction on a nano-scale, see *nano-machines*.

frictional electricity modern name triboelectricity, the separation of charge by rubbing. This was the first type of electricity known, and dates back to ancient times. Traditional ways to produce it were to rub glass with silk (making the glass positive and the silk negative), or amber (whose Greek name is *elektron*) with paper (the amber becomes negative). Nowadays, cellulose acetate is often used instead of glass and polyethylene in place of amber.

The early electrostatic machines (e.g. those of van der Graaf and Wimshurst) used the process to produce continuous currents. There are only a few uses of frictional electricity now other than in trivial contexts; rather, it tends to be a nuisance (as in causing shocks in rooms with carpets and sparking near fuel pumps, for instance).

Friedman, Jerome (1930–) physicist, joint winner (with Henry Kendall and Richard Taylor) of the 1990 Nobel Prize for work on the structure of nucleons leading to the quark model. Carried out in the 1960s, this involved the deep inelastic scattering of electrons produced by the Stanford Linear Accelerator.

fringes patterns of light and dark bands obtained by interference (or by that process in diffraction). If the input light is not monochromatic, the fringes are coloured. Fringing of this type occurs with any wave. For instance, in the case of sound it will produce loud and quiet regions.

front layer photo-cell type of barrier layer photo-cell in which the metal layer is a thin transparent film on the semiconductor surface. This is more sensitive to blue light than is the back-layer type.

frost temperature (also called **frost point**) the air temperature lower than 0 °C below which moisture in the air will condense as frost or 'freezing fog' rather than dew or mist

Froude number dimensionless number used to analyse the passage of a floating object through a fluid. It has several values and expressions; one is v^2/gl; here v is the relative speed of object and fluid, g the acceleration of free fall, and l the length of the object, at the water line or equivalent, in the direction of motion.

ftl travel faster than light travel

fuel substance which burns, or is used up in some other way, to produce useful energy. Organic fuels used in combustion (burning) may be natural (for instance, wood, the world's most common fuel); fossil (non-replaceable, such as coal) or manufactured (as town gas is from coal). Nuclear fuels may be naturally occurring, i.e. non-replaceable (as is uranium-235), or produced by breeding from non-replaceable nuclides; fusion holds out the hope of energy for ever, in that its fuels (such as hydrogen) are very common.

fuel cell device which offers the direct conversion of the chemical energy from a redox reaction to electricity. The reactants (fuel, such as hydrogen and oxygen) pass continuously into the cell, where the reaction takes place in the presence of a catalyst. A voltage results between two plates in the cell.

While a fuel cell can output high electric power smoothly and for as long as there is a fuel supply, its efficiency is somewhat less than that of a modern rechargeable (secondary voltaic) cell. The figure overleaf outlines the structure of a hydrogen–oxygen cell.

fuel element the smallest package of fuel in a nuclear reactor. A common arrangement has the fuel held in small spheres of ceramic (the elements) in a metal tube, the **fuel rod**. Several rods fit into each **fuel channel** in the core of the pile.

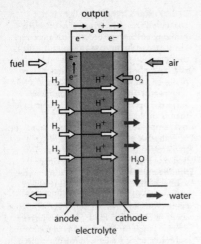

This typical hydrogen fuel cell uses a solid polymer electrolyte and porous carbon electrodes that contain a platinum catalyst.

fuel injection the central principle of diesel and some jet engines: a pump forces fuel into the hot compressed air, where it burns at once.

fuel rod See *fuel element*.

fuel value See *energy value*.

full radiator old term for a perfect black-body radiator, with **full radiation** being an old term for black body radiation

full-wave rectification a process whose input is an ac signal and whose output is full-wave rectified ac – a continuous (though not constant) dc. See *rectification* for such a circuit and the waveform.

function expression of the relation between two variables, members of sets y and x, written $y = f(x)$; each element x of one set maps to an element (value) y of the other. A graph is a common way to show how the two relate, with x plotted along the horizontal axis and y along the vertical.

function generator electronic circuit whose output signals may be set to various shapes and frequencies, in order, for instance, to test other circuits

fundamental a) basic as with *fundamental constants*
b) short for fundamental frequency (fundamental mode)

fundamental constants those constants of nature that appear in many diverse expressions and have no simpler structure. However, except for the gravitational constant (it seems), they all relate closely; thus the expression for the fine structure constant is $2 \pi Q_e^2 / h\, c$.

The table below shows the main fundamental constants. There are many others.

fundamental frequency the frequency of that component of a sound (or of any wave-form) which has the lowest frequency; the others are its over-tones.

fundamental interaction (or **fundamental**

constant	symbol	value
speed of light in empty space	c_0	$2.997\,925 \times 10^8$ m s^{-1}
Faraday constant	F	$9.648\,5 \times 10^4$ C mol^{-1}
gravitational constant	G	6.672×10^{-11} N m^{2}kg^{-2}
Planck constant	h	$6.626\,18 \times 10^{-34}$ J s
Boltzmann constant	k	$1.380\,66 \times 10^{-23}$ J K^{-1}
Avogadro constant	N_A	$6.022\,05 \times 10^{23}$ mol^{-1}
elementary charge	Q_e	$1.602\,19 \times 10^{-19}$ C
gas constant	R	8.314 J K^{-1} mol^{-1}
absolute zero of temperature	T_0	-273.15 °C
fine structure constant	α	$7.297\,351 \times 10^{-3}$
permittivity of empty space	ε_0	0 $8.854\,188 \times 10^{-12}$ Fm^{-1}
permeability of empty space	μ_0	0 $1.256\,637 \times 10^{-6}$ Hm^{-1}

channel	strength	range	theory	mediator
gravity	10^0	$\infty - 1/r^2$	geometrodynamics	graviton
weak	10^{25}	10^{-18} m $-$ c $1/r^6$	flavordynamics	W and Z
electromagnetic	10^{36}	$\infty - 1/r^2$	electrodynamics	photon
strong	10^{38}	$\infty - 1$	chromodynamics	gluon

force) channel (mechanism or route) by which two particles interact with each other that we cannot explain in terms of anything more fundamental. It is best not to use the term force here as that implies a classical, Newtonian view of how the particles interact – we now explain forces that act at a distance in terms of the exchange of particles and/or how space-time curves.

'Traditionally', there are four such channels. The table above shows these, with, for each, the relative strength and range, and the main current theory that tries to explain it with its mediating (exchange) particle.

Less traditionally, the thrust of research in recent decades has been to unify these interactions. With some success, electroweak theory tries to combine the electromagnetic and weak interactions. With less success, grand unified theory tries to combine that model with the strong interaction. There has been no success in bringing gravity into an overall model, though people working in quantum gravity have that main aim.

fundamental interval the range between two fixed temperatures on a scale (100 kelvin in the case of the absolute, thermo-dynamic, scale between the ice and steam temperatures). In the case of a specific type of thermometer, it is the difference between the actual readings (e.g. of resistance or voltage) at the two fixed temperatures.

fundamental mode the mode of vibration (e.g. half wave) of a system or object at the natural or fundamental frequency; the harmonics have higher frequencies

fundamental particles rare term for elementary particles

fundamental units rare term for the base units in a unit system

fuse a) old term for melt (see *melting*)
b) safety device in an electric circuit that melts if it becomes too hot (i.e. as a result of too high a current). A circuit breaker, now more common, has the same function – it is much more costly, but is more secure and can be reset.

fusion a) old term for melting; the actual meaning is joining together
b) process in which small nuclei combine (fuse) to make larger ones, with the release of useful binding energy as radiation and kinetic energy of the products. To make the nuclei approach closely enough for this to happen, they need the high energy associated with high temperature (hence the name sometimes used: thermonuclear reaction). Iron and nickel have the largest binding energy per nucleon, as the graph there shows – so fusing nuclei to the left of these has much potential. See *nuclear physics*.
The required temperature is tens of millions of kelvin – at which level it is very hard to contain a dense enough plasma for a useful length of time; as yet, therefore, despite much research, practical (controlled, efficient) **fusion power** in a practical **fusion reactor** has not yet been achieved. Best so far (2006) is the world's largest fusion reactor, JET, which has managed an energy 'profit' for several seconds.

The system's **fusion energy gain factor**, Q, is the ratio of the power output of a fusion system to the power needed to keep the plasma in a steady state; there is no unit. For the reactor to be economically viable needs a value of at least 20; current systems achieve up to 1 or 2 (and then only for short times). See also *ITER*, and see *nuclear power*. Fusion is the process that provides most of the energy output of stars; there the scale is huge.

See also *cold fusion*.

future history that has not happened yet. 'The study of the past is the key to the present' is a well known saying. In the field of physics, however, it is not easy to predict even the near future – however well we understand recent and current trends. There is always the possibility of new discoveries that will upset theory – as happened with high-temperature superconductivity and could have been the case with cold fusion. Astronomy continues to produce observations that are hard to explain and new elementary particles are always possible. On the other hand, the main aim of theoretical physicists is to produce a unified theory of fundamental effects. That was indeed the aim of Einstein and others not so far off a century ago.

The current picture of physics is therefore very different from that just over a century ago, at the end of the classical period. Then people believed that there was nothing left to discover or to explore: the work of physics was starting to involve just the design of better and better hardware to make more and more precise measurements.

g symbol for gram, 10^{-3} kg

G a) symbol for gauss, old name (until 1930) for the oersted, cgs unit of magnetic field strength, also outdated
b) unit prefix symbol for giga, 10^9 (a thousand million)
c) symbol for giorgi (name strongly pushed for kilogram when this became a base unit) or grave (name of the kilogram until 1795)

g a) symbol for energy level *degeneracy* (sense d), there being no unit
b) symbol for gravitational field strength, near the Earth's surface being 9.81 N kg^{-1} (or 9.81 m s^{-2})
c) symbol for Landé factor, or gyromagnetic ratio
d) symbol for the metric tensor, a symmetric tensor field used to measure distance in space
e) symbol (also χ) for osmotic constant (coefficient)
f) symbol for specific Gibbs function, or specific Gibbs free energy, unit: joule per kilogram, J kg^{-1}

G a) symbol for electrical conductance, unit: siemens, S ($= A^{-1}$)
b) symbol for Gibbs free energy function, unit: joule, J
c) symbol for gravitational constant
d) symbol for rigidity
e) symbol for shear modulus
f) symbol for weight (now rare: W much preferred) unit: newton, N

γ gamma, the third letter of the Greek alphabet, used as the name for:
a) the term $1 - \sqrt{(1 - v^2/c^2)}$ (or $1 - \sqrt{(1 - \beta^2)}$) that appears in some relativistic equations
b) a magnetic field strength of 10^{-9} tesla
c) high energy electromagnetic radiation (gamma radiation), or a quantum (photon) of this

GaAs gallium arsenide, a semiconductor

Gabor, Dennis (1900–1979) physicist who received the 1971 Nobel Prize for the invention of holography. He did this in the 1940s, but the use of the technique in practice had to wait twenty years, for the development of the laser.

gadolinium Gd rare earth, transitional metal in the lanthanide series, period 6, number 64, oxide first separated in 1880 by Jean de Marignac, who gave it the name from the ore gadolinite (though the pure metal was obtained only a century later). Relative density 7.9, melting and boiling temperatures 1310 °C and 3260 °C; most common stable isotope ^{158}Gd. Used in making compact discs and data storage chips, in mobile nuclear power units, and in various alloys.

Gaede molecular air pump 'vacuum' pump based on a rotating cylinder, able to work down to 0.1 Pa

gain a) of a circuit or system, the ratio of output power P (or, sometimes, voltage V or current I) to the input value. It is a measure of amplification (the effectiveness of an amplifier).

Being a ratio, gain has no unit. However, even if the output is less than the input (**fractional gain**, loss, or attenuation), the value usually appears in decibels (dB). The ratios are as follows:
$$10 \log (P_2/P_1)$$
$$20 \log (V_2/V_1)$$
$$20 \log (I_2/I_1)$$
A power gain of 3 dB is the same as a doubling in signal power; – 3 dB denotes halving. (The actual values are 3.010.)
b) common term for amplification in general, the amplitude of a signal at the output of a system compared to that at the input. So **gain control** means (output)

amplitude control and is the same as volume control in the case of an audio system.

galaxy massive set of stars, gas and dust, often a member of a cluster or group of galaxies that move together. The Sun is a star way out in an arm of a spiral galaxy, the 'Milky Way' being our edge-on view of hundreds of millions of other stars. 'Our' Galaxy is shaped like a disc about 100 000 light years across, and about 5000 light years thick at the centre (where there may be a supermassive black hole). The disc rotates with a period of around 10^8 years.

There are six main types of galaxy; they vary in size and shape. Dwarf galaxies (which is most of them, e.g. there are at least ten known in orbit round our Galaxy) can be elliptical, spiral or irregular, and the same applies to full sized galaxies. Our Galaxy is none of these, however, but a **barred galaxy** – these have a much stronger magfield than the others, strong enough to affect gas currents in it, and strongly aid star formation.

Galilean transformation equations equations (in fact, standard vector relationships) that relate the measures by one observer (1) of an object's position and motion to those of a second observer (2); the observers are moving at velocity v relative to each other along the x direction. The relations match the Newtonian view of motion, and still hold true for everyday cases. On the other hand, they do not conform with relativity (on the other hand, the Lorentz transformations do).

$$x_2 = x_1 - v\,t$$
$$y_2 = y_1$$
$$z_2 = z_1$$
$$t_2 = t_1$$

Galilei, Galileo (1564–1642) mathematician who did much practical research into physics, in particular mechanics and thermometry. Galileo was a mathematician, physicist, astronomer, and populariser of science. He is best known for his rejections of aspects of Aristotelian physics and of the Earth-centred Universe propounded by Ptolemy, and extended by Tycho Brahé. He found that a simple pendulum swings with a period that does not depend on amplitude (but only on length); that all masses have the same acceleration of free fall at a given place; and that projectiles move in parabolic paths under gravity.

In physics, there are a number of uses of the name **galileo** or **Galileo**:

a) symbol: gal – old unit of acceleration, one centimetre per second per second (cm s^{-2}). This is because Galileo was the first to measure this value, g, in the case of the Earth (981 gal).

b) a space probe that reached Jupiter in 1995, went into orbit round it, and fired a probe into its atmosphere

c) the European *satellite navigation* system

Galileo's improvement of the new refracting telescope led him to observations that indicated that the Copernican theory of the solar system was the correct view. (The observations included sunspots and solar rotation; the motion of the main moons of Jupiter; the Earth-like surface of the Moon; and the phases of Venus.) This view led to some conflict with the Church, though less than people often suppose.

The **Galilean telescope** uses a converging and a diverging lens to produce an upright image of only low magnification and very small field of view, but in an instrument that is small and light – this is still the design used in theatre glasses.

gallium Ga metal, element 31, predicted by Dmitri Mendeleyev in 1869 and found from spectra by Paul Lecoq de Boisbaudran in 1875, who named it from the Latin *Gallia* (= France) – though some believed he named it after *gallus*, Latin for Lecoq (rooster). Relative density 5.9, melting and boiling temperatures 30 °C (yes, it will melt in your hand) and about 2400 °C; most common isotope ^{69}Ga. Main uses: in mirrors, for doping semiconductors, in low temperature alloys (e.g. as a flux for solder), in some photo-cells, and in various types of thermometer.

gallium arsenide GaAs very widely used example of a III-V semiconductor (i.e. one with equal numbers of atoms from groups III and V). It has a near ideal gap width of 1.4 V, and GaAs chips are coming to replace those of silicon because:

a) they can work at higher temperatures, up to about 400 °C;

b) they can be smaller (so use less power and act some five or six times faster);

c) they can work at much higher frequencies – up to about 250 GHz;

d) they can link much better to microwave and fibre optic systems;

e) they can work at much higher radiation levels.

On the other hand, silicon is cheaper and stronger, has higher hole mobility, and has a 'built-in' high quality insulator, silicon dioxide.

gallon gal the fundamental unit of volume in the Imperial (foot-pound-second) unit system, the volume of ten pounds of pure water at 62 °F and standard pressure: 4.5461 litres (dm^3). The US gallon is about 0.83 Imperial gallons.

Galtizin pendulum massive horizontal pendulum (i.e. one with a nearly vertical axis of vibration) at the centre of the seismograph

Galton whistle high frequency whistle working on the basis of edge tones. The fundamental frequency (which depends on section area and length) can be supersonic (i.e. above 20 kHz). Polymath Francis Galton (1822–1911) used it in research on how mammal species differ in physiology (and also devised the concepts of finger printing and of eugenics).

galvanic cell old term for primary electric cell (see *electrochemical cell*) The name comes from **Luigi Galvani** (1737–1798) who discovered 'animal electricity' (as opposed to static electricity) in the 1780s; versions of the story about frogs' legs twitching during a lightning storm are well known.

galvanising process of coating iron or steel with zinc to prevent corrosion (rusting)

galvanomagnetism see *Hall effect*.

galvanometer meter able to react to very small values of voltage or current. There are many designs, most of those now in use being electronic rather than electromagnetic.

gamma camera complex detector used in, for instance, medical diagnosis, based on gamma-emitting tracers (radioactive isotopes). A number of photomultipliers amplify the light produced by gamma photons passing through a scintillator; a computer analyses the output of these to locate the source of the radiation and show it on a screen.

gamma decay or **gamma-ray transformation** process of nuclear de-excitation that outputs gamma radiation. In nature, the most common cause for a nucleus to be excited is its creation in this state – e.g. as a result of alpha or beta decay or fission. Thus gamma radiation very often appears together with α and β radiations.

gamma radiation high energy electromagnetic radiation, with energy above about 10 keV (though there's much overlap with hard x-rays). Seeing it as a high energy wave means it has high frequency (small wavelength). This

also explains the high penetrating power in matter. See *nuclear physics*.

gamma rays narrow beams (or sometimes photons) of gamma radiation that behave like other electromagnetic waves; for instance, the study of the **gamma-ray spectrum** output by a given source gives information about the nature of the source. About 3% of the gammas emitted as the result of a fission process are **delayed gamma rays**: they arise from the de-excitation of an intermediate meta-stable state (as opposed to **prompt gamma rays**).

Short **gamma-ray bursts** are the most intense forms of radiation known. Lasting for around a tenth of a second, but of very high energy, they may come from the collision of neutron stars (as may gravitational waves). A second type of gamma ray burst, lasting much longer (over about two seconds), appears when a super-nova becomes a black hole.

gamma-ray spectrometer device that measures the details of gamma spectra – the energy (i.e. radiation intensity) at each wavelength – and thus detects a number of elements. There are two types – small (e.g. hand-held) ones that sense scintillation (in most cases in thallium-doped sodium iodide), and those (e.g. in space craft and surface roamers) using a photo-diode (in most cases made of ultra-pure germanium); the latter need cooling so are more bulky. In both cases, the gammas sensed come from background radiation or as secondary emission from cosmic ray impacts.

gamma-ray transformation the same as *gamma decay*

Gamow barrier the potential barrier around a nucleus that makes it hard for a second nucleus to approach – the Coulomb barrier in the context of nuclear fusion in particular. The main work of **George Gamow** (1904–1968) was in nuclear and astrophysics, cosmology, genetics, and the popularisation of science. Gamow also devised the concept of tunnelling in this context; because of tunnelling, there is a range of temperatures at which fusion can occur in a given case, below the theoretical value – the **Gamow window**.

ganged circuits two or more separate circuits with a single control. The use of this control affects both/all the circuits at the same time in the same way.

gap the energy difference between the top of

the valence band of a substance and the bottom of its conduction band – see *band model*. The **gap width** is the measure of this energy difference – most often given, however, in (electron) volts rather than in joules: 1 eV is about 1.6×10^{-19} J. The gap width value is a crucial factor in how effective a substance is as a semiconductor – not good if it's too large or too small.

gas matter in such a state that the particles fill all the space available, however its volume and shape may change. A vapour is also in such a state, but it is at a temperature below the critical value of the substance concerned; this means that it will condense into a liquid if the pressure becomes high enough. See also *plasma*. Compared to liquids, gases tend to have low values of density, fluid friction, light absorption and rigidity, and to be easy to compress.

gas amplification or **gas multiplication** process in which the ions produced in a gas accelerate enough in a high electric field to produce further ions to form a chain reaction or avalanche. The result may be **gas breakdown**, in which a very heavy discharge disrupts the structure of the substance; see, for instance, *spark*.

gas constant, or **molar gas constant**, or **universal gas constant** R universal constant that appears in the ideal gas equation of state: $pV = RT$. Here p, V and T are the pressure, volume and (absolute) temperature of the one mole sample of ideal gas. The value of R is 8.314 35 J K^{-1} mol^{-1}. See also *Boltzmann constant*, the corresponding value for a single gas particle.

gas-cooled reactor and **advanced gas-cooled reactor** nuclear fission reactor types in which a gas is the coolant – less efficient than liquid types as far as concerns fuel use, but more efficient at generating electricity. See *nuclear power*.

gas discharge the effect of current in a gas. As a gas is normally an insulator, much energy may be required to produce a discharge. A graph of this would show how the 'striking' voltage – that needed to start a discharge – in a gas depends on the gas pressure. A gas *discharge tube* shows such effects, including at different pressures. See also *spark*.

gas-filled relay *thyratron*

gas laws statements that describe in equation form how the pressure, volume, and temperature of a mass of gas relate to each other. The gas may be ideal (perfect) or

real – see *ideal gas* and *equation of state* respectively.

gas multiplication same as *gas amplification*

gasoline US word for petrol

gas thermometer thermometer whose working substance is a fixed mass of gas, the behaviour of which depends on temperature. There are two types of design. In one, the gas sample remains at constant volume; in the other it is at constant pressure. While such thermometers are nowadays rare in practice, but remain the basis of the thermodynamic scale of temperature.

gas turbine engine in which hot gas turns a turbine to give power; the system is cheap, quick to start up, and efficient, so is in widespread use in transport and in power stations. There are internal and external combustion versions. See jet, often taken to mean the same, but in fact a wider range of types of engine.

gate a) the control electrode of a field-effect transistor (though some fets have more than one gate)

b) the basic logic circuit (element), the value of whose single output depends on the value(s) at the input(s). Truth tables show the output value for each possible set of inputs. These use the binary values 0 and 1, though in digital electronic and information technology systems the values are two levels of voltage, while in logic (Boolean algebra) they are True and False.

The two-input AND and OR gates shown in Appendix 3 are common (as are such gates with more than two inputs, which act in just the same way). On the other hand, the NOT gate has only one input. We show both the non-standard symbols (simple to use) and the standard ones (maybe simpler for non-English users). The truth tables are opposite.

The other gates also appear, again with both symbols and truth table and for two inputs only. They are the NAND (not AND), NOR, XOR (or EOR, exclusive-OR), EQ (equivalence).

The six two-input (strictly multi-input, recall) gates above provide all input/output patterns required of standard logic circuits. There are other possible and more complex patterns; working back from their required truth tables, adepts can design a combined set of basic gates to give any effect. Indeed, sets of NAND gates alone or of NOR gates alone can produce any desired effect: NAND

Fig. 1

inputs		output	inputs		output	input	output
I1	I2	o	I1	I2	o	I	o
0	0	0	0	0	0	0	1
0	1	0	0	1	1	1	0
1	0	0	1	0	1		
1	1	1	1	1	1		

Fig. 2

I1	I2	NAND	NOR	XOR	EQ
0	0	1	1	0	1
0	1	1	0	1	0
1	0	1	0	1	0
1	1	0	0	0	1

logic systems use only NAND gates, and NOR logic systems NOR only gates. While NAND logic often makes elements more complex than they would be with the use of basic gates, the resulting systems are cheaper to build.

Note that there are also three-state gates, certain types of buffer, with outputs of 0, 1 and high-impedance.

gauge a) instrument for measuring a quantity, mainly diameter/thickness and pressure **b)** condition that leads to symmetry (invariance) of the vector and/or scalar potentials of a field during a given mathematical mapping. **Gauge theories** describe **gauge interactions** – those that concern electromagnetism or the strong or weak nuclear forces – and have had great success. The theories also involve exchange particles – the **gauge bosons**: photons, vector bosons, and gluons.

gauss G (or, rarely, Gs) the unit of magnetic field strength in the cgs electromagnetic unit system. $1 \text{ G} = 10^{-4}$ tesla. A **gauss meter** is an undamped moving coil meter used to measure magnetic field strength.

Gaussian distribution the normal distribution, i.e. the frequency distribution in statistics that gives the normal 'bell-shaped' curve.

Gaussian eyepiece eyepiece of a telescope, with a small window at the side to let in light which shines on the cross wires needed for accurate measurements

Gaussian system of units unit system, now very rarely used, that combines features of the electromagnetic and electrostatic cgs systems

Gauss, Karl Friedrich (1777–1855) one of the world's greatest mathematicians, also an astronomer, who invented a number of useful instruments in both fields. His main work in physics is in the field of electromagnetism.

gauss meter an undamped moving coil meter, or an electronic meter, used to measure magfield strength

Gauss's theorem The total electric flux normal to any closed surface in an electric field equals the total charge enclosed. So, if the volume contains charge Q, the flux out of it is $4\pi\,Q$. There are similar versions for magnetic and other vector fields.

Gay-Lussac, Joseph (1778–1850) scientist whose work in various areas of physics and chemistry, and in particular on gas laws, has been of much value. His studies of gases sampled from high-altitude balloons led to the **Gay-Lussac law**: the volumes in which ideal gases chemically combine relate by simple whole numbers to each other and to the volume of the product (if a gas). All measures are at constant pressure and temperature. The same name also sometimes applies to Charles' law.

gear toothed wheel used to transfer rotation from one shaft to a second with low slippage and low friction. There is one gear on each shaft (with a third between them if the shafts are to turn the same way); there are very many designs.

Geiger, Hans (1882–1945) nuclear physicist who worked on radioactivity and cosmic rays under Ernest Rutherford.

The **Geiger counter** (or **Geiger–Müller counter**, as it is based on the **Geiger–Müller tube** he designed with a colleague) was the first automatic sensor and counter of ionising particles. In the tube, the trail of electrons and ions formed by a quantum of radiation multiply to produce a pulse of current in the circuit. It works over only a certain range of applied voltage; this is the

Geiger plateau (or **Geiger region**) of the tube. See *nuclear physics: detectors*.

Geiger law the range of an alpha particle varies directly with the cube of its speed.

$$R = k\ c^3$$

This holds for a fairly wide range of ranges.

Geiger–Nuttall relation an approximate relation between the range of an alpha particle and the lifetime of its parent. The modern version links the energy W of the particle to the decay constant λ of the parent.

$$\log \lambda = a + b \log W$$

Geiger plateau or **Geiger region** See *Geiger counter*. The **Geiger threshold** is the lowest voltage of the counter tube's Geiger region.

Geissler tube early gas *discharge tube*, that led to the neon tube as used for displays. The original device, designed by glass blower **Heinrich Geissler** (1814–1879) in 1858, contained air or ether at about 1% of atmospheric pressure and showed marked luminosity.

gel solid or semi-solid (jelly-like) colloid

Gell-Mann, Murray (1929–) physicist whose work on elementary particles won him the 1969 Nobel Prize. With Feynmann, he proposed the quark as the 'atom' of elementary particles.

generalised coordinates the smallest number of independent values ('coordinates' that can be numbers, vectors or tensors) needed to describe in full the state of a given system. See also *degree of freedom*. In a given system, a **generalised force** is the ratio of the energy transfer of all the forces acting to the change that results in a given generalised coordinate.

general theory Einstein's 1915 theory of relativity and gravitation

generation set of particles whose members differ only in mass and flavour. According to the standard model of matter, there are three generations, based on the three generations of quarks.

generator machine with a mechanical input and an electric output. The first such machines were based on electrostatic effects; examples are the van de Graaf and Wimshurst generators. It is normal now to use electromagnetic systems such as the alternator and dynamo shown, but see also *magnetohydrodynamics*.

de Gennes, Pierre-Gilles (1932–) mathematical physicist who was able to apply simple classical methods to such complex systems as liquid crystals and polymers – and had success in explaining the observations about order. He won the 1991 Nobel Prize for this.

geomagnetism the nature of the Earth's magnetic field, and its study. Until early in the last century, knowledge of the Earth's field and its measures at any point on or near the surface was crucial for navigation.

At any point P, the field has three measures:
a) B_h: the horizontal component of the field strength B;
b) δ, the angle of dip (or inclination): the angle between the field direction and the horizontal;
c) α, the declination: the angle between the horizontal field and the geographic meridian (true north-south line).

The actual field strength is B_h sec δ.

The Earth's field changes with time. In the days when the Earth's field was used for long-distance navigation and position-finding (as it rarely is now), knowledge of the details of these secular (slow and – in most cases – steady) changes was also required. The most important such change is the variation of declination with time. (Magnetic storms cause large but temporary changes.) The history of the changes in the Earth's field is well known from the study of the magnetism of rocks. The polarity of the Earth's field has switched many times at varying intervals of a few million years. During a switch, the field strength becomes very low. As the field provides some protection from the solar wind (see also *aurora*), field reversal is significant to life on Earth.

geometric attenuation the fall in strength of any non-parallel signal with distance as a result of the inverse square law. See also *attenuation*.

geometric optics the study and use of light rays in optical systems with no account taken of their wave nature. Thus refraction and reflection come into this field (while diffraction, polarisation and interference are aspects of physical, or wave, optics).

The basic assumption is that light travels in a straight line in a uniform medium (rectilinear propagation). Both reflection and refraction follow simple geometric laws; this means that it is easy to draw diagrams (ray sketches) to explore how a system affects incoming light rays. The concepts apply to any form of radiation, as long as the aspects of its transfer do not depend greatly on wave physics.

geometrodynamics a) loosely, a name for Einstein's general theory of relativity (i.e. of gravity) **b)** strictly, quantum theory that uses the graviton as mediator to try to explain gravity

geometry semi-jargon term for the arrangement of the items of equipment needed to carry out an experiment

geon theoretical particle of electromagnetic or gravitational energy held together by its own gravitational field. John Wheeler (1911–) first studied them in theory in 1955 (and gave them the name, for 'gravitational electro-magnetic entity'), but no geons have been observed.

geophysics study of the physics of the Earth and its features. It includes atmospheric physics, geomagnetism, hydrology and seismology in particular; it also relates closely to geology and astronomy.

geothermal gradient gentle increase of temperature with depth below the surface in the crust of the Earth. It occurs at any point on the Earth but varies with the thermal conductivity of the rock.

geothermal power using high pressure, high temperature water from under the ground in regions of high seismic activity to provide local heating and, in some cases, electricity. This is universal in Iceland, but there are some such systems in Britain, in Cornwall, at least for research. The water may come from underground reserves or be pumped down cold from the surface.

germanium Ge brittle crystal solid element with a diamond-like structure, a metalloid element, number 32, group 4, period 4, predicted by Dmitri Mendeleyev in 1869 and found in 1886 by Clemens Winkler, who named it after Germany. Relative density 5.3, melting and boiling temperatures 940 °C and 2830 °C; most common isotope ^{74}Ge. A major use used to be as a semiconductor, but that is now less common; the small band gap, however, gives it value in infrared optical systems and airport security. It is a phosphor and used in certain alloys.

getter substance which combines strongly with certain others, in particular with impurities in given systems. Thus getters help capture remaining gas particles in electron tubes and vacuum pump systems (for instance, barium, charcoal or titanium in this case) or impurities on the oxide surface of semiconductors (for instance, phosphorus).

GeV symbol for giga electron volt, an old unit of energy still widely used in nuclear physics and applied science. $1\text{ GeV} = 10^9\text{ eV} = 1.6 \times 10^{-10}\text{ J}$. The US term BeV (for billion electron volts) is also occasionally used and is the source of the name Bevatron for an accelerator.

g-factor g name sometimes used for the *Landé factor*

ghost faint image on a tv (or radar) screen, slightly shifted with respect to the main image, caused by reflection of the inbound signal from, for instance, a large building or steep hill. See also *spectral ghosts*.

GHz gigahertz, 10^9 Hz

Giacconi, Ricardo (1931–) astrophysicist whose work led to the discovery of cosmic x-ray sources, and to a share of 2002 Nobel Prize

Giaever, Ivar (1929–) physicist who shared the Nobel Prize in 1973 for his work on the tunnel effect

giant magnetoresistance large change in the electrical resistance of a thin film of magnetic material with change in applied magfield, caused by interacting electron spins. The effect, discovered in 1988, is two hundred times stronger than standard magneto-resistance. With great potential for very high density magnetic data storage, magfield sensing, and magnetic read and write memory (ram), all already in use, applying this effect is a very active branch of spintronics. See http://policy.iop.org/vproduction/v5.html for more details and updates.

Gibbs function or **Gibbs free energy** G unit: joule, J the thermodynamic function of a gas sample ('system') at temperature T, as

$$G = H - T\,S$$
$$= W_i + p\,V - T\,S$$

Here H and S are the enthalpy ($W_i + p\,V$) and entropy of the system.

If there is a reversible change in the system at constant temperature and pressure, the value of G changes by an amount equal to the energy transfer into or out of the system. The **specific Gibbs free energy** g is G per unit mass, unit: joule per kilogram. Compare with *Helmholtz function*.

Gibbs–Helmholtz equation relation for the Helmholtz free energy F (which is rather similar to the Gibbs free energy). It gives the internal energy W in terms of the variation of F with temperature T (volume being constant): $W = F - T(\delta F/\delta T)$

Gibbs theorem When two ideal gases taking up the same volume mix isothermally to give the same volume of mixture, there is no change of entropy.

giga- G prefix for 10^9, as in gigagram (10^9 g, 1000 tonnes)

gilbert Gb unit of magnetomotive force in the electromagnetic cgs system, equal to $10/4\pi$ ampere-turns, named to commemorate William Gilbert

Gilbert, William (1544–1603) 'the father of magnetism', royal doctor whose part-time research led to advances in the subjects of magnetic poles and the Earth's magnetism, and also electrostatics. He was the first to use the terms 'electricity' and 'electric' in their modern senses.

Ginzburg, Vitaly (1916–) joint winner of the 2003 Nobel Prize 'for pioneering contributions to the theory of superconductors and superfluids'

giorgi G widely supported alternative name for the kilogram, the base unit of mass in the modern system, *Appendix 1*. (A second such unit is the grave, G.) The use of such a name would overcome one of SI's last problems, that of having a base unit with a prefix.

Giorgi system said as 'Georgie', the Italian way, the unit system that led to SI (the one now in most widespread use in most parts of the world. It was based on the metre, kilogram, second and ampere (abbreviated to mksa). In fact, when **Giovanni Giorgi** (1871–1950), electrical theorist and engineer, first proposed the system in 1900, he used the ohm for the basic electric unit; the ampere replaced this fifty years later, just before the system was superseded by SI in 1954.

Gladstone–Dale law When the density ϱ of a substance changes (as a result of change of pressure or temperature), the refractive constant n changes so that

$$n - 1 = k\,\rho$$

Here k is a constant.

Note, however, that there is no relation between the density and refractivity of different media.

Glaser, Donald (1926–) inventor of the bubble chamber, for which he was awarded the 1960 Nobel Prize

Glashow, Sheldon (1932–) winner of the 1979 Nobel Prize for his work on unifying the weak and electromagnetic interactions. He has also explored the concepts of quantum chromodynamics and charm.

glass complex of metal oxides and silicates cooled from the liquid state without gaining any long-range (crystalline) order. In that a glass may flow (albeit with extremely high friction, i.e. slow rate) and has no constant melting temperature, it is common to view it as a supercooled liquid rather than as a solid. Even so, it is very stable thermodynamically; for instance, a glass can return to its original state after bombardment with high energy electrons. Most glasses are transparent to light, and accordingly have many practical uses. See also *optical glass*.

glass fibre hair-thin strand of highly pure glass used to carry signals on light waves; see *fibre optics*.

glassy alloys very strong alloys that resist corrosion, mainly of iron and phosphorus with carbon or nickel. The name refers to their lack of crystalline structure.

Glauber, Roy (1925–) mathematical physicist who shared the 2005 Nobel Prize for his work developing the quantum theory of optical coherence (laser light)

glide the relative motion between two crystal (atomic) planes in a solid during plastic deformation under stress. A **glide plane** is a plane in a solid along which a glide may take place if the stress is of the right type and direction.

gloss the ratio of the light reflected by a surface specularly (i.e. as by a mirror), to the total light reflected

glow discharge the production of light throughout the volume of a low pressure gas when it passes a current. See *gas discharge*.

gluon virtual vector gauge boson, exchange particle in the theory of quantum chromo-dynamics (qcd), which tries to explain the strong nuclear force, i.e. the force between quarks. (It 'glues' the particles together.) According to quantum chromodynamics, there are eight gluons, all of spin 1; as they can also interact with each other, qcd is very hard to analyse. The gluon has two spin states rather than the three of other vectors (reduced by gauge invariance).

Like quarks, in normal matter gluons can exist only inside hadrons. Some aspects of qcd predict that hadrons can exist which contain only gluons – these are **glueballs**;

exotic hadrons would contain real gluons (rather than the virtual ones of normal matter). At extreme temperatures and pressures, matter would be in the form of quark-gluon plasma – there being no hadrons, the gluons (and the quarks) would become free. Gluons were first observed in 1979, though strong evidence appeared ten years before then.

Goddard, Robert (1882–1945) engineer whose life's work was the development of rockets. His main firsts were flight of a liquid fuel rocket 1926, flight of a supersonic rocket 1935.

Goeppert-Mayer, Maria (1906–1972) physicist who shared the Nobel Prize in 1963 for research into the shell model of the nucleus

Golay cell electromagnetic radiation detector of particular value in that it works over an unusually large range of wavelengths, in fact from ultraviolet to microwave. At its centre is a small transparent cell of gas. The system measures the change of pressure in this cell that results from its absorption of radiation energy.

gold Au metal known and valued since ancient times (at least 5000 years ago), perhaps having been the first metal known, the name being that in Anglo-Saxon English; transitional element, number 79, in period 6. Relative density 19.3, melting and boiling temperatures 1060 °C and rather over 2800 °C; only natural isotope[197] Au. Because of its softness, it is common in dental fillings, decoration, coinage and jewellery; its high conductivity and freedom from corrosion make it of value in microelectronic circuit wiring.

gold-leaf electroscope classical device (Abraham Bennet 1786, scale added by Volta in 1787) able to show potential differences and therefore charge states, now never used except perhaps in elementary teaching. The central metal core is always at constant potential, and the metal leaves (scraps of very thin sheet) repel to an angle that depends on the potential.

goniometer device designed to measure angles between planes, in particular reflecting planes in a range of optical systems (including crystals)

governor mechanical device used to control the rate of rotation of a shaft by simple feedback. A traditional system has two masses mounted on a collar on the shaft: as the speed rises, so do the masses; this opens a valve and makes the shaft harder to turn.

G-parity quantum number involved with elementary particles of zero baryon number and strangeness, conserved only in strong nuclear interactions

grad the gradient of a scalar field S, the product of del Δ and S. Grad is a vector field, given by grad $S = \mathbf{i}\delta S/\delta x + \mathbf{j}\ \delta S/\delta y + \mathbf{k}\ \delta S/\delta z$. Here $\mathbf{i}, \mathbf{j}$ and $\mathbf{k}$ are the unit vectors.

grade old term that relates to degree in many contexts, not only that of the unit of angle or temperature. In Napoleonic times in the case of angle measurement, the name was given to the decimal degree, one-hundredth rather than one-ninetieth of a right angle. This usage still remains in some parts of Europe. The term also appears in 'centigrade'.

graded-base transistor same as drift transistor

gradient m; no unit the slope of a line given as the change of vertical value y per unit change of horizontal value (x). If the line is straight, we have $m = (y_2 - y_1)/(x_2 - x_1)$. For a curve, this becomes the rate of change of y with x at any given point: the derivative of y with respect to x, called dy/dx. For the gradient of a scalar field, see *grad*.

Graham's law of diffusion gases diffuse at a rate inversely proportional to the square root of density. This is strictly true only for diffusion through a hole whose diameter is no more than a tenth of the mean free path of the particles. **Thomas Graham** (1805–1869) was a chemist; he also worked on osmosis and discovered colloids.

gram or **gramme** g the basic unit of mass in the various cgs ('metric') systems, a thousandth of a kilogram. The definition then was the mass of a cubic centimetre of pure liquid water under standard conditions. **Gram-atom** and **gram-molecule** are old terms for mole.

grand unified theory gut single theory, not so far achieved, that can at a sweep explain simply all the basic aspects of modern physics: the theories of relativity and quantum physics, the four fundamental forces (gravity, electromagnetism, weak and strong nuclear), and elementary particles. With some success, electroweak theory tries to combine the electromagnetic and weak *fundamental interactions*. With less success, grand unified theory tries to combine that model with the strong interaction. There has been no success in drawing in gravity.

graph diagram (almost always in two dimensions) that shows how two or more interdependent variables relate. Spectra and energy level diagrams are one-dimensional graphs: there is only one variable in each case.

graphene a newly found set of *graphite* compounds with some novel properties

graphite soft black form of carbon that sublimes at 3660 °C. It has many practical uses because of its high electric and thermal conductivity and its ease of gliding. (This last makes it a common lubricant.) The cores of many types of fission reactor are built of graphite blocks – it is a good moderator.

First made in 2004, **graphene** is a form of graphite with some novel – and important – properties. It has a two-dimensional crystal (buckyballs, bucky tubes* and normal – pencil lead – graphite crystals having zero, one and three dimensions). The crystal is a sheet of carbon one atom thick that can grow only at the edges; also only at the edges can the carbon combine with other elements (as, for instance, in $C_{62}H_{20}$). Graphene is a better conductor of electricity than anything else known, except superconducting metals, because in it the electron *drift* speed is relativistic. This, combined with the fact that

*One can view nano-tubes (bucky-tubes) as rolled up and capped graphene sheets. In a major UK survey in 2006, study of these structures was voted the most exciting area of physics.

it does not react with gases, gives it great value for electronic and microelectronic conductors – some people expect graphene transistors and chips by 2008. The very high conductivity also means that, in graphene, the Hall effect is relativistic; this had never been found before. In 2006, a new material was developed, with interesting properties and many potential uses – sheets of graphene embedded in a polymer. Other possible uses include hydrogen storage and processing electron spins in spintronic circuits.

graticule a) fine mesh of thin wires or threads, or of lines on a glass sheet, set in the focal plane of the eyepiece of an optical instrument. It allows a more effective measurement of the image than does the simple cross-hair from which it derives.
b) the grid on the screen of an oscilloscope

grating surface that diffracts incident light during reflection (reflection grating) or transmission (transmission grating) to produce colour fringes and holographic effects. Gratings have many uses in the production of spectra, of security images in bank notes and credit cards, and of decorations. See also *diffraction grating*.

gravimeter device designed to measure the difference in gravitational field strengths between two points, nowadays also to give their absolute values

gravitation

universal long distance (but weak) interaction through which all masses attract each other. It is the force that holds the Universe together, the stars in the galaxies, and planets in stellar systems. On the other hand, it has negligible effect within matter, in comparison with the much stronger electromagnetic and nuclear interactions.

There have been many theories of gravitation, but none can yet explain the effect. Newton was the first to explore gravitation with any success, in around 1666. He used the concept to explain first the motion of the Moon around the Earth, and then Kepler's laws. Later, he applied his ideas to many effects, such as precession and tides – but he didn't explain it, other than speaking of 'action at a distance'.

In recent years, several highly sensitive explorations of the Earth's field have led to claims that Newton's theory does not equate exactly with practice (see, for instance, *fifth force*). So far, however, all such results have later turned out to be due to tiny variations in the field because of large buildings, distant mountains, rock density variations, and so on. Also, in recent years, there have been many attempts to describe and explain gravitation on the basis of modern theory. Einstein's (general relativity) view of it as a distortion of space-time is helpful; gravitation may also be a type of long-range Casimir effect.

Quantum gravity models involve either exchange of gravitons or quantisation of Einstein's curved space-time.

gravitational acceleration g the effect of the force experienced by a mass in a gravitational field. g depends only on the field strength. Near the Earth, the standard value is $9.806\ 65\ \text{m s}^{-2}$ (N kg^{-1}).

gravitational constant G universal constant, of value $6.6742 \times 10^{-11}\ \text{N m}^2\ \text{kg}^{-2}$, that appears in the expression for the force of gravity. Some hold that G is not constant, but changes with the expansion of the Universe (i.e. with time).

gravitational field region of space in which there is a gravitational force on a mass. Its **gravitational field strength** (symbol g) is given by the force on unit mass; the numerical value is the same as the acceleration that results if the mass is free to move (fall). Thus the standard value can be expressed as $9.806\ 65\ \text{N kg}^{-1}$.

gravitational force the force between two masses that results from gravity. Newton's law gives a value to this gravitational force F between two masses m_1 and m_2, distant r apart:

$$F = G\, m_1\, m_2 / r^2$$

G is the gravitational constant.

In the special case of the Earth (mass m_e and radius r), this leads to an expression for g:

$$g = G\, m_e / r^2$$

Using $F = m\, a$, we then have an expression for the weight W of an object: $W = m\, g$.

gravitational interaction the mutual force of gravity between two masses: Newton's third law of force states that the two forces are the same but in opposite directions.

gravitational lens massive object in space that affects the light passing to us from an object directly behind it, in a lens-like manner though very weakly. The effect is a result of the bending of the light waves by the gravitational field of the lens object, as predicted by Einstein's general theory of relativity. If the lens and the other object are exactly in line, the image produced is a ring (an Einstein ring, as he put figures onto the concept).

First proposed in 1925, the concept was taken forward by Einstein in 1936; the first case observed was in 1979. Hundreds of these lenses are now known (with about a dozen being rings); they give astronomers a great deal of information about both the lens object and the object behind it. For instance, lenses give us more details about very distant objects, such as their very large red-shifts, while in 2006 came the first claims to have seen dark matter by this method. Also, a careful study of lens effects may lead to the detection of gravitational waves.

People are now working on the design of a telescope using the edge of the Sun to form a gravitational lens. The sensor would need to be 550 times further from the Sun than the Earth is, but the magnification would be at least 100 million – so a small dish at that distance could resolve stellar details as small as 80 km.

gravitational mass that feature of an object that causes it to experience a force in a gravitational field. Some of the discussion in these entries assumes that an object's gravitational mass is the same as its inertial mass. While the latter concerns how the object resists acceleration under any force, it appears to be the case that the two are the same: see *mass*.

gravitational potential W, unit: joule, J the energy at a point in a gravitational field that would be transferred in taking a mass between that point and infinity. In the case of a spherically symmetric field around a large mass M, the potential at a point distant r from the centre of mass is $G\,M/r$. For a mass m at a distance h relative to (say) the Earth's surface, $W = m\,g\,h$.

gravitational radiation See *gravity wave*.

gravitational red-shift See *relativity*.

gravitational units units in mechanics in pre-SI systems that involved gravitational acceleration g. In the Imperial (foot-pound-second) system, for instance, there were two units of force: the poundal (pdl) gives a mass of one pound an acceleration of 1 foot per second per second; the poundforce (lbf) accelerates a mass of one pound by g, 32 ft s^{-2}.

gravitational wave something like a ripple in space-time, crucial required outcome of Einstein's general theory of relativity as radiating from a moving mass. Such radiation can come only from very large masses, and the many searches (started by Joe Weber thirty years ago) have so far had no clear success. The mass changes in a supernova should produce enough gravitational radiation for this to be detectable even at a great distance, and supernovas have produced many quasars and black holes. However, gravitational wave detectors built and run during recent decades have not produced a single agreed positive result. The nearest was in the late 1970s – the orbit of a certain binary pulsar is falling at just the rate – 7 mm/day – it would if giving out gravitational waves; the discovery gained Russell Hulse and Joseph Taylor the 1993 Nobel Prize.

This has become a highly active field of research and a number of new gravitational wave telescopes are working or about to start. Indeed, a world-wide system of more sensitive gravity 'telescopes' is now being set up, with the first (in Germany) starting full-time work in mid-2006. Each consists of two 3 km vacuum tubes at 90°; laser light from the junction sent along each pipe reflects back to and fro many times between suspended mirrors before passing to an interferometer. The system should be able to detect mirror movement smaller than 10^{-18} m, a thousand times more sensitive than existing techniques and of the order of a thousandth the proton diameter.

In theory, such waves radiate from all accelerating masses, but only very massive small sources (such as neutron stars) can produce enough energy in this form for us to detect. From mid-2006, there have been high hopes of sensing these waves with enough detail to be able to recognise the signatures of different types of event (for instance, a supernova, or merging black holes).

graviton exchange particle in some quantum theories of gravity (geometrodynamics), proposed to explain gravitational interactions in a way consistent with quantum physics. The graviton would have zero rest mass (as gravity has infinite range) and spin 2 (as gravity is a tensor field of the second rank); it would transfer between two interacting masses at the speed of light. The graviton, if it exists, has not yet been discovered; finding gravitational waves would be much simpler.

gravitino supersymmetric partner of the graviton, as needed by some supergravity theories (those that combine supersymmetry and general relativity). It would be a fermion of spin 3/2 and would be produced by the decay of a supersymmetric particle.

gravity the gravitational interaction of one mass with a second mass. The *force of gravity* of an object in a field is its weight. See also *centre of gravity*.

gravity balance same as gravimeter

gravity cell primary chemical cell whose two electrolytes differ enough in relative density (formerly known as specific gravity, see *density*).

gravity lens gravitational lens

gravity meter see *gravimeter*.

gravity wave

a) low amplitude wave in a liquid surface, or in the atmosphere, that is under the control of gravity and not of surface tension or cohesion

b) name sometimes used for gravitational wave (but best avoided)

gravity well distortion in space-time caused by a massive object. The 'depth' of the well relates the gravitational potential energy needed to escape from the object concerned, or the escape speed.

gray Gy   the unit of absorbed radiation energy dose, one joule per kilogram of tissue; the unit's meaning is the same as that of the *sievert*, the unit of dose equivalent – so one must use the correct one. The name comes from that of Hal Gray (1905–1965), the radiobiologist who invented radiobiology.

Great Wall a crumpled surface of thousands of galaxies, the surface being over 500 million light years across. It is the largest structure so far found in the Universe (though there are smaller ones); the Great Wall as yet defies explanation. The smoothness of the cosmic microwave background radiation implies that the Universe was smooth long after such a large structure must have formed.

greenhouse building, often of glass, designed to accept the inward short wavelength radiation from the Sun but not to pass out the long wavelength radiation from the ground and objects inside. Thus, the inside gains net energy and needs less artificial heating.

greenhouse effect process that takes place in the atmosphere, in which certain high level gases allow through the inward solar energy, but do not let pass all the energy radiated from the ground. This makes the Earth's surface some 18 °C warmer than would otherwise be the case. It is believed that the effect is much stronger on Venus and is becoming stronger on Earth as the atmosphere's content of methane and carbon dioxide rises. Both methane and carbon dioxide are effective **greenhouse gases**: they are highly transparent to high energy radiation and yet able to absorb (and therefore trap) low energy radiation well. An advantage of this is that the outer atmosphere, now cooler than it was, is shrinking – spacecraft in low Earth orbits now suffer less drag so survive longer.

Grenz rays long wavelength (low energy) x-rays formed in a **Grenz tube**, an x-ray tube having a window with very low absorptivity for such radiation. The maximum energy of Grenz rays is 15 keV (about 10^{-16} J); they were once in widespread use for treating skin problems.

grey body surface that gives off radiation at all wavelengths, with the same distribution as a black body at that temperature, but always with lower intensity

grid a) the high voltage electric power transfer system over a region. It links large scale power plants with each other to provide an even and reliable service to consumers throughout the region.

b) the control electrode of an electron (vacuum) tube. All such tubes with more than two electrodes (cathode and anode) have a control electrode between them; the voltage on this affects the flow of electrons through the tube. The **grid bias** is the mean voltage of the grid. A **grid leak** is a high resistance path between grid and cathode; its aim is to prevent the build up of charge on the grid (for this would change the voltage).

Gross, David (1941–) one of three joint winners of the 2004 Nobel Prize, for work on the theory of the strong interaction, in particular for discovering a new aspect

ground a) US term for earth

b) basic, lowest: for instance, the **ground state** of a system is the state with the lowest possible energy

c) relating to the surface of the Earth. A **ground effect vehicle** is a hovercraft, i.e. a craft that gains its lift from the ground rather than from wings (foils). **Ground waves** are

- waves from an earthquake that stay very close to the surface
- radio waves that pass from transmitter to distant receiver by bending along the Earth's surface.

group speed the speed of transfer of energy in a wave pulse. This may not be the same as the wave speed (phase speed) if the wave consists of components which differ in wavelength.

Gruneisen's law The ratio of a metal's expansivity to its specific thermal capacity is a constant that does not depend on temperature.

guard ring large, flat, metal sheet that surrounds, and is in the same plane as, the electrode it 'guards'. It gives a uniform electric field over the whole of that electrode, and prevents any confusing edge effects. There are many similar equivalent devices in modern systems, including thermal ones.

guard wires earthed single wires or nets below high-voltage cables. If a cable breaks, it is earthed by the guard wire before it reaches the ground – and so presents no hazard.

Guillaume, Charles (1861–1938) metrologist, Director of the International Bureau of Weights and Measures, who invented the nickel–steel alloy Invar, a substance with a very low expansivity and of much value for measures and instruments. For this he received the Nobel Prize in 1920.

Guillemin effect form of magnetostriction in which a bent ferromagnetic bar tends to straighten in an axial magnetic field

Gunn diode device that uses the Gunn effect to produce electric oscillations (ac) in the microwave region while working at constant voltage (dc). It also has applications in optoelectronics.

Gunn effect the *negative resistance* in a sample over a certain range of applied voltage, so the characteristic curve is as shown. Over this range, free electrons are passing to a higher band where they cannot move so freely, so are no longer available for conduction. Very few solids show the effect, discovered by John Gunn in the 1960s – cadmium sulphide and gallium arsenide are the most common ones that do.

GUT Grand Unified Theory

gyration rotation about an axis. See also *radius of gyration*.

gyrator electronic element, most common in microwave circuits, which reverses the phase of signals that pass one way but leaves as they are those that pass the other way. Thus, it may cause capacitance to appear as inductance, and the reverse.

gyrocompass continuously driven gyroscope with a feedback system to keep its axis along the (true) meridian (north-south line). The system does not depend on magnetic effects (which cause many problems to navigators), but it does not work very well close to the poles.

gyrodynamics the science and applied science of spinning systems, in particular those with some gyroscopic features

gyromagnetic effects effects that link an object's magnetisation and its spin. See, for instance, *Barnett effect, Einstein and de Haas effect* and *Larmor precession*.

gyromagnetic frequency See *Larmor frequency*.

gyromagnetic ratio g or γ, no unit the ratio of the magnetic moment of a particle to its angular momentum. In the case of an electron (charge Q_e and mass m), the orbital value is $Q_e/2\ m$; that for its spin is Q_e/m. It is the same as the *Landé factor*, or Landé g-factor, but applied to an elementary particle.

gyroscope invented by Foucault in 1852, a fast spinning flywheel mounted on a low friction axis and in gimbals so that the axis can be in any direction. Once the spin axis is set, it resists change. In the same way, a spinning top, for instance, remains upright until the speed of spin drops below a certain point; this is the effect of **gyroscopic inertia**. The effect is at the heart of the gyrocompass (see also *laser gyroscope*), and in many other stabilising systems (including bicycle wheels).

If an outside force is strong enough to cause the axis to move, the axis will move at right angles to the force. This effect is **gyroscopic precession**. While spinning tops have a long history, Foucault was the first to explore the effects scientifically, in the middle of the last century, and produce the modern design.

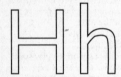

h non-SI unit prefix symbol for hecto-, 10^2

H a) symbol for henry, unit of inductance

 b) chemical symbol for hydrogen

h **a)** symbol for the Planck constant

 b) symbol for height (or altitude or depth)

 c) symbol for specific enthalpy

ħ symbol for the Dirac constant, $h/2\pi$

H **a)** symbol for enthalpy

 b) symbol for Hamiltonian function

 c) symbol for a 'heat function'

 d) symbol for the Hubble constant

 e) symbol for magnetic field strength

 f) symbol for degree of magnetisation (with subscript)

hadron one of a large group of elementary particles made up of the baryons and massive mesons (the pions and upward). Hadrons interact mainly by the strong nuclear force, but experience all three other forces to some extent.

hafnium Hf transitional element in period 6, a dense metal with proton number 72, after many clues first found in spectra in 1923 by Dirk Coster and György von Hevesy and named after Copenhagen (Latin: *Hafnia*). Relative density 13.3, melting and boiling temperatures 2230 °C and 4600 °C; most common isotope ^{180}Hf. Used in nuclear pile control rods and lamps, and as a scavenger for oxygen and nitrogen.

Hagen's formula See *Poiseuille formula*.

hahnium a) name (after Otto Hahn) for dubnium for a few years, until the Dubna team were able to prove they had created this element (number 105) first

 b) name (also after Otto Hahn) for hassium (element 108) until 1997

Hahn, Otto (1879–1968) nuclear chemist who won the 1944 Nobel Prize in chemistry for his work on nuclear fission. With Lise Meitner, he 'split the atom' by bombarding a uranium sample with neutrons to produce barium (and other products).

halation the faint glow that occurs around the spot on the screen of an oscilloscope. Its cause is internal reflections in the thick glass. A similar halo effect (hence the name) sometimes appears around the image of a bright object in a photo as a result of internal reflections in the camera lens.

half-cell one of the two electrodes of a chemical cell and the electrolyte that surrounds it. We can regard all cells as consisting of two half-cells combined.

half-life or **half-life period** $t_{1/2}$ the time needed for half of a sample of radioactive substance to decay. After two half-lives, only a quarter of the original sample is left; after three, there is only an eighth; and so on. Half-life relates to decay constant λ and to mean life τ thus:

$$t_{1/2} = 0.693/\lambda$$

$$t_{1/2} = 0.693\ \tau$$

The relationships follow from the fact that radioactive decay is exponential; they therefore apply to any such decay process (for instance, the decay of current in a CR circuit).

half-period zone circular band on a spherical wave front, used in the Fresnel analysis of diffraction. Its radius r is

$$r = \sqrt{(n\ d\ \lambda)}$$

Here λ is the wavelength of the radiation, d is the distance between the centre of the front and the observation point, and n (the order of the zone) is an integer.

Each zone is half a wavelength closer to the observation point than its neighbour is.

half-value thickness the thickness of an absorbing material which will cut input radiation to one half of its intensity. Rather than metre, the unit is sometimes the kilogram per square metre (mass per unit area).

half-wave dipole common type of aerial, often a straight wire. A passing radio wave of wavelength approximately twice the aerial length sets up resonance in the wire, with a current antinode at the centre and voltage antinodes at the ends. See also *dipole*.

half-wave plate slice of a birefringent substance (such as quartz), the thickness of which is designed to produce a 180° phase difference between the output ordinary and extraordinary rays; see *polarisation*.

half-wave rectifier or **half-wave circuit** circuit which passes only alternate half waves of an input ac signal and blocks the others; the output is a very 'lumpy' dc. This is the simplest form of rectifier – see *rectification*.

half-width one or other of the two measures of a spectral or resonance peak shown in the sketch

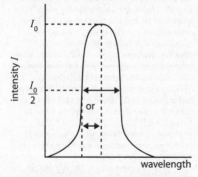

Hall constant R_H for a substance, the potential gradient in the Hall effect for unit field strength B and unit current density j

Hall effect discovered in 1879 by physicist **Edwin Hall** (1855–1938), the appearance of a voltage (the **Hall voltage** V_H) between the sides of a current-carrying conductor in a transverse magnetic field; this is an instance of the motor effect. Current, field, and Hall voltage should all be at 90° for greatest effect. If current and field are not at 90°, the **Hall angle** is the angle between the current and the net electric field (i.e. the resultant of the applied and Hall fields).

The size and direction of the voltage (which is useful only in semiconductors) gives

information about the sign and number of charge carriers. The relation (using the *Hall constant* R_H) is

$$V_H = R_H \, B \, j \, t$$

Here t is the thickness of the sample in the direction of the effect.

The effect also allows one to measure the mobility of the charge carriers – see *Hall mobility*.

The **quantum Hall effect** occurs at very small scales – as well as being forced to the edge of the conductor in a magfield, the conduction electrons must move in circular paths of given radii (energy levels). In the case of *graphene* only, there is a **relativistic Hall effect** – because graphene has a very high conductivity, the conduction electrons move at speeds close to that of light; this raises their masses.

Hall, John (1934–) physicist who shared half the 2005 Nobel Prize with Theodor Hänsch for their work that helped develop laser-based precision spectroscopy

Hall mobility the ratio of the Hall angle (see *Hall effect*) to the magnetic field strength involved

Hamiltonian or **Hamiltonian function** H
a) in classical physics, for a set of n particles of positions q and momenta p, the sum of the products of p and dq/dt minus the Lagrangian function L. L is the difference between the set's kinetic and potential energies. If L does not change with time, H is the total energy of the set – and then relates to the classical equations of motion.
b) in quantum physics, an operator which gives the equation of motion for a wave function

Hamilton's equations form of Lagrange's equations based on the momenta of particles in a set rather than on force; they are widely used in quantum physics. For each particle of position q and momentum p, the equations are as follows:

$$dp/dt = \delta H/\delta q$$
$$dq/dt = \delta H/\delta p$$

H is the total energy of the set.

Hamilton principle general version of the principle of least action: the time integral of the Lagrangian function of a conservative system has a stationary value

Hamilton, William (1805–1865) mathematician who, by the age of fifteen, had already carried out new work. While still an undergraduate, he became professor of astronomy at Trinity College, Dublin and the Irish Astronomer Royal. His main work in physics concerns the wave theory of light and the mathematics of mechanics.

Hänsch, Theodor (1941–) physicist who shared half the 2005 Nobel Prize with John Hall for their work that helped develop laser-based precision spectroscopy

hard not easy to affect or change. A **hard magnetic material**, such as many steels, is one that retains a degree of magnetisation in the absence of an outside field, even in a reverse field: see *hysteresis*.

hardness of a surface, the resistance to scratching or penetration. A common test is the Brinell test, which measures the load on a defined ball needed to produce a given dent. For **hardness scale**, see *Mohs scale*.

hard radiation ionising radiation of high energy (and therefore high penetrating power)

hard vacuum high vacuum, i.e. one with a particularly low pressure

hard vacuum tube discharge tube in which there is so little gas that electric conduction is almost solely by electrons (cathode rays); see *thyratron*.

Hare's hydrometer device able to measure the relative density of a liquid by balancing a column of it against a column of pure water

harmonic pure (sinusoidal) signal or tone whose frequency is a multiple of the fundamental frequency v_0 for the system in question. The fundamental itself is the first harmonic; the second harmonic has frequency $2v_0$; and so on. For a given signal or note, a Fourier series describes the overall complex effect as a full set of harmonics. (However, a given signal may contain non-harmonic frequencies as well.)

harmonic analyser device that takes a signal or note as input, and outputs the Fourier series; in other words it displays the values and relative strengths of the harmonics.

harmonic distortion See *distortion*.

harmonic motion regularly repeated set of actions or values that can break down into a set of sine waves (Fourier series); see *simple harmonic motion*.

Hartmann formula relation between the refractive constant n of a substance and the wavelength λ of an input radiation. There are various forms; below are the most general and the version for a prism.

$$n = a + b/(\lambda + c) \, d$$

For a prism:

$$dn/d\lambda = -e/(\lambda - f)^2$$

In both cases, a, b, c, d, e and f are constants.

Hartmann, Johannes (1865–1936) was a physicist and astronomer whose main work was in optics, lens evaluation, and wavelength standards.

Hartmann oscillator apparatus that generates high power acoustic (ultrasonic) waves in fluids, based on edge tones

hartree atomic (non-standard) unit of energy, the potential energy of an electron in the first shell of the Bohr atom. The value is 4.360×10^{-18} J (27.21 eV). Douglas Hartee (1897–1958) was a mathematician and physicist who developed numerical analysis for the needs of atomic physics.

hassium, Hs trans-uranic rare earth in the actinide series, period 7, proton number 108, first made in 1984 by Peter Armbruster and team, who gave it the name from the Latin *hassia* = Hesse, the German state where they worked. The most stable isotope is ^{269}Hs – whose half-life of less than ten seconds means little is known about the metal.

Hawking, Stephen (1942–) nuclear physicist whose major work has been in astrophysics and gravitation. Amongst other things, Hawking arrived at the concept of **Hawking radiation**, in 1974, as the process that would inevitably lead to the decay of mini black holes (should these exist). Though this radiation, caused by quantum tunnelling, is still unproved, the work was the start of the field of quantum gravity.

h-bar how to pronounce $\hbar$, the symbol for the Dirac constant

H-bomb hydrogen bomb – i.e. fusion bomb

head a) symbol h, unit: metre, m – the vertical height of a liquid column, giving a pressure at the bottom. The static pressure is the product of the head, the liquid density and the acceleration of free fall: $p = h \, \varrho \, g$.
b) the input end of a system. A **head amplifier** is a preamplifier placed at or near the input and used to boost the signal strength before transfer and further processing.
c) the front of an object. The **head wave** of an object moving at supersonic speed is the shock wave from its front.

health physics the branch of medical physics concerned with the health and safety of people working with, or otherwise exposed to, ionising radiations and neutrons. It includes biology, chemistry and ecology as well as physics.

hearing the sense by which animals gain information from sound waves (pressure waves); in most cases the ear is the specialised organ. A **hearing aid**, a device to assist hearing in impaired people, consists of a microphone and amplifier, with the output going to a tiny speaker in or behind the ear.

As with the other senses, it is convenient to define standards of perception; and in hearing, as in the others, few people fully meet the standard. **Hearing loss** is defined, at a given frequency, as the ratio of a person's threshold (minimum detectable power level) to the standard; the unit is the decibel.

heat

common name for thermal energy, that energy involved in the temperature and state changes of matter samples. Below are the main expressions for energy transfer ΔW (formerly Q) into and out of matter samples.

Changing the temperature of a sample by ΔT:

$$\Delta W = C\Delta T$$

C is the sample's **heat capacity** (thermal capacity is a better term), unit joule kelvin^{-1}, J K^{-1}; it is the product of its mass and the specific thermal capacity of the substance (c, unit J kg^{-1} K^{-1}).

Changing the state of a sample:

$$\Delta W = L$$

L is the sample's **latent heat**, or, better, latent thermal capacity, unit J. L is the product of the mass and the specific latent thermal capacity, l, whose unit is J kg^{-1}.

The word latent means hidden: during the change of state of a pure substance, no effect can be sensed as the temperature remains constant. (A former name for the energy of temperature change is **sensible heat**.) For the methods of thermal energy transfer, see *conduction, convection, evaporation,* and *radiation.* See also *insulation.*

Philosophically, there remains much confusion about heat and temperature and work and energy. Until the nineteenth century, most scientists believed in the caloric theory: they regarded heat as being the particles of a weightless, invisible fluid. The efforts of scientists such as Davy, Rumford, Joule and Helmholtz showed that heat and work are facets of the same thing; also Joseph Black exploited and explained how heat and temperature differ, over two hundred years ago.

There are very good reasons for not using the word 'heat' at all, except as a verb meaning to add energy to a sample to raise its temperature or to change its state. See also *thermal*.

heat engine system designed to produce mechanical energy (force) from the burning of a fuel; see *engine* and see also *Carnot cycle* and *Otto cycle*.

heat exchanger device designed for the efficient transfer of energy between a hot fluid and a cooler one. The hot fluid passes through a system of narrow metal pipes immersed in the cooler fluid.

heat function also called thermodynamic function, an expression for the total energy of a gas sample. See the specific types.

heat pipe device that allows the efficient net transfer of energy from a region at high temperature to one at a lower temperature. It contains a small volume of working fluid (often water) which boils at the hot end; the vapour diffuses to the other end, where it condenses, ready to diffuse back again through some kind of wick. Thus, there is energy transfer from the hot end to the cool end and no need for a pump – a continuous process (and a very efficient one), with the fluid cycling between states.

heat pump device for the transfer of thermal energy to a higher temperature region. The best-known example is the refrigerator.

heat shield an extra, dense surface layer on the front of a spacecraft designed to prevent over-heating and ablation of the craft itself from the friction of high speed entry into an atmosphere. The shield instead vaporises and sheds energy.

heat sink

a) system, useful in thermodynamic analysis, able to take in any amount of energy without change of temperature

b) lump of metal, often fitted with fins, attached to a temperature-sensitive object such as a chip, in order to conduct away and radiate excess energy

heat transfer the transfer of thermal energy ('*heat*'). See *conduction*, *convection*, *evaporation* and *radiation*, these being the main methods; and *insulation*, the attempt to reduce energy transfer by any of these means.

For any of the four methods of heat transfer, one can devise a **heat transfer coefficient** to describe the transfer rate. This is the rate of energy flow through unit area per unit temperature difference; the name is conductance in the case of conduction and emissivity for radiation. The unit is the watt per square metre per kelvin, $W\ m^{-2}\ K^{-1}$.

heating effect the temperature rise in a conductor carrying a current I against resistance R. The rate of transfer, p, is $I^2 R$. The effect is used in the very many kinds of electric '**heater**', in fuses and in incandescent lamps.

Heaviside layer the E-layer of the atmosphere

Heaviside, Oliver (1850–1925) electrical engineer, mathematician and physicist; his main work was in telegraphy and radio communications. He identified the Heaviside layer (E-layer) of the upper atmosphere.

heavy hydrogen common name for deuterium, ^{2}H

heavy ion in nuclear physics, any ion greater than about that of carbon. There is no clear definition – for instance, the heavy ion accelerators at Darmstadt work with ions from hydrogen to uranium.

heavy water deuterium oxide, D_2O, the common name for deuterated water, water in which most or all of the hydrogen atoms are replaced by deuterium atoms. (People also use the name for tritiated water, water containing a high fraction of T_2O, T being tritium.) Heavy water makes up about 0.014% of natural water, and has many uses, including as the moderator and coolant in the **heavy water reactor**, a type of nuclear reactor.

hecto- h unit prefix, best avoided, for 10^2, as in **hectare** – an area of 100 ares (one are being 100 square metres)

height h (or z) distance of a point above some reference level (for instance, mean sea level), a surface, or some other point. See also *barometric height*.

Heisenberg, Werner (1901–1976) physicist who won the 1932 Nobel Prize for his work (with Max Born) on the development of quantum physics (quantum mechanics). Heisenberg refused to accept any visual model of atoms, and instead used matrices to work out their behaviour and properties.

Heisenberg force the exchange of charge between nucleons. It is one of the three types of exchange force that may appear between nucleons.

Heisenberg's uncertainty principle an outcome of quantum theory stated by Heisenberg in 1927 – it is not possible to measure any pair of conjugate variables with such precision that the product of their errors is less than $h/2\pi$ (h, the Dirac constant, where h is the Planck constant). This is the

case because any measuring process involves an interaction which must alter the values to some extent.

Important pairs of such variables in the case of a particle are its energy and time (see, for instance, virtual particle), and its position and momentum. With, in the latter case, the uncertainties in position and momentum being Δq and Δp, the value $\Delta q \, \Delta p$ cannot be better than around h.

A crucial philosophical implication is that it is not therefore possible to have such perfect knowledge of a system that one can predict its future fully. (To do that was a great hope of classical physics.) See also *wave packet* for a wave view of a particle that gives a reason for this result. See also John *Bell*.

helical shaped like a helix, as in the case of a **helical spring** (often, wrongly, called spiral)

helicity the component of the spin of an elementary particle along its direction of motion: this may be either left- or right-handed.

helicon circularly polarised electromagnetic wave of low radio frequency that can transfer through a metal in a constant magnetic field. Helicons are of value for research into metal properties.

helion the nucleus of ^{3}He (as the alpha is the nucleus of ^{4}He)

helium He inert gas (group 0) in period 1, proton number 2, discovered in the spectrum of the Sun in 1868 by Pierre Janssen and then by Norman Lockyer, and the second most common element in the known Universe; the name comes from *Helios*, the Greek sun god (with the ending -ium as Lockyer – wrongly – thought this is a metal like sodium). Density 0.18 kg m^{-3}, melting and boiling temperatures 0.95 K (at at least 2.6 MPa) and 4.2 K; most common isotope ^{4}He.

On Earth, helium comes in abundance from the alpha decay of nuclides in rocks; it can sometimes be as much as 7% of natural gas. Helium has some most unusual features – see, for instance, *helium I and helium II* (the two liquid phases, both quantum fluids), and the *Joule–Thomson constant* – and its solid, liquid and gas phases have almost the same refractive constant.

Main uses: cryogenics (including cooling superconductors), giving lift to balloons and airships, high pressure breathing systems, and to protect such industrial processes as welding and growing semiconductor wafers

helium I and helium II two forms of the liquid of the isotope helium-4: helium I exists from the boiling temperature (4.22 K) down to 2.19 K (the lambda temperature); it is a common refrigerant. Helium II, the form existing below 2.19 K, is a quantum fluid: it has very high thermal conductivity and zero liquid friction, see superfluid. A **helium film** (Rollin film) will cover any surface partly in contact with helium II; indeed through it, an open container of the liquid will empty. Liquid helium-3 does not show any of this behaviour.

helix curve shaped like a line on the surface of a tube that is moving at constant speed parallel to the axis

Helmert's formula rule of thumb relationship that gives the value of g (the acceleration of free fall near the Earth's surface) in terms of latitude and altitude. This does not (and cannot) cope with local and anomalous variations.

helmholtz non-SI unit of dipole moment per unit area, 3.335×10^{-10} coulomb per metre

Helmholtz coils pair of identical coils used to provide a fairly uniform magfield in the region between them. The coils are parallel at a distance apart equal to their radius; they carry the same current in the same direction.

Helmholtz electric double layer concept devised by Helmholtz to explain the appearance of a voltage between two different surfaces in contact; he proposed a layer, one molecule thick, at the interface, with the molecules polarised to become positive on one side and negative on the other. Lenard later extended the theory.

Helmholtz free energy A (or F) unit: joule, J also known just as free energy, or as the **Helmholtz function**, the expression $W_i - T S$ for a system in thermodynamics; the 'system' is a sample of gas at constant volume. In words, the Helmholtz value for free energy (i.e. that available for transfer outside the system) is the system's internal energy (which is not available) less the product of temperature and entropy. Compare with *Gibbs free energy*.

During an isothermal change (one at constant temperature), the energy transfer dW is the same as the change of Helmholtz free energy, $dA : dW = dA = dW_i - T \, dS$.

The **specific Helmholtz free energy** a is A per unit mass, unit: joule per kilogram.

Helmholtz galvanometer galvanometer that uses a pair of Helmholtz coils to produce the central field, rather than a single coil as in the standard tangent galvanometer

Helmholtz, Hermann von (1821–1894) scientist whose work was important in very many fields – in particular, mathematics, physics and physiology. In physics, he invented various useful devices and systems as part of his research into acoustics, electrics, energy and the senses.

Helmholtz resonators set of bulbs of air that differ in size; the volume of the neck of each is much smaller than the contained volume. Each resonates at its own frequency to produce sound, when excited. The purpose of the set was to explore acoustics and sound systems, much like a set of tuning forks.

henry H the unit of inductance. It is the inductance of a single loop of wire when there is 1 V back emf when the current changes by one ampere per second. An alternative (and equivalent) definition is the inductance of a loop with unit flux linkage (one weber, Wb) when it passes unit current.
Thus $1 \text{ H} = 1 \text{ V}/(\text{A s}^{-1}) = 1 \text{ V s/A} = 1 \text{ Wb/A}$

Henry, Joseph (1797–1878) physicist whose best known work was in electromagnetism (in particular self-induction and sparking, though the electric motor and the relay are among his inventions). He discovered electromagnetic induction at the same time as, but independently of, Faraday. Henry also worked on telegraphy and the design of electromagnets. He found that sunspots are cooler than the surface around them, and he showed that sparking is often an alternating effect; this later led to the work of Maxwell and Hertz on radio communication.

Heron (Hero) of Alexandria (c.10 –c.75 ce) inventor and mathematician, best known for a design of a simple steam engine (that works but could take no load because of friction). Only his books remain – but Heron is a common name of the time, and historians date him anywhere in five centuries from about 250 bce to 250 ce.

hertz Hz named after Heinrich Hertz, the unit of frequency of periodic changes such as waves, the (cycle) per second. (The unit of a non-periodic repeated change, such as radio-active decay, is the (event) per second, s^{-1}.)

Hertz, Gustav (1887–1975) physicist who shared the 1925 Nobel Prize with James Franck for the experiment that confirmed the quantisation of atomic energy levels

Hertz, Heinrich (1857–1894) physicist, uncle of Gustav Hertz, student of Helmholtz and Kirchhoff. In 1886, he proved Maxwell's theories of electromagnetic radiation by a series of experiments in which he discovered radio waves (**Hertzian waves**). He then did much research on these waves.

Hertzian oscillator oscillator in which there is a small gap in the rod linking two capacitative spheres or plates. Sparking across the gap when the voltage is high enough involves an ac (see *Henry*); its frequency depends on the capacitance and inductance of the system. Radio waves of the same frequency radiate outwards; a similar spark gap a few metres away forms a simple detector. The Hertzian oscillator is still used occasionally. As a result of this work, radio waves are still sometimes called **Hertzian waves**.

Hertzsprung-Russell diagram fundamental graph of stellar brightness (luminosity or absolute magnitude) against surface temperature (spectral type or colour). Most stars appear somewhere in a broad band between bright hot giants and faint cool (red) dwarfs: this is the main sequence in which those stars spend at least 90% of their lives. Other types of star appear in other regions of the diagram; evolution to these regions is very rapid. Ejnar **Hertzsprung** (1873–1967) and Henry **Russell** (1877–1957) were astronomers who independently produced this high value tool in the early 1910s.

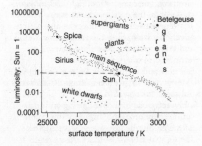

Hess, Victor (1883–1964) physicist who received a share of the 1936 Nobel Prize for

his work on cosmic rays. The **Hess diagram** is a graph of stellar luminosity against brightness (absolute magnitude); used mainly to analyse star clusters, it shows a curve for the stars of each spectral type.

hetero- prefix that denotes a difference (as opposed to homo-, that indicates sameness). Thus, a **heterodyne receiver** is a radio set whose output is the beat frequency signal that results from mixing the input with a radio frequency signal generated within the set.

heterodyne or **heterodyne receiver** type of radio receiver – see *beat receiver* and *hetero-*

heterogeneous mixed in type, or hybrid, as opposed to homogeneous. Thus, **heterogeneous multiplexing** is a form of multiplexing in which different signals pass through the system at different speeds.

heterogeneous radiation multichromatic radiation of a given type – i.e. it contains a number of frequencies.

heterogeneous reactor reactor in which the fuel is separate from the moderator

heterogeneous strain the strain on a substance under stress, where there is a non-linear relation between the coordinates of each point in it before and after the stress starts; compare with *homogeneous strain*.

heterojunction the interface between two crystals in contact of the same type (structure) but different substance, used in one form of semiconductor laser action

heteropolar generator generator in which the conductors carrying the induced voltage move through an effectively alternating magnetic field; this means that the voltage alternates. This is the standard type of alternator, ac generator.

Hewish, Anthony (1924–) astronomer who led the team that discovered pulsars in 1967 (though his student Jocelyn Burnell made the actual discovery). For this and for his other pioneering work in radio astronomy, he shared the 1974 Nobel Prize with Martin Ryle.

hexagonal system See *crystal system*.

hf high frequency

hidden variables a class of theories which try to explain the 'random' nature of quantum events by claiming that other variables – ones that we don't yet know about – control each such process.

Higgs boson or **Higgs particle** very massive boson, proposed as the exchange particle of the *Higgs field*. This appears in the Standard Model in some modern electroweak theories

Higgs field

(which try to unify the weak nuclear and electromagnetic interactions). The boson – or bosons, for there may be more than one – has not yet definitely been found; the mass is likely to lie over 1 TeV/c^2. See also *large hadron collider* and *wormhole*.

Theoretical physicist **Peter Higgs** (1929–) proposed the particle in 1963, as did others around the same time. The **higgsino** is the massive superpartner of the Higgs boson – see *supersymmetry*.

Higgs field type of fundamental field proposed by Peter Higgs and others in the late 1960s to provide elementary particles with mass. Though part of the electroweak theories, the field differs from others in being a scalar field (so its exchange particle, the Higgs boson, has zero spin); it also has a higher energy when zero than when non-zero (so the elementary particles must have gained their mass very early in the history of the Universe).

higher order derivative in calculus, derivative which follows from the differentiation of a derivative. The nth order derivative is dny/dxn, the derivative of the $(n-1)$th derivative.

high frequency hf in the radio wave frequency range 0.3–3 MHz

high-temperature gas-cooled reactor htgr design of gas-cooled reactor in which the coolant is helium gas

high-temperature superconductivity type II – even maybe type III – superconductivity, a major research field in the 1980s that has made little progress since then

high tension ht out-dated term for high voltage, 'tension' being an old word for voltage

Hilbert space multidimensional non-Euclidean space widely used in the analysis of wave functions. **David Hilbert** (1862–1943) was a mathematician who spent much time exploring the logic of the geometry of such spaces.

histogram graphical way to show frequency distribution of data grouped into classes. It is a bar chart that closely relates to the frequency polygon.

Hodgkin, Dorothy (1910–1994) crystallographer who received the 1964 Nobel Prize for chemistry for her work on the structure of vitamin B12. She had earlier been a pioneer in the use of x-ray crystallography, and later was the first person to use a computer to solve biochemical problems.

Hofstadter, Robert (1915–1990) physicist who worked on the size and structure of nucleons and was awarded (with Mössbauer) the 1961 Nobel Prize. He also worked on semi-conductor detectors and, more recently, in astronomy.

hole unfilled electron site in the band (energy level) model of a solid; an electron must have gained energy to escape from that site. In an electric field, holes migrate – in effect, by the motion of electrons the other way: this is **hole conduction**). It is the main process of conduction in p-type semiconductors, for these have more holes than free (conduction) electrons. The effect is rather like positron conduction, but the mobility of holes (as found by the Hall effect, for instance) is much less than that of free electrons or positrons. See also Dirac's hole theory.

holmium Ho rare earth from the lanthanide series, period 6, proton number 67, one of the seven elements found (1878, Jacques Soret and, a few months later by a different method, Per Cleve) in a rock in a quarry in Ytterby near Stockholm (hence the name). Relative density 8.8, melting and boiling temperatures 1470 °C and 2720 °C; only one isotope: ^{165}Ho. Holmium has the highest magnetic moment of any natural element – so it and its alloys and compounds have many uses in that field; it is also sometimes used in control rods in nuclear piles as it is a very good absorber of neutrons.

hologram photographically fixed interference pattern (see *holography*) that shows no obvious visual information, but in fact contains a vast amount of data. From it, one can reconstruct an exact three-dimensional image. It is common to use laser light both to produce the hologram and to reconstruct the image; however, there are natural light techniques as well. As the figure opposite shows:
a) to prepare the hologram, the plate is exposed both to direct laser radiation and to radiation reflected from the object, thus forming the interference pattern;
b) to view the hologram, light from the same laser is used to give a real three-dimensional image of the object.

holography method of recording an image of an object photographically by storing details of the amplitude, phase and wavelength of the light at all points in the image plane (rather than just amplitude and wavelength as in normal photography). Modern holography works with wave forms other than light

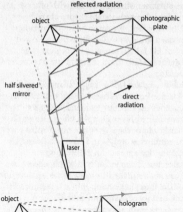

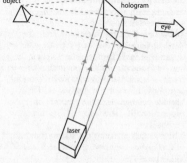

It is best – but not essential – to use laser light with holograms.

spinning polygonal mirror to steady the image formed by a lens.

One problem is that a single frame of a three-dimensional hologram of image area 100 mm square requires 25 GB of data, however. Initial work therefore had to sacrifice vertical parallax (i.e. the holographic effect in the vertical plane), halve the viewing angle from the normal 30°, cut the image size, and reduce the image resolution. Other research efforts have led to holographs of computer-designed images (valuable to architects, for instance) and holographic imaging of the tissues inside the body and structures deep inside aircraft bodies.

There are many other uses in science, applied science and the fields of engineering. However, the technique remains fiddly and costly.

homo- prefix that denotes uniform, of the same type; compare with *hetero-*

homogeneous made up of similar elements or components, or uniform in type and nature; compare with *heterogeneous*. Thus, **homogeneous multiplexing** is multiplexing where all the signals concerned transfer at the same rate.

homogeneous radiation monochromatic radiation (i.e. radiation having only one wavelength)

homogeneous reactor reactor in which fuel and moderator are finely mixed

homogeneous strain the strain shown by a sample under stress in which there is a linear relationship between the before and after coordinates of all points

homogeneous substance substance with the same chemical and physical properties at every point (like water, but not like wood)

homopolar generator a type of generator, such as the Faraday disc, in which the conductor(s) across which appear the induced voltage always move in a constant field. Compare with *heteropolar generator*.

Hooke, Robert (1635–1703) prime mover in the 'Golden Age' that involved many other well known scientists (such as Boyle, Newton and Wren) and the founding of the Royal Society of London (1663). His major works were in many fields, including, in physics, how to measure time better, observing the weather and measuring aspects of it, optics, gravity, and microscopy. His discoveries, inventions and activities covered vast areas of science and applied science, though (because of personal and political problems) he

alone; indeed, the concept, being based on interference, applies to any wave radiation.

Gabor in 1948 first proposed the theory of the technique, but it was not until the mid-1960s that laser technology made it feasible – lasers can provide the highly coherent stable light source needed.

Various techniques have been tried for making three-dimensional moving holographic images. One successful one announced early in the 1990s can produce such images floating in space, either in real time or from a recording (in both cases output from a supercomputer). The input laser beams pass through an acousto-optic crystal, one whose local refractivity changes as a result of an input sound wave. The sound wave carries the picture information, so the crystal adds the moving image to the beam. The laser beam reflects from a fast sometimes found it hard to get his work published. He invented the compound microscope, and it was his notion of an inverse square law for gravity that led to Newton's theory. **Hooke's law** – that an object's strain varies with its stress (up to the elastic limit) – followed his work in the 1670s on springs for clocks and watches. It applies to any elastic deformation under stress: the deformation (strain) is proportional to the cause.

horizontal component B_0 (or B_H) the horizontal component of the Earth's magnetic field: one of the three magnetic (or geomagnetic) elements of any point near the Earth's surface

hormesis unproved concept that very small doses of something harmful, such as ionising radiation, may be of value to living tissue

horn funnel-shaped aerial – see also *wave guide.*

horsepower hp old unit of power, now taken to be 746 watts, introduced by Watt to help sell his new steam engine in 1776

hot at a higher temperature than normal (the details depend on the situation)

hot-cathode tube electron tube ('valve') in which the energy given to the electrons is the thermal energy associated with the high temperature of the heated cathode

hot spot bright central region of an image (e.g. if projected), which can make viewing difficult

hot-wire relating to the physics of wires at high temperatures. There are many systems whose design depends on the characteristics of hot wires, as follows.

hot-wire instrument or **hot-wire meter** meter whose reading depends on the increase of length of a fine wire as a result of the heating effect of the current in it. A spring keeps the wire taut, so that its expansion can move the needle on the scale.

hot-wire microphone microphone with a fine heated wire at the centre. The variation of resistance of the wire with incoming sound waves gives a corresponding signal in the outside circuit.

hovercraft 'surface effect' vehicle that floats on a cushion of air inside its 'skirt', the air being kept at high pressure by a fan. The hovercraft is therefore not a true aircraft, as it does not gain lift from a foil: it rests on the air cushion on the surface, so can move with very low friction.

Hoyle, Fred (1915–2001) astronomer and populariser of science, who (with others) devised the steady state theory of the Universe (but, for some unknown reason, missed the 1983 Nobel Prize that followed)

ht (or **HT**) high tension, i.e. high voltage

htgr (or **HTGR**) high temperature gas cooled reactor

Hubble, Edwin (1889–1953) astronomer who was the first to realise (in 1923) that there are galaxies outside our own and to equate red-shift with the Doppler effect. He worked hard to have the Nobel Prize in physics cover astronomy as well – and died just before the committee agreed. The **Hubble space telescope** is named after him; conceived in 1946 and launched in 1990, it has become one of the most important optical telescopes in the history of astronomy; it will cease working soon after 2010. See http://www.spacetelescope.org/

Hubble law Galaxies move away from the Earth at velocities that vary directly with distance. It is far from certain whether the law – a foundation of the expanding Universe model – is valid; however, it makes a useful assumption for further discussion. The **Hubble constant** H is the constant of proportionality, about 10^{-17} s^{-1}, in the Hubble law.

Hubble radius or **Hubble distance** the radius of the known Universe, around 10^{27} m (10^{14} light years)

hue pure colour in the visible spectrum, relating to a single frequency of light. Compare with *tint.*

Hulse, Russell (1950–) research student of Joseph Taylor, with whom he discovered the first binary pulsar – of great value for exploring general relativity. The two won the 1993 Nobel Prize for this work.

hum ac signal induced in an electronic circuit by a nearby mains line. The **hum frequency** may be the same as the mains frequency or it may be an over-tone; this depends on the characteristics of the circuit.

humidity *d* unit: kilogram per cubic metre, kg m^{-3} measure of the content of water vapour in a sample of air. It is an absolute value. Normally, however, the measure used is *relative humidity*. *Hygrometers* measure either value. See also *dew temperature.*

hunting continual oscillation of a controlled quantity above and below the set value. Proper damping may cure this if it is a problem.

Huygens, Christiaan (1629–1693) Dutch

physicist and mathematician, who developed many of the ideas of Galileo in mechanics and astronomy, as well as working in his own areas, particularly that of optics. He invented the pendulum clock, proposed the wave theory of light, and discovered polarisation.

Huygens construction method of plotting the path of a wave front through an optical system. A wave front can be drawn given the previous one, as each point on the original front is regarded as a new source of radiation, i.e. of a new wavelet, moving out at the speed of the rays in the medium in question. The new wave front is tangential to all those wavelets.

Huygens eyepiece eyepiece consisting of two plano-convex lenses, set apart at a distance that depends on their focal distances. With some modification, it is still the basis of the eyepieces of telescopes and microscopes.

Huygens' principle the concept of wavelets arising from each point on a wave front. It is the basis of the *Huygens construction*.

hybrid combining two elements in a system. A **hybrid integrated circuit** is one that consists of two or more monolithic integrated circuits (i.e. ones on a single substrate) permanently linked in some way.

hydrated electron extremely reactive electron in a potential well in water, the result of secondary emission from water molecules by ionising radiation. The new ions group round the electron and trap it.

hydraulic powered by forces in a liquid trapped in an enclosed space. The parts that provide the power of many machines are hydraulic.

A motor moves a piston in an input (master) cylinder, linked by valves to the output (slave) cylinder in current use. The change of pressure in the output system cause its piston to move in turn.

The **hydraulic brake** and **hydraulic press** are good examples of the principle. In each case, the area of the input piston is much smaller than that at the output. If the pressure throughout the system is the same, this makes a very large output force compared to that input: this is a simple machine, with a high force ratio (mechanical advantage).

hydraulics the science and applied science of liquids under pressure. It includes hydrodynamics and hydrostatics in physics. Such aspects as irrigation are also sometimes in the list.

hydrodynamics old name for fluid dynamics –

the science and applied science of the motion of fluids (liquids and gases, perfect and real)

hydroelectric power station electric power station whose input energy comes from water held at a height by a dam and released as a fast stream to turn a turbine, which turns the generator shaft. Such systems, along with tidal flow stations, produce around a third of the world's electricity; they may affect the appearance of the landscape but (once built) cause no direct pollution.

hydrofoil object shaped to have a large lift force when moving through a liquid (in particular water). The same name applies to boats whose hulls rise out of the water (to reduce friction) when moving at speed because of the lift on the foils under the surface. An efficient hydrofoil craft has only the foils, struts, screw and rudder under water when moving at speed. See also *airfoil*.

hydrogen H at 90%, the most common element – a plasma – in the Universe, though a gas (the least dense gas) on Earth, discovered by Paracelsus in the early sixteenth century and first studied and recognised as an element by Henry Cavendish in 1766; the name means 'water-generator' (as water is H_2O) and the proton number is 1. Density 0.09 kg m^{-3}, melting and boiling temperatures 14 K and 20 K; most common isotope (99.99%) ^{1}H, but see *deuterium*, ^{2}H, also stable, and *tritium*, ^{3}H, radioactive. Hydrogen has many uses, e.g. in bubble chambers, as a fuel and as a refrigerant.

hydrogen bomb thermonuclear bomb, one that quickly releases the energy from the fusion of pairs of deuterium nuclei or of one nucleus each of deuterium and tritium; an atom (fission) bomb provides the high temperatures needed for the reaction. The use of a separate fission system to trigger the fusion process, with methods to direct its shock waves and radiation to the deuterium, was the secret of the early US success in the field. The true design was worked out in the USSR only after analysis of the fallout from US tests from 1952.

hydrogen electrode half-cell that puts hydrogen gas bubbles in contact with hydrogen ions (in a molar acid); the normal metal for the electrode is platinum. The value of the system's potential is taken to be zero, for reference against other half-cells.

hydrogen spectrum the line spectrum of

hydrogen gas, which consists of a number of series; all elements have such spectra, but that of hydrogen is the simplest as the H atom has only one electron.

The study of the Balmer series, the main series of lines in the visible region, led to a relation for wave number σ (the reciprocal of wave length). This was an early triumph for Bohr's quantum theory of the atom, for that had derived this relation from first principles. The general relation for all the spectral series (principal quantum number, $n = 1, 2, ...$) is in the box.

The principal quantum number, n, is the number of the energy level to and from which electron transitions take place to produce a given line. For the Balmer series, $n = 2$ – so the energy level involved is the first one above the ground state ($n = 1$). Also m is the set of integers greater than n (here 2); as m becomes larger, the lines become closer and closer.

$$\sigma = R \left(1/n^2 - 1/m^2 \right)$$

R is the Rydberg constant for hydrogen.

Bohr showed that he could express R in terms of a number of fundamental constants:

c_0	speed of light in free space
h	Planck constant
m_e	electron mass
Q_e	electron charge
ε_0	permittivity of empty space

He found that $R = m_e \, Q_e^4 / (8h^3 \, \varepsilon_0^2 \, c_0)$. The value of this, 1.1×10^7 m^{-1}, is within 0.05% of that found by experiment. Like the Balmer series ($n = 2$), the Lyman ($n = 1$) series includes lines in the visible region; $n = 3, 4, 5$ and 6 give the Paschen, Brackett, Pfund and Humphreys series, all in the infrared.

hydrometer instrument used to measure the density or relative density of a liquid

hydrophone device used to emit and/or collect sound waves under water. Hydrophones are of much use in echo-sounding, sonar and communications.

hydrostatics the study and use of forces and pressures in stationary fluids, i.e. those in equilibrium. It includes the forces and pressures on surfaces partly or fully submerged (immersed) in fluids; see, for instance, the *Archimedes' law*. The **hydrostatic equation** gives the pressure p at a depth h in a

fluid of mean density ϱ in a gravitational field of strength g: $p = h \, \varrho \, g$.

hygrometer device used to measure the humidity (water vapour content) of a sample of air or other gas, or its relative humidity. A **hygroscope** is a crude version, while a **hygrograph** gives a permanent record of values over time.

hyper-charge Y quantum number associated with elementary particles in the context of the strong interaction: the sum of the particle's baryon number and its flavours (strangeness, charm, bottom-ness and top-ness). Both the electromagnetic and strong nuclear interactions conserve hypercharge; weak interactions do not, though in this case there is a different type of hyper-charge. See also *hyperon*.

hyperfine structure the splitting of spectral lines on a very small scale as a result of such effects as isotopic and spin-dependent shifts

hypermetropia same as hyperopia: far or long sight, in which the eye can focus rays from a distant object only behind the retina. Converging lenses can compensate for this defect.

hypernucleus highly unstable nucleus formed when a lambda particle replaces a neutron

hyperon any baryon other than a nucleon with non-zero strangeness and a lifetime greater than about 10^{-23} s – in other words, one that is fairly stable under the strong nuclear force. The hyperon multiplet system consists of the lambda (a singlet), xi (doublet), sigma (triplet), xi (doublet) and omega (singlet) particles, and their antis. These all decay to (anti)nucleons by some weak process: yet high energy (strong) nuclear interactions also produce them in large numbers. The quantum number hyper-charge is used to explain this apparent anomaly: it is conserved in strong and electromagnetic interactions, but not in weak ones The quark model helps understand the hyperons.

hyperopia same as hypermetropia: far or long sight, in which the eye can focus rays from a distant object only behind the retina. Converging lenses can compensate for this defect.

hypersonic of an aircraft or flow speed, greater than about mach 5

hysteresis effect in a system that gives it some kind of memory of past states. The best known case is **magnetic hysteresis**, shown by magnetically 'hard' samples such as some steels; there is an electric analog, and

somewhat the same effects appear in certain cases of deformation (e.g. in car tyres).
Torsional hysteresis is an example of the last: the twisting strain of a wire passed often through a cycle of twisting (torsion) stress tends to lag behind the force.

To explore hysteresis, one passes the sample through a **hysteresis cycle**; this is a complete cycle of forward and reverse force which returns to the starting point. The result, when plotted on a suitable graph, is a **hysteresis loop**. That shown is the magnetic case, plotting applied field H against the sample's magnetisation B. Relevant measures on the graph are:
a) area shaded: the energy transfer in one cycle; the larger the area the 'harder' the material
b) B_s: the saturation magnetisation
c) H_s: the field needed to produce saturation
d) B_r: the remanence, the magnetisation that remains when H returns from Hs to zero
e) H_c: the coercive force, the reverse value of H needed to reduce the remanence to zero

The area of the loop is a measure of the energy absorbed in passing the sample through a cycle; see also *domain*. In practice, this energy raises the temperature of the sample: **hysteresis heating** (as in induction cookers) depends on the effect to provide highly controlled deposition of energy. On the other hand, it also leads to the **hysteresis loss** of useful input energy, a cause of inefficiency in electrical machines. The magnetic cores of such machines are therefore made of 'soft' materials in order to reduce this loss as much as possible.

Hz unit symbol for the hertz, the unit of frequency in the case of a periodic action, the cycle per second

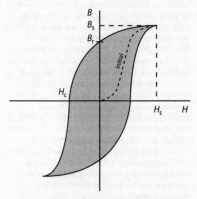

The hysteresis loop of a sample offers much information about the substance concerned.

i symbol for intrinsic semiconductor/semiconduction, in which the process does not depend on any impurity in the semiconductor

I chemical symbol for iodine, element 53

i **a)** symbol for the square root of −1 (solution to $i^2 + 1 = 0$). This is the basis of imaginary and complex numbers. An imaginary number has the form ni; a complex number is $m + ni$ where n and m are real numbers. In Argand space (i.e. in an *Argand diagram*), these two types of number appear respectively on the *i*-axis and between the *i* and real number axes.
b) symbol for the unit vector in the *x*-direction in Cartesian space

I **a)** symbol for action sense c), unit: joule-second, J s
b) symbol for electric current, unit: ampere, A
c) symbol for impulse, unit: newton-second, N s
d) symbol for intensity (e.g. luminous intensity, unit: candela, cd, and intensity of illumination (now called illuminance *E*), unit watt per square metre, W m^{-2})
e) symbol for moment of inertia, unit: kilogram metre squared, kg m^2
f) symbol for the nuclear spin quantum number

IAEA International Atomic Energy Authority, a UN quango founded in 1957 and based in Vienna, whose task is to promote the peaceful use of nuclear physics and limit the military use. The agency and its director general – Mohammed El Baradei – won the Nobel Prize for peace in 2005.

ic integrated circuit, a complete complex microelectronic circuit built into the surface of a chip of semiconductor

ice solid H$_2$O, frozen water, a slightly birefringent crystal with seven different phases (forms). Ice is highly unusual in being less dense than liquid water at the same temperature (and therefore floating on it); in fact the relative density of ice is 0.92. As a result, H$_2$O expands on freezing, with many consequences. This anomalous expansion of water is the result of strong hydrogen bonding.

An **ice calorimeter** is a device used to measure the specific latent thermal capacity of fusion (melting) of ice, or, by extension, that of any other solid. The **ice line** is a graph showing how the **ice temperature** (melting temperature of ice) depends on the pressure (see also *triple point*). By definition, the ice temperature is 0 °C exactly for pure water at standard pressure (101 325 Pa); however, this temperature is no longer one of the standards on the international temperature scale: the temperature of the triple point has replaced it.

Iceland spar mineral form of calcite, calcium carbonate, the first substance found to show double refraction. It was therefore widely used in the past to produce polarisers and analysers in the form of the Nicol prism. Polaroid now fills this role.

ideal perfect, and thus easy to describe and analyse in comparison with a real system of the same type. For instance, an **ideal crystal** has a structure with no defects or cracks; an **ideal gas** is one whose internal energy does not depend on volume and whose particles exert no force on each other. Neither of these exists in practice (though some real systems may be close): we define ideal systems for ease of analysis. For **ideal gas constant**, *R*, see *gas constant*.

idle component **a)** old name for the reactive component of the current or voltage in an ac circuit, the one that takes no power from the supply

b) a gear, wheel, or roller running between two similar parts to transfer motion from one part to the second with no change of speed or direction of motion

igfet insulated gate *field effect transistor*

ignition temperature temperature to which one must raise a workinq substance for it to burn in air (in the case of a fuel) or to start a fusion reaction (in the case of a nuclear fuel). In the former case, the term is sometimes flash temperature, or flash point.

illuminance E (or I) unit: lux, lx the power (rate of energy transfer) of visible radiation reaching unit area of a surface. The lux is the lumen per square metre.

illusion error of perception of reality, in most cases the result of the brain's incorrectly interpreting signals from the senses, in particular the sense of vision (**optical illusion**). In some cases, the illusion has a real cause: as with the bent appearance of a stick half submerged in water, and the change of perceived frequency of the sound of a passing car.

image the point in an optical system from which radiation from an object appears to come. This applies in all kinds of optical systems (indeed even sometimes to electronic circuits), not just those that handle visible radiation. We say an image is real if the radiation actually passes through the image plane (as in the case of the image on a screen from a slide projector); it is virtual if it does not (as with the image seen behind a driving mirror).

The main aim of the science of optics is to describe the image formed by a given system of a given object; we wish to know

- the **image distance** (from the centre, or pole, of the system)
- whether the image is real or virtual
- its size compared to that of the object (magnification)
- whether it is the same way up as the object (upright, or erect) or upside down (inverted).

In practice, the degree and types of aberration are important too, though at an elementary level we work only with ideal systems (such as very thin lenses).

Optics also involves the design of systems to produce required results. In the case of simple optical systems (such as a single mirror or thin lens), the image distance (v)

relates to the object distance (u) and the focal distance (f) thus:

$$1/v + 1/u = 1/f$$

Here we recall that in such a system we measure such a distance from the centre (or pole); we use the convention that a distance to a real point is positive and that to a virtual one is negative.

An **image converter** is a device designed to produce a visible image from invisible radiation (such as infrared, ultra-sound, or an electron beam). In a common design, the original (real) image falls on the surface of a photocathode; the photoelectrons produced give a visible image on a nearby fluorescent screen. An **image intensifier** is a system that raises the luminance (brightness) of a visible image; an **image orthicon** is a type of video camera tube.

Image processing is a major field of modern information technology, important for improving the images received from spacecraft (for instance); in some way or another, the user controls a computer program through a set of standard steps to increase brightness and contrast, edit small patches, and amend colour values. **Image separation** is the standard process of preparing a colour picture for printing: the result is several versions of the picture, as viewed in different lights or for printing in different inks; the standard versions are black, cyan, magenta and yellow.

A charged object distant d from a flat conducting surface experiences a force from the charge induced in the surface. The force is the same as that between the object and an image of it distant d behind the surface. Here, the force is sometimes called an **image force**, while **image potential** is the potential energy of the object as a result of the new field, given its charge Q and the permittivity of the medium ε:

$$W = Q^2/(16\pi \; \varepsilon \; d^2)$$

imaginary number number of the form $n\,i$, where i is $\sqrt{-1}$. See *i*.

imaging making, enhancing, manipulating and interpreting a visual image – still or movie – of something not usually seen (e.g. the surface of a distant planet or a live human heart). Now, it is normal to use a wide range of IT-based techniques for this kind of work – such as in spectroscopy and ultrasonography.

immersion objective an oil immersion lens – a way of raising the magnification of a compound microscope that involves putting the object and the front face of the object lens in oil

impact brief collision between two objects, with an impulsive force between them (in other words, it is the impulse of the force that is significant). People use **impact testing** to measure the hardness of a substance.

impact parameter b in Rutherford scattering (and, by extension, in all Coulomb collisions and the equivalent in other force field interactions), the closest distance between a passing particle and a target particle for there to be no interaction

impedance Z unit: ohm, Ω the opposition of a circuit to alternating current: the ratio of the applied alternating voltage to the alternating current that results. The term is now in more general use for the ratio of an alternating input or effect to the alternating output or result; see also *acoustic impedance.*

In all cases, impedance includes resistance and reactance terms and phase changes; see *electric circuit: ac circuit.* Where one circuit transfers a signal (or power) to a second, **impedance matching** is the process of ensuring maximum power transfer. In the case of a power transmission line, for instance, this requires the line impedance to equal both the output impedance of the source and the input impedance of the load; otherwise, reflections occur, with a loss of transfer efficiency. This is much the same as the process of blooming a lens.

imperial units the 'traditional European' system of units of measurement first started in a formal way in the Roman empire. The base units are the yard (yd, for length), the pound (lb, mass) and second (s, time).

Over the centuries, many local versions developed. These were standardised in Britain in 1824 and spread round the world through the British empire, in competition with the French metric system. ('Imperial' means to do with empire.) Imperial units are not coherent, logical, or easy to remember, however, and cause many problems when applied to science and engineering in particular. SI is slowly coming to replace them, therefore, even in daily life. See also *fps unit system.*

implosion process of bursting inwards, as opposed to explosion (bursting outwards). A container of low pressure gas may implode

on impact; gravitational implosion may be the process that turns a dying star into a black hole.

impulse unit: newton-second, Ns ($= kg\ m\ s^{-1}$) the product of a constant force F and the time t for which it acts, equal to the change of momentum Δp produced. An **impulsive force** is one of non-negligible size that acts for a very short time, as in an impact between cricket bat and ball. If the force is not constant, impulse is the integral of the force with time.

$$I = \Delta p = m\ \Delta\ v = F\ t = \int F\ dt$$

An **impulse generator** is a circuit designed to output a large but brief pulse of current; often this is by the charge and rapid discharge of a capacitor. In any electric or electronic circuit, **impulse noise** consists of sharp separate pulses (which produce clicks in a speaker, for instance). An **impulse current** or **voltage** is a current or voltage pulse: one with a very rapid rise time to the peak value and a fairly rapid decay thereafter.

incandescent hot enough to output a significant power in the form of visible radiation. The traditional electric lamp is of this kind: a current in a high resistance metal coil makes it 'white hot'. In other words, its output covers the whole visible spectrum to give the effect of white light. We call this an **incandescent lamp**; it is far from efficient, however, with only some 2.5% of the electric energy input able to appear as light (about 10 lumens per watt) in the case of a 100 W lamp.

incident inbound, input (or 'hitting') – an **incident ray** is one passing through one medium (vacuum or transparent substance) and hitting the surface of a second. The **angle of incidence** i is the angle between the incident ray and the normal, the perpendicular to the surface, at the **point of**

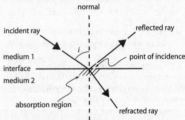

Three things can happen to a ray incident on the surface of a second medium.

incidence. Three things can then happen to the incident ray, as the sketch shows:

- it can bounce back into the first medium (*reflection*), and/or
- it can pass on into the second medium (*refraction*), and/or
- it can be absorbed (*absorption* c)) around the point of incidence.

inclination δ the angle of dip of the Earth's magnetic field at a given point on or near the surface: in other words, it is the angle between the field and the horizontal

inclined plane machine used to raise a load (of weight W) vertically by moving it up a flat slope with angle α. The input force is less than W (there is a 'mechanical advantage'; see *force ratio*) – in fact, the value is W sinα. On the other hand, the distance ratio is greater than 1: the distance moved by the load is greater than the height raised. The wedge and screw are very important forms of inclined plane; in the latter case, the plane wraps round a cylinder or cone.

inclinometer device used to measure magnetic inclination. The traditional design is the dip circle.

incoherent in the case of a beam of radiation, with no regular or uniform phase relationship between the waves. The fact that laser output is coherent gives it great value in many situations.

indeterminacy principle the same as uncertainty principle, the quantum physics concept that it is not possible to obtain perfect and complete knowledge of a physical system

index error the same as the zero error of a meter, in which it shows a non-zero reading with zero input. This is an important example of a non-random, or systematic, error.

indicator diagram graph of the pressure in (y-axis) and volume of (x-axis) the gas in a reciprocating engine's cylinder during a cycle of action. The area of the loop is the energy transfer during one cycle. See, for instance, *Carnot cycle*.

indicator tube small cathode ray tube whose screen image brightness or size varies with the input voltage. This type of voltmeter – the 'magic eye' – is now very rare.

indifferent equilibrium old name for neutral equilibrium

indium In group III, period 5 metal, proton number 49, discovered by Ferdinand Reich and Hieronymus Richter in 1863 by its spectrum – two close lines in the indigo region, hence the name. Relative density 7.3, melting and boiling temperatures 157 °C and 2080 °C; most common isotope ^{115}In, a beta emitter with half-life 440×10^{12} years. Its main use is in thin films in liquid crystal displays, though it also has a role in low melting temperature alloys and as a dopant.

---

induced

caused to appear or to happen without contact. See the various types of induction.

inductance or **self-inductance** L unit: henry, H the feature of a conductor which causes the magnetic field round it to oppose a change of the current in it. The source of the effect is a voltage induced between the ends; see *electromagnetic induction*. All conductors have some inductance, though it is often negligible in practice. The design of the inductor is to make the opposition as large as desired. The inductance of an inductor is the voltage induced between its ends when the current changes by one ampere per second.

Mutual inductance M is a similar feature in which the induced voltage between the ends of one conductor depends on the rate of change of current in a second conductor. See *voltage transformer*, a major use of the effect.

The unit of inductance (and mutual inductance) is the henry, H: the inductance giving 1 V with a rate of current change of 1 A s^{-1}.

induction

a) old name for the electric displacement at a point in an electric field

b) rare name for nuclear magnetic resonance

c) the appearance of an effect without contact from the cause. There are several cases in physics.

Electromagnetic induction is the voltage that appears between the ends of a conductor when the magnetic field cutting it changes. Faraday and Henry discovered this independently early in the nineteenth century. The **laws of induction** are as follows.

1 A voltage appears between the ends of a conductor in a changing magnetic field. (Faraday)

2 The size of the voltage depends on the rate of change. (Faraday)

3 The direction of the voltage is such as to oppose the change. (Lenz)

This last law relates to Le Chatelier's principle (and to the law of constant energy); see also *back emf*. The relation that results between the induced pd V and the change of magnetic field linkage Φ is

$$V = - L \, (\mathrm{d}\Phi \, / \mathrm{d}t)$$

Here L is the inductance of the conductor.

An **induction coil** is a type of transformer able to produce a near direct current, output at high voltage from a low voltage direct input. A trembler (make and break device as in an electric bell) switches the input current on and off again and again, causing it to change regularly. An **induction flowmeter** measures the rate of flow of a conducting fluid; there is a magnetic field across the channel or tube, and contacts pick up the induced voltage. Compare that with the Hall effect, a result of the inverse of induction, the motor effect.

There are various types of **induction heating** process in industry: eddy currents in a conducting object raise its temperature without raising that of its surroundings. **Induction motors** (asynchronous motors) are much more common than those with slip or split rings to carry current into the rotor. There are many designs; all consist of a fixed block (the stator) wound with current-carrying coils and a moving block (rotor) in which current is induced. There is a force on the latter as a result of the interaction between the fields of the two currents.

Electrostatic induction is the actual net movement of charge, or the polarisation of atoms, in a sample of matter in an electric field. The forces that result between the field and the induced charges may be large (as when a rubbed pen picks up small pieces of paper).

In much the same way, **magnetic induction** is the appearance of poles in a ferromagnetic sample in a magnetic field. Again, the resulting forces may be large: one example is picking up iron nails (with newly induced poles) with a magnet. This also explains how a sheet of paper covered with iron dust and tapped shows magfield patterns: induction causes the specks of dust to line up.

inductive concerned with electromagnetic induction. **Inductive coupling** is the transfer of a signal between two circuits by a process of mutual induction in a pair of coils (see *transformer*): changes in the current in one circuit induce a corresponding voltage in the second. An **inductive load** is an element in an ac circuit which has inductance and therefore tends to oppose change of circuit current. The opposition of the element is its **inductive reactance**, X_L; this relates to its inductance L and has the ohm (A) as unit. There is also a phase change as a result: the inductor tends to make the circuit current lag behind the applied voltage (by up to a quarter cycle). **Inductive tuning** is the process of changing the inductance in a circuit to change its resonant frequency: there is a variable inductor in the circuit for this purpose.

inductometer old name for variable inductor

inductor passive circuit element (in other words, one that needs no power) that affects the current as a result of electromagnetic induction. How well it can do so depends on its (self-)inductance, measured in henries, H. Even a straight conductor offers some inductance (opposition to change of current); we can increase this by forming the conductor into a coil, and by adding a magnetically soft core. The core may be laminated or made of ferrite dust to reduce the loss of energy caused by eddy currents. The most common design of **variable inductor** has a core which can move into and out of the coil.

Inductors are important elements in ac circuits; they offer ac resistance (called reactance) and also change the *phase* (sense a) between the applied alternating voltage and the ac in the circuit. See *electric circuit: ac circuits*.

industrial revolution the rapid changes of society and its systems that followed the widespread use of machines to replace muscle power. It is normal to describe this 'revolution' as having started in eighteenth century Britain. This is far from fair, however: machines causing social change have been common in many parts of the world from ancient times. Thus the windmill and water wheel were major sources of power throughout Europe in the middle ages, often enough to allow complex forms of mass production in large factories. On the other hand, eighteenth and nineteenth century Britain saw a very rapid growth in large factories and offices, and, as a result, a major move of people from rural areas to towns. Other countries in Europe followed quickly, and the related ideas of efficiency and public transport have spread throughout the world.

inelastic collision one in which the total kinetic energy after impact is less than that before – the constant ('coefficient') of restitution is less than 1. On an atomic or nuclear scale, the process is **inelastic scattering**.

inertance the imaginary component of acoustic impedance. It results from the inertia of the particles in the medium that carries the sound signal, and corresponds to inductance in the transfer of electric signals.

inert cell primary electric cell that works only after having water added. It therefore has a longer storage life than other types.

inert gas gas which takes part in very few reactions (or even none): it is a better name than noble gas for the family of helium, neon, argon, krypton, xenon and radon. Each such element has all electron shells full.

inertia the mass of an object in the context of Newton's first law of force: its tendency to resist acceleration (change of momentum). Given by the ratio of the net outside force to the acceleration produced, a better name is **inertial mass**; however, no test has yet shown any difference between it and gravitational mass. If, rather than force, our concern is with torque (turning effect), the corresponding measure is the object's **moment of inertia**, I. When several forces act on an object from outside, the **inertial force** is the equilibrant: the single force able to keep the object in balance.

In an **inertial system**, any observer sees that any object free of net outside forces moves with constant momentum. The observer in such a frame of reference is an **inertial observer**. If the whole system moves with constant acceleration, it is not inertial, and Newton's laws of force will not then appear to hold. It is normal to take the 'fixed' stars as the basis of an inertial frame of reference; **inertial guidance** systems (which involve accelerometers and gyroscopes) relate the position and speed of a vehicle to these stars at all times.

inflation expansion, often of a sample of a gas. **Cosmic inflation** is the very rapid – exponential – expansion of the Universe thought to have taken place at some time in the first picosecond after the big bang; the Universe may have become disc-shaped but otherwise remained almost isotropic and homogeneous.

information technology See IT.

infrared electromagnetic radiation close to the low energy (red) end of the visible spectrum, with frequencies in the range 5×10^{11}–7×10^{14} Hz (approximately). There are many uses for infrared beams to carry short range signals (for instance, for remote control units and intruder alarms). Infrared, in the range 1310–1550 nm, is also the norm (much more common than light) in fibre optic systems.

Infrared optics is much like that for use with light. However, many substances transparent to light (including glasses) do not pass infrared; lenses and prisms working in this region involve germanium, polyethylene, quartz, or rock salt (for instance). Detection is mainly by photo-cells, photographic film, and thermopiles.

Infrared radiation is still commonly called 'heat' (or thermal) radiation: most of the energy output by a surface below about 6000 K is in this range. It comes from the movement of charge on a molecular scale (for instance, molecular vibration and rotation). Air, like glasses, is opaque to infrared at most wavelengths; an **infrared window** is a range of wavelengths passed by the atmosphere, so of use in astronomy and satellite communications.

infra-sonic describes sound and other pressure waves whose frequency is too low to be heard. Such waves can have harmful effects on the human body.

inject a) to push a substance into a space, typically a shot of fuel into a cylinder of hot gas at the moment of ignition in the diesel engine cycle
b) to feed a signal into an electronic circuit

c) to transfer charge carriers into a semiconductor and so increase their density above the equilibrium level

in parallel describes elements in an electric circuit (or, by extension, other such systems) linked so that the flow splits to pass through them. See *parallel* and compare with *series.*

in phase of two or more waves of the same type passing the same way through the same region, there is a constant positive phase relationship between them. The **in-phase component** of an alternating current is the component in phase with the voltage between the ends of the resistive elements. See *electric circuits.*

input the energy, force, or signal applied to a system, or anything to do with this

in series of elements in an electric circuit (or, by extension, other such systems), linked so that the same flow passes through each of them in turn. See *series* and compare with *parallel.*

instantaneous value the value of a varying measure (such as velocity or current) at an instant in time. For a cyclic measure, plotting this against time gives a wave form whose maximum value is the amplitude.

insulated gate field effect transistor type of field effect transistor

insulation technique designed to prevent energy transfer between two places. Thus, the aim of **acoustic insulation** is to reduce the transfer of sound energy, for instance, with foams and other air-filled solids. **Electrical insulation** involves an electrical insulator (non-conductor) of adequately high resistance, as with the plastics or rubber sheath on a cable and around each core inside it. As well as high resistance, a good insulator has a high breakdown voltage and does not deteriorate with age; it must often also resist weather, animal attack, flame and fatigue. **Thermal insulation** techniques may be more complex as there are several possible methods of energy transfer: conduction, convection, evaporation and radiation. On the whole, however, a good electrical insulator is a good thermal insulator. See, for instance, *vacuum flask.*

insulator substance with low conductivity for the energy transfer in question. See *insulation.*

integrated circuit (ic) circuit whose elements (components) and links all lie in the surface of a small piece (chip) of semiconductor, in most cases silicon. Packed for protection into a case with pins that fit into sockets on printed circuit boards, chips are robust and cheap, use little power, need little cooling, and offer high speed working. They are the basis for a very wide range of modern hardware design, in audio, video, control and computing, for instance, and we treat them as single indivisible units when using and servicing them.

A **monolithic integrated circuit** is a single chip as described above; a **hybrid ic** consists of one or more monolithic ics mounted on a single substrate, or several elements similarly mounted and linked together. Thin film techniques may also be involved in the latter case.

The basic circuit elements of ics are diodes and transistors (the latter used as switches or gates) and linear amplifiers. The integration of several transistors into one chip in the early 1950s saw the start of the so-called third generation of information technology. The process was one of small scale integration (ssi); we have since moved through medium scale integration (msi) and large scale integration (lsi) to the current very large scale integration (vlsi), with millions to hundreds of millions of elements on a chip. The rate of growth of scale of integration goes on without slowing (though quantum limits may be a final barrier), towards giga scale integration (gsi); this involves thousands of millions of elements on a chip.

integrating circuit or **integrator** device, logic circuit or system whose output is the integral of the input with respect to time. The output may be a single value, as in an **integrating meter** (of which there are many kinds), or a wave form. In the latter case, if the circuit has a phase modulator after it, the whole becomes a frequency modulator; if it follows a frequency detector, it becomes a phase detector. See, for an example, *accelerometer.*

A simple integrator consists of a capacitor in the feedback loop of an amplifier. A **summing integrator** combines a summer (summing amplifier) and integrator so that the output is the sum of the integrals of the input with respect to time.

intensifier substance able to increase the contrast and density of a photographic image after processing. (Computer techniques of image enhancement are now more common instead, of course.) An **intensifying screen**, on the other hand, is used during exposure. It consists of a layer placed next to the film that intensifies the image produced. A fluorescent screen, for instance, outputs light when x-rays reach it; the light produced aids the image formation on the film.

intensity measure of the strength (often energy content or energy density) of some physical effect. In the case of a radiation, it is the rate of energy transfer (in other words, the power) of the radiation, per unit area: the unit is the watt per square metre, W m⁻². **Electric intensity**, on the other hand, is an old name for electric field strength; in the same way, **magnetic intensity** is a measure of magnetic field strength.

Luminous intensity I (or, as is common, just intensity) is the rate of visible light energy output by a source. The unit is the candela, cd; this is defined as the output, measured along the normal, of 1/600 000 m² of a black body surface at the temperature of freezing platinum at standard pressure. (That temperature is 2045 K.) One of the SI base units, the candela corresponds to 1.46 milliwatts per steradian (sr, unit solid angle). The unit of light radiation itself is the lumen (lm), the power transfer through a steradian from a point source of 1 cd luminous intensity; 1 lm = 1 cd sr. Lastly, the **intensity of illumination** (illuminance) of a surface has the unit lux (lx); this is the power of radiation reaching the surface when one lumen falls over one square metre.

There are similar concepts for all other types of radiation. In the case of sound waves, in particular, beam intensity I in a medium of mean density ϱ is

$$I = \tfrac{1}{2}\, \varrho\, c\, a^2\, v^2$$

Here, c is the speed of transfer and v is the frequency, while a is the amplitude.

In all such cases, intensity relates to the square of the amplitude.

The **intensity of magnetisation** of a sample, on the other hand, is the ratio of the magnetic moment to the volume (assuming uniform magnetisation). In a cathode ray tube, intensity modulation is the process whereby a voltage on the grid of the electron gun controls the brightness of the spot on the screen; it does so by changing the power of the electron beam on the basis of the signal input.

interaction transfer of energy between two objects (or particles) or between one object and a field. A force of some kind is involved: gravitational, electromagnetic, or weak or strong nuclear.

interelectrode capacitance the capacitance between a pair of electrodes in an electronic device. This may strongly affect how the device works, so needs to be considered during design.

interface the region of contact between two materials: between (solid or liquid) and (solid, liquid or gas)

interfacial angle the angle between a given pair of faces in a crystal structure. (Strictly, it is the angle between the normals to the two faces concerned.)

interference

a the effect on a wanted signal of unwanted signals, such as noise or cross-talk in a communications link or its power supply. The effect is to reduce the clarity of the wanted signal.

Natural interference follows lightning flashes and other atmospheric effects; human sources include mains hum, signals on nearby channels, ignition systems, and electrical equipment. Digital transfers are much more able to survive interference than analog ones. Other ways to reduce the effects of interference are, as appropriate, using directional aerials, careful siting of hardware, putting a filter in the mains supply and fitting **interference suppressors** to noisy systems.

b the effect produced by waves of the same type passing through the same space at the same time. At each point, the amplitude is the sum of the amplitudes of the component waves (taking sign into account); this is because it is not possible to have more than one value of displacement at a point. The waves emerge unchanged from the region of overlap, but may produce a static pattern in that space. Such an **interference pattern** (below) consists of a network of points of **constructive interference** (where the waves are in phase, so reinforce each other) and a network of points of **destructive interference** (where they are in anti-phase, so tend to cancel each other's effects).

Observed with any types of wave, the pattern may follow the process of diffraction: this is not itself interference, but simply a process that makes waves of the same type pass through the same space. An **interference figure**, on the other hand, is the pattern of coloured bands obtained from a crystal in converging polarised light between crossed polarisers. An **interference filter** is a filter that encourages certain wavelengths by constructive interference in a thin film, and bars others by destructive interference.

An interference pattern consists of a set of so-called **interference fringes** or **bands**. If the waves have the same single frequency, the bands are of high and low intensity that merge into each other (light and dark in the case of optical waves). The fringes are more complex if the wave trains cover a range of frequency; they show colour effects in the case of light, as happens with oil and soap films. See *superposition*.

For there to be permanent interference patterns, not only must the waves be of the same type and have the same intensity and frequency range. The angle of overlap must not be too great and the sources must be small. Also, the waves must be unpolarised (or polarised the same way) and coherent, having a constant phase relationship. Lasers are convenient coherent sources.

Historically, however, the only way to obtain coherent waves was to split the output of a single source into two beams. This can be by splitting the wavefront or by splitting the amplitude. The first method devised is that of Young's experiment (1801), this being the first definite proof that light is a wave form (but see *photoelectric effects*). Here, the single source shines on two close slits (division of wave front). The sketch shows the result. For other methods, see *Fresnel bi-prism* (division of wave front) and *Lloyd's mirror* (division of amplitude); see also *Newton's rings*, observed earlier but without recognition of significance, and *beats*, rather the same effect in the case of sound waves.

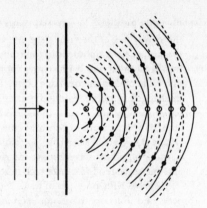

line of crest
- - - line of trough
● node giving destructive interference
○ anti-node giving constructive interference

The two-slit technique produces interference fringes by dividing the wave front from a single source.

For there to be constructive interference at a point, the difference in phase of the two (or more) waves must be zero. Where the waves come from a single source, this means that the path lengths must differ by some whole number multiple of the wavelength λ:

path difference = $n \lambda$ for constructive interference.

Destructive interference requires a phase difference of 180°, meaning:

path difference = $(n + \frac{1}{2}) \lambda$ for destructive interference.

In both cases, n is an integer, the 'order' of the fringe for constructive interference. At a distance L from two coherent sources l apart, this gives fringes whose separation is $\lambda L/l$. By such means, therefore, an interference effect allows us to measure wavelength: see *interferometry*.

interferometry the use of interference systems to obtain information about waves or about aspects of optical systems. Such techniques are now of value and importance in the case of all types of wave. See, for instance, *echelon*, *etalon* (both types of **interferometer**), and the *Michelson–Morley experiment*. **Radio interferometers** are sets of radio telescopes used to find the structure and direction of radio sources in the sky.

intermediate vector bosons the particles proposed as intermediary (exchange particle) in the case of weak nuclear force interactions. The electroweak theories of the 1960s predict three such massive particles, the two Ws, of opposite charge, and the neutral Z.

Beta decay is a typical weak interaction; we can express the common form of this process as the decay of a neutron into a proton, with release of an electron and an antineutrino $\bar{\nu}$:

$$n \rightarrow p^+ + e^- + \bar{\nu}$$

Rather than this simple decay, suggesting that a W^- (one of the intermediate vector bosons) is involved:

$$n \rightarrow p^+ + W^-$$
$$W^- \rightarrow e^- + \bar{\nu}$$

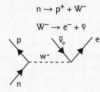

The high mass of the W means it takes part here only as a virtual particle: this means it is not easy to find. Indeed, it was not until 1983 that the three particles turned up in very high energy proton/anti-proton events using accelerators. However, there is much other experimental evidence for interactions like this, and more modern theories (such as those of quarks) serve only to strengthen the concept.

intermodulation the production of sum and difference frequencies by interference between similar waves

intermolecular force attraction between neutral atoms and between molecules – see *van der Waals force*. **Intermolecular potential** is the energy involved in separating two such particles to infinity against this attraction.

internal to do with aspects inside a system or device, rather than those that relate it to the outside world

Internal absorptance α_i measures how well a sample of substance can absorb radiation passing through it; it is the ratio of the power absorbed (between input and output) to the power input. Energy loss by scattering and reflection is not included. See *absorption*, and see also *internal transmittance*, below.

An **internal combustion** engine is one in which the power output comes from the place where the fuel burns; diesel, petrol and rocket engines are of this type, but a steam engine is not.

Internal conversion is a process of nuclear decay in the same class as gamma radiation: an excited nucleus loses energy with no change of nature (parent and product are isomers, in other words). Rather than emitting the energy as a gamma photon, however, the nucleus gives the energy to an electron in its electron cloud. The energy may be enough to eject the electron completely (it is then an Auger electron); if not, the electron returns to the ground state with the release of an x-photon. For a given excited nucleus, the **internal conversion constant** is the probability of internal conversion out of all available methods of decay.

The **internal energy** W_i (sometimes still given as U) of a sample of matter is the sum of the kinetic and potential energies of all the particles in it. The sample's **specific internal energy** is that value for unit mass. The first law of thermodynamics relates internal energy to energy input and output ('work done') in any change.

Internal forces are the forces between the particles of a sample (or the parts of an object), distinct from forces from outside; see *stress* and *strain*. One effect of internal forces in a solid is **internal friction**; analogous to fluid friction (viscosity) inside fluids, this causes, for instance, the damping of elastic waves.

The **internal resistance** r of an electric circuit is the resistance of the source; this affects the current in the circuit in the same way as the resistance of the rest of the circuit – see *source resistance*.

The **internal transmission density** of a sample of matter is an old name for its absorbance (energy absorbing power). In such cases, **internal transmittance** τ_i is the converse of *internal absorptance* (above):
$$\alpha_i + \tau_i = 1.$$
Also, absorbance (internal transmission density, above) is the logarithm of the reciprocal of internal transmittance:
absorbance = $\log_{10}(1/\upsilon_i)$.

The **internal work** of a sample of matter is an old name for the energy needed to separate all its particles to infinity. It relates, therefore, to intermolecular potential.

internal reflection See *total internal reflection*.

international agreed between all (or many) countries. (Universal is not so strong: it means used in all (or many) countries.) In the process of moving towards the SI unit agreement, many units came to have agreed definitions. They than carried the name 'international'; examples are the **international ampere** and the **international candle**. In much the same way, the **International Practical Temperature Scale** (IPTS) is the agreed set of fixed temperatures; first agreement was in 1968, though there have been amendments since. The **international system** itself was a precursor to SI, being a fully agreed set of electrical units.

International Thermonuclear Experimental Reactor See *ITER*.

interpolation estimating the value of a variable y at some point x between known values (y_1, x_1) and (y_2, x_2). This may be by using a graph or by a formula that depends on the rate of change of y with respect to x in that range.

intersecting storage ring approach to exploring high energy collisions between particle beams. The two beams go to high energy in a conventional accelerator and then transfer to pass in opposite directions round a 'storage ring'. This large circular tube contains a vacuum and uses magnets to keep the particles in orbit. In this case, the orbits are designed to intersect at only one place in the ring; there one can study how the particles interact.

interstellar extinction value A, unit: stellar magnitude how much the space dust between a star or other such object and the Earth absorbs and scatters the radiation used in the measurement of magnitude

interstitial found in a small crevice. An **interstitial atom** *or* **interstitial ion** is a 'foreign' (impurity) atom or ion in a crystal lattice that fits into a small space rather than taking up a lattice site; it is a point defect in the crystal, and may (or may not) produce a crystal defect on a larger scale.

interval the frequency difference between two waves, in particular between two notes in music. One may express this either logarithmically or as a ratio greater than 1. In the former case, a common unit is the milli-octave; the relation is $3322 \log_{10}(\nu_2/\nu_1)$. This gives the value 1000 when the waves are an octave apart, in other words when ν_2 is twice ν_1.

intrinsic inherent, to do with a system's own nature rather than any outside effect. **Intrinsic mobility** is the net mobility of

carriers in an *intrinsic semiconductor* (below); such a substance has an equal density of holes and electrons, the mobility of the holes being about a third of that of the electrons. The **intrinsic pressure** in a gas sample is the pressure drop near the surface because of the attraction between the particles; it is the term a/V^2 in the van der Waals equation.

The **intrinsic properties** of a substance are those, such as specific thermal capacity, which do not change much if the substance is impure or has an imperfect structure.

An **intrinsic semiconductor** (i-type) is a pure substance which behaves as a semiconductor without need for any impurity atoms. Pure silicon and pure germanium are of this kind; they have an adequately small energy gap in the band model to allow a certain density of hole–electron pairs at normal temperatures. Thus, there are equal numbers of these carriers in such a substance. Adding a suitable impurity ('dopant') will increase the density of one type of carrier only, making the substance an extrinsic (n-type or p-type) semiconductor.

invariant unchanging (i.e. constant) during some change in the system in question. An **invariant system** has no degrees of freedom.

inverse opposite. Thus, there are various inverse effects, opposite to those first explored. For instance, the Faraday effect is the rotation of the plane of polarisation of light as it passes through matter in a magfield. The **inverse Faraday effect** is the magnetisation of a medium by passing intense polarised light through it. The inverse Raman and Zeeman effects are also of importance in modern physics.

Inverse gain is the gain of a bi-polar transistor when used with the emitter as collector and the collector as emitter. Inverse gain is mostly less than normal gain.

The **inverse square** of a value x is the reciprocal of its square, $1/x_2$. An **inverse square law** applies when an effect varies with the reciprocal of the square of the distance from its cause or source. Examples are the intensity of radiation from a point source in a non-absorbing medium, and the forces of gravity and electromagnetism.

inversion process that takes place the opposite way to the norm. For instance, the anomalous behaviour of H_2O around 0–4 °C shows the reverse of the normal process of expansion with temperature rise. An **inversion layer** in the atmosphere is one whose temperature falls with distance from the ground.

For a thermocouple with one junction at a given low (or high) temperature, the **inversion temperature** is the high (or low) temperature of the other junction at which the net voltage in the circuit is zero. The term also applies to the temperature at which the Joule–Kelvin constant of a gas changes sign.

inverter a) electronic circuit whose output is the inverse of the input. In other words, if digital, it is a NOT gate, giving 1 for 0 input and 0 for 1 input; if it is an analog circuit, it introduces a 180° phase change to the input signal. An **inverting amplifier** is of the latter type, and amplifies the input as well.
b) direct current motor that drives an alternator, or any other system that converts input dc to ac (such as a power pack run from a car battery whose output is 240 V ac)

inverting prism prism used to make an upside-down image upright (or the converse). See right.

There are various ways to do this, a simple one being in the sketch. The prisms in a pair of prism binoculars invert, but also (their main function) fold the light path into a manageable length.

iodine I halogen, subliming solid at normal temperatures, group VII, period 5, number 53, discovered in 1811 by Bernard Courtois while working with seaweed ash, and named after its violet colour (Greek *iodes*) by Humphry Davy who identified it as an element in 1813. Relative density 4.9, melting and boiling temperatures 110 °C and 180 °C (though the solid also sublimes); only natural isotope ^{127}I.

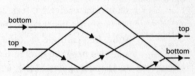

This inverting prism inverts without changing the direction of the rays.

ion

charged particle that consists of an atom or a group of atoms which has lost or gained negative electrons. For instance, in solution in water, the sodium chloride molecule, NaCl, ionises to give Na$^+$ and Cl$^-$ ions. We call these cations and anions: in an electric field, they move towards the cathode or anode respectively (but see *ionic atmosphere*). **Ion burn** is a mark in the phosphor coating of a cathode ray tube that results from the impact of high energy negative ions in the electron beam; see also *ion trap* (below).

An **ion drive** (or **ion thruster**) is a rocket engine proposed for long (e.g. interstellar) space trips. The propellant (perhaps liquid caesium) is turned into a vapour and then ionised; an electric field accelerates the cations to a high speed from the rear of the craft. The thrust this could produce in practice would not be high; all the same, keeping the engine going for a long time in empty space would produce very high speed. The approach can be very efficient because of the high charge-to-mass ratio of an ion, giving a high specific impulse. Ion drives are in use only in Earth orbit so far, though there have been some tests in outer space. A good source of up-to-date non-technical information is http://nmp.nasa.gov/ds1/tech/ionpropfaq.html

Ion implantation is a method of doping a semiconductor to produce the desired properties: the semiconductor is the target for a controlled beam of ions.

A common way to produce ions involves the use of ionising radiation; as the name implies, this can ionise the atoms and molecules it meets. An **ion pair**, cation and anion, or **ion–electron pair**, results from each such interaction (See also *plasma*.) For **ion microscope**, see *field ion microscope*. An **ion pump** can approach closer to vacuum than many other designs. When the gas is at a low pressure, a beam of ionising radiation (often high energy electrons) ionises the atoms that remain, and an electric field collects these. In much the same way, ions are produced in the ion gun (or **ion source**) of an ion accelerator (e.g. at Darmstadt): high energy electrons pass through the neutral gas and an electrode attracts the ions into the tube. In a cathode ray tube, there may be an **ion trap** to collect the massive negative ions in the electron beam before they cause damage to the screen (see above, *ion burn*).

ionic with properties that relate to, or depend on, a nature like that of an ion. For instance, the ions of an electrolyte each have an **ionic atmosphere**: each cation tends to have a number of anions around it, and each anion tends to have a number of cations in its 'atmosphere'. This has the effect of adding to the inertia of the ions in an electric field: it takes a while for each ion to break away from its atmosphere The Wien effect occurs when the field is so high that this inertia disappears.

In the solid state, an **ionic compound** has a regular array of anions and cations in its crystal lattice: the bonds are the electric forces between the ions. When molten or in solution it becomes a good electrolyte as complete ionisation takes place. (See *ion* for an example.) That is why electrolytes are such good conductors; the process of **ionic conduction** that takes place in them is very efficient. See *electrolysis* for the details. (Note that some people restrict the term ionic conduction to its appearance in solids only.)

Ionic mobility is the mobility of a given ion in a given situation, in other words its drift (terminal) speed per unit applied electric field; the unit is the $m^2 \, V^{-1} \, s^{-1}$ ($m \, s^{-1}$ per $V \, m^{-1}$). In an **ionic semiconductor** (not a common type of solid substance), the transfer of current by ions is greater than that by holes and/or electrons.

ionisation any process whose result is the appearance of ions. It needs input of energy to the particles of the substance. An

ionisation burst is a single pulse of current as output by an **ionisation chamber** (or **ionisation detector**). Here the energy comes from passing (ionising) radiation; this leaves a trail of ions in the gas in the chamber. A weak electric field causes the ions to drift toward the electrodes; the outside circuit can record, amplify, or count the pulses of current between these electrodes that follow the radiation events. For a major system that works with higher fields, see *Geiger–Müller tube*.

ionisation energy (or **ionisation potential**) the energy involved in ionising an atom or molecule by taking out a given electron; as the electrons in an atom's cloud are at different energy levels, each atom has a number of ionisation energies. The **first ionisation energy** is the energy transfer involved in removing the most weakly bound electron; the second, the second, and so on. Appearance energy (widely called appearance potential in industry, where it is the basis of a certain type of spectroscopy) is the ionisation energy (potential) of a given core electron.

In the case of an impurity semiconductor, **ionisation energy** is the energy needed to lift an electron from the valence band to an acceptor level, or from a donor level to the conduction band. The **ionisation temperature** of a substance is the temperature at which the mean energy of the particles equals the ionisation energy; above that temperature, the substance becomes a plasma. For **specific ionisation**, see ionising radiation.

ionising radiation any form of radiation able to transfer enough energy to an atom or molecule to ionise it. The specific ionisation of an ionising particle or photon is the number of ion pairs produced per unit length of path.

Radiation interacts with matter in various ways, not all of which are efficient at producing ions, even if the energy is high enough. Charged particles and high energy photons (in particular, of x and gamma radiations) interact directly with matter particles; these are **directly ionising radiation**. Others are **indirectly ionising radiations**: they can interact with matter, not to produce ions, but to produce other ionising radiations. Neutrons are an important example of this type.

ionosphere region of the Earth's atmosphere with an appreciable density of ions, the result of absorption of high energy ultraviolet from the Sun. The height is from about 60 km upwards, though this value (and the detailed structure of the region) varies with place and time. The main layers are the D-layer (day-time only), the E-layer ((Kennelly–)Heaviside layer), and the F (Appleton) layer.

All reflect radio waves to an extent, but not radio waves of all
wavelengths. Effective long-distance communication at wavelengths
not reflected by the ionosphere must use satellites.

ion propulsion the same as using an *ion drive*

IPTS the International Practical Temperature
Scale, the set of agreed fixed temperatures

ir infrared, the group of electromagnetic
radiations with frequencies in the range 5 ×
10^{11} to 7×10^{14} Hz (approximately)

iridescence the appearance of colour fringes
in thin films and butterfly wings (for
instance) as the result of interference. A
feature is that the fringe hues and patterns
vary with viewing angle – holograms are
the only way to capture the effect photographically. The root of the word is *Iris*,
Greek goddess of the rainbow.

iridium Ir metallic transitional element in
period 6, proton number 77, discovered by
Smithson Tennant in 1803, who named it
after the iridescence shown as it dissolves in
hydrochloric acid. Relative density 22.6
(perhaps the highest known), melting and
boiling temperatures 2410 °C and 4130 °C;
most common isotope ^{193}Ir. Uses: in various
alloys and polymer leds. The name also
belongs to a system of 77 (sic) communication
satellites launched in 1998 and bankrupt a
few months later.

iris diaphragm type of optical diaphragm
(variable stop – aperture – in an optical
system) that appears to work much like the
iris of the eye (whose function is the same). A
number of specially shaped thin leaves pivot
from points all round the edge; they turn
together one way or the other to reduce or
increase the central hole. The action of the
iris of the eye involves a ring muscle
(sphincter).

iron Fe transitional metal of period 4, proton
number 26, known since ancient times (by
definition since the start of the Iron Age –
some 7000 years ago, perhaps in India, when
people first started making, working and
using it); the name comes from the old
English *isarn* (perhaps from the Indo-
European *eyos*, = metal), while the symbol's
origin is its Latin name *ferrum*. Relative
density 7.9, melting and boiling temperatures
1540 °C and 2750 °C; most common isotope
^{56}Fe, which is very stable and formed inside

stars. Iron is by far the most widely used
metal (about 95% of all uses) – as such and in
the various alloys called steels – because of its
low cost and high strength.

iron loss the loss of useful energy (i.e. of
efficiency) in electrical machines caused by
energy transfer to the iron parts. The main
process involved is the induction of eddy
currents; forming the iron block from thin
strips (lamination) or powder largely cuts this
loss. See also *hysteresis*.

irradiance *E* unit: watt per square metre,
$W1\,m^{-2}$ the rate of energy transfer from
radiation to a surface, the general equivalent
of intensity of illumination for visible waves.
For the irradiance of the Earth by the Sun,
1 35 kW m^{-2}, see *solar constant*.

  Some people restrict the term to
electromagnetic radiation; there is, however,
no need to do this. Others give it as energy
per unit area, in J m^{-2}, but this is not correct.

irradiation exposure to ionising radiation in
any of a number of contexts. As a means of
treating or processing samples (perhaps for
medical or industrial reasons), this often
takes place in accelerators and nuclear piles;
special small irradiation chambers are also
common, supplied with suitable active
isotopes. Radiotherapy is a good instance of
this last approach.

irreversible change change during which the
system in question is not in equilibrium at
all times. Compare with *reversible change*.

irrotational flow fluid flow in which there is a
velocity potential; there is zero vorticity at all
points in the fluid. An **irrotational field** also
has no vorticity – it is a conservative field. An
irrotational wave is a pressure wave (for
instance, a sound wave); here too there is no
vorticity at any point.

is- prefix meaning the same, or constant (it
comes from iso-, which has the same
meaning). An **isallobar** is a line on the Earth's
surface joining points at which the rate of
change of air pressure is the same (see also
isobar). We describe a change as **isenthalpic** or
isentropic if, during it, enthalpy or entropy

remains constant. In the former case, during the change, the sum of the system's internal and external energies stays the same; in the latter, isentropic is the preferred word for adiabatic.

island of stability region of the periodic table in which some nuclides should be much more stable than their neighbours. The concept (first devised by Glenn Seaborg) follows from the shell model of the nucleus: when a nucleon shell is full, the nucleus has a 'magic number' of neutrons or protons – the nucleus would then be more stable as there is a large energy level gap before the next shell can start to fill.

The 'ocean of instability' starts after bismuth (element 83, though this is not a magic number) – nuclides after this are very or quite unstable in that they have half-lives less or far less than the life of the Universe. There is, however, an 'island of relative stability' at the end of the historical periodical table, peaking at $^{232}_{90}$Th (half-life 1.4 × 1010 years, whereas the other isotopes of thorium range from 75 000 years to 24 days), $^{235}_{92}$U (7 × 10^8 years) and $^{238}_{92}$U (4.5 × 109 years; the values for uranium's other isotopes ranging from 2.3 × 10^7 years to 70 years).

The magic numbers are not known for certain, though values of 82, 114, 120, 126 and 184 often appear. One of the most stable nuclides is of lead, $^{208}_{82}$Pb – which is 'doubly magic'. No other doubly magic nuclide has been found, though isotopes of singly magic $_{114}$Uuq (see *ununxium*) seem to be significantly more stable than their neighbours.

iso- prefix meaning the same, or constant (see also the short form *is-*). For instance, **isobars** are
a) lines on the Earth's surface joining points with equal air pressure
b) nuclides with the same nucleon number but which differ in proton number, as in the case of ^{3}H (tritium, hydrogen-3) and ^{3}He (helium-3). For **isobaric spin**, see *iso-spin* (below).

An **isochore** is a graph plotting two measures for two states of a system whose volume remains constant; the pressure–temperature graph of a gas that results from Charles' law is an isochore. **Isoclinals** or **isoclinics** are lines on the Earth's surface joining points of equal magnetic dip (i.e. inclination – see *Earth's magnetism*). **Iso-compounds** are chemical species with the same composition in terms of number and

type of atom, but that differ in structure; see also *isomers*, below. **Isodiapheres** are nuclides with the same neutron excess (isotopic number), this being the difference between neutron number and proton number.

An **isodynamic** line joins points on the Earth's surface with the same total magnetic field strength; these points do not necessarily have the same horizontal component as the angle of dip varies (see *Earth's magnetism*). On the Earth's surface too, an **isogam** joins points with the same gravitational field strength (the same value of *g*, the acceleration of free fall). Also when dealing with the Earth's magnetism, we speak of **isogonals** or **isogonics**: lines of equal value for the magnetic declination (variation). In the same context, three of the lines mentioned above (i.e. isoclinic, isodynamic and isogonic) are **isomagnetic lines**; these lines on the Earth's surface join points with the same values of one of the three characteristic measures of the planet's magnetic field.

isolated system closed system, one with no interaction with the outside world. This means there is no net force or net energy transfer from outside, so that the system's total angular momentum, energy, linear momentum and mass remain constant.

isolating transformer transformer used to prevent any direct link between a device or system and the outside, leaving inductive coupling as the only method of power transfer. This is mainly for safety reasons. Such transformers may also offer a change of voltage but this is not the prime function.

isolator device designed to allow microwaves through in one direction, but not in the other. In other words, it is a microwave valve.

Isomers are **a)** nuclides with the same proton and nucleon numbers (in other words, of the same isotope) that differ in energy state (i.e. degree of excitation); gamma decay causes transitions between them. An **isomeric state** is one with a life time long enough for it to be observed, i.e., it is meta-stable.
b) iso-compounds (above), chemical species with exactly the same atomic composition but which differ in structure and therefore in chemical and physical properties. **Optical isomers** differ in how they affect passing polarised light. They may, for instance, rotate the plane of polarisation to the right (dextrorotatory) or to the left (laevorotatory) when viewed from the output side.

An **isometric change** is an isochoric process (above) in the case of a gas, one of whose two measures in question is pressure; whatever the process, the gas volume stays the same. **Isomorphism** appears with certain crystals: the **isomorphs** have the same structure (they are **isostructural**) and similar chemical and physical properties. They thus form mixed crystals with ease. Note that **isomorphic** (or **isomorphous**) crystals are always isostructural, but isostructural crystals need not be isomorphs. An **isopiestic** change is one during which the sample pressure remains constant; the term usually refers to such changes where sample volume is under test.

Iso-spin (symbol I, or sometimes T or T_z) is a sub-set of flavour in particle physics – is one of the three internal quantum numbers used to describe and differentiate between the hadron class of elementary particles (the others are charm and strangeness); all three remain constant (are conserved) during strong nuclear interactions. For instance, on this picture, nucleons have iso-spin $1/2$: the proton's value is $-1/2$, while that for the neutron is $+1/2$. (Some people give the signs the other way round.) Heisenberg was the first to devise and use the concept; others extended it to nuclei as well as to particles (hence the alternative names – **isobaric spin** and **isotopic spin**). All the same, the measure is a mathematical tool, and has nothing to do with angular momentum. In modern analysis, one plots particles in an imaginary three-dimensional space, one of whose dimensions is iso-spin; we call this field **isospace**. **Iso-spin symmetry** is the same as charge independence, the principle that the strong nuclear force between two nucleons depends only on their angular momentum and spin.

The **principle of isostasy** is that, from the top of the Earth's mantle to the surface, any column of crust of a given cross-section has the same mass: whether the top of the column is ocean, lowland or mountain. Thus, we view the whole of the Earth's crust as floating on the mantle, in a sense, so that flows of the dense sima rock at the bottom of the crust will compensate for changes. **Isostructural** crystals have the same crystal structure ('habit'). On a surface or graph, an **isotherm** is a line joining points of equal temperature, while an **isothermal change** (or **isothermal process**) is one during which the

system's temperature remains constant. Compare with *adiabatic* (better called *isentropic*)in the case of work on gases.

For nuclei, **isotones** are species with the same neutron number while **isotopes** have the same proton number; in each case, the species differ in nucleon number. (They are *isobars*, see above, if nucleon number is the same.) **Isotope power** involves using the thermal energy produced in the decay of radio-isotopes. Thermocouples are the most common way to convert the output to a more useful form. The sources (called *radioisotope thermoelectric generators*, rtgs) are not nuclear reactors in the true sense, though people often call them by this name.

There are various techniques of **isotope separation**, for separating isotopes from each other. As their chemical properties are almost identical, using a chemical process for this is very hard (though it has been done); most techniques depend on the (still slight) differences in physical nature. The most important involve electromagnetic fields, centrifugation or diffusion; even here, in any case, the process needs to pass through perhaps hundreds of stages to obtain a useful yield. The **isotopic abundance** of an element is the composition of a sample in terms of its isotopes; the figures are given in relative terms, i.e., as a set of percentages. The isotopic abundance of most elements on the Earth depends little on source. See also *dating*. The **isotopic number** of a nuclide is an old name for its neutron excess: the difference between its neutron and proton numbers; see also *isodiasphere*, above. For **isotopic spin**, see *iso-spin*, above, and for **isotopic tracer**, see *tracer*.

Isotropy describes how a given property of a sample of substance is the same in any direction. A sample may be **isotropic** with regard to some properties, but anisotropic for others. Lastly, **isovolumic** means with the same or constant volume: a better name than isochoric (above).

i-spin same as iso-spin

IT information technology, the use for the good of society, of the ideas of science in the context of the handling, storage, processing and transfer of information (in the form of data and by electronic means)

ITER the International Thermonuclear Experimental Reactor, being built in southern France – the most costly international science project after the international space station. Planned to produce continuous power at a higher level than it uses by 2040, this research fusion reactor is hoped to lead by the end of the century to a viable source for 10–20% of the world's power.

iterative impedance the impedance between one pair of terminals of a simple four terminal circuit or network when the impedance of the load at the other pair is the same. Such a circuit has two pairs of terminals only, for input and output (ignoring any power supply leads). If the input iterative impedance equals that measured at the output pair, the circuit is symmetrical, and the value is its characteristic impedance.

i-type semiconductor intrinsic semiconductor, one whose properties do not depend on impurities

j symbol for the square root of –1 (*i* is more usual in physics), the 90° operator

J a) one symbol for the J/psi, or gypsy, particle
b) symbol for joule, the unit of energy

j a) symbol for electric current density, unit ampere per square metre A m^{-1}
b) symbol for impulse (*I* is much more usual)
c) symbol for the unit vector in the *y*-direction in Cartesian space

J a) symbol for magnetic polarisation (sense b), unit: tesla, T
b) symbol for the (now rarely used) 'mechanical equivalent of heat' (the Joule equivalent), 4.2 joules per calorie
c) symbol for moment of inertia (*I* is more usual), unit: kilogram metre squared, kg m^2
d) symbol for nuclear spin quantum number (*I* preferred)
e) symbol for rotational quantum number
f) symbol for second moment of mass

J/ψ particle which is an example of charmonium, the first such to be found – see *psi-particle*.

jamming deliberate interference with a communication signal which impedes access to the information

jansky rarely used unit of radiation energy density, equal to 10^{-26} W m^{-1} Hz^{-1}. This was of interest only in astronomy. The 1932 discovery by Karl Jansky (1905–1950) of extraterrestrial radio signals led to radio astronomy.

Jansky noise high frequency interference caused by radio signals from outside the solar system

Jan(s)sen, Zacharia (1580–1638) optician who seems to have been the first to develop a two-lens compound microscope system, some time between 1590 and 1609. The device was not, however, effective or publicised – Hooke

produced and used a good system some decades later.

Jeans, James (1877–1946) mathematician and physicist, best known for his work in astronomy and his popularisation of science

Jensen, Hans (1907–1973) physicist whose theory that the nucleus has a shell structure led to his Nobel Prize in 1963. He shared this with Wigner and Goeppert-Mayer, who devised the same theory independently.

JET Joint European Torus, the world's largest fusion reactor

jet a) the dense cone of particle tracks that results from the interaction of a very high energy particle (for instance in cosmic radiation) with a target nucleus
b) high speed flow of a fluid from a nozzle or small hole. **Jet tones** are the components of the variable hiss produced as a jet leaves its source (e.g. a gas jet from a nozzle in a fire or cooker); the cause is the vortices formed in the fluid stream at the edge of the hole.
c) engine that works on the basis of this: the jet carries mass in one direction, so there is a force (by Newton's third law) on the engine in

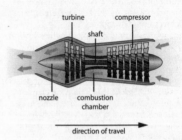

direction of travel →

The turbojet is a simple and efficient internal combustion engine.

the other direction. In the various designs (with all but the ram jet having a number of fans, or turbines), the jet is the engine's hot exhaust: a very high temperature mixture of air and burned fuel (the system is a form of internal combustion engine). Some boats and all rockets obtain their drive on the same principle, but rocket ships carry their own oxygen rather than trapping air as they fly. Engineers Hans von Ohain (1911–1988) and Frank Whittle (1907–1996) first devised this process of **jet propulsion**, in 1936–7, though the first jet planes to fly were over two years later. Their designs are the simplest, the **turbojet** shown.

The **ramjet** is a specialist jet used mainly in missiles. It is light and reliable, as there is no compressor and therefore no turbine – the engine forces air in through its forward motion, which must therefore be at high speed. The only moving parts are in the kerosene fuel pump to the injectors. This all makes it more like a rocket engine than a jet – but it is more efficient as it does not need to carry its own oxygen.

A **scramjet** – or supersonic combustion ramjet – is also very simple as it has no moving parts and takes all of the oxygen it needs to burn its hydrogen fuel from the air. However, scramjets do not work until they reach about five times the speed of sound – then the air passing through the engine becomes hot enough (by compression) for ignition. The rapid expansion of the exhaust gases that results creates the forward thrust. Most research into these engines takes place in the UK and Australia, the hope being to create engines which could power efficient air liners taking two hours to travel between London and Sydney.

jitter short-lived instability in a signal's phase or amplitude. Jitters cause, for instance, loss of synchronism and erratic motion in tv pictures for a few moments.

Johnson noise random noise in the signal in a conductor that results from the thermal motion of electrons. It is sometimes called thermal noise, but this should not imply any temperature effect.

Johnson–Rahbek effect 'Coulomb clamping', the force between layers of metal and semiconductor when there is a fairly high voltage between them. The effect is strong enough for use in a few contexts.

Joint European Torus (JET) the world's largest fusion reactor

Joliot-Curie, Irène (1897–1956) and **Jean Frédéric** (1900–1958) daughter and son-in-law of Marie and Pierre Curie, who shared the 1935 Nobel Prize for chemistry for their discovery of artificial radio-nuclides. They both moved into nuclear physics and nuclear power and died from radiation-induced cancer. See also *Curie, Irène Joliot-*.

Josephson effect the tunnelling of electrons without any applied voltage, as a super-current, through a very thin insulating barrier between two superconductors (a **Josephson junction**). Brian **Josephson** (1940–), then a student, predicted the effect in 1962 and demonstrated it the next year. He received the 1973 Nobel Prize for this work. The use of niobium junctions twenty years later removed the original practical problems and also gave extremely high switching speeds. There are now many uses, in data processing and storage, and in squids, for instance.

If there is a voltage across the junction, the leakage current alternates at a microwave frequency v that depends on its value V:

$$v = (2Q_e/h)\,V$$

Here h and Q_e are the fundamental Planck constant and electron charge.

The **inverse Josephson effect** appears when there is a microwave field across the junction: steadily raising the current through the junction makes the voltage increase in very precise steps. This effect, exact to within $1{:}10^{16}$, provides new laboratory standards of voltage for those who need such precision. See also *quantum Hall effect*, the two effects together giving highly precise values for the constants h and Q_e.

Josephson plasmon quasiparticle that results from quantising the oscillations of the free electron 'gas' of a metal – see *plasmon*.

joule J the unit of energy (and of 'work', an equivalent term, though in the past people did not think it was equivalent). The joule is the energy transfer involved when a force of one newton causes one metre displacement in the same direction. The main former metric units, the calorie and the erg, are around 4.2 J and 10^7 J respectively.

Joule, James (1818–1889) brewer who did much work on how 'work' and energy relate – the mechanical equivalent of heat – people then holding the belief that they did not relate at all. This was the start of thermodynamics.

Joule studied under John Dalton, and from 1852 worked in the field of thermodynamics with Thomson (later Lord Kelvin). A **Joule calorimeter** is an electrically heated calorimeter (device used to measure specific thermal capacities), while the **Joule cycle** is the reversible action of an engine using air as the working fluid.

Three concepts go by the name of **Joule effect** (see also *Joule's laws*, below):
a) the increase in temperature of a conductor when passing a current (as the result of resistance)
b) (or **Joule–Kelvin effect**) the change in temperature of a gas sample when it suffers an adiabatic expansion with no energy exchange with the outside, for instance, when it expands through a porous barrier into a region of lower pressure. The effect is the basis of an important approach to liquefying some gases.
c) (or **Joule magnetostriction**) the slight increase in the length of a ferromagnetic rod with an increasing magfield.

The **Joule equivalent** J is a name for the 'mechanical equivalent of heat' (also an old term, from the days when people did not realise that 'mechanical energy' ('work') and 'thermal energy' are the same thing). The value is about 4.2 joules per calorie. Showing this constant relationship in a number of cases involved Joule in much work; the concept led, however, to the modern views of energy mentioned above.

There are two **Joule's laws**:
a) The rate of thermal energy transfer from a wire of constant resistance varies with the square of the current in the wire. In other words, the energy transfer W in time t of a current I in a resistance R varies with the product $I^2 R t$. Using SI units, the constant of proportionality is 1, so that $W = I^2 R t$.
b) The internal energy of a sample of perfect (ideal) gas depends only on its temperature, and not on its volume. This sees a perfect gas as having no forces between the particles.

The **Joule–Thomson effect** is yet another name for the Joule(–Kelvin) effect noted above. For a gas at constant enthalpy, the **Joule–Thomson constant (coefficient)** is the rate of change of pressure with temperature. The value for helium is negative – so it rises

in temperature as it expands freely – down to its **Joule-Thomson temperature** of around 40 K at normal pressure, below which it behaves as a normal gas. The constant changes sign at a certain temperature in the case of most gases; that temperature is the Joule–Kelvin temperature (or inversion temperature) of the gas in question.

J/psi particle or 'gypsy' particle (from the sound of the full name), also sometimes called charmonium. It consists of a charmed quark bound to its anti-quark. This elementary resonance particle announced on 11 November 1974 by two teams (one calling it J and the other ψ) was seen both in proton-proton collisions and in high energy electron-positron collisions. It has a mass of some three proton masses and a lifetime of 10^{-20} seconds (which is quite long compared to those of other resonances that decay by the strong interaction). It was predicted as the combination of a charm quark and an anti-charm quark, these quarks being devised in attempts to unify electric and weak interactions (the electro-weak theories). The particle has zero values of charge, hypercharge and iso-spin.

junction the interface between two matter samples, in particular where significant electromagnetic effects result. (In fact, there is always at least a small voltage across such a junction.)

The **junction diode**, or **p–n junction**, is a junction between p-type and n-type semiconductor materials. This has the property of passing current only one way: it is a rectifier. The cause is the junction voltage that appears (as mentioned above), as a result of the diffusion across the junction of holes

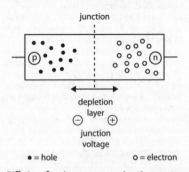

junction

depletion layer

junction voltage

● = hole   ○ = electron

Diffusion of carriers across a p-n junction causes a reverse voltage and a depletion layer.

and electrons. This process comes to an equilibrium situation as shown in the left-hand sketch. The diffusion allows holes from the p-region and electrons from the n-region to recombine. This forms
a) the voltage across the junction (the junction voltage or barrier), with the n-region positive and the p-region negative;
b) a region, called the depletion layer, across the junction where there are very few electrons and holes left.
With an outside voltage applied across the junction in the conducting direction (with a 'forward bias'), majority charge carriers of both signs can move and carry current. Reverse bias, on the other hand, increases the barrier voltage and the width of the depletion layer. Thus no charge can pass, except by the (very small) number of minority carriers. A typical diode passes only a few microamps when reverse biassed, but tens of milliamps with forward bias. The right-hand sketch (which also gives the standard symbol) shows the characteristic curve of the device. See also *diode* and *Zener diode*, the latter being a diode

in which a high enough reverse bias causes breakdown and an avalanche current.

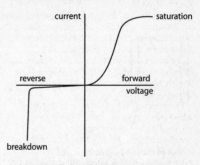

For **junction field effect transistor**, see *field effect transistor*. A **junction rectifier** is an old name for junction diode (above). A **junction transistor** is a normal (bi-polar) transistor. It consists, in structure and effect, of two junction diodes back to back, to make a p-n-p or n-p-n device.

k unit prefix symbol for kilo-, 10^3

K a) unit prefix symbol for Kilo-, 1024 in IT
b) unit symbol for kelvin, unit of temperature
c) symbol for the **K-meson** (kaon)
d) the electron orbit/shell with lowest energy in an atom, that can hold two electrons. **K-capture** is the main form of electron capture, a type of radioactivity process. The **K level** is the energy of the K shell; see also *absorption edge*.

k **a)** symbol for Boltzmann constant
b) symbol for chemical reaction rate constant
c) symbol for circular wave number
d) symbol for compressibility
e) symbol for radius of gyration
f) symbol for thermal conductivity (but λ preferred)
g) symbol for any unspecified constant

K **a)** symbol for bulk modulus
b) symbol for chemical reaction equilibrium constant
c) symbol for kinetic energy (but W_k preferred)
d) symbol for rotational quantum number (but J preferred)

Kaluza–Klein (KK) theory published in 1921–1926, an extension of general relativity to a five-dimensional space–time, by mathematician and physicist Theodor **Kaluza** (1885–1954) and Oskar Klein. The aim was to unify gravitation and electromagnetism, and the three-fold result is the Einstein field equations in 4D, the Maxwell equations (also 4D), and an extra field, the radion. After some years, people forgot Kaluza's idea in 1919 to work with general relativity in more than four dimensions, but it is now very important in string theory. (In fact, Gunnar Nordström first had the idea, but people forgot that too.)

Klein's contribution was that the theory's essential fifth dimension is real but curled tightly into a very small space showed that the Maxwell equations are still valid.

Kamerlingh-Onnes, Heike (1853–1926) physicist who discovered superconductivity and was the first to liquefy helium gas. He received the 1913 Nobel Prize for his work 'on the properties of matter at low temperatures'.

kaon hadron, one of four mesons (middle-weight elementary particles with the strangeness quantum number), with mass equivalent to around 496 MeV/c^2. There are positive and negative kaons (K^+ and its anti-, K^-), and a neutral kaon (K^0) with its antiparticle. The discovery of kaons in the output of cosmic radiation in 1947 (by George Rochester and Clifford Butler using a cloud chamber) marked the start of the profusion of elementary particles we since face. Kaons are short-lived (up to about 10^{-8} s) and decay into pions, the other main type of meson.

There are two neutral kaon forms, or eigenstates, with very different life-times (5×10^{-8} and 9×10^{-11} seconds), called K-long, K_L, and K-short, K_S, respectively. It is normal to view K^0 and its anti as composites of these two states. The discovery of CP violation in 1964 followed research into the decay of K_L.

Kapitza, Peter (1894–1984) physicist whose best known work is in low-temperature physics. He discovered superfluidity during his twenty years at Cambridge (where he first worked under Rutherford), and gained the 1978 Nobel Prize for his 'inventions and discoveries' in low-temperature physics.

Kastler, Alfred (1902–1984) scientist whose work on lasers ('optical methods of studying hertzian resonances in atoms') gained him the 1966 Nobel Prize

kathode original spelling of cathode, negative electrode or terminal

kc or **kcs** or **kc/s** old forms for kHz, i.e. kilohertz

K-capture the main form of electron capture, a type of beta decay (radioactivity process); a nuclear proton captures an electron from the K shell to produce a neutron. As the K shell is the closest to the nucleus of an atom, this is the most common type of electron capture.

keeper piece of 'soft' iron which, when placed between the poles of a horseshoe (U) magnet, much reduces the neutralising effect of the magnet's field. Thus, it helps the magnet retain its strength for a longer period. A pair of keepers, each linking a pair of unlike poles of two similar bar magnets, works in the same way.

Kellner eyepiece type of Ramsden eyepiece, common in prism binoculars. It corrects the aberrations of the original type.

kelvin K unit of temperature and temperature difference on the absolute (thermodynamic) scale. 0 K is absolute zero (−273.15 °C), and 1 K is the same temperature difference as 1 °C (1 degree C).

Kelvin, Lord See William *Thomson*. The **Kelvin balance** was a particularly accurate form of current balance, with a total of six coils rather than the usual two or four, while the **Kelvin double bridge** was a particularly accurate form of Wheatstone bridge for use with low resistances. (Neither is in much use now that electronic meters have such success.) For **Kelvin effects**, see *thermoelectricity*.

kelvin scale of temperature once common, but now rare, name for the absolute temperature scale (whose unit is the kelvin K)

Kendall, Henry (1926–1999) physicist, joint winner of the 1990 Nobel Prize for work on the structure of nucleons leading to the quark model. Carried out in the 1960s, this involved the deep inelastic scattering of electrons produced by the Stanford Linear Accelerator.

Kennelly–Heaviside layer the Heaviside (E) region of the ionosphere. Arthur Kennelly (1861–1939) was an engineer who proposed that the region exists at the same time as did Heaviside, but independently.

Kepler, Johannes (1571–1630) mathematician and astronomer who used the great set of highly accurate measures of the motion of the planets made by Tycho Brahe to prove that each planet moves round the Sun in an orbit that is an ellipse rather than a circle (as Copernicus had proposed). This was a crucial experimental development of Copernicus's heliocentric theory (which proposed circular orbits for the planets round the Sun, so did not tie well to practice). Kepler also devised the so-called astronomical telescope, a refractor with a converging eyepiece (rather than a diverging one as in Galileo's design); his other work in optics is important, and he came close to the law of refraction.

There are three **Kepler's laws** on this (stated for planets in the Solar System, but true for any object in free fall orbit in a gravitational field):
a) Each planet moves in an ellipse, with the Sun at one focus.
b) The line between planet and Sun covers equal areas in equal times (which means the planet moves faster the closer it is to the Sun).
c) The square of the year (period) of an orbit is proportional to the cube of half the major axis of the orbit.
Newton later used the laws to provide a strong basis for his new theory of gravitation.

kernel a) the core of an atom – nucleus plus filled shells – a positive ion
b) function of two (sometimes more) variables which occurs in an integral equation or transform. In nuclear engineering, the **diffusion kernel** is a function that relates the intensity of thermal radiation at one point in a finite homogeneous medium with that of neutron radiation at a second point.

Kerr cell fast optical shutter based on the **electro-optical Kerr effect**. This effect is the appearance of birefringence in a transparent, normally isotropic sample in an electric field; the sample then acts as a birefringent crystal with axis in the field direction. The Kerr cell consists of such a sample between two crossed Polaroids; no light passes until a strong field is switched on.

The **magneto-optical Kerr effect** describes how plane polarised light becomes elliptically polarised by reflection from the pole face of a magnet. The plane of polarisation must be the plane of incidence, or normal to it, for maximum effect.

Ketterle, Wolfgang (1957–) one of the three joint winners of the 2001 Nobel Prize, for being able to obtain Bose–Einstein condensates in 'dilute gases of alkali atoms', and for research into the properties of these condensates

keV symbol for kilo electron volt, 1000 eV, non-SI unit of energy still common in atomic and nuclear physics

key old name for push switch. It is still common in the case of certain designs (such as the Morse key, a tapper switch) and in keyboard or keypad, a set of labelled push buttons.

kg symbol for kilogram, base unit of mass

Kilby, Jack (1923–2005) joint winner of the 2000 Nobel Prize for his part in developing integrated circuits

killer impurity which cuts down luminescence in a phosphor, by trapping electrons

kilo- k unit prefix symbol for 1000. Thus, 1.5 kW is 1500 watts (the power usage of a lawn mower).

The **kilogram** (kg) is the base unit for mass, 1000 g. It is not strictly valid to use a unit multiple as a base unit (but the gram is too small for most purposes); the name giorgi, symbol G, for the kilogram had much support in SI's early years. The **kilowatt-hour**, kWh, is a unit of energy still sometimes used by electricity boards for convenience (and perhaps called a 'unit'). It is the energy used by a 1000 W device (for instance, a single bar electric fire or a pressing iron) in an hour, the same as 3.6 MJ.

In IT, computing in particular, **Kilo-** (often confusingly given with lower case initial) is the prefix for 'binary thousand', in other words for 2^{10}, 1024. Thus a Kilobyte, KB, is a thousand bytes, an important unit of data in store and in transfer.

kinematic concerned with the nature of motion. **Kinematic viscosity** is the (dynamic) viscosity (fluid friction) of a fluid divided by its density; the SI unit is the m² s⁻¹.

kinematics the study of the motion of objects in different situations (but without taking account of the causes of the motion, this being the province of kinetics). Its central measures are displacement, velocity (time rate of change of displacement), and acceleration (time rate of change of velocity). Therefore, the equations of motion lie at the heart of kinematics.

kinetic concerned with motion in general. We often find it convenient to discuss **kinetic energy** W_k (or K or T) as the energy of a moving object (the energy transfer in bringing it to rest in a given frame of reference); the value is

$$W = \tfrac{1}{2}\, m\, c^2$$

Here m is the mass of the object, and c is its speed in the frame of reference in question.

Kinetic equilibrium is the same as dynamic equilibrium (i.e. changing in detail but not overall). **Kinetic friction** is the friction between two surfaces when in relative motion. **Kinetic potential** is an old term for the Lagrangian function (the difference between the kinetic and potential energies of a system).

kinetic model, kinetic theory or **molecular kinetic theory** view of matter that describes physical properties in terms of particles and involves this set of statements:

a) All matter consists of particles.

b) The particles are always moving (i.e. they always have energy).

c) The energy of the particles relates to the sample's temperature.

d) The particles tend to attract each other.

The **kinetic theory of gases** goes further in order to explain in more detail how gases behave:

e) The particles are like hard smooth elastic points.

f) The volume of the particles of a gas is a negligible fraction of the volume of the gas sample.

g) When the particles collide, the time of interaction is negligible.

These ideas lead to these expressions for a perfect gas sample:

$$p = \tfrac{1}{3}\, \rho\, \overline{c^2}$$
$$RT = \tfrac{1}{3}\, n\, m\, \overline{c^2}$$

Here n is the number of particles, of mass m and mean square speed $\overline{c^2}$, in the sample; p is the pressure, T the temperature, and ρ the density of the sample; and R is the gas constant.

Kinetic theories depend on statistics: they deal with the bulk behaviour of large numbers of unpredictable particles (think of the uncertainty principle). Work in this area therefore often carries the name statistical physics.

kinetics the study of the effects of forces on the motion of objects. With kinematics, it forms dynamics. The concern of **chemical kinetics** is the rates and mechanisms of chemical reactions; of special interest is where a reaction does not follow the path predicted by thermodynamics.

Kirchhoff, Robert (1824–1887) chemist and physicist whose pioneer work in spectroscopy and electric circuits is of great value. In the former case, he explained the Fraunhofer lines in the Sun's spectrum and identified there a number of new elements. In the latter context, his rules for ohmic circuits (below) remain in wide use.

The **Kirchhoff formula** is an attempt to relate the vapour pressure p of a substance to its temperature T. The result has some limited validity:

$$\log_e p = a - b/T - c \log_e T$$

Here a, b and c are constants

The **Kirchhoff law** is that the emissivity of a point on a surface in a given direction equals its absorptance for radiation of the same wavelength coming from the same direction at the same temperature.

Kirchhoff's rules apply to ohmic electrical circuits and networks that carry steady direct currents:

1 The current entering any point equals the current leaving it (the law of constant current or charge).

2 The sum of the voltages taken round a complete closed path is zero.

We can reword the latter as:

2 The sum of the emfs in a complete closed loop of a circuit equals the sum of the products of current and resistance in the loop.

Kirchhoff's law The thermal radiation given off by a surface depends only on the temperature.

KK theory shorthand for the *Kaluza–Klein theory*

Klein, Oskar (1894–1977) mathematician and theoretical physicist, who developed the Kaluza–Klein (KK) theory with the idea that the essential fifth dimension is real but curled tightly into a very small space. He also worked towards

• the **Alfvén–Klein model** of the big bang;

• the **Klein fluid** (a view of matter in general relativity);

• the **Klein–Gordon equation** (a relativistic form of the Schrödinger equation used in the quantum theory of the nucleus); and

• the **Klein–Nishina formula** (for Compton scattering of x- and gamma-rays by an electron).

Klitzing, Klaus von (1943–) physicist whose 1985 Nobel Prize followed his research into the quantum Hall effect

klystron an electron beam tube designed to produce high power alternating output at the extremely high frequencies of microwaves; it is therefore a microwave source (or amplifier).

k-meson kaon, an elementary particle in the mass range between electrons and nucleons

knife edge sharp edge, often made of a hard mineral such as agate, used to act as the pivot (fulcrum) of some old types of balance

Knudsen flow flow of gas through a tube at a pressure so low that the mean free paths of the particles are much greater than the radius and length of the tube. Collisions between particles do not affect the flow therefore. The **Knudsen gauge** uses the principles involved to allow the user to measure very low pressures.

Kopp's law The specific thermal capacity of a compound is the sum of the atomic thermal capacities of the atoms in it.

Koshiba, Masatoshi (1926–) physical chemist who shared the 2002 Nobel Prize with Raymond Davis 'for pioneering contributions to astrophysics, in particular for the detection of cosmic neutrinos'

Kroemer, Herbert (1928–) shared the 2000 Nobel Prize with Zhores Alferov 'for developing semiconductor heterostructures used in high-speed electronics and opto-electronics'

krypton, Kr inert gas with no smell or colour, period 4, number 36, discovered in 1898 by William Ramsay and Morris Travers using the fractional distillation of air; the name comes from the Greek *kryptos*, meaning hidden. Density 3.7 kg m^{-3}, melting and boiling temperatures –157 °C and –152 °C; most common isotope ^{84}Kr. The main uses are in the krypton fluoride laser and artificial heart valve pneumatics.

Kundt tube glass tube that contains a scattering of fine powder. When there is a standing sound wave in the tube, the powder forms into heaps at the nodes. This allows an accurate measurement of the wavelength of the sound and, given the frequency, the speed of sound in the fluid in the tube.

Kusch, Polykarp (1911–1993) physicist whose shared Nobel Prize, in 1955, followed work on the magnetic properties of the electron

l unit symbol for litre, non-standard unit of volume, equivalent to 0.001 m³; the symbol L is preferred but not common in Europe

L a) symbol for lambert, old unit of luminance

b) preferred unit symbol for the litre, but common only in North America and Australasia

c) the second electron shell, with principal quantum number (n) value 2

l a) symbol for length

b) symbol for mean free path (but λ preferred), unit: metre, m

c) symbol for orbital angular momentum quantum number

L a) symbol for angular momentum

b) symbol for area (A much preferred)

c) symbol for Avogadro constant (N_A preferred)

d) symbol for inductance

e) symbol for Lagrangian function

f) symbol for luminance

g) symbol for mutual inductance (with subscript m, but M preferred)

h) symbol for orbital angular momentum quantum number

i) symbol for radiance

j) symbol for self-inductance

λ a) symbol for mean free path

b) symbol for thermal conductivity

c) symbol for wavelength (very often with a suffix for some specific value, such as λ_C for Compton wavelength)

Λ symbol for the lambda particle, the least massive hyperon

label easily detectable stable or radioactive nuclide used to investigate a process, such as a biochemical reaction. A **labelled compound** contains such a label. See also *tracer*.

laboratory system the inertial frame of reference of the current observer

ladder transmission line designed as a filter or network to offer a specified delay to passing signals. It consists of a sequence of identical four-terminal (quadric-pole) circuits that contain the appropriate series and parallel reactances.

laevogyric or **laevorot(at)ary** same as levo-, turning the plane of polarisation of passing radiation to the left when viewed from the front

lag a) to cover an object (such as a pipe or tank) with thermally insulating material (**lagging**), to reduce energy transfer to or from it

b) delay between sending a signal and its being received. For instance, using a satellite phone link introduces a noticeable lag of a fraction of a second compared to transfer by cable.

c) the constant time (delay) or angle (phase difference) by which an alternating measure follows behind some other alternating measure. For instance, in a purely capacitative ac circuit, the voltage between the ends of the capacitor lags behind the circuit current by a quarter of a cycle (90°). If the circuit is inductive, on the other hand, there is a **lagging current** (the current lags behind the voltage); the inductor is then a **lagging load**. Compare with *lead*, the reverse. See *hysteresis* for an example in mechanics.

Lagrange, Joseph (1736–1813) mathematician and astronomer who made many contributions to physics in both contexts. When he retired to France he was brought back to work on the new metric system.

The **Lagrange equations** are a set of second order differential equations of motion for a dynamic system, in effect a generalisation and expression of Newton's laws of force. There is one equation for each degree of freedom in the system. The **Lagrange function** L (often called the **Lagrangian**) for such a system is the difference between its total kinetic energy and total potential energy. **Lagrange's law** relates the object and image sizes (l_o and l_i), in media of refractive constants n_o and n_i respectively, to the angles α_o and α_i between the corresponding paraxial rays and the axis:

$$n_o \, l_o \, \alpha_o = n_i \, l_i \, \alpha_i$$

Lagrangian points (L-points) are particular solutions to the gravitational three-body problem. There are five such points in the plane of two massive objects at which other objects, of negligible mass, can remain in equilibrium. A number of people have explored the possibility of putting massive space stations at the Lagrangian points of the Earth–Moon system. There may, of course, be rocks there already (as is the case with the thousands of 'Trojan asteroids' in the Jupiter–Sun system); there are certainly faint clouds of dust at the two main Earth–Moon L-points.

Lamb, Willis (1913–) physicist whose work on the hydrogen spectrum gained him a share in the 1955 Nobel Prize. Part of that work led him to discover the **Lamb shift**. This is a shift (or broadening) of spectral lines as a result of the zero temperature (zero point) changes in the electromagnetic field around the source in question. The effect was first observed half a century ago in radar signals.

lambda λ or Λ the Greek letter ell, symbol for a number of measures in physics (in particular, for wavelength and thermal conductivity). A **lambda leak** (or superleak) is the escape of superfluid helium through holes too small for normal fluid particles. The **lambda particle** is the least massive hyperon (baryon with non-zero strangeness); it decays to a nucleon. It is a singlet, having only a neutral form (and a neutral antiparticle). The **lambda temperature** of a substance is the temperature at which some specific second order transition takes place. For instance, in the case of liquid helium, it is the temperature of the transition between the two phases, I and II, around 2.2 K.

lambert L old unit of luminance, the same as one lumen given out by one square centimetre of a surface

Lambert's law (Johann Lambert, mathematician, physicist and astronomer, 1728–1777)

a) The observed intensity of radiation from an element of a black body is proportional to the cosine of the angle between the direction of view and the normal to the element.
b) Also called Bouguer's law: Equal thicknesses of a medium absorb equal fractions of passing radiation. This means that the intensity I of radiation decays exponentially with depth x in the absorber:

$$I = I_0 \exp(-a\,x)$$

Here a is the medium's linear absorption constant for the radiation in question.

lamina thin flat sheet, taken in theory to have zero thickness. When a fluid transfers with **laminar flow**, we can view it as being in parallel sheets of constant speed. This is smooth (non-turbulent) flow; at some critical speed, the flow becomes turbulent: then the particles in a given lamina no longer all move at the same speed.
Lamination is the process of forming the iron core of an electric machine from thin sheets. The oxide layer on each surface is an effective electrical insulator when the sheets are clamped together. As there can no longer be large scale eddy currents flow, the approach much reduces the loss of useful energy to these.

lamp any artificial source of light (or other nearby radiation bands) used for experimental purposes (for instance, for testing or as an indicator). In most cases, lamps use electrical power and have small dimensions. There are many designs now, but three main classes: the filament (incandescent) lamp, the discharge (including fluorescent) lamp, and the laser/light emitting diode (led). See also arc and luminaire.

Lamy's rule (Bernard Lamy, mathematician, 1640–1715) More often called the sine rule of equilibrium: An object is in equilibrium when acted on by three coplanar forces F_1, F_2 and F_3 if:

$$F_1/\sin a_{23} = F_2/\sin a_{31} = F_3/\sin a_{12}$$

Here angle a_{23} is the angle between forces F_2 and F_3.

Land, Edwin (1909–1991) physicist who invented Polaroid in 1932. This was a great advance: previously people had to explore polarisation of light using bulky and costly Nicol prisms as polarisers and analysers.

233

Land set up the Polaroid Corporation in 1937 to make the most of his discovery, and also developed the 'instant' Polaroid **Land camera** in 1947. His research into colour vision is also of major importance.

Landau, Lev (1908–1968) physicist who explored many fields of quantum theory (including astrophysics, the nucleus, and solid state theory). His work in low temperature physics was of particular value; his theory of the superfluidity of liquid helium gained him the 1962 Nobel Prize.

Landé the **Landé factor** or **g-factor** (the old name for *gyromagnetic ratio* in the case of a particle): factor of proportionality that appears in expressions for the anomalous Zeeman effect. Here there is a fine structure splitting in energy levels in a magnetic field. The value of g depends on the orbital and spin angular momentum quantum numbers of the level in question; it ranges from 1 (pure orbital) to 2 (pure spin).

Langevin, Paul (1872–1946) physicist whose main work was in the structure of gases and magnetism; in particular his theory of diamagnetism is of much importance now.

langley out-dated unit of solar radiation density, approximately 42 kJ m^{-2}. The main areas of work of Samuel **Langley** (1834–1906) were in the design of flying machines and the relation between solar radiation and weather. He invented the bolometer in 1878.

Langmuir, Irving (1881–1957) chemist (Nobel Prize for chemistry, 1932) who developed the theory and practice of thin film physics and chemistry. The **Langmuir effect** is the ionisation of certain atoms in contact with certain hot surfaces. The atoms must have low ionisation energy and the surface must have a high surface energy for this to happen; however, the effect is in wide use in ion sources. Langmuir waves are oscillations in a plasma (e.g. the free electron 'gas' of a metal) that result from a disturbance – see also *plasmon*.

lanthanides the **lanthanum series** of elements, with proton numbers 58 (cerium) – 71 (lutetium), that follow lanthanum (57). They form a transition series within a transition series, and therefore have very similar chemical properties (as their outer electron shell structures are almost the same). The **lanthanide contraction** is the drop in atomic radius along the series.

lanthanum La transitional rare earth metal of the lanthanide series, period 6, with proton number 57, discovered in 1839 by Carl Mosander and named after the Greek *lanthano*, meaning 'to escape notice'. Relative density 6.2, melting and boiling temperatures 920 °C and 3500 °C; only stable isotope ^{139}La. Main uses are in carbon lamps and certain specialist alloys.

Laplace, Pierre (1749–1827) mathematician and astronomer who proposed the nebular hypothesis (for the formation of the solar system) we hold today. The **Laplace equation** is the basis of theories of potential; it gives the potential π at points in a field with zero net charge as

$$\delta^2\pi/\delta x^2 + \delta^2\pi/\delta y^2 + \delta^2\pi/\delta z^2 = 0$$

In terms of the **Laplace operator** (or **Laplacian**) div grad (∇^2):
$$\nabla^2\phi = 0$$

See also *Poisson equation*.

Last, the **Laplace transform** is a special case of the Fourier transform.

lapse rate the rate of change of temperature with height in the troposphere, the lowest region of the atmosphere, up to about 7 km (at the poles) or about 18 km (at the equator). This rate is not constant: it varies with latitude, with height and with time. Sometimes there is a temperature inversion: then the lapse rate is positive for a certain range of height; normally, though, temperature falls with height up to the tropopause, the junction with the stratosphere.

large electron-positron collider CERN's predecessor to the large hadron collider – see *lep*

large hadron collider (lhc) from early 2008, CERN's newest and the world's most powerful (circumference 27 km) particle accelerator. This is a proton synchrotron designed to explore the Standard Model of matter and the Higgs boson this depends on – see *nuclear physics*.

large scale integration (lsi) the integration of hundreds of thousands of microelectronic circuit elements in the surface of a single semiconductor chip

Larmor precession the gyromagnetic effect, the precession of the orbit of a charged particle in a magnetic field; it occurs in the direction of the field.

In the case of an orbital electron in a field,

the pulsatance (angular frequency) ω of the precession is $Q_e H/2 \, m_e \, c$; here Q_e and m_e are the charge and mass of the electron, H is the magnetic field strength, and c is the speed of light in empty space. The **Larmor precession frequency** (gyromagnetic frequency) is this value divided by 2π. See also *magnetic resonance* and the *Zeeman effect*.

laser

oscillator device able to produce an intense parallel beam of coherent near monochromatic radiation, in particular at a frequency in or near the visible region of the spectrum (see also *maser*). The name comes from the process involved, light amplification by stimulated emission of radiation, the process being such as to induce atoms to emit photons in phase with each other. Thus the output is coherent.

The first lasers produced a pulsed output from a crystal (for instance, of ruby). The active atoms are excited from the ground state (energy W_0), via an unstable state (energy W_2), to a meta-stable state (W_1), by an extremely intense and short lived flash of light; the process forms a population inversion that may last a large fraction of a second.

If the photon from one de-excitation passes close to a second excited atom, the latter decays as well (by stimulation); the two photons pass on together, in phase and parallel. While there is a great leakage of radiation from the sides of the crystal (efficiency is very low), the beam parallel to the crystal axis grows as it passes back and forth between the silvered ends. One of those ends is partly transparent, so the output pulse produced during the time that all the excited atoms decay to their ground states itself consists of a number of very brief pulses.

Modern lasers have a continuous output. Most consist of certain mixtures of gases in a tube; an electric discharge inputs energy at a constant rate high enough to maintain the population inversion. Lasers with a liquid dye as the active substance are tunable: the user can vary the output frequency over a certain range. Solid-state lasers no longer use the ruby laser system, except for very high output power (in which case – e.g., nuclear fusion research – neodymium glass lasers can be tens of metres long and deliver kilojoules per pulse); on the other hand, semiconductor lasers (below) and laser diodes (see *led*) can be tiny. A system that wholly differs from these is the *free electron laser*: laser action takes place in a very high energy electron beam.

Note that the output of no laser can be truly monochromatic, for reasons of spectral line broadening and the uncertainty principle. A typical bandwidth is 10^3–10^4 Hz: this is extremely narrow when compared to the frequency of light. Nor can the output beam be truly parallel, because of diffraction. See *stimulated Raman scattering*.

There are many uses for lasers (though for a long time people described the laser as a solution in search of a problem). The uses depend on one or more aspects of the high energy, high frequency, parallel, monochromatic, coherent output beam. They include surveying (see *lidar*, for instance), welding, cutting, spectroscopy, surgical techniques, guidance, communications (see further below), and holography.

The **semiconductor laser**, like any microelectronic device, offers very low size and power consumption as well as high reliability, sensitivity, and speed of action. The efficiency is important too: thus, some semiconductor lasers can convert 10% of the input power to laser light; this compares to 0.1% in the case of gas lasers working in the same region of the spectrum.

A semiconductor laser and detector pair has many uses; in, for instance, the compact disc read head (pickup), optical information processing, and fibre optics, all in some sense concerned with communication. Recent other uses are to monitor the growth of crystals and how particles behave in a liquid; in space, such experiments cost less than a tenth as much as those using conventional lasers. Also possible is **x-ray laser** action, there being various methods in use to attain it; see *x-ray*.

The **laser gyroscope** is in wide use in aircraft to aid navigation; it is cheaper, smaller and more reliable than a mechanical gyro (which it has almost completely replaced). In the spinning system, mirrors split and guide the laser beam both ways round a circle in each plane; movement round the axis of the circle leads to a change in frequency between the pair of beams, giving a beat effect at the detector. More modern versions use light fibres rather than mirrors to guide the beams.

latent hidden from the senses. In calorimetry, **latent energy** (still sometimes called **latent heat**) is the energy involved in a change of state or change of crystal form. During the energy transfer there is no temperature change to sense with a thermometer (or otherwise): hence the name 'hidden'. The **latent thermal capacity** of a sample is the energy involved in changing its state as specified (for instance, melting/freezing or from one crystal structure to a second); the unit is the joule. The **specific latent thermal capacity** of a substance is the latent thermal capacity of unit mass; the unit is the joule per kilogram, J kg^{-1}.

A **latent image** is one hidden in an exposed

photographic film before being developed to make it visible.

lateral to the side. **Lateral aberration** is a result of chromatic aberration; the images in different colours differ in size (across the axis). For **lateral inversion**, the left-right relation between optical object and image, see *perversion*. In an optical system, **(lateral) magnification** is the ratio of image size (height) to that of the object.

lattice a regular arrangement of points in two or three dimensions, as with the atoms of a crystal. In that particular case, **lattice conduction** is the transfer of thermal energy through a solid by the interaction of the vibrations of the fixed particles in the lattice (i.e. not by free electrons); a **lattice constant** or **lattice dimension** is the length of an edge of the unit cell. The **lattice energy** of a crystal is the energy needed to separate a pair of its ions to infinity at o K.

A **layer lattice** is a crystal structure in which all (or nearly all) the atoms or ions are in parallel planes, with none or few between these. Any substance with this structure (for instance, clays and graphite) shows easy slip between the planes, so is a good lubricant.

Laughlin, Robert (1950–) one of the three winners of the 1998 Nobel Prize, for their discovery of a 'new type of quantum fluid with fractionally charged oscillations'

law statement in science that has stood the test of time in that no one has so far been able to disprove it. It is never possible to prove a theory; all the same, many statements in physics have gained the status of law. There are, for instance, many **conservation laws**; each one states that during some given type of interaction, some given property of the system in question remains constant. In classical physics, the laws of constant charge, energy, mass and momentum (and others) are of this type; the same applies in modern physics to such measures as iso-spin, parity and strangeness.

Lawrence, Ernest (1901–1958) nuclear physicist who invented the cyclotron in 1929 and built the first working one two years later. This gained him the 1939 Nobel Prize. The trans-uranic element with proton number 103, the last of the actinide transition series, takes its name – lawrencium – after him.

lawrencium Lr transitional metal in the actinide series, period 7, with proton number 103, first prepared in 1961 by Albert Ghiorso's

team and named after Ernest Lawrence. Its properties are unknown as it is highly radioactive, the half-life of its most stable isotope, ^{262}Lr, being four hours; for the same reason, there are no uses.

laws of force (or **laws of motion**) See *Newton's laws*.

L-capture less common form of electron capture than K-capture, as it involves electrons from the less central L shell – in other words it is a form of beta decay in which a nuclear proton combines with an electron from the L-shell to form a neutron.

LC circuit circuit with impedance but no resistance, i.e. with some capacitance and/or inductance. Switching on a capacitative circuit causes a transient current peak as the voltage across the capacitor starts to grow to equal that supplied. Switching on an inductive circuit does much the same; the transient this time is the back emf that falls as the current grows to its new value.

lcd liquid crystal display, a type of low power visual display (in, for instance, watches, calculators and flat screens), in which the working substance is a nematic liquid crystal

LCR circuit ac circuit that contains some combination of inductance L, capacitance C and resistance R. Each of these offers a reactance, and the size and phase of the current in the circuit depend on their net effect – the circuit impedance – as well as on the applied voltage. See also *LC circuit*.

lead a) the constant time (delay) or angle (phase difference) by which an alternating measure changes in front of some other alternating measure. For instance, in a purely inductive ac circuit, the voltage between the ends of the inductor leads the circuit current by a quarter of a cycle (90°). If the circuit is capacitative, on the other hand, there is a **leading current** (the current leads the voltage); the capacitor is then a **leading load**. Lag is the reverse.

b) conductor used to feed power or a signal into a circuit or device

c) symbol Pb, soft, fairly inert, dense metal in group IV period 6 with proton number 82, known since ancient times (some 7000 years ago); the origin of the name is unclear, but the symbol comes from the Latin name *plumbum*. Relative density 11.3, melting and boiling temperatures 330 °C and 1740 °C; ^{208}Pb is the most common isotope, end of the thorium radioactive decay series. The metal's main uses are in accumulators, certain pigments, radiation shielding, lead glass, and sheathing for high voltage power cables; many older uses (e.g. plumbing, roofing, ballast, lead shot, and weights for fishing lines) are no longer approved – the metal is toxic.

The **lead equivalent** of a sample of potential radiation shielding material is the thickness of lead giving the same protection against the same radiation. The **lead/acid cell** is a very common secondary (rechargeable) electric cell. It has two interleaved sets of plates: spongy lead plates in series with the cathode and lead oxide plates connected to the anode; the electrolyte is sulphuric acid. The output of a fresh cell is 2.2 V on open circuit, falling to 2 V when it supplies a current.

leakage any low rate transfer from a store to the outside, normally not desired. Thus, in a magnetic circuit, the **leakage field** is that part of the flux (perhaps 20%) that escapes from the circuit so is of little or no use. In much the same way, the **leakage reactance** of a transformer is the reactance that results from the part of the field that cuts one coil only.

least action describes the path taken by a system between two states, when all the other possible paths involve a greater *action (sense c)*. Fermat's rule is a special case. In theory, the actual path may also have the greatest action, but this does not seem to happen.

least (potential) energy the state of a system in equilibrium; all the other states give it a greater total potential energy This relates to le Chatelier's rule.

least squares Legendre's approach to finding the line best fitting a set of points on a graph (or the corresponding equation). The best result is the one which has the smallest sum of the squares of the distances of the points from it.

least time as in Fermat's rule, describes the path of a ray through a system, all other paths taking a longer time. See also *least action.*

le Chatelier's rule If a system in equilibrium changes, the system reacts to minimise the effect. This follows from the law of constant energy; one of the various special cases in physics is Lenz's law of electromagnetic induction. Henri **le Chatelier** (1850–1936) was a chemist: the rule is of great value in dealing with chemical reactions.

Leclanché cell primary electric cell in very wide use. The original 'wet' form is now rare; this has a carbon rod for anode, a zinc cathode, and an electrolyte of ammonium chloride solution, to give around 1.5 V. The highly portable modern version, the dry cell, also gives that output; between the central carbon rod and the zinc case is a paste of ammonium chloride, other substances and gum.

led $\dashv\!\!\triangleright\!\!\vdash$ *light emitting diode,* small high efficiency light source, a semiconductor or other junction diode which emits a pure colour (narrow-band, but incoherent) when passing a current. The cause is a type of electroluminescence. The colour – in the range infrared to far ultraviolet (210 nm being the record at the time of writing), but often red – depends on the chemical nature (main band gap width) of the semiconductor used. Efficiency is not high, but still double that of an incandescent lamp. Shuji Nakamura received the second Millennium Technology Award, in 2006, for developing light emitting diodes of different colours, including white, and the blue laser, all sources of value in applications, including in less developed countries.

An **led display** is an oblong grid of closely packed leds, each able to switch under outside control to display static or moving patterns of text and/or pictures. An early small scale use of the led display was for calculators and watches; the leds give a much brighter image than lcds, but have too high a power demand.

The **laser diode** is an led designed to produce coherent monochromatic output; it involves injecting holes from the p-region into the n-region and electrons the other way. There are various types and many uses – see en.wikipedia.org/wiki/laserdiode; www.holoworld.com/holo/diode.html provides some help with using a laser pointer to produce holograms.

Lederman, Leon (1922–) Nobel Prize winner, 1988, shared with Steinberger for the development of the 'Standard Model' of particles as quarks and leptons

Leduc effect A temperature gradient appears across a metal sample at 90° to a magfield when there is a thermal energy flow through the sample. (The Nernst effect concerns the voltage that also appears.)

Lee, David (1931–) one of the three co-winners of the 1996 Nobel Prize, for their discovery of superfluidity in ^{3}He

Lee, Tsung-Dao (1926–) physicist who shared the 1957 Nobel Prize with his colleague Yang for the experiment that showed the violation of parity. In fact, this is widely called 'Mrs Wu's experiment', after Chien Wu: she actually carried it out; the result was a surprise to Lee and Yang.

Lees' disc method for finding the thermal conductivity of an insulator. A thin disc-shaped sample fits between two thick metal discs, both carrying thermometers, one of these being at a high temperature.

van Leeuwenhoek, Anton (1632–1723) microscopist who followed up the early work of Hooke to make many fundamental discoveries in biology with the new compound microscope

left-hand rule Fleming's motor rule: put the thumb and first two fingers of the left hand at 90° to each other. With the first finger in the direction of the magnetic field and the second in that of the current, the thumb gives the direction of the motor force on the current.

Legendre equation with x a complex variable:

$$(1 - x^2)\, d^2y/dx^2 - 2\, x\, dy/dx = n\,(n - 1)\, y = 0$$

The solutions to this second order differential equation are **Legendre functions** if the constant n is not an integer; they are **Legendre polynomials** if it is. The **associated Legendre equation** is much the same, and has much the same solution, but involves two constants rather than one. The equations appear often in advanced physics. **Adrien Legendre** (1752–1833) was a mathematician who first devised the method of least squares.

Leggett, Anthony (1938–) joint winner of the 2003 Nobel Prize 'for pioneering contributions to the theory of superconductors and superfluids'

Leibniz, Gottfried (1646–1716) philosopher and mathematician whose method of calculus (rather than that of Newton) is the one we use today. He was elected to the Royal Society in 1673 for his invention of an important calculating machine.

Leidenfrost effect (Johann Leidenfrost, doctor and theologian, 1715–1794) the 'dancing' of small liquid drops over a very hot surface as a result of an insulating cushion of vapour. As long as the surface temperature is above the **Leidenfrost temperature**, in turn well above the liquid's boiling temperature, the cushion prevents boiling. It is not certain that the

effect explains how people can 'walk on fire' but it does explain various other unsafe physics tricks.

Lemaître, Georges (1894–1966) physicist who first proposed the big bang theory of the origin of the Universe we hold today

Lenard, Philipp (1862–1947) physicist whose work on cosmic rays led him to the theory that the atom is mostly empty space: and to the 1905 Nobel Prize. He also carried out early tests on the photoelectric effect, being the first to show that electrons are produced. The **Lenard mass absorption law** is that the absorption of high energy electron radiation depends only on the density of the absorber.

length *l* unit: metre, m one of the chosen prime dimensions of SI and therefore of modern science. We define the unit – the metre, m – in terms of the speed of light in empty space.

Lenoir, Jean Étienne (1822–1900) electrician who invented the first successful internal combustion engine, in 1859, a two-stroke, single cylinder device running on town gas. Though used to drive a road vehicle a couple of years later, most of his engines made were for fixed use (e.g. for pumping). The Otto engine followed Lenoir's.

lens shaped piece of transparent material used to converge or diverge light passing through it by a set amount by refraction, or the corresponding structure or arrangement able to produce a similar effect in the case of other radiations. See, in particular, *dielectric lens, gravitational lens, magnetic lens* and *electron lens*. The shape involves the two surfaces on the axis having specified smooth curvature.

It is best to describe a lens by its effect on an input beam: **converging lenses** tend to bring this towards a focus (focal point), while **diverging lenses** have the opposite effect. It is assumed that the lens material is of higher refractive constant n than its surroundings; if the reverse is true, so are the effects.

Once it is made, the effect of a lens is fixed; we describe this in terms of its focal distance *f*. (However, note that the lens of the eye varies in power: a ring muscle makes it more or less converging by changing its shape. Also, a **liquid crystal lens** has a focal distance you can control, often over a very wide range.)

The value of the focal distance (often called focal length) of a lens depends on the refractive constant n of the lens material and the radii of curvature r_1 and r_2 of the two faces, as

$$1/f = (n - 1)(1/r_1 + 1/r_2)$$

where we assume the lens is for use in air or vacuum.

Viewed through a lens, light from an object at some distance, u, from the lens will appear to come from the image of the object; this is at distance v given by the **lens formula** as:

$$1/v = 1/f - 1/u$$

Throughout we use the convention that distances to real points are positive and those to virtual ones are negative. We also assume paraxial rays, ones close and nearly parallel to the axis. All the above, with suitable modification, applies to lenses handling any form of radiation.

All the above describes single (simple) thin lenses. A **compound lens** consists of two or more single lenses on the same axis: close, in contact or glued together. A **lens doublet** has two such lenses, a triplet three, and so on. See *compound lens*. In a **varifocal lens**, one can adjust the distance between the elements of a doublet to control the overall focal distance. On the other hand, a **zoom lens** adjusts in the same way to change the magnification without changing the focal distance See also *Fresnel lens*, a flat lens with the same effect as a spherical design.

Following the success of materials with graded refractivity in fibre optics, there is now a great deal of research into lenses made the same way. Should this succeed, effective low-mass lenses should become much cheaper. Some firms have claimed to be able to vary the refractive constant of lens material by up to 30% in a 10 mm layer.

Lenz's law the third law of electromagnetic induction: the direction of an induced voltage tends to oppose its cause. This follows from the law of constant energy and relates to le Chatelier's rule. Heinrich **Lenz** (1804–1865) was a physicist mainly working on the links between electricity and thermal energy.

lep CERN's 'large electron–positron collider', a ring 27 km round working from 1989 to 2000 able to produce and collide beams of electrons and beams of positrons with a total energy of rather over 100 GeV. Since lep closed, the space has been used for the large hadron collider, lhc. Among the triumphs of the lep are to provide support for various

aspects of the standard model of matter; lhc is to continue along this route, for instance finding the Higgs particle, which lep could only hint at.

lepton an elementary particle with spin $\frac{1}{2}$ (i.e. a fermion) that the strong nuclear force does not affect. In the standard model of matter, the lepton is the only fundamanetal particle other than the quark. The lepton group consists of electrons, muons, tau particles (these being the only leptons that decay into hadrons), the neutrinos that 'belong' to these three 'flavours', plus the antiparticles of all of them. The **lepton number** of a system is the number of leptons less that of anti-leptons; it is a conserved (constant) value (apart from a very small, and as yet unexplained, anomaly).

lever simple machine that consists of a long rigid rod and some form of pivot (or 'fulcrum', Latin for bedpost). The force and distance ratios depend on the distances between the pivot and the input and output forces (effort and load). Efficiency can be high, the main losses of input energy coming from friction, slippage, deformation of the rod, and the weight of the rod.

levogyric or **levorot(at)ory** of an optically active substance, turning the plane of polarisation of passing plane polarised radiation to the left (as viewed from the output side). The reverse is dextrogyric or dextrorotatory.

Leyden jar early capacitor, a glass jar lined in and out with metal foil. The name comes from the place in the Netherlands where the device first became popular as a store of charge.

lf low frequency, in the case of radio waves being in the range 3–30 kHz

lhc the large hadron collider, CERN's newest and largest particle accelerator

Lichtenberg figure complex star-like pattern formed by dust particles on the surface of an insulator at high potential

lidar light-based direction and ranging, a technique using narrow light beams to scan the environment and build up a picture. Most such current systems use lasers because of their highly parallel and pure output (though the beams can harm the eyes of animals in the way).

A major use of lidar is to assess and measure the concentration of pollutants in air over ranges of several kilometres. Remote, automatic monitors apply the technique; the wavelengths (infrared, visible, ultraviolet) used relate to the groups of pollutants that absorb the beam and thus reduce the reflected signal. Using a tunable laser, **differential absorption lidar** involves sending out in a given direction to a distant mirror pulses of absorbed and non-absorbed light. The difference in the reflected signals allows a precise measure of even very low concentrations of pollutants.

Other uses are in very precise surveying (for instance in forestry, seismology, astronomy and fusion physics), traffic speed sensing, and meteorology.

lifetime the same as mean life, a measure of rate of decay and similar in very many contexts

lift upward force on a foil moving through a stream of fluid. The cause is the difference in pressure between the upper and lower surfaces of the foil; its size depends also on the angle of attack, the angle at which the foil meets the flow.

light

band of electromagnetic radiation that gives a nervous sensation (vision) when it meets the retina of the eye. The wavelength range is of the order of 750 nm (red) to 400 nm (blue): in vision, wavelength relates to colour, see also *spectrum*. People's eyes vary in response to light, the visible range also falling somewhat with age.

Light radiates from surfaces at temperatures above around 900 K; below about 6000 K, however, most of the radiation is in the infrared region. Thus only about 2.5% of the radiation from a household filament lamp, working at 3000 K, is visible. 'Cold' light sources, such as fluorescent tubes, are more efficient in this respect. These produce light by specific transitions between electronic energy levels so have a line spectrum rather than a continuous one.

The nature of light was unclear for centuries, and there was controversy between those holding the particle view and the wave theorists since the end of the middle ages. Huygens showed how to explain the reflection and refraction of light in terms of wave radiation; Newton preferred the particle view ('corpuscular theory'). In 1801, the latter approach could not explain Young's interference experiments, and Fresnel soon proved the detail of wave interference mathematically.

By the end of the nineteenth century, however, people found they could not explain the facts of black body radiation and the photoelectric effects using the wave theory; Einstein and others developed the quantum theory to relate the two opposing views. The quantum of light is the photon.

We now accept that light radiation has both wave and particle (photon) natures. The relation $W = h \nu$ brings them together as required, W being photon (quantum) energy and ν wave frequency (h is the Planck constant).

Even so, we use the wave picture for most situations (in

particular for interference, diffraction, polarisation and scattering) and the photon model for black body radiation, photoelectricity, and such interactions with bulk matter as the Compton effect. The same dual approach applies in other sciences where light is involved, for instance in the processes of photosynthesis and photography.

A **light-emitting diode** (*led*) is a semiconductor junction diode which gives out light of a pure colour (often red) when passing a current. In use, the led is given a forward voltage (bias), and this applied voltage forces minority carriers across the junction; here they re-combine with majority carriers to release energy as light photons (and temperature rise). The process involved is electroluminescence. The wavelength of the output depends on the size of the energy gap in the semiconductor. The brightness of an led depends directly on the size of the current; too large a current will, however, overload the device and cause it to fail. It is normal, therefore, to mount a current limiting resistor in series with an led. See *led* the symbol.

Light exposure is a measure of the total energy received by unit area of a surface from visible (or other) radiation. It is the product of illuminance and illumination time; the unit is the lux second. For **light guide**, a thread of pure glass of graded refractive constant able to retain a light beam for a long distance, see *fibre optics*. A **light meter** gives a reading of illumination; it is some form of photo-cell.

light absorption the loss of certain wavelengths, and thus of brightness, as light passes through a transparent medium (such as a glass with metallic impurities, or even the purest glass of the best fibre). See also *absorption, photoelectric effects, scattering,* and *stellar spectral classes.*

light-emitting diode See *led* and under *light.* See also *organic light emitting diode* (oled).

lightning electrical discharge between cloud regions of opposite sign, or between the cloud base and the ground. The stroke itself, if visible, is a forking spark, hence the name **fork(ed) lightning**. If one sees only the reflection of the flash among the clouds, rather than the flash itself, the effect is **sheet lightning**. In either case, there are in fact several separate strokes in a flash; **ribbon lightning** is the effect of seeing those several strokes in turn when there is a very strong wind.

The energy output of the flash (tens of megajoules) is of light from the recombination of the ions formed by the discharge current. The current also raises the temperature of the channel very significantly; the pulse of sudden expansion causes the radiation of the crack of sound (rumbling being the result of reflections from the ground). A lightning flash also produces a strong radio energy pulse, detectable at some wavelengths right around the world in the form of a whistler.

The cause and nature of **ball lightning** are far from clear; indeed, many doubt that it exists. There are, however, lots of stories of a ball of light moving slowly and without sound through aircraft and buildings during a thunderstorm. After a few seconds, the ball disappears with a bang.

The **lightning conductor** is much better called **lightning rod** as it does not have the task of conducting the current of a lightning flash. (Indeed the high energy of a lightning stroke would melt the rod.) Its role is to discharge the charge that builds up on the Earth's surface under a highly charged 'thunder cloud'. The process involved is the action of points at the upper end, the lower end being earthed.

light year unit of distance on an astronomical scale, the distance travelled by light through empty space in a year, close to 10^{16} m. See also *parsec*, 3.26 light years.

limit the most extreme allowed value in some context. For instance, the **limit of proportionality** is that value of the deforming stress applied to a sample when the system is at the end of the linear part of the stress/strain graph (see elasticity).

limiter governer in an electric circuit, a device whose design is to keep some value (such as the circuit current or voltage) within a set range or below a set limit. For instance, a Zener diode is a voltage limiter.

limiting current the highest value of current a given electrolytic cell can pass. The factor leading to this is the rate of diffusion of the ions of the electrolyte in the field. This cannot be greater than a certain value; it depends in turn on their concentration and nature. See also *ionic atmosphere*.

limiting friction the greatest value of friction force able to appear between two surfaces. If there is no relative motion between the surfaces, we have a case of static friction; if there is sliding (or dynamic) friction. The limit value in the latter case is less than that in the former.

linac linear (particle) accelerator, one giving energy to particles as they pass through a long straight tube

Linde liquefaction (Carl von Linde, refrigeration engineer, 1842–1934) the process of turning a gas into a liquid by iso-thermal compression followed by expansion through a nozzle. This causes the temperature to fall by the Joule–Kelvin effect. The gas cooled this way lowers the temperature of the next batch in a cycle that continues until a liquid appears.

line

a straight one-dimensional figure (with no thickness therefore) defined by any two points and linking them. For **line defect**, see *defect*. The **line frequency** of a television signal is the number of screen lines broadcast (and displayed) per second; it is the product of the number of lines per frame and the frame frequency (the number of frames per second). A **line integral** is the integral along the line of a curve of any function that is continuous and single-valued along that curve. A **line of force** is the path in a force field followed by a defined point object affected by the force; it is of value to draw fields using such lines (though, unless the field is quantised, they have no reality). Then, such lines (which cannot, by definition, cross) show, by their direction at any point, the field direction and, by their density, the field strength. A **line of magnetic induction** is a name, still sometimes used, for the same concept in the case of magfields specifically.

  b single bright or dark image in a **line spectrum**. This is a spectrum that consists of a series of lines rather than a continuous band; each line relates to a specific transition between energy levels. For various reasons, no such spectral line (or other resonance) can be perfectly sharp (monochromatic); in other words, there are various causes of **line broadening** to make the line wider on a wavelength graph. These include the uncertainty principle; the physical nature of the crystals giving a line x-ray spectrum; and processes such as Doppler broadening, pressure broadening and Stark broadening. In such cases, the line width gives, in wavelength or frequency terms, its actual range of values; see *half width*, the usual measure, and see also *broadening*.

  c channel (path, not always sharply defined) for a signal or power supply. **Line noise** is the existence in a channel of unwanted signals (from, for instance, interference or crosstalk) that make the wanted signal harder to receive. **Line speed** is the rate of data transfer through a digital channel, often given in bits per second. In the case of a pair of power lines, **line voltage** is the voltage between the lines.

linear concerned with transfer in a straight line, or showing a straight line when plotted on a graph. The **linear absorption constant** of a sample of matter appears as α in this relation for its absorption of input radiation (though sometimes appears as a). An old name for absorptivity, this applies only when absorption is the only process of attenuation; see *linear attenuation constant*, below, and *Lambert's law*.

$$I = I_0 \exp(-\alpha l)$$

Here I_0 is the input intensity and I that after the radiation has travelled a distance l in the substance.

A **linear accelerator** (linac) is an accelerator that gives energy to charged particles in a beam as they pass along a long straight cylinder. In a common design, the track contains a series of tubes of increasing length, alternate ones being linked to one or other side of a radio frequency (rf) alternating voltage. A particle in the beam accelerates in the gap between two tubes; while it drifts through the next one, the voltages of all the tubes switch. Thus when the particle leaves each tube, it is again in an accelerating field. See *nuclear physics: particle accelerator*.

A **linear amplifier** is one the amplitude of whose output is directly proportional to that of the input (usually digital, i.e. pulsed).

The **linear attenuation constant** of a sample of matter appears as μ in this relation for its reduction of intensity I of input radiation by scattering and absorption. For cases when absorption is the only process of attenuation, see *linear absorption constant*, above.

$$I = I_0 \exp(-\mu l)$$

Here I_0 is the input intensity and I that after the radiation has travelled a distance l through the substance.

For **linear charge density**, see *charge density*. A **linear function** of two variable measures y and x gives a straight line graph. Its form is $y = mx + c$; m is the gradient (slope) of the line, and the constant c is where the line meets the y axis (i.e. it is the value of the y-intercept). In a **linear circuit**, there is a continuous linear relation between output and input – in other words, the output is always the same multiple of the input; this is therefore an analog circuit. The **linear energy transfer** (let) of a radiation passing through matter is the energy deposited in the medium by the radiation, per unit distance. It is the same as the linear stopping power (below) of the medium for that radiation. The **linear expansivity** α of a substance is the change in length Δl of a sample when the temperature T changes by 1 K (see expansion):

$$\Delta l = \alpha \, \Delta T$$

Linear extinction constant is an old name for linear attenuation constant (above). A **linear inverter** is an amplifier whose output is always the inverse (i.e. the NOT) of the input; see *inverter*. The **linear momentum** p of an object of mass m moving at velocity v is the product mv; see *momentum*. The law of **constant linear momentum** states that this does not change for any interaction within a closed system.

The **linear motor** was invented by Wheatstone in 1845, and developed over a century later into an efficient working system by Eric Laithwaite (1921–1997, engineer and populariser of science). It has the same form and action as a rotary induction motor, but the 'rotor' and 'stator' are both flat rather than cylindrical. As a result, the output is a straight line force rather than a torque. The design is used in some types of pump, vehicles (for instance, some hover trains), and in the rail (space) gun. In each case, the 'rotor' lies along the track and the 'stator' is the moving part. For **linear oscillation**, see *oscillation*.

Linear polarisation is a rare term for plane polarisation. **Linear programming** applies concepts of algebra and calculus to a set of unrelated variables in order to obtain the best solution to a problem that involves them. You express each variable as a linear function between set limits, and then relate these on a graph or by computer. The **linear stopping power** of a sample of matter is the energy loss of a charged particle passing through unit distance in it; the unit is $J\ m^{-1}$.

linkage a) set of rigid rods, linked to each other and constrained in how they can move, designed to transfer motion between two points (and perhaps amplify it by a lever effect as well)
b) the product of magnetic flux and the number of turns in the coil or circuit through which it passes. The unit is the maxwell turn.
c) the coupling between two circuits that allows energy transfer

liquefaction process of turning a gas into liquid form. There are various ways to do this, see, for instance, *cascade* and *Linde liquefaction*.

Lippmann, Gabriel (1845–1921) physicist who invented the first (but rather impractical) process of colour photography, in 1892, gaining the Nobel Prize as a result, in 1908. He also predicted, but could not find, the piezoelectric effect.

liquid state of condensed matter in which the attractive (cohesive) forces between the particles are enough to produce a surface but not enough to prevent continuous random movement. Therefore, liquids are fluids: they can flow, and they take the shape of the container. The particles are close, so there is short range order. Liquids are hardly compressible; in other words, the volume of a liquid sample is almost constant, however the pressure changes; see also *expansion*.

The **liquid drop model** of the nucleus views the nucleons as like the particles within a drop of liquid. Most relevant to massive nuclei, the model has a 'surface tension' view of the nuclear potential barrier and explains very well the process of fission. Compare it with the other nuclear structure models – see *nucleus, sense b)*. For **liquid manometer**, a device designed to measure the pressure of a gas, see *manometer* and (for work on air pressure) *barometer*.

liquid crystal Friedrich Reinitzer discovered the first liquid crystal in 1888, finding that the plant extract cholesteryl benzoate has a turbid phase of over 30 °C between the true crystal and true liquid phases. In that transition phase (the mesomorphic phase), a sample shows double refraction and interference effects. Since then, people have found this so-called liquid crystal phase in a number of substances and in two main forms. Indeed, they are quite common, soap solution (for instance) being well known for these effects.

A normal well-ordered crystal loses all long range order as it becomes a liquid at a well-defined temperature: in a liquid crystal phase, the substance, while still a liquid, has a significant degree of long range order. Magnified in polarised light, for instance, a liquid crystal shows a domain structure, with each domain having the molecules aligned in a single direction.

Thermotropic liquid crystals move into this phase by way of temperature change; lyotropic liquid crystals do so during concentration change. Thermotropic liquid crystals can show one or more of four types of phase:

- The nematic type (perhaps the most common) has long molecules that lie parallel in the crystal and liquid crystal domain.
- A special type of this is the chiral (or cholestric) liquid crystal; here the molecules also have a spiral form which twists in the liquid crystal domain.
- In a smectic liquid crystal, the long parallel molecules lie in sheets, so the domains are only two-dimensional.
- The discotic phase appears with disc-shaped molecule which stack in columns. A special type of this is the chiral (or cholestric) liquid crystal; here the molecules also have a spiral form which twists in the liquid crystal domain. In this case, the double refraction concerned – first proposed by Fresnel in 1866 but not found till 2006 – is of two oppositely circularly polarised rays.

Liquid crystal displays (lcds) mostly use nematic substances. In an electric field, a sample rotates the plane of polarisation of incident light to a greater or lesser degree. The sample lets through plane polarised light unless the applied field aligns its molecules differently. To display colour (though with some slowness of response), the light from each pixel passes through some combination of three filters: this allows each pixel to show black, blue, green, cyan, red, magenta, yellow or white.

The **liquid crystal lens** is a recent development, a lens whose focal distance changes with the voltage applied. Such a lens typically has a thin layer of liquid crystal in a glass sandwich with concentric rings of embedded control electrodes. Such lenses can often vary in focal distance over a very wide range and, within a few milliseconds, can switch between values needed, either automatically or under user control. Throughout the range, such a lens suffers

little scattering, few aberrations and little astigmatism. Some such lenses are as yet heavy, but they could be used in glasses (spectacles) which can adjust automatically to suit what the wearer looks at.

Lissajous' figures (Jules Lissajous, mathematician, 1822–1880) graphs of complex harmonic motion, the patterns obtained by combining two simple harmonic motions in different directions. Each pattern depends on the frequencies v, amplitudes A, and phase difference ($\Delta\phi$) between the two signals (as applied, for instance, to the two axes of an oscilloscope display). The two signals are $x_1 = A_1 \sin(2\pi v_1 t)$ and $x_2 = A_2 \sin(2\pi v_2 t + \Delta\phi)$.

lithium Li the least dense alkali metal, group 1, period 2, number 3, discovered in 1817 by chemistry student Johan Arfvedson and named after the Greek *lithos*, meaning 'rock'. Relative density 0.54, melting and boiling temperatures 180 °C and 1350 °C; most common isotope ^{7}Li. The metal's main uses are for thermal transfer (as it has the highest specific thermal capacity of any solid), in some dry cell types, as a flux, as a source of alphas (using the Cockcroft-Walton process), and in some ceramics.

litre L or l non-SI, but common, unit of volume: 1 dm^3, 1000 cm^3, 10^{-3} m^3. The original definition (the volume of 1 kg of pure water at 4 °C and standard air pressure) gives a value of 1000.028 cm^3, rather than 1000. The risk of confusion means one should not use the unit in precise experiments. Nor should one use the symbol, l, because of the danger of confusion with 1 (or I). However, the preferred symbol, L, is not yet common in Europe.

Litz wire fine stranded wire in which each strand is insulated. This offers a reduced skin effect when the wire carries high frequency signals.

Lloyd's mirror important nineteenth century (first, 1834) technique for giving interference fringes by division of wavefront. Light from a point source grazes the surface of a mirror to produce a point image close to the source. The direct and reflected beams form the two coherent beams needed for interference.

ln symbol for natural (Naperian or exponential) logarithm, one taken to base e (2.718 ...)

load a) old name for the output force of a machine, from the days when that really was a load to be lifted or pulled along: its name implies weight and, indeed, machines are still often used to lift heavy objects. On the other hand, they (for instance, levers, gears and pulleys) can and do apply force in any direction.

b) burden of particles in a fluid (for instance, a river) or solid. **Loaded concrete** is concrete with a high concentration of dispersed metal atoms to provide better nuclear shielding.

c) the circuit, element or material that receives and absorbs power from a source. Here **load impedance** is the impedance of the load, its overall opposition to ac. A **load line** is a line added to the set of characteristic curves of a circuit to relate circuit current to applied voltage for the actual load at the time. To load a communication channel is to add inductance in order to reduce amplitude distortion of the signal transferred.

lobe angular extent over which an aerial transmits or receives most strongly. A polar graph shows how signal strength varies with angle. The pattern is often complex, but there are always one or more lobes.

lodestone the mineral magnetite, Fe_3O_4, the first substance known to show magnetic effects. Suitably shaped pieces of the rock made to float in a bowl formed the first magnetic compasses; it is not clear when people first devised this aid to navigation or when they needed it (sailing away from land). The word 'lode' comes from the Old English for course, also used in earlier times in lodestar, the name for the North Star.

Lodge, Oliver (1851–1940) physicist who carried out much important early work on electromagnetic radiation. He invented an early detector for radio waves. (He also spent much effort on the paranormal, spiritualism in particular, and became a believer in the latter.)

log a) short for logarithm. In science, log x means the logarithm of x to base 10 (common logarithm); in mathematics it is the logarithm to base e (natural logarithm). In science we write the latter as ln x.

b) record of usage produced automatically by a system, as in the case of the log of phone calls. See also *data logging*, the automatic collection and storage of signals from sensors for later processing.

c) device used to measure the speed of a craft through water. Ships and large boats now use a pitot tube, as do aircraft.

logarithm the power to which a base must be raised to give a stated value. Thus 0.3010 is the logarithm of 2 to base 10 (written log 2 or $\log_{10} 2$); this is because $10^{0.3010} = 2$. See also *log* sense a). Before calculators and computers became common, people used tables of logarithms to help with calculations. These were first devised by John Napier (mathematician, 1550–1617), in around 1610; **Naperian logarithms**, named after him, are natural logarithms, those to base e (2.718 ...). Those to base 10 are **common logarithms**.

One can add, subtract, divide and multiply logarithms to obtain the products, quotients, roots and powers of numbers. The slide rule (also common until three decades ago) is a machine with logarithmic scales one can use at speed to give approximate values.

A **logarithmic curve** is the graph of a function of the form $y = a \log x + b$; a, and b are constants. Such relationships are common in physics. For instance, **logarithmic decrement** λ is a measure of the time rate of decay of an oscillation or radioactive sample. With A_n for the amplitude of the nth cycle, we have $\lambda = (\ln A_1 - \ln A_n)/(n - 1)$. People also use the term in much the same way to describe the scattering of neutrons by nuclei in a pile.

logic branch of mathematics which involves the analysis (with no regard to meaning) of the patterns of reasoning which follow from premises (input data) to conclusions (output). The binary digital mode of working of almost all IT systems is easy to describe by **Boolean logic**; we use the values 0 and 1 to describe the inputs, states and outputs. A **logic circuit** is the same as a gate, a controlled switch. This has (in most cases) one output, the value at which depends on the values at the one or more inputs. A **logic network** is a linked set of gates.

Lomonosov, Mikhail (1711–1765) scientist and philosopher; son of a Siberian fisher-man, he ran away to Moscow to obtain education, and became professor of chemistry at St Petersburg university. His main work was to tidy the writing and grammar of the Russian language, but he also devised important early statements of the particle structure of matter and the kinetic theory.

longitudinal along, in the direction of; compare with transverse, across.

Longitudinal aberrations in optical systems are the effects of aberrations measured along the axis of the system.

For an object moving at relativistic speed, the **longitudinal mass** m_l is the ratio of extra force to acceleration produced in the direction of the motion. It relates to the object's rest mass m_0 as follows.

$$m_l = m_0/(1 - \beta^2)^{3/2}$$

β is the ratio of the object's velocity to the speed of light in empty space.

The vibrations of a **longitudinal wave** are in the same direction as the energy transfer; pressure (including sound) waves are of this type.

long sight hyperopia, a defect of vision in which the point of nearest clear vision is more distant than for the perfect eye

long-tailed pair basis of the differential amplifier, a pair of identical junction transistors with their emitters linked and with a high resistance in the emitter line

loop a) single turn of conductor, or one turn of a coil. A **loop aerial** may have one or more loops in its flat coil; the main uses are for portable radio and tv sets and for direction finders.

b) any complete path through a circuit or network from one side of the supply to the other. See also *Kirchhoff's rules*.

Lorentz(–Fitzgerald) contraction reduction in the observed length l of an object moving in that direction with high velocity v, relative to the value l_0 that is obtained when the object is at rest in the observer's frame of reference.

$$l = l_0 \sqrt{(1 - v^2/c^2)}$$

Here c is the speed of light in empty space.

The names are those of the physicists Hendrik Lorentz and George Fitzgerald (1851–1901). It was Fitzgerald who proposed the contraction, in fact.

Lorentz used the ideas of Maxwell to explain reflection and refraction of light waves and went on to suggest his 'electron theory': that light involves the motion of electrons in a stationary electromagnetic ether. He was able to predict from this that the wavelength of light would change as it passes through a strong magfield. Zeeman showed this to be the case in 1896, and gained with Lorentz the 1902 Nobel Prize for this, the Zeeman effect. However, the results

of the Michelson–Morley experiment in 1887 did not agree with the electron theory, so Lorentz proposed time dilation. The expression for this is

$$t = t_0/\sqrt{(1 - v^2/c^2)}$$

with the symbols as used above.

He then linked this equation with that for contraction to give the *Lorentz transformations*. These relate the descriptions of a system in two frames of reference, one moving at constant high velocity relative to the other. The transformations led to Einstein's special theory of relativity (1905).

The **Lorentz force** is the force F on a charge Q moving with velocity v at an angle θ through a magnetic field B:

$$F = Q\,v\,B \sin \theta$$

A **Lorentz invariant** is a measure which does not change during the Lorentz transformations.

The **Lorentz–Lorenz formula** relates the refractive constant n of a gas to its density ϱ. The second name is that of Ludwig Lorenz, mathematician and physicist, who published the formula in 1869, the year before Lorentz independently did the same.

$$(n^2 - 1)/(n^2 + 1) = k\,\rho$$

where k is a constant

The **Lorentz unit** is a unit of frequency used to describe the spectral line splitting that follows the Zeeman effect (above).

Lorentz transformations transformations in space– that keep constant the space– interval between two events and also the origin of the measures. In effect, they describe how the space– coordinates of observers in different inertial reference frames relate. The mathematical basis for the special theory of relativity, Poincaré named them for Hendrik **Lorentz** (mathematician and physicist, 1853–1928; see also *Lorentz contraction*), one of many people to develop them in the 1880s and 1890s. Lorentz gained the 1902 Nobel Prize with Zeeman for their research into how magnetism affects radiation.

The space– coordinate systems in two frames, (x_1, y_1, z_1, t_1) and (x_2, y_2, z_2, t_2) moving relatively at velocity v in the x-direction, relate as in the box. Here γ is $1/\sqrt{(1 - v^2/c^2)}$, c being the speed of light in empty space. The second expression in the box shows the transformations in matrix form. β is $v/\sqrt{(c^2 - v^2)}$.

The correspondence principle is that when v is very small compared to c (i.e. when v/c tends to zero), the transformations reduce to the Galilean version.

$$x_2 = \gamma (x_1 - v\, t_1)$$

$$y_2 = y_1$$

$$z_2 = z_1$$

$$t_2 = \gamma (t_1 - v \times x_1 /c^2)$$

$$
\begin{bmatrix} x2 \\ y2 \\ z2 \\ c\,t2 \end{bmatrix}
=
\begin{bmatrix}
-\beta\gamma & \gamma & 0 & 0 \\
0 & 0 & 1 & 0 \\
0 & 0 & 0 & 1 \\
\gamma & -\beta\gamma & 0 & 0
\end{bmatrix}
\begin{bmatrix} x1 \\ y1 \\ z1 \\ c\,t1 \end{bmatrix}
$$

The transformations describe – but also predict. They predict time dilation (also a major outcome of the special theory) and can prove that electric and magnetic field forces are the same. The latter result comes when we apply the transformations to the Maxwell equations; indeed, to explore this, was why the Lorentz transformations were devised (though still based on ether thinking).

Loschmidt number L (Joseph Loschmidt, chemist and physicist, 1821–1895) correct name for the *Avogadro number*, the norm in some countries

loss in general, the transfer of energy that enters a system into some form in which it is of no use: in other words, a source of inefficiency. In the case of an electrical machine, for instance, it is common to speak of
- copper loss: the effect of resistance (transfer to electrical heating of the conductors), and
- iron losses: the energy involved in the core in domain switching (see *hysteresis*) and in transfer to eddy currents.

In much the same way, an ac supply to an inductor does not produce the 90° phase shift of a perfect inductor (see also *electric circuits*). The **loss angle** is the difference between the 90° phase angle of theory and the phase angle met in practice with a given inductor. A similar term applies to capacitive circuits. The **loss factor** of an insulator across which there is an ac supply is the product of relative permittivity and power factor. (Power factor is the ratio of the power used to that supplied.)

lossy involving high loss, in the case of a circuit or power line, for instance

loudness subjective sensation that relates to the intensity (energy density) of a sound wave that enters the ear-brain system. The **loudness level** of a given sound is the sound pressure level of a pure tone of defined frequency that seems to have the same loudness.

loudspeaker transducer (sensor) with input of sound waves and a corresponding electric output. See *speaker*.

Love wave horizontal circular or elliptical wave in a narrow fixed layer of elastic matter, an important type of earthquake wave. They are much the same as Rayleigh waves, but the latter are vertical.

lower sideband a sideband of an amplitude modulated carrier, the one with frequencies below that of the carrier

low frequency with a low value of frequency compared to some standard. In the case of radio waves, it covers the range 3–30 kHz.

low pass filter filter that passes low frequencies but blocks high ones

low tension old term for low voltage, below about 60 V (though the limit depends on context in fact)

low voltage safe voltage, less than around 60 V (though the limit depends on context in fact)

lsi large scale integration, building hundreds of thousands of circuit elements into a chip

lubrication use of a film at the interface between two relatively moving surfaces to reduce friction and wear. The **lubricant** (the substance used in the film) may or may not entirely separate the surfaces. If it does, we call the approach **fluid lubrication** (the lubricant may then be a liquid or a gas).

Lucretius (c.98–c.55 bce) poet and philosopher whose works include many dealing with science – with one, *De rerum natura* (On the nature of things), building on the atomic theory of matter. He was the last (but the greatest) person in ancient times to propound the atomic theory of matter. His other antireligious views led to these theories being suppressed.

Lüder band or **Luder band** band of local plastic deformation in a sample close to fracture. A common cause in a metal is welding.

lumen lm unit of intensity of visible light ('luminous flux'), the intensity output by a point source of one candela to a cone of solid angle one steradian. A point source of one candela whose output is the same in all directions emits 4π lumens (as 4π is the total solid angle around a point).

The light output of a standard 100 W incandescent lamp is around 1700 lumens.

luminaire special purpose lighting unit, especially one used for stage lighting

luminance L measure of the brightness of a source of light that is not a point. In a given direction, it is the intensity per unit area at 90° to that direction. The unit is the candela per square metre (cd/m^2), the former name for this being a nit. The **luminance signal** of a video signal is the part that carries the information about the brightness at each point of the display (the chrominance signal carries the colour information).

luminescence the output of radiation from a surface whose particles have absorbed energy from some non-thermal source and become excited. There are many types of luminescence, because there are many ways to excite the particles; the main ones are:

bioluminescence: chemiluminescence in living tissue

cathodoluminescence: energy source – cathode rays

chemiluminescence: energy source – a chemical reaction

electroluminescence: energy source – an electric field

ionoluminescence: energy source – an ion beam

photoluminescence: energy source – electromagnetic radiation

radioluminescence: energy source – nuclear radiation

sonoluminescence: energy source – ultrasonic waves

triboluminescence: energy source – friction (for instance, in grinding)

See also *phosphorescence*, luminescence which does not stop as soon as the source of excitation stops, and *fluorescence*, where it does.

luminosity the observed brightness or luminous intensity of a source of visible radiation. This is a subjective, somewhat imprecise measure. See *photometry* and see also *magnitude*.

luminous emitting visible radiation. The **luminous efficacy** K of a source, usually given for a set wavelength, is the ratio of the intensity, as perceived, to the actual intensity:

luminosity is a subjective measure. A related, but not identical, term is **luminous efficiency** χ, the ratio of light output to total (radiant) output.

The modern name for **luminous emittance** is luminous exitance (below). The **luminous energy** of a radiation is the product of its intensity and the time period of transfer; the unit is the lumen second, lm s. **Luminous exitance** M is the intensity of light radiated per unit area of the source in question.

Luminous flux is an old name for rate of flow of luminous energy. The unit is the lumen (lm), the power output in unit solid angle (one steradian) by a point source of **luminous intensity** (see also *magnitude* and *photometer*) one candela. This term, in turn, is the power emitted per unit solid angle (steradian); the unit is the candela, cd.

All these **luminous measures** concern the energy content of light as determined by a human observer; their physical (objective) counterparts use the adjective radiant rather than luminous.

Lummer–Brodhun cube device used as a photometer – it has two right-angled prisms to allow the observer to compare the intensity of a source with that of a standard. The observer sees a disc of light from one source, with round it a ring of light from the other. When the sources have the same brightness, the edge between ring and disc disappears.

lumped circuit circuit in which the energy storage and energy loss elements are lumped together into single small units of capacitance, inductance and resistance

lunar concerned with the Earth's Moon. A **lunar eclipse** is an eclipse observed when the Earth's shadow falls on the Moon's surface, this is able to happen only at times of Full Moon. To an observer in the shadow on the Moon, there is then an eclipse of the Sun.

The Birmingham **Lunar Society** (so called as it met at times of Full Moon so its members could see to go home) was Britain's major provincial learned society in the eighteenth century (at a time when the Royal Society was in decline). The members included Matthew Boulton, Erasmus Darwin, Joseph Priestley and Josiah Wedgwood; they came to be called lunaticks (from the name of the society).

lutetium Lu transitional rare earth in the

lanthanide series, period 6, with proton number 71, discovered by Georges Urbain in 1907, who named it from the Roman name for Paris. Relative density 9.8, melting and boiling temperatures 1650 °C and 3400 °C; most common isotope ^{175}Lu. Lutetium is very costly, so has few uses, and those are all in chemistry.

lux lx the unit of illuminance, one lumen per square metre

Lyman series first series of lines in the *hydrogen spectrum*, with quantum number 1

m **a)** unit symbol for metre, unit of length
b) as m_e, unit symbol for the electron mass
c) as such or as m_u, old unit symbol for atomic mass unit
d) symbol for minute (either sense, also min or ′)

M symbol for mega-, the unit prefix for 10^6

m **a)** as m_e, symbol for the electron mass
b) symbol for mass, unit: kilogram, kg
c) symbol for gradient, in particular the slope of a linear graph, no unit
d) symbol for magnetic moment, unit A m^2
e) symbol for magnetic quantum number, no unit
f) symbol for pole strength, unit: weber, Wb
g) symbol for relative stellar magnitude, no unit

M **a)** symbol for absolute stellar magnitude, no unit
b) symbol for bending moment, unit: newton metre, N m
c) symbol for luminous exitance (was luminous emittance), unit: lumen per square metre, lm m^{-2}
d) symbol for Mach number, an object's speed compared to that of sound in the same context, no unit
e) symbol for magnetic quantum number, no unit
f) symbol for magnification, no unit
g) symbol for (degree of) magnetisation, unit: ampere per metre, A m^{-1}
h) symbol for mass of primary object in a multi-object context
i) symbol for molecular mass (was molecular weight), no unit
j) symbol for moment (i.e. torque, symbol: T), unit newton metre, N m
k) symbol for mutual inductance, unit: henry, H

l) symbol for radiant exitance (was radiant emittance), unit: watt per square metre, W m^{-2}
m) symbol for relative molecular mass, no unit

μ **a)** unit prefix symbol for micro-, 10^{-6}
b) unit symbol for micron, old name for micrometre
c) symbol for the muon, a lepton

μ **a)** sometimes used symbol for absorption constant
b) sometimes used symbol for magnetic (dipole) moment
c) symbol for (relative) magnetic permeability (with μ_0 for that of empty space, the magnetic constant), unit: henry per metre, H m^{-1}
d) symbol for proper mass or rest mass, m_0 preferred
e) as μ_B, symbol for the Bohr magneton, and μ_n, symbol for the nuclear magneton, both being units of nuclear and atomic magnetic moment
f) sometimes used symbol for the arithmetical mean of unspecified values

Mach, Ernst (1838–1916) physicist and philosopher of science, whose philosophy led to his rejection of supernatural concepts – that is of ones that experience and experiment cannot confirm. Thus Mach, though an active and important scientist, did not believe in atoms. Mach's friend Einstein found his views helpful in getting away from Newtonian space–time and developing his theory of relativity but Mach did not believe in relativity either.

A major area of Mach's research was into the shock waves that come from an object moving faster than the speed of sound in the medium in question (i.e. with Mach number > 1). Such motion produces a high-pressure cone in the medium; the **Mach angle** of the motion is the half-angle of this **Mach cone**.

machine device designed to transfer power from one place (the input) to a second (the output). If that is its only function, its correct name is a mechanism. In most cases, however, the design is also such that with a small input force (effort) one can overcome a large output force (load). Such a machine has a 'mechanical advantage' (which is why people use it); we measure this by the force ratio: the output force divided by the input force. Hence Archimedes's well known statement: 'Give me somewhere to stand and I will move the Earth.' In fact, he'd also need a very long stick – maybe the one Mikhail Lermontov wrote of? 'O vanity! You are the lever with which Archimedes wanted to lift the earthly globe!' (*A hero of our time*, 1840)

Other designs have a force ratio less than 1, but benefit users by having a small input distance of motion to obtain a large output distance. The measure of this is the distance ratio (old name velocity ratio); the value is the ratio of the distance moved at the input to that moved at the output.

The efficiency of a machine is the force ratio divided by the distance ratio. In practice 100% efficiency (all energy input appears at the output in useful form) is not achievable in mechanical systems; friction between the moving parts, and moving the weights of these absorbs much input energy. See *lever* and *pulley* for typical machines in more detail; hydraulic belts, cogs and other drives are also mechanisms and/or simple machines.

machine tools powered tools, whatever their source of power, able to work (cut, drill, shape, smooth, etc.) wood metal samples much more finely and quickly than can hand-held tools

Mach number M the ratio of the speed of an object to the speed of sound in the medium concerned. The motion is sub-sonic if the value is less than 1, supersonic if it is greater (or, to some, greater than about 1.2), and hyper sonic if greater than 5. Some air flows are supersonic when the craft's speed is about 0.8 M; all of them are above about 1.2 (hence

the above confusion); the region 0.8–1.2 is the trans-sonic range.

Mach wave the shock wave produced by the front of an object moving at greater than Mach 1. The double bang heard from a supersonic plane is the effect of the shock waves from its front and its tail. See also *Cherenkov radiation*.

McLeod gauge mercury manometer (pressure meter) able to measure pressures down to about 0.001 Pa, important in vacuum physics and engineering. H G McLeod designed this in 1874.

McMillan, Edwin (1907–1991) physical chemist who discovered the first trans-uranic element (neptunium, nucleon number 93), in 1940, gaining the 1951 Nobel Prize for chemistry. He also developed the synchrotron, for which work he received the 1963 Atoms for Peace award.

macromolecule molecule, such as those of rubber and other polymers, which consists of a large number of identical monomers – or any other molecule (often organic), with a molecular mass over about 5000

magfield shorthand for magnetic force field

magic number one of a number of values of proton number, neutron number or nucleon number which may confer unusual stability in a nuclide. The values – peaks in the binding energy/nucleon number graph – are 2, 8, 20, 28, 50, 82 and 126. The shell model is a view of the nucleus which attempts to explain this.

maglev magnetic levitation

magnesium Mg alkaline earth metal, group III, period 3, number 12, isolated by Humphry Davy in 1808, who called it magnium after Magnesia, major source of a major salt – he didn't want people to confuse the name with manganese, but still the full name stuck in most languages. Relative density 1.7, melting and boiling temperatures 650 °C and 1090 °C; most common isotope ^{24}Mg. After steel and aluminium, magnesium is the most common structural metal, often as an alloy, including in cars, as it is strong and light and not too costly. It is also common in electronic circuits.

magnet

shaped solid object given a permanent magnetic field (see *magnetisation*) and thus able to attract a piece of iron. The domains are aligned (rather than being in closed loops) and so have a net external effect.

The only significant use of magnets until this century was as the needles of magnetic compasses. Since then, the strength per unit mass has doubled about every decade, so that permanent magnets have taken more and more of the roles of electromagnets (but users can control their effects little, if at all). The main developments over the last century have followed the appearance of ferromagnetic materials other than those based on iron.

A major breakthrough occurred in 1967, when Phillips produced a ferromagnetic alloy of cobalt and samarium. Since then, various useful alloys of iron or cobalt and a rare earth element have appeared. In particular, in 1983, simultaneously in the USA and Japan, the first magnetic alloy of iron, neodymium and boron was developed. Hard and costly to produce, quick to corrode, and with low Curie temperatures (80–320 °C), this now appears in many types – and much research goes on to improve the properties of these 'neodymium' magnets.

In the same vein, a major advance was the discovery in 1990 of an unusually effective organic magnetic material; several organic materials are already known to have magnetic properties, but this one (a derivative of benzene) is around 10 times stronger than expected. See also *magnetic material*.

magnetar name for a supermagnetic star – a neutron star with an extremely strong magfield

magnetism the properties of magnets (whether permanent or electro-) and their force fields, and the science and applied science of all these. The phenomenon is part of the electromagnetic interaction, one of the four types of interaction between matter particles: moving charge generates a magnetic field, while a changing magnetic field induces charges to move.

Magnetism has been known and used since ancient times as certain rocks (such as magnetite, called lodestone) have the property. The use of lodestone as the basis for the magnetic compass dates back at least some 2000 years, to early China, for instance. The scientific study of magnetism started in the 16th century, after people had come to some understanding of the principles of magnetic induction and magnetisation.

Electromagnetism, the production of a magnetic field by a current in a coil (in particular) has been explored since early in the nineteenth century; this was an essential step that led via radio waves to modern physics.

Electrons have magnetic properties as a result of their spin and motion in orbit, and all matter contains electrons. Thus, all matter has magnetism. The most important type, ferromagnetism, follows the grouping of the 'atomic magnets' into domains. Ferromagnetism is a property of relatively few substances; diamagnetism, ferrimagnetism and paramagnetism (and their anti forms) are weaker but more common. Indeed, if a substance shows no other property, it is diamagnetic: a magnetic field repels a sample of it. The figure on page 265 shows electron spins in some different types of magnetic material.

Modern uses of (electro-)magnetism range very widely – from small magnets to hold notes on fridge doors to massive machines, from magnetic media to store data in information technology systems to the superconducting magnets of accelerators and some motors. The magnetic compass, on the other hand, is being replaced by better aids to navigation, except in very small vehicles. See also *geomagnetism*, the Earth's magnetism.

magnetic concerned with a magnet or electromagnet and its properties in this context, including those of the Earth. A **magnetic amplifier** is a transductor, a device rather like a magnetically saturated transformer, set up so that a small change in current in one coil can produce a large change in that in a second. The Earth's **magnetic axis** is the axis of the dipole which provides the main part of the Earth's field. It does not coincide with the geographical axis (that of rotation); currently the angle between the two in the UK is a few degrees, but this varies greatly with time and place. The task of a **magnetic balance** is to measure the small magnetic forces between dia- and paramagnetics directly; the beam is a long bar magnet which the user can load to restore to equilibrium. A **magnetic bottle** consists of a number of magnetic field sources arranged so that the field at the centre can contain a plasma without its touching any physical container; the concept is important in fusion research. See also *magnetic mirror*, below.

Magnetic materials consist of tiny magnetic regions, called domains, each with a pair of poles. In some cases, it is possible to isolate the domains as permanent **magnetic bubbles**; two decades ago, there were hopes for **magnetic bubble storage**, with these domains in trains (series) of thousands held in loops in the surface of the substance, storing bits of data.

The concept of a **magnetic circuit** is a magnetic analog with an electric circuit. It is a closed path formed by magnetic lines of force; see also keeper. The table shows the corresponding measures with the magnetic symbols and units.

magnetic circuit	electric circuit
magnetomotive force, F (ampere turn)	electromotive force, E (V)
magnetic flux, Φ (Wb)	electric current, I (A)
reluctance, R (H^{-1})	resistance, R (Ω)
permeability, μ ($H\,m^{-1}$)	conductivity, σ ($\Omega^{-1}\,m^{-1}$)
permeance, $-$ (H)	conductance, G (Ω^{-1})

For a magnetic circuit, **magnetic conductance** is an old name for permeance. It is the reciprocal of reluctance, i.e., it is $\mu\,A/l$; here A is the section area of the circuit, and l is its length. The unit is the henry, H. The **magnetic compass** is a freely floating magnetised needle used to give the direction of the magnetic meridian (north-south line) for navigation; see also *lodestone*. The **magnetic constant** is an old name for the permeability of empty space μ_0, $4\,\pi \times 10^{-7}$ H m^{-1}.

Magnetic cooling is a common way to reduce the temperature of
a) a paramagnetic substance, able to achieve below 1 K: it involves isothermal magnetisation and then adiabatic demagnetisation;
b) a substance with net nuclear magnetic moment, giving around a microkelvin: isothermal moment alignment, then adiabatic demagnetisation.

The process of **magnetic crack detection** in magnetised materials involves dusting the surface with a fine magnetic powder. This tends to settle most in regions of high magnetic effect, such as in cracks (and at the edges of domains). The **magnetic dating** of magnetic objects and of rocks in the Earth's surface involves comparing their directions of magnetisation with the record of changes in the Earth's field; for instance, the magnetic poles switch every 5000 to 50 000 000 years. For **magnetic declination**, see *declination*. For **magnetic deflection**, see *deflection coils*. For **magnetic delay line**, see *delay line*. **Magnetic deviation** is the angle between the magnetic meridian and the direction of a compass, the difference being caused by, for instance, iron ore in a mountain or iron and steel in a ship. **Magnetic dip** is another element of the

Earth's field, the angle between the field and the horizontal; see *dip*.

A **magnetic dipole** is a close pair of north- and south-seeking poles, as in the case of a magnet. For **magnetic dipole moment**, see magnetic moment (below), a better term. The **magnetic disc** was, until this century, a cheap and convenient store of data for information technology systems such as computers, but optical and flash media are taking over. For **magnetic domain**, the basic unit of a ferromagnetic material, see *domain*, and also magnetic bubble (above). For **magnetic double refraction**, the birefringence that appears in a vapour in a strong field, see the *Voigt effect*. This is an example of a **magnetic effect**, an optical effect observed in a magnetic field; see also, for instance, the effects of *Faraday, Kerr* and *Zeeman*.

The **magnetic elements** are the three main measures of the Earth's field at a given place on or near the surface. Full knowledge of them allows one to fix one's position (given the necessary maps). See *declination, dip* (inclination) and *horizontal component* (of the Earth's field strength). The values all change slowly over the years, this being magnetic variation (below). In the same context, the Earth's **magnetic equator**, or aclinic, joins points on the Earth's surface at which the angle of dip ('inclination') is zero. The **magnetic energy** of a magnetised sample is the difference between its magnetic induction in a field and the strength of the field. The unit is the tesla ampere per metre ($T A m^{-1}$).

A **magnetic field** (or **magfield**) is a region of space in which a piece of iron would experience a force. At a given point, **magnetic field strength**, H in empty space and B in a space with a sample showing induced magnetisation, relates to the force on a unit pole at that point. The unit is the ampere per metre, $A m^{-1}$. It is common, though confusing, to describe fields in terms of **magnetic flux**, something like lines of magnetic force. The total flux through a surface is the product of the **magnetic flux density** ('induction', which relates to field strength) and the area of the surface taken normal to the field. The unit of flux is the weber, Wb; the unit of flux density is therefore the weber per square metre ($Wb m^{-2}$), called tesla, T. This is the normal unit of magfield strength; for instance, that of the Earth is on average about 3×10^{-5} T.

What is the largest possible magfield strength? For a long time, it was thought to be 10^{46} T, but a new calculation in 2006 reduced that to 10^{38} T. (This was based on 'pure' qed, i.e. without having to assume that monopoles exist.) The largest known fields (e.g. those around black holes) can be as high as 10^{13} T – but the new value for the theoretical limit is well below the 10^{44} T believed to exist around cosmic strings. On Earth, the largest continuous field achieved for a long time was 6 T – from a 200 tonne superconducting dipole at the Argonne National Laboratory near Chicago. The current value is around 40 T. Using a pulsed current can raise the value to over 70 T – as high as 850 T if the magnet is allowed to destroy itself. The site http://www.magnet.fsu.edu/focus/howstrong.html covers the range of Earth-bound values, from 850 T down to 3×10^{-10} T, the field of the human body.

Magnetic focusing involves the use of shaped magnetic fields to control beams of charged particles (much as in the magnetic bottle, above, and magnetic lens, below). The technique is common in cathode-ray tubes and electron microscopes (see *electron lens*) as well as in accelerators in general. The **magnetic hardness** of a substance is a qualitative term used to describe the size of a field needed to saturate a sample: 'soft' materials are easy to magnetise and also lose their energy with ease; 'hard' materials are not easy to magnetise, but retain the effect much longer. For **magnetic hysteresis**, which relates to this in detail, see *hysteresis*; it is the lagging of a sample's magnetisation behind an applied field. For **magnetic inclination**, see *dip*.

Magnetic induction is a name for magnetic flux density in the sense of the magnetic field strength arising from the magnetisation of a sample; see *induction*. **Magnetic intensity** is a synonym for magnetic field strength in *any* context; see *intensity*. A **magnetic interval** is the time between two successive reversals of the Earth's field; this is far from constant – ranging from 5000 to 50 000 000 years, as it does – as the Earth's core orientation is a chaotic system. **Magnetic leakage** is a cause of inefficiency in electromagnetic machines: the leakage of field from a core in such a way that it does not contribute to the useful output. A **magnetic lens** is a shaped magnetic field designed to focus a beam of charged particles, such as electrons, by a form of refraction. A typical design has an electromagnet with two ring-shaped pole pieces through which the beam passes. See magnetic focusing, above.

Magnetic levitation (**maglev**) is a method, using repulsive magnetic fields, of raising a vehicle a small distance (maybe 20 mm) from the ground to reduce friction. Various small-scale maglev train projects have started (and, in many cases, failed) over recent decades; Japan shows more interest than most. Indeed, they plan a 500 km/h maglev line between Tokyo and Osaka, and have so far a 20 km stretch of dual line in full scale use. See also *train*.

A **magnetic material** is a ferromagnetic material – one showing strong magnetism below its Curie temperature because of a domain structure. There are many magnetic materials now, as well as the traditional iron and some steels, cobalt and nickel: see *magnet*. On or near the Earth's surface, a **magnetic meridian** is the north–south line, the line along which a compass points. A **magnetic mirror** is a region of such high magnetic field strength that it reflects any charged particles that approach it. A magnetic bottle (above) uses this approach.

There are various kinds of so-called **magnetic moment**, and much confusion between them. However, in general, the magnetic moment m of a given source of magnetism is its strength. In a uniform magfield, there is a torque T on the source given by the (vector) product of the magnetic moment and the field strength B. ('Moment' is an out-dated word for torque; strictly, we should not use it here anyway, as magnetic moment is not a torque.)

$$T = m\,B$$

a) (effective definition: magnetic moment m; unit: weber metre, Wb m ($N\,m/A\,m^{-1} = kg\,m^3\,A^{-1}/\,s^{-2}$)) the torque on a source of magnetism per unit magfield strength when source and field are at 90°. It is a vector in the direction of the source axis.

b) (classical physics: magnetic dipole moment μ – also called magnetic moment; unit: weber metre, Wb m) the product of the pole strength of a magnetic dipole and the distance between the poles. It is a vector that points in the direction of the dipole axis.

c) (classical physics: *electromagnetic moment, m* – also called magnetic moment or magnetic dipole moment; unit: ampere square metre, A m²) the product of the current in a loop and the area of the loop. In the case of a coil of n turns, the value is n times greater. It is a vector that points in the direction given by the right hand grip rule (as area is a vector in that direction).

d) (quantum physics: magnetic dipole moment m – also called magnetic moment, atomic magnetic moment and electron magnetic moment; unit: Bohr magneton, μ_B) vector that relates to the spin of a charged particle and/or its orbital motion in a system. That's because such spin or orbital motion is an electric current, and such an electric current has the magnetic effect of a dipole; in this case, the source (spin or orbital motion) is quantised, therefore so is the moment: the Bohr magneton is a kind of quantum number as the moment has discrete values.

Because of the torque (turning force) on the magnetic source in a field, the source has magnetic potential energy. Its value is the scalar (dot) product of the magnetic moment and the field strength (it depends on the angle θ between the two vectors).

$$W_p = -m\,B = -m\,B\sin\theta$$

It follows from that, switching between parallel and antiparallel to the field involves an energy transfer of $2\,m\,B$.

A **magnetic monopole**, also called Dirac monopole (as it was first proposed by Dirac), is a particle with a single magnetic 'charge; it would be a lone north- or south-seeking pole. Some theories predict these in order to unify the electroweak and strong interactions, but none has ever been observed. While no modern theory makes the monopole impossible, some would need significant amendment were such a particle to be found. Indeed, there is no such thing as a magnetic pole – as in the case of line of force, it is a fiction that helps people describe what they observe. Even so, with m the symbol for pole strength (unit: weber, Wb), there is the usual inverse-square expression for the force F between two poles distance r apart in a medium of permeability μ.

$$F = m_1\,m_2/(\mu\,r^2)$$

Searches for monopoles, using either squids or accelerators, have produced only one fully positive result, in 1982. The lack of evidence implies monopoles, if they exist, are no more common than one for each 10^{30} nucleons, and the mass is no less than 600 GeV/c^2.

The process of **magnetic polarisation** is that of magnetisation by the appearance of poles in a sample (by the switching of domains). The degree of polarisation Q measures how far the process is towards saturation. A **magnetic pole** is a region in which a magnetic field is most concentrated. It has no more reality than that, but see magnetic monopole, above. The unit pole is the unit of **magnetic pole strength**; it is the strength of the pole which, placed one metre from a second one in empty space, exerts a force of one newton. Thus magnetic pole strength is the force on the pole in question when 1 m distant from a unit pole. The pole strength of a magnet is its magnetic moment divided by the distance between its poles. The unit is the weber, Wb.

The Earth's **magnetic poles** are the two points on its surface where its magnetic field is vertical. Its north-seeking pole is in the Antarctic (near the South Pole), while the south-seeking pole is near the North Pole. See **Earth's magnetism**. The **magnetic potential** in a magnetic circuit (above) is an old name for magnetomotive force (though this is hardly any better). In a magfield, there is meaning, however, in the term **magnetic potential difference**; this is the energy involved in moving a unit pole between the two points in question. That leads to a second meaning for the magnetic potential – at a point in a field, it is the energy transfer in moving a unit pole there from infinity. The **magnetic permeability** μ of a substance is a measure of how easy it is to carry a field; see *permeability*. For **magnetic quantum number** m, see *quantum numbers*.

Magnetic recording involves storing signals or data on or in the surface of a suitable medium for later access. Magnetic tape (below) and discs (above) are the most common forms: there are others, such as the ill fated bubble storage and the mram chips first sold in 2006. **Magnetic resistance** is an old name for reluctance, the opposition of a magnetic circuit (above) (or part of one) to a magnetic field. There are several forms of **magnetic resonance**, in which atoms or nuclei absorb radio frequency waves when in a magnetic field. See nuclear magnetic resonance in particular. **Magnetic rotation** is a name for the Faraday effect of a magnetic field on polarised light. A sample of magnetic material shows **magnetic saturation** when it can be magnetised no further: the domains are aligned as much as they can be; see also *hysteresis*. True saturation is the full alignment of all the particles' spins; this can appear only at absolute zero.

A **magnetic shell** is an imaginary dipole sheet with one side an n-pole and the other an s-pole. **Magnetic shielding** involves making the magnetic field in a region of space as low as possible (to prevent interference for instance). Putting the region within a box ('screen') of high permeability (i.e. ferromagnetic) material is the usual method; compare this with *Faraday cage*, the same kind of approach for electric fields. The material so strongly refracts the field, that little passes through it. A **magnetic shunt** is a moveable piece of iron near a permanent magnet; moved to various positions it varies the overall field strength to different degrees.

Magnetic storms are solar storms that greatly affect the Earth's field. Events called flares on the Sun's surface sometimes eject large streams of plasma that reach the Earth; the Earth's field focuses these streams. (See also *aurora*.) This can induce large currents in the atmosphere. The Earth's field rises by up to 0.25%, then falls to perhaps 0.5% below the normal value; it takes several days to return to normal. Nowadays, the effects of such events on radio communications and space travellers are far more significant than that on the Earth's field. For **magnetic susceptibility**, see *susceptibility*. **Magnetic tape** was a very common form of magnetic recording medium until early this century; optical and flash media are now taking over.

Magnetic polarisation is

a) the process by which a sample of magnetic material becomes magnetised: the domains line up more and more so their north-seeking poles more and more face the same way.

b) J, unit tesla, T – the degree to which a sample has become polarised, i.e. a measure of the alignment of the elementary magnetic moments. It is much the same as the sample's magnetisation, M, therefore; the two relate through $J = \mu_0 M$.

For **magnetic potential energy** (unit: joule, J) see under *magnetic moment* (above).

magnetic variation

a) an old term for magnetic declination

b) the variation in the Earth's magnetic field with time. There are three forms of magnetic variation in this sense:

- sudden, short-lived changes as a result of severe solar storms: see *magnetic storm*, above
- annual cycling, greatest at times of maximum solar activity
- secular, i.e. lengthy, change that has altered, for instance, the declination and inclination (dip) at London over the last three centuries from 0° and 74° to 8° and 66°

Magnetic viscosity is the delay in magnetisation of a substance after a field is suddenly applied, over and above that due to eddy currents. A **magnetic well** is a rarely used name for *magnetic bottle* (above).

magnetic vortex a magnetic dot with a curling field, a structure rather like a cone in the field at a magnetic surface in which the nearby magnetic moments of the atoms line up into concentric circles. Some people hope the concept may lead to magnetic data storage at densities as high as a gigabyte per square centimetre.

magnetisation the alignment, to some degree, of the domains of a magnetic sample, a result of an outside field. There are two main methods:

a) using a strong outside field, e.g. putting the sample into a coil which carries a large direct current;

b) using the Earth's field, e.g. banging the sample when it lies north–south.

The latter is not a practicable method for making a magnet; on the other hand, the effect of it is very common, and most steel objects (for instance, steel building frames and ships) are magnetised quite strongly as a result. People sometimes call this **intrinsic**

magnetisation, or **spontaneous magnetisation**. See *demagnetisation* for methods of removing magnetism.

As a measure, magnetisation is *M*, the degree of magnetisation, as in the box.

$$M = B/\mu - H$$

Here *B* is the total field strength and *H* the applied field strength, while μ is the permeability of the substance.

The **magnetisation curves** of a magnetic substance are those obtained by passing it through hysteresis cycles of varying total energy transfer (i.e. varying applied field strength). **Magnetisation energy** is the energy needed to saturate unit volume of a substance. For **remanent magnetisation**, see *remanence* – it is the magnetisation left in a sample after the magnetising force has become zero. **Spontaneous magnetisation** is the appearance of magnetism in a sample without any human effort: see *intrinsic magnetisation*, above.

magnetism the properties of magnets (whether permanent or electro-) and their force fields, and the science and applied science of all these. The phenomenon is part of the electromagnetic interaction, one of the four types of interaction between matter particles: moving charge generates a magnetic field, while a changing magnetic field induces charges to move.

Magnetism has been known and used since ancient times as certain rocks (such as magnetite, called lodestone) have the property. The use of lodestone as the basis for the magnetic compass dates back at least some 2000 years, to early China, for instance. The scientific study of magnetism started in the 16th century, after people had come to some under-standing of the principles of magnetic induction and magnetisation.

Electromagnetism, the production of a magnetic field by a current in a coil (in particular) has been explored since early in the nineteenth century; this was an essential step that led via radio waves to modern physics.

Electrons have magnetic properties as a result of their spin and motion in orbit, and all matter contains electrons. Thus, all matter has magnetism. The most important type, ferromagnetism, follows the grouping of the 'atomic magnets' into domains. Ferromagnetism is a property of relatively few substances; diamagnetism, ferrimagnetism and paramagnetism (and their anti forms) are weaker but more common. Indeed, if a substance shows no other property, it is diamagnetic: a magnetic field repels a sample of it. The figure shows electron spins in some different types of magnetic material.

The different types of magnetism relate to the different patterns of alignment of the elementary magnets.

Modern uses of (electro-)magnetism range very widely – from small magnets to hold notes on fridge doors to massive machines, from magnetic media to store data in information technology systems to the superconducting magnets of accelerators and some motors. The magnetic compass, on the other hand, is being replaced by better aids to navigation, except in very small vehicles. See also *geomagnetism*, the Earth's magnetism.

magneto simple generator of ac that consists of a magnet made to turn inside a coil. Electromagnetic induction causes a high voltage to appear between the ends of the coil.

magneto- prefix concerned with any aspect of magnetism, in particular where it relates to other aspects of science. Thus, **magneto-acoustics** is the science and applied science of the interaction of acoustic (sound) waves, particularly in a metal, with magnetic fields. See, for instance, *magneto-damping* (below). The **magneto-caloric effects** are the rise or fall in temperature of a sample during magnetisation or demagnetisation respectively. Adiabatic demagnetisation is an important method of cooling; it is fortunate that the temperature drop that follows a given degree of demagnetisation in a given substance is greater the lower the starting temperature. **Magnetodamping** is an aspect of magneto-acoustics (above): the more rapid decay of sound waves in a metal when it is in a strong magnetic field. The **magneto-elastic effects** describe the changes of the magnetic behaviour of a sample as a result of strain. For **magneto-electronics**, see *spintronics*, for which it is an alternative name.

Magnetoencephalography (meg) is a recently developed non-invasive technique for mapping brain activity. It has advantages over the more traditional electro-encephalographies (eeg systems) that use surface electrodes or implants. A superconducting quantum interference device (squid) can detect the magnetic effects of electric currents in the brain. See also *biomagnetism*. The meg system analyses the patterns of signals from an array of hundreds of squids to study neural activity inside the living brain. The magnetic signals sensed are a few femtotesla (10^{-15} T), so such systems need careful shielding from outside factors, such as the Earth's magfield. A **magnetograph** outputs a continuous record over time of the three elements of the Earth's magnetic field, as given by three sensors at the point in question.

Magnetohydrodynamics (mhd) is a silent and effective way to move a conducting fluid. A magfield across the sample drives the charged particles (by the motor effect), taking the fluid with them. Pumps based on mhd are fairly common for work with plasmas (see, for instance, *magnetic bottle*) and molten metals. Also, using mhd to make a jet of sea water (a good conductor) leads to a useful and quiet method of boat and ship propulsion; so far, however, this has not proved fully cost-effective. (The largest superconducting magnet of the 1980s – 180 t in mass – was built for this last purpose.) **Magnetohydrodynamic waves** are matter waves that appear in a conducting fluid in a magnetic field, an important feature of stars and of plasmas on Earth.

A **magnetometer** is a device designed to measure or compare magnetic field strengths. The once common **deflection magnetometer** (with at the centre a very small bar magnet fixed to a pointer over a scale), and **vibration magnetometer** (similar but allowed to vibrate with a period whose square is inversely proportional to the field strength) have been superseded by the Hall probe. **Magneto-motive force** (mmf) F_m is the line integral of any magnetic field over a closed path in that field; see also *magnetic circuit*. The unit is the ampere turn, A t. For **magneton**, see *Bohr magneton*, a fundamental constant: the magnetic moment due to the current associated with an electron of spin $h/4\pi$. For **magneto-optical effects**, see *magnetic effects* (the interactions of electromagnetic radiation with magnetic fields), and for **magnetopause**, see *magnetosphere* (below). **Magnetophotophoresis** is a form of photophoresis in which specks of dust in a gas move in spirals along an applied field in the presence of light.

Magnetoplasmadynamics is a name for *magnetohydrodynamics* (above), whether restricted to work with plasmas (as in astrophysics and nuclear fusion) or not.

The **magnetoresistive effect** is the change of resistivity of a substance in a magfield; see also the related *Hall effect* and *magnetostriction* (below). See also *mram* and *giant magneto-resistance*. The **magnetosphere** is the region around the Earth that contains the planet's magnetic field and the van Allen radiation belts that result from the interaction of this with the solar wind. The 'sphere' is shaped somewhat like a doughnut, with axis the Earth's magnetic axis; however, there is a kind of bow wave (the **magnetopause**) on the Sun side and a lengthy tail on the other side.

Magnetostatics is a branch of magnetic theory analogous to electrostatics; its basis is the magnetic pole, rather than the electric charge, and its main concern is with unchanging magnetic phenomena. Though the single pole is not known (see *magnetic monopole*), the approach has some value. **Magnetostriction** concerns the various effects of a ferromagnetic sample's magnetisation on its physical size and shape: for instance, a long rod contracts slightly in a magnetic field. The effects and their reverse are the basis of various high frequency vibrators (e.g. sound and ultrasonic sources) and detectors. Thus, a **magnetostrictive oscillator** is an iron rod in a coil fed with high frequency current. The rod vibrates along its length with the same frequency, and, if suitably cut, will resonate. Compare this with the *quartz oscillator*.

magnetron electron tube used to produce microwaves. The electrons given out by the hot cathode circle in tight orbits in the applied magnetic field and generate microwave oscillations in shaped cavities in the anode. Hence the name, still sometimes used, of **cavity magnetron**.

magnification the apparent size of an image compared with the actual size of the object, in particular in optical systems in the widest sense. It is normal to use angular magnification (or *magnifying power*) M for real, complex optical instruments.

> angular magnification =
> angle of the image at the eye/angle of the object at the unaided eye

For simple systems (e.g. with just a few thin elements), most common is **linear magnification** (or **lateral**, or **transverse**),

m (no unit); this is the ratio of the image height (distance of a given point from the axis) to the object height (distance of the same point from the axis), as

> linear magnification m
> = image height/object height
> = image distance/object distance

> (we measure each distance from the centre, or pole, of the system)

Simple geometry of paraxial rays gives the second ratio (which is much more simple to use in practice).

Useful magnification is the greatest magnification offered by a system which allows for the observer to detect extra detail; further magnification is pointless, and is called empty.

magnifying lens or **magnifying glass** simple

microscope, a single converging lens used to obtain a magnified upright virtual image of a close object (closer than the focal point). The greatest magnification obtained is around 10. See *lens*.

magnifying power of a single lens or mirror or of an optical system, the *angular magnification* obtained in a given case. The more precise definitions have the object at the near point, 250 mm from the eye, in the case of a microscope, and at the far point, infinity, in the case of a telescope.

magnitude a) the size of a pure number and, by extension, of any measure, whether dimensionless or not

b) *m or M*; no unit: specifically (particularly in astronomy, for stars and other objects in the sky) brightness, strictly at visual wavelengths only, but by extension in any band and/or when measured photographically or by some other device.

In most contexts, an object's **absolute magnitude** is of more value than the apparent value, for it relates to the object's actual luminosity. We define absolute magnitude M as what the object's apparent magnitude m would be if ten parsecs (32.6 light years) distant, rather than its actual distance r. The relationship between them also involves A, the interstellar extinction value for the wavelength concerned.

$$M = 5 + m - 5 \log_{10} r - A$$

Apparent magnitude, the actual perceived brightness in a clear sky, is much easier to estimate, but has much less meaning. Hipparchus, a Greek scientist, drew up the scale on which the present one is based, in around 120 bce. He classed all the stars into six groups, from 1 (the brightest) to 6 (just visible); over the centuries, the system was rationalised into three features:

- have exactly equal steps as far as concerns visual appearance;
- allow fractional values;
- allow values less than 1 and greater than 6.

A scale defining an object of apparent magnitude 1 to be a hundred times brighter than one of magnitude 6 leads from the first feature above (William Herschel in the early nineteenth century). In 1856, Norman Pogson proposed that this be made exact. That was agreed; it means that for each step in magnitude, brightness changes by the fifth root of 100, i.e. 2.512, the Pogson ratio. This also gives the Pogson scale which relates the apparent magnitudes m_1 and m_2 of two stars to their perceived brightnesses (luminous intensities) I_1 and I_2. It is a logarithmic scale (as is the case with most sensory stimuli in fact – e.g. see *decibel*).

$$m_1 - m_2 = 2.5 \log_{10} (I_2/I_1)$$

In practice, it is easier to assess absolute magnitude from a detailed knowledge of the temperature, size and structure of the object in any given case; we can then use the above absolute relation to estimate the distance. This is a very important technique of modern astronomy. Partly this follows the growing use of photographs and then photometers to allow people to measure luminous intensities rather than trying to assess them by eye. Also, the scale has grown – upwards to and past zero (defined as the visual output of an object of 2.65×10^{-6} lux) to cover the too large range of magnitudes of objects in the first class, and below six as telescopes showed fainter and fainter objects (down to about 30 in fact).

Also, various magnitude scales have appeared as people have devised systems working with different wavelengths – such as

- photoelectric (based on instruments with filters giving a narrow band-pass)
- UBV (the mean of such magnitudes measured in the ultraviolet, the blue and 'visual' yellow)
- bolometric (based on bolometers, such as thermopiles and resistance thermometers, which can measure the intensity of a wide range of radiations from the object in question)

magnon quantum of magnetic energy, as discussed in, for instance, the interaction of neutrons with waves of magnetic spin in matter. It is an analogue of the photon and phonon.

Magnox trade name, derived from magnesium oxide, for a range of magnesium alloys used to build the fuel containers in the first generation of British nuclear reactors. 'Magnox reactor' is thus a widely used term for a nuclear reactor.

Magnus effect sideways force on a spinning object moving through a fluid. It is the cause of the swerve of spinning balls and frisbees moving through air. Attempts have also been made to use the force to produce propulsion, as in the Flettner motor, a vertical spinning cylinder found on a ship.

main sequence star star lying on the main

band in the Hertzsprung–Russell diagram, in other words, one in the middle part of its life

Majoranna force the force of exchange of charge and spin between nucleons

majority carrier an electron in an n-type semiconductor or a hole in p-type. Compare with minority carrier. Intrinsic (i-type) semiconductors have equal numbers of holes and electrons.

make-and-break electromagnetic circuit element used to switch the direct supply off and on again at rapid rate. Electric bells and buzzers use this, as does the induction coil. When there is a current, an electromagnet attracts a piece of iron away from the contact of the switch; this breaks the current. A spring returns the iron to the contact, closing the switch again to re-start the cycle.

malleable soft enough to be able to be beaten very thin and into shapes (as in the case of most metals, in particular copper and gold)

Malter effect the high positive charge that develops in an insulated semiconductor layer as a result of secondary electron emission

Malus's law The intensity of light passed by a polariser/analyser unit is proportional to the square of the cosine of the angle between them. The name is of Étienne Malus (1775–1812), engineer and physicist, who found and explored polarisation by reflection in 1809.

manganese Mn transitional element in period 4, proton number 25, isolated in 1771 by Carl Scheele (who named it after Magnesia, the region in Greece where its main ore is found) and his assistant Johan Gahn. Relative density 7.2, melting and boiling temperatures 1240 °C and 1960 °C; only natural isotope ^{55}Mn. Main uses are in steels and alloys of other metals.

Manganin one of a range of copper–manganese–nickel alloys with high resistivity and low temperature coefficient of resistivity, widely used for 'resistance wire'

manometer device used to measure the pressure in a fluid. There are many designs that depend on usage and on whether the user wants an absolute or a relative reading.

mantissa the value part of a number in scientific standard form. In $a \times b^c$, e.g. 3.1 × 10^6, a (3.1) is the mantissa.

mantle the region between the Earth's crust and its core, also called the mesosphere. It is several thousand kilometres thick (but definitions vary).

Marconi, Gugliemo (1874–1937) physicist and engineer, pioneer of radio communication, basing it on the work of Heinrich Hertz. Starting this work in 1895, Marconi took out a patent and founded his Wireless Telegraph Company in 1896, setting up radio stations almost at once. In 1901, he was able to transmit a signal across the Atlantic (from Cornwall to Newfoundland), and received the Nobel Prize in 1909.

Marič, Mileva (1875–1948) mathematician, also trained in physics, who became Einstein's first wife (1903–1919). Some people believe she made major contributions to Einstein's early work; she certainly collaborated with him and others before she became pregnant by Einstein in 1901.

Mariotte, Edmé (1620–1684) physicist who also discovered the blind spot of the eye. **Mariotte's law** is the same as Boyle's law: they discovered the relation between gas pressure and volume independently.

Marx circuit system able to produce brief intense pulses of direct current. It involves charging a number of capacitors in parallel and then making them discharge in series.

mascon mass concentration, region of an object's crust whose mean rock density is greater than the object's norm, thus with a higher than normal gravitational field strength. The effect on the Moon was strong enough to crash spacecraft after a number of orbits until the data tables improved.

maser device able to produce a coherent, parallel, intense beam of microwave radiation. The action is either as an amplifier (of, for instance, very weak signals in astronomy), or as an oscillator (for use in atomic clocks). The name is an acronym for microwave amplifier (or amplification) by stimulated emission of radiation, though the m soon came to stand for molecular as the wavelength range lengthened through infrared to visible light. The laser's first name was **optical maser** because of this. In other words, the maser pre-dated the laser, though the theory, action and effects are much the same.

mask chemical layer placed for a short time on the upper surface of a wafer during the process of making one layer of a chip. The system deposits the new layer of junctions and links only through gaps in the mask; this is then removed.

Mason's hygrometer standard type of wet and dry bulb hygrometer (device to measure the relative humidity, or moisture content, of the air). The standard includes the air speed required for the reading: around 1 m/s.

mass basic measure of the amount of matter in an object. The unit is the kilogram, kg (but see giorgi). An object's mass appears in two contexts: gravitational (as in Newton's equation for that force), and inertial (as in mass = force/acceleration). The former comes in two versions: they depend on whether the mass in question is the larger or the smaller. If it is the smaller, we speak of the **passive gravitational mass** – a measure of the effect of a gravitational field on the object; if it is the larger, it is the **active gravitational mass**, the cause of the gravitational field in the region.

Inertial mass relates to the object's resistance to acceleration. It is theoretically distinct from gravitational mass, but (so far) no experiment has shown that they differ. The mass of an object in either context is therefore constant; however, there is an increase when the object's speed is high compared to that of light, c, see *relativity*). As a result, **relativistic mass** is the total mass of an object moving fast, with speed v. Compared to its rest mass, m_0, we have the expression

$$m = m_0 / \sqrt{(1 - v^2 / c^2)}$$

An object's **centre of mass** is the point at which we can think of its mass as being concentrated, when considering the action of forces; see *centroid*, and compare with *centre of gravity*. The **mass absorption constant** of a substance is a measure of how well it absorbs a given radiation (ignoring scattering and reflection processes); it is the same as linear absorption constant divided by density. The law of **mass action** relates the concentrations of reactants and products in a chemical reaction to the reaction rate and the equilibrium constant; see *activity*. In nuclear reactions, on the other hand, **mass defect** is the change of mass between starting nuclei and products; in particular, it is the mass equivalent to the binding energy of the nucleus or nuclei concerned. In other words, it is the difference between the mass of a nucleus and the sum of the masses of its nucleons. The equivalence mentioned there is **mass–energy equivalence**, the intimate relation between those two basic measures as given by special relativity theory. In particular, Einstein showed that any change in the energy of a system relates to an equivalent change of mass (and the other way round); the relation is

$$\delta W = \delta m \, c^2$$

Here δW is the energy change and δm that of mass, while c is the speed of light in empty space.

(But most people know it as $E = m \, c^2$).

We can also use the concept to view an object's mass as equivalent to the sum of its rest mass and its kinetic and potential energies. In cases where this is important, we should replace the laws of constant energy and of constant mass with the **law of constant mass–energy**: the sum of the masses and energies in a closed system is constant. See also *relativistic mass*, above: the extra mass of a fast moving object relates to its kinetic energy.

The **mass moment** I (or J) of an object is the product of its mass and the distance to some point, line, or plane. It is one measure of the distribution of an object's mass about an axis. In detail, mass moment depends on the situation: there are various mass moments for different contexts; see also *moment of inertia*.

The **mass number** A of an atom, nucleus or nuclide is the number of nucleons (which are massive) in the nucleus; it is the sum of proton number and neutron number. The name nucleon number is better; older even than mass number is atomic mass unit, amu, which was treated as a number too. The **mass reactance** of an acoustic system is that part of the acoustic reactance due to the inertia of the particles. In an electrical system, the **mass resistivity** of a sample (of, for instance, wire) is the product of mass and resistance, divided by the square of the length, unit A kg m^{-2}; see also *resistivity*.

A **mass spectrograph** and a **mass spectrometer** both allow the user to identify an atom, molecule, or compound through its mass number. The systems separate positive ions using electric or magnetic fields, or both. In the first case, the output is a record on photographic film; the second is electrical, using detectors (sensors). Both are approaches to **mass spectroscopy**, giving a **mass spectrum** whose variable is charge–to–mass ratio, and whose order depends on the degree of ionisation of the ions. The first such experiments were those of J J Thomson in the 1890s. A common use is to identify the relative abundances of isotopes in a sample (this can tell us something of the sample's source).

Last, the **mass stopping power** of an irradiated sample of matter is a measure of its effect on the energy of the radiation. It is the linear stopping power divided by the density.

massive having (non-zero) mass, or (in some contexts) having a relatively high mass. The **massive chiral fermion** is a virtual (quasi) particle believed to play a part in graphene – it has mass, but should not have, as particle physics theory predicts that chiral particles have no mass.

master oscillator stable high-frequency oscillator used to provide the carrier wave of a radio broadcast signal

matched much the same in relation to some stated property. Two linked parts of a circuit have **matched impedance** at a given frequency when there is maximum power transfer between them. **Matched transistors** (for instance) have very similar characteristics in some defined way, and so work well together in contexts where near identical properties are crucial.

material a) made of matter
b) a substance
The **physics of materials** concerns how a substance behaves under stress, for instance its elasticity. A **material radiation** may show a wave nature; however, it also shows a particle nature, and the particles have non-zero rest mass.

materials science the study of matter and its applications, crucial in pre-history (think of the iron and bronze ages). The concern of the materials scientist is the nature, structure and properties of a substance, while the applied science concerns its processing and performance. Thus, this science includes chemistry as well as physics, and feeds into various fields of technology and engineering.

Mather, John C (1946–) astrophysicist and joint winner, with George Smoot, of the 2006 Nobel prize for research into the cosmic microwave background radiation that provides such strong evidence for the Big Bang model of the origin of the Universe

Mathiessen's rule The product of a metal's resistivity and temperature coefficient of resistance is constant, however impure it may be.

matrix a) two-dimensional array of scalar elements, with m rows and n columns. The name of each element of the array A is its row and column coordinates, i and j, called subscripts whether written as such A_{ij} or otherwise: $A(i, j)$.
b) the fine grained background of a rock, e.g. of a sedimentary rock in which coarser stones and fossils lie
c) a uniform substance in which lie, for instance, unstable particles, as in the case of the pellets in a nuclear fuel rod

matrix mechanics the branch of mechanics, first explored by Heisenberg in 1925, that uses matrices (sense a) to represent measures such as energy and momentum. While it appeared separately from wave and quantum mechanics (though at around the same time), those three views of the world are closely linked and their expressions are exactly equivalent. In fact, the formulas of matrix mechanics were able to provide the first complete picture of quantum physics. The wave functions of wave mechanics are vectors in matrix mechanics; the relationships between the matrices that contain the vectors correspond to transitions between states and include the Planck constant of quantum mechanics. Schrödinger was the first to show how matrix and wave mechanics relate.

matter any collection of material particles, in other words, of ones with non-zero rest mass, as opposed to pure energy (such as that of electromagnetic radiation): in other words (but crudely) matter is anything made of atoms and/or subatomic particles. We can describe physics as the study of matter and energy and their interactions (including the inertia of lumps of matter and the way they interact gravitationally). See also *negative matter* and *states of matter*.

 Matter waves are particle radiations, but also show a wave nature (see, for instance, *de Broglie wave* and *electron optics*).

maximum-and-minimum thermometer device able to record the highest and lowest temperatures since last reset. There are many designs, as there are many uses (for instance, in agriculture, industry and research).

Maxwell, James Clerk (1831–1879)

widely adept physicist and mathematician whose main work was on the theory and laws of electromagnetism; his studies of electromagnetism and thermodynamics led to many of the crucial developments of the nineteenth century. The Edinburgh Royal Society published his method for drawing ovals when he was fifteen. A quarter of a century later, after holding chairs in physics at Aberdeen and London, he became the first professor of experimental physics at Cambridge, and set up the Cavendish laboratory.

maxwell Mx an old (cgs) unit of magnetic flux (field strength), equal to 10^{-8} weber

The **Maxwell–Boltzmann distribution** (or **Maxwell distribution** or **Maxwell distribution law**) gives the number of particles in a sample of a perfect gas having speeds in each range. If the total number of particles in the sample is n_0, the number n at temperature T with energy greater than some value W is

$$n = n_0 \exp(-W/R\,T)$$

Here R is the gas constant. The relation gives a characteristic energy distribution curve that depends on temperature. The graphs show this frequency distribution (or probability density distribution in the jargon).

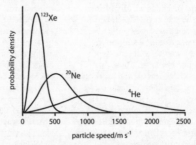

The mean, mode and median speeds of the particles of an ideal gas sample at a given temperature (here 25 °C) relate inversely to the particles' mass.

Based on classical statistics, the expressions below give various features of the frequency distribution of the speeds of the particles in a given case.

root mean square speed $= \sqrt{(3\, k\, T/m)}$
mean speed $= \sqrt{(8\, k\, T/\pi\, m)}$
maximum speed $= \sqrt{(2\, k\, T/m)}$

Here k is the Boltzmann constant, T the absolute temperature, and m the particle mass.

The **Maxwell bridge** was an early form of the Wheatstone bridge for work with inductances rather than with resistances. The **Maxwell colour triangle** puts red, green and blue at the corners of an equilateral triangle, and the complement of each at the centre of the opposite side. Then any colour is a point in the triangle. The chromaticity diagram is the modern version of this. (Like Dalton, Maxwell did useful work on colour vision too.) People view **Maxwell's demon** as sitting by a tiny door between two spaces, one full of gas and the other empty. The demon would open the door to let a gas particle pass only if it has high energy; thus, the demon could lower the temperature of the first space and raise that in the second. This would be against the second law of thermodynamics: but no one has ever proved that a demon-like (but dumb) system cannot exist (or that it can). A recent variation in which a demon controls a Szilard engine has been disproved, on the other hand. **Maxwell's distribution law** describes the Maxwell distribution, above.

The set of four **Maxwell equations** are perhaps his most important legacy; they formed the basis for much crucial later work, including the development of radio and aspects of modern physics (in particular, nuclear physics). None of them was new, however – but Maxwell brought them together, amended them in some cases, and integrated them into a view of electromagnetism as a single subject of physics study. Also, Maxwell would not recognise the set of four equations we now have – he published a set of twenty equations in twenty variables in 1865 and tried to revise them in 1873 (but his new work was not popular); the form we have now, using vector calculus, came from Heaviside and Gibbs in 1884, after Maxwell had died. Later, after the Michelson–Morley experiment led (eventually) to relativity, people worked to include them (successfully); indeed, the Kaluza–Klein theory (from 1919) of general relativity in a five-dimensional space–time leads straight to the Maxwell equations.

'Maxwell's equations' relate various features of the classical theory of electromagnetic fields, $\nabla\cdot$ being the div of the measure in question, and $\nabla\times$ being its curl. All the measures that are vectors appear in bold.

From the Faraday and Lenz laws of induction:

$$\nabla \times E = -\partial B/\partial t$$

From Gauss's law:

$$\nabla \cdot D = \rho$$

From Gauss's law for magnetism:

$$\nabla \cdot B = 0$$

From Ampère's law:

$$\nabla \times H = J + \partial D/\partial t$$

B	magfield strength (induction, flux density)	tesla (Wb m^{-2})
D	electric displacement	coulomb per square metre
E	electric field strength	volt per metre
H	magnetic field strength	amp per metre
J	current density from free charges	only amp per square metre
ρ	charge density (free charges only)	coulomb per cubic metre
t	time	second

Last in this context, Maxwell showed from his equations that the electromagnetic field effects described by them would transfer by perpendicular transverse waves of electric and magnetic field, and with a speed c_0 given by $1/\sqrt{(\varepsilon_0\,\mu_0)}$. The value given – around 310 Mm s^{-1} – was close to that then known for the speed of light in empty space, so Maxwell took it to be the same. This work therefore opened up a whole new field of physics theory and physics research as people (for instance, Hertz) tried to explore its significance. However, Maxwell did not live to see the result, the proof that his work was valid.

Maxwell's formula relates the refractive constant n of a substance to its relative permittivity ε_r. It applies as long as the substance is not ferromagnetic:

$$n = \sqrt{\varepsilon_r}$$

The **Maxwell–Wagner effect** is the build up of charge at the interfaces of substances that differ in conductivity and permittivity.

m/c common shorthand for 'machine'
mean any kind of average, but in particular the **arithmetic mean**, which for a set of n numbers is given by the sum divided by n. Often when a measure appears in an expression in physics, its mean value is what matters. Strictly, the symbol is a bar over the symbol, as in $\bar{x}$ and $\overline{c^2}$; people, however, often leave out the bar. The **mean deviation** of a data set is the mean of all the deviations (taking each as positive). The **geometric mean** of a set of n numbers is the nth root of their product; this is not much used in physics.

The **mean free path** λ is the average distance travelled by the particles of some kind of fluid (including the electron gas of a metal and the neutron gas of a reactor pile) between any, or specified, interactions (e.g. between captures and/or collisions). In the case of a uniform fluid, with n identical particles of radius r per cubic volume:

$$\lambda = 1/(\pi \, r^2 \, n)$$

Mean free path also relates to the fluid friction – or viscosity – η and the fluid's density ϱ:

$$\lambda = k \, \eta/\rho \, \bar{c}$$

Here $\bar{c}$ is the mean speed of the particles; κ is a constant of value around 0.5.

The **mean life** τ of an unstable object (such as a meta-stable atom, radioactive nucleus, or decaying elementary particle) is the mean time it lasts before decay. It is the time for an initial population to decay to $1/e$ in size. Mean life is the reciprocal of the decay constant, and is also $T^{1/}_{2}/0.693$, where $T^{1/}_{2}$ is the half-life concerned. (These relationships follow from the nature of exponential decay.)

The **mean square speed** $\bar{c^2}$ of a set of identical particles is the average of the squares of the speeds. For a gas it varies directly with absolute temperature T, as given by the Maxwell distribution:

$$\bar{c^2} = 3 \, k \, T/m$$

Here k is the Boltzmann constant and m is the mass of one particle. $\bar{c^2}$ also appears in the important kinetic theory relationship between the pressure p and density ϱ of the gas:

$$p = 1/3 \, \rho \, \bar{c^2}$$

mechanical concerned with machines and their features and aspects, in particular with the transfer of forces and energy through them. Thus **mechanical advantage** is an old name for the force ratio of a machine: the ratio of the output force (load moved) to the input force (effort applied); there is no unit. **Mechanical impedance**, at a point in a structure, is the ratio of the net outside force at the point to the velocity that results; it depends strongly on frequency, and at the resonant frequency is very low (which can mean a hazard in the case of many types of structure). The unit is the mechanical ohm.

mechanical equivalent of heat J the Joule equivalent (hence the symbol) – the constant relationship in transfers between thermal energy ('heat') and physical energy ('work') in a system, from the days before these were known to be the same, 4.18 joules per calorie. We do not now believe that 'forms of energy' differ so the concept has no further use. Some people still, however, use the corresponding term **mechanical equivalent of light** for light energy transfer, around 1.5×10^{-3} watts per lumen.

mechanics the study of forces and their effects on objects. If the forces cause no change of momentum in a given frame of reference, they are in equilibrium; this is the subfield of statics. Dynamics applies when there is a net outside force on an object so a change of momentum results. The concern of kinematics, lastly, is motion with no account taken of the cause.

mechanisation the process of moving towards the use of machines rather than muscle power. In a sense, this is the basis of the 'industrial revolution' of the last thousand or two years (since, for instance, water and wind mills became common). However, just as important are the use of methods of mass production and the social changes that result from large factories working this way. See also *automation* (where the machines have an element of automatic control) and *machine tools*.

mechanism simple machine – such as a lever or cog – whose only function is to transfer power from one place to a second

mechanomotive force the root mean square value of an alternating force

median type of average: for a set of n values in order, the value of the element at the middle of the set. If n is even, the median is half-way between the two central values.

medical physics physics applied to medicine, in particular to body part imaging and radiotherapy, major career areas. The main fields are relevant uses of IT and electronics and of radiation of all types (including radiation protection), but medical physics units (in hospitals, for clinical healthcare, and in higher education, for research and development) can deal with a wide range of systems. For a full list, with links to websites with full detail, see http://en.wikipedia.org/wiki/medicalphysics.

medium the nature of the space in which something happens or through which something passes. It may be material (solid, liquid, gas, plasma) or vacuum, and fields of any kind in it are often relevant. **Medium frequency** (mf) radio waves are in the wavelength range 100–1000 m (with frequencies 3 down to 0.3 MHz).

meg(a)- M unit prefix for million, 10^6, as in megahertz, MHz, 10^6 Hz, and megohm, MΩ, a million ohms

megaphone rigid cone of small solid angle used to amplify input sound waves by matching impedance

megatonne mt unit of energy output of nuclear weapons, equivalent to that of 10^6 t of chemical high explosive

Meissner effect the repulsion of all magnetic field from inside a superconductor: the sample becomes a perfect diamagnetic

Meitner, Lise (1878–1978) physicist who, with Otto Hahn in 1917, discovered element 91, protactinium. Alone, or with Hahn, or with her nephew Otto Frisch, she worked on the fission of uranium for twenty years. Indeed, she discovered the process and was able to predict the chain reaction.

meitnerium Mt transitional rare earth in the radioactive actinide series, period 7, with proton number 109, first made in 1982 by Gottfried Münzenberg and his team and named after Lise Meitner. The most stable isotope is ^{276}Mt, with half-life less than a second – so its properties are unknown and there are no uses.

melting change of state from solid to liquid, the converse of freezing. The name fusion is now rare. For a pure uniform substance at standard pressure, the state change occurs at a fixed **melting temperature**. A number of such temperatures are used as fixed temperatures on thermometry scales.

membrane very thin (strictly two-dimensional) barrier between two spaces that allows particles to pass through to some extent. It is semipermeable if particles can pass one way only.

memory the way some physical systems behave in a way that depends on their past history. See also *hysteresis*.

Mendeleev, Dmitri (1834–1907) chemist whose work on the properties of the elements led to his periodic table (1869): and to the realisation that atoms must have some structure to explain this. See also Lothar Meyer, who devised his version of the table independently of Mendeleev.

mendelevium Md radioactive transitional rare earth in the actinide series, period 7 and proton number 101, created in 1955 by a team led by Albert Ghiorso and named after Dmitry Mendeleev, who devised the periodic

table of elements. Most properties are unknown except for the melting temperature, which is 830 °C – as even the most stable isotope, ^{258}Md, has a short half-life (50 days).

meniscus the curve of a liquid surface near an interface, the effect of capillarity; see also *angle of contact*. A **meniscus lens** has one face concave and the other convex.

mercury Hg transitional metal, the only one liquid at room temperature, period 6, number 80, known since ancient times and called Mercury after the fast-moving planet (it is also still sometimes called quicksilver, but here 'quick' means 'alive' rather than 'fast-moving'). Relative density 13.6, melting and boiling temperatures –39 °C and 357 °C; most common isotope ^{202}Hg. There have been many uses, though, as the metal is toxic, the number is falling fast.

The **mercury barometer** (now rare; see *barometer*) uses the varying vertical height of the mercury column under a Torricellian vacuum in a glass tube to measure the air pressure. There are many designs, but all are bulky and costly as well as a hazard. There are also many types of **mercury cell**, an electric or electrolytic cell with one or both electrodes of mercury or amalgam (mercury alloy). As with the barometer, the **mercury in glass thermometer** is rare now; see *thermometer*.

There are also various **mercury switches**, often with the tube of mercury tilting to make contact and an inert gas to reduce arcing. The **mercury vapour lamp**, widely used as a spectral line source rich in ultraviolet, has mercury vapour at low pressure between mercury electrodes.

meridian the plane that contains the given point on the Earth's surface and the Earth's axis, and the great 'circle' of longitude that that plane makes at the surface. On the surface at the point in question, the meridian lies geographically north/south. A *meridian circle*, a device with a long history, consists of a small telescope lying in the meridian and used to measure the coordinates of objects in the sky. The old name is transit circle: one can use the device only when the observed object transits (crosses) the meridian. The Earth's *magnetic meridian* at a given point is the magnetically north/south line on the surface. The inclination, the angle between it and the geographical meridian, is one of the three elements of the Earth's magnetic field.

mesa transistor transistor design in which

the base region lies above the surface of the substrate (like a table = Spanish *mesa*)

mesfet metal-semiconductor field effect transistor, with, for instance, a layer of gallium arsenide on silicon

mesh network of lines joining nodes (junctions), in particular in electric circuits. A **mesh connection** in an electrical machine has such a form, the most common types being the delta and star.

meson member of a group of unstable particles, bosons with integral spin and odd parity that interact by the strong or the weak nuclear force, believed to be quark/anti-quark pairs in a cloud of virtual quarks and gluons. The name comes from their being intermediate in mass (meso = middle) between the electron (a lepton) and the nucleons (also hadrons). Yukawa first proposed that the strong nuclear force involves the exchange between the particles concerned of a new particle, the meson. Searches led first to the muon (= mu meson); the mass of this did not closely fit the theory, however, though is between that of the electron and that of the nucleon: the pion, found soon after, is the particle in question.

The pseudo-scalar mesons have zero spin, as the quark/anti-quark spins oppose each other; this family has eight or nine members: the pion triplet, the eta (which is its own anti), the kaon doublet, and the anti-kaon doublet.

The vector mesons have the quark/anti-quark spins in the same directions, so have a net unit spin themselves. Again, there are eight or nine members.

Mesonic atoms (or **mesic atoms**) are forms of exotic matter, with mesons instead of orbital electrons. Their energy levels provide much information about the strong interaction.

mesosphere
- the middle (= meso) layer in the Earth's crust, more often called the mantle
- the thin layer of the Earth's atmosphere above the ozone layer at the top of the stratosphere. Lying between about 48 and 53 km, the layer has an atmospheric temperature maximum, some 10 °C.

meta-centre the point in a floating object round which the object rolls and pitches. If the centre of mass lies beneath the meta-centre, the object is in stable equilibrium: after any roll or pitch it will tend to return to the upright state.

metal element (or mixture of elements) with a high density of free electrons. This gives it high electric and thermal conductivity as well as the typical high reflectivity ('lustre'). Good metals are also malleable and ductile (may be hammered thin and drawn into fine wires). Some 75% of elements are metals, though the bound between metals and non-metals is far from precise (see also *metalloid*).

A **metal oxide semiconductor** (mos) is a common type of transistor (strictly the substances which form its junction). A **metal rectifier** is a metal/semiconductor junction, the first type of junction found to be able to rectify. The common early design was the cat's whisker, a fine copper wire pressing on a crystal surface; radios using such a rectifier were 'crystal' sets.

metallicity of a star, the fraction of mass that consists of elements higher than helium, with population II stars having a much lower value than population I stars (because of the different types of cloud from which they formed)

metallising coating an insulating substrate with a thin conducting metal film, as in some forms of integrated circuit

metalloid semi-metal, element with some metal and some non-metal features. The group consists of boron, silicon, germanium, arsenic, antimony, selenium, and tellurium: these form a diagonal band down the periodic table; all are semiconductors.

meta-material a substance whose electro-magnetic properties follow from its structure rather than from its components. The structure must be smaller than the wavelength of the radiation used, in some cases much smaller. Opal is a meta material in the case of visible light – its structure is of very small silica spheres. Some artificial meta-materials have negative permittivity and permeability. This gives them a negative refractive constant – the refracted ray is on the same side of the normal as the incident ray, and higher frequencies have longer wavelengths. People are starting to find new uses for these new materials. For instance, a meta-material modulator developed in 2006 is the first that can switch terahertz signals efficiently. See also *cloaking*.

meta-physics the field of philosophy concerned with reality and existence, the natural and the supernatural. The name has in fact nothing to do with physics as such, coming from a librarian's putting a book of

Aristotle's philosophical essays after one on physics. The book gained the title meta-physics (= after physics). People often call Aristotle the father of this subject.

meta-stable of a system, in a state in which it may rest for an unusually long time before dropping to a lower energy state: it thus appears stable, but is not. Supercooled water (liquid H_2O below 0 °C) is meta-stable: it will remain liquid until some shock or speck of dust causes it to crystallise. See also *bubble chamber* and *cloud chamber*, both of which work with a meta-stable fluid in this sense. Meta-stable excited states of nuclei and atoms are not very rare; see, for instance, *laser*.

meteorology the study of aspects of the atmosphere concerned with climate and weather

meter device used to measure some aspect of the environment and give a direct reading (compare with *sensor*). See the many types. For **remote meter reading**, see *telemetry*.

The basic electric current meter (ammeter) is still the moving coil type (though electronic meters are fast gaining ground). The standard design is a milliammeter, with full scale deflection of the needle with, say, 10 mA dc. A rectifier can convert this to work with ac also, while shunts and multipliers change the range as follows. Here R_m is the resistance of the meter and I_m and V_m the (maximum) current and voltage for full-scale deflection of the needle. We have $V_m = I_m R_m$.

To change the basic meter to an ammeter able to read up to I, which is greater than I_m, use a shunt (bypass resistor, one mounted in parallel to the meter). The shunt current $I_s = I - I_m$; the voltage $V_s = V_m$ (as they are in parallel). The resistance of the shunt, R_s, is therefore:

$$R_s = V_s/I_s = V_m/(I - I_m)$$

In practice, this is a very low resistance: shunts must be thick metal bars.

To change the basic meter to a voltmeter able to read up to a value of V, which is greater than V_m, use a series resistor. Its current $I_s = I$; the voltage $V_s = V - V_m$. So the resistance of the series resistor is:

$$R_s = (V - V_m)/I_m$$

This is a high resistance.

In summary, we can change a basic meter into an ammeter using a suitable low

resistance in parallel, and into a voltmeter with a suitable high resistance in series. Indeed, we can also change the device into one that reads resistance (see *ohm-meter*) and power (see *watt-meter*). See also *multi-meter*.

method of mixtures approach to the measurement of the thermal properties of substances (e.g. specific thermal capacity) that involves mixing samples that differ in temperature. Done inside an insulated container (calorimeter), this makes the energy transfers easy to explore.

metre m the base unit of length, since 1983 defined in terms of the speed of light in empty space (previously in terms of a certain line in the krypton spectrum). The basis of the 'metric' system of units devised in Napoleonic France, this length was first defined as one ten-millionth of the distance through Paris between the North pole and the equator: but the surveyors made an error.

The **metre bridge** was, until recently, a common form of the Wheatstone bridge used to measure an unknown resistance; at its centre is a 1 m length of uniform resistance wire. The **metre-candle** is an old name for lux, the lumen per square metre, unit of illuminance. The **metre-kilogram-second system** of units was the most recent form of metric system that led to SI.

metric system the first rational system of units of measure, work starting in France in 1791. It is the basis of the modern SI. A **metric ton** is the name used for the tonne (1000 kg) in countries where metric systems are not common.

metrology the science and applied science of weights and measures, the aim being precise, useful, reproducible definitions of units within a fully integrated, rational system

Meyer, Lothar (1830–1895) chemist who published his independent periodical table of the elements in 1870, just after Mendeleev

mf medium frequency, of radio waves

mhd magnetohydrodynamics, a silent effective method of moving a conducting fluid

mho old name for siemens, the unit of conductance (reciprocal ohm, hence the word)

Michelson–Morley experiment crucial experiment in the early history of modern physics. It was carried out a number of times from 1881 by Albert Michelson (physicist, 1852–1931, Nobel Prize 1908 for the design and use of his precision optical instruments); Edward Morley (chemist 1838–1923) joined in the last, most sensitive, attempt, in 1887. The aim was to measure, by an interference method, the speed of the Earth through the fixed ether then thought to fill all space. Any such motion through the ether would have produced interference fringes between the light crossing the ether and that moving in the same direction. The experiment failed each time: this shows that there is no need for such an ether to carry electromagnetic radiation and thus that the speed of light is the same in all directions. See *Lorentz contraction*, the explanation for this result which led to the special theory of relativity.

not to scale

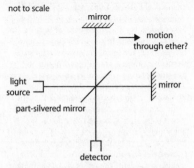

The experiment showed that light travels at the same speed whether across or along the supposed ether.

micro- a) prefix for a millionth (10^{-6}), symbol μ, as in micrometre, μm, 10^{-6} m
b) prefix meaning small in general. Thus, a **microbalance** is able to measure very small masses, a **micro black hole** is one on an atomic scale, which various theories predict – see *mini black hole* – and a **microcalorimeter** is able to measure very small thermal energy transfers.

Work with very small currents involves the **microchip** (now more often called *chip*), with its complex set of **microcircuits**, and leads to such modern hardware as the small personal computer (once often called **microcomputer** or **micro**). All this concerns **microelectronics**, the science and applied science of very small circuits that carry very small currents (no more than a few milliamps, or much less), and of the elements that make up such circuits. Now it tends to be restricted to microcircuits, i.e., those on chips, and in

particular to the design and use of microprocessors.

Microfluidics is a new field of fluidics in which the channels concerned are no more than a few tens of micrometres across.
Microgravity is the name for the effectively 'weightless' conditions in a spacecraft in free fall. Without the intrusive effects of full gravity, physical (and chemical) effects are often much easier to analyse. For instance, there are no convection and buoyancy. In fact, the contents of a spacecraft in free fall are not fully 'weightless': they attract each other slightly, while spacecraft movements that follow changes in solar radiation and people's actions may produce pulses as high as 1% of the force of gravity on Earth.

Free fall in spacecraft is not the only way to achieve microgravity, however. In the early days of space research, people achieved the effect by putting planes in long parabolic paths; this could give microgravity for up to half a minute. Japan is now building a **microgravity well**: 720 m deep in total, this has a free-fall zone 500 m long, giving 10 s of the effect very cheaply. (The other 220 m are for braking.)

A **micrometer** is able to measure very small lengths and distances (and changes in these), within around a micrometre. The various designs use a vernier, with or without a screw.

micron μ old name for micrometre, 10^{-6} m (1 μm)
microphone or **mike** sensor for sound waves, i.e. a device with sound (pressure) wave input and a corresponding electric output. There are very many designs, the main ones being in the list.
a) capacitor microphone: the waves vibrate one plate of a charged air capacitor, the other being fixed – it needs a power source (except in the case of the electret version), but provides very high quality and is the norm in studio work
b) carbon microphone: the pressure waves vary the resistance through a box of loosely packed carbon grains, as in the early phone mouthpiece; this type needs a power source and has a very low quality output – but it is robust and able to amplify
c) crystal ('electrostatic') microphone: which depends on the piezoelectric effect in a piece of quartz or Rochelle salt – it works best in contact with the source and is often reversible (so can also be a small speaker)

d) 'dynamic' (moving coil) type: with currents induced in the coil as it vibrates in a magnetic field, this being the most sensitive and accurate, but probably the most costly
e) ribbon: much the same, except the conductor is a metal foil rather than a coil – this makes the device strongly directional (bi-lobed, in fact); it has to be very fragile, however, to work well at low frequencies.

microprocessor complex microcircuit (i.e. one on a chip or on a set of chips) that carries out the functions of the processor of an IT system; in other words, it contains a control unit (and clock), an arithmetic and logic unit, and the necessary registers and links to main store and peripherals. The first one was on a set of four chips, produced by Intel in 1971.

microscope device able to give an enlarged visual image of a small close object. The **simple microscope** is a single converging lens, able to give magnification up to around 10. For higher magnification, we use a **compound microscope**; here there are many designs, all using two or (many) more lenses along the axis of the system. The sketch shows the principle.

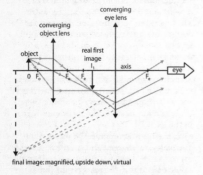

final image: magnified, upside down, virtual

The compound microscope gives an upside down, distant, enlarged, virtual image of a close object.

Modern designers are moving away from the traditional tube, folding the optics into smaller and smaller volumes and thus making the system more and more portable and easy to use. One such type also uses fibre optics to feed light onto rather than through the object; this thus need no longer be transparent.

In **microscopy**, working with microscopes, the limit on resolution (detail definition) is the wavelength of the radiation used to scan or view the target. Thus, conventional light microscopy cannot resolve details below a few hundred nanometres. One way round this may be to squeeze excitons (hole-electron pairs) through holes as small as 1 nm. On the other side, they decay into light to produce a light beam the size of the hole. Such a visible light beam can sometimes allow optical microscopy down to molecular levels; the name is **molecular exciton microscopy**. Another approach to using light for work below the wavelength of light involves the tunneling of photons; often this uses the intense light of synchrotron radiation.

Using radiations of shorter wavelength than visible light allows higher resolution than conventional optical microscopes. On the other hand, the radiation cannot be seen, so needs conversion (e.g. using a fluorescent screen). See, for instance, *electron microscope*.

microtomography scanning an object to provide internal detail on a small scale – see *x-ray microtomography*.

microwave high frequency electromagnetic radiation in the spectrum between the radio waves used in tv and the infrared region. They include the radiations used for radar and microwave cooking (below) as well as those used in communications links. The frequencies are of the order of 10^9–10^{12} Hz, with wavelengths from around 0.5 m down to 1 mm. See also *electron paramagnetic resonance* (a major type of *microwave spectroscopy*), and *maser*, a microwave oscillator. A **microwave cooker**, also widely used for heating samples in the lab, works at a frequency at which the protons in water resonate strongly. The gain in energy that results causes a rapid rise in temperature.

The cosmic **microwave background radiation** comes uniformly from all directions, with an equivalent temperature of just below 3 K and a red-shift factor of 1500. Formed some 300 000 years after the big bang the radiation's exact degree of uniformity is currently a matter of much research. Through its slight variations, it provides information about the start of the Universe. Those variations arise from the peaks and troughs of sound waves in the early Universe (when matter was very dense) that arose from the initial cosmic inflation soon after the big bang. See also http://physicsweb.org/article/world/19/5/5 on the polarisation of the cosmic microwave background which opens up a new window on the Universe when it was just 10^{-35} s old.

migration the same as *diffusion* sense b), the net movement of matter particles as a result of their random motion under some force. The term is common in the case of the movement of ions during electrolysis and of dopant particles during semiconductor doping, for instance.

mil unit of length still sometimes used in engineering, a thousandth of an inch

Millennium Technology Award applied science equivalent of the Nobel Prize, offered every two years in the sum of €1 000 000. The second winner (2006) is Shuji Nakamura, for developing light-emitting diodes of different colours (including white) and the blue laser, all sources of value in applications, including in less developed countries.

milli- m unit prefix for a thousandth, 10^{-3}, as in millimetre (mm), 0.001 m. A **milliammeter** is an ammeter able to measure currents in the range 0.5–10 mA, a sensitive current meter in other words.

Milli-charged particles have charges far smaller (perhaps a thousand times smaller) than that of the electron; some theories have shown that these objects may exist, though searches for them have been fruitless as yet. If they exist, their masses would be small, but not zero – in the 10 keV/c^2 to 200 MeV/c^2 range.

Millikan, Robert (1868–1953) physicist whose 'oil drop experiment' to measure the charge of an electron gained him the 1923 Nobel Prize. He also worked on the photoelectric effect (proving the Einstein relation and finding a close value for the Planck constant) and on cosmic rays (which he named).

The **Millikan oil drop experiment** involves keeping a small charged drop of oil in equilibrium under gravity in an electric field. The electric force depends on the charge on the drop; the common factor in the many values of that is the electron charge.

mini black hole common name for micro black hole, i.e. a very small one, one perhaps on an atomic scale, which various theories predict – see **black hole**.

minimum deviation D_{min} unit: degree the smallest angle by which a ray passing through an optical system can change direction. See *prism* and *Abbe prism*.

Minkowski space the mathematical context of the Universe of Einstein's special relativity, with three physical dimensions and one for time. The model allows one to describe the motion of an object as a curve in this space. From this, the concept arose of curved space–time in what we call a gravitational field.

minority carrier an electron in a p-type semiconductor or a hole in n-type. Compare with *majority carrier*, a hole and an electron respectively. Intrinsic (i-type) semiconductors have equal numbers of holes and electrons.

minute a) non-SI unit of time, sixty seconds **b)** unit of angle, also sixty seconds, a sixtieth of a degree

mirage illusion that results from the internal reflection of light in a layer of warm air next to a surface (for example, hot ground or a hot wall). Images of distant objects appear as if reflected by a layer of water (the image of the sky) at the surface.

mirror optical element that works by reflecting the radiation in question in a certain designed way. We can best describe mirrors by their effects on incident parallel beams of radiation. Plane mirrors reflect these as parallel beams; concave mirrors converge the beams; convex mirrors diverge them. See also *spherical aberration* and the *caustic curve*. The power to converge or diverge depends on the focal distance f of the mirror, the distance between the focal point F and the pole P; in fact, optical power, p (unit: dioptre D), is $1/f$.

A **mirror image** is a shape identical to a second, except with structure reversed as if viewed through a mirror: for instance, left and right hands. See *parity* for this in the case of elementary particles. Some people believe that dark matter may consist of **mirror** matter – just like normal matter, but with every particle replaced by its **mirror particle**. As yet (2007), however, no one agrees what a mirror proton (for instance) is, and no one has shown that mirror matter exists. In the case of nuclei, we speak of **mirror nuclides**; these are pairs of isobars (species with the same nucleon number), but the proton number of one is the neutron number of the other, and the converse. Parent and product nuclei in beta decay may be mirrors.

mismatch situation in which two linked circuits do not have the same impedance, so that power transfer between them is not efficient

mixing the same as oscillation sense b), the conversion – in a beam – of particles into their antiparticles and, in at least some cases, back again

mks unit system the metre-kilogram-second system of units, sometimes called the Giorgi system. It derived from the metric system called cgs and led to SI in the 1970s. The **mksA** system is the same, but with the ampere A as a fourth base unit. There were two systems of electromagnetic measures: the electromagnetic system which gave the value unity to the permeability of free space, and the electrostatic system which gave the value unity to the permittivity of free space.

mmf (or **MMF**) *magnetomotive force*

mmHg millimetre of mercury, old unit of pressure. It comes from the use of mercury manometers and barometers to measure pressure; in these, the measure was the height of a mercury column. 1 mmHg is around 132 pascal.

mobility the drift speed of a specified type of charge carrier per unit applied electric field. Thus, in the case of a semiconductor, we talk of electron mobility and hole mobility; the values are not the same and use of the Hall effect can measure them.

mode a) type of average of the values of a set of numbers: the most common value, that at the peak of the frequency distribution. If the distribution has two peaks, it is **bi-modal**; if more than two, it is **multi-modal**.
b) style of action. For instance, the fundamental and the other harmonics are the modes of oscillation of the system in question.

moderator substance used in the cores of nuclear reactors to reduce the speed of fast neutrons to allow their capture by fissile nuclei. The aim is for the neutrons to have 'thermal' energies, i.e. to be moving at around 2000 m s^{-1}, when they should be in thermal equilibrium with the rest of the pile. Graphite has been common for this purpose; systems that use water or heavy water as moderator also employ that for cooling.

modulation process of changing some aspect of a (radio, for instance) wave (the 'carrier', or **modulating wave**) in order to transfer information (the 'signal', or **modulating wave**). Shown are the two main types, modulating (= changing in a regular way), first, the carrier wave amplitude (strength) and, second, its frequency (or wavelength); there are other styles, such as phase modulation. The sketch shows analog

modulation, but there is the same kind of process with digital signals.

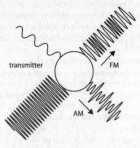

To carry a signal, we can modulate a carrier radio wave (lower left) by changing either its amplitude or its frequency to suit the carried signal (upper left).

There are also various styles of **pulse modulation**, in which a digital signal carries analog information. **Pulse code modulation** (pcm) is the most common. At set intervals, the modulator reads the size of the analog signal and codes it as a binary number. The signal then transfers as a set of binary numbers.

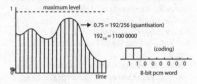

For a digital wave to carry an analog signal, we can use a technique like pulse code radiation.

In each case, the overall data transfer system is as in the next sketch, the transfer being by electromagnetic waves in space, air, wave guides or fibres. A **modulator** is any device in any system for mixing the two signals; at the far end, a demodulator separates the two and removes the carrier. (The word 'modem' means modulator and demodulator.)

This is a generic sketch of any modulation/demodulation process.

modulus a) old term for absolute value, in particular for the value (size) of a vector. In the case of a vector in Argand (complex) space – one given by $a + bi$ (i being $\sqrt{-1}$) – the modulus is $\sqrt{(a^2 + b^2)}$.

b) old term for constant of proportionality. For **modulus of elasticity** E, the ratio of stress to strain in an elastic system (now called elastic constant), see *elasticity* and *Young*.

Mohs scale a scale of the hardness of substances, devised by mineralogist **Friedrich Mohs** (1773–1839). It is non-quantitative but in common use. Each natural substance (mineral) in the list can scratch those above it but not those below it. The list is, from soft to hard:

talc
halite (rock salt)
calcspar
fluorspar
apatite
felspar
quartz
topaz
corundum
diamond

There have been various attempts to extend the list to include modern materials and to give a degree of quantitativeness. A common version includes five extra minerals between talc and diamond, and that version is now semi-quantitative.

Moiré fringes pattern of light and dark regions obtained when two similar regular sets of lines or dots overlap at a slight angle. The effect makes a useful model of interference fringes and has some uses in industry. The name is that of a type of fabric (silk, in the early days) which looks 'rippled' like this. There are many good animations of moiré fringes on the web; my favourite is www.mathematik.com/Moire/

mol the mole, the SI unit of amount of substance, 6.022×10^{23} particles

molality m the number of moles of solute to 1 kg of solvent, a measure of the concentration of a solution. If the solution is dilute, molality is very close to molarity. This measure is of major value in such contexts as the effect of concentration on freezing temperature. A **molal solution** has one mole of solute in one kilogram of solvent.

molar concerned with molarity in general, and standing for 'per mole' in particular (as 'specific' stands for 'per kilogram'). Thus, a molar value of a measure is the value for one mole of the substance in question.

The **molar fraction** x of a solution (or

mixture) is the ratio of the number of moles of solute (one substance) to the number of moles of solvent (the other(s)). People often give this as a percentage, in which case the name may be **molar percent**.

The **molar gas constant** R is the standard (universal) gas constant. It appears in the ideal gas equation for one mole of any gas:

$$pV = RT$$

Here p, V and T are the pressure, volume and temperature of the molar sample.

The **molar latent thermal capacity** of a substance is the latent thermal capacity ('latent heat') of one mole: the energy involved in changing the state, as specified, of one mole. In the same way, the **molar thermal capacity** is the thermal capacity of one mole: the energy involved in changing the temperature of one mole by one kelvin. A **molar solution** has one mole of solute in a litre of solvent (see *molarity*). The **molar volume** of a substance is the volume of one mole under stated conditions; for an ideal gas at standard temperature and pressure it is 0.0224 m^3 (22.4 litre).

molar absorptivity (once called extinction coefficient) ε for a given pure substance at a given wavelength, a measure of how much one mole absorbs compared to the input

molarity M unit: mole per cubic decimetre, mol dm^{-3} common term for molar concentration, the concentration of a solution given as the number of moles of solute in one litre of solution. A molar solution has a molarity of 1.0 – one mole of solute per litre of solvent.

mole mol the SI base unit of amount of substance. A mole of a substance is the amount whose mass in grams equals the relative molecular (or atomic) mass (former names: molecular and atomic weights). It contains the same number of particles (molecules or atoms) as 12 g (0.012 kg) of carbon-12, ^{12}C. That number is the Avogadro number, 6.022×10^{23} (to be more precise: $6.022\ 141 \times 10^{23}$).

We can therefore extend the concept to any kind of particle rather than just those of elements and compounds. For instance, a mole of photons is 6.022×10^{23} photons. For **mole fraction**, see *molar fraction*. See also *molar* (= per mole), in particular molar solution and molar volume.

molecular concerned with a single molecule

or with a view of matter on the scale of molecules. A **molecular beam** is a narrow beam of neutral molecules in high vacuum; the molecules are far enough apart that we can ignore their interactions. Such beams have value in a number of areas of research. A **molecular compound** is a weakly bound set of two or more molecules, while a **molecular diagram** shows the structure of a molecule, with its bond lengths and angles and its free charges. For **molecular exciton microscopy**, see *microscopy*.

The **molecular extinction constant** of a substance (in particular a liquid) is an absolute measure of its ability to absorb light, see the *Beer–Lambert law*. The **molecular flow** of a gas is rather like that of a molecular beam (above); here too we can ignore the interactions between the molecules of the gas: but this time the gas is in a pipe or tube so that interactions between the gas molecules and its surface are significant. A **molecular gauge** is a device whose design allows one to measure very low pressures, in the order of around 0.1 Pa down to 10 μPa. Its action involves a measure of the fluid friction (viscosity) of the gas that remains: at low pressures, this depends only on pressure.

A **molecular orbital** is an orbital (electron cloud, or wave function of given energy) of a molecule: it is much like that of a single atom, but, in general more complex (for example, with localised and non-localised regions). A **molecular pump** is a pump for work at very low pressures, while a **molecular spectrum** is the complex band spectrum of a substance (normally a gas) in molecule form. A **molecular shell** is a group of molecular orbitals (above) with the same principal quantum number. The **molecular weight** of a substance is an old name for its relative molecular mass.

Molecular wire uses zeolites to package electrical conductors of molecular scale. The molecules are long ones of certain polymers, able to conduct well along the axes. Zeolites, porous minerals with well-defined channels, form a good basis for growing these molecular crystals as desired.

molecule a single neutral atom, or a neutral group of atoms linked by some type of chemical bond. It is the smallest unit of a substance with the properties of that substance. However, metals and some covalently bound solids do not have molecules as such: we can view any sample as

a single giant molecule (not the same as macromolecule, though here too the concept of a molecule is of little value). On the other hand, some people view ions and free radicals as molecules.

The symbol for a molecule is its molecular formula: a list of the symbols for the atoms involved, with the number of each (other than 1) as a subscript. Thus the molecules of N_2, nitrogen, each have two nitrogen atoms. Given the relative atomic masses of the atoms, we can add to find the relative molecular mass of the substance. There are various methods of analysis able to find the formula of a molecule; in modern times, these include spectroscopy and the diffraction of electrons and x-rays.

molybdenum Mo transitional metal in period 5, proton number 42, discovered by Carl Scheele in 1778 and named after a major ore. Relative density 10, melting and boiling temperatures 2610 °C and 4600 °C; most common isotope ^{96}Mo.

moment T old name for torque, the turning effect of a force. It is the product of the force and the perpendicular (shortest) distance from the turning point; the unit is the newton metre, N m. See also moment of a general vector, below. The **magnetic moment** m of a magnet, current carrying coil or spinning electron measures the strength of its magnetism; see *magnetic moment*. For **mass moment**, see *mass*.

The **moment of inertia** I of an object is the rotational analog of its mass. In other words, it is the object's resistance to angular acceleration. Its value for a given object, relative to an axis about which it may turn, is the sum of the values of $m\,r^2$ for all its particles; here m is the mass of the particle and r the radius of its arc if the object turns (i.e. the distance to the axis). The table shows some important cases for axes normal to the object concerned (mass m) through its centre of mass; see *parallel axes* for other cases.

object	dimension	moment of inertia I
thin ring	radius r	$m\,r^2$
thin disc	radius r	$\frac{1}{2}\,m\,r^2$
solid sphere	radius r	$2/5\,m\,r^2$
thin rod	length l	$m\,l^2/12$

We define the **moment of a vector** other than force in much the same way as for force. In full, the moment (turning effect) of a vector about an axis is the product of the size of the vector and its shortest distance from the axis; this is a scalar, the scalar product. The moment about a point is a vector product, however – it is the product of the vector and the distance between the point in question and any convenient stated point on the line of the vector. There are cases when these moments are of use, in particular when dealing with (as well as force) velocity and momentum.

moment of momentum an old term for angular momentum

momentum p the product of the mass of an object and its velocity: $p = m\,v$. This is **linear momentum**; see also *angular momentum* (the rotational analogue), the product of the object's moment of inertia and its angular velocity. Both are conserved measures: they are constant for closed systems. *Newton's laws of force* explore the **law of constant momentum** in the linear case; there are direct analogues for rotation. **Momentum space** is the mathematical space in which we plot the graphs of the *band model* of solids.

monatomic with molecules each of which consists of one atom only. Thus, helium (He) is a monatomic gas, while hydrogen (H_2) and hydrogen chloride (HCl) are diatomic.

monochord single stretched string (made of wire or some other elastic substance) used as a sound source (indeed as a musical instrument in some cultures). Most models of sonometer are monochords.

monochromatic having a single colour or (in general for a wave) frequency or (for any radiation) energy. It is not possible to produce any radiation of truly a single energy, even with a laser: even the lines of an ideal line spectrum have some width. The half width of the radiation peak, as shown, is a common measure of spread; the half width range contains around 90% of the energy. See also *line broadening*, for factors which make a theoretically monochromatic line even less sharp.

monochromator device for producing monochromatic radiation from a poly-chromatic source. In the case of light, for instance, we can use a prism or grating to disperse the light and a slit to select a narrow frequency (colour) band.

monoclinic system a type of crystal system (structure)

monolithic of one single piece. A **monolithic circuit** is an integrated circuit on a single chip (rather than on several).

monopole a) also called Dirac monopole, a single north- or south-seeking pole; see *magnetic monopole*.
b) type of radio aerial, such as used on cars. It is half a dipole, with the other half replaced by a large earthed sheet of metal; the image of the half-dipole in the sheet has the same effect as the missing pole.

mono-stable having only one stable state, compared to bi-stable, for instance. A **mono-stable circuit** is an electronic circuit, most often a multivibrator, with one stable state that can switch to a second state for only a short time on receipt of a trigger pulse. Its output is therefore a pulse of known short length. Other functions of the circuit include pulse delay in logic systems.

Moore's law the 1965 prediction of Gordon Moore (co-founder, and later chairman, of Intel, the chip-maker) that the number of transistors on a chip would double every year. In 1975, he changed 'every year' to 'every two years' – and, so far, the industry has been able to keep up; indeed, in the mid-2000s the doubling period is around a year and a half. The 'law' extends to all measures of computer power, such as processing speed and main store size.

Morley, Edward (1838–1923) chemist who worked with physicist Michelson on the crucial Michelson–Morley test for the ether

mos metal oxide semiconductor, basis of a very widely used class of transistor called metal oxide semiconductor or insulated gate field effect transistors (mosfet or igfet). In a field effect transistor (fet), the input electrode (gate) links capacitatively to the channel, between the source and the drain as shown. Thus, the device controls the current in the channel by voltage rather than by current (as in the normal – bi-polar – transistor). This makes fets fast acting and gives them a low power demand.

Moseley's law Lines in the x-ray spectrum of an element have frequencies that depend on the proton number of the element. For the lines from the same transition (for example, the K lines) in a set of elements, the square root of the frequency varies with proton number. The graph of frequency against proton number is a **Moseley graph**. Henry Moseley (1887–1915) was a physicist who worked with Rutherford. He showed that many of the properties of an element depend on its proton number.

mosfet mos field effect transistor

Mössbauer effect emission of gamma photons of sharply defined energies from certain crystals during gamma decay. In most cases, the gamma decay product nuclei recoil away from the photons, so the photons have a broad range of frequency (energy). Each Mössbauer gamma decay involves the recoil's being taken up by the whole crystal lattice rather than by one nucleus; this means we can ignore the recoil. As the energy of the gamma photons in these cases is sharply defined, it is easy for the substance to re-absorb them. Related effects provide a strong test for general relativity. Physicist Rudolf Mössbauer (1929–) shared the 1961 Nobel Prize for his work on gamma ray absorption.

motion change in the position of an object relative to some frame. The motion may be in a straight line (linear), rotation about an axis inside the object, revolution (rotation about any axis): or some combination of these. For the **laws of motion**, see *Newton's laws of force*; see also *relative motion*.

motor machine designed to produce motion, i.e. force, from some input energy, normally electric. The standard design of a simple electric motor has a current-carrying coil in a magfield; there is a turning force on the coil and therefore on its shaft, as a result of the **motor effect** (Faraday, 1831). The motor effect is the force on a current in a magfield; see also *cathode ray oscilloscope*, *Hall effect* and *moving coil meter*, for instance.

In that simple design of motor, the magnet or electromagnet that forms the field is fixed: hence the name stator. The coil, its core, shaft, and electric contacts all rotate – they form the rotor. The coil needs an alternating current so it can keep turning in the same sense. At its ends there are therefore two separate metal rings to feed in current from an ac supply; if the supply is direct (dc), there is a single ring split into two half-circles instead.

The **induction motor** used for heavy duty work with an ac supply is in fact the most common of the hundreds of designs of electric motor. The field comes from a number of coils, each fed with current out of phase from the next to produce a rotating field. The rotor now consists of a set of copper bars round the shaft; currents induced in the bars by the changing field interact with the field to produce a torque, in the same way as before. An important type of induction motor is the **linear motor**. See also *stepper motor*, common in control circuits.

Modern non-standard motor designs use various other approaches. For instance, **piezo-motors** (able to offer particularly high torque at low speed) use the converse of the piezo-electric effect in some way, for instance, in a set of ceramic sectors that form the stator disc. Each sector in turn receives an electric signal which deforms it into a bump. The travelling wave of bumps that passes round the stator drags the rotor into motion.

Mottelson, Ben (1926–) physicist whose work on the nucleus gained him a share of the 1975 Nobel Prize

Mott insulator a substance the band model of matter predicts is a conductor, but it is not. Cases first appeared in 1937; Neville Mott explained the effect as the result of unexpected interaction between the conduction electrons of the substance.

Mott, Neville (1905–1996) physicist whose main work has been in the field of solid state studies. He received the 1977 Nobel Prize for work on the electronic structure of magnetic systems.

moving coil of some types of electric device, one in which a current-carrying coil is free to rotate in a magfield. In the case of the **moving coil meter** the coil tries to rotate against a spring. The angle it turns thus depends on the strength and direction of the current. The shapes of the permanent magnet pole pieces and of the fixed iron core make the field radial.

The **moving coil microphone** depends on electromagnetic induction rather than the motor effect (the two effects are converse). An input sound wave vibrates the coil to and fro in a permanent magfield; the induced current that passes to the outside circuit follows the shape of the wave. See *microphone*. On the other hand, the **moving coil speaker**, which may have just the same structure, *does* use the motor effect: an input electric signal makes the coil vibrate to match, giving sound waves.

moving-iron instrument electric meter in which the current passes through a fixed coil. Inside the coil are two small blocks of iron, one fixed and the other free to move against a spring; the latter carries the pointer of the meter. The current in the coil magnetises the two pieces of iron which then repel. Compared to the moving coil meter, this is simple and works with ac or dc inputs, while the design is robust: on the other hand, the scale is not linear.

MP3 Short for MPEG-1 Audio Layer 3 (an international standard since 1991), a widespread digital audio encoding/compression and recording/replay system. Although it is a standard, there are various sub-systems on the encoding side. See http://en.wikipedia.org/wiki/Mp3 for details, including the discussion http://en.wikipedia.org/wiki/Talk:MP3 about its accuracy.

mram non-volatile magneto-resistive read and write chips first sold in 2006 and believed (by the maker) to be likely to replace flash core and back-up data storage

M-theory sometimes called U-theory, and once called the theory of strings, the master theory (first proposed in 1995) that would combine the five superstring theories by way of their dualities. It is common to see the situation as like the separation of the theories of electricity and magnetism until united by Maxwell – and it is common to expect that M-theory too will need new maths. (For instance, it involves eleven dimensions, though sometimes looks like ten dimensions only.)

Müller, Karl Alexander (1927–) solid state physicist and Nobel Prize winner, 1987, sharing this with Bednorz for their joint discovery of the first 'high temperature' superconductor

multi- prefix that denotes applicability to a number of objects rather than to just one. A **multi-channel analyser**, for instance, has a number of outputs based on a chosen variable of the single input. The variable is frequency or pulse height in two very common types: here, over time, the outputs provide some kind of frequency or pulse height spectrum from the input signal. A **multi-meter** is a simple digital or moving-coil meter in a box fitted with a pair of probes. A number of switches allow it to give readings for current, voltage and resistance in various ranges, and for ac or dc as appropriate. See *meter* for how this happens in the case of ac/dc, current and voltage, and *ohm-meter* for the approach used for resistance measurement.

A **multiplet** is a group of objects with some common factor. Examples are.
a) a compound lens: a set of several lenses on the same axis, perhaps even glued together, to produce the desired effect with a minimum of aberration
b) a group of elementary particles, such as the leptons, which have the same set of basic properties but differ in one or two quantum numbers
c) a group of close, related spectral lines, and the group of energy levels (that have the same orbital quantum number but differ in spin quantum number) from which they arise: see *multiplicity*.

A **multi-stable circuit**, such as a **multivibrator**, is strictly bi-stable: it has two stable states between which it can switch (rather than the one of a mono-stable).

multiplexing treating a number of signals so they can pass through a single channel at the same time. There are many methods. In **frequency division multiplexing** (fdm), each signal modulates a different frequency band in the whole band width available. **Time division multiplexing** (tdm) involves sharing the (digital) signals between different time slots.

Clearly, the channel's band width must be at least as great as the sum of those of the signals carried: the higher the band width (the 'frequency' of the channel), the more signals it can carry. Hence the modern use of microwaves and of optical fibres rather than of radio for high volume voice, data and video traffic.

multiplication constant/factor k of a chain reaction, the ratio of the rate of production of neutrons to the rate of their 'loss' by absorption and leakage. If $k < 1$, the reaction is subcritical (not able to sustain itself, but see *natural reactor*); $k = 1$ makes it just critical (as in a fission reactor); with $k > 1$, the reaction is supercritical, explosive.

multiplicity of an energy level system (and of the line spectrum that results), the value of $2S + 1$; S is the spin quantum number for the system for given orbital quantum number L. It is the number of ways to fit the electrons into the levels available.

multi-wire proportional chamber See *nuclear physics: detectors*.

muon (or, in the past, **mu-meson**) elementary particle, a lepton, 207 times more massive than the electron, discovered in 1937 by Anderson. The name comes from mu-meson, but the particle is not a meson (the pion, pi-meson, discovered soon after, is). The muon is closely related to the electron (and with the tau particle, makes up the lepton family); indeed, an early name for the muon was 'heavy electron'. Thus, muons decay into electrons (and positive, or anti-, muons decay into positrons), while an electron/positron pair can annihilate and produce a muon/anti-muon pair. Each too is associated with a neutrino/antineutrino pair: the **muon neutrino** is not the same as the electron neutrino. For **muonium**, an atom with a positive muon instead of a proton or with a negative muon instead of an electron, see *exotic matter*.

muonium Mu (or μ^+e^-) type of atom with a muon as its nucleus and an outside electron. It is not the same as a **muon atom**: a proton with a μ^- outside. See **exotic atoms** for both.

musical scale ordered sequence by frequency of pure sounds (notes) used in music, with defined steps (intervals) between those notes. In most of the many scales explored in human history, the sequence repeats every octave, there being a doubling of frequency over that interval. The **diatonic scale**, with eight notes (five tones and three semi-tones) to the octave, is the most common in the Western music of recent centuries.

mutual of an interaction between two objects, reciprocal (i.e. shared or two-way). The **mutual capacitance** of two capacitors is a measure of how charging one affects the other: it is the charge induced on one for a given change of voltage of the other. The **mutual conductance**, g_m, of an amplifier is the change of output current divided by the small change of input voltage causing it. **Mutual induction** is the electromagnetic induction of a voltage in one conductor by a change of the current in a second; the voltage transformer uses this to great effect. The quantitative measure of the effect is **mutual inductance** M: the ratio of the induced voltage to the rate of change of current causing it. The unit is the henry, H.

myopia short- or near-sight, a defect of vision which prevents one from clearly seeing objects further away than perhaps a few metres. The eyeball is extra long, so that rays from distant objects come to a focus in front of the retina. A diverging lens can correct the problem. Laser treatment involves burning away part of the front of the cornea to reduce the length of the eyeball.

n a) unit symbol for nano-, 10^{-9}

b) symbol for negative, as a type of semiconductor and its main charge carrier (electrons)

c) symbol for neutron

d) symbol for north-seeking pole ('north pole' in day-to-day language)

N a) unit symbol for newton, unit of force

b) symbol for the gas nitrogen, element 7

n a) symbol for amount of substance (number of moles)

b) symbol for number of items in a sample (in general or in statistics)

c) symbol for number of moles

d) symbol for number of turns (loops) in a coil

e) symbol for principal quantum number, which denotes energy level

f) symbol for refractive constant (refractive index, sometimes μ)

g) symbol for rotational frequency, unit revolution per second, rev s^{-1}

N a) symbol for the neutron number of a nuclide

b) symbol for number of particles

v a) symbol for frequency (often f), unit: hertz, Hz

b) symbol for revolution rate, unit: (revolution) per second, s^{-1}.

Nakamura, Shuji (1954–) electronics engineer, second winner (2006) of the Millennium Technology Award, for developing light-emitting diodes of different colours (including white) and the blue laser, all sources of value in applications, including in less developed countries

NAND ⊐▢▷– = NOT AND, type of gate, with an output of 1 unless all inputs are 1, as in the truth table

inputs		output
I_1	I_2	O
0	0	1
0	1	1
1	0	1
1	1	0

nano- a) unit prefix for 10^{-9}, symbol: n, as in nanometre, nm, 10^{-9} m

b) in common speech, very, very small; more scientifically in size something like a few nanometres. This is the order of size of **nanoparticles** – such as buckyballs and **nano-tubes** – and **nano-structures** (like **nano-wires**)

nanoelectronics (also sometimes called **nanotechnology**) extension of microelectronics to smaller scales to allow Moore's law to remain valid, with **nano-circuits** built on **nano-tubes** (see *graphite*) and, in due course perhaps, graphene. The circuits concerned can be extremely small, for instance 10% or 20% of the width of a human hair, which makes them extremely fast. At the time of writing, nanoelectronics remains a research field, and only in the spring of 2006 the first transistor built on a nano-tube appeared (http://news.bbc.co.uk/1/hi/sci/tech/4839088.stm); a year later people had started using nano-tubes attached to chips as heat sinks. Also at the time of writing (2007), any possible dangers of breathing in nanoparticles remain unknown.

There are also various designs of **nano-machines**, for instance a **nano-motor** that uses a single 'chiral' molecule powered by

Brownian motion. Friction can be a problem in such cases, though; to overcome this, people are working on the use of electric fields and on vibration.

After nano-tubes, the most important physics research area in 2006 is the structure and use of **nano-wires**. These are conductors of the order of 10^{-9} m in size – lines of repeating atoms (metal, such as nickel, or semiconductor, such as silicon) or molecules (organic, such as dna, or not, e.g. $Mo_6S_9\_xIx$) that may one day have a crucial role in nano-circuits. At this scale, quantum effects have a great impact. See also *quantum dot* and *quantum wire*.

Napier or **Neper, John** (1550–1617) mathematician who invented logarithms early in the seventeenth century. **Naperian logarithms** are 'natural logarithms', those taken to base e (2.718 ...) rather than to base 10. The name of the neper, unit of ratio, also comes from that of Napier.

natural as found in the world or Universe rather than being artificial, made by people. The **natural abundance** C (no unit) of a given isotope (nuclear species) is the fraction of atoms of that species (constant neutron number) in a natural sample compared to all atoms of that element (constant proton number). It is usual to express it as a percentage. It was once thought that the value does not depend on the source of the sample; this is now known not to be true in all cases. See, for instance, *natural reactor* (below), and also *abundance*.

Natural convection is the appearance of convective cooling in the absence of a draught (it being forced convection if there is a draught). The **natural frequency** of oscillation of a system is the one observed when the system is not forced to vibrate any other way: it is the same as the fundamental frequency. **Natural logarithms** are logarithms to base e (2.718) rather than to base 10. **Natural philosophy** is an old name for science – in particular physics – in various languages.

The highest known levels of **natural radioactivity** on Earth are on a hill in Brazil and a town in Iran; someone living there would gain an annual dose of a quarter of a sievert (hundreds of times the average British dose). A **natural reactor** is a body of uranium ore showing clear evidence that in the past a chain reaction was sustained (though a subcritical one). The best known such

deposits are in Gabon, the first being found in 1972 – now sixteen are known in three bodies of uranium ore; all show a much lower than normal ratio (abundance) of uranium-235, ^{235}U; they would have been working one to two billion years ago. At the time the rock was laid down the abundance in it of ^{235}U was well over the 3% needed to sustain fission: and the ore was rich enough in uranium for an intermittent chain reaction to last some half million years with ground water as moderator. See also *enrichment*.

near infrared the part of the infrared region of the spectrum close to the visible region. Some people use a similar term for ultraviolet: **near ultraviolet**.

near point the point closest to an eye from which the eye can focus light without undue strain. In the case of the standard ('normal') human eye, for instance (e.g. in the design of optical systems), we take the distance to this to be 250 mm. A short-sighted (**near-sighted**) person has a closer near point than that, and a closer far point than infinity; on the other hand, a long-sighted person cannot see clearly such close objects. See also *accommodation*.

necking the decrease of thickness of a stretching wire (for instance) when under tension past the elastic limit: it is the physical sign of plastic deformation, a stage that may lead to fracture. See also *Poisson ratio*.

Néel temperature the temperature at which the susceptibility of an antiferromagnetic substance is greatest. Above this value there is no antiferromagnetic behaviour: thermal motion becomes too great for the magnetic order to remain. It is the analog of the Curie temperature of a ferromagnetic. Louis **Néel** (1904–2000) gained the 1970 Nobel Prize for his work on magnetic materials and the Earth's field.

negative associated with a value less than zero, and/or with behaviour the opposite of the norm, and/or with a charge of the same sign as that of a negative electron, a fundamental particle. For **negative feedback**, see *feedback*. In a *discharge tube*, the **negative glow** is a bright region next to the Crookes space, the main part of the glowing region close to the cathode. A **negative lens** is a diverging lens (one with a negative focal distance and power, using the real is positive convention).

Unlike antimatter, **negative matter** has the opposite mass to normal ('positive') matter. It could exist somewhere in the Universe – i.e. no current theory forbids it – but has not (yet) been found. Rather, some current theories require it, and imply that positive matter attracts both positive matter and negative matter, while negative matter repels both. This leads to the possibility of a drive with a chunk of negative matter in a positive spacecraft: both would accelerate together in the same direction. (Energy and linear momentum are conserved, as a moving negative matter object has negative kinetic energy and momentum.) When negative and positive matter combine, the effect is nullification, i.e. no energy output (rather than annihilation as with matter and antimatter).

A **negative mirror** is a diverging mirror (one with a negative focal distance and power, using the real is positive convention). **Negative pole** is a rare term for south-seeking magnetic pole. **Negative proton** is now a rare name for anti-proton.

Samples of some substances, and some types of device, show **negative resistance** over a certain input voltage range: as the voltage rises through the range, the current falls. The cause is the transfer of conduction electrons to a higher band in which they find it harder to move. The first transistor (the point contact device of Bardeen and Brattain) showed the property; for a while, it was used to make a single transistor flip-flop for the first third-generation computers. The negative resistance Gunn effect is common in semiconductors such as gallium arsenide and indium phosphide. Here, a plot of current passed against applied voltage shows negative slope in the Gunn region; this means negative resistance if we define resistance as the ratio of the change of voltage to the change of current. On the other hand, the static resistance, the ratio of current passed to applied voltage, is positive at any point.

It is also possible to have a **negative specific thermal capacity**, as with a saturated vapour in certain circumstances: one needs to extract energy in order to raise the temperature. Also, there is some value in the concept of **negative temperature** on the absolute scale. This does not, however, refer to temperatures lower on that scale than absolute zero (0 K, –273 °C): rather it is the state of a nuclear spin system in a magnetic field for which a small energy input leads to a fall in entropy. Some people also use the term for a population inversion: where there are more atoms in a high energy level than in the ground state.

negaton or **negatron** names offered for the negative (normal) electron for a few years after the antielectron was found. People first called the latter a positron (or positon); others then proposed to rename the electron by symmetry.

nematic common type of liquid crystal, widely used in liquid crystal displays because of the effect of its long molecules on plane polarised light

neodymium Nd transitional rare earth in the lanthanide series, period 6, with proton number 60, isolated in 1885 by Carl von Welsbach and named 'the new twin' (compare with *praseodymium*); it has two allotropes. Relative density 7, melting and boiling temperatures 1020 °C and 3070 °C; most common isotope ^{142}Nd. Main uses are for colouring glasses (including astronomical lenses for filters) and in magnetic alloys.

neon Ne inert gas in group 0 and period 2, proton number 10, first obtained in 1898 by William Ramsay and named after Greek *neos*, = new. Density 0.9 kg m^{-3}, melting and boiling temperatures –249 °C and –246 °C; most common isotope ^{20}Ne. Main uses are in discharge tubes, tv and vacuum tubes, cryogenics, and (with helium) lasers.

neon lamp a) discharge tube filled with low pressure neon, giving an intense red/orange light. The light from neon's discharge is the most intense of all the inert gases, and people sometimes call any advertising discharge tube a 'neon lamp' or **neon tube**. **b)** very small neon discharge tube, normally called just 'neon'. It glows red above a certain voltage (typically in the range 150–250 V for early neons, but below 90 V now). Its main uses are as an indicator lamp with a very low power demand and as a voltage stabiliser. The latter use follows the fact that when the discharge starts, the voltage between the ends stays constant over a wide range of currents.

neper Np unit of difference on a natural logarithmic scale (whereas the more common bel and decibel (dB) use a common, base 10, logarithmic scale). The difference between two currents, I_1 and I_2, for instance, is ln (I_2/I_1) Np. 1 Np = 8.69 dB. The name comes from an alternative spelling of John Napier.

neptunium Np transitional metal in the

radioactive actinide series, period 7, with proton number 93, first produced in 1940 by Edwin McMillan and Philip Abelson, and (being the first element after uranium) named from the first planet after Uranus. Relative density 20, melting and boiling temperatures 640 °C and 3900 °C; most stable isotope ^{237}Np, alpha-emitter with half-life 2×10^6 years. Neptunium is fissionable, but (as far as is known) has not been used in nuclear piles or weapons; ^{237}Np, by neutron absorption, is the major source of ^{238}Pu, a very important nuclear fuel for spacecraft.

The **neptunium series** is one of the four major radioactive series (decay chains), the only one not found in nature (because the first few species in it, which include neptunium-237, ^{237}Np, have short half-lives); the members of the series have mass numbers of $(4n + 1)$, n being a whole number.

Nernst, Walther (1864–1941) chemist whose award of the 1920 Nobel Prize for chemistry followed his discovery of the third law of thermodynamics. The **Nernst calorimeter** is a special design of calorimeter that involves electrical heating of the metal sample whose specific thermal capacity is needed. The **Nernst effect** is the appearance of a voltage between the sides of a metal sample in a magfield when there is a temperature gradient in the metal. For best effect, the energy flow and the field are at 90° to each other; so too will then be the voltage. This is the thermal analog to the Hall effect. The

Nernst energy theorem is a (rare) name for the third law of thermodynamics. The **Nernst glower** is an infrared lamp whose output is close to that of a black body when the conditions are right.

network a set of electric elements linked by conducting paths. The network becomes a circuit only if there is a source of electrical energy and a complete path for current between the two sides of the source. However, people often use the word to mean an electric circuit (or, by analogy, circuit in other contexts) – one in which there is a number of loops (complete paths between the two sides of the supply) and nodes (junctions between loops at which the current splits). The concern of **network analysis** is to find the current supplied or that at any part of a given network, given the supply voltage (and frequency if ac, and the relevant oppositions (resistances and reactances). See *Kirchhoff's rules*, major tools in this area.

Neumann's law the relation between the voltage V induced and the rate of change of magnetic field at a conductor that follows from the law of electromagnetic induction. If the field change comes from a change in electric current I, the relation is of the form shown.

$$V = -L \, dI/dt$$

Here L is the inductance concerned (unit henry H).

with no net effect, in particular no net charge. A **neutral current** (or neutral weak current) is a view of the weak interaction between leptons and hadrons where there is no change of lepton charge. (People sometimes use 'currents' to describe, or even explain, these reactions in general. They are the flow of exchange force particles.)

When an object or system in **neutral equilibrium** is disturbed, it stays in the new state, tending neither to return nor to move further; see *equilibrium*. A **neutral filter** for light is a grey filter: it cuts all passing wavelengths by much the same fraction. A **neutral point** in the force field around several sources is a place where there is no net force. The sketch shows a typical simple case between magnetic poles of the same type, using the lines of force model.

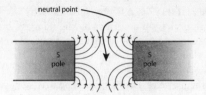

neutral point

S pole

S pole

The **neutral temperature** of a thermocouple is the highest temperature at which one can easily use it. The electric output rises with temperature rise up to that value, and then starts to fall. For **neutral weak current**, see *neutral current* (above).

neutralino one of a set of particles (wimps) given by supersymmetry theory, made up of superpartners of 'normal' particles. The set includes various mixtures (eigenstates) of the superpartners of the Higgs particle (higgsino), the photon (photino), and the Z boson (zino). The mass of the lightest neutralino would be in the range 30–5000 GeV/c^2 and could well be a major part of dark matter.

neutrino 'neutral little one', elementary particle proposed by Pauli in 1930 (with extension by Fermi in 1933) to explain the details of beta decay. If beta decay involves only the two particles in fact observed (product nucleus and electron), the laws of constant energy and momentum would fail. Pauli's neutrino avoids that problem; it would have zero (or very low) mass and spin $\frac{1}{2}$, travel at the speed of light, and carry off energy and linear and angular momentum.

Neutrinos are very common, being produced in a large fraction of nuclear interactions and decays; however, they are very hard to detect and to study as they barely interact with matter. (A light-year thick block of lead would absorb, i.e. stop, only half the neutrinos in a beam.) It was not until 1956 that such a particle was observed. There are three 'flavours' of neutrino – a neutrino (and antineutrino) for each lepton (electron, muon, tau), each with rest mass equivalent to no more than something in the range of 4 to 22 eV/c^2. (Some data obtained in 1995 seemed to propose a fourth type of neutrino; however, in 2007, other data disproved that.) Neutrinos appear in many modern theories, and many modern theories try to explain the observational problems. The biggest of these problems is that the observed intensity of neutrino flow past the Earth is much smaller than the standard theories can predict. One explanation is that a neutrino in transit oscillates between the three flavours; people have found such oscillation in the case of some other particles.

Whether or not neutrinos have mass is the subject of much

current dispute; if they do, it is very small indeed. The standard model of matter relies on neutrinos having no mass, but some recent experiments have implied that this is not correct. Those experiments claim to detect neutrino oscillation (or mixing), a change from one flavour to a second (and probably back again) that depends on the particles' having mass. The first such claim was in 1998; one in early 2006 gives a value for the neutrino mass of 10^{-7} times the electron mass. See http://neutrinooscillation.org, a site which also includes much useful material on neutrinos generally.

The world's largest **neutrino detector** is 'IceCube', being built in the Antarctic ice cap, 2006. It consists of $1 \, km^3$ of ice and should detect many tens of neutrinos a day (all three types): fifty times more than earlier systems. An array of thousands of photo-multipliers has the task of detecting the Cherenkov radiation produced. Other large detectors are in deep mines on land and on the ocean bed (using sea water for shielding).

There is a great urgency about such neutrino research. That is because all detectors so far have picked up a far smaller number of neutrinos than all theories expect, as noted above. Thus, the longest such experiment (and one of the first), Davis's $380 \, m^3$ of perchlorethylene in a Dakota gold mine, found during its twenty years only a third as many electron neutrinos from the Sun as predicted. Also, the 1989–1990 trial of a detector using 30 t of liquid gallium found none at all, though should have been more sensitive than Davis's. By the mid 2000s, theorists still have not been able to explain these results, hence the need for better results.

The **sterile neutrino**, not yet found, is the main constituent of dark matter in some theories. It is 'sterile' as the weak nuclear force has no effect on it (unlike 'normal' neutrinos) – gravity is the only force that affects it, as is the case with dark matter in general. Sterile neutrinos also explain the lack of balance between matter and antimatter; the great speed at which stars formed in the early Universe (despite the lack of atomic hydrogen then); and the way some supernovas expel pulsars with great energy.

neutron fundamental elementary particle, found in the nuclei of all matter other than normal hydrogen. It has no electric charge, spin $\frac{1}{2}$, and a mass of about 1.009 mass units, slightly more than that of the proton (the other nucleon). The neutron interacts by the strong nuclear force, so is a hadron (like the mesons, the proton, and other baryons).

The free neutron is not stable, having a half-life around 612 s; the form of (weak) beta decay concerned is to the proton (which is stable), electron, and anti-electron neutrino, as in the box. Within a nucleus, however, neutrons can remain in existence for ever by the continual exchange of pions with protons:

$$n \rightarrow p^+ + e^- + \bar{\nu}_e$$

When the proton was found, in 1919, people could at last accept Rutherford's 1911 model of the atom as having a tiny massive positive core. It was then thought that nuclei contain enough electrons to make the difference between the electron number and the proton (mass) number. Problems with this theory grew, and by 1930 the idea that there was a neutral nucleon had become common. Chadwick discovered this particle, now the neutron, in 1932.

neutron absorption of a nucleus, taking in a free neutron (from the 'neutron gas' in a pile, for instance). For the nucleus to capture a neutron rather than have it bounce away means the neutron must have no more energy than a certain limit. Once the nucleus has absorbed a neutron, its neutron and mass numbers increase by 1 – it is a new nuclide, which is likely to decay. ^{235}U becomes ^{236}U, which decays by fission, with the release of more neutrons: this is the basis of the chain reaction. ^{238}U become ^{239}U, whose beta decay leads to ^{239}Pu, itself a fission fuel like ^{235}U.

The **neutron bomb** is a 'clean' fusion (hydrogen) bomb designed as an improved nuclear weapon. Its main output is an intense burst of high energy neutrons, lethal to all life and electronics in the area, but there would be little physical damage and little fallout. **Neutron capture** is the absorption of a neutron by a nucleus; the most common result is that the new nucleus decays by the emission of a gamma photon (though decay by alpha, proton or neutron emission happens in some cases). For **neutron capture therapy**, see *neutron radiotherapy* (below). As a beam of neutrons (like a beam of any elementary particle) has wave properties, **neutron diffraction** is feasible; indeed, it is an important technique for research into crystal structure. It is much like electron and x-ray diffraction in practice.

The **neutron excess** of a nucleus (old name: isotopic number) is the difference between the neutron number and the proton number; in almost all cases, this is positive. Isodiaspheres are nuclides with the same neutron excess. **Neutron flux (density)** is a measure of the intensity of neutron radiation in a space (e.g. inside a reactor). It is the product of the number of free neutrons per cubic metre and their mean speed; the unit is the neutron per square metre per second: thus, one can view the measure as a type of power. The **neutron number** N of an atom, isotope or nucleus is the number of neutrons in the nucleus; it is the difference between the nucleon number A and the proton number Z.

Neutron radiotherapy is an alternative to x-ray therapy; each attempts to treat medical problems (such as a cancer) using a beam of high-energy radiation. Each too (as well as others of the same type) is imprecise and can damage healthy tissue significantly. **Neutron capture therapy** overcomes this major problem. First devised in 1936, but only now coming into practice, the technique involves the take-up by the target tissue of some such nuclide as boron-10, followed by irradiation with a low-energy neutron beam. The boron absorbs neutrons to form boron-11; the radioactive decay of this produces harmless lithium-7 and alpha particles. These have a range of five and nine micrometres compared to a typical cell size of ten micrometres; thus, it is easy to destroy the target cells with very little damage to healthy tissue.

A **neutron star** is, according to many theories, the result of the supernova collapse of a fairly massive star. It would contain a mass of about 1.6 suns in a ball a few kilometres across, and consist entirely of neutrons formed as the very large gravity forces protons and electrons together. It is therefore hugely dense and also has a very strong magfield, perhaps 10^{11} T (10^{16} times stronger than that of the Earth). When two neutron stars collide, the event provides (it is believed) the source of the most highly energetic radiation known as a result of the even more intense magfields produced – the short gamma ray bursts that last only around a tenth of a second but involve very large amounts of energy. There are thousands of pulsars known. People believe they are neutron stars that produce a beam of radiation rather than a uniform 'glow'. If the beam crosses the Earth as the star spins, we observe a pulsar. Pulsars may spin at up to 2000 times a second; quakes in the neutron core as the object settles can change that value suddenly.

At least one theory overcomes some problems of the view of these stars as built only of neutrons by postulating instead that the collapse produces a more stable strange star (that is, one built of strange matter, a roughly equal mix of up, down and strange quarks).

new Cartesian convention at one time a very popular convention for the signs given to measures in optical systems. Now the 'real is positive; virtual is negative' system is much more common. As the name implies, the former relates to the use of Cartesian coordinates: values are positive if to the right of the centre of the system and/or above the axis; in other cases, they are negative.

Newcomen, Thomas (1663–1729) engineer, inventor of the first practical steam engine (around 1710). This was a stationary machine mainly used for pumping water from mines.

newton N

the unit of force, the force able to accelerate one kilogram by one metre per second per second: $1\,\text{N} = 1\,\text{kg m s}^{-2}$. A **newton-meter** is a force-meter, such as a spring balance.

Newton, Isaac (1642–1727) physicist and mathematician (and philosopher and theologian, alchemist and politician). He is most remembered for developing a form of calculus (but see *Leibnitz*), the particle theory of light, a working theory of gravity (see also *Kepler*), and a good general concept of force. His output in very many fields was prolific, in fact: though many of his ideas were wrong and thus held up progress.

Newtonian derived by Newton or from some aspect of Newton's work. The strain in a **Newtonian fluid** (as opposed to a non-Newtonian fluid) is proportional to stress and to time; the fluid friction (viscosity) is the constant involved. A **Newtonian force** is the name sometimes given to any force which shows an inverse square law. The basis of **Newtonian mechanics** is the set of laws of force (below, also called laws of motion).

A **Newtonian telescope** is a simple reflecting system. The first such to appear, it is still the basis of many instruments. A large converging (now parabolic) mirror collects the light rather than the object lens of the refractor; a small flat mirror at an angle to the axis directs the converged beam to a lens at the side of the tube.

Newton's attributed to Newton or to some aspect of his work. **Newton's cradle** is a well known desk toy that shows well the laws of constant momentum and constant energy; alas, Newton didn't have one on his desk – actor Simon Prebble invented and named it in 1967. See also *Bose gas*. **Newton's law of cooling** is that an object loses energy at a rate proportional to the difference between its temperature and that of the surroundings. (This is true for natural convection only if the temperature difference is small, but is more generally correct for forced convection, as in a breeze.)

We can express **Newton's law of gravitation** as

$$F = G\, m_1\, m_2/r^2$$

Here F is the force of gravity between masses m_1 and m_2 distance r apart, and G is the gravitational constant.

The law survived unchanged for almost three centuries, until Einstein made some small changes; it is still fully valid in normal ('classical') contexts. **Newton's law of fluid friction** states that the opposition to the relative motion between a fluid and a solid is proportional to the square of the relative speed, the section area of contact, and the density of the fluid. See *viscosity*.

Newton's laws of force (or **laws of motion**, 1687) in modern terms are as follows:

a) An object's momentum is constant unless a net force acts on it from outside.

b) The net outside force is proportional to the rate of change of momentum produced.

c) If object A applies a force on object B, B applies the same force on A.

In fact, Galileo was the source of the first law which, like the third, is a statement of the law of constant momentum. The second law becomes:

$$F = m\,a$$

The net outside force = the object's mass × its acceleration.

Because of this, the unit of force, the newton N, is 1 kg m s^{-2}.

One can observe **Newton's rings** by looking down on a large converging lens resting on a mirror. The light and dark rings centred on the point of contact are, in fact, interference fringes (obtained by division of amplitude). Newton did not, however, recognise the effect as a proof of the wave nature of light: that was left to Thomas Young two centuries later. The effect is just like that of the air wedge: in this case, we have

$$l^2 = n\,r\,\lambda$$

Here n takes the values 0, 1, 2 ... for the corresponding dark rings, distance l from the centre, r is the radius of curvature of the face of the lens in contact with the mirror, and λ is the wavelength of the light.

NiCd cell the nickel–cadmium cell
nickel N transitional metal in period 4, element 28, known since ancient times and named from Kupfernickel, = devil's copper (as silver miners found the ore got in the way of their work). Relative density 9, melting and boiling temperatures 1450 °C and 2730 °C; most common isotope ^{58}Ni. Nickel is ferromagnetic; its main use is in alloys.
nickel–cadmium and **nickel–iron cells** important secondary (rechargeable) cells which are much lighter than the corresponding lead–acid cells. NiCd cells are in very wide use nowadays, in the dry form for portable electronics items, torches, toys and so on; they match the main dry primary (alkaline) cells (such as AAA, AA ...) but are also important in portable phones and electric tools, emergency lighting systems, and so on. Wet versions power some electric cars and the starter motors of planes.

The NiFe cell was an important secondary, first developed by Edison, and giving around 1.3 V. The plates are steel, the anode bearing a mixture of nickel and nickel hydrate, and the cathode being iron oxide; the electrolyte is potassium hydroxide. These cells have rather gone out of favour, but as they are robust and reliable, people are looking at them again for use in such contexts as back-up for renewable power systems and electric vehicles.

Nicol prism until the development of Polaroid™, the main way to produce polarised light. Indeed, the Nicol is better in that it absorbs less light. On the other hand, it is bulky, fragile and costly. A crystal of calcite – which is birefringent (doubly refracting) – is cut with a 68° angle, cleaved along the optic axis, and glued with Canada balsam. The extraordinary ray passes through the system, but the Canada balsam layer totally reflects the ordinary ray. As a result, the output is plane polarised light. This leaves the block parallel to the input because of the 68° angle.

Niépce, Joseph (1765–1833) inventor, the first person (1826) to produce a permanent photograph; he later became a partner of Daguerre. His process involves an exposure of many hours on an asphalt layer on a pewter plate.

NiFe cell the nickel–iron cell

night glasses prism binoculars with very large object lenses, so able to collect more light than usual and thus to produce an image of a very dim object. Modern systems for **night vision** amplify even very dim signals, but often also work on other radiations than visible light (e.g. infrared) and use a fluorescent screen to make the image visible.

niobium Nb transitional metal in period 5, proton number 41, discovered by Charles Hatchett in 1801 (who gave it the name columbium Cb, still often used) and re-discovered by Heinrich Rose in 1846 (who named it after Niobe, daughter of Tantalus, as its ore has the name tantalite). Relative density 8.6, melting and boiling temperatures 2470 °C and 4740 °C; only natural isotope ^{93}Nb. Used in various alloys, including for jewellery (because of its blue colour) and in nuclear industries (as it has a very low neutron capture cross-section).

nit now rare name for the candela per square metre, SI unit of luminance

nitrogen N main gas of the Earth's atmosphere (and once called phlogisticated air), group V, period 2, proton number 7, first found and studied by Daniel Rutherford in 1772; there have been many names proposed, that in English meaning 'soda producer' (soda being a name for sodium nitrate, which

has many uses). Density 1.25 kg m^{-3}, melting and boiling temperatures –210 °C and –196 °C; most common isotope ^{14}N. Main uses are to provide an inert atmosphere (e.g. for electronic circuits during manufacture, explosives, high voltage systems, and food stuffs) and to fill tyres.

nmr nuclear magnetic resonance, a technique for internal imaging, especially in medicine

Nobel, Alfred (1833–1896) inventor of dynamite and gelignite, a pacifist who became very bitter about the military uses of his explosives. He left most of his money to a fund for the annual award (since 1901) of **Nobel Prizes** in various areas of science and applied science (later also peace). This book provides notes on the Nobel laureates (prize winners) in physics to 2006 (by alphabetical order in the body of the book and also in *Appendix 4*), and on those physicists gaining prizes in other fields gained or where the work is relevant to us in some other way. The prizes are big in money terms as well as in honour: the 2006 Nobel Prize in physics was worth £700 000 (€1 000 000 or $1 300 000).

nobelium No transitional rare earth from the radioactive actinide series, period 7, with proton number 102, first made in 1958 by Albert Ghiorso and colleagues; the name (from Alfred Nobel) was given by a second team who claimed – wrongly, it turned out, to have made it the year before. Few properties are known: even the most stable isotope, ^{259}No, has a half-life of less than an hour.

nodal line line joining a set of nodes (points of minimum amplitude) in an interference fringe pattern

nodal points in an optical system, the two conjugate points on the axis (one each side of the centre of the system) where angular magnification is 1

node a) junction or place of overlap, e.g. in a circuit or on a graph or chart
b) point of minimum (perhaps zero) amplitude in an interference pattern or standing wave, half-way between a pair of antinodes: for instance, dark in the case of light; quiet in the case of sound (for instance, in this case, at the closed end of a sounding air pipe); no signal in a resonant circuit; minimum particle probability in an electron cloud

noise sound that is not wanted, or, by extension, any unwanted component of any kind of signal. Thus, there may be electrical noise in a circuit and radio noise in a communications channel. The **noise factor** of a circuit is the ratio of the signal-to-noise ratio at the output to that at the input. See also *interference*.

Noise cancellation can be much better at cutting down the ill effects of unwanted sound than acoustic insulation. It involves generating a sound signal of much the same amplitude but exactly opposite phase to the noise. A microphone near the point in question picks up all sound waves; an electronic circuit filters out those that present no problem; a speaker outputs the anti-sound of the rest. See *active noise control*, and for **noise control**.

no-load in the case of usage of an electrical machine, with no output but clearly defined inputs, used for specification (as in the no-load input impedance). Here, as in other contexts, load means output.

non- prefix for absence or exclusion. Thus, of a pair of variables, **non-commutative** means not commuting in the case of a given commuting operation – multiplying numbers is commutative (as $A \times B = B * A$), but dividing is not (as $A/B \neq B/A$). Again, **non-degenerate matter** is in a normal state, so that normal laws apply (see *degenerate*).

In a **non-Euclidean geometry** there can be none (Riemann) or more than one (Lobachevski) line through a given point parallel to a given line; in a Euclidean world there is always one and only one such line. Gauss was, in fact, the first person to explore this field, but he did not publish his work. A **non-inductive** device, circuit or system is one with no inductance (inductance can cause problems during switch on and off).

Nonlinear behaviour does not show as a straight line on a graph, so is less easy to describe than linear behaviour. For instance, **nonlinear distortion** appears in circuits which affect a signal by an amount that depends on the instantaneous value of (say) amplitude. **Nonlinear optics** is the optics (study of behaviour) of light of such intensity that the electric fields concerned affect the medium in a nonlinear (non-proportional) way to produce unusual effects (such as refractive constant that varies with intensity; the generation of various harmonics to the input beam; Raman scattering, etc.). For **nonlinear oscillation**, see *oscillation*.

A **non-metal** shows none of the features of a metal; see also *metalloid*. There are some seventeen non-metal elements, at the top right-hand corner of the periodic table. They tend to form anions. A **non-Newtonian fluid** has two mixed phases, and, as a result, fluid friction (viscosity) that depends on the two concentrations and on the rate of flow. A **non-ohmic** circuit or circuit element does not behave as described by Ohm's law (in other words, the current observed does not vary directly with the voltage applied). A **non-reactive** circuit or element offers no reactance to ac (i.e. it has no capacitance or inductance); with a **non-reactive load**, there is no current lead or lag in the circuit.

no parallax with a lack of apparent relative motion between two viewed objects as the observer moves: this must be because they are at the same distance from the observer. See *parallax*. If one can obtain no parallax between a virtual image and a real marker (such as a pencil held behind a converging lens used to magnify), the two are at the same distance from the eye.

NOR ⊐⟩∘ NOT-OR type of gate, with an output only if no input is 1; see the truth table.

inputs		output
I_1	I_2	O
0	0	1
0	1	0
1	0	0
1	1	0

Nordström, Gunnar (1881–1923) theoretical physicist whose theory of gravitation appeared before, and competed with, Einstein's general theory of relativity. His idea, which people forgot but was later developed by Kaluza and is now crucial for string theory, was to add an extra dimension to our 'normal' 4D space–time view.

normal a) usual or standard. Thus, the position of **normal adjustment** of a telescope is to have the image at infinity (relaxed eye); in the case of a microscope, normal adjustment brings the image to the viewer's near point. In other words, normal adjustment is the setting of an optical system

so that the image is where one would view the object with the unaided eye: many other adjustments are possible in practice, some perhaps being preferred by a given user. The **normal distribution** is the frequency distribution of a data set giving a bell-shaped curve; see Gaussian distribution. **Normal temperature and pressure** ntp is the old name for standard temperature and pressure, stp: 0 °C and 101 325 Pa.
b) perpendicular (line at 90° to a surface at some stated point, such as the point of incidence of a ray at a surface). A **normal ray** arrives at the surface along the normal (so reflects straight back on its path or enters without any bending due to refraction).

normalisation a) heating steel above a certain temperature and letting it cool, in order to relieve internal strains
b) tweaking a finite function so that the area between curve and x axis becomes 1
c) process that converts numeric data to a standard form, in most cases to give maximum precision

north-seeking pole n the pole of a magnet or electromagnet that tends to swing to point roughly towards the North Pole of the Earth (the North Star, Polaris, in the sky). The Earth's magnetic North Pole is therefore south-seeking (as poles of different kinds attract each other). The common way to write north-seeking pole is n-pole.

NOT gate —▷○— single input gate whose output is the converse (reverse) of its input

input	output
l1	0
0	1
1	0

note musical sound described in terms of pitch (the subjective analogue of frequency) and perhaps volume (loudness, i.e. amplitude), length (duration), and timbre (quality, i.e. mix of harmonics)

nova explosion of the outer layers of certain types of star: highly energetic, but not destructive (whereas a supernova is the death of a star). The normal pattern (but see below) is an increase in brightness by 10^2–10^6 over a few days. This corresponds to a surface temperature of around a hundred million degrees. After this rapid growth in brightness is a decline to the original state that may take years. After the nova is no longer visible to the eye, however, high radiation levels remain for a long time at shorter wavelengths. The star loses perhaps 0.01% of its mass each time a nova event happens (which may be as often as every few tens of years, weeks in a few cases). However, if the nova is one star of a double system, it may gain a lot of mass from its partner during the event. For **recurrent nova**, see *white dwarf*.

Noyce, Robert (1928–1990) electronic engineer, inventor of the silicon chip, and founder of both Fairchild and Intel, major manufacturers of integrated circuits

n-p-n transistor common type of bi-polar transistor, with regions of n-type semiconductor each side of the p-type base

n-pole standard shorthand for north-seeking pole

ntp normal temperature and pressure, now known as standard temperature and pressure, stp

n-type extrinsic semiconductor material with more free electrons than holes, and how such a substance behaves. Current passes through it mainly as a flow of negative charge (free electrons), the minority carriers being positive holes. See also *donor*.

concerned with the nucleus of an atom. This is a massive but very small positively charged object. To other positive objects outside, it presents a repulsive force, and therefore a **nuclear barrier** against approach. It is normal to quote barrier height in energy terms, often electron volts (eV), there being about 10^{20} eV in a joule. A **nuclear cell** is a small electric cell (wrongly called by some a reactor) that gains its power from the decay of a radioisotope; such cells are common in spacecraft and for polar systems. **Nuclear charge** is the absolute (positive) charge of a nucleus, the product of its proton number and the proton charge. For **nuclear emulsion**, historically a most important radiation detector with some specific uses still, see *emulsion*. For **nuclear energy**, obtained from the reorganisation (fission, decay, fusion) of nuclei, see *nuclear power*, below. We often express **nuclear energy change**, the energy output of a nuclear reaction, in electron volts (eV), indeed in most cases in MeV, there being some 10^{14} MeV in a joule. For **nuclear energy levels**, e the *shell model* of the nucleus. **Nuclear engineering** is the applied science of gaining energy from changes in nuclei, as in nuclear power. **Nuclear fission** is the breakdown of a large nucleus into two roughly equal smaller ones, with the release of some neutrons and energy. See *fission* and *nuclear power*.

There are two forces that specifically affect nuclear particles: the weak nuclear force and the strong nuclear force. The term **nuclear force** on its own means the latter. This is very strong on the nuclear scale of around 10^{-15} m (one femto-metre, fm), and is some 137 times larger than the electric repulsion between protons. Thus, it is the force that holds neutrons and protons together in a nucleus. If the energy involved is not great enough, as in the case of many larger nuclei, there is an instability which may lead to fission (splitting) or some other form of radioactive decay (mainly alpha). On the other hand, small nuclei may become more stable if they combine; this is the process of **nuclear fusion**: see *fusion* and also *binding energy*. **Nuclear heat of reaction** is an old term for *nuclear energy change* (above). For **nuclear isomers**, nuclei with the same numbers of protons and neutrons but different energy states and behaviour, see *isomer*.

Nuclear induction, more often called **nuclear magnetic resonance** (nmr), is a technique for internal imaging, specially in medicine (where it is very common). The system scans the target in a strong alternating magnetic field, using, at the moment, large, massive and costly magnets; different elements resonate at different frequencies. In particular, it is easy to detect the protons in water; that's the value for medicine. The technique is effective for the study of soft organs (the brain, heart, etc.): safer and quicker than using x-rays, and less destructive or imprecise than other methods. In a recent development, it has also become possible to assess muscle damage, as injured muscle tissue appears differently on the display: the water content and distribution even weeks after hard exercise differ from the norm. In another, scanning with a laser and checking how the scanned surface rotates the plane of polarisation, allows real-time working in two dimensions.

The **nuclear magneton** μ_n is a fundamental constant, the unit of **nuclear magnetic moment**, around 5.05×10^{-27} J T^{-1}. That is $1/1837$ the value of the Bohr (electron) magneton μ_B. These magnetons relate as

$$\mu_n = \mu_B \, (m_e/m_p)$$

Here m_e and m_p are the masses of electron and proton.

See *Bohr magneton*.

Nuclear medicine is a branch of health physics, the concern of which is the use of ionising radiation for diagnosis and treatment. See *tracer* for a major field of work in this context. For **nuclear models**, views of the structure of the detail of nuclear structure, see *nucleus*.

For **nuclear spin** and **nuclear spin quantum number** I (or J), see *spin* (a form of angular momentum).

Nuclear warfare is the military use of **nuclear weapons** (there are civil uses of these too, in large scale engineering, and potentially as a space drive). Such weapons are explosive devices based on a fast developing chain reaction in fissile material, in so-called atom bombs (but the reactions involved are nuclear and not atomic). Fission weapons release far more energy in a small space and time than chemical high explosive; however, the radioactive fallout is a major long-term hazard. Fusion weapons, the so-called hydrogen and neutron bombs, does not produce fallout as such. However, to achieve the very high temperatures required for fusion often involves a fission bomb as a trigger. Research in this area started in several countries in around 1939; the US Manhattan project led to the Hiroshima and Nagasaki fission bombs that ended the second world war. Within far less than a decade of that, both the US and the USSR had fusion weapons; hence the nuclear arms race of the cold war that lasted over thirty years.

By **nuclear waste** we mean by-products of the nuclear industries, the hazardous radioactive nuclides with intense activity and/or half-lives measured in thousands or millions of years. There have been many attempts to design secure methods to dispose of those that are of no further use. The most practical – but very costly and still not fully secure – is to leave them in concrete boxes deep in stable rock formations. (Study of the natural reactors in Gabon shows that active nuclei have diffused only a few centimetres in over a billion years.) An early idea of Carlo Rubbia (the energy amplifier) was to induce further transmutations in such waste until the products are no longer harmful. A Japanese team started to re-develop this method in 2006 by using a sub-critical nuclear core; rather than a chain reaction, a proton accelerator with a neutron-rich target is the source of the neutrons needed. As well as being able to produce nuclear waste with much less potential hazard, the design should produce more energy than it consumes; on the other hand, some people suspect that some of the dangers may be too great.

On the other hand, also in 2006, there were claims of a way to speed up alpha decay so that waste would become harmless in well under a century. The method involves embedding the source in a metal and cooling to a few degrees kelvin. Many people are still to be convinced. For the initial report, see http://physicsweb.org/article/news/10/7/13

that field of physics (practical and mathematical) that involves the study of nuclei, and of their properties, particles and interactions. It was born in 1911, when Rutherford was the first to propose that atoms have a 'planetary' structure of central massive positive nucleus with electrons in orbit around it; this followed his work on alpha particle deflection. The applied science of nuclear physics (with detectors, accelerators, nuclear piles) is neither straight forward nor cheap. The quest for cheap energy (and weapons) has caused the subject to gain massive funds over the decades.

There is little of much interest as regards the physics of stable nuclei. However, nuclei that are not stable tend to break down into two or more simpler objects. The main concern of nuclear physics is this 'radioactive breakdown' of a nucleus to a lower energy state.

Here we look first at radioactivity (which includes fission – see also *nuclear power*), then turn to artificial radioactivity (where people become active in causing nuclear breakdown rather than just watching it), and last look at aspects of nuclear technology – the radiation detector and the particle accelerator.

Perhaps the main concern of modern nuclear physics is the nature of subatomic particles and how they behave. Notes on these appear in this book under the particles' names and as part of the various theories that try to link them all together.

nuclear breakdown the decay of a nucleus that is not stable into two or more simpler objects (taking 'objects' in the widest sense). There are various types of decay, but as a nucleus that is not stable must become more stable during such a breakdown, there is always energy transfer to outside. See *binding energy*.

alpha decay a form of nuclear fission in which a nucleus of X splits into a nucleus of Y and an alpha (which is a nucleus of helium); those two fission products carry away excess energy (as kinetic energy), called **alpha decay energy**, W.

$$AZX \rightarrow \,_{A-4}^{Z-2}Y + \alpha\,(^4_2H) + \text{energy, } W$$

The half-lives of alpha-active nuclides vary from 10^{-7} to 10^{+10} years; most are much longer lived than beta- and gamma-emitters. Alpha-activity provides evidence for the liquid drop and shell models of the nucleus; the tunnel effect in wave mechanics explains it.

alpha radiation(α radiation) one of the three common forms of radioactive radiation, being made up of alphas (alpha particles). It carries away a great deal of kinetic energy (several million electron volts per particle). However, as the alpha's mass is large, the speed is not high – perhaps a million km s^{-1}; this means that, as the rate of ionisation of atoms along the path is very large, the radiation is short range and easily absorbed by matter. Alpha radiation may have a range of a few centimetres in air, but even a thin sheet of paper or living tissue (such as your skin) will absorb it.

beta decay a form of radioactivity, being a weak nuclear reaction with half-life of the order of several minutes. During the process, a nucleus emits a beta particle β; its nucleon number remains the same, but its proton number increases by 1. The process involves the decay of a nuclear neutron (n) into a proton (p), an electron (β^- or e$^-$, the **beta particle**) and an anti-neutrino ($\bar{\nu}$), with the release of (kinetic) energy W. The maximum energy of the beta in the case of the decay of an isolated neutron is 1.25×10^{-13} J (0.78 MeV).

The example below is the beta decay of carbon 14 (this is **beta-**

active, which means that a sample gives out betas; for a major use, see *carbon dating*).

$$n \rightarrow p + e^- + \bar{v} + W$$
$$^{14}C \rightarrow {}^{14}N + e^- + \bar{v} + W$$

A sample of any such beta-active substance outputs a flow of **beta rays**, a fairly short range radiation that needs care in detection.

While the neutron decay mode outlined above is the most common type of beta decay, it is not the only one. Others (found, however, only with artificial isotopes) are: as follows, with the reactions in the next box.

a) positron emission, with the decay of a proton to give an antielectron (β^+ or e^+) and a neutrino (v); and

b) (orbital) electron capture, equivalent to positron emission, the electron coming in most cases from the K shell, i.e. K-capture (though L-capture sometimes happens too).

a) $\qquad\qquad\qquad\qquad p \rightarrow n + e^+ + v + W$

b) $\qquad\qquad\qquad\qquad p + e_K^- \rightarrow n + v + W$

In this latter case there is no actual beta radiation; however electromagnetic energy emission results from the fall of an outer electron into the gap in the inner shell. The nuclear recoil that follows the neutrino-emission is also detectable.

beta-ray spectroscopy the capture and analysis of beta radiation in terms of the range of beta-particle energy: the maximum range helps identify the source. Note that a given decay does not produce beta particles with a fixed energy: the energy removed by the neutrino varies. Indeed, beta energy ranges led Pauli in 1931 to predict the existence of the neutrino, though it was not actually found until 1955 (by Cowan and Reines). The neutrino is also needed to keep the system's angular momentum (spin) constant during decay, the other particles all having half a unit of spin.

gamma decay or **gamma-ray transformation** process of nuclear de-excitation that outputs gamma radiation. In nature, the most common cause for a nucleus to be excited is its creation in this state – e.g. as a result of alpha or beta decay or fission. Thus gamma radiation very often appears together with α or β radiations.

gamma radiation high energy electromagnetic radiation, with photon energy above about 10 keV (though there's much overlap with hard x-rays). Viewing it as a high energy wave means it has high frequency (small wavelength). This also explains the high penetrating power in matter. However, it is often even more useful to consider the radiation as made up of 'packets', each of energy W. Here $W = h\nu$ on the basis of the quantum theory; the packets, or quanta, are gamma photons.

gamma rays narrow wave-like beams (or sometimes beams of photons) of gamma radiation that behave like other electro-magnetic waves; for instance, the study of the **gamma-ray spectrum** output by a source gives information about the nature of the source. About 3% of the gammas emitted as the result of a fission process are **delayed gamma rays**: they arise from the de-excitation of an intermediate meta-stable state (as opposed to **prompt gamma rays**)

Short **gamma ray bursts** are the most intense forms of radiation known. Lasting for around a tenth of a second, but of very high energy, they may come from the collision of neutron stars (as may gravitational waves). Whatever the cause, they tend to occur only in small, irregular galaxies that are low in metals.

fission the splitting of a fissile nucleus into two (or in a very few cases more) large nuclei – called **fission fragments** or **fission products** – and a small number of neutrons; the total nuclear binding energy falls, with the excess transferred by the fragments and the neutrons.

Fission is a form of radioactivity: a fissile nucleus splits this way on its own or after it absorbs a slow (thermal) neutron; if this latter can take place, we call the process **induced fission** and the nucleus **fissionable**.

The best known fissionable nuclide is uranium-235; one common mode of fission is in the box. ^{235}U is fissile too, with 0.25 spontaneous fissions per second per kilogram.

$$^{235}U + {}^1n \rightarrow {}^{143}Ba + {}^{90}Kr + 3{}^1n + \text{around 30 pJ}$$

chain reaction cascade of neutron-induced nuclear fissions, each of which gives further neutrons. The neutron-induced fission of uranium-235 can output up to two or three child neutrons (the number depends on the input energy and the fission mode). Those neutrons may escape from the uranium sample, or non-fissile nuclei may absorb them; however, if one or more (on average) causes a new fission, we have a chain reaction.

The design of a fission reactor aims to have exactly one new net fission produced per fission. If the branching ratio is much greater than one, the chain becomes an explosive avalanche as in the case of a fission bomb.

In other words, where there can be a chain reaction, the energy made available by fission can be used in nuclear power piles and atom bombs. The former are **fission reactors**; there are many types of these – see *nuclear power* for the main ones.

Spontaneous fission, a true form of radioactivity, is very common among the trans-uranic elements. The process is just the same as induced fission, except it needs no input neutron. Neutron sources contain such nuclides (often californium-252, half-life just over 2.5 years, branch ratio 3%). Uranium-238 is fissile in this sense too, with 7 fissions s^{-1} kg^{-1} – see *fission-track dating*.

artificial radioactivity radioactivity not observed in nature, i.e. a result of human activity. In most cases, this is because the nuclides concerned have decayed in nature since their creation because of their short half-lives. For **nuclear reaction**, see *reaction*. See also *natural reactor*, a reactor which ran for many thousand years in the past in a uranium ore in Africa and left remains we can still explore.

Fusion is in effect the reverse of fission, a process in which small nuclei combine (fuse) to make larger ones, with the release of useful binding energy as radiation and kinetic energy of the products. To make the nuclei approach closely enough for this to happen, they need the high energy associated with high temperature (hence the name sometimes used: **thermonuclear reaction**). Iron and nickel have the largest binding energy per nucleon, as the graph there shows – so fusing nuclei to the left of these has much potential.

The required temperature is tens of millions of kelvin – at which level it is very hard to contain a dense enough plasma for a useful length of time; as yet, therefore, despite much research, practical (controlled, efficient) **fusion power** in a practical **fusion reactor** has not yet been achieved. Best so far (2007) is the world's largest fusion reactor, JET, which has managed an energy 'profit' for several seconds. See *nuclear power*. Fusion is the process that provides most of the energy output of stars; there the scale is huge. See also *cold fusion*.

A **detector** is a system used to detect, and perhaps to count, and/or measure aspects of, high energy radiations (including those of subatomic particles). A range of such systems appeared in the early decades of the last century when nuclear physics started – the first was the highly effective, but in use highly tedious, spinthariscope (William Crookes, 1903). Modern detectors have many other uses in science and technology outside of nuclear physics: in astronomy (special types of telescope); biomedicine (see, for instance, *positron emission tomography*); engineering; materials

science; and so on. Most modern detectors are very large and very
costly.

The early bubble chambers, cloud chambers and emulsion
detectors are position (i.e. track) detectors – they amplify the tiny
signals from a passing ionising particle or photon into a visible
track. As a result, the ionisation appears as a trail made from the
electric charges and excited atoms in a neutral working substance.
In the more modern scintillation detectors, the recombination of
the ions and/or the de-excitation of the atoms produce visible
photons the system can detect and amplify using a photomultiplier.

Other modern units collect the charges produced by
ionisation, rather than letting them recombine. In this class are the
avalanche detector, charge-coupled device, Geiger–Müller tube,
ionisation detector (see also below), proportional chamber (see also
below), semiconductor detector and spark chamber. See under the
individual headings. In this group, perhaps the simplest involves a
single pulse of current – the output of an **ionisation chamber** (or
ionisation detector). Here the energy comes from passing (ionising)
radiation; this leaves a trail of ions in the gas in the chamber.
A weak electric field causes the ions to drift toward the electrodes;
the outside circuit can record, amplify, or count the pulses of
current between these electrodes that follow the radiation events.
The semiconductor counter is much the same, except the 'working
fluid' is a solid; the others in this family are all gas-filled.

The **Geiger counter** (or **Geiger–Müller counter**, as it is based
on the **Geiger–Müller tube** he designed with a colleague) was the
first automatic sensor and counter of ionising particles. In the tube,
the trail of electrons and ions formed by a quantum of radiation
multiply to produce a pulse of current in the circuit. Suitable
multiplication takes place over only a certain range of applied
voltage; this is the **Geiger plateau** (or **Geiger region**) of the tube. In
this range, the size of the output pulse does not depend on how
much ionisation the original quantum caused.

A **proportional chamber** or **proportional counter** is a gas-
filled particle detector working in the proportional region (below)
of the discharge curve of the gas used. The size of an output pulse
therefore varies with the amount of ionisation caused by a passing
particle. As a result, we can feed the output to a multi-channel
pulse height analyser to obtain a spectrum of the radiation in the
chamber: or just count the pulses in the case of the counter.

Rather than working with the simple chamber just described,
the norm is to use the **multi-wire proportional chamber** – called

wire chamber in daily life, and devised in 1968. Set up rather like a spark chamber, this contains thousands of parallel fine wires perhaps 1 mm apart and held at about 2 kV, each linked to preamplifiers and the detecting circuit. This allows the user to determine where in the chamber a particular event occurs. Each wire causes an avalanche near it as it accelerates the electrons produced by the input radiation. The multi-wire system has a number of versions, such as the more precise '**drift chamber**' and a pair of crossed drift chambers; each version improves the system's precision at detecting and finding the details and paths of passing charged particles.

particle accelerator device designed to accelerate charged particles to a high speed relative to the speed of light in empty space. In other words, it is a system designed to transfer high energy to charged elementary particles or ions for use in research into nuclear physics, to make nuclides, and to treat certain cancers, for instance. In most cases, the energy transfer takes place through electric fields, with magnets used to guide and focus the particle beam. Such systems also need

- a source of the input particles for the beam;
- high vacuum to keep unwanted scattering low;
- targets (which may be a second beam) for the interactions; and
- detectors (above) to assess the electromagnetic and particle radiations produced.

Most accelerators work with electrons and/or protons and/or their antis, though some produce beams of more massive ions. In most cases, these output beams are not for use in themselves, but to interact with solid targets (or with other beams) to produce more exotic particles. The purpose of a few high-energy electron accelerators is to produce synchrotron radiation.

There are two main styles of accelerator, linear and cyclic. In the linear accelerator (linac) – which has become quite rare except for feeding beams to cyclic systems – the beam passes along a straight tube, perhaps for several kilometres, accelerated by electrodes of increasing size. In the other case (such as the cyclotron), the beam passes hundreds or thousands of times through the same cyclic accelerating system, gaining energy in each cycle. It is normal to measure the energy of the particles in the output beam, rather than their speed – the particles often travel so close to the speed of light that their mass increase is very large. Throughout this account we use the electronvolt, eV, as the energy unit, the joule being far too large. The output of the world's most powerful accelerator in 2006, the tevatron, can be as high as 1 TeV,

10^{12} eV – but that is just a few microjoules, the energy of perhaps a small insect in flight. (1 eV = 1.6×10^{-19} J.) Very crudely, however, we can say that 1 MeV – the output energy of the very first machines – is around a pico-joule.

Cockroft and Walton received the 1951 Nobel Prize for their design and use of the first accelerator (1932). They passed protons through 0.5–0.7 MV, giving them energies of 0.5–0.7 MeV. Cockcroft and Walton fired the protons at a lithium target to produce alpha particles; this was the first nuclear reaction to involve accelerated particles, the output providing the first proof of Einstein's mass-energy relation. Before that, the highest energy particles people could study came from cosmic rays – unpredictable, to say the least. Even so, although they are very rare, the highest energy primary cosmic rays – at a joule, or even more – still far surpass anything produced on Earth. (Even so, almost all nuclear physics research between 1919 and 1932 used natural alpha particles; the energy of these can be as high as 8 MeV.)

Oxford's Professor of accelerator physics, Philip Burrows, said in 2006 that 'studying interactions using a particle accelerator is like watching two people successfully collide peas fired from pea-shooters when one is on the Sun and the other is on Jupiter.' Even so, accelerator energies and beam intensities are now high enough for users to find quite a few events of value each year (even one a week). For instance, the hadron-electron ring near Hamburg has enough energy for the electrons to collide with the quarks inside protons ...

Now follow the main types of accelerator, in alphabetical order.

betatron a cyclic accelerator for electrons in which the energy transfer to the particles involves a process of induction rather than the more usual electric field acceleration. In other words, the electrons gain energy as they move through an increasing magnetic field. First designed in 1940, the circular beam tube is the secondary of a huge voltage transformer. The field is alternating, between specially shaped pole pieces; electrons enter the chamber when the field is just rising from zero, and leave at its maximum value, several hundred thousand orbits later. The system, now rare, could produce beams of a few hundred megavolts and use these as such or to produce high energy x-rays.

bevatron type of weak focus proton synchrotron (a cyclic type). The name comes from BeV (billion electron volts), an old term for GeV. Used from 1954 for forty years, it had the early triumph of first showing the anti-proton (1954).

cyclic accelerator accelerator in which the charged particles (in most cases, positive ions) in the beam pass again and again through the same (or much the same) path in a vacuum. During each cycle, they gain energy. Most modern accelerators are cyclic, rather than being linear – linacs are very costly at high energy.

cyclotron the classic example of a cyclic accelerator, first devised by Lawrence and others in 1930. It works on the principle of **cyclotron resonance**: each particle gains energy from the alternating electric field through which it passes as it crosses the gap between the dees (hollow D-shaped sections). In the simplest case, a radio-frequency signal between the dees provides the accelerating field. Inside a dee, each particle 'coasts' (while the electric field changes). The magnetic field from two poles, above and below the dees, produces the circular path. Each pulse of acceleration causes the particle to move in an arc of slightly larger radius; during a number of cycles, it thus passes from the source at the centre to the target at the edge.

The relativistic increase of the mass of the particle with energy (speed) makes the higher energy versions more complex. The magnetic field needs to be focused, but it is most important to adjust the applied frequency quickly during the whole action. This last feature, without which the particles start to arrive at the gap badly out of step, gave rise to the design of the **synchrocyclotron** (see also below). This type of cyclotron (first built in 1946 but now rare) was able to accelerate charged particles to relativistic speeds. The frequency of the alternating electric field between the dees falls with time, as the relativistic mass increase of the particles makes them take longer to cycle through the system. The current design is the **isochronous cyclotron**. As its name may imply, it has constant frequency, but deals with the relativistic mass increase by having a magfield that becomes stronger with distance from the centre. See also *synchrotron* (below).

Since that time, many designs of accelerator have come into use for work with different beam particles, and giving higher and higher currents and energies. Users of modern systems justify the enormous building and running costs of modern machines (like telescopes, good examples of 'big science') on the grounds of the wealth of knowledge they give us. Typical is the electron/positron collider, a cyclic system able to produce, store, and let collide beams of electrons and of positrons. Europe's LEP (large electron-positron collider) was the most powerful machine of this type for fifteen years; its main ring is 27 km across and has 4600 magnets to focus the beams.. In the ring, four bunches of electrons and four of positrons cycle in opposite directions for hours, colliding every few microseconds. LEP was used from 1989 to 2000 to give beam energies of 50 GeV.

Diamond Light Source UK's biggest investment in pure science since the 1970s, a research facility planned to start work at the end of 2006 and designed to explore *synchrotron radiation* (below). The radiation comes from passing a beam of 3 GeV (30×10^{-9} J) electrons from a linear accelerator and a synchrotron into a storage ring where they lose energy by synchrotron radiation (mainly at x-ray wavelengths) in the strong fields of large magnets. Seven 'beamlines' channel the radiation into research areas (though the plan is to raise the number to about forty over the years). Planned research projects are in such fields as biotechnology, materials science, atomic structure and medicine.

intersecting storage ring approach to exploring high energy collisions between particle beams. The two beams go to high energy in conventional accelerators (often linacs) and then transfer to pass in opposite directions round a 'storage ring'. This large circular tube contains a vacuum and uses magnets to keep the particles in orbit. In this case, the orbits are designed to intersect at very few places in the ring; there the users can study how the particles interact.

large hadron collider (lhc) In 2007, Europe's Large Hadron Collider (LHC) took over the world's top energy spot. It lives in the 27 km system built for the lep and uses superconducting magnets to contain the beam. Designed to explore the standard model of matter and the Higgs boson, its output is two beams of bunches of protons moving at 99.999 999% of the speed of light; the two beams produce for analysis 800 million collisions a second, 14 TeV each. If the LHC raises more questions than it answers, the high energy physics community will start to lobby for a further step up in accelerator power, from about 2010.

linear accelerator (linac) an accelerator that gives energy to charged particles in a beam as they pass along a long straight cylinder. In a common design, the track contains a series of tubes of increasing length, alternate ones being linked to one or other side of a radio frequency (rf) alternating voltage. A particle in the beam accelerates in the gap between two tubes; while it drifts through the next one, the voltages of all the tubes switch. Thus when the particle leaves each tube, it is again in an accelerating field. In a different design, there is a travelling radio frequency electromagnetic wave in a wave guide. The particle accelerates on the back of the electric component of the wave as the frequency rises. In both cases, the particles form bunches as they pass through the system.

synchrocyclotron form of cyclotron (cyclic accelerator) able to accelerate charged particles to relativistic speeds. The frequency of the alternating electric field between the dees falls with time, as the relativistic mass increase of the particles makes them take longer to cycle through the system.

synchrotron cyclic particle accelerator in which the path of the charged particles has almost constant radius (whereas that in a cyclotron increases in radius with energy). As in a cyclotron, the energy to accelerate the particles comes from a high frequency alternating electric field. In this case, however, the applied magnetic field strength rises with particle energy.

Synchroton radiation is the electromagnetic radiation output by a charged particle moving at relativistic speed in a curve (as in the magnetic field of a synchrotron). There are important astronomical sources too (as there are many strong magnetic fields in the Universe). People use synchrotron radiation to form wafer surface designs with very high precision. The radiation ranges from infrared to x, and the intensity is many times greater than any natural or artificial radiation; it is highly parallel too. Indeed, synchrotron radiation is the most common source (but a bulky and costly one) of high energy x-rays for chip-making and other such purposes. Some firms produce 'table-top' synchrotrons designed as a source of such radiation – perhaps a metre across and costing a million pounds.

tandem generator an approach to the use of the van der Graaf generator which doubles the energy of the output beam. The input is of negative ions which accelerate to the energy available; the system then 'strips away' (removes) electrons from the ions to produce positive ions, which accelerate once again through the whole energy range.

Tevatron in the early years of the 21st century, one of the world's most powerful accelerators. First running in 1983, this synchrotron is able to accelerate beams of protons and anti-protons to energies as high as 1 TeV (10^{12} eV, just over a microjoule). Its main claim to fame is finding the top quark in 1995.

the industry built on the quest to obtain energy effectively, efficiently and economically – and safely – from nuclear sources. There are two main sources – fission and fusion (see *nuclear physics*). Also, the energy taken away from radioactive nuclei by the particles and photons produced in their decay is a useful source of power on a small scale. Many such systems exist, often with thermocouples to use the high temperatures produced to give an electric current. Designed on scales from very small heart pacemakers to units with masses of hundreds of kilograms in satellites and other remote places, the systems often take the name of nuclear reactor. See, rather, *radioisotope thermoelectric generators*, rtgs.

Nuclear power is quite common around the world; in 2000 fission plants produced about a third of centrally generated electricity. It raises its own green concerns, but does not contribute as much to global warming (the greenhouse effect) as other sources of mains electricity. Britain was the first country to have a nuclear power programme running (starting with Calder Hall, which fed power to the grid from 1956 to 2003, though its main purpose in the early years was to make plutonium for weapons). Britain now (2007) has detailed plans to re-start the programme.

fission power It is best to keep the term nuclear reactor for the core of the **nuclear power plant** (in a ship) or of the **nuclear power station**. Here the energy comes from the fission of such species as uranium-235. Again, the energy (see *binding energy* for its source) is in the form of the photons and the motion of the particles that leave the decay; again, it causes high temperatures in the structure. Now, however, a coolant (cooling fluid) removes the excess energy; the coolant cycles through a heat exchanger, the water in which makes high pressure, high temperature steam to drive turbines.

The core of the nuclear power plant, then, is the **nuclear reactor**; here takes place the controlled (steady and safe) chain reaction that keeps a constant supply of energy output. The core, heavily shielded against harmful escape of radiation, contains and allows access to

- the nuclear fuel, often uranium-235 as pellets of oxide in tubes called fuel rods;
- the coolant, passing through many small channels between input and high temperature output, to remove the output energy;
- the moderator (sometimes the same as the coolant), designed

to slow the neutrons produced in each fission to low enough energy ('thermal' energy) to cause further fissions;

- control rods, neutron absorbers which can move in and out to keep the neutron density within the desired range – for moment-to-moment control and for safety. The advanced heavy water reactor has a great tank of borated H_2O above it for safety: if something goes wrong, the water drops down and stops the reaction. (Borated H_2O is a very good neutron absorber.)

The first artificial controlled reactor started up in Chicago in 1942, three years after the concept of a chain reaction (observed by Fermi in 1934) became clear. (See, however, *natural reactor*.) Now there are many common designs of fission reactor that produce large scale electricity cheaply, cleanly, safely and reliably; there are useful by-products too. As a result, in some countries such systems output much of their electricity. However, there are problems that make assessment and development much harder – these include correct methods of costing; the disposal of wastes and of retired reactors; safety in action; uncertainty about future fuel resources; and social problems that result from the decline of traditional methods.

working for fission power

advanced gas-cooled reactor (agr, or AGR) type of thermal (fission) reactor, the second generation built in Britain. The moderator is graphite and the coolant is carbon dioxide (with the heat exchanger inside the pressure vessel).

boiling water reactor (bwr) reactor in which water acts as both moderator and coolant; the water boils when in contact with the fuel rods in the core. Therefore the output energy leaves the core in high pressure, high temperature steam.

breeder reactor design that uses fissile and fertile material as fuel to produce more fissile material. (All reactors produce some more fuel; the ratio is enough to make spent fuel re-processing worth while.) Making 1.2 times as much fissile material as is used is common; the limit in theory is 1.8. If the ratio is greater than 1.01, the reactor needs little more fuel in the future – it fuels itself, in effect. There are two main designs – the **thermal breeder reactor**, in which the future fuel is thorium, and the **fast breeder reactor**, in which it is depleted uranium.

Calder Hall Britain's – and the world's – first nuclear power station, a Magnox type, opened in 1956 and closed in 2003

heavy water deuterium oxide, D_2O, the common name for deuterated water, water in which most or all of the hydrogen atoms are replaced by deuterium atoms. Heavy water makes up

about 0.014% of natural water; its uses include acting as both the
moderator and coolant in the **heavy water reactors**. These are more
efficient than light water (H_2O) reactors as H_2O strongly absorbs
neutrons (so the fuel must be enriched). The main design of this
type is Canada's **pressurised heavy water reactor** system also in
wide use in India. Still on the drawing board (in India) is the
advanced heavy water reactor; this pressurised type is also novel in
having a thorium-based fuel, a *heterogeneous* (below) moderator, and
a tank of borated H_2O above the core for safety.

heterogeneous
1 of a reactor, one in which the fuel is separate from the
moderator. On the other hand, a homogenous reactor has the
fuel and moderator mixed
2 of a reactor, one with more than a single type of moderator. For
instance, the advanced heavy water reactor uses carbon powder
in the fuel rods (80% of the moderator by volume) and the heavy
water also used as coolant (20%).

high-temperature gas-cooled reactor (htgr) design of gas-cooled
reactor in which the coolant is helium gas

homogeneous reactor one in which fuel, coolant and moderator are
finely mixed – the fuel is dissolved as a salt in H_2O or D_2O. There
are few of these machines working now, though there are many
great advantages of the design – corrosion problems are very severe.
A typical design needs only a few hundred grams of uranium.

liquid metal fast breeder reactor not yet in widespread use, a design
of fast breeder reactor cooled by liquid metal (e.g. lead, as in some
Russian submarines, or sodium); much the same design uses
molten sodium chloride. Such reactors do not need a moderator as
the chain reaction works with fast neutrons.

Magnox trade name (derived from magnesium oxide), for a range of
magnesium alloys used to build the fuel containers in the first
generation of British nuclear reactors. 'Magnox reactor' is thus a
widely used term for a nuclear reactor.

nuclear waste disposal a problem that first became serious sixty
years ago, when people started to enrich (process) large volumes of
natural uranium to fuel nuclear physics research reactors, atom
bombs, and then power reactors. The British Committee on
Radioactive Waste Management reported four leading burial
options in 2005, and in 2006 recommended deep disposal. This
means burying the waste permanently in chambers as much as
2 km below the surface in areas of very stable geology. Meanwhile,
Britain continues to store radioactive waste in temporary stores –
just above or just below ground (but out of contact with the

biosphere). In 2006, there were nearly 500 000 cubic metres of such waste, perhaps as much as 10 million tonnes. See http://news.bbc.co.uk/1/hi/sci/tech/4949096.stm

pressurised-water reactor (pwr) reactor in which the coolant is H_2O under high pressure. This also acts as the moderator and has some reflection effect. The pressure is high enough to keep the liquid from boiling at an operating temperature of perhaps 500 °C.

fusion power As a result of the problems with fission power, for decades there has been much research into nuclear fusion as a source of energy. In nuclear fusion reactions, nuclei combine to produce net output of useful energy (see *nuclear physics*). In more detail, fusion is a process in which small nuclei combine (fuse) to make larger ones, with the release of useful binding energy as radiation and kinetic energy of the fusion products. (By this process, stars produce energy through most of their lifetimes.) Iron and nickel have the largest *binding energy* per nucleon, as the graph there shows – so fusing nuclei to the left of these has much potential.

To make the nuclei approach closely enough for them to join – to break through the Coulomb barrier – they need the high energy associated with high temperature (hence the name sometimes used: **thermonuclear reaction**). The required temperature is tens of millions of kelvin – at which level it is very hard to contain a dense enough plasma for a useful length of time; as yet, therefore, despite much research, practical (controlled, efficient) fusion power in a practical **fusion reactor** has not yet been achieved. Best so far (2007) is the world's largest fusion reactor, JET, which has managed an energy 'profit' for several seconds.

The **tokamak** is a major design of fusion reactor, developed in USSR over a number of decades, and now the most common used for research around the world. (In 1968, a tokamak reached over 10 MK, far better than anything else in the world, so the design became the main one used.) The plasma is confined to a torus-shaped (doughnut-shaped) tube in some form of magnetic bottle (such as theta pinch and zeta pinch); it is raised in temperature to the several million degrees needed for fusion by making the plasma the secondary of a transformer.

Fusion power should be very cheap, very safe, and very 'green', though it's taking a long time to achieve. All the same, there *has* been progress, and many people are certain that practical commercial systems will appear in the next two decades. See also *cold fusion*.

nucleation process that leads to a stable particle of condensed matter in a less condensed phase (for instance, a crystal in a fluid or a liquid droplet in a vapour) The nucleus may be a small particle of the same substance (for instance, a seed crystal), a reactive impurity particle, or a catalyst.

nucleon elementary particle in the baryon (massive particle) class of the hadrons (those involved in strong interactions). There are two nucleons, the positive proton and the slightly more massive neutron, as well as their antiparticles; Heisenberg was the first to suggest that these are different states of the same particle.

All nuclei contain protons; all other than 'normal' hydrogen (in other words hydrogen-1 as opposed to heavy hydrogen) contain neutrons as well, in most cases in slightly larger numbers. Hence the name of nucleon. Nucleons interact by the exchange of pions (Yukawa particles, 1935); pi mesons exist in the nucleus only as virtual particles, however – see *nucleor* – and in any event are much less massive than nucleons. Therefore the mass of a nucleus (and, it follows, that of an atom) is due almost solely to the nucleons. Taking the mass of each nucleon as 1 (atomic mass unit), the **nucleon number** (symbol A) of a species (the number of nucleons) gives its mass relative to that of ^{1}H.

nucleon annihilation See *annihilation*.

nucleon number A formerly called atomic mass unit, amu, and then mass number, the number of nucleons in a given nucleus. This gives the nuclear mass in terms of ^{1}H.

nucleonics name sometimes used for nuclear physics

nucleor the core of a nucleon in the model of that particle as a core and a cloud of virtual pions

nucleus a) small speck of matter that acts as the basis for the nucleation of a crystal in a fluid or of a droplet in a vapour

b) the massive (see *nucleon number*) but very small central positive region of an atom or simple ion. The atom or ion is a large cloud of negative electrons around this core. The radius of a nucleus is the distance at which there is a very rapid fall in the density of the matter concerned (nuclei have no surface as such). Very crudely, it is the product of a constant called r_0 (about 1.3×10^{-15} m) and the cube root of the nucleon number A:

$$r \approx r_0\, A^{1/3}$$

That value leads to the nuclear volume, and hence to the density:, which is around 10^{15} kg m^{-3}. See also *neutron star*.

There are various models of nuclear structure, each with at least some variations, and each well able to explain certain aspects of the behaviour of nuclei: but not others. None is perfect, therefore. The main models are:

a) liquid drop, now less often discussed, but giving a very neat picture of fission

b) shell, good for nuclear spectra and the magic numbers, and the parent of

c) generalised Hartree–Fock (whose concern is for potentials between nucleons as well), and of

d) Fermi's gas model (which is able to reduce problems associated with the surface of the nucleus)

e) unified model with features of both liquid drop and Hartree–Fock

nuclide species of nucleus with a certain number of protons (the proton number Z) and of neutrons (the neutron number N). (The common name is isotope, but that is not fully helpful.) We write each nuclide with the chemical symbol (which depends on Z), often giving the value of Z again as a subscript in front of the symbol. Also in front of the symbol, but as a superscript, is the value of A, the nucleon number. A second method uses the full chemical name or its symbol followed by the nucleon number. Thus:

$^1_1$H (hydrogen-1, H-1) is normal (light) hydrogen: one proton, no neutrons, and

$^{235}_{92}$U is uranium-235 (U-235), with 92 protons and 143 neutrons.

If a nuclide is excited (in other words, if it has energy above the ground state and is fairly stable), this, to some, is not the same nuclide; it is, in any event, an isomer. We write such a nuclide with a star (*) after the chemical symbol; if the excited state is meta-stable, we use ^{m} rather than *. The latter is the case of ^{235}U^m, whose half-life is 26 minutes. In either case, the release of a gamma photon marks the decay of the nuclide back to the ground state.

null zero in value as a result of balance, applied to

a) any method of measuring that involves coming to a balance, and therefore a zero reading of some kind.

b) any point within any field pattern at which the field strength is zero.

nullification the result of combining normal

and negative matter, there being no energy output when this happens (unlike the case of normal and antimatter)

numerical aperture symbol NA, no unit measure of the resolving power of a microscope. The value is the product of the sine of half the angle of view of the object lens and the refractive constant of the medium that contains the lens. As the refractive constant of oils is at least 1.5, while that of air is (in effect) 1.0, using an oil immersion lens raises the microscope's resolving power.

nutation the vibration of the axis of a spinning object round its mean direction. This is a feature of the motions of a spinning top, of a gyroscope, and of the Earth (for instance). Compare it with *precession*, much the same effect, but the result of an applied torque.

Nyquist law The rate of change of the power of the thermal noise in a resistor with signal frequency is proportional to the absolute temperature. The constant involved is the Boltzmann constant, k.

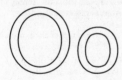

O a) symbol for oxygen, element 8
b) symbol for the output of a system or circuit

O symbol for object distance (normally u)

ω symbol for any of the omega mesons

ω a) symbol for angular velocity, unit rad s^{-1}
b) non-standard symbol for frequency, unit Hz

Ω a) symbol for solid angle, unit: steradian, sr
b) symbol for the density parameter of the Universe, the ratio of the density of the Universe to the critical density

Ω a) symbol for ohm, unit of resistance, reactance and impedance (and, in the past, upside-down, for the mho, the ohm^{-1}, now called siemens S)
b) symbol for any of the omega baryons

oasm old rationalised metric system of units based on ohm, ampere, second, metre

object the place where the light that enters an optical system comes from; its image is where it seems to come from. See also *virtual object*. In an optical system with more than one element, the **object lens** or **object mirror** is the single lens or lens assembly closest to the object or mirror furthest from it. Its functions are some or all of those in the list – so the design of this element is a major factor in the efficiency and effectiveness of the system as a whole. The object lens/mirror's function is
a) to collect a large amount of input light from the object
b) to converge (sometimes diverge) that light for further processing
c) to correct for the aberrations of the rest of the system.

   The **object plane** is the plane centred on the object that is at 90° to the axis of the system. The **object space** is the whole region

through which light passes between object and object lens/mirror.

objective still quite common name for the object lens of a system

oblate of a spheroid, with one of the axes shorter than the other two equal ones; compare with *prolate*. The spinning Earth is oblate about the axis of spin: that axis is rather over 100 km shorter than the other two.

observable any physical value that one can measure and therefore describe in some expression. The expression may be in matrix or wave function form in the case of modern physics.

obtuse of an angle, in the range between 90° and 180°

Occam, William of (c. 1290–1349) philosopher whose principle of simplicity is of great value in science. Called **Occam's razor**, this suggests that the simplest theory that fits the facts is the one most likely to be correct.

occlusion the way a solid mass, or its outside surface, retains a layer of fluid particles. See also *adhesion*.

occultation the eclipse (hiding) of a moon, planet, star or galaxy by a closer, apparently larger one, as the latter passes in front of the first, from the point of view of the observer. Occultations can give very precise information about the structure and motion of each of the objects concerned.

occupation number the mean number of particles in a given quantum state, normally used for each of a set of states such as orbitals

octave interval on a scale of music between two notes, the frequency of one being twice that of the other. By extension, the term applies to any two signals whose frequencies relate this way. See, for instance, *harmonics*.

octet set of eight similar objects or systems.
The word applies to
a) the set of eight electrons in the outer shell
of some atoms;
b) a group of eight elementary particles with
similar quantum numbers that forms a
symmetrical pattern in iso-space (see *eightfold
way* and *unitary symmetry*). See, for examples,
meson.

ocular old name for eye lens (the lens of an
optical system closest to the eye)

odd–even nucleus one with an odd proton
number and an even neutron number. Such
a nucleus is more stable than an odd–odd
nucleus, but less stable than an even–even
one. These points relate to magic numbers
and the nuclear shell model.

oersted Oe old unit of magnetic field
strength. 1 Oe = 79.6 $(250/\pi)$ A m^{-1}. The
name comes from that of the physicist.

Ørsted, Hans Christian (1777–1851) physicist
called 'the father of electromagnetism' – he
was the first to show that a current has a
magnetic effect (in 1820, though after years
of looking), from its effect on a compass
needle.

Ohm, Georg (1787–1854) teacher whose
research into the new science of electric
current led to **Ohm's law** in 1827. This states
that the current in a metal sample at constant
temperature varies with the potential
difference between its ends:

$$I \propto V$$

This leads to

$$I = V/R$$

R is the resistance of the sample, unit ohm Ω.

It was some ten years before most people
came to accept the statement. During that
time, Ohm suffered a great deal of mockery
and hardship (for instance, he lost his job).
At last, however, he became a professor, at
Munich, and could carry out further research
(including into acoustics).

ohm Ω **a)** the unit of electrical resistance. It
is the resistance between two points of a pure
conductor when 1 V between them gives a
current of 1 A between them.
b) by extension, the unit of electrical
reactance and impedance, respectively the
opposition to ac. By further extension, the
acoustic ohm is the unit of acoustic
impedance, and the **mechanical ohm** is that
of mechanical impedance.

ohmic of a conductor or circuit element,

one for which Ohm's law is valid; compare
this with *non-ohmic*. An **ohmic contact** is a
junction between two ohmic conductors
that is itself ohmic. The **ohmic loss** from a
current in a circuit or element is the energy
loss W due to the temperature rise in the
resistive components:

$$W = I^2 R t$$

Here I is the current concerned, R the total
resistance, and t the time for which the
current passes.

ohm-meter electrical meter able to give the
value of the resistance (unit: ohm) of a circuit
element or of a complete circuit or section.
An internal source in series with the meter
produces a voltage between its terminals. See
also *multi-meter*, a current meter designed to
work on a number of ranges of current,
voltage and resistance.

oil substance that dissolves in ether but not in
water, and is greasy to the touch. For the **oil
drop experiment** that led to a value for the
charge of the electron, see *Millikan*. An **oil-
immersion lens** is a microscope object lens
used with oil in the object space between its
front face and the object. The high refractive
constant of the object space raises the
resolving power of the system.

Oklo place in Gabon, West Africa, near which
was found the first natural reactor, one active
over 1700 million years ago; sixteen are now
known in that area – and nowhere else. The
Oklo phenomenon is the name for such a
natural reactor process.

Olbers, Heinrich (1758–1840) astronomer who
discovered many early asteroids ('minor'
planets) and carried out much other solar
system research and development. The
Olbers paradox demands that in a uniform
infinite Universe (as was then the model), the
night sky would be as bright as that in day
time, and uniformly so. The paradox is that
this is not the case. Olbers explained the
problem by proposing clouds of absorbing
matter between the stars; these were later
found – but the finite age of the Universe and
the red-shift due to its expansion are the true
solutions.

oled organic light-emitting diode

omega Ω **a)** elementary particle in the
hyperon class of baryons, made of three
strange quarks, and with rest mass
1672 MeV/c^2. Discovered in 1964, the negative
omega minus particle completed an octet

proposed by the unitary symmetry theory; the event was therefore hailed as a great advance, the particle's properties being very close to those proposed. In particular, it has a hyper-charge of −2: with its anti-, the only particle to do so. In the same family is the **charmed omega**, Ω^0_c, mass 2698 MeV/c^2; this has a charm quark instead of one of the three strange quarks.
b) *meson* of spin 1, a mix of the up/anti-up and down/anti-down quark pairs
c) an omega particle made use of in the *Star Trek* series, and able to disrupt subspace (hyper-space) and thus prevent warp travel. This omega particle is a molecule, and is as real as subspace and warp travel.

Onnes, Kamerlingh (1853–1926) low temperature physicist whose 1913 Nobel Prize was for 'his investigations into the properties of matter at low temperatures which led ... to the production of liquid helium'

Onsager, Lars (1903–1976) mathematician and chemist who gained the 1968 Nobel Prize in chemistry for research into irreversible thermodynamic processes. The **Onsager formula** relates the relative permittivity of a fluid to its relative permittivity for high frequencies and its molecular dipole moment.

opacity the ability of a substance to absorb incident radiation of some kind, the converse of transparency. As a measure, it is the ratio of the input radiant power to that output, the reciprocal of transmittance. See also *opaque*.

opalescence a) the colour fringes observed with certain minerals (such as opal) as a result of interference in very thin surface films
b) the similar appearance of light reflected from small suspended particles in a fluid

opamp operational amplifier, a signal processing chip

opaque of high opacity, in other words able to pass little or no incident radiation. Note, though, that a substance opaque to one radiation may be highly transparent to a second. Thus, glass passes visible light well, but is opaque to ultraviolet and to most thermal radiations. Compare with *translucent*. An **opaque projector** (now rare, former name epi-scope) gives an image on a screen of an opaque object. The object variably reflects light from an intense beam into the focusing lens.

open circuit circuit not able to pass a current as the conducting path is not complete: the switch may be open or there may be a break in a conductor (for instance).

opera glass(es) also called theatre glass(es), type of binocular in which each telescope is of the Galilean design. The device is compact because of the design, but offers only low magnifying power.

operating point the pair of voltage and current values for the normal use of a given transistor

operational amplifier or **opamp** integrated circuit that works on an input signal to produce a desired output. The first opamps were high performance dc amplifiers for use in analog computers: they could add, subtract, integrate and differentiate their input(s) to form the output. Modern versions do that and much more, in digital as well as analog systems.

The perfect opamp has infinite gain, linearity, band width and input impedance, and zero size, price, noise level, output impedance and temperature drift. No opamp is perfect, of course, but many very good systems now exist. The sketch shows the basic standard symbol in a circuit using this device as an inverting amplifier.

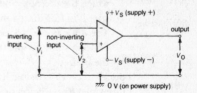

operator symbol for, and concept of, an operation – such as those in arithmetic (+, −, / and so on) and differential operators (like del or nabla, ∇)

ophthalmoscope optical instrument that allows one to examine the internal structure of the inner eye. Its powerful lamp and lens combine with the eye's own optical system to give very high magnification of, for instance, the retina.

Oppenheimer, Robert (1904–1967) nuclear physicist who headed the Manhattan project at Los Alamos from 1943 to 1945 (when he resigned) to develop the US fission bomb: but fought against the extension to the fusion bomb after the second world war. His views prevented his continuing in sensitive research. However, he was deeply involved in the development of nuclear power, and carried out much important early work on the theory of cosmic rays and black holes.

The **Oppenheimer-Phillips process** is a nuclear reaction between a deuteron and a nucleus in which the latter strips the neutron from the former.

optic

a) to do with visible light and, by extension, infrared and ultraviolet (these being nearest to it in the spectrum). The **optic-acoustical effect** is the appearance of waves of pressure (and perhaps therefore of sound) in a light radiation absorbing gas when the input light radiation varies. It results from the repeated warming and cooling of layers of the gas. The **optic sign** of a birefringent crystal is negative if the refractive constant for the ordinary ray is greater than that for the extraordinary ray, and positive if it is less. See also *optic axis*, c) below.

b) old slang name for human eye

c) to do with the eye and, by extension, with other optical systems. The **optic axis** of any such system is the principal axis, the line linking the centres of the lenses and/or mirrors. People also use the term for the direction through a birefringent crystal in which the ordinary and extraordinary rays travel with the same speed; in this direction there is therefore no double refraction. The **optic nerve** carries sense signals from the eye to the brain (indeed, some people think of it as the link between two parts of the brain.

optical to do with visible light and, by extension, infrared and ultraviolet (these being nearest to it in the spectrum). An **optical activator** is a substance which, when added in tiny amounts to a phosphor, causes the phosphor to give out the desired colour during phosphorescence. **Optical activity** is the rotation of the plane of polarisation of input plane polarised light by certain crystals and compounds (such as solutions of chiral molecules like sugars; crystals with rotated planes such as quartz; and spin-polarised gases). The molecules (or atoms) of an optically active substance have an asymmetric structure which can exist in left- and right-handed mirror image forms (these being optical isomers, with the property of chirality). The dextrorotary form turns the plane clockwise (viewed from the output side); the laevorotary form turns it anticlockwise. The angle of turning depends on the concentration of the active substance and the thickness of the sample.

The **optical axis** of any optical system is the principal axis (see *optic*, sense c)), the line linking the **optical centres** of the lenses and/or mirrors. Sometimes called the pole, the optical centre of a lens is the point in it through which rays pass without change of direction; in the case of a mirror, it is the geometric centre of the active part of the surface. **Optical character reading** (ocr) is a method of machine-readable input to IT systems in which a light source/sensor unit (or an array of these in a scanner) detects the pattern of light and dark on a sheet; suitable software interprets the output signal. For **optical communication**, see *fibre optics*; for **optical condenser**, see *condenser* sense b); and for **optical crystal,** see *optical molecule*, below.

The **optical density**

a) of a sample (sometimes called extinction) is a measure of its absorbance. In fact it is the logarithm of its absorptance (in other words, of the reciprocal of its transmittance). A useful relation is $\log_{10} I_0 / I$; here I_0 is the input intensity, while I is that output.

b) of a substance is a term (rarely used now) for its refractivity.

There is no relation between optical density (however defined) and physical density.

Optical discs store signals or data as pits in the surface of a suitable medium, a laser and photo-cell in the optical read head gaining access to the data on demand. Most optical discs (such as compact and video discs) are read only (you cannot record data on them). The **optical depth** or **optical distance** d travelled by a ray in a medium is the product of the refractive constant n of the medium concerned and the true distance l:

$$d = n \, l$$

Where the optical path involves different media, the optical distance is the sum of these products. In either case, it is the distance the ray would travel in empty space in the same time as it passes through the medium or system. See also *Fermat*.

The **optical elements** of an optical system (see below) are the parts that affect the path of input light, such as lenses, mirrors, prisms. The **optical extinction constant** of a substance is the extinction per unit distance in it. An **optical fibre** is a single fine strand of very pure (very highly transparent) glass, treated so that an input light signal cannot escape (see *fibre optics*). Widely used in engineering, an **optical flat** is an extremely flat (within 10^{-6} m) sheet of glass or quartz. Interference fringes obtained when a test sample is in contact with it provide a measure of the flatness of that sample; there are also uses in spectrometry. For **optical force**, see *optical molecule*, below. An **optical glass** is a glass with suitable optical properties and made in such a way as to keep to a minimum defects such as bubbles, strains, and colour centres. In other words, it is highly homogeneous. See crown and flint for examples (and the table below), and also *fibre optics*. For **optical illusion**, see *illusion*; for **optical isomer**, see *optical activity*, above, and for **optical length**, see *optical path*, below.

glass	density/kg m^{-3}	refractive constant
crown	2500–2600	1.51–1.61
flint	2700–4800	1.53–1.75

An **optical lever** is a mirror in an optical system in which reflection causes the output change of direction of a ray to be double the input angle, the angle by which the mirror turns. It thus allows one to measure very small displacements or angles of twist (e.g. in the suspension of a meter needle): the angle doubles, and also the greater the distance to the screen, the greater the movement of the spot of light produced.

The refractive constant for ordinary rays in an **optically negative** substance is greater than that for extraordinary rays; this is not the norm: most substances are **optically positive** in this sense. See also *optic sign*. **Optical maser** is an old name for laser.

An **optical molecule** is an artefact: a set of two or more tiny plastics spheres held together in stable 'molecular' form in a light beam. People first noted in the 1970s the power of a light beam to trap and support small particles (in that case bacteria); more recently, various teams have explored the optical force that can hold small particles stably. It seems to arise from interference effects between the radiation scattered by the particles and the radiation in the beam: when the particles are a whole number of wavelengths apart, constructive interference produces (low-energy) potential wells which hold the particles. As yet, only pairs of particles have been held together in this way; workers hope, however, to build larger structures of this type, to be called optical crystals.

The **optical path** of a ray through a medium or optical system is its actual track; people also give the name to the distance the ray would travel in empty space in the same time as it passes through the system. See *optical distance* (above) for the method of calculation. Here **optical path difference** Δd relates to phase difference $\Delta\phi$ thus:

$$\Delta\phi = 2\,\pi\Delta d/\lambda$$

λ is the wavelength of the radiation concerned.

An **optical processor** is a chip where all the data transfers are on light waves, rather than involving electric currents. There is no need for fibres to carry the light the small distances concerned, and, as light waves pass through each other without effect (unlike electric currents), optical processing has the potential of being much cheaper and faster than microelectronic systems. The logic circuits involve laser (or led) light processed by liquid crystals. The first working systems of this kind were announced in 1990, though research in the field has a long history; there are still problems, e.g. to do with the choice of crystal. See also *optoelectronics*.

The process of **optical pumping** uses intense light pulses to raise the energies of the active atoms of a laser substance to form a population inversion. **Optical pyrometers** measure high temperatures (those of 'red heat' and above, where there is visible light output) with fair precision and accuracy. The disappearing filament type consists of a length of fine wire in an inert gas in a glass cell; the user can vary the current to change the colour of the wire as it glows. You view the hot surface through the cell and adjust the current until you can no longer see the wire; this is then at the same temperature as the surface. It is possible to calibrate the scale of a current meter in the circuit to show temperature. Modern systems with silicon sensors and suitable electronics are coming to replace all these thermometers.

Optical rotation is the rotation of the plane of polarisation of plane polarised radiation by an active substance; see *optical activity*, above. An **optical system** uses light as a tool (often for viewing objects) or as a method of data transfer. It consists of one or more optical elements, such as glass blocks, fibres, lenses, mirrors and prisms, to change the paths of input light rays as required. An **optical telescope** works with input light from distant objects to form a magnified image. The two main types are the refractor (with an object lens) and the reflector (with an object mirror); there are, however, very many designs of each, and many ways to use the image produced. Astronomers also now employ telescopes that work with most other forms of radiation. Not all radiations pass readily through the atmosphere, however: for telescopes placed on the surface of the Earth (even at high altitude), there are only certain 'windows' (frequency/wavelength or energy bands) that provide good input intensities. The atmosphere's **optical window**, for instance, includes all visible light and also ultraviolet down to around 200 nm. The ozone layer absorbs shorter ultraviolet waves.

optics the science and applied science of light and vision, and, by extension, those that deal with at least some other radiations. The concern of **geometric optics** is how rays behave in the many different types of simple and complex optical systems; **physical optics** handles the nature of light and aspects such as colour and wave nature (including diffraction and interference). In the latter case, **nonlinear optics** is the optics of such intense coherent radiation (as from a laser) that the electric polarisation of the medium is important.

optoacoustics the study of phenomena that link light and sound waves. The **optoacoustic effect** is the absorption by a substance of pulses of light to produce pulses of high temperature and therefore raised pressure, giving sound waves. **Optoacoustic spectroscopy** uses this effect to give information about the molecular structure of the substance and about any impurities.

optoelectronics the whole field of the science and applied science of optical data transfer: the efficient production of the light carrier waves and their modulation by the signals in question; the transfer methods themselves; and the detection and separation of the two at the other end. By the early 1990s, the term had come to include work with the whole non-radio part of the electromagnetic spectrum. Integrated optoelectronic gallium arsenide chips can include photodevices (cells, filters, and led lasers, and waveguides) linked with electronic ones in the surface. See also *fibre optics* and *optical processor*.

optofluidics the development of configurable optical systems, by applying optics concepts in microfluidic systems, giving highly compact, integrated devices

OR gate ⎬⟩⎯ gate whose output is 1 only if at least one input is 1

inputs		output
I_1	I_2	O
0	0	0
0	1	1
1	0	1
1	1	1

orbit path (usually closed and cyclic) of an object that moves round a second because of some attractive force between them. The orbits of astronomical objects because of gravitational forces are well defined and stable (see, for instance, *Kepler's laws*). On the other hand, in the case of the motion of charged particles in, for instance, nuclei, atoms, and magnetic fields, the uncertainty principle means the orbits are smeared into clouds, and we can discuss only probabilities of speed and position.

In the case of gravitational orbits, the paths (if closed) are elliptical, though they *could* be perfect circles. Other possible non-closed orbits are parabolas and hyperbolas. For a circular orbit of an object of small mass, the **orbit speed** c_o depends only on the mass M of the central object (the 'primary') and the radius r of the orbit:

$$c_o = \sqrt{(G\,M/r)}$$

Here G is the gravitational constant.

We also have. $c_o = c_e/\sqrt{2}$
c_e being the escape speed, and

$$c_o = \sqrt{(rg)}$$

with g being the gravitational field strength at radius r.

The **orbit time** (period T) is

$$2\pi/r \sqrt{(r_o{}^3/g)}$$

Here r_o is the radius of the primary (the Earth or other planet, for instance).

We have a range of paths, including true (cyclic) orbits, for different values of projection speed c from a point P above the atmosphere:

a) $c \ll c_o$: object falls toward the ground in a parabola
b) $c < c_o$: object moves through part of an ellipse until it reaches the ground
c) $c = c_o$: object orbits in a circle
d) $c_o < c < c_e$: object orbits in an ellipse
e) $c = c_e$: object escapes on a parabolic path
f) $c > c_e$: object escapes on a hyperbolic path.
Last, if the masses of the two objects are much the same, the objects orbit each other, moving round a common centre; we must then amend all the above relations.

orbital a) concerned with the orbit of one object round a second. Thus, an **orbital electron** is one in the electron cloud of an ion, atom or molecule; **orbital electron capture** is a form of beta decay in which the nucleus absorbs an orbital electron. For (gravitational) **orbital speed**, see *orbit speed*.
b) the wave function that describes the motion of an electron in the cloud of an atom or molecule: it gives the probability of finding the electron at each point in the space. No more than two electrons can share a given orbital (and they will have opposite spins). We can define each orbital in terms of three quantum numbers: for the energy level (size), angular momentum (shape), and orientation of the 'orbit'. Spectroscopy provides the information concerned. The symbol *for* the **orbital angular momentum quantum number** or **azimuthal quantum number** is l (or L). This is the angular momentum of a particle or group of particles that move in an orbit, or appear to do so. It is normal to express it in units of h, – that is, the Planck constant h divided by 2π.

order a) an integer (whole number n or m) associated with a given fringe in an interference or diffraction pattern. In interference, a high energy/amplitude (bright, for instance) fringe occurs for a path difference of $n\,\lambda$ between the waves of wavelength λ concerned; minima appear at $(n + \frac{1}{2})\,\lambda$. **First order fringes** correspond to $n = 1$; second order to $n = 2$; and so on. In the same way, the spectra from a diffraction grating differ in order, the zeroth order being the undiffracted state (straight through, therefore no diffraction, therefore no spectrum).
b) measure of departure from randomness. An **ordered phase,** as in the case of a crystal, involves the particles concerned being in defined preferred sites in the lattice; liquids

show short range order only (but see *liquid crystal*); there is no order in the particles of a gas

ordinary ray in double refraction by a birefringent crystal, the output ray that obeys Snell's law. An input ray splits into two; these differ in the speed of transfer through the substance. The **ordinary refractive constant** describes the ordinary ray, as it follows Snell's law; this does not apply to the extraordinary ray. The two rays are plane polarised at 90° to each other. See, for instance, *Nicol prism*.

ordinate the distance of a point in Cartesian space from the x-axis, its y coordinate in other words

organic light-emitting diode oled light source built from a layer of an organic semiconductor – current in this raises free electrons to a higher energy level from which they release photons as they return. This is electroluminescence. Doping the semiconductor with the right mix allows the release of light of any colour. In the past, the dopants have been phosphors – but these are inefficient and also break down quickly. Mixing those with certain fluorescers overcomes those problems, leading the way to light sources as bright as light bulbs, but using less current and lasting for ten or more thousands of hours – and even made as thin sheets to cover walls and ceilings (for instance). By early 2006, oleds had started to appear in the displays of such things as cellphones, radio sets, and MP3 players. Prototypes of displays as much as a metre across exist, as well as of lighting sheets.

orrery mechanical model, not to scale, of the main planets and moons of the solar system. The Earl of Orrery was the patron of the inventor, George Graham (1673–1751), whose first design appeared around 1710.

orthicon type of video camera tube in which an electron beam scans the array of photo-cells on which the image forms. In the **image orthicon**, the first image is on a surface that emits photoelectrons to a storage plate; the electron beam then scans that plate.

ortho- prefix that implies right (i.e. correct), or at a right angle (90°). An **orthochromatic film** is a black and white film that is sensitive to light at both ends of the spectrum, or even one whose spectral response is much the same as that of the eye. **Orthogonal** lines are at 90°. The nuclear spins of each of the two nuclei in a molecule of **orthohydrogen** are in the same direction. For **orthorhombic crystal**, see *crystal system*.

oscillating Universe model of the Universe, not now often discussed, in which the size of the whole Universe varies cyclically, but without catastrophe at any point in the cycle. Some views of the big bang model suppose that the bang is the catastrophic start of an ever repeated cycle rather like this.

oscillation

a) the same as vibration, a cyclic energy variation in a system (for instance, mechanical, electric, or particulate), or a single such cycle itself. A **damped oscillation** shows a curve of amplitude falling quickly against time; see *damping*. In a **forced oscillation**, the frequency is set from outside the system, and differs from the system's natural frequency. Then, when the frequency is at the natural frequency of the system, resonance occurs. A **free oscillation** (or free vibration) of an object is when it vibrates with no outside force acting; it then vibrates in its natural mode.

We can describe any oscillation by way of a differential equation (see, for instance, *simple harmonic motion*). The equation of a **linear oscillation** is linear (like a straight line, with no power terms); that for a **nonlinear oscillation** is nonlinear.

b) the change of a sub-atomic particle from one type to another, and (according to some theories and evidence) back again, as with

- the change of a kaon to an anti-kaon (which relates to the discovery of CP-violation);
- that of a neutrino from one flavour (electron, muon or tau) to a second (which relates to neutrino mass); and
- that between B-mesons and anti-B-mesons (which relates to matter/antimatter balance and occurs at frequencies as high as 3×10^{12} Hz).

oscillator system in which free oscillation can occur with little energy input. It can be mechanical, electric, or particulate (e.g. molecular, atomic or nuclear). The designs of many electronic circuits aim to oscillate; the purpose of each is to output an undamped (constant amplitude) signal of given shape and frequency, while the input is dc.

oscillograph device whose output is a plot on paper of the variation of some physical or electrical input. People once used the word for oscilloscope as well.

oscilloscope device whose output is a visual display (on a screen) of the variation of some physical or electrical input. See *cathode-ray oscilloscope*, though there are other designs.

osmium Os transitional metal, the densest known (relative density 22.6), period 6, proton number 76, identified in 1803 by Smithson Tennant and named from 'osmi', Greek for stink (osmium oxide has a strong – and toxic – odour). Melting and boiling temperatures 3030 °C and 5020 °C; most common isotope ^{190}Os. Used in hard wearing alloys, and, until replaced by tungsten, lamp filaments.

osmosis the free diffusion of a solvent through a semipermeable membrane (or 'skin') from a region of higher solute concentration to one of lower concentration. Thomas Graham first studied the process, coining its name in 1858. See also *van't Hoff*, who put it on a quantitative basis.

Electric osmosis is the same process, caused or accelerated by a current in the solution. **Isotopic osmosis** uses such a process to separate isotopes (for instance of liquid helium). See also *osmotic*, including for **reverse osmosis**.

osmotic concerned with the process of osmosis, the flow of solvent through a semipermeable membrane. An **osmotic constant** g (or ϕ) is a factor introduced into the equation of ideal osmosis to cover a given non-ideal case. The **osmotic pressure** π of a given solution is the pressure one must exert in the output side of the membrane to reduce the osmotic flow to zero. Higher pressures cause a solvent flow the other way – this being reverse osmosis. The osmotic pressure of sea water is 2.7 MPa; desalination plants use reverse pressures of about twice that to produce pure water from sea water.

ossicles the three small bones of the middle ear that transmit the vibrations of sound waves from the drum to the inner ear. Their names come from their shapes – hammer (malleus), anvil (incus) and stirrup (stapes). They are the smallest bones of the human body, being around 18 mm in total in the case of an adult male (though they are close to full size at birth).

Ostwald, Friedrich (1853–1932) chemist, called by many the founder of physical chemistry. He gained the 1909 Nobel Prize in chemistry for his work on catalysis (an aspect, for instance, of the crucial **Ostwald process** for making nitric acid). The **Ostwald viscometer** is a device used to measure the viscosity

(liquid friction) of a liquid. A capillary tube links two bulbs at different heights; the user measures the time taken for the liquid to flow from the upper bulb to the lower.

Otto, Nikolaus (1832–1891) engineer who designed (in 1876) the first fully successful internal combustion engine. This quickly replaced the steam engine (an external combustion system) in many applications, and led to the development of motor cars. The **Otto cycle** (compare with *Carnot cycle* and *Diesel cycle*) is the thermodynamic cycle of spark ignition (petrol) engines. The stages of the cycle ('strokes') in this case are

a) input of fuel/air mixture at air pressure

b) adiabatic compression (pressure)

c) ignition causing high temperature and pressure at constant volume

d) adiabatic expansion (power), and

e) exhaust at constant volume.

out of phase of two or more cyclic processes (e.g. waves), not peaking together (at the same time or in the same place). If they are fully (180°) out of phase, there is destructive interference at that point. See *phase*.

output the energy taken from a system, or the place or time at which it is taken. The **output impedance** of a circuit or other system is the impedance offered to it by the output (load). An **output transformer** may sit between such a circuit and its load in order to match impedances to provide a more efficient energy transfer.

over-current such a large current drawn from the output of a circuit that there is an over-load event

over-damping such a high level of *damping* of the oscillation of a system that the system comes to rest very quickly with no oscillation at all

over-load situation in which the current drawn from a circuit is higher than the maximum design value, as a result of too low an output resistance or impedance (for instance a short circuit). **Over-load protection** in the circuit may involve a fuse, or a circuit-breaker (electromagnetic automatic switch).

over-shoot to go above a system's design maximum value during the rapid rise of some measure, as (for instance) in a pulse. This is likely to lead to oscillation; see also *damping*.

overtone component of a note with higher frequency (and, in most cases, lower amplitude) than the fundamental. The fundamental and overtones of a note are its harmonics.

oxygen O second most common gas in the Earth's atmosphere, and most common element in the crust, a gas in group VI, period 2, with proton number 8, first isolated in 1771 by Carl Scheele – but his publisher's neglect meant that Joseph Priestley's later success (1774) receives most credit; the name means 'acid generator'. Density 1.4 kg m^{-3}, melting and boiling temperatures –220 °C and –180 °C; most common isotope ^{16}O. Oxygen, being essential to life, comes in tanks for use at altitude and in hospitals; it is the common oxidant in liquid fuel rockets.

oxygen temperature the fixed temperature on the international standard scale before the most recent (1990) given by the boiling temperature of liquid oxygen at standard pressure, 90.188 K

ozone triatomic form (allotrope) of oxygen: O_3, a highly toxic blue gas with a strong smell. It is toxic because it is a very strong oxidising agent.

The **ozone layer** in the atmosphere is a region between 30 and 40 km high in which the concentration of ozone (formed from oxygen by solar ultraviolet radiation) is high. 'Holes' in this layer occur as a result of air pollution: in them the concentration falls so much at certain times of the year that high intensities of ultraviolet can reach the surface of the Earth. Such 'holes' form over the Arctic and Antarctic regions in the spring time, though more clearly in odd numbered years. The minimum concentration recorded so far was around a third of the normal level; the average is around 50% of normal.

As well as the 'holes', over the past twenty years there has been a steady general loss of ozone from the ozone layer. The rate of decline is around 3% per decade. In 2006, however, there were reports that the Antarctic 'hole' has stopped growing and may start to reduce over the next decades.

p a) unit prefix symbol for pico-, 10^{-12}

b) symbol for proton

c) symbol for extrinsic semiconductor whose main charge carriers are positive 'holes'

P a) symbol for phosphorus, element 15

b) unit prefix symbol for peta-, 10^{15}

ᵖ a) symbol for (molecular) electric dipole moment

b) symbol for linear momentum, unit: kilogram metre per second, kg m s^{-1}

c) symbol for pressure, unit: Pascal, Pa (= newton per square metre, N m^{-2})

P a) symbol for electric polarisation

b) symbol for power, unit: watt, W

c) symbol for probability

π symbol for the ratio of a circle's circumference to its diameter, 3.14159 ...

Pa symbol for pascal, the unit of pressure, 1 N m^{-2}

packing fraction the ratio $(m - A)/A$ for a nuclide of mass m (in atomic mass units) and nucleon number (mass number) A. This is an expression of mass defect, being positive (indicating instability) for both light and heavy nuclides, and negative in the middle range.

pad an attenuator with a fixed attenuation

pair creation or **pair production** the formation of a particle and its antiparticle as the result of

a) the interaction of a photon or high energy particle with a strong field (such as that of a nucleus), or

b) the de-excitation of an excited nucleus (this being **internal pair production**).

The most common example is the appearance of an electron/positron pair from a gamma photon in a nuclear field. (The process needs a field in order to allow conservation of momentum and energy.) Pair production was an important piece of evidence in support of Dirac's anti-electron (hole in the electron sea) theory.

pal(a)eomagnetism the study of the Earth's magnetism in past ages as shown in rocks laid down at different periods. One outcome is that the direction of the Earth's field has 'flipped' many times.

palladium Pd transitional metal in period 5, number 46, discovered in 1802 by William Wollaston and named after the second minor planet Pallas (as Olbers discovered this at much the same time). Relative density 12, melting and boiling temperatures 1550 °C and 3140 °C; most common isotope ^{106}Pd. Mainly used in catalytic converters, in certain alloys, and in jewellery.

panchromatic film black and white film that is sensitive throughout the visible spectrum

parabolic shaped like a parabola, or with a cross-section of that shape. A parabolic section surface reflects a parallel beam through a single focal point; conversely, it forms a parallel beam from rays reaching it from that point, the focus. Thus, a **parabolic dish** is an aerial able to receive low power radio signals from a certain direction, or to transmit a parallel beam (narrow-cast rather than broadcast); the parabolic shape allows use of the whole surface of the dish.

parahydrogen form of hydrogen gas in which the nuclear spins of the two atoms in each molecule are antiparallel. Compare with orthohydrogen, in which the two spins are parallel.

parallax the apparent change in the relative position of two objects that differ in distances when viewed from different points. The stereoscopic effect of our binocular vision is a result of parallax.

When there is **no parallax**, the objects are at the same distance. The 'no parallax' method is of great value in finding the position of a virtual image: move a real object to and fro beside it until there is no parallax between them.

parallel of straight lines, in the same direction and always the same distance apart. More broadly, it means side by side or in step. In particular, the word describes elements in an electric circuit arranged so that the circuit current splits to pass through them; it later re-joins. The voltage between the ends of each such element is the same.

In the case of resistors in parallel, the size of the current in each is in inverse proportion to its resistance, Rn. That leads to this relation for the resistance R of the section.

$$1/R = 1/R1 \div 1/R_2 + 1/R_3 + \ \dots$$

For a set of capacitors in parallel, the section capacitance C is the sum of the individual capacitances:

$$C = C_1 + C_2 + C_3 + \ \dots$$

An object has **parallel forces** acting on it if we can show them as parallel lines; an example is the weight of a balloon and the upward force ('upthrust') on it from the air. We can then find the resultant by simple addition, taking sign into account. There may also be a net torque (turning effect), if the parallel forces don't pass through the same point: use the principle of torques (moments) to find this. Much the same thinking applies to any type of vector.

parallelogram rule Show two vectors acting at a point as the two adjacent sides of a parallelogram drawn to scale: the diagonal between them gives their resultant. If the vectors are V_1 and V_2, an angle A apart, the resultant V is

$$V = \sqrt{(V_1{}^2 + V_2{}^2 + 2\ V_1\ V_2 \cos A)}$$

The angle B between V and V_1 is $\sin^{-1}(V_2/V) \sin A$. It is most common to apply these concepts to finding the resultant of two or more nonparallel forces or velocities. See also *parallel*.

parallel plate capacitor See *capacitor*.

paramagnetism the magnetic behaviour of a substance with a small positive susceptibility (and, therefore, a relative permeability of value just over 1); these two measures often vary inversely with temperature as described by the curie law. A **paramagnetic** substance has unpaired electron spins and thus a net

dipole moment; in a magnetic field, a sample will concentrate (and therefore increase) the field as the spins align with the field. As a result, the sample itself lines up parallel with the field (hence the name).

Above its Curie temperature, a ferromagnetic substance loses its domain structure; then only paramagnetic behaviour remains. The same applies to an antiferromagnetic substance above its Néel temperature. One can therefore view ferromagnetism and antiferromagnetism as special cases of paramagnetism. See also *Pauli paramagnetism*.

parameter one of the set of variable measures that describe a system, but which is kept constant during a particular study of the system. The temperature of a gas sample is a parameter in iso-thermal work, for instance. A **parametric amplifier** is a type of amplifier whose action depends on 'pumping' (varying sinusoidally) the value of some parameter (such as the capacitance of a capacitor in the circuit).

parasitic unwanted in a given context. Thus, in a fission chain reaction, **parasitic capture** is the capture of a neutron by a nucleus without any fission as a result. A **parasitic oscillation** may appear at the output of an amplifier or oscillator: it is an unwanted oscillation caused by stray reactances (and also a cause of inefficiency, in that it takes input energy from the system). A **parasitic voltage** is any unwanted signal in a circuit.

paraxial parallel and close to the principal axis of an optical system. The strict definition concerns rays for which $\sin \alpha = \alpha$ applies closely; here α is the angle in radians of such a ray to the axis at the point in question (e.g. at a surface). Only **paraxial rays** meeting a spherical reflecting or refracting surface then pass (or appear to pass) through the focal point of the surface: the others suffer spherical aberration.

parent in a chain, the nucleus that decays into a second nucleus (sometimes called the child, or daughter)

parity property of a wave function in quantum mechanics which describes how the function behaves when one reflects all the physical coordinates through the origin (to form a kind of mirror image). The wave function has even (or positive) parity if this transformation keeps it the same; the parity is odd (or negative) if the sign of the function changes. Taking those two cases as $+ i$ and $- i$

respectively lets us view parity as a quantum number. Parity is not conserved (does not stay constant) in weak interactions (as Mrs Wu's experiment shows).

parsec a standard (but not SI) unit of distance on astronomical scales. It is the distance (3.258 light years, 3.086×10^{16} m) at which one astronomical unit (150 000 km) subtends an angle of one second ($1''$). In other words, it is the distance of an object with a parallax of $2''$ when viewed from opposite sides of the Earth's orbit.

partial a) any pure (single frequency) element in a complex signal or note. Its frequency may (or may not) be above that of the fundamental, and may (or may not) relate simply to that of the fundamental. For instance, over-tones are partials (but partials are not over-tones). See also *harmonic*.
b) not complete: as a **partial eclipse** involves less of a shadow than a total eclipse. The **partial pressure** of a gas in a mixture of gases is the pressure that the gas would exert in just the same situation if the other gases were not present. See *Dalton's law of partial pressures*.

partial derivative a derivative which may be of value when the function is of two or more independent variables. In the case of $y = f(x, y)$ for instance, there are two partial derivatives. Obtain each by differentiating with respect to one variable while keeping the other(s) constant; in this case, this gives $\delta y/\delta x$ and $\delta y/\delta z$. See also *derivative*.

particle small piece of matter, where 'smallness' depends on context. In physics, we use the word mainly for ions, atoms and molecules, and for their constituents: so a **particle accelerator** is one which can take some such (charged) particles to very high energies. See also *elementary particle* (subatomic particle), and, for (elementary) **particle symmetry** (a mathematical relation between sets of similar particles), see *unitary symmetry*.

In chemistry and applied science, on the other hand, a 'particle' may be an assembly of molecules, such as a speck of dust or a droplet.

particle accelerator device designed to accelerate charged particles to a high speed relative to the speed of light in empty space. See *nuclear physics* and see also *photon accelerator*.

particle physics also called high energy physics, the study of the elementary 'building

blocks' of matter and radiation and how the react with each other. The concept that matter consists of elementary particles ('atoms') is more than 2500 years old; however, the modern subject concentrates on subatomic particles and has done so since the late nineteenth century. In the late twentieth century, there had been hundreds of particles found or specified; the Standard Model of physics explains these as combinations of a much smaller number.

partition function Q (or Z) in thermodynamics, a statistical expression for the distribution of particles in the different energy states in a group. See *equipartition*.

parton old name for quark

Pascal, Blaise (1623–1662) philosopher, physicist and mathematician whose most well known work in physics concerns air pressure. He was also one of the founders of statistics (and his early work in mathematics also led to the calculus). Pascal's experiments on air pressure in around 1646 are why we now use his name for the unit of pressure; amongst other things, he explored how air pressure varies with height above sea level. In related work, in what we now call hydrostatics and hydrodynamics, he came to **Pascal's law**. This is that the pressure in a sample of fluid in equilibrium is constant at all points. This is the basis of hydraulics.

pascal Pa the unit of pressure and of stress, with 1 Pa = 1 newton per square metre, $N\ m^{-2}$. The pressure of the air at sea level is around 100 000 Pa. There have been many other units of pressure in the past, some of which are still in use in certain contexts.

Paschen, Friedrich (1865–1947) physicist who did much important early work in spectroscopy. The **Paschen-Back effect** results in a variation of the patterns of the Zeeman effect (spectral line splitting) at very large field strengths. A hyperfine structure appears, due to the separate quantisation of the electron orbital and spin angular momenta of the substance. At even greater field strengths, the Zeeman effect predominates once more.

The **Paschen law** is that the electrical breakdown of a gas between parallel plate electrodes occurs at a voltage that depends only on the plate separation and the gas pressure. The **Paschen series** in the hydrogen line spectrum is the third such series of lines

$(n = 3)$, after those explored by Lyman and Balmer ($n = 1$ and 2 respectively).

pass-band range (band) of frequencies allowed through to the output of an electronic filter circuit. This passes waves of the designed frequency range, but blocks all others.

passive a) of a meter or other measuring device, does not change the value measured **b)** of an electronic component or sub-circuit, contains or demands no source of energy. This means it cannot produce power or amplify a signal. Examples of passive elements are resistors, capacitors and diodes, and sub-circuits that contain any of these elements only. Compare with *active*. A **passive aerial** is a parabolic dish.

Paul, Wolfgang (1913–1993) physicist who developed the ion trap technique for working with mass spectroscopy and shared the 1989 Nobel Prize for that with Hans Dehmelt

Pauli, Wolfgang (1900–1958) Nobel Prize winner (1945) as a result of the **Pauli exclusion principle**. This is that no two fermions in any system may have the same set of values of the quantum numbers concerned. The concept applies in particular to the electrons in an atom, where the quantum numbers are n, l, m and s, and to the nucleons in a nucleus.

  Pauli paramagnetism is a form of paramagnetism where there is extremely small positive susceptibility. It appears in a few metals (such as the alkali metals) as a result of the magnetic moments of the free electrons.

p-brane or **brane** object that appears in string, superstring, and M-theories, where p is its number of dimensions. A 0-brane is a particle with no dimensions, a 1-brane is a string, a 2-brane is a membrane, and so on. There are many other types.

pcb printed circuit(board), a straight forward way to mass produce circuits with places for individual elements and their connectors

pcm (or **PCM**) pulse code modulation, a form of modulation in which the system converts each value of the analog input into a sequence of pulses that code as 1s and 0s

pd potential difference, or voltage, unit: volt, V

pdm pulse duration (or width) modulation, a form of modulation in which the system converts the analog input into a sequence of pulses that vary in length

peak maximum, in particular of the value (displacement) of a cyclic measure (such as a wave or other harmonic motion). If the measure is cyclic, the peak value is a quite common way to describe it. Much more common, though, is the root mean square value. The **peak factor** of the system is the ratio of the peak value (e.g. amplitude) to the root mean square value. If the vibration is sinusoidal, the value of this is $\sqrt{2}$, i.e. 1.414 ...

Peltier, Jean (1785–1845) scientist who discovered the thermoelectric effect that bears his name (see below). In that case, the **Peltier constant** for a given pair of metals is the power transfer to temperature change per unit current. **Peltier cooling** uses the effect to cool a volume of space. Clearly, therefore, the **Peltier effect** is the change of temperature of the junction between two different metals when there is a current across the junction. See *thermoelectricity* for more on this and related effects.

pencil narrow beam (bundle of rays) of radiation, parallel or nearly so

pendulum hanging object that vibrates (moves to and fro) in a regular periodic motion. The main types are as follows, but there are many others.

- **simple pendulum:** the object, called a bob, is small (in theory a point); the suspension (thread) has negligible mass compared to that of the bob; the point of suspension is fixed. The bob swings to and fro under gravity after an initial displacement. If the amplitude is not too large, the period T is

$$T = 2\pi \sqrt{(l/g)}$$

Here l is the length of the pendulum (the distance from pivot to bob) and g is the gravitational field strength.

See also *Foucault pendulum*, a simple pendulum with a very massive bob and a very long thread.

- **cycloidal pendulum:** simple pendulum in which a pair of cycloid shaped surfaces constrains the motion of the fixed end of the thread. This makes the period T quite independent of amplitude.

- **conical pendulum:** simple pendulum whose bob moves in a small horizontal circle. This makes the thread trace out the surface of a cone. The above relation for the period T applies, as long as the circle is indeed small.

- **compound pendulum:** in this case, it is a rigid massive object that swings to and fro under gravity (for instance, the pendulum of a 'grandfather' clock). The above relation applies again, but, instead of l, we have $(L^2 + k^2)/L$. Here L is the distance between pivot and the object's centre of mass, while k is the radius of gyration about an axis through the centre of mass. (If the object's mass is m and its moment of inertia is I, $k = \sqrt{I/m}$.) In each case listed, the motion is simple harmonic. Also, in each case other than the first, the pendulum's equivalent length is the length of the simple pendulum with the same period. If the vibration is at the fundamental frequency, one needs little input energy to each cycle to maintain the motion for ever. Hence the use of pendulums for time keeping.

pentatonic scale in music, scale with five notes to an octave (interval over which the frequency doubles)

penumbra region of partial shadow: where the light source is partly, but not fully, hidden from view. See also *partial eclipse*, the view of the source from within the penumbra.

Penzias, Arno (1933–) astrophysicist and radio astronomer who shared the Nobel Prize in 1978 with Robert Wilson for the discovery of the Universe's microwave background

perfect having ideal (and often far from real) nature, behaviour, properties. Thus, a **perfect crystal** has no defects, including from thermal motion. A **perfect fluid** has no fluid friction (resistance to motion or to change of shape, viscosity). A **perfect gas** is an ideal gas, a collection of perfectly elastic masses of zero size, which exert no forces on each other except when they collide. Last, a **perfect radiator** is a radiator with some output at all wavelengths, i.e. a black body.

perigee the point in the orbit of the Moon round the Earth at which the Moon is closest to the Earth (apogee is the furthest point). In the orbit of a planet around the Sun, **perihelion** is the closest point (and aphelion the furthest).

period T unit: second, s the time for one full cycle of a cyclic (harmonic, periodic) motion, such as a wave

periodic showing some form of cyclic repeated behaviour. A **periodic motion** is one which regularly passes through a set of exactly repeated states (though the term allows a small change in amplitude). The **periodic table** classes the chemical elements in order of proton number as a table that shows the repetition of certain properties (in particular, valency) down each column (group); there is also a gentle change from element to element across each row (period). This description relates to the way the electrons enter the shells.

peripheral away from the centre, near the edge. A **peripheral (unit)** of an IT system is any part of the hardware outside the central processor: in other words, any item of hardware used for data input, output or backing storage. **Peripheral vision** is at the edge of the visual field: images are away from the centre of the retina; they do not offer much detail or colour, though people's peripheral vision of moving objects is highly effective.

periscope optical instrument with a bent axis, in particular one rather in the shape of a Z. It allows the user to obtain images of objects past some barrier. The simplest device has two plane reflecting surfaces as shown, either mirrors or prisms (using internal reflection in the latter case). It is common to add lenses to make the system a Z-shaped telescope.

Perl, Martin (1927–) joint winner of the 1995 Nobel Prize, for his discovery of the tau lepton

permalloys series of heat treated alloys of about 20% iron and about 80% nickel, often with other elements, designed to have very high permeability, very low magnetostriction and low resistivity – therefore being good for making magnets and cores

permanent unchanging, or hard to change. A **permanent dipole** is one whose positive and negative electric charge centres are always apart, as in the case of certain kinds of molecule. Compare with *electric induction*, where the separation of the centres appears only in an electric field. **Permanent gas** is an old term for a gas whose critical temperature is below around 0 °C. It is a gas, in other words, which one cannot condense by pressure alone: it needs a lowered temperature. See also *vapour*. For **permanent magnet**, a magnetised 'hard' ferromagnetic sample, see *magnet*. The **permanent set** of a sample of substance is an internal strain which remains after the stress has reduced to zero.

permeability μ unit: henry per metre, H m^{-1} the strength of the magnetic field B induced in a sample of substance, per unit applied outside field H.

337

$$\mu = B/H$$

The permeability of empty space has a value of $4\pi \times 10^{-7}$ H m^{-1}; given the symbol μ_0, this sometimes carries the name magnetic constant. We refer to it when we define the **relative permeability** μ_r of a substance. A constant in most cases, this is the magnetisation of the sample compared to the magnetisation of empty space for the same value of H. This term has no unit.

$$\mu_r = \mu/\mu_0$$

If the value of μ_r is less than 1, the substance is diamagnetic; if greater than 1, it is paramagnetic; if much greater than 1, it is ferromagnetic (but now the value depends on H). μ_r also relates to the susceptibility χ of the substance thus:

$$\mu_r = 1 + \chi$$

The **differential permeability** of a substance is the slope of its hysteresis curve.

permeance λ the magnetic conductance of a sample of substance, the reciprocal of its reluctance. See also *magnetic circuit*.

permittivity ε unit: farad per metre, F m^{-1} the strength of the electric field D induced in a sample of substance per unit applied outside field E.

$$\varepsilon = D/E$$

The permittivity of empty space has a value of 8.854×10^{-12} F m^{-1}; given the symbol ε_0, this sometimes carries the name electric constant. We refer to it when we define the **relative permittivity** (former name: dielectric constant) ε_r of a substance. This is the electric polarisation of the sample compared to the polarisation of empty space for the same value of E, as in the box.

$$\varepsilon_r = \varepsilon/\varepsilon_0$$

An alternative (but equivalent) definition of relative permittivity is the ratio of the capacitance of a capacitor with the substance filling the space between the plates to the capacitance of the same capacitor with empty space between the plates. Relative permittivity has no unit; its value is always greater than 1, and may be as high as several thousand in the case of some ferroelectric materials.

perpetual motion feature of a machine with efficiency 100%: once set working, it will go on for ever without slowing. Such a machine, should it be possible, would be of no value: to obtain useful energy from it requires an efficiency greater than 100%. That would give us a perpetual motion machine of the first kind, not allowed by the first law of thermodynamics (the law of constant energy). A perpetual motion system of the second kind does not fail for this reason, but because it offends the second law of thermodynamics: it would for ever be able to take energy from a low grade reservoir (such as the sea) and output useful (high grade) energy. The third kind of such machine is that described in the first sentence above: one with 100% efficiency exactly; this is not impossible in theory, but would require zero friction (as in a superconductor).

Perrin, Jean Baptiste (1870–1942) chemist who gained the 1926 Nobel Prize in physics for using Brownian motion to obtain a value for the Avogadro number

persistence the afterglow of (for instance) the screen of a cathode ray tube and of the low pressure gas of a discharge tube for some time after excitation ceases. See *phosphorescence*.

personal constant fixed correction one can apply to the readings obtained by a given worker using a given analog meter, to improve accuracy. This relates to systematic error.

persorption such a high level of absorption of a gas by a solid sample that the result is a solid solution

perturbation small change in some value in a system as the result of some factor outside the system. A passing object may perturb the orbit of a second object around a third, for instance; such perturbations have been of value in finding new members of the solar system. The **perturbation method** is an approach to solving a complex problem by solving a simpler version first: and then exploring the effect of small changes (perturbations) on that version in order to approach the real case. **Perturbation theory** describes how to do this, and applies to modern as well as to classical problems.

perversion lateral inversion: the apparent left–right reversal of reflected images. This does not lead to a paradox as some people claim: it follows from the fact that for each point of the object the image lies directly behind a plane mirror, opposite that point, and the same distance from the mirror.

pet positron emission tomography, a new medical scanning technique fast growing in importance

Petit, Alexis (1791–1820) physicist: see *Dulong and Petit*.

petrol (US: gasoline) fuel used in some types of internal combustion engine (and as a solvent). For the **petrol cycle** used in the **petrol engine**, see *Otto cycle*.

Petzval field curvature a name for the optical aberration of field curvature

*p*H measure of the acidity of a solution in water, the logarithm (to base 10) of the hydrogen ion activity (roughly the same as concentration). At the centre of the scale is pure liquid H_2O, with a *p*H value of 7; acid solutions have values between 0 and 7, while alkalis (basic solutions) have *p*H greater than 7 (up to 14).

phase ϕ

a) the fraction of a cycle passed by an oscillating system (for instance a pendulum or wave), given as an angle; this is the **phase angle** (with a whole cycle being 2π radians or 360°). The reference point from which we measure is normally when the amplitude is 0 and passing from negative to positive. An example is the alternating voltage supply to a purely inductive circuit and the ac in the circuit. The phase of each wave changes from point to point; however, there is always a **phase difference** $\Delta\phi$ between them – 90° (a quarter cycle) in this case. The *x*-axis here is the **phase axis**; the whole thing would be just the same with the more common time axis.

A **phase discriminator** is an electronic circuit whose output signal amplitude reflects the variations of the phase of the signal at the input. As a form of modulation, **phase modulation** is much the same as frequency modulation. A **phase shift** is a sudden change in the phase of one cyclic measure, or in the phase difference between two such measures. It can also be the result of 'tweaking' the phase angle of an ac system to improve the power factor. The **phase velocity** (or the wave velocity) of a wave is the speed at which a given part of the cycle (for instance, a certain peak) travels through the medium. This may not be the same as the wave's group velocity when the wave is in a dispersive medium.

b) how the particles in a sample of matter aggregate (clump or link with each other). In simple cases, this is the same as the state of the sample (solid, liquid, gas, and, perhaps, plasma). In more complex cases – such as alloys, with more than one component – there can be many more phases. In this sense, we use **phase diagrams** to describe the possibilities in such cases. Diagrams of this type show by lines those sets of points of equilibrium between phases (i.e. the points of **phase equilibrium**) that obtain for the system or set of systems in question. Isothermal curves are of this type; see, for instance, under *Andrew's experiments* and *saturated vapour pressure*.

c) the angle between the origin of an *Argand diagram* and a given point in the space around it. We can describe any such point by its coordinates a and b along the real and imaginary axes respectively. In other words, we can describe it as $(a + ib)$. (Here i is $\sqrt{-1}$.) On the other hand, we can describe it using the equivalent of polar coordinates: by its distance r from the origin (where $r = \sqrt{(a^2 + b^2)}$) and by its phase angle ϕ from the positive real axis (where $\phi = \tan^{-1}(b/a)$.

Because of this, we often call an Argand diagram a picture in **phase space**. See also *phasor*.

d) description of the fraction of the surface of a planet or moon that is illuminated by the Sun from the point of view of an observer. We express this as a decimal fraction of the whole visible surface, or by terms such as (in the case of the phases of the Moon) New Moon (phase close to 0) and Full Moon (phase close to 1).

phase contrast imaging method of raising the resolution of images that depends on detecting the phase change along the path of rays from different parts of the image. Used with medical x-rays, for instance, the technique lets the user see details of soft tissues the standard absorption system cannot detect – the different tissues differently affect the speed, and thus the phase, of the rays.

phasor a rotating vector in an Argand diagram (model of phase space) that represents some cyclic measure, such as the supply voltage in an ac circuit. The projection of the vector on one of the axes corresponds to the measure's instantaneous value of amplitude (the value from moment to moment) in the cycle. The phase angle between two such measures is the angle between their phasors. For a sketch showing this, see *electric circuit: ac circuits: Argand diagram*.

Phillips, William (1948–) one of the three joint winners of the 1997 Nobel Prize, for developing ways to work with laser light to trap and cool atoms

phlogiston theory developed by alchemist Johann Becher (1635–1682) and disproved by Antoine Lavoisier in the 1770s, that any combustible substance contains phlogiston and frees it when the substance burns (to leave a 'calx')

phon p old unit of the perceived loudness level L of a sound. A sound is n phons loud if it sounds as loud as a tone of 1 kHz frequency whose intensity (pressure level) is one decibel above the threshold of hearing (10^{-12} W m^{-2}). The normal unit now is the dB(A).

phonon in a rigid, elastic crystal lattice, the quantum of vibration, sound or thermal energy (named from the context of sound waves by analogy to the photon). The phonon is a boson with zero spin. It is sometimes of value to describe the thermal motion of atoms as phonons passing through the substance; also they are part of the thermal and electrical conduction processes and likely to play a part in high-temperature superconductivity. The energy W of a phonon of frequency v is

$$W = h v$$

Here h is the Planck constant.

phosphor substance that shows luminescence, either on its own or with a small amount of an activator. Luminescence here includes the appearance of output radiation outside the visible spectrum, and also includes fluorescence and **phosphorescence**. This latter is luminescence where there is radiation output that continues more than a microsecond after the source of excitation has stopped. Indeed, some phosphors in this class glow for hours or even days longer than this. See also *after-glow*. An activator (or optical activator or accelerator) is a substance which, when added in tiny amounts to a phosphor, causes the phosphor to give out the desired colour during phosphorescence.

phosphorus P non-metal with three allotropes, group V, period 3, element 15, discovered by amateur alchemist Henning Brand in 1669 and named after its phosphorescent features (Greek 'fosforos' = light carrying). Relative density 1.8–2.2, melting and boiling temperatures 44 °C and 280 °C; only natural isotope ^{31}P. There are various military and metallurgical uses.

phot symbol: ph old unit of illumination, the name for the lumen per square centimetre (10 000 lux)

photino the massive superpartner of the photon – see *supersymmetry*.

---

photo-

prefix concerning light energy (input, process, transfer and output) and, by extension, aspects of infrared as well as of other forms of electromagnetic radiation with more energy than the visible spectrum.

In the process of **photocatalysis**, light input accelerates (positive catalysis), or retards (negative), a chemical action. See also *photo-chemistry*, below. The **photocathode** of a photo-cell (below) is the electrode from which photons appear; in most cases, this is kept at a negative potential (as the name implies).

photo-cell device in which input light (in the broad sense) causes the transfer of electrons. These cells have many uses: including as solar cells – on Earth (though they are too costly for very wide use here), or in space.

There are various types. In the early days of work in this area, a photo-cell was a photoelectric cell: an electron tube in which the photoelectric effect causes a current between two electrodes in a tube of vacuum or low pressure gas. Most common now is the photoconductive cell or photoresistor; here the two electrodes (cathode and anode) have a piece of semiconductor between them. The current is very small in the absence of light: but input light causes free electrons and holes to appear, thus raising the conductivity and the current. Photoconductive cells work in the infrared region as well as at higher energies. The photo-diode and phototransistor are somewhat like this in principle too, while the output of photovoltaic cells (the most efficient type at the moment, hence their use as solar cells) is a voltage rather than a current. A cadmium sulphide cell is a photoconductive photo-cell with much higher sensitivity and much wider frequency range than the selenium type. Used in light exposure meters, infrared security sensors and street lamps, for instance, it consists of a sample of cadmium sulphide between two electrodes, one being transparent.

The highest efficiency of silicon-based photovoltaic solar cells is with 'stacked systems', able to work with a broad range of sunlight wavelengths, which have figures close to 40%. (Other approaches use optical techniques, such as mirrors, to concentrate the input radiation 100-fold or more.) Commercial units are much less efficient, somewhat below 20%; a typical one consists of perhaps a hundred 100 mm × 100 mm modules in series, each able to generate 3.5 A at 0.5 V in full tropical noon sunlight. The power to mass ratio can be as much as $1 \, kW \, kg^{-1}$. For space use, silicon cells have the problem of degrading quickly, so gallium arsenide is coming to be a more common base. See also *solar power*.

Photochemistry is the study and applied science of chemical systems in which input light causes or catalyses (see *photocatalysis*) chemical action. Here the concern is with cases in which a single molecule absorbs each input photon (unlike radiation chemistry, where it is the overall energy absorbed that is important). Einstein's law of **photochemical equivalence** is that there is a simple relation between the yield of photochemical product and the number of photons absorbed. This remains true even if some secondary process raises the yield, or a reverse process makes it less.

Photochromism is the reversible change of colour of a sample of substance exposed to high energy light or ultraviolet. **Photo-conduction** involves input light's raising the conductivity of a sample of substance. The absorbed light frees electrons, which add to the conductivity. If the substance is a semiconductor, conducting holes may appear as well. Various designs of **photoconductive cell** make use of this process; see *photo-cell*, above. **Photodetachment** is the release of an electron from a negative ion when this absorbs a photon.

A **photo-diode** is a reverse biased semiconductor diode (p-n or p-i-n) in which photons absorbed in the central depletion layer cause free electron/hole pairs to appear. The applied bias (voltage) collects a fraction of these before they combine again. See also *photo-cell* (above), *phototransistor* (below), and *semiconductor detector*.

Photodisintegration is a nuclear change caused by an absorbed photon; see *photonuclear reaction*, below. **Photoelasticity** is the double refraction that appears in some substances when under stress. Such substances include glass, Cellophane and Perspex; passing polarised white light through stressed samples leads to a display of coloured fringes which gives information about the stress pattern. Working in this way with glass or Perspex engineering objects or models is a test technique of great value. **Photoelectric absorption** is the way a surface completely absorbs a photon (i.e. it takes in all the photon energy) in one form or other of photoelectricity. For **photoelectric cell**, the type of photo-cell based on electron conduction in a vacuum, see *photo-cell* (above). For **photoelectric effects** and **photoelectric equation**, see *photoelectricity*, below.

photoelectricity the general name of any process in which light and/ or similar radiation absorbed by a substance leads to electric changes. The first such effect discovered (by Hertz in 1887) was **photoemission**: the release of electrons from a zinc surface irradiated by ultraviolet. In 1899, Thomson and Lenard measured the specific charge of the output radiation Hertz had found, and identified it as a flow of electrons. There was much research in the field around that time, and the **laws of photoelectricity** appeared (all for monochromatic light):

a) The output photocurrent (flow of photoelectrons) is proportional to the intensity of illumination of the surface.

b) There is no time delay between the start of the light radiation and the start of the emission, however faint the light.

c) The maximum energy of photoelectrons output from a given surface depends only on the frequency of the light (it does not depend on its brightness, for instance).

c) Below a frequency that depends on the surface (the cut-off, or threshold, frequency), there is no emission, however bright the light.

The model of light at that time was the wave theory. However, a wave model cannot explain these observations (laws). In 1900, Planck proposed the quantisation of generated light in an attempt to explain problems with the nature of black body radiation. Einstein developed these ideas in 1905 to form the **Einstein theory of photoelectricity**; he proposed the quantisation of absorbed light (and therefore of light in transfer). His view of photoemission involved the light photons giving their energy to electrons near the surface; it led to **Einstein's photoelectric equation**:

input energy = output energy

$$h\nu = W + \tfrac{1}{2} m_e \nu_{max}^2$$

Here h is the Planck constant, ν the frequency of the input light, W the energy function of the surface (the energy involved in removing an electron from it), m_e the electron mass, and ν_{max} the maximum escape speed of a *photoelectron* (an electron released in photoemission).

We also have an expression for W, the energy function of the surface, in terms of the cut-off frequency ν_0:

$$W = h\nu_0$$

Once Millikan had verified the Einstein equation by experiment, the concept of the photon as a wave-particle became secure. This is not a classical physics concept: before the start of the twentieth century, one could not view a radiation as having both wave and particle properties. Being able to explain photoelectricity was therefore one of the factors that led to the change from classical to modern physics.

All the above theory applies to photoemission in particular. However, the broad concept is just as valid for the other forms of photoelectricity. The broad concept is that, in a photoelectric interaction, a single photon gives all its energy ($h\nu$) to a single electron (whether the electron escapes from its bonds or not). The other main forms of photoelectricity (also used in photo-cells of various designs) are

a) **internal photoelectricity**, i.e. the photoconductive effect: photon energy absorbed inside a sample frees electrons or electron/hole pairs; illumination can raise conductivity by 10^4 or more.

b) the photovoltaic effect: a voltage appears between the two faces of an irradiated sample.

The technique of **photoelectron spectroscopy** involves study of the energies of photoelectrons in order to gain information about energy levels in the source atoms. **Photoemission** is the common term for photoelectric emission, the release of electrons from an irradiated surface. **Photoferroelectricity** involves working with the images formed on a special ceramic thin film surface to which a voltage is applied. There are two types, but in both cases illumination changes the polarisation of a ferroelectric sample.

Photofission is nuclear fission induced by the absorption of a high energy gamma photon.

Photography involves obtaining a (fairly) permanent image in the form of a pattern of chemical change in an emulsion (or other substance); in most cases the image is latent (hidden) and needs to be developed and fixed in some way. **Colour photography** uses layers of emulsion sensitive to the primary colours red, green and blue. See also *camera*. **Digital photography** involves having the image in electronic form rather than chemical form. This allows for ease of storage, processing and transfer.

Photoionisation is ionisation induced by the absorption of photons, while **photolithography** is a technique for the transfer of a circuit pattern to the surface of a wafer in the process of making chips. **Photoluminescence** is a form of luminescence induced by absorbed photons, and **photolysis** is chemical breakdown induced by light. The purpose of **photometers** (of which there are many designs) is to measure the intensity of a radiation, or the brightness of a source (e.g. the magnitude of a star); this approach is central to the science and applied science of **photometry**, whose concern is such measurement. **Photomicrography** involves making photographic images of small objects: it is a mix of photography and microscopy, in other words, with cameras fitted to microscopes in effect.

The purpose of a **photomultiplier** is greatly to provide a strong signal from very low light levels. Most modern designs – working with perhaps 2 kV – can produce tens of millions of photoelectrons for a single input photon. A photocathode absorbs that input photon, and emits a photoelectron. This accelerates across to the first 'dynode', where, by secondary emission, it can release many more electrons. The process repeats through a number of dynode stages to produce an output current pulse. See also *avalanche diode*, a particularly effective and compact type.

The **photon** is the quantum of electromagnetic radiation, with energy $h\nu$, h being the Planck constant and ν the wave frequency. That photons exist shows the wave/particle duality of this radiation (as with others); they have momentum $h\nu/c$ (c being the speed of light). This massless particle is a boson with unit spin. The virtual photon is the exchange particle for the Coulomb force; virtual photons can travel at any speed, rather than just c. A **photon accelerator** is a device or system designed to raise the frequency of laser light pulses that may also one day come to be used to accelerate charged particles. The approach with the most promise in 2006 is to pass two laser beams together through a plasma; one forms a wave in the plasma and the other's pulses 'surf' on the wave and gain higher frequency, energy, and group velocity. People have reached frequencies in the GHz range using this technique.

A **photonic crystal** is an insulating nano-structure with a periodicity that affects passing electromagnetic waves in much the same way as a semiconductor affects electric current – there is a kind of band structure with allowed and forbidden levels. The process involved is a form of diffraction in a solid structure with the right set of electrical or magnetic features. Opal is a natural photonic material (though does not have a large band gap); artificial ones are grown from colloids in most cases. Photonic crystals can control the transfer of light on a nano-scale, but are (at the time of writing) still too new for there to be many clear, confirmed uses.

Photoneutron emission is the release of neutrons from nuclei when these absorb photons of gamma with greater energy than the neutron binding energy. *Photofission* (above) is one such process, as nuclear fission always involves the release of free neutrons. Both of these processes are **photonuclear reactions**: nuclear changes of some kind induced by absorbed photons. Several other types appear above.

The movement of small suspended particles (for instance of dust in air) as a result of radiation is the process of **photophoresis**. The path of the motion that results is in the form of a helix (spiral); the axis of the helix is in the direction of the light (unless an electric or magnetic field modifies the process). See *electrophoresis*. In normal light levels, **photopic vision** is vision in which the cones of the retina are the main light receptors (as opposed to scotopic vision). The sensitive surface used in the process of photolithography mentioned above is **photoresist**. A **photosensor** is a light sensor, the same as a *photo-cell* (above) in practice. A **phototransistor** is a type of transistor sensitive to electromagnetic radiation. Radiation input to its emitter causes more charge carriers to appear in the base, and thus raises the collector (output) current. The effect is of a photo-diode with an amplifier.

Phototropy is another name for *photochromism* (above). The **photovoltaic effect** is the appearance of a voltage across the interface (junction) between two substances in contact, in the presence of light. The photo-diode is an example of a photo-cell of this type; some people call them barrier layer photo-cells, as there is a barrier at the junction. Electrons freed within the cell can pass one way through the barrier, but not the other; this forms the voltage concerned.

photonics the technology that applies the physics of optoelectronics to systems that work with photons rather as electronic systems work with electrons; the frequency range of the systems concerned is 300 GHz to about 10 THz. See also *polaritonics*, the technology between photonics and electronics, working at a few hundred tera-hertz.

physical concerned with material things and with energy: in other words with the matters of interest in physics. The concern of **physical chemistry** is how the physical and chemical structures of matter and their properties and changes relate to each other; it uses the concepts of physics to explain the findings of chemistry. **Physical geography** deals with the natural structures of the Earth (such as mountains and valleys) and *their* properties and changes; it relates closely to geology and certain aspects of astronomy. **Physical optics** is the behaviour of light and similar radiations, in particular where their wave nature is concerned; see *optics* for how this differs from geometrical optics. The **physical sciences** are those concerned with energy and non-living matter – physics, chemistry, geology and astronomy, and their close relations; so the term covers the whole field of natural science other than biology.

physics the study of matter and energy, and of their forms, properties, and interactions. It is an observational, experimental science, but one in which theory (for instance, philosophy and mathematics) plays a major role in meeting its aim. This is to explain all the features of the non-living world in simple, unified terms (though biology, in so far as it develops from chemistry, also depends on the laws of physics). Physics is the basis of all fields of applied science and engineering as well.

Originally, the subject was natural science (science = knowledge), in other words the study of the non-living aspects of nature. In particular, it was a branch of philosophy (and some people still call the subject natural philosophy). The work of Galileo, Huygens

and Newton put it on a solid experimental (or observational) and mathematical basis, and the subject became what we would now call **classical physics**.

By the late nineteenth century, almost all of the phenomena treated seemed open to full explanation and description. However, people then discovered cathode rays and radio-activity which they could not explain; also, they were not able to explain and describe black body radiation and photoelectricity with any success at all. In 1900, Planck came up with the idea of quantisation in the former case (thus solving the so-called ultraviolet catastrophe); in 1905, Einstein used that central concept to explain the photoelectric effect. Both theories see light and similar radiations as having wave and particle aspects at the same time; we can therefore no longer call this subject classical physics: this work was the start of **modern physics**.

The aim of people doing research in the field now, a century later, is to describe, explain, and unify all the many types of phenomena we now know (and, of course, others that may appear in due course). Links with astronomy and chemistry are very strong at the basic levels at which these people work. We can still handle many aspects of mechanics and geometrical optics as classical physics; most of the other sub-fields of physics need modern concepts for a full treatment. Those sub-fields are acoustics, electricity and electronics, magnetism, nuclear physics, physical optics, quantum physics, relativity, and thermodynamics.

physisorption the form of adsorption that involves van der Waals forces

pick-up strictly any transducer in general – but people on the whole restrict the term to a 'read head' able to detect a signal stored on a disc and to convert that to an electric output.

pi π a) the ratio of the circumference (perimeter) of a circle to its diameter. The value of π, 3.14159 ..., is irrational, never ending: it has now been worked out to tens of millions of decimal places. π relates to the other fundamental irrational number, e, by Euler's formula, as in the box.

$$e^{i\pi} = -1$$

Here i is $\sqrt{-1}$.

b) the **pi-meson** – the intermediary (exchange) particle of the strong nuclear force. See *pion*.

c) the **pi network** – a linkage of two parallel circuit elements with a series element between.

d) name for the **pi theorem** (or Buckingham theorem) of dimensional analysis: one can always write a function of physical measures as a relation between non-dimensional combinations of the corresponding physical quantities. The function is valid if the dimensions balance. This is the usual approach to dimensional analysis but can be extended to much more complex cases.

pico- p unit prefix for 10^{-12}, a millionth of a millionth, as in a picofarad, pF, 10^{-12} F

piezoelectric axis a direction within a piezoelectric substance along which tension or compression will produce a voltage. This concerns the **piezoelectric effect**: the appearance of opposite charges at the opposite ends of an asymmetric crystal (for instance, one of quartz) when the crystal is in tension or compression. The voltage that results varies directly with the strain; its sign depends on whether the strain is tensile or compressive. Research into pyroelectricity led to the discovery of the effect in 1880 by Pierre Curie and his brother Jacques; it has many uses, in particular in transducers for such systems as sonars, the record player pick-up, microphones, viscometers, elastometers, and many types of electronic circuit.

The converse effect has the same name: it is the change of shape of such a crystal caused by a voltage between its ends. It is a special case of electrostriction. An alternating field applied to such a crystal causes it to vibrate; if the frequency of the ac has the right value, the crystal will resonate. It is then a **piezoelectric oscillator**, widely used as the tuned circuit of various electronic devices (such as the 'quartz clock'), as the frequency output is highly stable. It can act either as a standard frequency source or as a frequency stabiliser.

piezomagnetism the converse of magnetostriction, in other words the effects of strain on the susceptibility and remanence of a ferromagnetic sample

pile a) old word for a battery of electric cells
b) now rare term for the central part of a nuclear fission plant, the core that contains the nuclear fuel and in which the chain reaction occurs

pinch effect a constriction in the cross-section area of a tubular conductor, of a liquid conductor, or of a plasma when it carries a very large current. The effect is the result of the motor effect interaction between the current and its own magnetic field. In the case of plasmas used in nuclear fusion research, the effect is one way people are exploring to hold the very hot plasma safely in its solid container (compare with *magnetic bottle*).

pin cushion distortion distortion in an optical system that causes the sides of the image of a square to become concave. See *aberration* for a sketch.

p-i-n diode p-n diode with a narrow region of intrinsic semiconductor as the depletion layer between the n-region and the p-region. Such diodes have many uses as modulators and switches in microwave systems (the greatest working frequency in 2006 being about 1 GHz).

pin-hole camera photographic camera or *camera obscura* in which there is no lens and therefore only a very small aperture. While the image that results is very faint, it shows no aberration. The radius of the hole giving the best effects is 0.85 $\sqrt{(l\,\lambda)}$; here l is the distance between the pin-hole and the plate or screen, and λ is the mean wavelength of the light.

pink noise part-way between white noise and red noise (Brownian noise) – noise whose intensity varies inversely with frequency

pion π elementary particle, the meson (with, therefore zero spin) mainly involved as an exchange particle in the case of the strong nuclear force. There are three types, the neutral pion (π_0) and those with positive and negative charge; the first has an antiparticle, while the last two are the antiparticle of each other. All have very short lives, in the range 10^{-16}–10^{-8} seconds. Their masses are around 170 electron masses.

In 1935, Yukawa proposed the concept of exchange particles for the fundamental nuclear force (and for other forces); his theory was able to predict the nature of the 'meson' (meaning middle mass particle, i.e., one lying in mass between electron and nucleon) in great detail. It was a disappointment therefore when Anderson discovered a middle mass particle in cosmic ray showers the next year: it did not match Yukawa's very well at all. In fact, Powell discovered the Yukawa particle in 1947 (also in cosmic ray showers): we now call Anderson's particle the mu-meson, or muon; it is in a different class from the pions.

piston short cylinder that moves to and fro (reciprocates) inside the cylinder of a pump or engine; it drives, or is driven by, a fluid under pressure. A **piston gauge** (or **free piston gauge**) is a pressure meter for work with fluids at high pressure: the fluid pressure pushes at one side of a piston in a cylinder; the user measures the force needed to keep the piston in the zero position.

pitch the frequency aspect of a sound as perceived by the human hearing system, with high pitched sounds having high frequencies

pitchblendes series of minerals that form the world's most important sources of uranium (in the form of uranium oxides). The main mines are in Czechoslovakia (from where

Marie Curie obtained her working material), Zaire and North America. Pitchblendes also contain useful ratios of radium, polonium (both found by the Curies), and thorium.

Pitot tube widely used meter for fluid speed. Such meters are common on the outsides of boats and aircraft, for instance, to measure their speed. One open end of a bent tube points into the fluid stream; at the other end there is some kind of manometer. Henri Pitot (1695–1771) devised this design in 1732.

The reading is of the total fluid pressure: static as well as dynamic. Most modern designs measure the former component as well as the total, and thus give the user a direct reading just of the pressure due to the speed (and therefore a reading of the speed itself).

planar of the widely used process for making **planar semiconductor devices** (such as diodes, transistors and chips). The silicon substrate on which are grown and built oxide and other layers (see *wafer*) is a plane surface that provides mechanical strength as well as the desired electrical properties.

planck

name proposed for the SI unit of action, the joule second

Planck, Max (1858–1947) physicist, most noted for his making the first steps towards the quantum theory (which, with relativity, forms the basis of modern, i.e. post-nineteenth century, physics). As a student of Kirchhoff and Helmholtz (and successor to the latter's chair in Berlin), in his early research he followed the steps of Clausius in thermodynamics; he moved from there to the study of black body radiation and then the problem of the ultraviolet catastrophe (see *Planck formula*, below, for what this is).

Planck's quantum theory of 1900 had great success in overcoming that problem. In a sentence, this proposed that a measure – in this case the energy of radiation – could hold only certain values, rather than all values in a range. Planck viewed his approach as a mathematical trick only, however (one with no reality, in other words); he did not change his mind after Einstein's success in extending the concept to photoelectricity in 1905, or after the Bohr model of the atom in 1913 which depends on the same concept. All the same, the work gained Planck the 1918 Nobel Prize.

The **Planck constant** h, appears in many fields of physics (and other sciences) as the constant of quantisation. Planck's theory of 1900 saw radiation as produced in jumps of energy, each of value h v, where v is the frequency of the radiation concerned. Here h is the fundamental unit of *action* (sense c), with value 6.6256×10^{-34} joule seconds. A common form in which h appears in practice is $h/2\pi$, so common that it has its own symbol: h. See, for more on this, *quantum Hall effect*, *quantum mechanics* and the *uncertainty principle*. For **Planck distance**, the quantum of distance, see *Planck length* (below).

People's early work on black body radiation was able to lead to mathematical expressions which fairly well fit the energy distribution of radiation at long wavelengths. At shorter and shorter wavelengths, on the other hand, there was more and more divergence between theory and practice (hence the term 'ultraviolet catastrophe'). The **Planck formula** was the first to fit observation at all wavelengths. This tells us that the power output per unit area of a black surface at wavelength λ is as in

$$P\lambda = 2\,\pi\,h\,c^{3/} \left(\lambda^5 \exp(h\,c/k\,T\,\lambda) - 1\right)$$

Here k is the Boltzmann constant, c the speed of light, and T the absolute temperature of the surface. We call hc^3 the 'first radiation constant', and hc/k the 'second radiation constant'.

For typical energy/wavelength graphs, see *black body*; the catastrophe was that these graphs didn't peak but rose faster and faster with falling wavelength.

The formula is the basis of the **Planck law** (or **Planck radiation law**), and involves the quantisation of the process that produces the radiation; thus, as noted above, it is the basis of the crucial quantum theory.

Within classical thermodynamics, the **Planck function** Y of a substance is an energy function; it relates to the Gibbs function G of the substance and to the temperature T thus:

$$Y = -G/T$$

The **Planck length** l_P (or Planck distance) is the quantum of distance, about 1.6×10^{-35} m.

$$l_P = \sqrt{(h\,G/c^3)}$$

Here h is the Planck constant h, divided by $2\,\pi$, G is the gravitational constant, and c is the speed of light in empty space.

The Planck length is one of the three 'natural' **Planck units** – the others being the quanta for mass and for time. Thay are all natural (says Planck) in that they are all functions of the fundamental constants *h*, *G*, and *c*, as above.

plane flat surface, one such that all points of the straight line between any two points in the surface lie in the surface too. The **complex plane** is that defined by the axis of real numbers and an axis of imaginary numbers at 90° to it; each off-axis point is a complex number. See *Argand diagram*. The **plane of flotation** of an object floating in a liquid is the plane in the object that corresponds to the liquid surface.

The **plane of polarisation** of plane polarised radiation (below) is (strictly) the plane along a ray that contains the ray and the magnetic field vibrations. The original definition applied to plane polarisation by reflection: the term then was for the plane that contains the incident and reflected waves; this happens to contain also the magnetic wave. The electric wave vibrates in a plane at 90° to this; unfortunately, many people call this the plane of polarisation (though plane of vibration is a better phrase in this case). Unpolarised radiation has the magnetic (and therefore the electric) vibrations evenly spread in random directions; **plane polarisation** involves some process – such as reflection – that keeps them each to a single plane. The radiation is then **plane polarised**. For **plane of vibration**, see plane of polarisation, above. See also *polarisation*.

planemo a planetary mass object that does not orbit a star. A few tens are known, with masses of the order of that of Jupiter; of these few tens, two found in 2006 form a pair in orbit round each other.

planet massive solid object (perhaps with a gassy atmosphere and a liquid core) in orbit round a star (such as the Sun). A planet gains most of its energy from its star, but some energy (the amount depends on the planet's mass) flows from the core to the surface, being the result of nuclear decays. If the mass is so great that nuclear reactions produce enough energy for the object to be a glowing sphere of gas, it is a star, and not a planet. If the mass is too small for gravity to form the object into a sphere (or close to one), it is not a planet but a **minor planet**. Using these

definitions, plus a vague one about sweeping its area clear of debris, the International Astronomical Union looked again at the solar system in 2006 and put Pluto in a new group – it became a **dwarf planet**, joined by a few other objects at much the same distance, and also by Ceres, the largest of the minor planets. A dwarf planet is larger than a minor planet, but not massive enough to clear other objects from its orbit. (That last is the new criterion on which the 2006 decision depended; it is controversial and hard to apply in practice, and it is possible the IAU may re-visit the whole process in 2009.)

We call a planet that orbits some other star an **exo-planet** (or **extra-solar planet**). After much searching, the first of these turned up in the 1990s, with over 200 known at the time of writing (mid-2007), the least massive being below ten Earth masses. The current view is that perhaps ten per cent of Sun-like stars have planets.

Astronomers long held that planets form from a *planetary nebula* (sense a), proto-planetary disc (proplyd for short) – there are many now known, with the first being found in 1984. This high density disc of gas (and, later, dust and specks of ice) rotates round a proto-star. If the stellar wind doesn't blow the disc away and the star doesn't absorb it, the disc starts to condense into small centres called **planetesimals**, and may in the end form a **planetary system** like ours.

planetary electron out-moded term for an electron bonded to the nucleus of an atom. The term comes from the Rutherford 'solar system' model of the atom; it implies a clear precise orbit, which is not the case here – see *electron cloud* or *shell*.

planetary nebula
a) old name for protoplanetary disc, the disc-shaped gas and dust cloud round a very young star from which planets may form
b) the spherical gas cloud round a dying star – from which planets do not form – given off from the outer layer of a red giant

plano- prefix for flat, like a plane surface. A **plano-concave lens** has one flat face and one concave face (so is diverging: being thinner in the middle than at the edge); a **plano-convex lens** is a converging lens with a plane face and a convex face.

plasma to some, the fourth state of matter, a gas with so much energy that its atoms have become almost fully ionised. The plasma thus consists of nuclei, ions and electrons: all are charged particles (though the sample as a whole remains neutral). The gas of stars and the matter within a discharge tube are plasmas; the temperature in the former case may be high enough for controlled fusion to take place if the gas is hydrogen.

The problem to workers in the field of artificial fusion is to contain the plasma (i.e. without having it touch the solid walls of the container) for a time long enough for fusion to appear with useful energy output at a sensible efficiency.

A **solid state plasma** is the free electron gas' of a metal or the similar assembly of electrons and holes in a semiconductor. A **plasma display** is a flat screen system using the light radiation caused by a current in a low pressure gas. **Plasma waves** are oscillations in a plasma that result from a disturbance – the same as Langmuir waves. A **plasma torch** is a device able to heat a gas electrically to such a high temperature that the jet can melt steel and create chemical compounds. A **plasmatron** is a discharge tube with a current between the hot cathode and the anode; an input signal can modulate the current by its effect on the area of the plasma.

plasmon quasiparticle that results from quantising the oscillations of a plasma (i.e. the Langmuir waves in it), in particular those of the free electron 'gas' of a metal. This is a **Josephson plasmon**. A **surface plasmon** stays in the metal surface and has an even stronger effect on the metal's optical properties. The frequency of these waves is very high – maybe 100 THz – so there are high hopes of using plasmons for data transfer inside chips as well as for actually making chips. The applied science of plasmons in such contexts is **plasmonics**.

plastic of a solid, easy to mould. This is because the bonds between the atoms or ions of the solid are weak enough to allow slippage between them when they are under stress; the result is the **plastic deformation** or **plastic flow** of the sample. After the stress returns to zero, the sample stays in its new

state, as opposed to the case of an elastic solid, where it does not.

A **plastics** substance (a 'plastic' in common language) is one made from a polymer. Rubber, bitumen (as in asphalt and tar), and some resins are among the few natural examples; most known are artificial, almost all being made from oils. A **plastic magnet** is a plastics substance (polymer) with ferro-Magnetic properties. As well as being an insulator, its value lies in being able to form uniform magnetic thin films (important for sensors and perhaps in such devices as speakers). The first such substance was announced in 1990: it is ferromagnetic below –180 °C (i.e. at the temperature of liquid hydrogen); the first working at room temperature appeared in 2004.

plasticity that property of a substance that allows one easily to deform it permanently. See *plastic*.

plastometer device designed to measure the plasticity or flow of solid samples

plate a) the anode of an electron tube ('valve') **b)** very large solid chunk of the Earth's crust, floating on the mantle, and making up a continent or subcontinent, or a large part of one. **Plate tectonics** explains continental drift by the slow but steady movement of these plates over the mantle during geological time because of convection currents.
c) rare term for printed circuit board

platinum Pt transitional metal in period 5, number 78, used in jewellery since ancient times in North Africa and South America, but first studied in the 1740s by William Brownrigg; its name comes from the Spanish, meaning 'little silver'. Relative density 21, melting and boiling temperatures 1770 °C and 3800 °C; most common isotope ^{195}Pt. Platinum is common for electrical contacts, electrodes, alloys, and lab equipment.

platinum resistance thermometer thermometer at whose centre is a coil of platinum wire; as the resistance of this depends on temperature, it is easy to use it as a fast and accurate meter. See also *bolometer*.

pleochroism the result of the different degrees of absorption by a birefringent substance of the ordinary and extraordinary rays, a form of dichroism. A **pleochroic halo** is a multi-coloured ring system within mica (for instance), the result of stopping an alpha particle.

plutonium Pu transitional radioactive, rare

earth metal in the actinide series, period 7, proton number 94, created in 1941 by Glenn Seaborg and team and named after the dwarf *planet* **Pluto** to complete the sequence uranium (Uranus), neptunium (Neptune) and plutonium. Relative density 20, melting and boiling temperatures 640 °C and 3230 °C; most stable isotope ^{239}Pu, half-life 24 000 years. Produced in breeder reactors, this is the main fissile element in nuclear weapons; 238Pu is very suitable for radioisotope generators (as in spacecraft, but also, for a time, in heart pacemakers).

p-n diode junction between p-type and n-type semiconductor material. This is a rectifier (diode) in that it conducts far better one way than the other. See *junction diode*.

pneumatics the study and use of the dynamic properties of gas samples, in particular their pressures. A **pneumatic circuit** is an approach to the control of machines by way of switching pressures between gas tubes. Its action is much like that of an electric circuit in many ways. Compare it with *hydraulics*, the same but with liquids (the volume of whose samples cannot change with pressure). **Pneumatic systems** (widely used in robots, for instance) are cheap, quiet and reliable.

pmos positive channel metal oxide semiconductor. See *mos* and *transistor*.

p-n-p transistor transistor with p-type emitter and collector, and an n-type base between these

Pockels effect the linear electro-optic effect, in other words the Kerr effect as occurs in a piezoelectric crystal. Here a strong outside electric field changes the crystal's refractivity and makes it birefringent. The birefringence varies with the field, whereas in the Kerr effect it varies with the square of the field.

Pogson, Norman (1829–1891) astronomer who proposed, in 1856, the logarithmic **Pogson scale** of apparent magnitude still used, based on the **Pogson ratio** (about 2.5) between integers on the scale

point geometrical entity with no dimension, in practice a very small area close to a circle in shape. For **point contact diode** (e.g. cat's whisker), see *diode*; this was the basis of the first (1948) transistor, the **point contact transistor**. A **point contact electrometer** is a device able to measure currents as low as 10^{-18} A – so able to count individual electrons. See also *ammeter*. In a crystal lattice, a **point defect** may be a vacant site, an

impurity atom, or a crystal atom not in its normal place; see *defect*. The value of a **point function** depends on the position of the point concerned in some kind of field. A **point source** of radiation is one so small that the wave fronts produced are exactly spherical; we can think of all radiation sources as points from a large enough distance (as is the case with all the stars other than the Sun).

poise P cgs (i.e. out-moded) unit of dynamic viscosity (fluid friction), with a value of 0.1 pascal second (0.1 Pa s)

poiseuille Pl name proposed for the pascal second, the SI unit of dynamic viscosity

Poiseuille flow the laminar (steady, uniform) flow of a fluid, flow with no turbulence. A sketch of the speeds of the layers (laminas) at different distances from the container walls has a parabolic front; the speed is constant in any layer between a wall (zero speed) and the centre (greatest speed). Jean-Louis Poiseuille (1799–1869) did a lot of work in this area.

Poiseuille formula or Hagen's formula formula, based on Newton's laws of fluid friction, that relates the rate of flow, dV/dt, through a pipe of radius r and length l, to the fluid's friction (viscosity, ξ) and the pressure difference Δp between the ends of the tube:

$$dV/dt = \pi\ r^4\ \Delta p/8\ \xi\ l$$

poison a substance which reduces the efficiency of a process. The term is most common in nuclear reactor physics, applied phosphor science, and photo-emission.

Poisson, Siméon-Denis (1781–1840) mathematician who became a peer for his research in mathematical physics. The **Poisson distribution** is the normal distribution for discrete events. It is a frequency distribution formula that applies to cases where the chance of occurrence of a given event is small (for instance, the radioactive decay of one nucleus in a sample), the average rate is known, and the rate does not depend on time.

The **Poisson equation** has a number of forms, for different aspects of physics. In the case of static electricity, it describes the electric potential V that results from a given distribution of net charge. Where there is no charge, or all charges are paired (no net charge), the equation simplifies to the Laplace equation.

$$\Delta^2 V = -\rho/\varepsilon$$

Here ρ is the charge density (coulombs per cubic metre) and ε is the permittivity of the medium. Δ^2 is the Laplace operator (div grad) for partial differentiation.

The **Poisson ratio** is an elastic constant for an object under tension (e.g. when necking); it is the ratio of the lateral strain (the sample's contraction, $\Delta A/A$) to the longitudinal strain (its extension, $\Delta l/l$). If the sample's total volume remains unchanged, the value is 0.5; in practice (where the volume tends to rise), the ratio is less than that.

polar concerned with either the two 'poles' (points of opposite property) that appear in a number of situations in physics, or with a single point (viewed in this context as a single pole). Thus at the ends of a **polar axis** in a crystal, the crystal properties differ somewhat; such an axis is an axis of rotation that contains no centre of symmetry. A **polar bond** is an ionic bond. For **polar coordinates**, see *coordinates*. A **polar diagram**, or **polar graph**, is a graph with a point origin and no axes; it shows the value of the measure in question as it varies with angle round the origin.

polarimetry using a **polarimeter** of some type to measure the optical activity of a sample. Optical activity shows as the rotation of the plane of polarisation of light passing through. The basic polarimeter consists of one fixed and one rotatable sheet of Polaroid, initially set crossed so that no light passes. With the sample between the Polaroids, one needs to turn the second Polaroid to cut out the light. The angle of turn is the angle of rotation (or activity) of the sample. This gives information about the nature and concentration of the active sample, and the technique is widely used by chemists and engineers. The most sensitive types of polarimeter use interference effects.

polarisation some process whose result is a directional aspect to some property of some sample of some substance. The main cases are as follows.

a) in an electric cell – a chemical or physical action at the electrodes that prevents their most efficient use, thus making the output voltage fall lower than it should be. A cause may be gas bubbles coating an electrode. The cell needs a depolariser to counter the effect.
b) in an electrolytic cell – **electrolytic polarisation** – chemical action that leads to a voltage (called back emf) that tends to oppose (and thus reduce) the one applied
c) in an insulator ('dielectric') – the distortion of the neutral atoms in an electric field to give each one a positive and a negative end. This is because the electron cloud tends to displace one way, while the displacement of the nucleus or core is the other way. The sample itself as a whole therefore gains a positive face and a negative face. The result is that it now has a dipole moment, the value of this per unit volume being the quantitative polarisation (**electric polarisation** P) of the sample.
d) in a ferromagnetic sample – the same as magnetic induction, causing an n-pole and an s-pole to appear at its opposite ends in a magfield
e) in an elastic medium, such as around an earthquake in the Earth's crust – the appearance of horizontal (S_H) and vertical (S_V) shear waves (S waves, or shake waves) much like polarised light waves (below, g))
f) in a beam of matter particles with non-zero spin – the alignment of the spins in a single direction as a result of scattering (for instance)
g) in a beam of electromagnetic radiation the restriction of the plane of the electric field vibrations to a single direction. (The magnetic field vibrations lie in a plane at 90° to this.) As we can view such a beam of radiation as made up of photons (which have non-zero spin), we can call this a version of e)

If the electric field waves lie in the same plane all along the radiation path, the radiation is **plane polarised**. **Circular polarisation** appears when the plane of polarisation rotates at a constant rate along the path. **Elliptical polarisation** is much the same, except the rotation traces an ellipse. We can view the former as two plane polarised waves of equal amplitude at 90°; elliptically polarised radiation then has two such waves that differ in amplitude.

There are various ways to polarise light. The first found and explored, early in the nineteenth century, is **polarisation by reflection**. Light reflected from any surface is partly polarised (Malus, 1809). The angle of incidence is the Brewster angle (or **angle of polarisation**, or polarising angle). Light incident at this angle (about 57° in the case of glass) reflects almost fully polarised, with the vibrations parallel to the surface; the refracted light is plane-polarised at 90° to that. In 1812, Brewster showed that:

$$i_B = \tan^{-1} n$$

Here n is the refractive constant of the medium that reflects the light.

In Malus' experiment to explore the effect, the brightness of the output beam from the first glass sheet was compared while turning the second glass sheet about an axis to give various angles. In this kind of context, the first glass sheet met by the light is a **polariser** (it can polarise an input beam); the second is an analyser (we can use it to explore the polarisation).

After the work of Malus and Brewster, people tried other ways to produce polarised light. They found that crystals with the feature of double refraction (birefringence) split an input ray into two: the 'ordinary' ray (which follows Snell's law for the substance) and the 'extraordinary' ray (which does not). These rays are both plane polarised, but in planes at 90° to each other. Nicol prisms, made from such crystals (often of calcite), gained great success as polarisers and analysers; however, they are bulky, costly and fragile. The artificial substance Polaroid is much better, so is now very widely used in this context (though is less efficient). Polaroid is a dichroic substance (as is the mineral tourmaline, also once common for this purpose).

Polarisation is the only property with which we can distinguish between transverse and longitudinal waves: it is possible to polarise transverse waves but not longitudinal ones, as Young showed in 1817. All electromagnetic radiations are transverse, so can show the effect, therefore; indeed radio and microwaves are polarised when first produced, as are the light waves from a laser. In the case of the polarisation of electromagnetic radiation, the phenomenon mainly results from the interaction of the electric field component of the wave with matter. Other causes apply to other forms of transverse wave.

See also *Kerr effect, optical activity, photoelasticity* and *solar compass*.

polariser device used to plane polarise input electromagnetic radiation, as in a polarimeter

polarising angle name for the Brewster angle, or angle of polarisation: the angle of incidence at which polarisation by reflection is most complete

polariscope simple type of polarimeter, used to measure optical rotation

polariton a quasiparticle that can transfer energy through certain samples of matter; it is the basis of **polaritonics**, a technology partway between very high frequency electronics and photonics. This rightly implies that a polariton is a mixture of a dipole (such as an electron or a phonon or other exciton) and a photon. A feature of a polaritonic substance is that the speed of light through it depends on the frequency. The frequencies of polaritonic systems are hundreds of terahertz, THz.

The **polariton laser** is a major application of the field. Lasers of this type developed in 2007 can work at room temperature at a tenth of the power of conventional solid-state lasers.

polarity state of an object or system in which some measure has opposite value or sign at two different points. This applies not just to a magnetised ferromagnetic sample (a magnet, with an n-pole and an s-pole), but (for instance) to electric circuits or elements with a positive 'pole' (point or terminal) and a negative one.

Polaroid ™ synthetic sheet material able partially to plane polarise input light. It consists of aligned long polarising crystals (e.g. of iodoquinine sulphate) in a plastics (nitrocellulose, for instance) base. This structure makes the substance cheap and robust. On the other hand, it is less efficient as a polariser than some systems, and absorbs quite a lot of the input light. There are many uses, the most common being in anti-glare sun glasses. The first patent was in 1929.

The **Polaroid camera** (instant camera or Land camera, after Edwin Land, its inventor, 1947, who founded the Polaroid Corporation in 1937) uses a film pack which contains its own developer and paper as well as the light sensitive emulsion. After exposure, rollers force these together, to develop the image and print it on the paper within a number of seconds. There have been versions for making moving pictures, including in 3D.

pole a) (better: centre or optical centre) the geometric centre of a reflecting or refracting surface. In the case of a lens, we can take it to be the geometric centre of the common section of the two surfaces, though there is little difference unless the curvature is large. All rays through the pole of a thin lens suffer no deviation; all rays reaching the pole of a mirror follow the laws of reflection with the axis being the normal. In geometric optics, we measure all relevant distances from the pole of each element.

b) electrode in an electric or electrolytic cell and, by extension, a terminal of a source of electricity or switch. The word is now rare in both senses, however.

c) one of the two points – called north and south – where the Earth's spin axis passes through the Earth's surface. By ancient extension, the word also applies to one of the two points – called celestial north (at the moment, near Polaris, the Pole Star or North Star) and celestial south (which is not near any bright star).

d) region of a field, in particular the magnetic field in a magnet or electromagnet, where the field strength is greatest. See also *dipole* and *multi-pole*. Such magnetic poles always appear to come in pairs, called north seeking (n-pole) and south seeking (s-pole); however, see *monopole*. When using a magnetic compass, we find that at most parts of the Earth's surface, the direction to the nearest magnetic pole (the line of the magnetic meridian) is close to the direction to the geographical pole (the line of the geographic meridian).

The **pole face** of a magnet or electromagnet is the end surface through which passes the useful field. In other words, it is the surface that faces the region of interest. A **pole piece** in an electric machine is a shaped piece of ferromagnetic material fixed to a pole face to make the output field as strong as possible and of the desired form. The **pole strength** (m, unit: weber, Wb) of a pole is the force on it from a unit pole a metre away in empty space. A unit pole repels a similar unit pole a metre away in empty space with a force of one newton.

e) for **Regge pole**, see *Regge model*.

Politzer, David (1949–) one of three joint winners of the 2004 Nobel Prize, for work on the theory of the strong interaction, in particular for discovering a new aspect

polonium Po radioactive metalloid in group VI, period 6, with proton number 84, discovered in 1898 by Marie Curie, who named it after the Latin name for Poland. Relative density 9, melting and boiling temperatures 250 °C and 960 °C; most stable

isotope (of the record 33 (2006)) ^{207}Po, half-life 5.8 h.

poly- prefix that implies plural, more than one in number (or, in some cases, many rather than just a few). Thus, a **polyatomic** molecule consists of two or more atoms. In this context, more specific terms are diatomic (with two atoms) and triatomic (with three). Note that some people reserve polyatomic for molecules with three or more atoms, or even for those with four or more (i.e. more than triatomic). **Polychromatic** describes radiation with a number of different frequency or energy components, whether of similar intensity or not. Compare this with *monochromatic*, in which case too the 'chrom' refers to colour (equivalent to frequency in the case of light waves).

A **polymer** is a chemical compound whose molecule consists of a large number of identical small units called monomers. Many plastics are of this nature, as are proteins, rubber, and some minerals. See also *macromolecules*, of which polymers are a class. Note that, while electrically conducting polymers are quite common, in 2006 appeared such a polymer with a number of other features of a metal; for details, see http://physicsweb.org/article/news/10/5/1

Polymorphism is a feature of some substances that exist in a number of forms that differ in structure (maybe at different temperatures and/or pressures); see also *isomer*. A **polynomial** is a mathematical expression which is the sum of two or more terms; these may or may not be similar in form. A **polyphase system** is an electric network in which there are two or more ac supplies of the same frequency that differ in phase. See *three-phase* for a very common approach to power supply in practice.

In the late 1960s, there was much research into **poly-water**, a form of liquid H_2O thought to exist as a polymer (above); with freezing temperature of -40 °C and boiling temperature of 150–500 °C, as well as relative density as much as 1.4, those who believed in it thought it differs greatly from normal water. After a few years, people agreed that poly-water is normal water, but with a fairly high concentration of silicates dissolved from the glass in which it is made or other impurities.

Popper, Karl (1902–1994) philosopher of science, whose ideas on the nature of science and scientific research have had much influence. In particular, Popper's view that one can never prove a theory means it is important to test for its falsity rather than for its correctness. Further, a theory that it is impossible to disprove may not be a scientific theory anyway; for instance, the theory that God created the Universe is a matter of faith rather than of science.

population I/II stars way to class all stars we now observe (though there is actually a smooth spectrum) with population II stars having formed first, so having a much lower value of metallicity than population I stars. The latter formed from clouds of hydrogen (and helium) and produced the higher elements during their main sequence phase. Those higher elements passed into the gas clouds – by stellar mass loss of various types up to supernova events – from which the younger stars formed.

population III stars very massive stars believed to have appeared in the early Universe at about the same time as quasars. Unlike quasars, no one has (yet) observed such an object; however, they may lie within distant blue galaxies, and these *have* been seen.

population inversion situation in which, at a normal temperature, there are more particles in a system at a high energy level than in the ground state. Some people view this as a form of negative temperature. Reaching a population inversion is an essential aspect of the efficient production of laser output. Here the upper energy level is meta-stable: particles can accumulate in it as a result of excitation without dropping back to the ground state too quickly.

Porro prism, or **Porro–Abbe prism** also (wrongly) called roof prism – see *Abbe–Porro prism.*

posistor thermistor with a positive temperature constant of resistance

position point or location at which some-thing exists or at which we make some measurement. Position is not an absolute term: we must always relate it to an arbitrary zero (the origin); it is normal to use some kind of coordinate system for this purpose. In some such systems, a **position vector** *r* is the vector (defined in direction and in size, i.e. length) of the point from the origin concerned. Other such special terms appear in astronomy and navigation, for instance.

positive a) associated with a charge opposite

to that of an electron (which we define to have a negative charge). The **positive column** is the main glowing region in a discharge tube; when the tube is long, as for lighting, this region almost fills the whole length between cathode and anode. A **positive electron** is an electron with a positive charge; in other words, it is the positron, the 'normal' electron's antiparticle. The **positive glow** of a discharge is a name for the *positive column* (above). The **positive pole** of an electric circuit is an old name for the positive terminal of its supply. **Positive rays** (also called canal rays) are streams of positive ions, in a discharge in particular. The term is now rare, since the nature of this radiation has been made clear.
b) associated with a value greater than zero, as opposed to negative (with a value less than zero). For **positive feedback**, see *feedback.* A **positive lens** or **positive mirror** has a focusing power greater than zero: it will converge an input beam.
c) associated with the higher of two (or more) voltage levels. Thus, a digital electronic system uses **positive logic** if we call the higher voltage pulses 1s and the lower levels 0s.

positron the antiparticle of the 'normal' (negative) electron; its positive charge gives it its name. Anderson discovered the particle in the cloud chamber tracks of cosmic radiation in 1932: some electron-like tracks curved the other way in an applied magnetic field. This was just a few years after Dirac predicted the existence of such a particle (see, for instance, *antimatter* and *hole*).

Positrons also appear as the result of some types of beta decay, or in electron/positron pair production from the interaction of a photon with a nucleus. As a positron is annihilated when it meets a negative electron, its lifetime in practice is short; it is, however, a stable particle.

positron emission tomography pet body scanning technique of fast growing importance in medicine. The patient receives a positron-emitting isotope used as a tracer (often ^{18}F); the scanner, a circular array of detectors round the patient, detects the pair of gamma photons produced by each positron decay. A computer analyses the signals and plots the site of decay of each positron. Doctors use the technique to survey blood flows, in particular those to certain organs or to tumours, and the distribution by

357

the body of a drug, for instance. It is possible to label oxygen and most drugs with a positron-emitter.

positronium quasi-stable pairing of positron and electron as an exotic atom (an analog of the hydrogen atom). The half-life is around 10^{-7} s when the spins of the two particles are parallel (**orthopositronium**); when they are antiparallel (**parapositronium**), it is a thousand times shorter. In either case, annihilation destroys the pairing; mainly, the output is three gamma photons in the former case and two in the latter, the total output energy being 1022 keV. A **positronium molecule** is a briefly stable combination of two electrons and two positrons, in the analog of the hydrogen molecule.

Post Office box practical form of Wheatstone bridge (or potentiometer) once very widely used as a kind of multi-meter. The user – often a Post Office phone engineer in the days when the PO ran most of Britain's phone systems – would fit keys into any of a large number of sockets to allow any combination of resistance to switch into the test circuit.

pot nickname for *potentiometer* (senses a and b) within certain groups of electrical and electronic engineers

potassium K alkali metal in group I, period 4, with proton number 19, prepared by Humphry Davy in 1807 and named after pot-ash (potassium carbonate), a product of burning plants in iron pots. Relative density 0.86, melting and boiling temperatures 63 °C and 760 °C; most common isotope ^{39}K. Used in making glass and, as an alloy with sodium, as a thermal conductor.

potassium–argon dating in geology, a technique of *dating* rocks that depends on the radioactive decay of potassium-40 to argon-40. As the half-life is so long (some 1.28×10^9 years), one can use the approach to date back close to the birth of the Earth.

potential V unit: volt, V at a point in an electric field, the energy involved in the transfer of a unit charge between infinity and that point. In some cases, it is of more value to use the ground state rather than infinity as the reference level. The unit is the joule/coulomb, called the volt V. We can define the term in just the same way in the case of other types of force field. For a gravitational field, for instance, the unit is the joule/kilogram. We also use the term for the state of a system rather than for a point; the reference level

may be either the ground state or infinity, again.

Therefore, though much of the rest of this discussion concerns **electric potential** (the potential at a point in an electric field), it applies on the whole just as much to the other fields and to the states of systems. We need to reword the definition and rename the unit for each case, of course. Lastly, note that it is common to use the term potential where strictly potential difference or potential energy (below) applies.

A **potential barrier** is the region of high electric potential energy round the nucleus of an atom; this affects the passage of charged particles (either as they try to escape or as they approach). Other names are Coulomb barrier (as this is for an electric field), Gamow barrier (as Gamow was one of the first to explore its effects in detail), or nuclear barrier. A **potential depression**, on the other hand, is a region of low potential energy; see *potential well*, an important special case, below.

Rather than potential, it is often of far more value to discuss the **potential difference** V (or U), pd for short in this book, between two points in the field, or that between two points in a circuit (which is an electric field distributor). Again, we define this as energy per unit charge transferred, using the relation

$$V = W/Q$$

Here W is the energy involved in the transfer and Q is the charge transferred.

The unit is the joule per coulomb, called volt V (hence the common name of voltage for pd). The pd between a given pair of points does not depend on the path taken between them.

Specific to electric circuits, the **potential divider** is an arrangement of resistance that allows one to share an applied pd between two sub-circuits, or to apply a fraction of it to one sub-circuit. See *voltage divider*.

In effect, the full name for potential and potential difference as here described is **potential energy** W (or V or U). This is the energy of an object or system that relates to its position or state, the *energy* that could potentially transfer from the object or system. There are many special cases of potential energy (pe), for the different types of force field and the different types of energy state.

Here are some – but we could argue that all energy is potential energy.

a) chemical pe: the energy involved in the chemical change between two states

b) deformation (elastic) pe: that involved in the strain within the elastic limit of a sample of solid

c) gravitational pe: the energy of an object of mass m at height h above the reference level in a gravitational field of strength g, given by mgh

d) pressure pe: the energy of a fluid sample because it is under high pressure (in truth the form of elastic pe that applies to fluids rather than to solids)

   Potential flow is a name for the smooth (laminar, Poiseuille) flow of a fluid. A **potential function** is either

 • a harmonic function which satisfies the Laplace equation $\Delta^2 V = 0$, which describes how potentials in a field behave,

or

 • for a conservative force system, a scalar function which satisfies the relation $F = - - \Delta V$.

The **potential gradient** at a point in a field is the rate of change of potential with distance at that point, distance being taken in the direction of maximum gradient. In the electric case, it is a measure of the field strength, $E = dV/dx$ – the unit being the volt per metre. The **potential scattering** of a wave or particle by a nucleus is a process we can describe as reflection by a hard sphere. A **potential transformer** is a rare name for the voltage transformer.

   A **potential well** in a force field is a region of low potential (energy W), with high gradient at the edge, and perhaps with higher potential than usual just outside the edge. We can view the nucleus in this way. The nuclear energy well (as relates to a nearby neutron, for instance) and the well for nuclear and Coulomb (electric) energies (as for a proton, for instance). The second curve is in fact the well for a nucleus with high proton number; those with low proton number have deeper wells with a much flatter base.

potentiometer a) common (but unfortunate) name for a variable resistor (often called pot for short). This usage follows, one assumes, from the common use of the device at the centre of a voltage divider (*potential divider*), and its slight resemblance in some designs to some designs of potentiometer, sense c).

b) fairly common (but also unfortunate) name for a variable resistor used to produce a variable voltage divider effect, as in the volume control of an audio system. Some people in the field use the term, or at least the short form 'pot', for any variable resistor.

c) device used to measure electric potential difference (pd, i.e. voltage) – the true meaning – by balancing two potentials to give zero reading on a sensitive current meter. The metre bridge was perhaps the most common form until modern meters appeared.

pound lb unit of mass in the 'imperial' (foot, pound, second) system, defined as about 0.454 kg. (The **metric pound**, fairly common in some countries of Europe, is 0.5 kg.) In the same system, the **poundal** (pdl) was one of the units of force: the force able to accelerate one pound by one foot per second per second. The other unit was the **poundforce** (lbf), the force able to accelerate one pound by g, 32 feet per second per second.

powder photography technique of x-ray crystal diffraction analysis in which the sample is a fine powder. It is common to rotate this sample in the x-ray beam to produce an output pattern that consists of a set of concentric circles.

Powell, Cecil (1903–1969) physicist who gained the 1950 Nobel Prize for developing ways to photograph particle tracks, and thence being the first to see traces of the pion

power P **a)** the number of times to multiply a value by itself, the number being the index or exponent. Thus in 10^3 ($10 \times 10 \times 10$, i.e. 1000), the power (or index or exponent) is 3.

b) the ability of an optical element or system to converge input radiation to a focus. It is the reciprocal of the focal distance, and the unit is the dioptre, D (= metre^{-1}). As with focal distance, a negative value shows that the element or system will diverge input radiation – an image is virtual.

c) the rate of energy transfer from or to a system. In other words, it is the energy transfer per second; the unit is the watt, W (= joule per second). (The early non-SI units called **horse power** (hp, first used by Watt in the 1780s) are around 740 W, except for **boiler horse power**, 9.81 kW.)

   The relation used to find power depends on context, for instance as follows.

 • electric power – with V, I and R the (direct) voltage, current and resistance involved:

$$P = V\,I = I^2\,R = V^2/\,R$$

In ac circuits, V and I are not normally in phase. One must then involve the *power factor* (below) of the system, and use the unit volt amp(ere) (VA) rather than the watt.

• mechanical power – with force F and v the velocity of the object:

$$P = F\,v$$

The relation for rotational motion is equivalent.

The gain or **power amplification** of an amplifier is the ratio of the output power to that input; in logic too, this is much the same as efficiency. There are similar definitions for other such systems (such as transducers). A **power amplifier** is an amplifier with gain (power amplification, above) greater than 1; its input may be a preamplifier, but its output goes direct to the next stage, the system's output transducer (rather than being amplified again).

In an ac circuit, the **power component** is the component of the current that is in phase with the applied voltage; a more common name is active current (as opposed to reactive current, the component that is 90° out of phase with the voltage). In the same context, the circuit's **power factor** is the cosine of ϕ, the phase angle (phase difference) between current I and voltage V. The power of a dc circuit is $V\,I$ (unit: watt, W); that of an ac circuit is $V\,I\cos\phi$ (unit: volt amp, VA); hence the name: $\cos\phi$ is the factor by which one multiplies dc power to obtain the ac power value. See also *active power*.

Power line transfer is the transfer of signals (which may include control signals and data) through a power cable rather than through a special channel. This is quite common on a domestic scale (such as for intercomms and baby alarms) as well as for communication through the national power grid of a country. A **power reactor** is a nuclear reactor whose prime function is to produce electricity for the community (rather than to produce isotopes or to act as a research tool). A **power supply** (or **power supply unit**, psu) is an electronic system with an electrical power output; it either acts independently of mains power, or converts mains power to a form suited to the device. In the former case, it may be no more than an electric cell; in the latter case, it is likely to have a transformer (to produce the desired voltage level), a

rectifier (to convert the ac to half wave or full wave dc), and a smoothing capacitor (to convert that to a dc with only a small ripple). See http://www.smpstech.com/tutorial/to1int.htm#SMPSDEF for plenty of examples.

The design of a **power transistor** allows it to handle more than about a watt, perhaps as much as several hundred watts. It needs some form of cooling for this reason.

Poynting, John (1852–1914) physicist whose research covered many fields, in particular electromagnetic radiation and its effects and gravitation. The *Textbook of Physics* he wrote with J J Thomson was a classic undergraduate work for many decades. The **Poynting theorem** is that the power transfer from electromagnetic radiation varies with the product of the electric and magnetic amplitudes (field strengths), E_0 and B_0 respectively. For an electromagnetic wave, the **Poynting vector** S gives, in size and strength, the wave's power transfer per unit area; the unit is the watt per square metre, and the value is $\frac{1}{2}\,E_0\,B_0$. The vector's direction is that of the energy transfer, forward at 90° to the electric and magnetic vectors (which are themselves at 90°).

praseodymium Pr transitional rare earth in the lanthanide series, period 6, proton number 59, prepared by Carl von Welsbach in 1885 and given its name from the Greek for 'green twin' (compare with neodymium, new twin). Relative density 6.8, melting and boiling temperatures 930 °C and 3500 °C; most common stable isotope ^{142}Pr, half-life 19 h.

preamplifier amplifier whose output is a second amplifier, either another preamplifier, or the main (power) amplifier in the system. The aim of the approach (rather than using a power amplifier with higher gain) is to keep the noise level low. Another name is head amplifier, from the fact that a common site for a preamplifier in practice is close to the source of the signal (the 'head' of the system, like the head of a river).

precession the rotation of the axis of spin of an object round a second axis. We can see the effect in spinning tops and gyroscopes; it also applies to the Earth's spin (due to torque from the gravity of Sun and Moon) and to orbits. The cause is always a **precessional torque**, with the spin axis moving at 90° to the torque direction.

precision often confused with accuracy, the

number of valid significant figures in a result or in the reading of a meter – 'my height is 2.345 m' is precise but not accurate; 'my height is 1.8 m' is accurate but not precise. A **precision laser** has a tightly focused beam which allows use for high precision cutting, and so on. In making chips, for instance, precision lasers help work the surface in very fine detail (to high resolution).

pre-emphasis method of selective amplification of a signal in order to reduce the noise level: the modulation factor at high frequencies is made to differ from that at low frequencies. De-emphasis is the reverse process at the far end of the channel, and emphasis is the whole technique (used in audio digital recording and fm radio).

presbyopia the appearance of far sightedness in older eyes. One's near point moves further away as the lens becomes less flexible with age.

pressure p unit: pascal (Pa), the newton per square metre the force at 90° to a surface (real or otherwise) per unit surface area:

$$p = F/A$$

As well as the pascal, there are many still quite common non-SI units, such as:
- the atmosphere (atm) – the standard ('normal') pressure of the air at the surface of the Earth, 101 325 Pa
- the bar: almost the same, at 10^5 Pa (and the millibar, 100 Pa)
- the millimetre of mercury (mmHg): the pressure of one millimetre of mercury (see below for the relation), 133 Pa
- the torr (from Torricelli): a special name, now rarely used, for mmHg

People often design objects to make the pressure from a force high (as at the point of a drawing pin) or low (as with the tyres of a tractor). Here, respectively, the contact area is made low or high, and we can use the concept of *force ratio* to give the effectiveness of the design.

The pressure at a point within a fluid is an important concept, though is not easy to define in practice unless one brings in an imaginary surface at the point in question. The most useful expression is

$$p = d \rho g$$

Here d is the depth of the point within a fluid of mean density ρ in a gravitational field of mean strength g. Rather than d, we

sometimes use h (for height, or for 'head' (below)).

This relation is equivalent to the first, in that the force in question is the weight of fluid in a 1 m² column above the level concerned. It is simple to apply that relation to a point in a liquid: less so for a gas (where, also, a variation can applies in the case of air pressure). Also, many forms of manometer (pressure meter) and barometer (air pressure meter) used to depend on the relation.

The pressure at a point in a fluid at rest
- is the same in all directions (else fluid would move to make it so);
- acts at 90° to any surface (the same applies);
- does not depend on the shape of any container.

See also *buoyancy, centre of pressure, gas laws, hydraulics* and *pneumatics*.

Pressure broadening is one cause of the broadening (increased width) of the lines of the line spectrum of a gas. This effect also involves a slight shift in the central wavelength, and follows the increased number of collisions between the gas particles at higher pressures. A **pressure cooker** (autoclave) prevents the vapour from the cooking foods from escaping. As a result, the pressure rises to perhaps twice atmospheric; this raises the boiling temperature, in practice to around 115 °C. The result is much more rapid cooking.

A **pressure gauge** is a pressure meter (manometer), there being many designs (some mentioned in this book). Primary gauges (which measure pressure directly) include the liquid manometer and the free piston gauge; secondary ones (which measure the effect of pressure) include the Bourdon gauge. There is a large group of such meters designed to work at very low gas pressures. In a liquid, **pressure head** is the height of a column of the liquid (often water, sea water, or mercury) giving a pressure same as that stated. (Hence the pressure unit mmHg.) Indeed, a head (column) of liquid is a quite common way to produce a desired pressure in a fluid sample or on a solid surface.

Prévost exchange theory A body radiates the same power at a given temperature, whatever the temperatures around it. This leads to the concept of energy flow towards a state of thermal equilibrium, an important aspect of thermodynamics. **Pierre Prévost** (physicist and philosopher, 1751–1839) published the concept in about 1800.

Prigogine, Ilya (1917–2003) physical chemist whose research into non-equilibrium thermodynamics led to his 1977 Nobel Prize for chemistry

primary of the first rank (i.e. most important), or coming first in time. One cannot recharge a **primary cell**; a secondary cell is rechargeable, on the other hand. A set of **primary colours** has three members, depending on the context. There is an infinite number of sets of three primary colours (hues) of light; the only criterion is that you cannot produce any one of the three by mixing the other two in any way. Common primary hues are red, green, and blue, or their complements: cyan, magenta, and yellow. One can mix three such hues to produce the effect of white light (see *chromaticity*). The primary pigments (dyes or paints) are red, yellow, and blue. See also *colour separation* and *colour mixing*.

  Primary electrons are those absorbed by a surface to produce an output of secondary electrons. Much the same applies to **primary ionisation**, the ionisation produced by an ionising particle or ray (and not that produced by the ions formed). In cosmic rays too, **primary radiation** is that reaching the upper layers of the atmosphere; there it can produce showers of secondary particles and radiation. By extension, a primary radiation is any radiation as output by its source, before significant interaction with matter. The **primary winding** of a transformer is the one that carries the input current (the secondary gives the output).

principal main. The **principal axis** of an optical element or system is the line normal to its optical centre(s), or pole(s). See also *axis*. The **principal focal points** of the element or system lie on this (see *focal point*). The **principal plane** of a doubly refracting crystal contains the ordinary ray and the optic axis. For the **principal quantum number** n, see *quantum numbers*. In an optical system, the **principal points** are the two conjugate points on the axis where lateral magnification is 1.

principle fundamental law (but not provable any more than is any law in science)

printed circuit board (pcb) flat plastics (insulating) board which carries a pattern of conductors etched ('printed') as tracks, and sockets for the pins of chips and other elements. See also *surface mounting*. The first

pcbs were the radio proximity fuses used in British anti-aircraft shells in 1944, though the concept was first demonstrated in 1936.

prism block with two opposite parallel identical faces. Made of transparent material, these have many important roles in optics. Very often, those two faces are triangles, though the **penta-prism**, with five-sided faces, is used in certain types of camera view finders.

  The main uses of prisms are
  • to disperse radiation by refraction, to produce a spectrum of wavelengths over a range of output angles;
  • to change the direction of input radiation by one or more internal reflections.

The **prism binocular** uses a pair of *Abbe-Porro prisms* in each arm to fold the processed light into a short length, and also to make the otherwise upside down image the correct way up.

probability P value that tells us the chance of one particular outcome of an unpredictable event. It is common to express this as the ratio of the number of cases (expected or observed) of the particular outcome to the number of cases of all possible outcomes. The concept is of great importance in modern physics. Thus, the **probability density distribution** or **probability density function** of a radiation or electron cloud gives the probability at each point of finding a particle. It is $|\phi|^2$, ϕ being the Schrödinger wave equation concerned.

probe device able to obtain a signal (energy) from some space while causing as little disturbance in the space as possible

Proca equations set of equations, analogs to the Maxwell equations, but for particles of spin 1

process set of one or more actions carried out on an input (raw material such as light or data, for instance) to give an output that has more value to the user. A **processor** is a device with the task of doing this.

programmable uni-junction transistor type of thyristor, with a region of negative resistance in its characteristic curve

progressive wave travelling wave, one in which the whole periodic displacement moves through the medium. Compare it with *standing wave*, one that is restricted to a certain space.

projectile object moving freely (i.e. in free fall) in a field, having been given some initial momentum. What follows applies to gravitational fields, but there are analogues for any other type of field.

The object starts at velocity v at an angle Θ to the horizontal. If Θ is 90°, the motion is linear in the vertical direction; the equations of motion then apply at once, with the acceleration being g. In all other cases, we must keep separate the vertical and horizontal components of the motion. In the vertical direction, again the equations of motion apply; in the horizontal direction, there is no force, so the velocity is constant. The path of the projectile is thus an arc of a parabola.

If we ignore the effects of air friction, we have

- greatest height $h = v^2 \sin^2 \Theta / 2g$
- time to reach greatest height $t = v \sin \Theta / g$
- horizontal range $R = v^2 \sin^2 \Theta / g$

In practice, we must take account of friction, wind and the variations of these and even sometimes of gravity over the path of the object. All this involves the science and applied science of ballistics, brought to a fine art by hundreds of years of (pre-rocket) weapon design.

projector optical system designed to place an image on a screen of some kind of object. There are various types; they differ in

- the size of object;
- the size of the audience (e.g. personal microscope type versus cinema projector);
- whether the object is real or in the form of data (data projector);
- whether it is flat or has a three-dimensional nature; and
- whether it is opaque (episcope) or transparent (diascope or overhead projector, slide/strip/film projector).

In all types working with a real object, light from the source passes through a condensing (concentrating) system to the object, and thence to a lens unit that can focus a large, bright, real image on the screen.

Prokhorov, Aleksandr (1916–2002) physicist who shared the 1964 Nobel Prize for work in quantum electronics that led to the maser

promethium Pm transitional rare earth in the lanthanide series, period 6, element 61, created in 1942 by Chien Wu and colleagues, and named after Prometheus who (in the Greek myth) gave fire to the human race. Relative density 7, melting and boiling temperatures 1050 °C and 2500 °C; most stable isotope ^{145}Pm, half-life 18 years (decay by electron capture). There are uses in

thickness gauges, long lasting light sources, lasers and nuclear cells.

prompt neutron neutron produced in a reactor as a result of the main type of fission (e.g. of a ^{235}U nucleus) rather than by the decay of a fission product nucleus (which happens after some delay)

proof plane small conducting disc at the end of an insulating handle, now rare. One used it to sample the charge at a small part of a surface, transferring the charge it collects to some kind of electroscope for study.

propagation transfer, of energy in some form. The **propagation coefficient** or **propagation constant** P of a transmission system or line is the natural logarithm of the ratio of the input strength to the output strength. The usual unit is the neper. The **propagation loss** of an electromagnetic radiation beam is its energy loss per unit distance: by absorption, divergence and scattering. This is an important measure in work with radio.

proplyd short for protoplanetary disc, the common modern term for planetary nebula

proportional of two measures that vary with each other: two measures are directly proportional if doubling one doubles the other. Two measures are **inversely proportional** if halving one doubles the other. The relations concerned are

$$y \propto x \quad \text{i.e.} \quad y = k\,x$$
$$y \propto 1/x \quad \text{i.e.} \quad y = k/x$$

with the two measures being y and x. Here k is some constant.

Graphs of y against x and of y against $1/x$ respectively are straight lines through the origin.

A **proportional chamber** or **proportional counter** is a particle detector working in the proportional region (below) of the discharge curve of the gas used. The size of an output pulse therefore varies with the ionisation caused by a passing particle. As a result, we can feed the output to a multi-channel pulse height analyser to obtain a spectrum of the radiation in the chamber: or just count the pulses in the case of the counter.

The **proportional limit** is that point on the elasticity curve of a solid sample at which the deformation becomes no longer proportional to the stress. This point is not the same as elastic limit, though it may be very close in some cases.

The **proportional region** of the discharge

curve of a gas is that part at which the current passed through the gas varies directly with the voltage between the electrodes. It lies between an initial curve and the main (Geiger) plateau. See *spark*.

protactinium Pa transitional, radioactive metal in the actinide series, period 7, proton number 91, found in 1913 by Kasimir Fajans, and named 'first actinium' as it decays into actinium. Relative density 15, melting and boiling temperatures 1570 °C and 4000 °C; most common stable isotope ^{231}Pa, half-life 32 000 years.

protective relay circuit breaker that 'trips' (switches open) when the current through it passes the design greatest value. It acts like a fuse, therefore (and for the same reasons), but after use one can reset it.

proto- prefix that denotes first (or early), as in prototype, an early model or the first stage of implementing a design. **Protoplanetary disc** (proplyd for short) is the common modern term for planetary nebula. A **proto-star** is one in the first equilibrium stage after birth and gravitational collapse.

proton p elementary particle, a nucleon like the neutron. It has unit positive charge (i.e. the same as that of an electron but of opposite sign: $+1.602\ 192 \times 10^{-19}$ C) and a rest mass of 1.673×10^{-27}kg (somewhat less than that of the unstable proton). It is a baryon (massive particle) with ½ spin and iso-spin values; as it is the least massive particle with baryon number 1, it is highly stable (for baryon number is constant in all interactions). The proton, being the ion of normal hydrogen, is also of much importance to chemists, in particular in solution in water. See *hydrogen*.

  Proton decay is a rare process of radio-active breakdown: the unstable nucleus emits a proton and loses 1 from its nucleon and proton number values. The **proton magneton**, a measure of magnetic moment, is a less common (but more accurate) name for nuclear magneton. A **proton microscope** is much like an electron microscope, but works with a proton beam rather than with an electron beam; this gives improved resolving power. The **proton number** of a nuclide is the number of protons in the nucleus; this is the same as the old atomic number that distinguished the chemical elements from each other.

  The **proton–proton chain reaction** is a series of nuclear fusion reactions that produce helium from hydrogen. There is very significant transfer of energy to radiation; people believe the reaction is the basis of solar energy (and, therefore, of the output of most other stars) – it becomes efficient at temperatures above about 10 MK. There are three possible routes from ^{1}H (hydrogen-1, in other words, the proton p) to ^{4}He (helium-4, i.e. the alpha particle α). The main one is in the box. At each stage, there is release of energy to the outside in the photons (and a little as kinetic energy of the massive products). The net energy transfer to outside of the three steps shown to produce one alpha is 27 MeV (about 3×10^{13} J); an alpha is 0.7% less massive than the four protons, this being its mass defect.

1	$p + p \rightarrow\ ^2H + \nu + e^+$ (very slow)
	$\nu + e- \rightarrow 2\ \gamma$
	(instantaneous; electron from shell)
2	$p +\ ^2H \rightarrow\ ^3He + \gamma$
3	$^3He +\ ^3He \rightarrow \alpha + p + p$

The other particles involved are the deuteron 2H and the triton 3H; ν is the electron neutrino and γ is a gamma photon.

For **proton resonance**, see nuclear magnetic resonance. The **proton synchrotron** is a cyclic proton accelerator (see *synchrotron*) able to achieve very high energies. The machine is important as the proton beam is of high value as a nuclear probe and a treatment for cancer (though people believe an anti-proton beam would be four times better).

pseudo-Goldstone bosons (pgbs) group of eight particles, as yet undiscovered, needed by apparent breaking symmetry in quantum chromodynamics. See *axion*, one of this group of particles.

psi particle ψ one of the two names for the J/psi, or gypsy, particle

psu power supply unit, system which either works independently of mains power to give a dc output, or converts mains power to a dc form suited to the device supplied

Ptolemy (Claudius Ptolemeus) (*c.*90–*c.*168) mathematician, astronomer and geographer, perhaps based in Alexandria, whose book, the *Almagest*, summarised the whole of the Greek astronomical knowledge of the time. Ptolemy's geocentric (Earth-centred) Universe, a polished version of earlier theories of this type, came to be widely accepted until the end of the Middle Ages (see *Copernicus*), despite the views of the

Pythagorean group. Ptolemy also carried out some useful work in physics.

p-type denoting an extrinsic semiconductor with more free 'holes' than electrons. As we can think of a hole as a positive charge carrier, such a substance passes current mainly as a flow of positive charge. Hence the name ('p' for positive) See also *acceptor*, the type of impurity that causes this behaviour.

pulley simple machine that transfers input power through the tension in a 'string' (rope, cable, chain) wound over one or more wheels. There are two types of single pulley system and two types of multiple system. In practice, friction will reduce the force ratio and the string's stretching will increase the distance ratio.

A block and tackle (invented by Archimedes) is a pulley system with one or more fixed large pulleys and one or more moving large pulleys.

pulling frequency pulling – the tendency of a second oscillator to synchronise (draw the frequency of) a main oscillator's output towards that of its own

pulsar a point source of continuous pulsed radio radiation in the sky, with similarly pulsed light output in a few cases, and pulsed gamma and/or x-rays in others. Jocelyn Bell Burnell found the first pulsar in 1967 (gaining her supervisor Anthony Hewish the 1974 Nobel Prize); now we know many hundred of these sources. They come in three classes – the class depends on how the object obtains its huge energy output: slowing fast rotation, gravity, and falling magfield.

There is no standard, agreed theory of how pulsars arise and decay. Most people, perhaps, believe that a pulsar is a neutron star formed by a supernova. When a fairly massive star runs out of nuclear fuel, the core collapses suddenly under gravity. The burst of neutrinos and the gravitational energy output by this supernova collapse ejects the star's outer layers. The massive core that remains becomes so compact (perhaps 25 km across) that its electrons and protons combine to form neutrons (it becomes a neutron star, in other words). These neutrons may in turn compact further a few seconds later, with a new burst of neutrinos, to produce strange matter (roughly equal numbers of up, down and strange quarks): a strange star.

Whether neutron or strange star, the core now rotates at high speed (because of the law of constant angular momentum plus contribution from captured orbiting matter). If a jet of radiation from one point of the surface sweeps past the Earth as the object spins, this will produce for us the pulsar effect. The fastest pulsar yet discovered (2006) rotates at around a thousand rev s^{-1}. As a pulsar loses energy from radiation, its spin rate slows. In some cases, young pulsars in particular, 'glitches' occur every few years (or, in at least one case, weeks): there is a sudden increase in pulse (i.e. spin) rate, and/or a total break. People believe these glitches follow quakes in the structure, causing it to contract further.

To the pulsar effect mentioned above may be added small changes in frequency if the pulsar is part of a binary (two-star) system. Their analysis was a new way to confirm general relativity (the spin-orbit coupling effect and gravitational waves). The most interesting case in this context is of a binary pulsar some 2000 light years away – both stars involved are pulsars, their orbit period is 2.4 hours and they eclipse each other as they orbit. There are many tests possible in such as case – see http://news.bbc.co.uk/1/hi/sci/tech/5356910.stm – and those so far tried confirm general relativity very well. As yet, however, there is no sign of gravitational waves from this source.

pulsatance ϕ unit: hertz, Hz angular frequency, the number of complete rotations of a system in a second. A wave or simple harmonic motion of frequency ν has a pulsatance of $2\pi\nu$.

pulsating star star with cyclically varying output (brightness). There are many types, some known a long time, of which the pulsar is perhaps of the most interesting.

pulse a brief change in the value of an electric current or force (for instance). The sudden disturbance transfers as a transient wave. The task of a **pulse amplifier** is to increase the value of an input pulse (and sometimes to shape it as required). A **pulse amplitude analyser** is the same as *pulse height analyser* (below).

Viewing a long series of pulses as a digital carrier signal leads to various methods of modulation (changing it so it carries information or data). The main ones are pulse amplitude modulation (changing the pulse heights)

- **pulse code modulation** (in which pulse groups of os and 1s represent the codes

for the values transferred – see below) and
- **pulse duration modulation** (changing the lengths of the pulses).
- **Pulse height modulation** is a name for pulse amplitude modulation, while
- **pulse width modulation** is the same as pulse duration modulation.

Any or all of those may bear the name **pulse modulation**.

A **pulse generator** is an oscillator circuit whose output is a train of pulses of set shape, size and frequency; see *astable*. A **pulse height analyser** records (counts) an input pulse only if its amplitude (height) is within a set range. A multi-channel system can, in effect, sort input pulses into the right ranges and record the total in each range during a given period. This is of great value in analysing the signals from particle detectors in nuclear physics.

A major advantage of digital data transfers as opposed to analog ones is that the noise level in the output can be made very low. An analog circuit must amplify all frequencies (signal and noise) in much the same way; the digital equivalent is the regenerator. This outputs a perfectly shaped (noiseless) pulse for each noisy input pulse; the process is **pulse regeneration**.

pulse code modulation (pcm) a widely used system for digitising an analog (e.g. audio or video) signal to allow transfer of the data or information with far less noise. The first use was for time division multiplexing telegraph signals. Here are the main steps in the process, which takes place (as shown vertically below) in an analog to digital converter (adc) ...

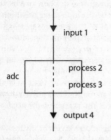

1 input to adc – analog signal such as that shown in the first sketch with frequency a few hundred hertz
2 process in adc – several thousand times a second (or much more often for video) the system measures the input's

amplitude and converts the value to a digital (stepped) one. In the second sketch, we have a four-bit system, so the sixteen (4^2) allowed values lie between 0 (0000) and 15 (1111) – the sketch shows the digitised signal by the dotted line.
3 simultaneous process in adc – the system converts each digital level to a binary value (as above, 0 → 0000 and 15 → 1111 – as in the table.
4 output from adc – series of pulses that stands for that set of binary values for the signal levels (the format shown is amplitude modulation)

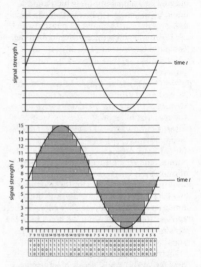

In practice, sampling is more frequent than a few kHz, and each output 'word' has more bits than four (eight may be enough for audio). The output signal from the adc goes into the data transfer channel concerned; if other signals join it in the same channel, it will need some form of multiplexing.

pump machine for one of these tasks:
1 raising a mass of liquid to a higher level (as from a well or for fire fighting)
2 adding energy to a mass of fluid (for instance, by raising its pressure)
3 moving a mass of fluid from one place to a second
4 removing the gas from a space to approach a vacuum (see below)

Note that 'fluid' here includes a fluidised solid in particle form.

The Archimedes screw was one of the first lift pumps, still in use, though first devised around 250 bce. Piston pumps (now the most common type) started to appear in the next few centuries; they developed steadily throughout the Middle Ages. The steam engine, to some the breakthrough of the industrial revolution, was aimed at powering pumps to extract water from mines. There are many modern designs, often of very high efficiency. See also *hydraulics* and *magnetohydrodynamics*. For moving water, the main types are displacement pumps (which trap a volume of liquid and force it into the output); centrifugal pumps (which move a liquid by the reverse action to that of a water mill or water turbine); and jet pumps (which use flowing high pressure water to drag the wanted water from a space).

Microscopic pumps are of value for cooling electronic chips. One such design works by electrohydrodynamics (so has no moving parts), has a volume of 3 μl, and can move 20 ml of fluid per minute. **Vacuum pump systems** have, in most cases, two stages. First the rotary backing pump exhausts the gas to the air and can work down to about a pascal; after this a set of one or more mercury filter pumps removes more gas, perhaps down to a micropascal.

punch-through old name for the avalanche breakdown of a semiconductor junction. See also *Zener*.

pupil the actual aperture of the eye, adjustable in size by the action of the iris (a ring muscle). The pupil appears black as the inner eye reflects very little light (its function is to absorb light, after all). The task of a **pupilometer** is to measure the size and shape of the pupil in detail. See also *entrance pupil* and *exit pupil*, the effective apertures of an optical instrument.

Purcell, Edward (1912–1997) physicist who shared the 1952 Nobel Prize with Bloch, for his discovery of nuclear magnetic resonance

pure note or **pure tone** monochromatic sound, of one frequency only (i.e. with no over-tones)

push-pull of a system whose output consists of two signals or vibrations in anti-phase. This needs two matched sub-circuits. In a **push-pull amplifier**, the two outputs come together to produce a single wave-form; the effect is to reduce distortion and inefficiency significantly.

pwr, or **PWR** pressurised water reactor, reactor using high pressure water as a coolant and moderator

pyr(o)- concerned with high temperature (or, sometimes, high thermal energy). The design of a **pyranometer** is to measure the total solar radiation reaching a surface; that of a **pyrheliometer** is to measure the total energy content of solar radiation.

Pyroelectricity is the appearance of opposite charges on the opposite faces of a sample (of an asymmetric crystal such as tourmaline) with temperature change; **pyrolysis** is the chemical breakdown of a substance at a high temperature.

A **pyrometer** is a thermometer for use over a range of high temperatures. The group includes the platinum resistance thermometer and some thermocouples; the **optical pyrometer** has the extra advantage that you can use it at a distance from the light source. There are several types of optical pyrometer, the disappearing filament design being the most common. Modern systems with silicon sensors and suitable electronics are certainly to replace all of these. **Pyrometry** is the science and applied science of high temperature measurement.

Pythagoras (*c.*570–*c.*500 bce) philosopher among whose group's ideas were the Sun-centred Universe, harmony (including the 'music of the spheres', mathematical harmony in the heavens), and the belief that mathematics can describe all aspects of the natural world. To some, Pythagoras was also the first known experimental scientist: there is some evidence that he carried out experiments on sound sources, using the monochord (sonometer, which he invented) as a standard of pitch.

While Euclid's writings (c 300 bce) give the earliest known statement of **Pythagoras' theorem**, there is some evidence that this was well known a long time before Pythagoras. The theorem is that the square of the length of the longest side of a right-angled triangle equals the sum of the squares of the lengths of the other two sides.

q symbol for generic coordinate

Q symbol for Q value/factor (quality), most senses

q **a)** old symbol for quantity of energy (W preferred), in particular of thermal energy; unit joule, J (at the time calorie, cal)
b) symbol for shear stress, unit: N m⁻²

Q **a)** symbol for electric charge
b) symbol for fluid flow rate, unit: cubic metre per second, m³ s⁻¹
c) symbol for partition function
d) old symbol for quantity of energy (W preferred), in particular of thermal energy; unit joule, J (at the time calorie, cal)
e) symbol for reactive power, unit: watt. W

q band a microwave band used in radar, with frequencies in the range 36–46 GHz

qcd the theory of quantum chromodynamics, a major modern theory of the strong nuclear force

qed the theory of quantum electrodynamics, the theory of how electrons interact with an electromagnetic field

Q factor or **Q value** **a)** measure of the energy balance of a vibrating system. Thus, for a resonant circuit, it is 2π times the ratio of energy stored to energy 'lost' (wasted); there are similar relations in a number of other contexts of energy transfer through systems that vibrate.
b) measure of the efficiency of a reactive element or circuit. It gives the effect of the resistance on the oscillation produced. There are a number of definitions for slightly different situations; not all are consistent with each other. In this case, the best way to define the Q-factor of an LCR circuit is

$$Q = 1/R \sqrt{(L/C)}$$

When the circuit is tuned (i.e. working at its resonant frequency), there is a peak in the

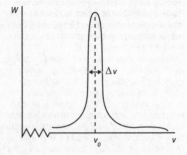

energy transfer graph. Q is the sharpness of the peak – the efficiency of the circuit; this is the ratio of the central frequency v_0 to the half-width Δv.
c) name for fusion energy gain factor, the ratio of the power output of a fusion system to the power needed to keep the plasma in a steady state
d) the binding energy difference between parent nuclide(s) and child nuclide(s)
e) in statistics, the false discovery rate when trying to assess significance or correlation
f) in health physics (radiation), a conversion factor between absorbed dose (unit: gray) and equivalent dose (i.e. effective dose, unit: sivert) that depends on the type of radiation in question

qft quantum field theory, the view that any particle is the quantum of some kind of field

Q_e/m_e the ratio of the electron's charge to its mass, more often (though less correctly) given as e/m

qso quasi-stellar object, old name for quasar

quadrature state of being out of phase by 90° – two such signals are 'in quadrature'

quadripole circuit or network with two input terminals and two output terminals only (other than power supply terminals). A **symmetric quadripole** circuit has the same electrical nature if one swaps input and output (i.e. reverses its action).

quadrupole pair of electric or magnetic dipoles that we can think of as a single object. That describes a special case: people also use the term for any system in which the distribution of charges or poles is more complex than that of a single dipole. In any such case, one can define a **quadrupole moment**; this adds to the system's monopole and dipole moments to produce the net effect observed. (However complex the system, there is rarely any need to go to a higher multipole moment to describe it.) **Quadrupole radiation** is the radiation from a vibrating quadrupole.

quality a) the feature of a sound (sometimes called timbre) that depends on its spectrum. (That is the pattern of the harmonics – i.e. single frequencies – present in the sound, and their relative strengths.) It is how the same note played on two different instruments will differ. See also *over-tone*.
b) the absence of distortion in the sound output of an audio system (with high fidelity meaning high quality). By extension, the term applies in the same way to the output of any electronic system.

quality factor
a) of a system: its Q factor or Q value (Q stands for quality here)
b) of an ionising radiation which presents a hazard: the factor by which one must multiply its energy content to obtain the dose

---

quanta

the plural of quantum, the smallest amount of a physical quantity that can exist. Thus, photons are quanta, of electromagnetic radiation.

quantisation
a) decision to assume that a given measure is quantised rather than being continuous as classical physics would have it
b) existence within a system of one or more measures which must be quantised rather than being continuous

quantised of a measure whose value is discrete rather than continuous: there can be only certain values in the allowed range rather than any values. The smallest allowed step in a given case is the quantum concerned. For instance, to explain the photoelectric effect, Einstein had to assume that the input radiation is quantised. It may be that this applies to all fundamental measures on a small enough scale.

quantising error loss of precision that follows from a process of analog-to-digital conversion. The input analog signal is continuous: it can have any value between zero and its maximum. On the other hand, the output digital signal can have only 256 different values in the case of an eight-bit word, for instance: from 0000 0000 to 1111 1111. Thus the conversion of input to output cannot be exact for many input values. An everyday example follows from asking for people's ages to the nearest year: this is a 'digital' measure that can rarely lead to an exact answer.

quantum (plural quanta) the smallest amount of a physical quantity that can exist. This concept is crucial to modern physics: classical physics assumes all measures are continuous rather than quantised. The first case known was the quantisation of electric charge: the realisation that one cannot have a charge smaller than that on the electron. Agreed, it seems that quarks have fractional charges – see quantum chromodynamics, below – but none has been observed (hence the theory of confinement), and, in any event, theory tells us that quarks always combine to give particles with integral charge.

The **quantum barrier** is the ultimate limit to miniaturisation, e.g. in making chips. Some effects of quantum theory (below) tell us that data transfers will not be consistent and reliable if they involve currents smaller than a certain size in channels separated by insulation thinner than a certain size. For instance, some electrons can 'tunnel' through an insulator if it is too thin, as in the Josephson effect. For **quantum bit**, see *qubit*. **Quantum chemistry** involves the use of quantum theory concepts and theories in the study of molecules and their interactions.

Perhaps most people in physics believe that the theory of **quantum chromodynamics** (qcd) is close to being a correct view of the nature of the strong nuclear force and of the particles (in particular, nucleons) that interact this way. It is the quantum theory of the fundamental interaction, with the gluon as mediator, and explains the results of many experiments by viewing the strong particles as made up of the fundamental fermions (spin $\frac{1}{2}$ particles) called quarks, plus gluons. We can say that the proton consists of two 'up' quarks (charge $\frac{2}{3}$) and one 'down' quark (charge $-\frac{1}{3}$); the neutron has two 'down' quarks and one 'up' quark. And so on ... however, the Δ^{++} strong particle has spin $\frac{3}{2}$ and charge +2 (as the name shows). Therefore, it must consist, on the evidence given above, of three 'up' quarks. However, that breaks the Heisenberg uncertainty principle and, in any event, causes problems with other properties of the Δ^{++}. The qcd theory therefore goes on to propose that we describe each type of quark on the basis of 'colour charge' (hence the name of the theory): with each one having quantum numbers for 'red', 'yellow' and 'blue'. (These names do not relate at all to what we know of as colour.) The theory goes much further than that, but remains able still to explain all current observations in this field. Compare it with *quantum electrodynamics* (below).

The fairly new field of **quantum computing** involves using such concepts as wave equation superposition and *quantum entanglement* (below) to handle and process data. Quantum computing systems – none yet exist – would use the qubit as the unit of data rather than the bit. A **quantum defect** is the difference between the total quantum number of some atoms, as given by spectral data, and the nearest integer. A **quantum discontinuity** is the discrete input or output of energy involved in a quantum jump (change) in a system. A **quantum dot** is a nano-crystal of semiconductor, 2–10 nm across, that contains 100–1000 electrons, holes or electron-hole pairs (excitons). Because of quantum effects, very important on this scale, the dot has a set of discrete energy levels – for that reason, some people call it an artificial atom. Unlike an atom, dots are easy to work with – to study by spectroscopy, for instance – and there are hopes to use them in nano-circuits and for data storage (see *qubit*). The **quantum efficiency** of a photon-induced change (for instance, in photochemistry or the photoelectric effect) is the number of events that follow the absorption of a single photon (quantum of radiation).

Quantum electrodynamics (qed) is the fundamental theory of electromagnetic interactions; it uses the photon as mediator to try to explain the electromagnetic exchange force. It appears to have full success in being able to explain the behaviour of photons, electrons and muons (these three being the quanta of electro-magnetic fields); it also seems to be the basis of all forces between atoms and molecules, and of electromagnetic fields on all scales from nuclear to astronomical. All the same, the trend in the physics in recent decades has been to combine (unify) this theory with that of the weak nuclear interaction. Electro-weak theories are the result. While none of these has yet met with great success, it remains the case that qed is an excellent structure on scales larger than around 10^{-18} m. Quantum electrodynamics has a long history: Dirac, Heisenberg, and Pauli started work on the quantisation of Maxwell's equations in the late 1920s. The need for someone to do this was the acceptance of the existence of the photon, the quantum of electromagnetic radiation: Maxwell's equations, being classical, assume that this radiation is continuous.

Quantum electronics concerns quantised aspects of changes in atoms and molecules and in the interactions of these with electromagnetic and pressure waves. The study of lasers and various resonance phenomena needs such an approach.

The phenomenon of **quantum entanglement** appears when the quantum states of two quite separate systems (and that includes chunks of 'empty space') are not separate – they are entangled, i.e., linked, in some way. Changing the state of one will then instantly change the state of the other. However, this does not mean we can now transfer information in the normal sense through space faster than light; even so, there are a number of major uses of the process on the horizon – such as developments in quantum computing. This last is because entanglement has also been observed in the case of a string of qubits, the units of data in this kind of computing.

A **quantum field theory** (qft) views a field as quantised into an assembly of particles; the total energy, charge, and so on of the field is the sum of those of the particles. From the first such theories in the mid-1920s, it took twenty years to reach the first one to have any success – quantum electrodynamics. Note that such a theory also views any particle as the quantum of some field.

A rare term for the model that uses the W and Z bosons as mediators to try to explain the weak nuclear exchange force is **quantum flavo(u)rdynamics**. A **quantum fluid** is
a) a group of valence electrons moving together when in the fermionic condensate phase;
b) such a group of electrons at very low temperature and very high pressure;
c) any superfluid.

A field theory of **quantum gravity** has been an aim for several decades, with quantum wormholes a major modern aspect. If all space contains a high density of quantum wormholes, as some theories predict, these objects would play an important part in any theory of quantum gravity. This is because they would modify quantum mechanics and the behaviour of matter. Quantum gravity covers all post-Einstein models of gravitation – they involve either exchange of gravitons or the quantisation of Einstein's curved space–time.

Klaus von Klitzing gained a Nobel Prize for physics in 1965 for his discovery of the **quantum Hall effect**. Like the Josephson effect, this results from the quantised behaviour and energy of electrons in electric and magnetic fields. The effect appears when electron flow is essentially two-dimensional in a very thin semiconductor film at a temperature near o K. It leads to a set of extremely precisely constant ratios of Hall pd V_H to current I:

$$V_H/I = (1/n)(h/Q_e^2) \qquad n = 1, 2, 3, \ \ldots$$

Here h is the Planck constant, and Q_e is the electron charge.

The process allows high accuracy resistance standards (as the ratio V/I is a resistance). It also lets us obtain high accuracy measures of h/Qe^2; when combined with the high accuracy values of $h/2Q_e$ got from the Josephson effect, these led to much more precise values than before of these two fundamental constants.

In a continuous x-ray spectrum, the **quantum limit** is either
a) the shortest wavelength in the spectrum, or
b) the energy associated with that wavelength.

However, people use the term quantum limit very widely to mean some small scale limit to some kind of work that may (or may not) relate to quantum noise or the uncertainty principle. In other words, it if often a phrase without real meaning, just as are **quantum jump** and **quantum leap**.

Quantum mechanics – a subset of quantum field theory (above) and also called **quantum physics** – is the mathematical expression of the quantum theory (below). In essence, it is a statistical theory that deals with the probabilities of events and energies in almost any area of classical physics at an atomic scale or smaller. Matrix and wave mechanics are both equivalent mathematical structures. **Relativistic quantum mechanics** is the extension by Dirac to include the effects of relativity; as we can view quantum theory and relativity as the two main planks of modern physics, this theory is its true basis. So-called **quantum microscopes** depend on the quantum tunnel effect. An important type is the photon scanning tunnel microscope; this allows optical imaging of non-conducting (e.g. living) objects with resolutions below the wavelength of light.

Quantum theory (below) and quantum mechanics (above) let us describe any atomic or molecular system in terms of a set of **quantum numbers**; each number can take half-integral and/or integral values (sometimes including zero). The unit in most cases is h, the Planck constant h divided by 2π. Each number gives the value of a conserved quantity in the system. Also, each number relates to an eigenfunction of the Schrödinger wave equation for the system. The main quantum numbers are as follows.

1 *angular quantum number*: for any angular momentum (e.g. that of a spin or orbit), the same as the orbital quantum number (7 below)

2 *azimuthal quantum number*: same as orbital angular momentum quantum number

3 *inner quantum number*: for the total angular momentum of a single atomic electron (symbol j) or for that of all the atomic electrons in the system (symbol J)

4 *internal quantum numbers*: those used to describe the hadrons (mesons and baryons: strong particles), in particular their iso-spin, strangeness and charm

5 *magnetic quantum number*: for the component of any angular

momentum in a magnetic field, symbol m; m has the values $-I, -I + 1 \ldots 0 \ldots I - 1, I$ (I is the orbital quantum number)

6 *molecular quantum number*: the molecular equivalent of an atom's orbital angular momentum quantum number, symbol λ; in this case, σ, π, δ ... stand for the electrons with molecular quantum numbers 1, 2, 3 ... (by analogy with the corresponding s, p, d ... of atomic electrons)

7 *orbital (angular momentum) quantum number*: symbol l, for the orbital angular momentum of an atomic electron, or the rotational angular momentum of a simple diatomic molecule; $I = 0, 1, 2 \ldots n - 1$, with those values matching the s, p, d ... notation

8 *principal quantum number*: symbol n, for the energy levels (shells) of an atomic electron system, where n depends only on distance from the nucleus

9 *rotational quantum number*: r, that for the rotation of a molecule

10 *spin quantum number*: s, for the intrinsic (spin) angular momentum of a simple or complex particle (such as a nucleon or nucleus); its values are $-\frac{1}{2}$ or $+\frac{1}{2}$. Note that in quantum physics, 'spin' does not mean rotation

11 *total quantum number*: rare name for the principal quantum number (8 above).

We can also express other quantised measures of these physical systems in terms of quantum numbers. Such measures include charge (as noted above), baryon number, lepton number, and the various types of parity. All are conserved in certain interactions.

We can then go on to say that we can fully describe the **quantum state** of any atomic and molecular system in terms of its set of quantum number values. Thus, the hydrogen atom in the ground state has one electron in the K shell; the quantum numbers of this are: $n = 1$, $l = 0$, $m = 0$, $s = 0$.

Quantum statistics in turn goes on to explore the distribution of the elementary particles of a system among the various quantum states (quantised energy levels, for instance). There are two main types:

1 Fermi–Dirac statistics, that applies to fermions (particles of half integral angular momentum); we must therefore use Pauli's exclusion principle. The number n_S of particles of a given type in a state S of energy W_S at temperature T is

$$n_S = 1/(\exp(W_S/k\,T) + 1)$$

(k is the Boltzmann constant.)

2 Bose–Einstein statistics – for bosons (particles with integral spin), where the exclusion principle does not apply. Here we have

$$n_S = 1/(\exp(W_S/k\,T) - 1)$$

The quantum theory

The quantum theory grew from the start of the twentieth century in order to explain a number of phenomena which classical physics could not handle. First, in 1900, Planck devised the concept of quantisation; as a result, he had great success in describing in mathematical terms the distribution of black body radiation with wavelength. He had to assume that when electromagnetic radiation appears from some source, there are energy changes in the source given by $W = h\,v$. Here h is a constant we now call the Planck constant, while v is the frequency of the radiation output.

Next, in 1905, Einstein applied the concept to photo-electricity, again with great success; in this case, it is the absorption of electromagnetic radiation that is quantised. It therefore follows, people assumed, that if electromagnetic radiation is quantised when it appears and quantised when it disappears, it is quantised all the time: electromagnetic radiation thus has both wave and particle properties.

The next step was Rutherford's discovery (1911) that the atom consists of a small but very massive positive core with a group of much less massive electrons around it. Classical physics predicts that accelerating charges (as would be electrons in orbit) would lose energy by the emission of electromagnetic radiation. Clearly, this does not happen (as atoms are stable): so in 1913 Bohr proposed the quantisation of the orbits, with radiation input or output only when there are jumps between orbits (excitation and de-excitation) and jumps between orbits and outside the atom (ionisation and neutralisation). Once more, the quantum picture had great success, though it was not able fully to explain the observed details of the spectra that arise from these jumps. Quantum mechanics arose in the 1920s to provide the more detailed mathematical basis needed for this – until then, there was no actual justification for quantum theories: they simply explained the facts.

Indeed, even now, a hundred years later, it is not clear

what the physical reality of the quantum theory and its concepts is; however, it may be possible to explain much of the theory on the basis of random but consistent interactions between particles and the zero-point fields they generate at all times.

Other major stages in the development of quantum theory are:

1920   The Stern–Gerlach experiment shows the quantisation of particle spin.

1924   de Broglie: quantum theory of particles based on their wave nature.

1925   Heisenberg: matrix mechanics; Schrödinger: wave mechanics. Schrödinger later shows that these are the same in essence, both being views of quantum mechanics.

1927   Heisenberg: the uncertainty principle
        Bohr and Heisenberg work on the 'Copenhagen interpretation' of quantum mechanics.
        The Davisson–Germer experiment shows electron diffraction.
        Dirac starts to unify quantum mechanics and special relativity.
        Quantum chemistry starts, with the quantisation of the covalent bond in H_2.
        Attempts start to devise quantum field theories, by Dirac and Pauli in particular.

1948   The qed theory is fully developed, after twenty years.

1963   The qcd theory starts, with quarks proposed to explain the eightfold way model of hadrons.

1971   Electro-weak theory confirmed, having come from several points independently over eight years.

The **quantum voltage** of a quantum of energy is the voltage through which one must accelerate an electron in order to give it that energy. A **quantum well** is a potential well with only two dimensions: it traps particles so they can move in only two dimensions. Quantum physics still applies, so that, in particular, the well has a set of discrete energy levels. The same is true of a **quantum wire** – a linear quantum dot, therefore one that confines the charge carriers to a single dimension. For **quantum wormhole**, see *wormhole*, this being a link between two different points in space–time, either on a scale of metres or far smaller than a nucleus. Both types are theoretical objects, with, as yet, no observational proof.

The **quantum Zeno effect** is an almost paradoxical result of the behaviour of a particle's wave function. Each time the particle is observed (for instance, in a detector), the wave function collapses back into the form of the newly created particle. As a result, however short the particle's half-life period, continuous observation of it means it will never decay. See *Zeno effect*.

quantum encryption a range of techniques for secure information transfer that depend on quantum theory, in particular entanglement and uncertainty. One aspect is the transfer of a key, without which it would not be possible to decode a coded message. The record distance (2006) for sending a quantum key is over 180 km. A second aspect depends on the uncertainty principle to detect whether someone has intercepted a signal.

quark (in English, pronounce this to rhyme with squawk) in current theory, the most fundamental massive subatomic particle, making up all hadrons (nucleons and mesons); according to the standard model of matter, the only other fundamental particle is the less massive lepton. Only quarks act through all four of the fundamental forces; their other main property is confinement (only the top quark doesn't show this because of its very short lifetime).

As with other particles, we use quantum numbers to describe and define quarks. There is much the same list as this for anti-quarks.

a) All quarks have a baryon number B of value $\frac{1}{3}$.

b) All quarks have a lepton number L of 0.

c) All quarks have an electric charge number Q, either $+\frac{2}{3}$ (called up-type quarks) or $-\frac{1}{3}$ (down-type).

d) All quarks have a weak iso-spin I, of value either $+\frac{1}{2}$ (for up-type quarks) or $-\frac{1}{2}$ (down-type).

e) Each pair of quarks with iso-spin $+\frac{1}{2}$ and $-\frac{1}{2}$ forms a 'generation'.

f) There are three generations of quarks (as is the case with leptons, as, indeed, the standard model of matter expects); therefore, there are six quarks, with these flavours:

up-type – up, charm, and top/truth (in order of mass)

down-type – down, strange, and bottom/beauty (ditto)

g) Each flavour comes in three 'colours' (red, yellow and blue).

See also *quantum chromodynamics* (qcd) and *unitary symmetry*, the theories that build on all this, and the *J/psi* (a particle). According to those two theories, nucleons consist of three quarks bound together, while the structure of a meson is of a quark and an anti-quark bound together. See *gluon* for the binding particle.

No free quark has yet been found for sure (though some experiments of the last decade imply they produce top/anti-top pairs or lone top quarks). As a result, many people believe they cannot exist alone in nature (that being the property of confinement). All the same, there is strong evidence for all those mentioned other than the top quark, and particle pairing theories imply that that too must exist. Theory and practice also imply that the top quark has a larger mass than the bottom quark, which is the most massive of the five others (at rather over 4 GeV/c^2, 4.5 times the mass of the proton). The top quark's mass is around 170 GeV/c^2; this should soon make it more accessible to accelerators – on the other hand, it has a very short life-time.

The **quark model** was first proposed in 1964, independently by Gell-Mann and Zweig; it was Gell-Mann who gave the particles the name we now use. It comes from a passage in *Finnegan's Wake*. For **quark number**, see *baryon number*. The strong nuclear force is what binds quarks into nucleons (as well as nucleons into nuclei). It is not yet known why quarks seem never to exist on their own; nor is it known why the quarks in a nucleon have combined rest masses so much less than a nucleon mass. The FAIR system being built at Darmstadt will explore these questions, by firing anti-proton beams into hydrogen (proton targets) at the correct energy to produce charm quarks. See also the *CKM matrix*, a 3 × 3 matrix that shows the nine possible weak interaction quark decays. And see http://pdg.lbl.gov/quarkdance/ for what quarks get up to in their time off.

quarkonium a meson with no (net) flavour that consists of a quark and its anti-quark. One type is **charmonium** – the J/ω is an example; in each of these cases the quarks are charm quarks.

quarter-wave plate (or **quarter-wave retardation plate**) thin slice of a birefringent crystal, often quartz. This is cut parallel to the optic axis and of such a thickness that the ordinary and extraordinary rays leave with a quarter-wave (i.e. 90°) phase difference. (This happens because those rays travel at different speeds through the crystal) See also *quartz wedge*, a variable retardation plate.

quartet set of four related objects, for instance set of four close spectral lines from the same source

quartz common form of silica (silicon dioxide, SiO_2), transparent when pure ('rock crystal'), and forming hexagonal prisms as it grows. It is the main component of sand and other minerals, and forms many gem stones. Quartz is birefringent, so has some uses in polarisation systems. Also it is piezoelectric and electrostrictive, and is therefore the basis of the **quartz clock** (and watch) systems as well as of many other types of precise oscillator. Here, an electronic oscillation of around 100 kHz makes the crystal vibrate at its resonant frequency. Feedback allows very high precision of the system's output frequency; this system divides to give seconds.

The **quartz–iodine lamp** is a tungsten filament lamp with a quartz bulb and an atmosphere of iodine vapour. It works effectively at a high temperature, and runs at a lower voltage than other filament lamps (so its life time is longer without loss of brightness). For **quartz oscillator**, see *quartz clock*, above. A **quartz wedge** is a retardation plate which one can slide across a light beam to vary the phase difference introduced.

quasar from quasi-stellar (radio) object (once called qso), an astronomical point source of intense radiation, often in the radio regions.

Many tens of thousands of quasars are now known (the first being discovered in the mid-1960s); most workers believe they are the highly active cores of new galaxies of stars. Quasars are the most radiant objects known (perhaps 10^{40} W – i.e. a greater output than our galaxy). The output is such that we can observe quasars at great distances: indeed, most have high red shift values (above 4, the highest by 2006 being 6.4). Thus quasars can act as markers for the distant (i.e. early) Universe, the 6.4 quasar being at least 13 thousand million light years distant, and therefore formed not much more than 1000 million years after the big bang.

quasi-crystal type of diffraction grating with no periodicity in its structure (whereas a crystal – which diffracts X-rays – has a periodic, repeating, structure). Metal films with holes in a 'random pattern' are quasi-crystals that diffract light in useful ways.

quasi-particle same as *virtual particle*

qubit (or quantum bit) pronounced cubit, a quantum dot with only two possible states, the actual state being a superposition of these. The probabilities for the two states sum to 1; measuring the state will collapse the qubit to one or the other. For use in a quantum computer, one must therefore measure the state of each qubit many times to obtain the distribution of the results.

quenching a) use of some method to prevent some intrusive process, or to cut off such a process quickly when started. In a Geiger tube, for instance, a halogen gas acts as a **quenching agent**; it stops the discharge formed by an ionising particle very quickly: then the tube is receptive once again to further radiation.
b) plunging a hot metal sample into cold water (or other liquid) to cool it quickly and thus fix its crystalline structure

quiescent current the current in an active circuit when the input is zero

Quincke's tube gas in glass resonance tube of variable length, used, for instance, to measure the speed of sound in the enclosed gas

quintal metric unit of mass, 100 kg

r a) symbol for roentgen, unit of ionising radiation intensity

b) symbol for right (as opposed to left, often as a subscript)

R a) symbol for ohm used by some people without access to Ω

b) symbol for Réaumur, everyday temperature scale used still in some parts of Europe

c) circuit symbol label for resistor

r a) symbol for correlation coefficient

b symbol for cylindrical or spherical or radial polar coordinate

c) symbol for molecular position

d) symbol for nuclear radius

e) symbol for position vector

f) symbol for radius, unit: metre, m

g) symbol for radius vector

h) symbol for electric source resistance (internal resistance), unit: ohm, Ω

R a) symbol for Hall coefficient (as R_H)

b) symbol for molar gas constant, about 8.3 joule per kelvin per mole, J K^{-1} mol^{-1}

c) symbol for radiance, unit: watt, W (L preferred)

d) symbol for range, unit: metre, m

e) symbol for the reactivity of a nuclear reactor

f) symbol for reflectance, unit: percent or decibel, dB

g) symbol for reluctance, unit: ampere turns per weber, A t Wb^{-1} (= turns per henry, t H^{-1})

h) symbol for resistance, unit: ohm, A

i) symbol for Reynolds number (as R_e)

j) symbol for Rydberg constant

k) symbol for universal gas constant (Boltzmann constant), about 1.4×10^{-23} joule per kelvin, J K^{-1}

ρ a) symbol for density, unit: kilogram per cubic metre, kg m^{-3}

b) symbol for radial polar coordinate

c) symbol for reflectance, unit: percent or decibel, dB (R preferred)

d) symbol for resistivity, unit: ohm metre, Ω m

Rabi, Isidor (1898–1988) physicist whose research on the magnetic properties of atoms and molecules led to the maser and the atomic clock: and to his 1944 Nobel Prize

rad a) symbol for the radian, unit of angle (also ^{c})

b) (symbol sometimes rd) non-SI unit of absorbed radiation dose, 10^{-2} joule per kilogram of body tissue

radar band of microwave radio waves in the spectrum of electromagnetic radiation with a length range of 100–10 mm (frequency range 10^9–10^{10} Hz), both sets of figures being approximate. The name comes from radio detection and ranging, the use of pulses of such waves to scan the environment; detection of the pulses reflected by objects gives their direction and range.

German scientists explored the concept from 1904 for use with shipping, but used too high a frequency for sky scanning. The first patents on the 'use of television with aircraft' came from John Logie Baird in 1926, major developments following in Britain a decade later (Robert Watson-Watt, physicist, 1892–1973).

While humans do not seem to be able to detect radar microwaves directly, it may be that the low frequency hum some people hear when near a powerful radar transmitter is a real result of the radiation's interacting with the hair cells of the cochlea (inner ear: see *resonance theory of hearing*). There is also uncertainty as to the degree of hazard associated with modern powerful radars: the very high electric fields near them may have physiological effects.

Radar astronomy involves the use of radar techniques to measure the distance, and map the surface, of nearby astronomical objects in the solar system – it is a branch of radio astronomy.

radial concerned with measures that depend on the distance from a point (as with radius). **Radial acceleration**, unit: metre per second per second, m s^{-2}, is the component of an object's acceleration along the radius of the curve in which it is moving. See also *angular acceleration*. **Radial symmetry** is symmetry about a point or single axis. **Radial velocity** is the component of a velocity along the line of sight of the observer (towards or away).

radian unit of angle, a supplementary SI unit (i.e. it is one that we do not need – as it has no dimension – but find useful). It is the angle between two radii of a circle which cut off the fraction $1/2\pi$ of the circumference. (The length of that arc of the circumference is then equal to each of those radii of the circle.) In other words, there are 2π radians in a circle (360°); 1 radian (symbol rad, or sometimes ^{c}) = 57.3°.

radiance or **radiancy** L (or R) the radiant intensity (energy content) of a source of radiation per unit area (taken at a right angle to the radiation). The unit is the watt per steradian per square metre, W sr^{-1} m^{-2}. Usage is not consistent, however: some people take it as the radiation power in total, and others the power per unit solid angle (in watts, or in watts per steradian respectively).

radiant radiated, or associated with radiation (often of light). The **radiant absorptance** (of a surface is the total power of a radiation absorbed by a surface as a fraction of that reaching the surface; **radiant absorptivity** is the radiant absorptance of unit area (per square metre, m^{-2}). The **radiant efficiency** η of a source is the ratio of the power radiated to the power input (unit: none, or percent). The source's **radiant emittance**, now called **radiant exitance**, M, is the radiated power output per unit area of surface; the unit is the watt per square metre. At the receiving end, the **radiant exposure** H is the radiant energy received by unit area of a surface; unit: joule per square metre. In either context (i.e. emission or absorption), and in transfer, **radiant flux** is an old term for *radiant power* (below): the energy involved per second. Also, **radiant heat** is an old name for infrared radiation (thermal radiation).

The **radiant intensity** I of a source or of a radiation is the power per unit solid angle, in watts per steradian (W sr^{-1}); some people also take it per unit area (see radiance) The **radiant power** P of a radiation (in production, in transfer, or on reaching a surface) is the energy involved per second; the unit is the watt. The **radiant reflectance** of a surface is the power reflected as a fraction of the radiant power reaching the surface, per unit area in the case of **radiant reflectivity**. Its **radiant transmittance** is the power that passes through the surface as a fraction of that reaching it.

radiation a) the transfer of thermal energy by a wave process (as compared with convection, conduction and evaporation). Here the waves are electromagnetic radiation, mainly of infrared.
b) energy output as moving matter particles or photons by a radioactive decay
c) ionising radiation – a biological hazard as ionisation can damage living cells, in particular their growth and reproduction. Therefore, **radiation absorption** is the effect in the body of ionising radiation. See *absorption* sense b) and the *gray*, the unit of dose of absorbed radiation energy. **Radiation protection** includes all techniques designed to avoid or reduce radiation damage in living tissues or in other materials. Amongst those techniques are those that limit exposure as well as those designed to treat overexposure. The symptoms of **radiation sickness** in people – the result of overexposure to ionising radiation – include listlessness, nausea, and vomiting some hours after exposure, and hair loss later. Anemia, internal bleeding and various malignant growths may develop and lead to death. The most common cause of radiation sickness is the use of radiation for therapy (see *radiotherapy*); see also *Chernobyl*.
d) in general, the emission of energy from a source, and the energy itself while in transit. The radiation may be a wave form or appear as moving particles: or it may have features of both of these in different contexts (as, in particular, is the case with the electromagnetic wave/particles, or quanta, called photons).

There are various **radiation belts**, zones of high energy charged particles very high in the Earth's atmosphere. See the *van Allen belts* for details of the two main ones. **Radiation broadening** is the most basic cause of the broadening of spectral lines from complete monochromaticity (their increased width from perfect sharpness, in other words). Its cause is the less than perfect sharpness of the energy levels involved, because of the finite life times of states other than the ground state. The concern of **radiation chemistry** is the various effects of ionising radiation on chemical structures and reactions; this includes photochemistry and photography at all wavelengths, as well as radiobiology – radiation induced mutations and other effects on living matter. Note that the subject is not the same as radiochemistry. For the **radiation constants** that appear in the **radiation formula**, see the *Planck radiation law*.

The **radiation impedance** of a surface as it vibrates in a medium is the fraction of the absorbed power that the surface radiates as sound (pressure) waves. The **radiation length** of an ionising particle in a medium is the mean length of the path over which its energy will fall to $1/e$ of its value as a result of collisions. The **radiation loss** of such a particle is its transfer of energy to 'Bremsstrahlung' (braking radiation). A **radiation monitor** is a detector that will cause an alarm if the radiation level near it goes above a set value. The concern of **radiation physics** is the interaction between ionising radiation and matter.

Radiation pressure is the pressure that radiation reaching a surface will exert. The cause is the momentum carried by the radiation. If the angle of incidence is zero (the radiation arrives along the normal), the value equals the radiation's energy density. At the Earth's surface, electromagnetic radiation pressure is negligible (around 10^{-5} Pa); in space, on the other hand, it causes the gas and dust 'tails' of a comet (and is a form of space craft drive of potentially great value), while in stars it prevents the outer layers from imploding under gravity. Sound waves too can exert a significant radiation pressure, as night club dancers know; all microphones rely on that.

A **radiation pyrometer** is a thermometer for working with a high temperature surface; it measures the radiation output of the surface in some way. See optical pyrometer for an important example. **Radiation resistance**, a measure of the sound wave power output by a surface, is the same as *radiation impedance* (above). The various **radiation units** are becquerel, curie, dose, gray, rad, rem, roentgen and sievert.

radiation-less annihilation See *annihilation*.

radiative associated with the output of radiation. Thus, **radiative capture** is the capture by a nucleus of a particle from outside, with the immediate escape of a photon. An example is electron capture, which is a type of beta radiation. A **radiative collision** is an interaction between two charged particles in which some kinetic energy transfers to 'Bremsstrahlung' (braking radiation); the most important case is the interaction between a high energy electron and a nucleus. Within a nucleus, **radiative transitions** are decays down the energy level ladder with the release of gamma radiation.

radical a group of tightly bound atoms that can transfer as whole from molecule to molecule during a chemical reaction, or exist on its own in ion form

radio

a) a broad range of low energy radiations within the electro-magnetic spectrum, widely used for communications and including the microwave (radar) regions. For convenience, we divide the radio region into bands.

Band 1 is the infra-sonic region, with 2 and 3 for sonic frequency waves, and 4 for ultra-sonics. We use band 6 for amplitude modulated (am) radio signals, and 7 is for frequency modulated (fm) signals and tv. The higher bands are of microwaves: bands 10 and 11 are the radar bands, with sub-bands p, l, s, x, q, v, and w. This brings us to 100 GHz, where the radio spectrum merges into the infrared region. Clearly, therefore, the term **radio frequency** has little meaning; all the same, people use it to refer to electromagnetic waves in the range 10^6–10^{11} Hz – so-called **radio waves**. (For some reason, some people call them air waves, which means nothing – unless it means sound waves.)

b) the use of radio waves to communicate information and data. All radio transmissions involve transmit and receive aerials (US: antennas). These are conductors held parallel to the plane of polarisation of the waves; a straight aerial works best if its length is about half the length of the waves concerned.

In the transmitter, a piezoelectric oscillator generates a steady radio frequency carrier wave. After amplification, the system modulates it (see *modulation*) with the signal that carries the information or data concerned. The modulated carrier is amplified to a high enough power for radiation from the aerial. The output radio wave weakens quickly: the inverse square law applies in cases of broadcast radio. Thus, the distant receive aerial can react to only a tiny fraction of the output power (along with other signals and also background noise).

At the receive end, a circuit tuned to the desired carrier frequency outputs only that frequency to the demodulation ('detection') and amplifier systems; see *tuned circuit*. There may need to be a number of amplification stages before the signal that carries the information or data is strong enough for use at the speaker. See also *beat oscillator* and *heterodyne*.

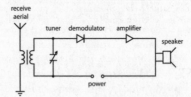

At the receive end, sub-circuits convert the weak radio signal into loud sound at the speaker.

People use hf (short wave) radio to communicate over long distances as it is convenient (despite some distortion). A transmission can bounce a number of times between the ionosphere and the Earth's surface, so can follow the curve of the Earth. Transmitters and receivers are cheap, small and light and use little power; they can work with speech, digital data and still video, and are in wide use in transport. On the other hand, the ionosphere is far from uniform in time and space (including height), so short wave radio transfers cannot work 24 h a day between two given points at a given frequency.

At frequencies below about 30 MHz, the ionosphere tends to absorb waves rather than reflecting them; at higher frequencies, it tends to transmit them: thus using the ionosphere for long distance radio transfers has a very small window. The window changes quite a lot each day (which is why short-wave stations transmit at very different frequencies at different times). Communication at higher frequencies still (which includes tv) is line of sight only therefore: communications satellites are the only method for long distance transfer of such signals through air or space.

The main use of elf (extremely low frequency, i.e. very long wave) radio was, in the past, underwater communication: water is transparent to such waves (and to very few others in the electromagnetic spectrum). On the other hand, such a long wavelength means very large aerials; the communication rate is very low too: only a few bits per second.

Maxwell was the first to predict (in the mid-1860s) the possibility of radio waves, as a result of the work that led to the Maxwell equations for electromagnetic radiation. Hertz reached success in generating and receiving them in 1887. In 1894, Lodge sent the first signal by radio; voice transfer followed in 1900; and Marconi made the first trans-Atlantic broadcast the next year. Amplification remained a very severe problem until the triode valve (an electron tube able to work as an oscillator or as an amplifier) was invented in 1906; that allowed public broadcasts to start after the first world war. The next major development in the industry was the invention of the transistor in 1948: this was a highly efficient and cheap replacement for the triode.

A **radio altimeter** is a device used in aircraft to find height above ground or sea. It is much like a down pointing radar. **Radio astronomy** is the study of the radiations in the radio spectrum that come to us (including by reflection radar astronomy) from elsewhere in the Universe. The main instruments used are the various forms of *radio telescope* (below). In 1932, Karl Jansky traced to the centre of the galaxy the interference (noise) in a telephone system on which he was working. The first serious radio astronomer was Grote Reber – he worked in his back garden from about 1937 with a 9.5 m radio telescope. After the second world war, people worked with larger and larger instruments and were able to build up pictures of the radio sky in more and more detail; radio astronomy soon became big science. (The first, 66 m, Jodrell Bank instrument started work in 1947 – and it was very big science for those days.) The field allows the effective study of very distant galaxies on the one hand, and the remains of supernovas on the other; see also the *microwave background radiation, pulsar* and *quasar.*

Radio communication involves the use for telecommunication of radio waves broadcast or beamed (narrow cast) through a medium, i.e., with no wires or other wave guides. Though, as noted above, the term **radio frequency** is vague, to say the least, it is in wide use. For instance, **radio frequency heating** is a form of induction heating, a process of raising the temperature of metals by induced eddy currents from an alternating field, with alternating frequencies above around 25 kHz. Some cookers work on this basis, and there are many uses in industry.

The **radio interferometer** is an important type of radio telescope. It involves linking the output signals of two or more radio telescope dishes and exploring the interference patterns the system detects. This allows very high accuracy in fixing the position of a **radio source**, a source in the sky of waves in the radio region. Such sources include **radio stars**: point sources of radiation (some of which are very distant galaxies). The Jodrell bank instruments are part of a national array of seven and have therefore a much higher resolution (at 5 GHz as good as the Hubble telescope at optical frequencies).

The standard **radio telescope** is a kind of receive aerial, a parabolic mirror for radio waves, made of sheet metal, or (on the larger scales) a metal grid; the best instruments have a surface accurate within a fraction of a millimetre. This works much as does the optical mirror of a reflecting light telescope, but has a radio energy detector as the actual sensor. As radio wavelengths are much longer than optical wavelengths, a radio telescope must be far bigger than an optical reflector to achieve the same resolution (level of detail). On the other hand, even with poor resolution, people have learned a great deal from the study of the radio sky. All the same, techniques such as *radio interferometry* (above) improve the precision of the output greatly. Another such is aperture synthesis, where the system combines signals gathered from radio telescopes at the ends of very long base lines to give much of the effect of a single telescope of that diameter. The largest base line available is the Earth's orbit round the Sun: by taking readings six months apart, one can to some extent simulate a radio telescope with a 300 million km dish. This action is much like parallax.

A major problem for radio astronomers is that the Earth's atmosphere absorbs many radio wavelengths: there are only a few **radio windows** through which we can observe. As happened in 1990 with optical astronomy, the use of radio telescopes in orbit (i.e. outside the atmosphere) by-passed the problem; they are small – a few tens of metres across – but use interferometry and aperture synthesis very well.

a) concerned with radiation, in particular that associated with radioactive decay. Thus **radioactivation analysis** is a technique of chemical analysis which involves irradiating the test substance and identifying and measuring the radioactive nuclides produced. The irradiation often uses neutrons (see *radioactive collision*, below). See *activation analysis*.

The nuclei of a **radioactive** substance are unstable, and, as a result, subject to random decay. See *radioactivity*, below. The **radioactive age** of a rock is the age as fixed by some technique of radioactive dating; the term is not common. A **radioactive chain** is the same as a *radioactive series* (below). **Radioactive collision** is an old name for the nuclear (n, γ) reaction: that in which a nucleus absorbs a neutron and loses energy in the form of a gamma photon. The gamma output is characteristic of the nucleus concerned; this therefore gives us a useful technique of analysis: see *activation analysis*. For **radioactive dating**, the use of knowledge of radioactive decays in, for instance, the remains of living organisms and rocks to find their age, see *dating*.

A **radioactive series** (or chain) is a sequence of nuclides, each formed by the decay of the previous one. All these series end in a stable nuclide. There are four such series, whose members have mass numbers given by

$4n$ the thorium series, from ^{232}Th to ^{208}Pb
$4n - 1$ the actinium series, from ^{235}U to ^{207}Pb
$4n - 2$ the uranium series, from ^{238}U to ^{206}Pb
$4n - 3$ the neptunium series, from ^{241}Pu via ^{237}Np to ^{209}Bi

In each series, most of the main decays involve the release of alpha particles (hence the factor of 4 above); beta- and gamma-decays do not change a nuclide's mass number. The output of alphas (helium nuclei) and the decay to stable lead or bismuth explains why many ores have a high concentration of lead and helium. Note that the neptunium series does not exist in nature: the nuclides near the top of it all have short half-lives. See *actinium series* for an illustration in detail.

A **radioactive tracer** is a radioactive nuclide used as a tracer to explore the flows of fluids and molecules within living tissues, for instance.

Radioactivity is the random (apparently spontaneous) decay of an unstable nucleus, with the release of excess energy (see binding energy) as some form of radiation. The main types of process involved are as follows, given with their effects on the proton number Z and nucleon number A of the starting ('parent') nucleus.

• alpha decay: output of He++ (nuclei); Z falls by 2, A by 4.
• beta decay: output of electrons e; Z rises (e–) or falls (e+) by 1, A unchanged. A form of e^+ decay is electron capture, the electrons concerned being in the shells out side the nucleus – see, for instance, *K capture*.
• gamma decay: output of photons; Z and A unchanged.
• fission: nucleus 'splits'; output of neutrons as well as the two child nuclei.
• proton decay (rare): output of protons, Z and A fall by 1.
Radioactivity of any given nuclide takes place at a rate that does not vary with temperature, or the chemical or physical state of a sample: it depends only on concentration. However, in 2006 a team claimed to be able to reduce hugely the half-life of an alpha emitter by embedding it in a metal and cooling the metal to a few degrees kelvin. If true, this would be a great help for those trying to deal with nuclear waste. The report – http://physicsweb.org/article/news/10/7/13 – followed work on a form of cold fusion.

Artificial radioactivity is that observed in isotopes formed in, for instance, reactors (in other words, isotopes that do not exist in nature); these include all the isotopes of all the transuranic elements. **Natural radioactivity**, on the other hand, is the activity of isotopes which do exist in nature. The **radioactivity level** A of a source is the power output, defined in a suitable way for each type of source. The other measure of an active substance relates to this: it is how quickly the activity falls off, see half-life and the other measures noted there.

The concern of **radiobiology** is the effect on living cells, tissue, and organisms of ionising and other radiations. For **radiocarbon dating**, using ^{14}C as a method of finding the age of once living material, see dating. **Radiochemistry** involves the use of radioisotopes (i.e. radioactive species) for chemical research, in particular analysis. It includes the preparation and production of radioisotopes and compounds that include them, and is not the same as radiation chemistry. See also *tracer*. A **radiogenic** substance is one that results from radioactive decay (is generated by radioactivity, in other words); the term is not common.

A **radioisotope** of an element is an isotope that is radioactive. If the activity is great enough, the source rises in temperature; the **radioisotope thermoelectric generator** (rtg) puts this to use – in other words, it uses the thermal energy produced in the decay of a radioisotope. Thermocouples are the most common way to convert the rtg output to a more useful form. The sources are not nuclear reactors in the true sense, though people often call them by this name. While the rate of production of such energy falls off with time (because of half-life), an isotope power source can be cheap, compact, reliable, and long-lived. They are common in spacecraft and automatic equipment in remote places: for instance, the 1989–2003 Galileo mission to Jupiter involved two rtgs; each started with 8 kg of plutonium-238 set in ceramic (for safety) and produced 570 W at launch, falling by 0.6 W/month.

Radiometric dating is an old name for dating by radioactive means. A **radionuclide** is a nuclear species that is unstable, so decays by radioactivity.

b) concerned with radiation in general. A **radiobalance** allows one to measure the power of input radiation; the system – of which there is a number of designs – balances the energy absorbed against the Peltier temperature change in a thermocouple.

A **radiograph** is a shadow photograph made from the radiation (e.g. x-rays) that passes through an object; the radiation may be ionising or otherwise. **Radiographic stereometry** involves building up a three dimensional picture of the object from radiographs made at different angles; the modern extension of this is tomography (a form of scanning). The field concerned with all these is **radiography**; this is the applied science of making an image (on film or on screen) of the rays that pass through the structure of a semi-opaque sample – in industry as well as in medicine. On the other hand, **radiology** *is* a branch of medicine (though there are many other uses) – the general science and applied science of high energy ionising radiations, and their use for diagnosis and treatment (e.g. in this case, *radiotherapy*, below).

A **radiolucent** substance is fairly or highly transparent to a given radiation (compare with *radioopaque*, below). **Radioluminescence** is the emission of visible light from a radioactive sample; in this form of luminescence the energy involved comes from radioactive decay inside the sample itself. Tritium gas in a fluorescent tube has long been a common source, but modern solid state methods (binding the tritium with the phosphor) produce much brighter light. **Radiolysis** is the (chemical) breakdown of the molecules of a substance by ionising radiation; a major case is the breakdown of water (e.g. in spent nuclear fuel ponds) to produce hydrogen. The products may include other molecules (perhaps excited ones), ions or excited atoms. **Pulse radiolysis** uses very short pulses of very intense rays to allow the study of very fast reactions.

The function of a **radiometer** is to measure (or at least detect) electromagnetic radiation; there are many designs for the various regions of the spectrum. Some depend on the **radiometer effect**, the force on a surface in a low pressure gas that results from its uneven temperature rise in radiation. The **Crookes' radiometer** is in this group: a set of four vanes, each with one shiny surface and one blackened surface, on a low friction mounting in low pressure air. The set rotates as the dark faces, being warmer, cause incident gas particles to rebound with higher energies than those meeting the shiny faces.

A **radiomicrometer** (much better called microradiometer) is a highly sensitive radiometer. The current from the input energy at a thermocouple junction passes through a meter able to work with exceptionally low values. A **radio(-o)paque** substance blocks given input radiation so this cannot pass through a sample; compare with *radiolucent*, above. A **radiosensitive** living cell is unusually open to damage by ionising radiation. **Radiotherapy** involves the use in the treatment of disease of high energy radiation from a beam (e.g. output by an x-ray tube or particle accelerator), or from lower energy radiation from isotopes placed within the body. There are various techniques in each case, all designed to minimise the damage to healthy cells and tissue. **Radiotransparent** is much the same as *radiolucent* (above).

c) concerned with electromagnetic radiation in the radio region of the spectrum. A **radiogoniometer** is a type of radio direction finder, a device that gives the user information about the direction of the source of input radio waves. **Radiolocation** is an old name for radar techniques. A **radiosonde** is a small freely floating balloon used to carry instruments and a radio transmitter high into the atmosphere; the transmitter returns the data gathered by the instruments. A **radiospectroscope** is a system for the display of the spectrum of wavelengths (energies) of radio waves absorbed by an aerial.

radion the size of the fifth dimension in quantum field theories with a fifth dimension (e.g. the Kaluza-Klein theory) as a function of the normal four of space–time

radium Ra radioactive, alkaline earth metal in group II, period 7, with proton number 88, first found in pitchblende (a uranium ore) and isolated by Marie and Pierre Curie in 1898 and named after its high radioactivity. Relative density 5, melting and boiling temperatures 700 °C and 1140 °C; most common stable isotope ^{226}Ra, half-life 1600 years.

radius r **a)** straight line between the centre of a circle and its edge (circumference) **b)** the length of that line, half the diameter and $1/\pi$ the circumference of the circle. The **radius of curvature** of a surface, such as the surface of a 'spherical' lens or mirror, is the radius of the sphere of which the surface is part. The **radius of gyration** k of a rigid object (mass m) from a given axis is the distance from that axis a point mass m would be to have the same moment of inertia (I) as the object:

$$k = \sqrt{(I/m)}$$

The **radius vector** r of a curve in space is the vector that links the general point on the curve to the origin of a set of coordinates.

radix the number base ('root') of a number system. It is the number of distinct digits 0 ... n in the system, n being radix minus 1 (9 in the case of the denary system).

radon Rn radioactive inert gas, group 0, in period 6, with element number 86, discovered in 1899 by Ernest Rutherford as an 'emanation' (gas given off) of thorium and by Pierre and Marie Curie as an emanation of radium; their 'thoron' and 'radon' are in fact both isotopes of element 86. Melting and boiling temperatures –71 °C and –62 °C; most stable isotope (of twenty) ^{222}Rn, half-life 4 days. Radon escapes ground rocks like granite (e.g. in Derbyshire and Devon) and builds up in caves, mines, buildings and drinking water – this may cause thousands of lung cancer deaths a year in the EU. On the other hand, some people believe that small levels are of positive benefit to health: this is a type of hormesis.

rail gun an electromagnetic method for firing metallic projectiles at high speed, proposed as a means for sending objects into Moon orbit from the Moon's surface (for instance). One approach is to use a homo-polar generator to transfer several megajoules at, say, 750 kA to the two metal rails (out and return) in the barrel. The conducting projectile ('armature') on the rails experiences a motor force (see *linear motor*) able to accelerate it to several km s^{-1}.

rainbow coloured band across the sky, that may appear when one views rain or spray with one's back to the shining Sun. A ray of sunlight reaching a water drop can **a)** pass straight through (deviation 0°); **b)** be reflected straight back (deviation 180°); or **c)** be refracted, totally internally reflected once, and then escape, being refracted again. The sketch overleaf shows the possibilities. It shows that the (c) rays all deviate by at least 138°, but that a large proportion come out at close to that angle. That applies to red light; the angle for blue is 139°: so the bow has red on the outside and the spectrum to blue inside that. Outside the red edge, the sky is much darker than inside the bow, as much less light comes from those directions.

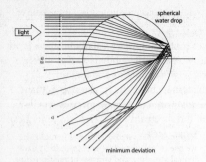

light

spherical
water drop

minimum deviation

All this applies to the primary bow. A secondary bow results from two internal reflections of (c) rays, the angle concerned here being 128°. The secondary bow therefore lies above the other. Also, that classical explanation does not tie up fully with experience: because of complex interference effects and the fact that rain drops are rarely spherical.

Rainwater, Leo (1917–1986) physicist whose work on the structure of non-symmetric nuclei and the liquid drop model led to his sharing the 1975 Nobel Prize

Raman, Chandrasekhara (1888–1970) scientist who became professor of physics at Calcutta university in 1917. He discovered the effect that bears his name and received a Nobel Prize for it in 1930. He left Calcutta a few years later and set up various excellent institutes and departments around the country. His main interest throughout has been wave physics, starting with (for instance) music and interference.

The **Raman effect** (or **Raman scattering**) follows this interest: Raman proposed it in 1922 and observed it in 1928. It is the change of frequency of monochromatic light scattered by molecular matter, the interaction of photons in an intense beam with optical phonons. The effect is weak (it involves perhaps 1 in 10^7 photons), so only the later advent of lasers made it of much interest: but then led to the important technique of **Raman spectroscopy**. This follows from the fact that the range of wavelengths in the scattered beam is characteristic of the medium concerned. The **hyper Raman effect** and the **inverse Raman effect** are two of the four non-natural variations of the effect that appear as the result of the very intense electric fields (at least 10^9 V m^{-1}) in a giant pulsed laser beam. The former case needs

much stronger fields than that – and when these are in place, high overtones of the input frequency appear. In the latter case – not really an inverse – a number of overtone lines appears as a result of the illumination of the sample by a high energy continuous beam at the same time. See also *stimulated Raman scattering*; here the new lines are under-tones.

ramjet engine See *jet engines*.

Ramsay, William (1852–1916) Chemist who discovered the noble gases in the 1890s and received the 1904 Nobel Prize for chemistry. He'd worked on this field to an extent with Lord Rayleigh, who received the 1904 Nobel Prize for physics for the discovery of argon.

Ramsey, Norman (1915–) physicist who shared the 1989 Nobel Prize for his work on the hydrogen maser and related atomic clocks

r & d research and development, the work needed to bring a new product to market

random without any plan or defined order, based on chance only. **Random noise** is the same as true noise, signals that are unpredictable in frequency and time and interfere with a wanted signal. A **random number** is a value produced entirely unpredictably. There are many random processes in physics, in particular at small scales. (Random numbers produced by IT systems are not random, though appear so in almost all cases.) A **random walk** is a view of transport phenomena such as diffusion; stochastic analysis (to use the statistical name) can give the most probable distance from a starting point after a number of moves in random directions. The best known case is the movement of a speck in a fluid due to Brownian motion.

range a) the distance W travelled by an ionising particle to the point where it no longer has enough energy for ionisation. The value depends on the type and energy of the particle, and the nature of the medium (in particular its stopping power). **Range straggling** is the variation in the range of a number of particles of the same type and energy in the same medium. In some cases, the effect is extremely small.

b) the distance over which a force that does not follow the inverse square law is significant. For the strong nuclear force, the range is around 10^{-15} m, much the same as the nuclear radius.

c) the distance R from its firing point at which a projectile reaches the ground
d) the distance from an observer of an observed object – radar is 'radio detection and ranging'. There are also various types of optical **range finder** (a device – based on sonar, radar, or a laser – for measuring the distance to an object), such as those used with cameras.

Rankine, William (1820–1872) engineer who (with Clausius and Kelvin) founded the science of thermodynamics, though his fields of research were many. The **Rankine balance** allows one to compare the susceptibilities of weakly diamagnetic and paramagnetic materials. The **Rankine cycle** is a theoretically ideal cycle of action for a steam engine, ideal in being reversible. It is much the same as the Carnot cycle, but has a separate boiler (steam generator) and condenser. The **Rankine temperature scale**, now rarely used, is the absolute scale based on the Fahrenheit degree. Absolute zero on this scale is at −459.671 °F.

Raoult, François-Marie (1830–1901) Chemist who did much new work on solutions. **Raoult's law** relates the vapour pressure p_s over a solution to that p_0 of solvent alone: p_s is the product of p_0 and the ratio of the number of moles of solute to the number of moles of solution. He also showed that making any 1% solution (by moles) would reduce the solvent's freezing temperature by 0.63 °C.

rare earth old name for the oxide ('earth') of one of sixteen elements found in certain ores. They are not particularly rare either. See *Ytterby*, source of a rock that contains seven of the sixteen.

rarefaction reducing the pressure of a gas, making it 'rarer'

ratio relationship between two numbers x and y, shown as x: y (or as a fraction x/y). See also *proportion*. The **force ratio** of a machine is the ratio of the output force ('load') to the input force ('effort'). The **turns ratio** of a voltage transformer is the ratio of the number of turns in the input coil to the number in the output coil. The **voltage ratio** of the transformer is the ratio of the output voltage (when there is no load) to the input voltage. For a perfect (i.e. lossless) single phase transformer, the two ratios have the same value.

rational a) able to be written as the ratio of two polynomials. This is the case with a

rational function – its formula is the ratio of two polynomials. Indeed, a polynomial is a rational function too.
b) able to be written as the ratio of two whole numbers, as is the case with each **rational number** (any positive or negative whole number or fraction, plus zero). There is an infinity of rational numbers, and an infinity of irrational numbers; together they form the set of real numbers.
c) reasonable or consistent with logic
Rational unit systems (such as the main one we use in physics – SI – see *Appendix 1*) have logical structure and consistent relationships. Thus in SI, relationships that involve circular or spherical symmetry include factors of 2π and 4π respectively.

ray very narrow beam of radiation, of light or one of its near neighbours. If we know how certain rays behave in a given optical system, we can build up a **ray diagram** or **ray sketch** (by the process of **ray tracing**) from which we can deduce the site and nature of the image of any object at any point. This is important in the analysis and design of optical systems. The main rays used in these diagrams are as follows; in all cases, the ray must travel very close to the system axis.
a) lens (taking account of whether this is converging or diverging):
- ray through the pole passes straight on (no deviation)
- ray through a focal point leaves the lens parallel to the axis
- ray parallel to the axis leaves in the direction of a focal point
- ray through one point 2F (which is twice the focal distance from the pole) leaves in the direction of the other point 2F
b) mirror (similarly):
- ray to the pole (centre) comes off at the same angle to the axis
- ray through the focal point leaves parallel to the axis
- ray parallel to the axis leaves in the direction of the focal point
- ray through the centre of curvature returns through the centre of curvature
c) prism: use the laws of refraction to find the paths of rays.
There are a number of ray sketches in this dictionary.

Rayleigh, Lord (John Strutt, 1842–1919) physicist whose work on the inert gases in air led to his Nobel Prize in 1904. The same year, William Ramsay gained the Nobel Prize in chemistry for the discovery of the inert gases in air; sources differ as to which of the two suggested the idea.

The **Rayleigh criterion** is for just resolving two points or spectral lines using an optical system: the central maximum of the image of one should fall on the first minimum of the image of the other. See also *Abbe criterion*. The **Rayleigh disc** gives a sensitive measure of fluid speed. The disc hangs in the stream; the torque on it varies with the fluid velocity (or its root mean square if the flow is alternating, as in the case of a sound wave).

The **Rayleigh–Jeans law** gave the spectral distribution of energy in black body radiation as a function of temperature. It had success only at long wavelengths; at short wavelengths, its predictions varied wildly from practice: the ultraviolet catastrophe. The Planck radiation law, based on quantisation, overcame the problem. The **Rayleigh law** is that for low levels of magnetisation, the energy transfer in a hysteresis cycle varies with the cube of the magnetic induction. The **Rayleigh scattering** of a radiation is that by objects which are small compared to its wavelength: the intensity of the scattered radiation varies with the fourth power of the wavelength. The concept explains the reds and blues of sky colour. In an earthquake or similar, some of the output energy transfer is by **Rayleigh waves**; these are surface waves that involve a complex vertical circular or elliptical motion in the direction of energy transfer, rather like liquid surface waves.

RC circuit circuit which combines capacitance and resistance – the capacitative reactance X_C is $1/(2 \pi \nu C)$, and the voltage leads the current (by 90° if R is zero).

rd symbol sometimes used for the rad, a unit of absorbed radiation dose, 0.01 J kg^{-1}

reactance X measure of how much a circuit element will oppose electric current, the ac equivalent of resistance. It is the ratio of the root mean square voltage to the rms current; the unit is the ohm, χ. For each of the three types of reactive circuit element, the reactance is

capacitor of capacitance C $X_C = 1/2 \, \pi/\nu \, C$
inductor of inductance L $X_L = 2 \, \pi/\nu \, L$
resistor of resistance R $X_R = R$

ν is the frequency of the supply, and $\mathbf{j}$ the square root of -1.

In the first two cases, the value i appears in order to indicate the phase difference between voltage and current; there is no phase difference (and no frequency dependence) in the case of a pure resistor. See also *effective resistance* and see *ac circuits*.

A **reactance coil** is the same as an inductor. **Reactance drop** is the drop in voltage between the ends of a reactance in an ac circuit; its value is the product of current and reactance.

reaction a) old name for a force that appears in response to a given force, under Newton's third law of force. (Newton said something like this – in Latin – 'To an action [= force] there is always an equal opposing reaction.') The word is still common as the name of the force that

- is the third-law twin (symbol R) of an object's weight, when the object lies on a surface, or
- gives a jet or rocket drive its thrust.

b) chemical interaction between elements and/or compounds that involves a change. The speed of the reaction is the *reaction rate k.*

c) interaction between nuclei, or between a nucleus and some other particle, that causes a change in a nucleus. By extension, many people include radioactive decay as a nuclear reaction. A nuclear reaction is endo-ergic if the nucleus absorbs energy during the change, and exo-ergic otherwise Rutherford carried out the first artificial nuclear reaction, in 1919; he allowed alpha particles (helium nuclei) from a radium source to interact with nitrogen gas, as shown.

$$^{14}_{7}N + {}^{4}_{2}He \rightarrow {}^{17}_{8}O + {}^{1}_{1}H$$

We can also write this as the ^{14}N (α, p) ^{17}O reaction.

reactive concerned with reactance. Thus, in an ac circuit, the **reactive component** of current (or voltage) is the component of the circuit current (or voltage) that is 90° out of phase with the voltage (or current). (Active component is the part that is in phase.) In that case, **reactive current** is the reactive component of the current, and reactive voltage is the reactive component of the voltage. A **reactive load** is any circuit, sub-circuit, or element in which there is a phase difference between the applied (input) voltage and the current that results (output) – it is a load with reactance, in other words. The **reactive power** P (or Q) of an ac circuit is the

rate of energy transfer, given by the product
of circuit voltage, current and power factor
(cosine of the phase difference between
current and voltage). For **reactive voltage**, see
reactive current (above).

reactivity ρ measure of how close a nuclear
reactor is to criticality. A value of zero applies
to the case when the reactor is just critical;
negative and positive values relate to sub-
and super-criticality, respectively.

reactor a) vessel that contains a nuclear
reaction (**nuclear reactor**) or chemical
reaction. See also **natural reactor** in the
former case.
b) circuit element which offers reactance
to ac

real a) of any number which has no
imaginary component (whereas a complex
number does)
b) actual rather than apparent – thus the real
depth of a pool of water is how deep it
actually is (as perceived when you're
standing in it, say), while the apparent depth
is how deep it seems to be when viewed from
above.
c) of any aspect of an optical system through
which rays pass (rather than appear to pass,
which is virtual). Thus a **real image** is one
through which the rays that form it pass, so
one can project it on a screen. The **real is
positive convention** is a system by which one
assigns signs to measures in optical systems:
we take measures for a real object or image
(etc) as positive, and those to virtual ones as
negative. For **real object**, see virtual object.

Réaumur scale scale of temperature, still
sometimes used in parts of rural continental
Europe. It has 0° R for the ice temperature
and 80° R for the steam temperature.

receive aerial the aerial through which a radio
signal passes into a 'receiver' at the output
end of its channel

reciprocal of a number n, the number $m = 1/n$.
In other words, the product of a number and
its reciprocal is 1. This applies to real and
complex numbers, but there is, however, no
meaning to the reciprocal of 0.

reciprocating machine machine in which
there is some kind of to and fro motion,
often of a piston in a cylinder (as in
reciprocating engines and pumps)

recombination the joining again, after
being formed, of pairs of electrons and holes
in a semiconductor, or of electrons and
ions in any medium. In the latter case, the
recombination constant of such a process

is the recombination rate divided by the
square of the density of ions.

rectangular coordinate system the system of
giving the position (coordinates) of a point in
space in terms of three axial distances x, y, z,
from a defined origin

rectification process of taking an alternating
input and forming a corresponding direct
output, as with a rectifier

rectifier

device or system with an alternating input
and a corresponding direct output. In other
words, a rectifier allows current through it in
only one direction (the direction shown by
the arrow in the symbol above). Most
rectifiers now are semiconductor junction
diodes, though other systems (such as the
metal/semiconductor junction) carried out
the same task in earlier times. Indeed the
word diode comes from electron tube applied
science, the two-electrode electron tube, or
valve, being a rectifier; also the word diode
now widely means rectifier in general and
a valve lets something through only one
way.

A single diode has a half wave output (it
passes alternate half cycles with no current
between them). Linking two or more diodes
can produce a full wave rectifier. The sketch
shows a simple full wave 'bridge' circuit, with
an output capacitor able to smooth the waves
to a ripple on a level direct current output.
Below it are sketch graphs against time for
a) the input – an alternating voltage
b) the direct current in a single diode fed
with such a voltage – half-wave rectified ac
c) the output of the circuit shown before (or
without) the capacitor – full-wave rectified ac

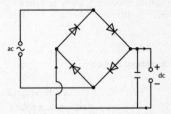

This 'bridge' rectifier can produce a very smooth
dc output.

d) the final output of the circuit shown – full-wave rectified ac smoothed to a rippling dc

A **rectifier meter** is a direct current meter

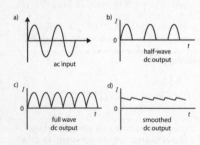

a) ac input

b) half-wave dc output

c) full wave dc output

d) smoothed dc output

fitted with a rectifier so it can also work with ac. **Rectifier photocell** is an old name for the barrier layer (photovoltaic) photo-cell.

rectilinear propagation the way a radiation travels through a uniform non-scattering medium without deviation (change of direction). Diffraction makes this never quite true in practice.

red colour associated with low energy, low temperature visible radiation. A **red giant** is a normal star near the end of its life, as its nuclear fuel runs out; its diameter may increase hundreds of times while it adjusts.

The **red shift** is the result of the Doppler effect on electromagnetic radiation from a source moving away from the observer. The radiation shifts to a longer wavelength: towards the red in the case of blue or yellow light. (The corresponding blue shift appears when the source moves towards the observer.) As the effect depends on the source speed relative to the speed of light, it is easy to observe only when the source speed is high – as is the case with radiation from distant stars and galaxies. We can also observe the **Einstein red shift** (or **gravitational red shift**) in radiation coming towards us from a highly massive source; see *relativity*.

Most theorists accept the speed implication of observed red shifts and use results to give direct values of source speeds. The outcome is the belief that the higher the source speed (i.e. the higher the shift), the more distant is the source – a crucial aspect of the theory of the expanding Universe. In turn, this relates to the age of the source since the big bang. On the other hand, new explanations for astronomical red shifts appear from time to time, and it is not always easy to refute these. For instance, research in the late 1980s shows that red shifts can follow partial polarisation of the radiation or its scattering by media (e.g. space close to quasars) whose optical properties vary at random in time and space.

The highest red shift observed in an astronomical object (2006) is 6.4 for a quasar and 7.0 for a galaxy; these correspond to distance of over 13 000 million light years, i.e. to the object's age of about 1.5 billion years. Such large shift values are starting to concern astronomers who use the high degree of uniformity of the cosmic background microwave radiation (shift factor 1500) to deduce that the Universe was uniform for a very long time after the big bang. Should shifts reach a value of 10 – and that of a recently found lensed galaxy may do so – current theories would have to change.

red noise also called Brownian noise as it relates to the noise of Brownian motion, noise whose intensity varies inversely with the square of frequency

red shift Doppler shift of electromagnetic radiation to a lower frequency (i.e. towards the red end of the visible spectrum), when viewed from a point towards which the source is moving. This phenomenon is relevant mainly in astronomy. See *red*, and compare with *blue shift*.

reference tone standard sound of fixed known frequency and amplitude (pitch and loudness)

reflectance R (or ρ)

for a surface, the fraction of incident radiation it reflects. This may apply either to a specified radiation type and energy (often then called reflectivity), or to a wide spectrum; in any event, it may depend on wavelength, and perhaps on angle of incidence, as well as on the nature of the surface itself. The reflection density of the surface is the logarithm of the reciprocal of the reflectance. See also *reflectivity* and *albedo*, the reflectance of the ground or of a planet or moon.

reflecting of an optical system which uses one or more mirrors for major tasks. A **reflecting microscope** uses a mirror rather than a lens as its objective; this means it can focus a wider range of wavelengths to the same point. A **reflecting telescope** is an astronomical telescope with a large mirror for its objective; this is far cheaper and easier to build and use than a converging lens of similar size.

reflection

a) the way rays in a medium meeting the surface of a second medium bounce back, to stay in the first medium. While this is something that happens at the surface, the rays in fact enter to a certain depth; see *reflectivity*. See also *echo* (the effect of reflection of sound waves), *image* and *mirror*, and compare with *absorption* and *refraction*.

In any given case, the **angle of reflection** is the angle between a reflected ray and the normal (perpendicular to the surface) at the point at which the ray leaves the surface. Rarely used, the **reflection density** of a surface is the logarithm to base 10 of the reciprocal of the reflectance, $\varrho - \log_{10}(1/\varrho)$. Some people call the reflectance of a surface its **reflection factor**.

Whenever a surface reflects a ray, the so-called **laws of reflection** apply. These are:

a) the reflected ray is in the same plane as the incident ray and the normal;

b) the angle of reflection r equals the angle of incidence i.

If the surface is not smooth, each ray in a parallel beam may come off at a different angle to the surface. This is **diffuse reflection**: but each ray follows those laws, even though the output does not appear to be uniform; see also *diffuse reflection*. On the other hand, if the surface is smooth, all the rays of a parallel beam come off parallel: this is **regular reflection**, which some people call **specular reflection**, from the Latin word for mirror. For **total internal reflection**, which happens when a ray meets a surface of a substance of lower optical density at an angle greater than the critical value, see *total internal reflection*.

b) the way part or all of a signal or current in a circuit section 'bounces back' from the interface with a second section. The **reflection coefficient** Γ of a transmission line, for instance, is a measure of how much this happens in a given case; or the reflection may be from a break or other fault in the line. In any event, it depends on the impedances of the two circuit sections – these must match for best transfer.

reflectivity the reflectance of a sample of substance so thick that making it thicker does not affect the reflectance. Strictly, reflectance and reflectivity are the same, though the latter more often refers to reflection at a given wavelength.

reflector

a) word for mirror

b) device designed efficiently to reflect all input rays back in the direction from which they came. See, for instance, *corner cube*.

c) short name for reflecting telescope, as used in astronomy, there being a large number of designs. In each case, the task of the main mirror is like that of the object lens of a refracting telescope: to collect as much light from the viewed object as possible and converge it towards the eye lens.

d) part of the shielding of a nuclear pile, the function being to reflect neutrons back into the system for use

refracting angle the angle between the two refracting surfaces of a prism

refracting telescope telescope which uses a lens to collect light from the object viewed and converge it. A reflecting telescope has a mirror for this purpose. The sketch shows a typical simple system; see also *Galilean telescope* – this has a diverging rather than a converging eye lens. As both object and image are at infinity, we cannot use the normal expression for magnification; instead, this is the ratio of angle *b* to angle *a*.

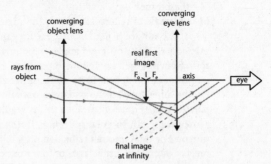

Both input and output rays are parallel; however, they are not at the same angle to the axis.

refraction

a) the change of direction of an electric field across the boundary between two insulators. The ratio of the tangents of the angles on the two sides equals the ratio of the relative permittivities.

b) the potential change of direction (deviation) of radiation when it passes from one medium into a second with a change of speed. There is no deviation if a ray enters the surface along the normal (the line at 90° to the surface at the point of incidence). The **angle of refraction**, as shown, is the angle between the refracted ray and the normal. In any case of refraction, the **laws of refraction** apply:

1) the refracted ray is in the same plane as the incident ray and the normal;

2) the sine of the angle of refraction r divided by the sine of the angle of incidence i is constant for a given pair of media and for rays of a given energy (e.g. wavelength).

That second statement is Snell's law; the reciprocal of the ratio described is the refractive constant n for the transfer concerned. **Diffuse refraction** occurs when the surface is not smooth, and/ or the second medium is not transparent (see *diffuse refraction* for details); if it is smooth, we have **regular refraction** (compare with the two cases for reflection).

refractive constant (or **refractive index**) n, no unit for a given pair of media and a ray of given energy, the ratio of the sine of the angle of incidence to the sine of the angle of refraction:

$$n = \sin i / \sin r$$

A stricter definition gives the refractive constant $_1n_2$ between medium 1 and medium 2 in terms of the speeds c of the radiation in the two media:

$$_1n_2 = c_1 / c_2$$

Then

$$_1n_2 = \sin i_1 / \sin i_2$$

Here i_1 and i_2 are the angles between the rays in the two media and the normal.

However, all these are for relative refractive constant, a value that involves two mediums. The most fundamental definition is for the absolute refractive constant of a medium:

$$n = \sqrt{(\varepsilon_r \, \mu_r)}$$

For the medium concerned, ε_r and μ_r are its relative permittivity and relative permeability.

This follows from the fundamental definition of the speed of light in terms of those two measures. Some artificial meta materials have negative values for permittivity and permeability. Some people argue that this gives them a negative refractive constant – the refracted ray would be on the same side of the normal as the incident ray, and higher frequencies would have longer wavelengths.

The value of the constant depends not only on the two media but on the energy of the radiation. In the case of electromagnetic rays, that is wavelength (or frequency). It is common to give the values of the D-lines of the sodium spectrum for refraction of visible light.

Other expressions for refractive constant appear in special cases, as follows.

a) For a sequence of refractions at a sequence of interfaces:

$$_1n_3 = {_1n_2} \times {_2n_3} \text{ (and so on)}$$

b) For total internal reflection, where c is the critical angle:

$$n = 1/\sin c$$

c) For the real (true) and apparent depths of a transparent medium (like a glass block or a pool):

$$n = \text{real depth/apparent depth}$$

d) For a substance which absorbs most of the input energy (such as a metal in the case of light):

$$n_c = n (1 - i k)$$

Here we call n_c the complex refractive constant (as i is $\sqrt{(-1)}$); k is the absorption constant for the medium.

e) For the material of a prism of refracting angle a:

$$n = \sin ((a + d_{min})/2)/\sin (a/2)$$

The angle d_{min} is the angle of minimum deviation for the glass of the prism and the wavelength concerned.

See also the *Gladstone–Dale law* and *refractivity*.

refractivity N measure of the ability of a medium to bend (refract) rays that enter its surface. A common definition relates the measure to the refractive constant n and the density ϱ of the medium, as in the box.

$$N = (1/\rho) \times (n^2 - 1)/(n^2 + 1)$$

refractometer device able to measure how well a medium refracts input radiation (in particular, light). There are many designs, each based on one or other of the relations given for refractive constant.

refractor

a) word for lens

b) short name for refracting telescope, especially as used in astronomy, there being a number of designs. In each case, the task of the main lens is like that of the object mirror of a reflecting telescope: to collect as much light from the viewed object as possible and converge it towards the eye lens.

refractory able to survive high temperature without physical or chemical change, so of value for thermal insulation and for building high temperature structures such as ovens

refresh to apply a suitable signal to ensure stored data does not vanish. Thus the contents of most screen displays needs a new scan tens of times a second. The process involved is **refreshing**, based on a **refresh circuit**, while the number of times a second it takes place is the system's **refresh rate**.

refrigerant the working fluid that cycles round a refrigerator system, cooling one area and giving out the energy elsewhere

refrigerator device for cooling a space and its contents. A pump passes the volatile working fluid (refrigerant) in a closed loop round the system. In liquid form, it passes through a narrow opening and expands; this lowers its temperature as it turns to vapour. Later in the cycle, the pump compresses the vapour into a liquid once more; at that stage latent energy passes to the outside. The **ideal refrigerator** is a heat engine working backwards. See also *adiabatic cooling* and *heat pipe*.

regelation the freezing again of water produced from ice under pressure, just by removing the pressure

regenerative involving some kind of feedback to reduce energy usage. Many designs of electric vehicle, for instance, use **regenerative braking**; to slow the vehicle, a switch converts the electric motor to a generator: this transfers energy from the motion to re-charge the cells. See also *rheostat braking*. The process of **regenerative cooling** appears in such systems as the Lindé method of liquefying gases; expansion cools the compressed gas, which in turn cools the compressed gas that follows

it. Some types of vehicle (such as some buses that start and stop often) have a **regenerative flywheel** – this stores energy for use in acceleration and takes it back during braking.

regenerator a) heat exchanger that absorbs energy from a waste output fluid (e.g. gases from a furnace or cooling system) for use elsewhere

b) the equivalent of an analog repeater in a digital communications channel. A repeater simply amplifies its input signal to form the output, with all the noise amplified too; a regenerator constructs it from fresh again. In most cases, a digital (pulsed) signal is binary, having only two values (high and low). Even after the signal passes through a channel with a great deal of noise (interference), it is not hard to tell from instant to instant whether the signal is high or low. The output of the regenerator consists of a number of clean pulses, each generated on the basis of the input. This is the main reason why digital data transfers over very large distances gain very little noise.

Regge model mathematical model – devised by the physicist Tullio Regge (1931–) – of the scattering of high energy elementary particles. It involves the now widely used **Regge calculus** to see such scattering as a process that involves the exchange of **Regge poles** rather than simple exchange particles; the spin of a pole depends on its effective mass. A graph shows how spin depends on mass, the line of the graph being a **Regge trajectory**. Discrete points on each such line represent resonance particles, **reggeons**.

Regnault, Henri (1810–1878) chemist whose most well known work was on the physical properties of gases. **Regnault's apparatus** is a U-tube of special shape used to find the absolute thermal expansivity L of a liquid. When the two arms differ in temperature, the liquid levels differ by a height from which one can find L. The **Regnault hygrometer** is a type of dew temperature hygrometer (device able to measure the relative humidity, or moisture content, of a sample of air).

regular reflection reflection of a beam of radiation from a smooth surface. If the input beam is parallel, so is the reflected beam. Compare with the *diffuse reflection* of a

beam from a rough surface. **Regular refraction** is much the same – the effect of a transparent medium rather than a translucent one.

Reines, Frederick (1918–1998) joint winner of the 1995 Nobel Prize, for being the first to detect the neutrino

rejector (circuit) ac circuit working at the frequency at which its impedance is greatest. The impedance is then the **rejector impedance**, and the circuit blocks input signals.

---

relative

not absolute in value, but compared to some standard. It is often more convenient to discuss a relative value than an absolute one; this is because relative values make it easy to compare measures and easy to see changes. No relative measure has a unit; they are all ratios, by definition.

The **relative activity level** a, of a radioactive substance is the ratio of its activity to that of some standard. The **relative aperture** of an optical element or system is its f-number. The **relative atomic mass** A (once 'atomic weight') of a nuclide or element is its mass in atomic mass units; the mass of a nucleon is 1 unit on that scale. The formal definition is the ratio of the mean mass of an atom to $\frac{1}{12}$ of the mass of a ^{12}C atom.

The **relative density** d (once 'specific gravity') of a substance is the number of times denser it is than liquid H_2O. One can use a thin walled **relative density bottle** to find this in the case of a liquid sample. The **relative humidity** of a gas sample (in particular one of air) is its actual moisture content as a fraction of what the sample could hold at that temperature and pressure. It is normal to give this fraction as a percentage. Compare it with *absolute humidity*.

The **relative molecular mass** M (once 'molecular weight') of a compound is the ratio of the mean mass of a molecule to $\frac{1}{12}$ of the mass of a ^{12}C atom. The **relative permeability** μ_r of a material is its absolute permeability compared to that of empty space; this is a measure of its magnetic nature. In much the same way, we measure the electric nature of a substance by its **relative permittivity** ε_r (once 'dielectric constant'). This is the number of times greater is its absolute permittivity than that of empty space. In practice, it is the ratio of the capacitance of a capacitor with the substance between the plates to that of the same capacitor with air between the plates.

The **relative velocity** of one moving object compared to a second is the difference between their velocities in a stated direction. If neither of the two objects is moving at relativistic speed, their relative velocity $_1v_2$ is the vector difference of their two velocities:

$$_1v_2 = v_1 - v_2$$

See also *relativity*.

relativistic associated with the effects of relativity, in particular with an object's motion at a speed close to that of light in empty space (c_0 – or, often, c – close to 300 000 km s^{-1}). Such high speed does not affect all the properties of the object: **relativistic invariance** applies to those which do not change with speed.

For the **relativistic Hall effect** (e.g. as observed in graphene), see *Hall effect*.

The **relativistic mass** m_r of an object moving at such a speed (v) is greater than its rest mass m_0 in the frame of reference of the viewer; see the box. The concept is not of great value – for instance, an object's relativistic mass depends on the frame of reference in which it is observed, and one can't replace m_0 with m_r in all contexts to change relationships from classical to relativistic.

$$m_r = m_0/\sqrt{(1 - v^2/c^2)}$$

A **relativistic object** or **relativistic particle** is in a **relativistic state**; in other words, its speed relative to the observer is close enough to c_0 to affect some observed properties. We call such a high speed **relativistic speed** or **relativistic velocity**.

relativity a theory that relates matter to space and time and also explores the fundamental nature of these. Einstein's **'special' theory of relativity** (1905) states that observers moving at constant speed relative to each other (called 'inertial' observers) have the same view of the Universe. In particular, they find the same value, almost 300 000 km s^{-1}, for the speed c_0 of light in empty space. A list of the main predictions of this statement follows; people have observed all of them in practice. Some apply to a 'relativistic object' (one moving at a speed v which is a large fraction of c); in such cases, the factor β may appear:

$$\varepsilon = \sqrt{(1 - v^2/c^2)}$$

Here are the predictions of special relativity:

a) The length l_r of a relativistic object is less than its rest length l_0 – this change being the Fitzgerald contraction.

$$l_r = l_0\beta$$

b) The mass m_r of a relativistic object is greater than its rest mass m_0.

$$m_r = m_0 / \beta$$

c) As a result of (b), mass and energy are equivalent, both the same as mass-energy. A change of mass Δm is associated with a change of energy ΔW. This appears in the concept of mass defect in nuclear fission and nuclear fusion.

$$\Delta W = \Delta m \, c^2$$

d) Time moves more slowly for a relativistic object than for the same object at rest, from the viewpoint of an observer; this effect is time dilation (and it relates to the twin paradox).

$$t_r = t_0 / \beta$$

e) As a result of (d), events that take place at the same time in one frame of reference may not take place at the same time in a relativistic frame; this is the concept of simultaneity.

f) The simple relation for relative motion does not apply for relativistic objects. The relative velocity $_1v_2$ of objects 1 and 2 moving at velocities v_1 and v_2 (taken in the same direction) is:

$$_1v_2 = (v_1 - v_2)/(1 - v_1 \, v_2/c^2)$$

g) As a result of several of the above, it is not possible to accelerate an object to, or past, the speed c. At that speed, the object would have infinite mass (as well as zero length and rate of passage of time).

Minkowski's mathematical approach to the concepts of special relativity led to the view of time as being 'the fourth dimension' of the Universe: this lets us describe the motion of an object by way of a curve in four-dimensional space–time. See also the *Lorentz transformations*, equations which link the motion as observed in different frames.

In 1915, Einstein published the **general theory of relativity**; this concerns accelerations as well as velocities. In particular, it deals with relative motion between accelerated ('non-inertial') frames of reference. The theory predicts that acceleration and gravitation are equivalent and one cannot distinguish between them; it also predicts that space curves in a gravitational field. These mean respectively that

a) inertial mass and gravitational mass are the same; and
b) light bends in a gravitational field: its energy rises (blue shift) as

it falls through a field and falls (red shift) as it rises out of a field.

Again, practice shows these predictions to be valid. In particular, astronomical observation is important: general relativistic effects do not differ much from those of Newtonian gravity. One predicted effect, one soon to be observed, was the precession of the orbit of the planet Mercury. Another astronomical test follows the study of single and double pulsars: the only explanation for the pattern of their output is precession as a result of spin-orbit coupling or orbit-orbit coupling respectively. This too is an effect of general relativity.

With quantum theory, relativity is the basis of modern physics. The main aim of modern physics now is to unify these two theories and to see all fields as, like gravity, aspects of the geometry of space–time.

relaxation any process in which a system takes a significant length of time to react to a change. There are many cases in physics where return to equilibrium does not at once follow a change (such as an applied stress). For instance, the state of a dielectric sample (insulator sample) changes with the outside field – however, the sample does not change as soon as the field changes, but 'relaxes' exponentially and almost gently into its new state. (Relaxation time – defined below – is almost zero in the case of metals.) At the other extreme, relaxation describes the gentle reaction to changes on an astronomical scale – such as energy exchange between stars to minimise total energy after the collapse of the core of a star cluster.

The **relaxation heating** of a sample of matter below 1 K allows one to supply a controlled quantity of energy. It involves applying steady and cyclic magnetic fields. The **relaxation length** of a measure which falls exponentially with distance is the distance over which the value falls to $1/e$. This applies, for instance, to the ways an electron beam interacts with a plasma, an elastic sample reacts to a change in stress, a building settles after an earth tremor, and even the pattern of strain in a helicopter will relax after a change in outside forces (e.g. wind pressure or cable tension). A **relaxation oscillator** is a circuit with a saw tooth output. In much the same way as with relaxation

length, the **relaxation time**, of a value that falls exponentially with time, is the time for the value to drop to $1/e$. This last relates to (for instance) **dielectric relaxation**.

relay electromagnetic or electronic device, switching whose input switches the output. In most cases, the output ('switched') current is greater than the input ('switching') current, this being of great value in control systems. (For instance, a power station control staff member can flip a switch that carries a very small current in order to control a lethal current of hundreds of amps.) Electromagnetic relays can handle large output currents but are slow in action. The thyristor is the basis of the most common electronic relays: small and fast, but not so well able to handle such large currents.

reluctance R (or S) unit: henry per metre, H m^{-1} in a magnetic circuit, the analog of resistance: the ratio of the magnetomotive force applied to the magnetic flux that results in the circuit

reluctivity of a substance, the reciprocal of its magnetic permeability – a measure of how hard it is to carry a magnetic force field (magnetic flux)

rem unit of dose of radiation, originally an abbreviation of 'roentgen equivalent [for a] man', one hundredth of the sievert (the SI unit)

remanence in the magnetic hysteresis cycle of a substance, the degree of magnetisation that remains when the magnetising field taking a sample to saturation becomes zero again

remote control control from a distance of the action of a system. The control signals pass through an electric circuit or by electromagnetic waves (such as radio, infrared, or light).

repeater electronic circuit in a transmission system (for instance, a phone network) with the task of amplifying an input analog signal to form the output. A major problem compared with the regenerator used in digital signal transfer is that the repeater amplifies input noise as much as the input signal. Thus, the quality of an analog signal worsens with distance.

research the choice and use of suitable techniques to obtain new knowledge, to develop existing knowledge, or to explore ideas. **Applied research** has the task of finding answers to specific practical problems (so is a form of technology); on the other hand, **fundamental research** (or **pure research**) has no specific application in mind. In practice, the bounds between these two can be very vague; often, for instance, someone working on a given real life problem makes a new scientific discovery – or someone who discovers a new effect can at once see how people's lives could improve from using it in practice.

residual a positive or negative difference between two numbers, in statistics a measure of the error in a result

residual magnetism less formal – and less precise – name for remanence, the degree of magnetisation that remains when the magnetising field taking a sample to saturation becomes zero

resilience measure of the elastic potential energy stored in a deformed sample of substance. It is the energy involved in deforming unit volume of the substance to its elastic limit; the unit is the joule per cubic metre.

---

resist

to oppose something, in physics some kind of flow. See below. Also, **photoresist** (sometimes called just resist) is some substance used in making the patterns of the layers of chips. After exposure to the light through the mask, acid can remove only part of the substance; the rest resists the acid. (There are two types of photoresist – the part that resists either was exposed to the light or was not exposed to it.)

resistance the opposition of a sample of substance to some specified change or effect. In particular, the **electrical resistance**, R, of a sample of substance is the ratio of the direct pd V between the ends of the sample to the current I, through it. The unit is the ohm, Ω:

$$R = V/I$$

In the case of solid metals, the cause of this resistance is the collisions between the free electrons that carry the current and the fixed atomic cores and crystal defects of the sample. See also *Ohm's law, impedance, negative resistance, resistivity,* and *superconductivity.*

There are similar concepts to electrical resistance in, for instance, the transfer of sound energy through a medium, the transfer of mechanical energy between the parts of a system, the

flow of a fluid through a tube, and the transfer of thermal energy.
See also *reluctance*, the name for magnetic resistance.

In the case of a non-ohmic conductor or sub-circuit (that is, a
conductor or sub-circuit whose electrical resistance varies with
voltage), it is of more value to use **differential resistance** (or
incremental resistance) rather than simple resistance as defined
above. The value of this form is the ratio of a small change of
voltage, ΔV, to the associated change of current, ΔI:

$$R = \Delta V/\Delta I$$

A **resistance box**, now rarely (if ever) used, is a set of resistors with
different resistances of known values in a box so that one can
switch any series combination into a given circuit. It is, in other
words, a variable resistor whose resistance can be set to a specific
value. A **resistance gauge** is a meter for high values of fluid pressure
which depends on the effect of the pressure on the resistance of a
metal sample. A **resistance strain gauge** measures the strain in an
object by way of its effect on the resistance of a wire (or set of wires)
on the surface of the object.

The action of the **resistance thermometer** depends on the
effect of temperature on the resistance of a metal sample (in most
cases with this in the form of a wire). The most common type has a
coil of fine platinum wire wound on a mica former; the coil's
resistance R_Θ at Celsius temperature Θ is as follows.

$$R_\Theta = R_0 \left(1 + \alpha\theta + \beta\theta^2\right)$$

Here R_0 is the resistance at 0 °C, and α and β are constants; except in
high precision work, one can ignore the β term.

Resistance wire, such as Nichrome, has a high resistance per unit
length and other well defined electrical properties, so is good for
making heating elements and wire-wound resistors of desired
resistance values.

resistivity ρ unit: ohm metre, Ω m former name 'specific resistance',
a measure of the tendency of a substance to oppose electric current.
It sets the resistance R of a sample (for instance, a wire) of the
substance of length l and section area A as

$$R = \rho \, l/A$$

At normal temperatures, the resistivities of different types of
substance fall into broad ranges as in the table. (See also *temperature
coefficient of resistance* – the resistivities of metals rise with
temperature rise, while those of other materials fall with
temperature.)

substance	resistivity/ς m
conductor	$10^{-8}-10^{-6}$
semiconductor	$10^{-7}-{}^1\!o^{-6}$
insulator	$10^{-6}-10^{15}$

resistor electric circuit element whose value in a circuit is a result of its electrical resistance alone. The resistance of a resistor may be fixed, or the device may be a **variable resistor** (one whose value the user can change).

There are many types of resistor; they vary in

a) how they're made – the most common designs being a coil of wire (wire wound resistor), a cylinder of ceramic material, a film of carbon, or (mainly within a chip) a length of semiconductor

b) power rating, the greatest electrical power (product of voltage between the ends and current passed) at which they can work continuously

c) accuracy – how close the resistance of a sample lies to the stated value.

The resistance of a resistor depends to an extent on temperature, though most modern types have fairly constant values. Off-the-shelf resistors tend to lie within the range 0.1 ohm to 10^9 ohm.

resistor-resistor logic a way to connect logic gates as inputs in a logic circuit using a resistor for each gate to give a specified logic output. The technique is simple, but needs the resistors to be accurate.

resistor-transistor logic a way to connect a transistor and resistor to each input of a logic circuit such that the system amplifies and inverts (NOTs) the desired output. Another transistor used as a NOT gate at the output will reverse the inversion if the design doesn't need this.

resolution a) an indication for an optical element or system of the smallest angular or linear distance between two objects or two spectral lines for which one can observe separate images. See *resolving power*, this being the quantitative measure.
b) the same indication for any other type of system whose output depends on separating inputs by way of some physical measure (for instance, mass in the case of a mass spectroscope, and energy in the case of a particle detector). See *resolving power* for the quantitative measure.
c) process of replacing a single vector with a pair of vectors in stated directions (in most cases at 90°) that gives the same effects. Those two vectors are a pair of components of the first vector; that, in turn, is the resultant of the two.

resolving power quantitative measure of *resolution*, senses a) and b). There are many definitions depending on context. Thus, for an optical element or system, resolving power is the angle at the eye between two objects that leads to just separable images. Scanning across the image would then give a graph of image intensity with angle. The Rayleigh criterion for successful resolution is that the first dark diffraction fringe of one image is at the same place as the central bright fringe of the other.

This concept leads in practice to this relation for resolving power: 0.61 λ/D; here λ is the wavelength concerned, and D is the aperture of the system. See also *Abbe criterion*. The resolving power of the human eye is about 10^{-3} – it can just distinguish between two points 100 μm apart at the standard near-point distance of 250 mm.

There are other cases where the task of some system is to separate images that are close in terms of some physical measure (such as wavelength, time or mass). For most of these, we use for resolving power the relation $X/\Delta X$. Here ΔX is the minimum difference in the measure that the system can detect, while X is the actual value of that measure.

The **chromatic resolving power** of a spectroscope is much the same – it is a measure of how close two wavelengths can be while still being separated.

resonance a) an excited state of an elementary particle, with a life time of the order of 10^{-23} s. A graph of some such nuclear process as the capture cross-section (probability) σ of particles by a target may show one or more peaks around certain energies. In that energy range, therefore, there is a higher than normal probability of the kind of event the cross-section describes. The **resonance cross-section** is the cross-section for that event at the peak in the graph, while the **resonance energy** is the energy of the particles in the beam at that value. Where this peaking effect occurs, we may call the process **resonance scattering**.

It is reasonable to associate a resonance peak with the production of a **resonance particle**. The theory is that, at a peak, the interaction is able to produce a particle whose mass equals the total relativistic energy of the two interacting particles concerned. Analysis of the situation and the graph allows one to state other properties of these resonance particles; indeed, in some cases, one can actually observe them, despite

their short life times. Hundreds of resonance particles are known to exist, excited states of mesons and baryons.
b) form of fluorescence in which the output radiation has the same frequency (energy) as that input. See *laser* for an important example.
c) increase of the amplitude of the vibration of a system when an alternating input force or signal is close to the natural frequency of the system. There are many cases in physics: in particular for mechanical systems (including sound sources); for ac electric circuits; and for the bonds of nuclei, atoms and molecules.

According to the **resonance theory of hearing**, each of the tens of thousands of hairs in the cochlea of the human ear has a different resonant frequency. Each frequency in an input sound wave therefore excites only a small set of hairs; these send an appropriate signal through the auditory nerve to the brain. Resonance of the brain is one theory for the supersonic hearing (in some cases, up to 120 kHz) of some people.
resonant able to show resonance (in any sense) in the right conditions. A **resonant cavity** is a space which will resonate strongly when fed with a sound or electromagnetic wave at or near a certain frequency. A **resonant circuit** is an electric circuit that contains reactance as well as resistance, so gives a strong output signal for an input at or near a certain frequency. Tuning circuits in broadcast receivers are of this type, for instance. In both cases, the 'certain frequency' concerned is the system's **resonant frequency** (in other words, its natural frequency).

The resonant frequency v_r of an ac circuit is

$$v_r = 1/(2 \pi L C)$$

This involves the circuit's inductance L and capacitance C, but not its resistance.

One can design the system to resonate at any frequency over a wide range, by tweaking the values involved.

See *sound sources* for the values of resonant frequency of acoustic systems.

A **resonant line** is a cable an integral number of quarter wavelengths long and with open- or short-circuited ends; used in aerial and oscillator circuits, it helps keep the output stable at the design frequency.
rest state of an object which has zero speed in

the frame of reference of the observer. The object's **rest energy** is the energy equivalent to its rest mass (below). If the rest mass is m_0, the rest energy W_0 is

$$W_0 = m_0 c^2$$

Here c is the speed of light in empty space.

An object moving relative to an observer has a greater mass than its **rest mass**; this effect is significant only if the speed is a large fraction of c: see *relativity*.

restitution the return of an object to its initial state after some such temporary process as elastic deformation. (Restitution means return to, or restoring, in this case to a starting state.) For the impact between two objects, the value of the **restitution constant** (or **coefficient of restitution**) shows the extent of elasticity in the impact. The constant is the relative velocity between the objects after impact divided by their relative velocity before impact. Here we ignore sign, and we take components of the velocities if the impact is not along the line of the two centres of mass. The table shows the ranges of values and the corresponding type of impact.

value	impact
> 1	superelastic
1	perfectly elastic
< 1	inelastic
0	plastic

resultant a) of two or more vectors, the single vector with the same effect. See *parallelogram rule* for how to find the size and direction of the resultant of vectors.
b) of two or more pure tones (monochromatic sound waves), the combination tone the hearer detects

retardation deceleration, slowing down. A **retardation plate** has the function of introducing a given path difference between the ordinary and extraordinary rays passing through it. It is made of a birefringent substance, cut to a suitable thickness. See, for instance, *quarter wave plate*.

retentivity old name for remanence, the degree of magnetisation that remains when the magnetising field taking a sample to saturation becomes zero

retina the light-sensitive layer of the eye. It contains rod and cone cells, receptors (sensors) for light intensity and colour respectively; these link to nerve cells that feed output signals to the optic nerve that passes to the brain. On the retina, the fovea is the most sensitive region; the blind spot, on the other hand, has no receptors. (However, the blind spot is small, the image from the other eye covers it, and the brain fills in any missing detail.)

retro- looking, or going, back. In the case of a planet, for instance, **retrograde** motion is in the opposite direction to the norm. A **retro-reflector** is a device designed to reflect input radiation back along its path, for instance, a corner cube. A **retro-rocket** is one designed to slow down a space craft.

reverberation the persistence of sound waves at a given point in a space after the sound source has stopped; the effect is due to reflections (echoes) and, in some cases, also to resonance. The reverberation of a room is a major acoustic feature, important in the design of studios, teaching spaces and concert halls, for instance. A **reverberation chamber** is a room designed for a high (but often variable) degree of reverberation for the purpose of acoustic research. Compare with *anechoic chamber*, or dead room, one with near zero reverberation.

reverse bias voltage applied to an element or circuit in the direction (**reverse direction**) of greater resistance, as opposed to forward bias. The current that results is the **reverse current**; in the case of a diode or transistor, this is very small.

reversibility a) principle that if a ray passes between two points A and B in an optical system, a ray between B and A can follow the exact same path in the other direction.
b) feature of a thermodynamic change or process that allows the system to return to its initial state. Such a change or process is **reversible** and involves the system's being in equilibrium at each point. A **reversible engine** is a heat engine which can (in theory, but not in practice) pass through its cycle of actions in either direction; see *Carnot* and *Rankine*.

Reynolds, Osborne (1842–1912) engineer best known now for his work in fluid mechanics. His improvements to the design of centrifugal pumps, while (the first) professor of engineering at Manchester University, were of great importance.

Reynolds' law gives an expression for the

pressure needed to maintain fluid flow at constant speed through a given tube. The **Reynolds number**, Re or N_R, describes the flow of a given fluid (density ϱ and friction χ) at a given speed c past a given solid surface (for instance, that of a tube). Its value is

$$Re = \rho\ c\ l/\chi$$

Here l is a characteristic length for the solid surface.

Re relates to the inertial and friction forces concerned; a high value implies low stability of flow.

rf radio frequency, in the range 10^6–10^{11} hertz (Hz). An **rf modulator** is a device or circuit able to modulate a carrier wave in this frequency range: in other words, to cause it to carry a useful signal.

rhenium Re transitional metal in period 6, with proton number 75, first isolated in 1908 by Masataka Ogawa, but not recognised as rhenium; Walter Noddack and colleagues found it again – the last natural element found – in 1925 and called it after the River Rhine. Relative density 21, melting and boiling temperatures 3200 °C (the second highest) and 5600 °C; most common natural isotope ^{187}Re, half-life 4×10^{10} years. The metal has many very high temperature uses, despite its cost.

rheo- prefix that implies flow or current. **Rheology** is the branch of physics concerned with friction between surfaces and with the flow of fluids and plastics. In the last case, the design of a **rheometer** allows one to measure the plastic flow (deformation) of a sample by giving readings of stress and strain over time.

A **rheostat** is an old name for variable resistor, in other words a device one can adjust to control the current (charge flow) in an electric circuit. A few designs of electric vehicle still use **rheostat braking** while slowing down: a switch converts the motor to a generator, and the energy output raises the temperature of a rheostat. Compare this with regenerative braking, where the energy output re-charges the driving cells.

rhodium Rh transitional metal in period 5 with proton number 45, discovered by William Wollaston in 1803 and named from the Greek 'rodon' = rose, because of the pink colour of its salts. Relative density 12, melting and boiling temperatures 1960 °C and 3700 °C; only stable isotope ^{103}Rh. It is mainly

used to harden platinum and palladium alloys, for contacts and electrodes, and in scientific and optical equipment.

rhombohedral system a type of crystal system

ribbon microphone type of microphone in which input sound waves vibrate a small metal ribbon between the poles of a magnet. Electromagnetic induction causes a varying current output.

Richards, Theodore (1868–1928) chemist who obtained accurate values for the relative atomic masses of well over half the natural elements. His work gave evidence for the isotope theory, and led to his Nobel Prize for chemistry in 1914.

Richardson equation the Dushmann equation that relates the density of thermionic electrons output by a surface to its temperature. Sir Owen **Richardson** (physicist, 1879–1959) received the 1928 Nobel Prize for this and other work on thermionic emission of electrons, crucial in the early development of electronics in practice.

Richter, Burton (1931–) physicist who discovered the psi resonance (mass 3095 units), the first particle found that contains a charmed quark. He shared the 1976 Nobel Prize for this work.

Richter scale numerical scale of earthquake intensity based on that devised in 1935 by Charles Richter (seismologist, 1900–1985). The values on the non-linear scale of 1 to 10 range from 1–3 (tremors that only special sensors can detect) to 8–10. Any quake over about 8 on the scale gives total destruction around the epicentre. The strongest recorded earthquake occurred in Chile in 1960 and measured 9.5 on the scale; there have been five others of 9 and over in the past 500 years. The values arise from the output amplitude and frequency of the surface waves as measured by instruments; it is, however, common to relate them to environmental effects as observed by people in the region. Thus a 'rather strong' quake, 4.5–4.8 on the scale, will wake most sleeping people.

Riemann, Georg (1826–1866) mathematician whose major work was the development of the mathematics of various multi-dimensional, curved, non-Euclidean spaces. The overall theory concerned is **Riemannian geometry**. Minkowski applied its ideas to describe the space–time of Einstein's relativity. Some people therefore describe that model of the Universe as **Riemannian space–time**.

Righi effect the thermal Hall effect, the appearance at 90° to a magfield of a temperature difference between the edges of a metal strip as it conducts thermal energy

right-hand rules a) Fleming's dynamo rule for electromagnetic induction: hold the thumb and first two fingers of the right hand at right angles to each other – they show you the directions of the motion (thumb), magnetic field (first finger), and induced current (centre finger).
b) the right-hand grip rule: with your right hand, grip a conductor that carries a current in the direction of the current; your fingers show the direction of the induced magfield.

rigid describes an object whose structure (size and shape) does not change whatever force acts on it from outside. No object is fully rigid: all deform to some extent.

rigidity modulus G rare name for the shear constant of elasticity of an object

ripple small amplitude wave, such as that imposed on a direct current at the output of a smoothed full-wave rectifier. Only surface tension controls the motion of ripples on the surface of water. A **ripple tank** is a device used to project the effects of water surface waves onto a screen to show how waves behave in different contexts.

rise time the time for a pulse to reach its greatest value from zero

rms root mean square, a way to describe measures of an alternating signal or system

robot programmable mechanical device or machine, from the Czech word for 'work'. The term first appeared in a 1921 play by Karel Čapek. **Robotics** is the study and applied science of robot design and usage, particularly nowadays the application of concepts of machine intelligence (such as machine vision and problem solving).

A **robot scientist** is a computer running an appropriate expert system, software which stores human expertise in statement form, applies that to new situations, and learns better and better how to make relevant decisions. Systems like this help to handle the vast amounts of data now generated by science, to pick up patterns, and to develop hypotheses and experiments which could test these. This approach is of most use in biology (where one can view cells and systems of cells as information processors) and medicine (for instance, the automatic checking of mammograms); even so, high energy physics is one of many areas where it is having at least some success.

rocket engine of a vehicle that stores both its fuel and its oxidant, allows these to mix in its combustion chamber, and obtains forward thrust (force) from the high energy exhaust gases that leave through a nozzle at the rear. As the system does not need a supply of air (unlike the case of a jet engine), it is effective outside the atmosphere. Some people also use the term for nuclear powered space craft, where the exhaust is a high energy jet output from a reactor.

rod photosensitive cell in the retina of the eye whose output relates to the input light intensity (energy content or brightness)

Rømer, Olaus (or **Ole**) (1644–1710) engineer, mathematician and astronomer whose achievements include the first measurement of the speed of light in empty space c. He first observed that the eclipses of the satellites of the planet Jupiter occur later than expected when the Earth is furthest from Jupiter; he then used the data to obtain a value for c. His result, around 214 000 km s^{-1}, is of the right order bearing in mind that people did not then have a good knowledge of the distances in the Solar System.

roentgen r unit of dose of x- or gamma-rays, now replaced by the coulomb per kilogram: $1 \text{ r} = 2.58 \times 10^{-4}$ C kg^{-1}. Another unit that relates to this is the **roentgen equivalent man** (see *rem*).

roentgenium Rg transitional, trans-actinide metal in period 7, proton number 111, prepared in 1994 by Sigurd Hofmann and team and named after Wilhelm Röntgen. No properties known: its most stable isotope, ^{280}Rg, has a half-life of less than four seconds.

Röntgen (or **Roentgen**), **Wilhelm** (1845–1923) the first person to receive a Nobel Prize in physics, in 1901, this being for his discovery of x-rays (x being his name, because he did not know what they were: many other people called them **roentgen rays** for a long time). The 1895 discovery was an accident: Röntgen was working on the new cathode radiation (a source of x-rays) when he noticed a phosphor screen across the room fluorescing.

Rohrer, Heinrich (1933–) physicist whose first scanning tunneling electron microscope appeared in 1980. He shared the Nobel Prize in 1986 for this work.

roll motion of an object (such as a ship or aircraft) about the long (fore and aft) axis. It is common to use stabilisers to reduce this in the case of ships.

rolling friction the friction of a surface rolling over a second. This friction is very low in most cases (hence the use of ball bearings, for instance), but it is not zero.

root mean square (rms) the effective value of an alternating measure. It is the square root of the mean of the squares of the instantaneous values. If the alternating measure is sine wave in form, the rms value is 0.707 times the peak value.

When dealing with alternating circuits and signals, it is normal to use rms values for voltage and current. This is because, in those cases, the rms value is equal to the direct voltage or current which has the same effect.

In statistics, the **root mean square deviation**, found in the same way, is a measure of the average difference between two sets of numbers. In the case of the motion of the particles of a fluid, their **root mean square velocity** relates to temperature. This too is a statistical device.

rot common name for the rotation of a vector. This is the same as its *curl*.

rotameter device used to measure the rate of flow of a fluid. It consists of a float in a vertical graduated glass tube; the height of the float depends on the rate of flow.

rotation a) for a vector or vector field, the same as its *curl*

b) turning through an angle, perhaps as a continuous motion as the Earth rotates around its axis (see *circular motion* and the tables under *rotational*). See also *optical*

rotation, turning the plane of polarisation of polarised radiation through an angle. A **rotation photograph** is a form of x-ray crystal image formed by turning the crystal round an axis at 90° to the beam during exposure.

rotational concerned with rotation (either sense) Thus

a) a **rotational field** is a vector field with a non-zero *curl*

b) the **rotational frequency** n of a rotational motion is the number of cycles in a second. For examples of **rotational motion**, see *circular motion* and *simple harmonic motion*, and refer to the tables below. The **rotational quantum number** J (or K) of a system is the same as its spin quantum number.

The first part of the table below gives rotational analogues of linear measures, with symbols and units. The second gives the equations of rotational motion.

rotor originally the turning part of a rotating electric machine, such as a generator or motor, compared to the stator. There are, however, designs in which this distinction is less clear (for instance, the linear motor). The rotor of a standard generator is the coil and associated assembly in which the induced voltage appears; in the case of a standard motor, it is the coil and/or assembly which experiences the useful force.

rtg radioisotope thermoelectric generator, a power source (widely used in satellites, buoys and other remote systems) based on the decay of an active isotope and using a thermopile to remove the energy

Rubbia, Carlo (1934–) nuclear physicist who

linear measure	rotational measure
displacement (s): m	angular displacement (θ): rad
velocity (v): m s^{-1}	angular velocity (ω): rad s^{-1}
acceleration (a): m s^{-2}	angular acceleration (α): rad s^{-2}
mass (m): kg	moment of inertia (I): kg m^2
momentum (p): kg m s^{-1}	angular momentum (L): kg m^2 s^{-1}
force (F): N	torque (T): N m
$\omega_2 = \omega_1 + \alpha t$	(missing θ)
$\theta = ((\omega_1 + \omega_2)/2) t$	(missing α)
$\theta = \omega_1 t + \frac{1}{2} \alpha t^2$	(missing $\omega 2$)
$\theta = \omega_2 t - \frac{1}{2} \alpha t^2$	(missing $\omega 1$)
$g v_2{}^2 = \omega_1{}^2 + 2 \alpha \theta$	(missing time t)

shared the 1986 Nobel Prize with van der Meer for their discovery the year before (a record short gap between event and prize) – they found the W and Z particles of the electroweak theory.

rubidium Rb alkali metal, group I, in period 5, with proton number 37, first found from its spectrum, in 1861 by Robert Bunsen and Gustav Kirchhoff and named because of its two ruby red lines. Relative density 1.5, melting and boiling temperatures 39 °C and 690 °C; most common isotope ^{85}Rb. Many minor uses, e.g. in electronics and in atomic clocks; see also *Bose gas*.

rubidium–strontium dating form of dating of rocks based on the radioactive decay of rubidium-87 (half-life 5×10^{11} years) into stable strontium-87

ruby laser the first type of laser to work (Ted Maiman, 1960); it fired 1 ms pulses of coherent deep red light from a ruby crystal.

Rumford, Count (1753–1814) scientist whose real name was Benjamin Thompson

Ruska, Ernst (1906–1988) pioneer in electron microscopy and Nobel Prize winner, 1986

Russell–Saunders coupling form of nuclear and atomic coupling, that between the resultant of all the orbital angular momenta of the system's particles and the resultant of all their spins

ruthenium Ru transitional metal in period 5, proton number 44, found after much work by Karl Klaus in 1844 and named after the Latin for Western Russia, where the ore came from. Relative density 12.4, melting and boiling temperatures 2300 °C and 3900 °C; most common isotope ^{102}Ru. Used to harden platinum and palladium and to make titanium resist corrosion, and in advanced high temperature single crystal superalloys for aerospace use.

Rutherford, Ernest, Baron (1871–1937) great physics researcher, whose first work was with J J Thomson on magnetism, radio and x-rays. He carried out a wide range of work into radioactivity and nuclear physics. With Frederick Soddy (1877–1956, 1921 Nobel Prize in chemistry for work on radioactivity), he devised the laws of natural radioactive decay

and transmutation and explored alpha, beta and gamma rays; with Hans Geiger, he worked on alpha rays, including alpha scattering, and developed his crucial model of atomic structure; he then went on to artificial transmutation. Rutherford carried out the first artificial nuclear disintegration in 1919; this involved the conversion of nitrogen to oxygen nuclei in an alpha particle beam. For his early work on radioactivity, Rutherford gained the 1908 Nobel Prize for chemistry for showing that alpha particles are helium nuclei.

Rutherford scattering is the name sometimes given to the Coulomb (electrostatic) scattering of a passing positive particle by a nucleus. The term follows Rutherford's discovery of the effect in 1909; it led to his nuclear model of 1911.

rutherfordium Rf transitional trans-actinide rare-earth in Period 7, proton number 104, perhaps created in 1964 by Georgi Flerov and his team and for sure created in 1969 by Albert Ghiorso and his team and named after Ernest Rutherford. The most stable isotope is ^{265}Rf, whose half-life is 13 h; its properties are not known and there are no known uses.

Rydberg, Johannes (1854–1919) physicist who carried out much important early quantitative work on spectra. A **Rydberg atom** is one with one or more electrons having very high values of the principal quantum number; this gives the atom some features well worth study. The **Rydberg constant** R appears in expressions giving the wavelengths of lines in the hydrogen spectrum. Its value depends on the element concerned, but is of the order of 1.097×10^7 per metre. The **Rydberg series** is the general expression for the lines of the hydrogen spectrum. The **Rydberg spectrum** is an ultraviolet section of the absorption spectrum of a gas with many fine lines, and thus of value in research.

Ryle, Martin (1918–1984) radio astronomer who shared with Anthony Hewish the 1974 Nobel Prize. This was for his exhaustive study of the radio sky which led to the rejection of the steady state theory of the Universe (for the big bang theory).

Ss

s a) symbol for second, the standard unit of time

b) symbol for the 'sharp' spectrum type, azimuthal quantum number $I = 1$

c) symbol for source, terminal of a field effect transistor

d) symbol for south-seeking (for the pole of a magnet)

e) symbol for the strange quark

S a) symbol for geographic south pole and the direction towards it

b) symbol for shear (shake) wave, transverse type of body wave (e.g. from earthquakes)

c) unit symbol for siemens, unit of admittance, of conductance, and of susceptance, equivalent to Ω^{-1}

d) unit symbol for stokes, unit of fluid friction (viscosity)

e) symbol for the strangeness quantum number

f) symbol for sulphur, element 16

g) symbol for switch in circuit diagrams

h) symbol for symmetric group (as in unitary symmetry, SU)

s a) in the case of a circle, symbol for arc length

b) symbol for displacement (vector) and distance (scalar), unit: metre, m

c) symbol for (electron) spin quantum number

d) symbol for specific entropy, unit: joule per kilogram per kelvin

e) symbol for standard deviation (σ preferred)

S a) symbol for apparent power

b) symbol for area and surface area (A preferred in both cases), unit: square metre, m^2

c) symbol for entropy, unit: joule per kelvin

d) symbol for the Poynting vector, which gives the energy flux of an electromagnetic field

e) symbol for reluctance (R preferred)

Σ a) symbol for the set of sigma particles (which are hyperons)

b) symbol for the sum of a set of similar terms

σ a) symbol for electrical conductivity, unit: per ohm per metre, $\Omega^{-1}\,m^{-1}$

b) symbol for nuclear reaction cross-section (probability), unit: square metre (or barn)

c) symbol for standard deviation

d) symbol for the Stefan–Boltzmann constant

saccharimeter type of polarimeter designed particularly to find the concentration of sugar solutions

safety factor the ratio of a structure's breaking stress (strength) to the greatest stress its designers expect it to have to withstand. Factors of three or more are common.

safety lamp a design of oil lamp for use in mines where there is a danger of methane explosion from the naked flame. A metal gauze around the flame allows in the air needed for burning, but conducts thermal energy enough to prevent the escape of any methane flame. The most famous designs were those of George Stephenson and Humphry Davy.

safety valve valve in the surface of a container of high pressure fluid, held closed by a weight or spring. If the pressure becomes too great, the valve opens. This is safer than letting the container explode.

sag same as brownout, a sudden fall in power supply voltage lasting for some seconds

Saint Elmo's fire old name for glow (or brush) discharge. Named after Saint Erasmus (patron of sailors), the discharge appears at sharp points (such as on ships' masts and rigging)

in an electrical storm. The discharge can reduce the chance of a lightning strike.

saltation the irregular motion of solid particles along a bed in a fluid flow. These particles are too large to form a suspension; indeed, they may include large rocks in the case of a river in flood.

Sakharov, Andrei Dimitrievitch (1921–1989) nuclear physicist. Twenty years in disgrace, with ten in internal exile, he is most remembered now for his political influence on the USSR. Sometimes called 'the parent of the Russian H-bomb', Sakharov contributed greatly to the understanding of thermonuclear reactions in general (for instance, in reactors and stars). A 1950 paper anticipated the Tokamak fusion reactor design, and other important contributions lie in the fields of cosmology and gravity. In particular, he proposed in the 1970s that gravity may not be a fundamental force but an effect of other fields and/or of zero-point effects. The theory still receives much attention.

Sakharov's fall into disgrace started with his urging Khruschev to halt the testing of nuclear weapons (because of fallout); this campaign led to his receiving the 1975 Nobel peace prize.

Salam, Abdus (1926–1996) theoretical physicist whose successful electroweak theory, unifying the models of the weak and electromagnetic forces, gained him a share of the 1979 Nobel Prize.

salt compound formed when an acid neutralises a base. Most salts are ionic compounds that make good electrolytes when in liquid form or in solution.

samarium Sm transitional rare earth in the lanthanide series, period 6, element number 62, found from its spectrum in 1879 by Paul de Boisbaudran, who named it after its ore. Relative density 7.5, melting and boiling temperatures 1075 °C and 1790 °C; most common isotope ^{152}Sm.

sample a) to select a subset of a population for detailed statistical analysis. People do this if the population is too large to analyse as a whole. For the results to be valid for the whole group, however, the sample chosen must fully represent the group.
b) to select certain instantaneous values of a signal rather than them all. The set of values that results can represent the signal fairly well. See, for instance, *pulse code modulation*.

saser sound amplification by the stimulated emission of radiation, the sonic equivalent of laser action, with phonon output rather than photon output. The first prototype system (2006) passes a current through a superlattice of semiconductor layers and vibrates it at the same time. With a coherent output sound beam in the nanometre region, the system should have uses in scanning and quantum computing.

satellite a) weak line found close to a strong line in a spectrum. It arises from less common events or species (such as a rare isotope).
b) object in orbit around a much more massive object. Thus Pluto is a satellite of the Sun (though no more than a dwarf planet, unlike the Earth), and Charon is a satellite of Pluto. An **artificial satellite** is a spacecraft in free fall in orbit round the Earth (or round the Sun or any other object in the solar system). To reach such an orbit round the Earth, rockets give the satellite a speed relative to the Earth of around 8 km s^{-1}. Once in orbit, as there is very little friction, the craft needs no more motive power than for the occasional adjustment.

A **communications satellite** has the task of aiding communications between points on the Earth's surface (or between other space-craft and/or the Moon). The standard approach is to put the satellite in a geo-stationary orbit – one at such a distance (some 36 000 km above the equator) that it takes the same time to orbit the Earth as the Earth takes to rotate once on its axis. Thus, the satellite always appears to be at the same point in the sky. It detects and amplifies signals sent in a narrow beam to it, and sends them back in a narrow or wide beam, in the latter case to cover a fairly large area of (footprint on) the surface. There, a small aerial (dish or flat like a squarial) detects the signal and feeds it to an amplifier and receiver. This was first done in 1972, though proposed by Arthur C Clarke in 1945.

Satellites in close orbit – the first being Telstar, launched 1962 – on the other hand, cannot be geostationary. Aerials therefore have to track them as they pass across the sky, though the stronger signals need much less amplification. Thus the various world-wide cellphone networks use 70 or more satellites in orbits a few hundred kilometres high; this means there will always be at least one satellite in view from any part of the surface.

Neither system is ideal, so the concept of the **Molniya satellite** provides a good halfway house. These satellites – designed for telecommunications in high latitudes – have 12-hour polar orbits that are so shaped that they spend most of their time in sight of Russia (their home land).

Survey satellites have the task of observing the Earth's surface, for military purposes or for surveying work. As they can fix where objects on the ground are within a metre or so, they have become of great value for **satellite navigation**. The sensor – in a boat, plane or car, say – works out where it is from the signals of a number of such satellites and compares it to a map in its data store. The US military runs the current world-wide system, GPS (29 satellites; 15 m resolution on the ground), and the Russian military runs the GloNASS system (24 satellites; 60 m). Europe planned its own – Galileo (30 satellites; 4 m) – to be working by 2010 and to provide a very high intensity service.

Satellite signals have many uses outside telecommunications and scanning. Measuring the phase difference between two points on the Earth's surface over time, for instance, allows seismologists automatically to measure movements of a few millimetres.

saturable reactor inductor (with inductive reactance) whose core responds in a non-linear fashion to the magnetic field of the coil, in particular as regards its magnetic saturation: the design is such as the core saturates at and above a certain coil current. The inductance then changes and controls the current. Magnetic amplifiers use saturable reactors too.

saturated able to accept no more of something, the state concerned being saturation. A **saturated colour** is one right on the edge of the chromaticity diagram: it contains no white component at all. Working with the **saturated mode** of an electronic device involves using it over a flat portion of its characteristic curve. A **saturated solution** can accept no more solute at the given temperature and pressure. A **saturated vapour** is a vapour in dynamic equilibrium with the liquid form; each second as many particles enter the vapour phase from the surface as condense back into liquid. The pressure of the vapour is then the **saturated vapour pressure** of the sample; for a given substance, this depends only on temperature.

saturation equilibrium state of being

saturated, able to accept no more of something. The **saturation current** of an electronic device is the value of the current above which increase of applied voltage does not change the current. It is, in other words, the value of the plateau current in the characteristic curve. The **saturation voltage** is the voltage at the start of the plateau. **Magnetic saturation** is the state of a ferromagnetic substance at which it can be magnetised no more strongly; see *hysteresis*. See also *supersaturation*.

Savart, Felix (1791–1841) physicist whose work includes aspects of sound, polarisation of light, and the Biot–Savart law

Savery, Thomas (c1650–1715) engineer who patented an important early steam engine in 1698. The patent covered Newcomen's system of 1712, and the two became partners to develop and market the latter.

sawtooth of a signal a graph of whose variation with time looks like the edge of a rip saw – with, in each cycle, a slow steady rise ('sweep') followed by a fast straight fall ('flyback'). Such signals, used for instance to provide the time base of oscilloscopes, appear at the output of such circuits as the relaxation oscillator.

saxion the scalar superpartner of the axion, according to supersymmetric theory

scalar a measure with no direction aspect (whereas a vector has direction as well as size). Speed is a scalar (but velocity is a vector); mass, resistance, and temperature are other examples of scalars. A **scalar field** is a region of space which we can fully describe by a scalar value at each point; for instance, the temperature of the surface of an oven is a scalar field. The **scalar product** $a \cdot b$ of two vectors a and b is a scalar (on the other hand, a vector product is a product of vectors which is itself a vector). For instance, the energy transfer W involved when a force F displaces an object by a distance s is a scalar product: F and s are vectors, but $W = F \cdot s$ is a scalar.

scaler a circuit whose output consists of pulses, each of which indicates a set number of input pulses. In other words, the output is a scaled down version of the input. A scaler consists of one or more **scaling circuits** that have the same function; a **binary scaling circuit** gives one output pulse for each two input pulses, while a **denary scaling circuit** has a 1:10 ratio. Scalers are of value as event counters, in association with (for instance) particle detectors.

scandium Sc transitional metal in period 4, proton number 21, one of the elements predicted by Dmitri Mendeleev in 1860, found in 1879 by Lars Nilson and named for Scandinavia, source of its only ore. Relative density 3.0, melting and boiling temperatures 1500 °C and 2800 °C; only stable isotope ^{45}Sc. Main uses in lamps as a source of sun-like light and in high performance alloys.

scanning passing through a region (of one, two, or three dimensions) point by point in some linear fashion in order to sense or study or change some aspect of the points. In general, a **scanner** is the hardware used, as in, for instance, the read head that forms the image of the contents of a sheet of paper in a copier, fax machine or scanning machine (also called scanner). Various types of radiation detector are the basis of scanners in industry, research and medicine. **Scanning coil** is a rare name for an electromagnetic deflection coil of a cathode ray tube.

A **scanning electron microscope** is a very important type of electron microscope, and there are many designs. An electron beam scans the surface of the object. This causes secondary emission of electrons; at each point, the number of electrons produced depends on the geometry and nature of the surface. An anode passes the secondary electron output to an electron multiplier to produce the highly amplified current used to form the image. Typical magnifications are tens of thousands of diameters, resolution being a few nanometres. Those values are not as good as those of the transmission electron microscope, but the object does not need to be in a very thin slice. On the other hand, there are **scanning transmission electron microscopes** which achieve the best of each approach. Also, the He$^+$ probe of the scanning helium ion microscope gives a much sharper image with very little damage of the surface.

A **scanning tunnelling electron microscope** (stem) can not only provide images of individual atoms in special circumstances (with a magnification of several million), but also manipulate them (e.g. to start a specific chemical reaction). Its action depends on the quantum tunnel effect. A team from IBM showed the latter use tellingly, early in 1990, by spending over 20 h moving 35 xenon atoms on a surface into an image of the company name.

scattering the deflection and/or loss of energy of the particles or quanta of a radiation (any type) to the particles of a medium or to those in a second beam. The size and nature of the effect depend on the energy involved, in most cases, and give a form of reflection on an atomic scale. Many of the colour effects observed of small particles (such as the atmosphere, dust, and so on) result from the dependence of scattering on wavelength (light energy). **Elastic scattering** is when there is no energy change in the scattering process; see *restitution* for other types.

There are many special names for the scattering process observed in specific contexts. For instance, **Coulomb scattering** is the scattering of positive radiation by nuclei (which have a Coulomb, or electrostatic, field). Other names for this are **nuclear scattering** and **Rutherford scattering** (the latter after its discoverer). **Rayleigh scattering** is the scattering of radiation by particles which are very small compared to the dimensions of the radiation particles or quanta; it is the cause of the scattering of sunlight in the atmosphere to produce blue sky and red sunset. For **resonance scattering**, see *resonance*. **Thomson scattering** is the scattering of photons by free electrons. In this case, the **scattering cross-section** σ (see also below) is

$$\sigma = (8 \pi r_e^2)/3$$

Here r_e is the radius of the electron.

The **scattering amplitude** a of a substance measures the scattering power of its atoms for a given radiation. It is not quite the same as cross-section (which is a probability), but the two relate as

$$a = \sqrt{(\sigma/4\pi)}$$

σ is the scattering cross-section of the process (see above).

Schawlow, Arthur (1921–1999) physicist whose work with Bloembergen on laser spectroscopy led to their sharing the 1981 Nobel Prize

schematic block diagram showing the logic of a system, used in particular in the US as meaning circuit diagram. It shows the system's elements and how they inter-relate and interact. **Schematic symbols** are those used in such diagrams.

Schiller layer surface layer which gives pretty colour effects as a result of diffraction and scattering of light by a regular array of bumps or holes. We can see this effect with certain crystals, butterfly wings and petals.

schlieren (from the German for 'streaks', and therefore sometimes spelled **Schlieren**) streaks one can observe inside a transparent medium as a result of variations of density (and therefore of refractive constant). This is the cause of the shimmer of air above a fire and the streaks in a water/ethanol mixture, for instance. **Schlieren photography** is a technique for the study of such thermal effects, shock waves, and turbulent flow using a high-speed camera to detect and record the streaks.

Schmidt lens large flat lens, with low converging power around the centre and low diverging power near the edges. Placed in front of a spherical converging mirror, this corrects for the mirror's spherical aberration, so provides a large, yet cheap and effective, telescope mirror. Such **Schmidt telescopes** are of great importance in astronomy. Bernhard Schmidt (1879–1935) was the optician who invented the system in 1931.

Schmitt trigger bi-stable circuit whose output is a flat pulse whatever the input (as long as this is above a certain value)

schorl European name for the mineral tourmaline (whose name is Sri Lankan), the mineral being important for its polarising effect

Schottky, Walter (1886–1976) one of the first electronic engineers, a physicist whose inventions include one of the first triode valve types of vacuum tube (1915), the tetrode (1919) and a version of the superheterodyne system (used in wireless). His theory of the Schottky effect (next paragraph) appeared in 1938.

The **Schottky barrier** is a distorted potential barrier at the surface of a solid that results from an outside electric field (and any image potential that may appear). The main effect is that the energy function falls, so that electron emission (of whatever cause) rises. In a **Schottky diode** the metal–semiconductor boundary acts as a rectifier but with differences that follow the existence of the Schottky barrier. In particular, the diode does not store charge when forward biased. This makes it very fast acting. The **Schottky effect** is the passage of electrons from a solid surface over the Schottky barrier rather than by tunnelling through it.

A **Schottky defect** is a defect in a solid sample of an ionic crystal that involves an empty site in the crystal lattice; it is the equivalent of a hole in a semiconductor. In a crystal of NaCl, for instance, a Schottky defect could be the hole left by a Na^+ ion that escaped its site, or that from a Cl^- ion.

Schottky noise, also called shot noise, is the variation in current in an electronic device as a result of small random changes in the rate of transfer of current carriers.

Schrieffer, John (1931–) physicist who shared the 1972 Nobel Prize with Bardeen and Cooper for the first theory of super-conductivity to have any success. That is the BCS theory, from the initials of the three.

Schrödinger, Erwin (1887–1961) physicist whose fundamental work in quantum mechanics gained him (with Paul Dirac) the 1933 Nobel Prize.

Foremost in that work was the **Schrödinger wave equation** (1926), the basis of his wave mechanics approach to modern physics. The equation describes how a particle moves in a force field. Wave mechanics views that particle (mass m) in transfer as having a wave form, of wave function ϕ (a complex function of position and time). The most general, time-independent, form of the Schrödinger equation is as follows.

$$\Delta^2\phi + (2\ m/h^2)\ (W - W_p)\ \phi = 0$$

Here Δ^2 is the Laplacian operator and h is the Planck constant h divided by 2π. W is the total energy of the particle and W_p is its potential energy.

This all relates to quantum physics in that the equation has only certain discrete solutions, these being for certain values (called eigenvalues) of W. In other words, the concept includes quantisation.

Schwartz, Melvin (1932–2006) one of the winners of the 1988 Nobel Prize, sharing it with Lederman and Steinberger for developing a way to generate beams of neutrinos. They also showed that neutrinos can exist in more than one 'flavour' and discovered the muon neutrino.

Schwarzschild radius r_s the radius of the event horizon around a black hole, also the radius of any object with mass below which it becomes a black hole. This is a result of general relativity, which predicts that no light can escape from a black hole (hence the name). The value of the radius for an object of mass m is

$$r_s = 2m \, G/c^2$$

Here c is the speed of light in empty space and G is the gravitational constant.

The concepts were first proposed in 1915 in a paper by physicist and astronomer Karl Schwarzschild (1873–1916) from Einstein's new general relativity.

Schwarzschild also produced another paper before he died (on the Eastern front), giving the first solutions (the **Schwarzschild metric**) to the general theory.

Schwinger, Julian (1918–1994) theoretical nuclear physicist who shared the 1965 Nobel Prize with Feynmann and Tomonaga for his work on the development of quantum electrodynamics. The **Schwinger effect** concerns the forming of an electron-positron pair by a photon-photon scattering event.

scintillation brief flash of luminescence that follows the absorption by a substance (a phosphor **scintillator**) of a quantum of ionising radiation. When a pair of ions recombines, the energy transferred appears as a photon of light. This provides a visual method to detect ionising quanta. (The spinthariscope was the first approach used in nuclear physics.) In a **scintillation counter**, the output photons are changed to electrons and amplified in a photomultiplier; an electronic circuit counts the output pulses. A **scintillation detector** combines one or more scintillating crystals surrounded by an array of many photomultipliers. This is highly efficient in that it gives much information about the path and energy of detected particles and photons; on the other hand, it is complex electronically. A variation is the Auger camera used to surround a patient in positron emission tomography radiology.

sclera or **sclerotic** the hard white outer layer of the eyeball

scotopic vision 'night vision', vision at low light levels. Here, as opposed to photopic vision, the rod cells of the retina are the main light receptors and there are no colour sensations.

scr silicon-controlled rectifier, or thyristor, a diode whose forward current depends on the current input through the gate (a third electrode)

scramjet engine a type of jet engine

screen visual display. Screens include the cathode ray tube types (for instance, in radar displays and older tv tube and computer system monitors), flat screens (such as liquid crystal display and plasma display), and reflective projection screens.

screw type of machine, related to the inclined plane and, in practice, to the second class of lever. While the force ratio (ratio of the output force to the input force) can be very high, efficiency tends to be low because of friction. The distance ratio of a screw is $2\pi r/p$ (where r is the radius of the screw and p is its pitch – the distance moved by the screw in one complete turn). A **screw dislocation** is a type of crystal defect in which a given plane in the lattice rotates and advances so that it looks like the thread of a screw.

seaborgium Sg transitional rare earth in the trans-actinide group, period 7, with proton number 106, made at the same time by a team led by Albert Ghiorso and one based at Russia's Dubna in 1974, but not confirmed till 1993; the name comes from Glenn Seaborg, long a member of Ghiorso's team. The most stable isotope, ^{271}Sg, has a half-life less than 150 seconds, so we know nothing of the element's properties.

search coil small conducting coil used as a probe in a system of measuring magnetic field strength

Searle's bar apparatus designed to measure the thermal conductivity of a good conductor. The substance under test is in the form of a long lagged rod (the bar); a coil heats this at one end and a steady flow of water cools it at the other. The power output is $m \, c_w \, \Delta T/t$. Here m is the mass of cooling water, of specific thermal capacity c_w, passed in time t; and ΔT is the steady state temperature difference between the rod's ends. The power output also equals $k \, A \, \Delta T/l$ where k is the conductivity, and A and l are the cross-section area and length of the rod.

second s (though sec is in wide daily use as the symbol) the unit of time, one of the base measures in the SI unit scheme (and most others). We define it in terms of the frequency of the radiation involved in the jump between two hyper-fine levels in the atom of caesium-133. The frequency is a little over 9.192 GHz, so that a second is just over 9 192 million cycles.

secondary of lesser importance than a primary aspect, or concerned with output as opposed to input. A **secondary cell**

(sometimes called accumulator) is one that the user can re-charge (whereas a primary cell, once used, has no further value). The **secondary coil** of a voltage transformer is the one that carries the output. **Secondary colours** are those formed by mixing pairs of primary colours.

A **secondary electron** is an electron released from a surface by taking energy from a photon or other particle (perhaps itself an electron) that enters the surface. This process of release of electrons from a surface by irradiation is **secondary emission**; it includes photoemission. In the case of cosmic rays, **secondary radiation** is the radiation produced when the original (primary) rays enter the Earth's atmosphere; while the primary radiation consists mainly of high energy positive particles, the secondary shower may contain many other particles and photons. In general, then, secondary radiation is the output of the interaction of any primary (input) radiation with matter.

In measurement, a **secondary standard** is a copy of the actual standard, or a unit whose value in terms of the primary unit is known precisely. For **secondary waves**, or **secondary wavelets**, see the *Huygens principle*. The **secondary winding** of a transformer is at the output, the same as the *secondary coil* (above).

Seebeck effect the appearance of a voltage, and as a result a current if there is a loop, when the two junctions of two pieces of different metals (or semiconductors) differ in temperature. This is the main photoelectric effect, found by Thomas Seebeck (physicist, 1770–1831) in 1821. The Peltier effect is the reverse effect, and the thermocouple is a major use.

Segrè, Emilio (1905–1989) physicist who discovered technetium (the first artificial element) in 1937. He shared the 1959 Nobel Prize for the discovery of the anti-proton in 1955. A **Segrè chart** is a plot of all known nuclides on axes of proton number (x) and neutron number (y).

seismograph device able to detect seismic waves, such as from earthquakes, with high sensitivity, and to produce a chart on paper and/or screen as output. The basis of any such system is a massive pendulum with highly sensitive motion detection, or a set of three of these at right angles. **Seismology** is the study of seismic effects, in other words, the vibrations in the Earth that result from earthquakes, eruptions, large explosions and

such events. It provides much information about the structure of the Earth on a variety of scales (so is of value in prospecting for minerals, for instance).

selection rule any rule that bans certain transitions between nuclear, atomic or molecular states. Such rules result from the conservation of physical measures. Despite the rules, some banned transitions may occur in certain cases (but with low probability).

selective observed only in certain ('select') cases. For instance, a substance may show **selective absorption** to light radiation, absorbing some wavelengths but not others; it is then a pigment, as in a surface or colour filter. In much the same way, **selective fading** is the fading of radio signals observed as a result of the absorption of some wavelengths by the atmosphere. Some people get by the problem by using two radio sets, with aerials a quarter wave apart.

selenium Se non-metal with several allotropes, in group VI period 4, and with proton number 34, identified in 1817 by Jakob Berzelius and named after Selene, minor Greek goddess of the Moon, as the element is close to tellurium (which is named after the Earth). Relative density 4.8, melting and boiling temperatures 217 °C and 685 °C; most common isotope ^{80}Se. Its photoelectric features make it widely used in photo-cells and photography; it is also used in rectifiers and to de-colour and colour glass.

selenium cell type of rectifier photo-cell with a layer of selenium on a metal base and a very thin layer of gold over it. The gold conducts current, but is thin enough to allow light to pass through.

self- prefix for a property that results from an interaction inside an object or system. Thus, **self-absorption** is the (total or partial) absorption of radiation by the source of the radiation; an example is how the outer layers of the sun absorb some light from lower down. The **self-capacitance** of an inductor or resistor is the capacitance it happens to have as a by-product of its structure; in analysis we can replace that capacitance by a capacitor in parallel with the device. **Self-energy** is
a) the same as rest energy (the Einsteinian energy equivalent of rest mass), or
b) the Helmholtz free energy that follows the polarisation of a dielectric sample.

The magnetic field of a **self-excited** generator is taken from the machine's output rather than from a separate source; there must be enough remanence in the core to provide a field when the machine first starts working. A self-excited oscillator will start to oscillate on switch-on with no external alternating trigger. The **self-inductance** L of an inductor or other circuit element is its own inductance, independent of any interaction with other elements. The process of **self-induction** is the appearance of a voltage between the ends of such an element when there is a changing current in it.

semi- prefix that implies half value, or, more widely, partial. A **semiconductor** is a substance that can conduct electric current to a small extent (much more than does an insulator, but far less well than a metal); see *resistivity* for values. The use of semiconductor systems has led to a revolution in electronics since the 1940s. There are some eighty semiconductor materials in common use, all with their pros and cons. Cheap, strong silicon (with its 'built-in' high quality insulator, silicon dioxide) is in widest use, followed by germanium and then by *gallium arsenide*.

At normal temperatures, a number of elements are semiconducting, and many insulators join the group at higher temperatures when there is enough energy for electrons to become free. A substance with semiconducting features at normal temperatures (such as silicon, germanium and gallium arsenide) is an **intrinsic semiconductor**. Such a substance carries current by the flow in opposite directions of an equal number of free electrons and missing electrons ('holes').

Adding the right impurity to an intrinsic semiconductor makes it conduct more efficiently (but still much less well than a conductor); it is now an **extrinsic semiconductor**. Its behaviour follows the release of charge carriers into the lattice from impurity atoms. If the impurity atoms are donors the charge carriers are electrons and the material is n-type; if they are acceptors the charge carriers are holes in a p-type material. So-called donor impurities add to the density of free electrons in the sample; most charge carriers are therefore now negative electrons, and the substance is an **n-type semiconductor**. On the other hand, adding an acceptor impurity in effect adds positive holes, making holes the majority charge carrier and the substance a **p-type semiconductor**.

In either case, the nature of the base substance, and the amount and type of impurity, allow the semiconductor designer to select the electrical properties of the product quite precisely. Silicon and germanium are the most common intrinsic semiconductors used in diodes and transistors; others have special roles such as that of gallium arsenide in the case of light-emitting diodes (leds). There are also a number of organic semiconductors (solid organic compounds which offer electron conduction that rises with temperature). See also *band model* and *chip*.

A suitable semiconductor sample is the basis of the **semiconductor counter** and **semiconductor detector**; This is a piece of semiconductor, or a p-n diode, that reacts to passing quanta of ionising radiation with a change in conductivity. When this absorbs ionising radiation, its conductivity rises as the result of freed electron-hole pairs. The pairs soon recombine, so an applied voltage leads to pulses of output current that allow the user to detect the radiation or to count its quanta.

A **semiconductor diode** is a junction between two types of semiconductor (but see *point contact diode*), often p-type and n-type. It has two electrodes (contacts); its functions relate to the fact that it passes current one way only – see *diode*. A **semiconductor (diode) laser** is a solid state laser: highly efficient, and of great importance in fibre optics and optoelectronics systems as well as in laser and compact disc readers. It is, in effect, a type of led, a junction diode that gives off light as electrons and holes combine. In this case, though, the light is of one wavelength only (or very closely so), and the waves are coherent as with any other laser. See also *scintillation*.

A **semi-metal** is a poor conductor of electricity and thermal energy, but with a clear metallic nature. The class consists of antimony, arsenic, bismuth, selenium, and tellurium; in each case, there is a very small gap between the valence and conductions bands. See also *metalloid*. A **semipermeable membrane** is a thin skin through which some types of substance can pass much more freely than others; see, in particular, *osmosis*. A **semitone** is the smallest interval in Western music, with a frequency interval of around $2^{1/12}$.

sensation level measure, in decibels, of the

intensity (loudness) of a sound wave compared to the minimum audible level

senses the means which report to an organism information about the world outside (**external senses**, such as sight and smell) and about its internal state (**internal senses**, such as pain and the sense of self (proprioception)). In the case of higher animals, all sense organs, whether localised (as is the ear) or generalised (the skin) contain specialised detector (sensor) cells with nerve cells to link these to the nervous system.

sensitive with high sensitivity. A **sensitive flame** significantly changes height and/or shape as sound waves pass through it. The **sensitive time** of a detector is the time period during which it can detect quanta of ionising radiation. In much the same way, its **sensitive volume** is the space through which it can detect most efficiently.

sensitivity a measure of the smallest input a sense organ, sensor, or measuring device can detect. More strictly, it is the change of the output of the device for unit change of input.

sensor a transducer with an electric output that relates to its input of some other form of energy. There are many types of sensor, each able to output a signal as a result of some kind of input: such as sound waves, light, strain, acidity, magnetic field strength, movement, etc. Sensors are crucial in process control and appear in many types of input device in IT. There are also many specialised sensor types for less common uses – such as silicon carbide sensors for extreme temperatures (from inside volcanoes to the surface of Mars) and smart polymers that can detect tiny amounts of specific substances (e.g. water in aircraft fuel).

separation energy the energy involved in the transfer of a nucleon from the nucleus of a given isotope

series a) sequence, as, for instance, with a radioactive series of isotopes
b) way to link elements in a circuit so that the same current passes through each in turn – 'in series'. A **series-wound machine** (generator or motor) has the field coil(s) in series with the main coil.

servo(mechanism) automatic control system with power amplification and a degree of feedback of output to input

sextant once crucial aid to navigation, a device invented in 1730 and used for over two centuries, with which the user can measure the height of an object above the horizon.

The most common use was to measure the height of the sun above the horizon at noon, in order to help fix where the user is on or above the Earth's surface. The standard design has a 60° metal frame, hence the name. Satellite navigation systems (and the Earth-bound systems that came first) make this technique of little value.

shadow area or space that does not receive radiation (in most cases, light) from a given source, because some object is in the way. A point source gives a sharply defined shadow region, called the umbra. If the source is large, on the other hand, the umbra has a region of partial shadow round it, the penumbra. The amount of radiation reaching the penumbra ranges from almost 100% (hardly any shadow) to nearly 0% (near umbra, total shadow). However, diffraction of radiation round the edge of the object can lead to a more complex shadow structure. See also *eclipse*, though here refraction of radiation through an atmosphere can even more confuse the simple model.

A **shadow mask** is a solid sheet with holes in it used in some types of colour tv set crt to allow the electron beam(s) to reach precisely correct points on the screen.

Shannon, Claude (1916–2001) mathematician who carried out important work in various fields of IT, communications theory in particular. His **Shannon model** of a communications system (1948) remains of great value today. This views the system with
a) a communication channel linking an information source at the transmit end and an information sink at the receive end;
b) the basic aim of having an exact reproduction at the receive end of the message that leaves the source;
c) problems with noise from some other source (specific or general) along the channel; and
d) a way to encode the source message before transfer to produce a signal that suffers least from noise, with the other end being able to decode the message to reverse the process.

shear deformation which we can view as the sliding of parallel planes of a sample over each other (so that, in section, a rectangle becomes a parallelogram). It is always a component of bending deformations. The **shear constant** or **shear modulus** G of an elastic sample is the tangential force needed to give unit shear (angular deformation);

much the same in its context as elastic constant, it is sometimes called the rigidity constant of the substance. The sample's **shear strength** is the shear stress needed to pass the shear elastic limit, with **shear stress** g being the effect of the four shear forces on an object that produce shear; the unit is newton per square metre, N m^{-2}.

A **shear thickening fluid** (stf) is a non-Newtonian liquid whose viscosity rises hugely when under shear stress. A household example is a suspension of corn flour in water – very liquid until stirred when it at once becomes near solid. There are many uses for stfs, for instance, in making bullet-proof fabric for clothing. A **shear wave** is the same as a transverse wave in an elastic substance, in other words one in which the displacement is at 90° to the direction of energy transfer; we call it a shake wave or S wave in the context of an earthquake.

shell group of electrons in an atom, or space for such a group, all of which have the same value for the principal quantum number n. The shell that acts as if closest to the nucleus is the K shell ($n = 1$); it can contain two electrons only (with opposite spin quantum numbers). The higher shells, with $n = 2, 3, 4$ ..., are L, M, N ... . These have sub-shells, called s, p, d, f, g ..., filled as dictated by the exclusion principle. Early models of the atom saw each shell as much like the orbit of a planet in the solar system. In fact, however, the electrons smear out in space as given by their wave functions, or orbitals; these show the probabilities of finding each of them at particular points in space. The model relates to quantum theory in that we can assign energy levels to each shell and sub-shell.

The **shell model** of the nucleus involves much the same view of the nucleons, as far as concerns the wave functions for the particles within the nuclear energy well and their quantum numbers and energy levels. The main support for this model comes from the 'magic numbers'.

Shenstone effect the large rise in photo-emission from certain types of surface as a result of an electric current in the body of the sample

shf super high (radio) frequency, the so-called decimetre wave region, the range being 300–3000 MHz

shielding a) putting sensitive hardware into some sort of Faraday cage (for instance, a wire mesh around a cathode ray tube, and metal

screening on cables). The aim of this is to reduce
1) the power of signals that leak from the hardware, and
2) the effect of outside electric fields.
b) doing the same to exclude magnetic fields from a space
c) surrounding the core of a reactor, or any other source of high intensity radiation, with absorbent material (often concrete, steel, and/or lead)

shm simple harmonic motion, a cyclic (or periodic) oscillation we can describe in terms of a sine wave or phasor

shock electric shock, the physiological effect of electric current in the tissues of the body. A **shock wave** is a brief sound (pressure) wave with a very sharp leading edge that results when a high pressure region overtakes a low pressure region. This happens when an object with a sharp leading edge moves at high speed relative to a fluid. Explosions can cause shock waves in solids as well as in fluids.

shock absorber or **damper** device or system designed to increase damping, in most cases to just below critical. An electrical analogue is the resistor in an oscillating (LC) circuit. A car's shock absorber – dashpot in US – has a piston that bears the load and whose head can move within a cylinder of viscous oil; this system often works in tandem with a sprig (which also resists displacement). A less common system has a magnet for the piston head and an aluminium tube for the cylinder – the effect of this is eddy current damping. In the future, smart liquid systems may appear.

Shockley, William (1910–1989) physicist who shared the 1956 Nobel Prize with Bardeen and Brattain for their work on the development of the transistor. He also started Silicon Valley.

short circuit linking two points in a circuit with a very low resistance path, so that most of the current by-passes the load in the circuit. The circuit current is larger than normal as a result, if not much larger, and this can lead to a hazard which fuses and circuit breakers should prevent.

short gamma ray bursts See *gamma rays.*

short sight myopia, very common defect of vision that means the eye cannot focus clearly on objects further away than a short distance (rather than being able to focus on objects up to an infinite distance). The cause is an over-long eyeball. The usual method of correction involves a diverging spectacle lens of suitable power; however, more and more people are trying laser treatment, which smoothly removes the most curved outer layer of the cornea.

shot noise name for Schottky noise, though Schottky called it Schrot noise, meaning (the patter of) bird shot. It is the effect of statistically random variations in the current in a conductor or output by a surface (for instance).

shower same as avalanche or cascade, process that grows with time or distance. A **cosmic ray shower** is a brief but intense beam of secondary radiation of mixed types formed when one high energy cosmic ray particle interacts with a molecule high in the atmosphere.

Shull, Clifford (1915–2001) physicist who shared the 1994 Nobel Prize, for his work on developing techniques of neutron diffraction

shunt in a circuit, name for an object linked in parallel with a given circuit element. The most common instance of this usage is for a resistor put in parallel with a meter to allow it to act as an ammeter for higher currents than the design range. In a **shunt-wound machine** (generator or motor), the field coil(s) are in parallel with the rotor coil.

shutter movable plate of absorbing material used to cut off or reduce a beam of radiation, as in a camera

SI acronym for the Système International d'Unités, the international system of units. After some two decades, this is now almost universal in science (though much less so in applied science, engineering, the US and daily life). See *Appendix 1* for full details.

sidebands the frequency bands above and below the main band of a carrier wave; they appear as the result of an amplitude or angle modulation process. Each sideband consists of a number of **side frequencies**. When a pure signal of frequency v modulates a carrier of frequency v_c, two single side frequencies appear, at $v_c - v$ and $v_c + v$. When the carried signal covers the range v_1 to v_2, the lower sideband has the range $v_c - v_1$ to $v_c - v_2$; the upper sideband starts at $v_c + v_1$ and goes up to $v_c + v_2$.

Sidebands also follow frequency and phase modulation, but in a more complex way. In all cases, however, it is possible to filter out at least some sideband signal without loss of the carried information; this cuts down the overall signal band width, so makes transfer cheaper and simpler. See *single sideband transfer*.

Siegbahn, Kai (1918–) nuclear physicist who shared the 1981 Nobel Prize for his work on high-resolution electron beam spectroscopy

Siegbahn, Manne (1886–1978) physicist whose crucial early work on x-ray spectroscopy gained him the 1924 Nobel Prize. He had also produced a theory of x-ray production that linked well with the then new quantum theory.

siemens S the unit of electric conductance, acceptance and susceptance, the reciprocal ohm (Ω^{-1}, once called mho). The name is that of a German family of engineers, the work of later members of which concentrated on electromagnetic machines and telegraphy.

sievert Sv the joule per kilogram, $J\ kg^{-1}$, unit of radiation dose equivalent. Do not confuse it with the gray (also $J\ kg^{-1}$), the unit of absorbed radiation dose – the equivalent dose is the product of the absorbed dose and a Q factor (that relates to the type of radiation absorbed) and a factor that relates to all relevant aspects of the body being irradiated and the exposure time.

sigma particle one of the hyperons, a baryon with slightly more mass than the nucleon, there being three forms: positive, neutral and negative. Each has spin $\frac{1}{2}$ and an antiparticle.

signal a message (or data) in transfer in some coded form, described by the Shannon model for the channel concerned. Any signal is a set of values, analog or digital, of some quantity that represents amplitude against time. In most cases, the quantity is an electric measure. A **signal generator** is an electronic circuit whose design is to output an alternating voltage signal of set frequency (and perhaps shape). This output is not a true signal in that there is no information content. On the other hand, it is easy to modulate the output of most such systems so that it does carry some test signal at least. **Signal processing** involves using hardware and/or software to accept a signal and output a more useful form. Examples are filtering the signal to remove noise, and picture processing to give a clearer image.

During transfer, as the Shannon model shows, processes of interference add noise to the initial signal sent from the source. At the far end, the **signal-to-noise ratio** (symbol: s/n, no unit) gives a measure of how strong the wanted signal is compared to the added noise: it tells us how much the latter competes with the former – if the ratio is too low, the risk of at least some loss of information is too high. As a ratio (in this case, of energies or powers), strictly there is no unit; all the same, it is common to use the decibel. Digital signals suffer much less from output noise than do analog signals, as it is much easier to separate a digital signal from the noise.

sign convention an agreed approach to giving signs to the distance measures involved in the action of an optical system. The **new Cartesian convention** relates closely to the Cartesian coordinate system for graphs: it treats distances right of and/or above the pole as positive; those left and/or below are negative. This is easy to learn and use – but has the major drawback that the formulas of optics then depend on the situation. As a result, the **real is positive convention** (which does not lead to that problem) is now much more common (and is the one used in this book). Here we take as positive the distances associated with real objects and so on, and as negative those associated with virtual ones.

significance a) measure of how likely a given outcome could *not* be the result of chance **b)** estimate or measure of the probability that a given measure lies outside a group of such measures whose distribution is known

silicon Si metalloid in group IV, period 3, with proton number 14, first made by Jakob Berzelius in 1824 and named after Latin 'silex' = flint (a hard form of silicon oxide, silica). Relative density 2.3, melting and boiling temperatures 1410 °C and 3250 °C; most common isotope ^{28}Si. Silicon is of great importance in the semiconductor and microelectronics industries, though the main uses are in light, strong alloys with aluminium and to produce silicones.

After oxygen, silicon is the most common chemical element on Earth. It is very common in rocks (and therefore sand) in the forms of silicon oxide (called silica) and of silicates (more complex compounds). Silicon is a semiconductor very widely used as the basis of modern diodes, transistors and chips. A **silicon cell** is the old name for a barrier layer photo-cell. A **silicon chip** is a tiny square of very pure silicon with a number of layers of patterns of impurity which build up in the surface a complex circuit of many thousands (even millions) of circuit elements.

A **silicon-controlled rectifier** (scr), or thyristor, is a diode whose forward current depends on the current input through the gate, a third electrode. A **silicon microstrip** is a form of charge coupled device (ccd) with cells in an array of linear strips, used as a particle detector. This is faster than the ccd, and gives more precise information; on the other hand, the electronics (with an amplifier for each strip) is much more complex in practice.

Silicon Valley the southern part of the San Francisco bay area of California, roughly centred on San Jose, so named because of the large number of high tech firms working there. The term was first used (by a journalist) in 1971.

silver Ag transitional metal in period 5, number 47, known in ancient times (at least 7000 years ago); the name comes from old Saxon 'siluvar', while the Latin word, 'argentum', gives us the symbol. Relative density 10.5, melting and boiling temperatures 960 °C and 2200 °C; most common isotope ^{107}Ag. The main use is as a precious metal (e.g. for coinage and jewellery), but silver is very important in electrical and electronic products, some solders and some types of cell.

sima class of rock in the Earth's crust – mainly silicon and magnesium

simple open to very straightforward description and analysis – as opposed to compound, which tends to be made up of simple components. Thus, **simple harmonic motion** (shm) is any motion that one can show as a single sine wave, while compound harmonic motion is the sum of a set of sine waves. Examples of shm include the way a simple pendulum oscillates (vibrates) or a source of pure sound (as in music: hence the name harmonic), and the variation involved in a simple wave motion. Some such type of motion appears when any system, taken away from its equilibrium state, experiences a restoring force that varies with the displacement from that state. This leads to the relation

$$m\, d^2x/dt^2 = -\,k\,x$$

Here x is the displacement of mass m, with d^2x/dt^2 the acceleration; k is a constant, the sign showing that the direction of the force is towards the centre (in the opposite direction to x).

During shm, there is a periodic interchange of energy between potential (position) and kinetic (motion) forms, the sum of the two being constant (as long as there is no damping). Here are other useful relations:

$$T = 1/v$$
$$T = 2\,\pi/\omega$$

Both of these follow from their definitions, with T the period, v the frequency, and ω the pulsatance of the motion.

$$x = x_0 \sin \omega\, t$$
$$dx/dt = +/-\,\omega\,\sqrt{(x_0{}^2 - x^2)}$$
$$d^2x/dt^2 = -\,\omega^2 x$$

Here x_0 is the amplitude (maximum displacement) of the motion.

We can describe motion round an arc (rather than along a line) in just the same way, with angle θ and amplitude θ_0 rather than x and x_0 throughout the above. In other words, various oscillations are simple harmonic. The table below describes some in outline.

A **simple microscope** has only one lens, whereas a compound microscope has at least two. The simple microscope is the single magnifying lens in other words. For **simple pendulum**, see *simple harmonic motion* (above) and pendulum.

simulation type of hardware or computer applications software that models some real life system or process. A simulation lets the user study the system or process cheaply and quickly, and with less risk (compared to trying to explore the real thing); a simulation can also be much simpler than real life, in that one can involve fewer factors in it.

simultaneity principle two events in two frames of reference that appear to an observer to happen at the same moment may not in fact be simultaneous. This is because the information concerned cannot travel faster than the speed of light. It is therefore possible for one observer to see event A before event B, while a second observer sees B before A. However, it is not possible for any observer to see an effect before a cause.

sine for an angle in a right angled triangle, the ratio of the side opposite to the angle divided by the hypotenuse of the triangle; any angle that size has the same sine value. A **sinusoidal** graph has the same form as that obtained when one plots the sine of an angle against the size of the angle. In such a case, the wave has a form such as in

$$x = x_0 \sin\omega t$$

Here, x is the displacement at a given instant t, while x_0 is the maximum displacement (amplitude); ω is the pulsatance, the product of 2π and the frequency v.

A **sine wave** is a wave (a form of energy transfer) which, when suitably plotted on a graph, has a sinusoidal shape. A pure sound wave in air is sinusoidal when plotted as air pressure at a point along the wave against time (or at a given time against distance from the source). Also we can plot some aspect of simple harmonic motion against time to produce the same shape of graph.

single phase type of alternating electricity circuit in which the input (supply) voltage is a simple sine wave. Compare this with multi-phase and, in particular, *three-phase* supply.

single sideband transfer method of signal transfer by modulated carrier wave which involves transmitting only one of the two sidebands the system outputs. This makes transfer simpler and cheaper; on the other hand, the circuit at the receive end must

motion of ...	period T	symbols
simple pendulum	$2\,\pi\,\sqrt{(l/g)}$	l: pendulum length
mass on spring	$2\,\pi\,\sqrt{(m/k)}$	k: spring constant
floating object	$2\,\pi\,\sqrt{(h/g)}$	h: immersion when at rest
liquid in U-tube	$2\,\pi\,\sqrt{(l/2\,g)}$	l: column length

replace the missing sideband before decoding the signal.

singlet any system with only one element or component, as opposed to a doublet (with two elements) or, in general, a multiplet. The term is most common for a spectral line with no doubling or for a single lens rather than a compound one.

singularity a) a point in a mathematical space where a function (or its graph) goes off to infinity. For instance, look at the function $y = 1/x$ – the graph shows a positive asymptote to the y axis as $x \rightarrow 0$ from the positive side, and a negative one as $x \rightarrow 0$ from the negative side. So we cannot define y when $x = 0$ – it is both $+\infty$ and $-\infty$. Here, then, there is a singularity at $x = 0$. A case in physics is the bouncing of a ball on the floor – theory tells us that the ball loses the same fraction of energy in each bounce: so (as with Zeno's paradoxes) the ball can never come to rest. The singularity is that, of course, the ball does come to rest. There is also a case in robotics – the straighter a robotic arm, the stronger the force even the weakest motor can exert: so in theory, when the arm is almost straight, it should straighten on its own.
b) a state of a changing system after which one cannot predict the next state – the next change leads to a state of chaos. The motion of a tidal wave (like Britain's Severn bore) is like this, though here there are many singularities.
c) a region of curved space–time for which the curvature is infinite, as for a black hole (within which the gravitational field strength and density of matter are infinite), a white hole, and a wormhole.

sink a negative source of energy (or of a fluid), in other words a point in a system at which the energy (or fluid) seems to be absorbed

sinusoidal able to be drawn as a sine wave (of x against t), and thus being of the form $x = x_0 \sin \pi \nu t$

siphon device using air pressure to transfer a liquid over a barrier to a lower level. The usual design involves an inverted U-tube with arms that differ in length; the tube must be full of the liquid before the siphon action can begin.

siren source of intense impure sound waves of variable base frequency. A common design involves passing high pressure air through holes or slots in a fast turning disc.

skin a) outer layer or surface of a sample of

substance. The **skin effect** is the passage of electric current through the outer layers only of a conductor when the supply is at high frequency: as the frequency rises, the depth of the active skin falls. The cause is internal self-inductance, and the result is an increase in resistance because of the reduced effective section area. It is therefore common to use tubes to conduct very high frequency currents. **Skin friction** is the resistance to motion of an object through a fluid; it follows the high shear in the layers of fluid next to the skin of the object.
b) or diaphragm, a 2D source of sound waves, as in the case of a drum (and ear drum)

skip distance the distance from a radio transmitter between the furthest range of ground waves and the nearest point able to receive sky waves. There may also, however, be surface waves, so if a skip distance exists there may still be reception at that range.

sky wave radio wave that reaches a receiver after reflection by the ionosphere – as opposed to ground wave, a wave that passes to the receiver parallel to the ground, and surface wave, one that hugs the surface.

slide projector once called diascope, system that projects on to a screen an image of a transparent object (the slide). This gives a brighter image than an opaque projector, in which light reflects from the surface of an opaque object.

sliding friction the effect of friction between two surfaces sliding over each other

slip one of the three main ways in which a metal sample may deform: there is slippage between layers of the sample, each many planes of atoms thick. The results may include necking and/or **slip bands**, parallel dark bands visible on the surface as a result of the 'staircase' effect there.
   to slide over a surface, rather than, for instance, rolling. A **slip ring** is a method of leading current between a turning coil and a fixed circuit, being joined between the ends of the former. A sliding contact, such as a carbon 'brush' presses on the ring as it turns.

slope resistance between a given pair of electrodes in a system, a very small change of applied voltage divided by the small change of current that results. One must normally specify this for a given value of applied voltage.

slow light media thin films able to reduce the group speed of light to as little as a fortieth.

There are other ways to produce slow light, some able to reduce the group speed to less than one per cent. However, the slower the light, the smaller its bandwidth.

slow neutron low energy neutron, in particular in a reactor. In most cases, the term is the same as thermal neutron.

slug non-SI unit of mass, the mass which accelerates by one foot per second per second when a force of one poundforce acts on it. A **metre slug**, also non-SI, accelerates by one metre per second per second when the force is one kilogramforce.

smart having – or seeming to have – a certain degree of intelligence, though not alive. The usage is most common in information technology.

In physics, a **smart liquid** is an electro-rheological liquid, one that becomes a stiff jelly in an electric field, or one which becomes near-solid when stressed (see *shear thickening fluid*). Though people have worked on the former effect since the 1930s, only in recent years has understanding become enough to lead to applications. A smart liquid consists of a suspension of very small particles (around a micrometre across) in an electrically insulating liquid; starch in transformer oil is a common DIY combination. In a field of the order of 3 MV m^{-1}, the particles polarise and form strong links parallel to the field. The strength of the links and thus of the 'jelly' then depends on the field strength; it is reversible, so allows a high degree of control with no moving parts.

Possible uses are static ones (for instance, safety catches and circuit testers), dynamic ones (like clutches), and in systems to provide selective damping of unwanted vibration (for instance, to act as shock absorbers in cars). There is great enthusiasm to apply the concepts to robots and office hardware (for instance, copiers), and some people see smart fluids leading to a new 'machine age'.

Smart liquids are a subset of **smart materials**, of which there are many, some in common use. The common feature is that one can control one or more properties of any such material through a stimulus from outside; the stimuli one can use are many – they include concentration of a solution (and that includes *p*H), electric field, humidity, magfield, pressure, radiation, and temperature. There are almost as many properties that can change as a result in

given cases. For instance, chromogenic samples show a colour change in response to changing chemical, electrical, optical and thermal stimuli; piezoelectric samples produce a voltage when under stress; and samples made of a shape memory alloy differ in shape at different temperatures.

Smart polymers are also a subset of smart materials. There are many of these, able to sense a range of aspects of the environment and thus form the basis of sensors and switches.

smectic of a type of liquid crystal, having the long parallel molecules lying in sheets, so the domains are only two-dimensional

Smoot, George F (1945–) astrophysicist and joint winner, with John Mather, of the 2006 Nobel prize for research into the cosmic microwave background radiation that provides such strong evidence for the Big Bang model of the origin of the Universe

smoothing cutting down variations in the case of an electric power signal (doing so in the case of movement and other contexts is damping). In particular, it is normal to add a large value **smoothing capacitor** in parallel with the output of a *rectifier* to form a **smoothing circuit**. The capacitor absorbs charge when the current rises, and lets it back into the system when the current falls; as a result the variations have a smaller amplitude, and form a 'ripple' rather than a wave. The larger the capacitance, the smaller is the amplitude of the ripple. A second approach involves the use of an inductor in series with the output.

s/n symbol for signal to noise ratio, a measure of the lack of interference in the signal's transfer

Snell, Willebrord (1580–1626) mathematician – whose family name was Snel; Snell comes from the Latin form, Snellius – who made the first major theoretical physics discovery; he also made the first measure of the Earth's radius and the value of π since ancient times. **Snell's law** (1621) – the second law of refraction – is that, when a light ray passes from one substance to a second, the ratio of the sines of the angles between the ray and the normal on the two sides of the surface is constant. The constant is the refractive constant for light of a given wavelength passing between the two given substances.

snr abbreviation for signal to noise ratio, a measure of the lack of interference in the signal's transfer

soap film thin, fairly long lived, sheet of soap solution (a liquid with a high surface tension) in air used to explore interference effects and mathematical shapes of minimum energy. The interference fringes observed depend on the thickness of the film and on whether one views reflected or transmitted light. A bubble is a film made into the shape of a sphere.

Soddy, Frederick (1877–1956) chemist who worked with Ramsey to discover helium. His concept of isotopes (1913), which followed work with Rutherford on radioactive decay, gained him the 1921 Nobel Prize for chemistry. The **Soddy–Fajans rule** is that alpha emission from a nuclide causes the nucleon number to fall by 2, while beta emission causes it to rise by 1.

sodium Na alkali metal, sixth most common element on Earth, group I, period 3, number 11, found in 1807 by Humphry Davy and named after soda (sodium carbonate, known since ancient times); the symbol comes from the alchemists' 'natron', from the Hebrew 'neter' = soda. Relative density 0.97, melting and boiling temperatures 98 °C and 880 °C; only stable isotope ^{23}Na. Main uses are in a few alloys, in the highly efficient sodium vapour lamps, and as a thermal transfer liquid in some reactor and engine designs.

sodium D line a very common standard for visible light – see *D line*.

sodium-line reversal the change of the sodium spectrum D lines from bright (emission spectrum) to dark (absorption spectrum) when the temperature of the sodium light source is the same as the temperature of an intervening sodium vapour sample. This allows one to measure the temperature of the sample.

soft involving little energy, when compared with a 'hard' sample. Thus, a soft ferro-magnetic material (such as **soft iron**) is easy to magnetise, but also loses its magnetisation easily; the coercivity is no more than a couple of hundred ampere per metre. A **soft radiation** has little penetrating power. A **soft vacuum** has a pressure above around 10^{-2} Pa, while a hard vacuum has a lower pressure than this. In a **soft vacuum tube**, ionisation of the low pressure gas in the tube affects how the gas behaves.

solar to do with the Sun and its energy output, the **solar system** being the Sun and the planets, dwarf planets, minor planets, moons, comets, dust etc that move with it through space.

A **solar cell** is a type of photo-cell whose aim is to obtain energy from solar radiation, either on the Earth's surface or in space. The history of the applied science goes back to 1876, date of the first photovoltaic cell of Adams and Day (though Henri Becquerel's father, Alexandre, discovered the effect forty years before). Such devices are far from efficient: it took sixty years to reach 1% efficiency and another two decades to achieve 6%. Then the 'space age' started, giving great impetus to solar cell research. This concentrates on materials and on methods of focusing the energy onto them. The highest efficiency reached by 2006 is 40%; however, highly efficient cells that cost a great deal to produce are not cost-efficient – most systems in use on the Earth's surface have efficiencies in the range of 10–20%. To increase effective efficiency also involves such techniques as concentrating the input radiation onto just the cell itself (rather than onto its support and wiring). A recent method uses a bloomed Fresnel lens to focus the light from a large area onto a cell. Even with simple silicon cells this technique can give effective efficiencies over 20%. The largest solar power station being built (2006) is in southern Portugal; designed to produce 11 MW, it consists of 52 000 photovoltaic panels steered always to point to the sun during daylight. But trapping solar energy by direct absorption in solid and liquid materials is just as important a research area as those described; such cells are far cheaper than ones that use semiconductors.

The **solar constant** is the solar power reaching unit area at the mean distance of the Earth from the Sun; its value is 1.39 kW m^{-2}. This is a tiny fraction of the total output of some 3×10^{26} W: and then much of the energy that reaches the Earth is reflected or absorbed in the atmosphere. During a **solar eclipse**, the shadow of the Moon crosses the Earth's surface; someone in the shadow will see only part of the Sun's disc (partial eclipse) or even none (total eclipse). For **solar energy** or **solar power**, see *solar cell*, above.

The **solar wind** is a stream of charged particles (mainly electrons and protons) that leaves the Sun's surface to form the Sun's outer atmosphere, or corona. At the distance of the Earth's orbit, the mean speed of the solar wind particles is around 150 km s^{-1}; 'gusts' (as a result of flares, in particular) may reach four times this, perhaps once every solar cycle. When the wind interacts with the Earth's magnetic field, it causes the radiation belts; its interaction with the upper atmosphere causes the aurora. The wind also causes the 'tails' of a comet and may well soon provide an efficient but slow form of space propulsion.

solarimeter same as pyranometer, a device able to measure the total solar power input at a surface

solenoid a) coil of wire with an iron core that is free to move along the axis of the coil as the field changes. Various devices use this effect. **b)** wire coil whose length is significantly greater than its radius. The magnetic field outside, when the coil carries a current, is much like that of a bar magnet. The field inside the coil tends to $\mu_r \mu_0 n I$; $\mu_r \mu_0$ is the permeability of its core, n is its number of turns per metre, and I is the current carried. A **solenoidal field** is a vector field with no divergence.
c) any coil of wire

solid state of matter in which cohesion between the particles is so high that samples tend to retain their shape. A true solid has a crystal structure and a single melting temperature (rather than a range); amorphous solids (such as glasses) have neither: in effect, they are supercooled liquids, though there is still some order at longer ranges than in true liquids. A **Hookean solid** is one whose samples follow Hooke's law with no hysteresis when deformed.

solid angle Ω angle in three dimensions, such as formed by the pointed surface of a cone. If we take the point of the cone to be at the centre of a sphere, the solid angle is the ratio of the area of its surface on the sphere to the square of the radius r of the sphere. The unit is the steradian (sr), the solid angle giving an end surface of area r^2. The solid angle at the centre of the whole sphere is thus 4π sr.

solidification freezing, the change of phase (state) of a sample from liquid to solid. The **solidification curve** of a substance is the line in a pressure/volume space (phase diagram, or eutectic) that separates the solid and liquid phases.

solid state concerned with the structure, properties and features of solid samples. A **solid state detector** is a piece of semiconductor, or a p-n diode, that reacts to passing quanta of ionising radiation with a change in conductivity. The old name was

semiconductor detector. It is a **solid state device**: one whose action depends on the behaviour of a solid sample. The branch of physics concerned here is **solid state physics**; the main branch of 'condensed matter physics', this deals in particular with crystals and their electrical, magnetic, mechanical, optical and thermal features – such as photoelectricity, semiconductivity, and superconductivity. A **solid state relay** is a relay using only solid-state elements: such as a transistor, thyristor or triac.

solution a mixture of two or more substances that is homogeneous on a molecular scale. Most solutions are of one or more solid substances or gases in a liquid; **solid/solid solutions** (such as alloys) exist too. The **solute** is the dissolved substance (recall that there may be more than one); the **solvent** is the bulk substance. The best liquid solvents have polar molecules and a high relative permittivity; water is the most effective known (partly because it is unusual in having such a high boiling temperature and low rate of evaporation for the size and nature of its molecules).

Sommerfeld, Arnold (1868–1951) physicist and mathematician whose main research was into the quantum theory of spectra. The **Sommerfeld constant**, $1/137$, is a name for the fine structure constant. **Sommerfeld waves** are electromagnetic waves that travel along the interface between a conductor and an insulator

son rare name for the product of some process or reaction, also called daughter or child

sonar acronym for sound navigation and ranging, technique for detecting and finding the position of underwater objects, such as shoals of fish and submarines, or the sea bed (in that case, the echo sounder works in much the same way). A sonar system outputs pulses of high frequency sound waves and analyses the echoes that return. There is growing concern about the effects of sonar on marine mammals (e.g. disorientation or decompression sickness may result, leading to beachings) – so bans are coming to be widespread.

sonic boom the sharp pulse of sound observed from the shock wave produced when an object travels faster than sound in air or other gas. This applies mainly to supersonic aircraft, but also, for instance, to the crack of the end of a whip.

sonogram the output of an ultra-sound or saser scan

sonometer (or, in music, monochord) single stretched 'string' (often a wire) whose length and tension the user can vary to allow the exploration of the system as a sound source. It is normal to mount the string between two sharp supports ('knife edges') on a sounding board; the former allows the length to be precisely set, while the latter amplifies the output. Pythagoras may have been the first to use a form of this research device, as he worked towards his concepts of harmony, harmonics, and the 'music of the spheres'.

sorption old name for adsorption

sound

a) the sensation produced when hearing waves of sound, sense c)

b) any type of pressure wave – including that of sense c) as well as sub-sonic and supersonic waves and P-type earthquake waves – in an elastic substance. Most of the concepts of sense c) apply in the other cases too.

c) pressure wave in an elastic substance of such a frequency (in the range of about 20 Hz to 20 kHz) that it can excite the sense of human hearing. (Note, however, that it seems some people can sense supersonic waves as high as 100 kHz.) The energy involved in the transfer causes the particles of the substance to oscillate to and fro (in other words in the longitudinal direction, along the line of transfer). In the case of a simple (pure) sound wave, the oscillation is a simple harmonic motion; a graph of the displacement of any one particle with time, or of all particles along the line at a given moment, is therefore a sine wave: so too are the corresponding graphs of the pressure in the substance. We can therefore describe sound as a longitudinal wave motion. It is therefore possible to reflect, refract and diffract sound waves.

We can also discuss a given sound wave in terms of its

- fundamental, or base, frequency (which relates to perceived pitch);
- amplitude (giving intensity, i.e., energy content, which relates to loudness); and
- structure in terms of frequency spectrum (quality or timbre).

The speed c of a sound wave depends on the medium and its characteristics (such as temperature). For sound in air the value is around 330 m s^{-1}; the figures for water and steel are about 1500 m s^{-1} and 5000 m s^{-1}. In fact, the speed c of sound in a medium is

$$c = \sqrt{(E/\rho)}$$

Here E is the medium's relevant elastic constant and ρ its density.

The elastic constants concerned are: its Young constant for a solid, its bulk constant for a liquid, and its adiabatic bulk constant for a gas. For a gas, in particular, we have

$$c = \sqrt{(\gamma p/\rho)}, \text{ or}$$
$$c = \sqrt{(\gamma R\, T/M)}$$

Here T is the ratio of the molar thermal capacities, p and T the gas pressure and temperature, and M its molar mass, with R the gas constant.

Sound absorption is the loss of energy from sound radiation passing through matter, the result of various processes – see *absorption* c). The **sound absorption constant** of a substance (medium) or of a surface is the ratio of the sound energy or power it absorbs to the sound energy or power input. The **sound energy reflection constant** of a surface is the ratio of the sound energy or power it reflects to that input. In both cases, the value depends on frequency. The **sound barrier** describes the extra energy an object needs to accelerate it to a speed above that of sound in a given medium; shock wave and sonic boom, for instance, relate to this.

As well as the ear, **sound detectors** include the various types of microphone. **Sound flux** is an old term for the power (rate of energy transfer) of a sound wave. Sometimes the transfer of sound energy into a given space is a nuisance; there are therefore various techniques of **sound insulation**, most involve absorption as in the case of an anechoic chamber (dead room) – but see *anti-sound*.

The **sound intensity** of a sound wave is the power transfer through unit area at 90° to the wave. The unit is the watt per square metre; its value is $p^2/\varrho\, c$, where ϱ is the density of the substance and c the speed of sound in it, while p is the wave's **sound pressure**. This last is the instantaneous value of the pressure due to the sound wave, above the rest pressure of the medium. As with any simple harmonic displacement, the mean value is zero – therefore, we take the **sound pressure level** to be twenty times the logarithm of the ratio of its root mean square value to some reference value; the unit of this measure is the decibel, dB. The most common reference level used in this context is 20 µPa (root-mean-square); this around the threshold of human hearing (the sound of a mosquito about three metres away). Taking this approach is not just useful in itself – perceived sound level seems to follow a logarithmic curve like this as well. Because of that, it is how most **sound level meters** show what they sense. There are various types of these; however, almost all give a root-mean-square reading using the dB(A) – both features are for the sake of tradition, not because they provide better data.

There are various techniques of **sound recording** too, making a permanent record of a sound wave for later reconstruction. The vinyl (plastic) 'gramophone' record (disc) goes back to Edison's phonograph system of 1877; the audio compact disc has now taken its place, while the MP3 solid-state system is moving in fast. Magnetic, optical and digital recording systems like these are common in the tape recorder, motion picture sound track and digital camera respectively. In any such system, the original sound forms the input to a microphone and amplifier, whose output - often with compression in the digital case – provides the type of signal required for storage and/or transfer. Later playback involves a suitable detector, decompressor (maybe), amplifier and speaker.

There are also many types of **sound source**. Each one is an object of one, two or three dimensions able to vibrate (see also resonance) in certain ways to produce pressure waves in the medium around it. Each one has a sequence of simple modes of vibration – fundamental (base) and over-tones – so can output a series of harmonics.

source where something has come from – as opposed to a sink – in particular the fluid or energy input to a system, such as a circuit. In that last case, it is often useful to discuss the **source impedance** z (for an ac circuit) or **source resistance** r (dc circuit). The value of this affects the current in the circuit in much the same way as does the circuit's impedance (resistance). It is normal to add source and circuit impedance (resistance) to obtain the whole effective impedance (resistance) that will oppose the current in the circuit.

source follower much the same as a *cathode follower* (an audio sub-circuit), but based on field effect transistors rather than on vacuum tubes

south(-seeking) symbol: s for a bar magnet, of the pole that tends to point to the geographical south if the magnet is free to turn. The normal symbol is s (with S for a geographical south pole, as in the case of the Earth).

space a bounded or unbounded region with (to the mathematician) any number of dimensions of any type. **Empty space** is a region of physical space in which there is no real matter, though it may (and in most cases

does) contain energy, fields, and virtual matter. A **space charge** inside an electronic device (such as a semiconductor sample or electron (vacuum) tube) is any region where there is a net charge density. This will form a potential barrier to the passage of current. For space–time – a view of 3D space and time as similar components of a 4D Universe – see *relativity*. A **space wave** is a radio wave received after one or more reflections from the surface of the Earth.

spallation a) the loss of fragments from an object's surface when stressed (as the surface of some objects crumble when you drop them)

b) the plume of rocks, dust, vapour and maybe gas that appears when a projectile or meteorite hits the ground. It is, therefore, also the loss of such parts of a surface in space due to long exposure to cosmic radiation – maybe a cause of future failure of old thin spacecraft parts, and a way to date moon rocks.

c) the significant reduction in the nucleon number (and, maybe, proton number) of a nucleus after a reaction in which it gives out several neutrons, alphas and/or betas.

Spallation is, in other words, a nuclear decay with an unusually large output of particles; this is almost always the result of bombardment by high energy radiation. The main case is in cosmic radiation interactions; however, using a heavy metal target in an accelerator can produce strong (but costly) pulses of neutrons (in the case of mercury, between twenty and thirty neutrons per impact).

d) a feature of cosmic rays themselves – the higher than expected ratios of light nuclei implies spallation from carbon, oxygen and nitrogen in the source of the rays or on their journey through space.

spark high energy (but often short lived) discharge in a gas or other insulator, or across the surface of an insulating solid. See *spark discharge*, below. A **spark chamber** is a detector of ionising radiation which allows the user to view or to photograph the tracks of radiation quanta. Now not much used except when searching for very rare events, it consists of two interleaved sets of thin, close,high voltage electrodes in a transparent insulating medium (often an inert gas). The ionisation along the track of a passing quantum causes a spark between each pair of electrodes, if triggered by a second detector outside.

A **spark counter** has much the same design. In this case, however, there is no attempt to view the sparks; rather, an electronic circuit counts the pulses of current that they cause.

A **spark discharge** takes place only if there is a high enough electric field in the medium concerned, the medium being a gas in most cases of interest. When the gas is at normal (or higher pressure), sparking is an intermittent (not continuous) process. This is because each spark discharge, a large scale transfer (avalanche) of electrons and ions, much reduces the electric field strength: in fact, to below the threshold value that can support sparking. It therefore takes some time for the field strength to build up past the threshold once more. This is also the case with lightning, sparking in the atmosphere on a massive scale.

However, as the gas pressure falls (as when one removes gas from a discharge tube), the threshold field value falls too. It thus becomes easier for a discharge to take place. For a given rate of supply of energy to the electrodes, therefore, the sparking frequency rises. Also, the sparks become broader; in

other words, the section area of the **spark channel**, the tube of ions and electrons (plasma) that carries the discharge in the medium, becomes larger.

If the pressure falls enough, the typical intermittent energetic sparking action is replaced by the appearance of one or more broad glowing streamers between the electrodes, even if the voltage (or electric field strength) is quite low. At even lower pressures still, a continuous glow fills the space in the tube between the electrodes.

Closer examination shows that the glow in the lower pressure discharge tube has a quite complex structure. Indeed, at even lower pressures than those above, the main glow region starts to break up into a number of striations (stripes). This is the situation shown in the figure for *discharge tube*; by this stage, one needs once more a somewhat higher voltage to maintain the discharge.

Taking still more gas from the discharge tube brings it closer and closer to a vacuum tube. The light output is the result of the recombination of ions and electrons in the gas (hence the colour output depends on the gas). Thus, as the pressure falls even further, the glow disappears: the dark spaces of the striations merge. There are now, of course, few ions and electrons to carry the charge that forms the discharge current, and, once again, we need a very high voltage (electric field strength) between the electrodes to maintain the current. The graph of discharge voltage against tube pressure shows this clearly. When the tube in effect contains a vacuum, the conduction of current is by electrons that leave the cathode and pass at high speed straight through to the anode. Cathode rays was the name first given to these high energy electrons; x-rays form the radiation output when cathode rays meet matter (such as the glass of the tube).

A **spark gap** is an arrangement of two electrodes with a high enough voltage between them for a spark to pass. A **spark over** is a spark discharge between two points in a circuit that appears other than by design. In **spark photography**, a spark provides the illumination required: this is very intense but very brief, so can 'freeze' quite fast motion in the dark. A **spark plug** is a spark gap used in the cylinder of a petrol engine; at the correct point of the working cycle, a spark across the gap provides the energy to ignite the fuel vapour/air mixture. The **sparking potential** of a spark gap is the minimum voltage between the electrodes at which spark discharge will take place.

A **spark spectrum** (or arc spectrum) is the spectrum of the light (and ultraviolet) output by a spark or arc; as the light comes from the re-joining of ions and electrons, the spectrum depends on the substance in the spark gap. In fact, there is so much concentrated energy in a spark or arc that irons can become further ionised (e.g. $Ne > Ne^+ > Ne^{++}$); this gives a second spark spectrum, in most cases in the ultraviolet – and Ne^{+++} can form too, to give a third spark spectrum.

speaker common name for loudspeaker, a transducer whose output is a sound wave image of its input electric or electronic signal

special theory Einstein's first theory of relativity (1905), concerned with relativity between frames of reference moving at constant relative speeds. The general theory (1915) concerns relatively accelerated frames of reference, on the other hand.

specific defined for unit mass of the substance in question. However, some terms still remain in which 'specific' relates to some other aspect of the system than mass. Thus, the **specific acoustic impedance** of a medium that carries a sound wave is the impedance of unit area.

The **specific activity** (symbol A) of a radioactive source is its activity per unit mass. The **specific charge** of a charged particle is its charge per unit mass: in other words, the ratio Q/m, taken as a fundamental property of the particle in some situations (as its of great value for identification). The use of the electron's specific charge is most common: Q_e/m_e, widely called e/m. The **specific entropy** s of a substance is its entropy per unit mass, while **specific gravity** is an old term for relative density, the density of a substance compared to that of water. (Also, a **specific gravity bottle** is a thin walled glass bottle designed to measure this last with ease, in the case of a liquid sample.) **Specific heat capacity** is an old term for specific thermal capacity. The **specific humidity** of a sample of moist air is the absolute humidity (moisture content) of unit mass. **Specific inductive capacity** is an old name for the relative permittivity of a substance. The **specific ionisation** of an ionising particle or photon is the number of ion pairs produced per unit length of path (at a stated pressure if passing through a gas).

A given substance has a number of **specific latent thermal capacities** l (unit: joule per kilogram, $J\ kg^{-1}$). Each is the energy involved in changing the state of unit mass of the substance: the different values relate to the different state changes that may take place, such as fusion (melting) or sublimation, and the different conditions under which a given change occurs. For such a state change, we have

$$W = m\ l$$

The energy transfer involved in state change is W, for a sample of mass m of a substance with specific thermal capacity l.

For **specific reluctance**, see *reluctivity* (the modern name), and for **specific resistance**, see *resistivity* for the same reason.

The **specific thermal capacity** c of a substance is the energy involved in changing the temperature of unit mass by one degree; the unit is the joule per kilogram per kelvin ($J\ kg^{-1}\ K^{-1}$). The actual value depends on the temperature as well as the substance, and over the decades various people have tried to produce theories and laws to explain this. The energy transfer W involved in changing the temperature of a sample of mass m by ΔT degrees is

$$W = m\ c\ \Delta T$$

Also, in the case of a gas sample, there is an infinite number of specific thermal capacities at a given temperature; these reflect the possible ways the temperature change can affect the volume and pressure of the sample. We therefore define two **principal specific thermal capacities** in this case, one (c_p) for a sample kept at constant pressure, and the other (c_V) for work at constant volume. The ratio of these, c_p/c_V, called γ, appears quite often in relations for gas behaviour; so too does the difference, $c_p - c_V$, the energy involved in expanding unit mass of the gas. For a perfect gas, in fact, we have the relation

$$c_p - c_V = R$$

R being the gas constant.

See also *negative specific thermal capacity*.

Also for a perfect gas, for any adiabatic change, $p\ V^\gamma$ has a constant value.

Last, the *specific volume* v of a substance is the volume of unit mass; this is the reciprocal of the density.

plural of spectrum, a one- or two-dimensional graph of some
quantity

spectral concerned with a spectrum, a one- or two-dimensional
graph of some quantity, in particular (in physics) of energy or of
some measure (such as mass or wavelength) that relates to this.
Spectral bands rather than discrete lines occur in some spectra;
a band may have a sharp edge or be more diffuse. The spectra of
stars depend in particular on surface temperature; it is therefore convenient to describe stars by **spectral class**, as in the
Hertzsprung-Russell diagram. A **spectral doublet** is a pair of close
lines in a line spectrum.

The **spectral emissivity** of a surface is the power radiated at a
given wavelength compared to that output by a black body at the
same temperature. It is quite common to define other specific
measures in the same way, in other words for a given wavelength.
Spectral ghosts are lines in a line spectrum that appear as the
result of some fault in the optical system rather then being output
by the source. The **spectral hardening** of a radiation passing
through a medium is the increase in its mean energy as a result of
absorption of low energy components by the medium.

The atomic **spectral notation** is a system for giving symbols to
energy levels so that their differences indicate the details of the
spectral lines that follow transitions between the levels. In the full
system, we describe each energy level in the form $(nl, {}^mL_j)$, giving,
in other words the level's structure and its fine structure. The table
shows the meanings of the symbols used there.

n	total quantum number (1, 2, ...)
l	azimuthal quantum number (coded s, p, d, f, ...)
m	multiplicity (1, 2, ...)
L	orbital quantum number (coded S, P, etc as above)
j	total angular momentum quantum number (half-integral and integral values)

Thus $(1s, {}^2S_{1/2})$ describes the ground state of the hydrogen atom. The
full system is often too unwieldy: more common, then, is the short
form, which leaves out the symbols for the fine structure. Then (1^2S)
describes the ground state of the hydrogen atom, with (2^2P) for the
next level, and $(1^2S - 2^2P)$ being the transition that gives the first
line of the Lyman series of the hydrogen spectrum. There is a

similar, but much more complex, system for giving symbols to molecular states for the same purpose.

The Lyman series mentioned above is an example of a **spectral series**: a sequence of lines in a line spectrum that relate in some way, in particular to a set of energy level changes that involve a given energy level. Such a series has a limit, given in most cases by setting n to infinity. The **spectral type** of a star is the same as its spectral class (above).

spectrograph system that gives a permanent record of the details of a spectrum. As well as the source of the spectrum itself, the system includes suitable sub-systems for

a) producing a parallel beam of input radiation (a so-called collimator – this was an early misprint for collineator, device for making rays collinear, i.e. parallel);

b) dispersing (spreading) the radiation into directions that depend on the measure of the spectrum in use (for instance, in the case of light spectra formed by wavelength or frequency, a prism or grating); and

c) recording the output (for instance, a photographic plate).

spectrometer system that allows the user to measure the details of a spectrum. The layout is much the same as that of a spectrograph, but with a suitable technique for measuring the directions of the components of the dispersed beam rather than simply recording the output. See also *mass spectrometer*, a device in this class used to measure the masses of the nuclides in a mixture of isotopes.

spectrophotometer type of spectrometer working on visible light (or on radiations close to that) with a photocell or photometer at the output

spectroscope simple type of spectrograph or spectrometer which lets the user view an optical (visible light) spectrum directly through a telescope, and perhaps to make measurements on the image. A **direct vision spectroscope** is even simpler, having a single straight tube of optical elements (which include a grating) through which one views the source.

spectroscopy the whole science and applied science of producing, viewing, recording, measuring, and interpreting spectra of any kind. Spectroscopy is of crucial value in many areas of science, especially optics, nuclear physics, chemistry, and astronomy. This is because a given spectrum is highly specific to its source, so it offers much power to analysts. The systems involved in practice depend on the nature of the input radiation and on its energy band (energy range).

spectrum plural spectra, a one- or two-dimensional graph of some quantity, in particular (in physics) of energy or of some measure (such as mass or wavelength) that relates to this. It is common to use the word also for the actual distribution rather than for the graph (which may be a visual display, a photograph, or the output of an electronic circuit).

Any spectrum in this context follows the existence of some form of discrete energy states in matter, each with its own energy level. This is most clearly the case with the **electromagnetic spectrum**, the very long spectrum by wavelength (frequency, energy) of electromagnetic radiation. For convenience, the bands within this particular spectrum carry different names; these are to some extent historical, but tend to relate to the specific nature and scale of their sources, their uses, and/or their detection methods. In order of increasing energy content and frequency (number of waves per second), in other words in decreasing order of wavelength and scale of source, the main bands are as follows. Note, however, that there is often significant overlap between the regions.

a) radio: involves aerials for sending and reception, for instance of the radio and television signals used in communication

b) microwave: involves very small aerials or other special sources, used for radar and point to point (narrow cast) communications in particular

c) infrared: the thermal ('heat') radiation produced by molecular changes, of wide use in fibre optics and remote control systems

d) visible light: waves the human eye can detect, with their own, well known spectrum (observed by the normal eye as hundreds of hues ranging in colour from red to blue); the sources in these cases are transitions between outer energy levels in atoms

e) ultravioet: produced by inner atomic energy level changes

f) x-rays: ditto, but with even more energy content

g) gamma-rays γ: the result of nuclear interactions and changes

The table shows the approximate frequency and wavelength ranges of these bands.

radiation	frequency/Hz	wavelength/m
radio	$3 - 3 \times 10^9$	$100 \times 10^6 - 10^{-1}$
microwave	$3 \times 10^9 - 300 \times 10^9$	$10^{-1} - 10^{-3}$
infrared	$300 \times 10^9 - 400 \times 10^{12}$	$10^{-3} - 700 \times 10^{-9}$
light	$400 \times 10^{12} - 800 \times 10^{12}$	$700 \times 10^{-9} - 400 \times 10^{-9}$
ultraviolet	$800 \times 10^{12} - 30 \times 10^{15}$	$400 \times 10^{-9} - 10 \times 10^{-9}$
x	$30 \times 10^{15} - 3 \times 10^{19}$	$10 \times 10^{-9} - 10 \times 10^{-12}$
γ	$3 \times 10^{19} -$	$10 \times 10^{-12} -$

The details of the electromagnetic spectra produced from given sources depend not just on scale but on the nature of the substance and its physical state (including its temperature). The source of a **band spectrum** (such as a compound in gas form) consists of independent molecules; the spectrum over a broad frequency range is a very large number of very close lines, giving the appearance of a band. On the other hand, a simple source (such as an element in gas form, in which the individual radiators are independent atoms) produces a **line spectrum**: a sequence of lines of well defined frequency; see, for instance, *hydrogen spectrum* and *Balmer*.

These two types of spectra are entirely characteristic of their sources, so are of major value in analysis. There is a third type of spectrum, however, where this is not the case. A **continuous spectrum** – like that of the Sun – involves sources whose radiators are not independent of each other: as in solid or liquid, or a gas under pressure. In this case, the output is of all frequencies in a wide range. See also *black body*.

We can view the spectrum of any source (whether it be a line, band or continuous spectrum) in either of two ways. If the source has a lower energy density than the background, as with a gas cloud in front of a star cluster, we observe a dark spectrum on a bright background. This is an **absorption spectrum**: the source in question selectively absorbs radiation from the background (and re-emits it in all directions), thus cutting out some wavelengths. See *Fraunhofer lines*, these being dark lines on the continuous spectrum of sunlight, due to absorption by the cooler gases of the Sun's and Earth's atmospheres. On the other hand, an **emission spectrum** is a bright spectrum on a dark background (and correspondingly in the case of non-visible radiations). Here the source has a high energy density and we work with the radiation it actually emits.

specular reflection the regular reflection of radiation from a surface that is smooth on the scale of the wavelengths concerned. (A **speculum** is an old name for mirror.)

speed c unit: metre per second, $m\ s^{-1}$ rate of change of position: the distance moved in unit time. This measure is a scalar whose vector form is velocity; the unit of both is the metre per second. We use the same symbol – or c_0 – for the special case of the **speed of light** in empty space, almost $3 \times 10^8\ m\ s^{-1}$, as this appears in many contexts in physics. We use c for the speed of light either because this

is constant and c is a symbol for constant, or from the Lain word celeritas, meaning swiftness. In any event, the symbol is now in wide use for speed to show that it differs from velocity, v.

The text above about the speed of light applies to the 'group speed' of a light wave, the speed of net energy transfer. When light passes through matter, the group speed, c, falls by an amount that depends on the refractive constant. (Some thin films – 'slow light media' – can reduce the group speed of light to a fortieth or even less. On the other hand, the high energy of laser beams can make the group velocity much higher than c_0 – the record (2006) is 300 times greater, in caesium.) Even so, the speed of the photons in any light wave in any substance is always c_0 – because in any substance there's free space between the particles. Photon scattering and absorption (maybe with re-emission) slow down the group speed.

sphere surface produced by turning a circle by 180° around any diameter. The surface area is $4 \pi r^2$; the volume is $\frac{1}{3} \pi r^3$: in each case, r is the radius of the sphere. A **sphere gap** is a spark gap between two metal sphere electrodes.

spherical shaped like a section of a sphere, or relating in some other way to a sphere. **Spherical aberration** is a type of aberration caused by an optical system in which the image of a point object is a disc. For the **spherical coordinate** r of a point in space, and for **spherical coordinates** in general, see *coordinates*. The surface of a **spherical lens** is a section of a sphere; because of spherical aberration (above), it is best to use only the central portion. The same applies to a **spherical mirror**, whose surface is also a section of sphere (but see *Schmidt lens*).

spherometer device used to measure the curvature of a spherical surface. The ends of its three legs are at the same level; the user adjusts the setting of a central screw so that its tip just touches the surface, and can then read the radius from a vernier scale.

sphygmomanometer sphyg for short, a mechanical or electronic device designed to measure the arterial blood pressure range of the human body

spike sudden brief voltage peak in an electric power supply, perhaps rising to several thousand volts. So-called uninterruptible power supplies should prevent spikes, and also the slower surges, from causing damage to electronic systems.

spin property of an elementary particle by which it acts as if spinning on an axis: in other words, it has an intrinsic angular momentum (apart from that due to any orbital motion) and, as a result, a net magnetic moment (see also *magneton*). Because of this, in a magnetic field, the particles align at an angle to the direction of the field and precess around that direction. The angle of alignment is quantised so that the component of the spin angular

momentum in the field direction is $m_s\, h/2\, \pi$, where h is the Planck constant. Despite all that, there is no physical reality at all to a particle's spin.

In that expression, m_s is the magnetic quantum number. This relates to s, the **spin quantum number** of the particle thus: m_s can take any of the values s, $s - 1$ ... 0 ... $-s$. The values of s for an electron are $\pm \frac{1}{2}$; thus m_s in this case can be $+ \frac{1}{2}$ or $- \frac{1}{2}$. In fact, most elementary particles are either fermions (with half-integral spin values like the electron) or bosons (with integral spins). The total spin S of a set of particles is the vector sum of the spins of the particles. In particular, the net spin of a nucleus is the resultant of the spins of its nucleons.

Spin flip is the reversal of the direction of spin of a particle. A **spin glass** is an alloy (not a glass in fact) of a magnetic element in a non-magnetic metal; at low temperature, the spins of the magnetic particles set in random directions, rather than being ordered (as in a ferromagnetic substance for instance). **Spin-orbit coupling** is a coupling (interaction) between the intrinsic (spin) and orbital angular momenta of a particle or object in orbit. For **spin resonance**, see *electron spin resonance*.

spinthariscope early form of scintillation detector of ionising radiation, with which the user viewed by microscope – and analysed by eye – the individual scintillations produced in a thin zinc sulphide layer. The development of effective photo-cells and photomultipliers allowed this very tedious task to be automated.

spintronics spin-based electronics, also called magneto-electronics, the fairly new – twenty-first century – field of electronics based on electron spin (which relates to magnetism) rather than on electron charge. The transistor is the basis of any form of electronics, as it is the building block of gates and all logic circuits. However, the transistor does not work well at very small scale because of 'quantum' (theory) effects, in particular the uncertainty principle. Also, tiny transistors suffer more and more from charge leakage and need power for them to retain their charge states. Spintronic gates, built on a pilot basis in the first years of this century, do not have any of these problems. The same applies to spintronic data storage, one system announced in 2007 being able to read the spin states of individual atoms.

See http://policy.iop.org/vproduction/
v5.html and *giant magnetoresistance*.

s-pole standard form for south-seeking
magnetic pole (whereas an S-pole is a
geographical south pole)

spontaneous fission the fission of a nucleus
that takes place without any known cause:
as compared with, for instance, a fission
induced by the absorption of a neutron

spreading the dispersal of a liquid as a film
over a surface, rather than its staying in
droplet form. A **spreading agent** is an
additive that helps this to take place by
reducing the surface tension of the liquid.
See also *wetting* and *angle of contact*.

spring device which deforms under load as
described by Hooke's law: in other words, the
deformation is proportional to the deforming
force, and, when the force is removed, the
system returns to its original state. The most
common types are the helical spring (a wire
shaped in a helix, like a Slinky), spiral spring
(with the wire in a flat spiral, as in many
clocks and watches), and leaf spring (with
metal strips, perhaps curved, clamped at one
or more points, as in vehicle suspension
systems). A **spring balance** is a type of force-
meter (or newton-meter) that consists of a
helical spring whose extension gives a
reading of the force applied. A common use
is for weighing objects.

spur a) junction in a network between the
main line and a smaller side line: or the side
line itself
b) a side track leading from the main track
of an ionising particle in (for instance) a
cloud chamber, caused by a low energy
secondary electron. See also *delta rays*.

sputtering the evaporation during electric
discharge, or the ejection during
bombardment, of atoms from a solid surface.
The process is often used to coat a second
surface with a thin layer of the atoms.

square wave train (sequence) of rectangular
pulses whose length is the same as the
distance between them. A square wave is the
sum of a series of odd harmonic sine waves
of suitable amplitudes.

squarial common name for a small (for
example, 300 mm) square radio receiving
aerial sensitive and accurate enough for use
with domestic satellite receivers. With various
advantages over the conventional parabolic
dish, this consists of an array of very small
printed horn aerials whose outputs link to a
single amplifier and frequency converter. At
the frequencies concerned (for example,
12 GHz), the wavelength is only a few tens of
millimetres; this means that the lengths of
the paths from the horns to the amplifier
must be very accurate to ensure the signals
are in phase. Squarials are less common now
than they were for that reason.

squegging oscillator oscillator whose output
amplitude rises from zero to a peak, then
falls to zero again – it quenches itself

squid superconducting quantum interference
device, a superconducting ring with a thin
central region, the current through which
shows high sensitivity to changes in an axial
magnetic field. First designed in 1964, it is an
application of the Josephson effect (1962).
Other uses that follow from this are to
measure very small changes of current or
voltage.

The ring may be a few millimetres across,
so that, if it is made of a high temperature
superconductor (cooled with liquid nitrogen
rather than with liquid helium), the probe
can be hand-held and still handle magnetic
field changes a billionth of the size of the
Earth's field. Biological research and medical
uses are therefore common, for instance in
checking the magnetic field generated by the
heart and brain. Thin film rather than bulk
squid rings are the aim of current research.

sr unit symbol for steradian, the unit of solid
angle

stability measure of how hard it is to disturb
an object or system from its equilibrium
state. In statics, stability depends on the
effect on the centre of mass of a small
displacement: the equilibrium state is
a) stable if the centre of mass rises
b) neutral if it stays at the same level
c) unstable if the disturbance lowers it.
In other areas of physics (such as electronics
and spectroscopy), there are also meta-stable
states, ones in which the system is unstable
but can still remain for a while before
dropping out.

stable equilibrium such a state of equilibrium
of an object or system that, if something
disturbs it, the object or system tends to
return to that original state

standard of any system generally accepted, on
a local or wider basis, as the one all should
use. In physics, there are many such
standards, as measures depend so much on
the situation. Thus, the **standard atmosphere**
is a standard pressure of 101 325 pascal, as
widely used in thermal physics.

A **standard cell** is one whose output voltage is defined to high precision, the most widely used for the purpose being the Weston cadmium cell. It is not hard for a given lab or workshop to obtain or build a standard cell; this can then act as the basis for checking the accuracy of their meters and so on. In statistics, the **standard deviation**, σ (or s), of a set of readings measures the dispersion of the set from the mean value; it is the square root of the variance, and is the square root of the mean of the squares of the deviations of the readings from the mean. A **standard electrode** is a half cell used to measure electrode potentials. The hydrogen standard electrode is the main one, though the calomel (mercurous chloride) system is quite common. A **standard illuminant** is a defined source of light used in colorimetry. The **standard near-point distance** of the eye is 250 mm: the distance to the point of closest clear vision for the 'normal' eye. **Standard pressure** is 101 325 pascal, as noted above, while **standard temperature** is 0 °C in physics (but often 15 °C in chemistry). In physics, then, **standard temperature and pressure**, stp, is 0 °C and 101 325 Pa.

standard model of matter a quantum field theory, first set up in 1970, that tries to describe and bring together all the fundamental forces and particles known, while keeping account of relativity and the quantum theory. The main gap at the time of writing is the lack of success in bring gravity into the model; however, it does link the electroweak theory – which relates the electromagnetic and weak nuclear inter-actions – and quantum chromodynamics which deals with the strong interaction. Also, the model has been able to predict a number of particles before they were found, and to describe them closely – these are the charm and top quarks, the W and Z bosons, and the gluons.

As regards particles, the model classes these into two groups – the first is the fermions (with half-integral spin) that make up matter and obey the Pauli exclusion principle. There are three generations of these, with eight particles in each – only the first generation members are stable enough to play a role in normal matter (they include the electron family and the up and down quark/anti-quark pairs which make up the nucleons; the particles in the other two generations decay quickly into the first generation. We can also class the fermions into leptons and hadrons; the former have no colour, while the latter do, being made of quarks.

The second group of particles consists of the exchange particles called bosons (with integral spin) that carry forces and do not follow the exclusion principle.

The bosons and the interactions they mediate appear in the table at the foot of the page.

Despite the success of the standard model, it cannot deal with gravitation, predict the masses of many particles in any logical, unified way, or explain either the lack of balance in the Universe between matter and antimatter or how inflation occurs. Since 1974, a large number of grand unified theories (GUTs) have tried to deal with these gaps – they have had no more success, so much research goes on into tweaking the standard model. For instance, the standard model cannot explain massless neutrino oscillation – nor can the GUTs – so now the standard model allows neutrinos to have a tiny mass. The large hadron collider (lhc) should support many aspects of the standard

interaction	boson group	SU* theory
gravitational (to a tiny extent)	Higgs bosons (not true bosons, though, and not yet found)	–
gravitational (to no extent)	gravitons (also never yet seen)	–
electromagnetic	photons	$SU(2) \times U(1)$
weak nuclear	W and Z particles	$SU(2) \times U(1)$
strong nuclear	gluons	$SU(3)$

* SU is unitary symmetry, which includes the 'eightfold way' and has had great success in setting up 'periodic tables' for matter particles.

model, as did its predecessor, the large electron-positron collider (lep).

standing wave a constant pattern of nodes and antinodes set up in a medium by the superposition (interference) of a progressive wave in one direction with the same wave reflected back from some boundary. In practice, reflection is rarely perfect, so that the amplitudes of the two waves are not often quite the same. The nodes are then points of minimum rather than zero amplitude. In a given case, the **standing wave ratio** is the ratio of the amplitude at an antinode to that at a node. The various pictures of *sound sources* show the one-dimensional standing waves concerned as dotted lines.

star a) method of linking three or more sub-circuits (for instance, the three single-phase circuits of a three-phase system).
b) massive but compact glowing ball of plasma held together by its own gravity and able to produce energy through nuclear fusion, a series of thermonuclear reactions in its core. The sun is a good example (though it is a little less massive than most). We class each star by its

- age – it evolves through a sequence of stages, mainly in the Hertzsprung–Russell diagram:
- birth: as it starts to condense from a disturbance in a molecular cloud (one of high density compared to most of space, but still a hard vacuum in our terms); it next begins to collapse under gravity and gain energy from that process – density and temperature rise
- proto-star stage: the first equilibrium stage (about 8% of the mass of the Sun, a few tens of million years), as the star contracts gently but the temperature stays stable; many proto-stars have a *planetary nebula* (sense a)) from which planets may later condense
- main sequence (where most stars spend 90% or more of their lives): throughout this time, as it passes from being a hot, blue giant, a star takes energy from the hydrogen-helium fusion reaction in its dense core
- red giant stage: once the core has lost its hydrogen, the star begins to collapse until fusion can re-start in a shell round the core; this makes the outer layers expand and cool: at the same time, the core contracts and heats up until it can sustain helium-based fusion reactions (in the case

of massive stars forming elements as far as iron)
- white dwarf stage (back on the main sequence after the loss of the outer layers of gas to form a *planetary nebula*, sense b), which then drifts away): further collapse and cooling over a few millions years
- death as a cooling red dwarf
- make-up in terms of chemical elements – the 'metallicity', the fraction of mass that consists of elements higher than helium being the main factor; population II stars have a much lower value (as little as a thirtieth) than population I stars (because of the different ages, and therefore types, of cloud from which they formed)
- mass (from which follow its luminosity and the length of time it spends on the main sequence) – the most massive stars (some 150 times more massive than the Sun) live only a few millions years, while the least massive ones (less than 10% of solar mass) may last a million million years
- spectral type or class, there being the classes given by the mnemonic 'Oh, Be A Fine Girl (Guy) – Kiss Me Right Now ... Smack!', though the last three are very small; each class has sub-classes that range from 0 – hottest – to 9 – coolest. The main classes are
- O: blue (surface temperature over 30 000 K) giants (radius about 15 times the solar radius r_s) – He$^+$ lines dominant in the spectrum
- B: blue-white (30 000–10 000 K) stars (about 7 r_s) – He lines dominant, and no He$^+$ showing
- A: white (10 000–7500 K) stars (about 2.5 r_s) – H lines dominant (and some metal ion lines)
- F: yellow-white (7500–6000 K) stars (about 1.3 r_s) – weaker H lines and stronger metal ion lines
- G: yellow (6000–5000 K) stars, including the Sun (which is a bit smaller than the mean of 1.1 r_s) – Ca$^+$ dominant
- K: orange (5000–3500 K) stars (about 0.9 r_s) – metal lines dominant
- M: red (3500–2000 K) non-giant stars (about 0.4 r_s) – molecular bands dominant

Stark, Johannes (1874–1957) physicist whose 1919 Nobel Prize followed his 1913 discovery of the **Stark effect**. This is the effect on spectral lines of a strong electric field across the source. The effect may be line displacement and/or line splitting. Explaining the effect in terms of quantum physics was a triumph for that theory. The Zeeman effect is much the same, but involves a magnetic field.

The **Stark–Einstein equation** gives the energy W absorbed per mole in a photochemical reaction:

$$W = N h v$$

N is the Avogadro number, h is the Planck constant, and v is the frequency of the absorbed radiation.

state **a)** the condition of a particle or system, as described by the values of all relevant properties. That includes the quantum numbers, which are the most relevant at the smallest scales.
b) the condition of a sample of matter, as described by the interrelationships between its particles. The **states of matter** are the main phases in which matter normally exists: solid, liquid, gas and (to some) plasma. A solid sample tends to be fixed in shape and size, and is hard to deform; this is because the particles concerned bond strongly. The others are fluids in that they have little resistance to flow: liquid particles bond weakly, enough to cause a surface (see also *surface tension*), but there is no interaction between the particles of gases and plasma, both of which therefore fill the space available and are easy to compress. The particles of a plasma are not the same as those of the other states: rather they are the ions and electrons caused by the breakdown of those particles; this is why some people do not call plasma the fourth state of matter. See also *kinetic model* and the state changes (such as melting) concerned.
static **a)** electronic noise that interferes with radio signals, caused by lightning and nearby sources such as motors switching on and off, computer activity, and ignition systems
b) common name for charge built up by friction between clothing and carpet (for instance) and for the small shocks that can result
c) unchanging or unmoving, or associated with something that is unchanging or unmoving – as opposed, in most cases, to dynamic. The **static characteristics** of an electronic circuit or element are the characteristic curves obtained by plotting dc rather than ac values. **Static electricity**

concerns the effects of electric charges at rest (whereas the concern of current electricity is charge in net motion); *electrostatics* is its study. **Static friction** is the friction (resistance to motion) between two surfaces in contact that are at rest – so it relates to the force needed to start them moving. The **static pressure** at a depth in a fluid is the pressure due to the weight of that depth of fluid rather than to any motion of or in the fluid; see *hydrostatics*. **Static tube** is a name for the Pitot tube; it is not a very common name.
statics the study of forces and torques in equilibrium, and therefore of objects and systems whose momentum and angular momentum do not change. (If either does change, the field is dynamics.) Hydrostatics is the study of fluid samples in equilibrium.
stationary much the same as static, or associated with the stationary value of some function. A **stationary orbit** is an old term for the geosynchronous orbit of a satellite above the equator, an orbit which causes the craft to go round the Earth at the same angular speed as the Earth rotates beneath it; this keeps the craft at the same point in the sky from the point of view of a given observer. A **stationary state** is a state of an atom (for instance) allowed by the quantum theory, as opposed to a forbidden state. A **stationary wave** is the same as a standing wave, the latter being the more common name.
statistical mechanics or **statistical physics** the use of statistics to predict the behaviour and properties of matter in bulk from a knowledge of the behaviour, properties and interactions of the particles concerned. Two major fields at atomic scales and below are Bose–Einstein statistics and Fermi–Dirac statistics; also, in nuclear physics, radio-activity is a statistical process. At atomic scales and above, the techniques are of great value in thermodynamics. Thus, kinetic theories depend on statistics: they deal with the bulk behaviour of large numbers of unpredictable particles (think of the uncertainty principle).
statistics the branch of mathematics that deals with the manipulation of data in numeric form. The concern of **analytical statistics** is the ways in which one may collect or sample data. **Descriptive statistics** involves the collation, classification, processing and presentation of data once collected. The aim of the latter field is to produce a mathematical model of the distribution of

the data; this also involves an understanding of probability theory.

stator originally the non-rotating part of a rotating electrical machine, that part producing the magnetic field. The term now applies more generally to the part of an electrical machine that is not the rotor.

steady state Universe an old view that the Universe has always existed in the current form, rather than having evolved from (for instance) a big bang. If a steady state Universe is expanding, matter must be created throughout it all the time in order to maintain the density. This requires creation at the rate of about 10^{-40} kg per cubic metre per second. However, evidence from astronomy appears to show that the density of the Universe is not constant. Hermann Bondi, Fred Hoyle and Thomas Gold devised the steady state theory of the Universe in 1948. This was more widely accepted than the big bang until the discovery of the cosmic background radiation.

steam the vapour phase of H_2O at or above the boiling temperature at the pressure concerned.

As a given mass of steam has some 1700 times the volume of the liquid, its potential for motive power has long been recognised. Thus, the **steam engine** was the first working type of heat engine; the first effective design appeared in 1698 (Thomas Savery, though Hero and others produced design some 1900 years before). The traditional steam engine, as used for pumping and for locomotives (tractors, road rollers, and rail engines), is a reciprocating external combustion engine: energy input to a boiler produces steam that is led to a cylinder where it drives a piston. This is of low efficiency, however, and has now been superseded by the *steam turbine* (below) and the various internal combustion engines.

The **steam line** is an old name for the line between the liquid and vapour phases on a graph of saturated vapour pressure against temperature in the case of H_2O. The **steam temperature** was a fixed temperature on the Celsius scale – 100 °C as given by the boiling temperature of pure H_2O at standard air pressure (101 325 pascal); its value is now 99.9839 °C. A **steam turbine** is a rotating steam engine, modern forms being highly efficient. High pressure, high temperature steam from a boiler passes through a series of turbine fans and forces these to rotate as

it loses energy. First developed in the 1880s by the engineer Charles Parsons, the steam turbine is widely used in ships and electricity generating stations.

Stefan, Joseph (1835–1893) physicist, mathematician and poet ('There always something will remain/That we shall not know, but why?'). The **Stefan constant** appears in **Stefan's law** (or the **Stefan–Boltzmann law**): the total power output by unit area of the surface of a black body is proportional to the fourth power of its temperature. The Stefan constant σ is the constant of proportionality, 5.7×10^{-8} W m^{-1} K^{-4}. This is a fundamental constant.

Steinberger, Jack (1921–) physicist who helped develop neutrino beam technology and discovered the muon neutrino. He shared the 1988 Nobel Prize with Lederman and Schwartz for this.

stellar relating to a star or stars. Thus **stellar spectra** are the spectra of the radiation from stars. See *star* for the main **stellar spectral classes**.

stem scanning tunnelling electron microscope, one able not only to view individual atoms in certain contexts, but to move them from place to place on a surface

step-down transformer voltage transformer whose output voltage is less than the input value. A **step-up transformer** has the reverse effect. The output/input voltage ratio equals the ratio of the numbers of turns in the output and input coils.

Stephenson, George (1781–1848) engineer whose achievements include

a) an efficient stationary steam engine (1812)

b) a steam-driven locomotive, the Blucher (1814), used to haul coal for a mine, and the first to use flanged wheels

c) a successful safety lamp for use in mines (1815)

d) the Darlington–Stockton railway (the first public railway, 1825)

e) the Rocket engine which from 1829 served the 65 km Manchester – Liverpool line at up to 50 km h^{-1}.

George's son Robert (1803–1859) was involved in the last two projects, but is best known for his bridges. On some later projects he worked with Brunel.

stepper motor a digital electric motor, one that turns at constant speed in steps of a given angle (rather than continuously). Stepper motors are of great value in many automated control systems, including robotics.

steradian sr the unit of solid angle – a supplementary SI unit (i.e. it is one that we do not need – as it has no dimension – but find useful). It is the solid angle which subtends an area r^2 at the surface of a sphere of radius r. As the area of a sphere is $4\pi r^2$, there are 4π steradians in a sphere (and there cannot be a larger solid angle than that).

stereo- prefix indicating a solid three-dimensional nature or appearance. A **stereoscope** is a device with which each eye views one of a special pair of pictures, in such a way as to obtain a 3D effect. The two pictures may be photos taken a few centimetres (or further, even much further in the case of very large objects) apart, as if through two eyes. A **stereoscopic microscope** has an optical system for each eye, so gives a 3D image of the object viewed. **Stereoscopic vision** is normal human 3D vision: the brain fuses the two images from the two eyes to produce a 3D effect.

Stern–Gerlach experiment the first demonstration (1922) that the angular momentum of the electron is quantised. Since tried with other particles (with the same success), the test showed that an electron beam passed through a shaped, highly non-homogeneous magnetic field has a line spectrum rather than a band. Classically, a particle with angular momentum ('spin') should have any value between the two extremes – so in this context would have a band spectrum; Stern and Gerlach found only two lines in the electron spectrum, showing there are only two spin states. Physicist Otto Stern (1888–1969) gained the 1943 Nobel Prize for his work on the magnetic properties of atoms and on the magnetic moment of the proton; his colleague for this experiment was Walter Gerlach (1889–1979).

Stevens, John (1749–1838) engineer whose achievements include.

a) the first steam boat with a screw drive (1802)

b) the first steam boat for use at sea (1809)

c) the first steam locomotive outside of UK (1825)

John's son Robert (1787–1856) worked on the above; he also invented the modern form of locomotive rail/sleeper system (1830).

stiffness k unit: newton per metre, N m^{-1} the force needed to produce unit deformation in

an elastic object or system, the reciprocal of its elasticity (also called compliance) if it is a single substance

stilb an old unit of luminance, the candela per square centimetre

stimulated emission the de-excitation of an atom, with the emission of a photon, caused by the passage nearby of an identical photon. The two photons then travel on parallel and in phase. This is an important aspect of laser action.

Stimulated Raman scattering is a rather similar process. Within a suitable sample, it allows the wavelength of radiation passing through to switch to a more useful value. For instance, a tube of high pressure methane will switch a 1.06 μm laser beam to 1.54 μm; the output cannot then cause eye damage. The process is one of the four variations of the spontaneous Raman effect and produces one or more harmonic under-tones of the inbound wave pulses.

Stirling, Robert (1790–1878) priest and inventor, whose **Stirling cycle** has long been the basis of a potentially highly efficient engine, the **Stirling engine**. Indeed, it competed with the steam engine until the electric motor and Otto's internal combustion engine superseded both at the end of the nineteenth century. Now, research into the engine, an external combustion type with a piston for use on land and in satellites, has grown a great deal; a number of types have been designed since the 1980s.

Used in reverse, driven by another engine to produce a heat pump effect, Stirling first designed the cycle in 1816 and built the first working model in 1818. That version of the cycle was, and remains, a heat pump that is of interest for refrigeration. A major problem in either version, however, has been to keep the working fluid in the system. This is even more so if the fluid is hydrogen gas (which is the most efficient otherwise.

The system consists of two opposing pistons in a single long cylinder with the working substance between them; it drives the pistons at 90° out of phase. Between the pistons is a heat exchanger, called a regenerator matrix. The sketches show the heat pump cycle; much the same principles apply to that for the Stirling engine. There is a good animation and description of the latter at http://auto.howstuffworks.com/stirling-engine.htm; however, this does not show or mention the regenerator matrix.

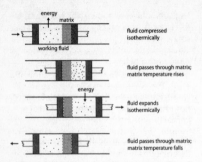

fluid compressed isothermically

fluid passes through matrix; matrix temperature rises

fluid expands isothermically

fluid passes through matrix; matrix temperature falls

stm scanning tunnel microscope, a microscope based on quantum tunnelling effects

stochastic process one that random walk statistics can explain, as in the case of Brownian motion. It is any process that depends on the laws of probability for random events. In particular, the probability of a given event at a given moment depends on the random events at earlier moments.

Störmer, Horst (1949–) one of the three winners of the 1998 Nobel Prize, for their discovery of a 'new type of quantum fluid with fractionally charged oscillations'

stokes St an old unit for kinematic viscosity (fluid friction), 10^{-4} m^2 s^{-1}

Stokes, George (1819–1903) mathematical physicist who was the first to use spectroscopy to analyse the light from the Sun and other stars, and the first to identify x-rays as electromagnetic radiation. **Stokes' law** concerns the fluid friction on a sphere passing through a fluid: the force that opposes motion is

$$F = 6 \pi r \chi c$$

Here r is the radius of the sphere and c its speed through a fluid of friction (viscosity) χ.

At the terminal speed, when c becomes constant, this force is the same as the driving force (e.g. the weight of the sphere if falling).

The **Stokes law of fluorescence** is that the wavelength of the radiation output in fluorescence is always greater than that input (in other words, it has less energy); however, the 'law' is not always valid if the target atoms are not in the ground state because of stimulated emission. The Stokes shift is the wavelength difference between input and output quanta.

stone old (Imperial) unit of mass, 14 pounds (6.36 kg)

stop an artificial aperture in an optical system, such as an iris diaphragm (which is a variable aperture, or **variable stop**). In optics and in electronics, the **stop band** of a filter is the range of frequencies (frequency band) which the filter blocks; it is the reverse of pass band. The **stop number** of an optical system or element is the same as its f-number – the focal distance divided by the diameter of the entrance pupil.

stopping power a measure of the ability of a substance to reduce the kinetic energy of particle radiation passing through it. The usual expression is the energy drop per unit distance, $-dW/dl$; this gives the unit joule per metre, J m^{-1}. The **relative stopping power** of a substance is its stopping power compared to that of some standard.

storage cell the same as a secondary cell or accumulator (one that the user can re-charge). A **storage ring** is a circular tube, with magnets all around the circumference to produce a uniform field, able to store a beam of charged particles output from an accelerator.

stp (or **STP**) standard temperature and pressure, in physics 0 °C and 101 325 Pa

strain ε the deformation of a sample per unit original dimension, as a result of a stress. The normal forms are **compressive strain** and **tensile strain**, i.e. when squashing or stretching the sample – both are the corresponding change of length of the sample per unit length:

$$\varepsilon = \Delta l / l$$

There is no unit. **Bulk strain** is the change of volume per unit volume, while **shear strain** equals the angular deformation produced during shear action.

A **strain gauge** is a device designed to measure the strain at a point (in most cases, in a surface). A common form measures the change of resistance of a wire bonded to the surface, as strain in the surface causes the wire to stretch; there are, however, very many other designs.

strange (in the early days) of a particle with a surprisingly long life time or (now) of one that contains a strange quark. Strangeness, S, is the corresponding quantum number. **Strange matter** would consist of roughly equal parts of up, down and strange quarks. Proposed in 1984, the theory of this holds that under very high pressures and temperatures (as during the big bang and supernovas), the quarks could link to form low energy (i.e. stable) 'nuggets' (sometimes called **strangelets**) of strange matter. Indeed, some people believe that supernovas produce **strange stars** (ones made of strange matter) rather than neutron stars. It is true that a strange star could spin as fast as 2000 rev s^{-1} to produce the effect of a very fast pulsar (the record is 716 rev s^{-1}, 2006) while speeds above 1500 rev s^{-1} would make a neutron star break up.

strangeness S quantum number, conserved in strong and electromagnetic interactions but not in weak ones, which relates to the nature of strange particles (such as hyperons). Particles which are not strange have a strangeness of zero. Strangeness is conserved in the strong and electromagnetic inter-actions, but not the weak interaction. A light particle that contains a strange quark cannot decay by the strong interaction, so has a long lifetime.

stratosphere the layer in the atmosphere from about 20 to about 50 km in altitude. It includes the ozone layer, an important absorber of solar ultraviolet radiation. The **stratopause** is the upper boundary of the stratosphere, where it merges into the mesosphere.

stray accidental but still perhaps significant. Thus the **stray capacitance** of a circuit is the capacitance associated with the circuit's junctions and electrodes. In circuits, a **stray voltage** can 'leak' from one part to another and sometimes cause interference. (In the US, the same name seems to be used for static as a cause of accident or, on the road, as a cause of motion sickness.)

streamer the appearance of a spark discharge in a gas at a fairly low pressure: a continuous (or near continuous) wriggling broad glowing discharge. The now very rare **streamer chamber** is the corresponding (i.e. low pressure) form of spark chamber.

streaming the net flow of a fluid induced by sound waves in it. A **streaming potential** appears between the ends of the pores when one forces an electrolytic fluid through a porous solid; it is the reverse of electro-osmosis.

streamline a line in a flowing fluid whose tangent at any point gives the direction of flow at the point. One can see the overall pattern of the flow from a set of streamlines.

Streamline flow is the same as laminar flow (in other words, steady rather than turbulent flow): the streamlines are all continuous and very close to parallel. **Streamlining** is the design of an object to reduce as much as possible the drag on it as it passes through a fluid. The most effective shape for subsonic speeds differs from that at supersonic speeds, however, because of the shock wave that appears in the latter case.

strength the greatest stress a solid sample can take before it fractures. It is normal to describe this in terms of the type of stress involved (for instance, shear strength for shear stress, tensile strength for tensile stress, and so on).

stress σ unit: newton per square metre, $N\ m^{-2}$ the force in an object per unit section area. It is normal to describe this in terms of the type of deformation (strain) that could result: thus tensile, compressive, shear stress. Pressure is the same as compressive stress; it has the same unit (newton per square metre, but called pascal). **Stress analysis** is the mathematical description of the stress throughout a sample, or the use of polarised light (for instance) to explore this. In practice, a stressed sample often suffers a complex pattern of **stress components** rather than being purely under tension, say. For **stress/strain curve**, see *elasticity*.

stretch marks marks that appear on the surface of samples of certain metals as a result of strain past the elastic limit. Lüder bands are an example.

stretched string an elastic string or wire under tension, in effect a one-dimensional sound source. The name is sonometer in physics and monochord in music (in which context a guitar, for instance, is a stringed instrument).

strings or **cosmic strings** long thin tubes of mass–energy, of radius perhaps as small as the Planck length (far below that of an atom) but with huge mass per unit length, deemed by some theories to criss-cross the Universe. Strings, or knots of (or loops in) strings, are proposed in order to explain the missing matter problem and as 'condensation nuclei' for galaxies. **String theories** were the many attempts like this of the 1980s and 1990s to combine the two main theories of modern physics (quantum physics and relativity); they tend to imply a Universe with as many as ten or eleven dimensions. Superstring theory is a more recent development – but again there are many versions. In neither case is there any evidence as yet (2006), though if strings exist, most string theories would say that so then do monopoles – and these have not been found. A good site on this whole topic is http://www.damtp.cam.ac.uk/user/gr/public/cshome.html. See also *causal dynamical triangulation*, a new theory of the universe that has already had some greater success than string theories.

stripping or **stripping reaction** nuclear reaction in which a compound particle passing close to a nucleus loses one of its components to the nucleus. An important case is the D-P reaction, in which a moving deuteron loses its neutron to a massive nucleus while passing, with the proton going on much as before.

strobe (or **stroboscope**) device or system which allows one to view a scene or object only at regular intervals. A common form produces a sequence of intense brief flashes of light. **Stroboscopic photography** uses the technique to obtain a series of 'snapshots' of a moving object in a single frame for later analysis.

strong force the stronger of the two nuclear forces, leading to a **strong interaction** between hadrons (baryons and mesons). While strong, this force has only a short range, of about 10^{-15} m. It is common to describe strong interactions in terms of the interchange of a virtual hadron between the particles concerned. A large number of quantum numbers must be conserved (constant) during such strong interactions and decays. Described by quantum chromodynamics, at the most fundamental level it involves interactions between gluons, quarks and anti-quarks, the gluons being the exchange particles concerned.

strontium Sr alkaline earth metal, group II, period 5, proton number 38, first isolated by Humphry Davy in 1808 and named after Strontian in Scotland (in whose lead mine people found the ore). Relative density 2.6, melting and boiling temperatures 780 °C and 1400 °C; most common stable isotope 88Sr. Used in some radioisotope thermoelectric generators, in some ferrite magnets, and for radiotherapy.

Strutt, John (1842–1919) became Lord *Rayleigh* in 1872, before he did much of the physics research for which he became famous and for which he gained the 1904 Nobel Prize

sub- prefix indicating a low, maybe unusually

low, level or value. Thus **subatomic** describes something smaller than an atom or able to form part of an atom, such as a **subatomic particle**, or elementary particle. A **subcarrier** is a radio frequency wave that carries a signal by modulation, but is itself carried by a full carrier – this is an early form of multiplexing (used to carry, for instance, stereo). It allows a single carrier to convey a number of subcarriers each with its own signal. Special decoders in the receivers would separate these signals and allow the user to choose between them.

A **subcritical** value lies below the critical value concerned. In particular, the term describes a chain reaction whose multiplication factor is less than 1, so that the chain reaction does not sustain itself. An atom bomb contains a number of sub-critical lumps, plus a system to force them together at the moment of detonation. The frequency of a **subharmonic** is an integral submultiple of the fundamental (base or natural) frequency of the system concerned.

sublimation the change of phase of a substance directly between solid and vapour, in other words without the formation of a liquid. This can occur with any substance below its triple point (the only pressure/temperature context in which solid, liquid and vapour phases can exist in equilibrium together). The process of freeze drying foods involves sublimation.

sub-millimetre band region of the electromagnetic spectrum between very short radio waves (the millimetric waves) and the far infrared. The wavelength range is 1 mm down to 0.2 mm (both values approximate).

Subsonic speeds are speeds below the speed of sound in the medium concerned.

subtraction process of the colour mixing of dyes (pigments) rather than of lights (which is additive mixing rather than **subtractive mixing**). The colour of the light reflected depends on the absorption (in other words, subtraction) of certain input colours by the dyes in the mixture. Much the same applies for light transmitted through a set of colour filters.

SU groups groups of elementary particles as described by one of the grand unified theories of unitary symmetry

sulphur (US: sulfur) S non-metal with three allotropes, in group VI, period 3, with proton number 16, known since ancient times (e.g. as brimstone, including in medicine), Latin name 'sulpur'. Relative density 2.1, melting and boiling temperatures 115 °C and 445 °C; most common isotope ^{32}S.

sum tone a perceived combination tone (sound wave) of frequency given by the sum of the frequencies of its components

Sun the star at the centre, and with 99.9% of the mass, of the solar system of planets and other objects. The Sun is a rather small but generally quite normal star (of type G); it consists mainly of 70% hydrogen and 28% helium. The visible disc is the surface of the photosphere, 1.4×10^6 m in diameter, whose temperature is almost 6000 K (though the core temperature is some 20 MK). The corona, partly visible in a total eclipse, is a much thinner gas outside the photosphere; the temperature of this can be as high as 1 MK, and it extends far past the orbit of the Earth (in the form of the solar wind). A **sundial** is an ancient form of shadow clock that usually shows local time (though one can calibrate it to standard time).

Sunspots are cooler, therefore darker, regions on the surface of the photosphere, sometimes visible with the (protected) naked eye (and noted in reports dating back since ancient times). Their exact nature is not clear, but they tend to come in pairs or groups and relate to intense solar activity (e.g. auroras), and most people assume they are storms. The temperature at the centre of such a spot may be as low as 5000 K. The **sunspot cycle** is a steady cycle of solar activity that peaks about every 11 years; there are other, less obvious, solar cycles. These have some effects on the Earth, as there is slightly more solar radiation at times of few sunspots. (During the Maunder minimum, 1645–1715, when there were very few sunspots, the Earth's mean temperature fell by about 1 °C. This was the time of Europe's 'little ice age'.) A new sunspot model (early 2006) predicted that the next cycle would start in late 2007 or early 2008 and be 30–50% stronger than the last.

super-

prefix for a high, perhaps unusually high, value or level. A **super-alloy** is an alloy designed for rugged high temperature use, most often based on cobalt, nickel or nickel-iron, with the main applications in aerospace (e.g. turbine blades and rocket engines); the first generation appeared in the 1940s, working up to around 700 °C. The fourth generation, single crystal super-alloys, mostly include ruthenium and can work up to about 1100 °C. See http://www.msm.cam.ac.uk/phase-trans/2003/nickel.html for a major resource on this topic from Professor Harry Bhadeshia of Cambridge University.

For **superantiferromagnetism**, see *supermagnetism*, below. A **super-atom** is a Bose–Einstein condensate – tens or hundreds of atoms linked at a very low temperature so they act as one.

superconductivity the passage of an electric current through a **superconductor**, a substance which offers zero resistance. Discovered by Kamerlingh-Onnes in 1911; it was little more than a curiosity for a long time, because of the extreme conditions involved. The phenomenon occurs in many metals and alloys, and in silicon, at very low temperature (no higher than a few K). A superconductor is also a perfectly diamagnetic substance. The cause of superconductivity is believed to be the pairing of electrons in such a way as to create an energy gap, or tunnel, in the energy band spectrum. There are two types of superconductivity – Type I in all elemental superconductors except carbon nano-tubes, niobium, technetium and vanadium, and Type II in other cases. See also *Meissner effect*.

Superconductivity disappears at the transition (or critical) temperature, or below this if the outside magnetic field is greater than a critical value. A type I superconductor has a very precise transition temperature and critical field strengths; type II substances show more gentle change. In the former case, the critical field strength at temperature T is proportional to $1 - (T/T_c)^2$, T_c being the critical temperature.

By another process (known since 1986), there is also superconductivity in a number of other – now complex (therefore Type II) – substances at higher temperatures. Those **high-temperature superconductors** known at the time of writing lie in the range 35–138 K (–135 °C). Their uses include applications in electromagnetism (such as highly efficient electromagnets and rotating machines), and in telecommunications (where, for instance, a superconducting resonator has a Q-factor tens of times better than with a normal approach). At low frequencies in particular, a superconducting device may have a hundredth the volume of a traditional system. Also, the squid, superconducting quantum interference device, is of great value for magnetic field measurement, for instance. Many people believe that, if there will ever be room temperature superconductivity – which could be of great value – it will be a third process in a third group of materials (maybe carbon-based ones).

Most current research in the field is into methods of making and working with high-temperature superconducting materials. The aims are to increase critical temperatures and sample strength and homogeneity, and to form samples into various useful shapes (rings, rods, wire, thin film, etc.). Also, the smoother the surface the better able a sample is to retain its properties at radio frequencies.

Work on organic superconductors is also developing; the strongest hopes of recent decades are that an organic substance superconducting at room temperatures is possible. As yet, however, the record is 12 K (a certain types of nano-tube). A second important research area is to develop 'tunable' superconductors: ones whose superconductivity one can adjust by way of the molecular structure. But maybe this is all a dead end? – research into high-temperature superconductivity peaked in about 1990 and has fallen all the time since then – there has been no real breakthrough for almost two decades.

Supercooling is taking the temperature of a substance below that at which some change would normally occur; in particular, a supercooled liquid is liquid below the normal freezing temperature. This tends to require high purity and no vibration (to be supercooled is to be in a meta-stable state); see, however, *glass*, and compare with *superheating* (below). A **supercurrent** is one to whose flow there is no resistance – in the cases of superconduction and Josephson flow in electricity and superfluidity in fluid flow. A **supercritical** sample of fissile material has a mass above the critical value; it can show a branching (explosive) chain reaction – in other words, its multiplication factor is greater than 1. The **superdense theory** is a name for the big bang theory of the start of the Universe; at the moment of the big bang, the Universe was a point of infinite density – and throughout the first few seconds after the big bang, the Universe was far denser than a black dwarf. For **superelastic collision**, see *restitution*.

Something **superficial** is concerned with the surface of a sample rather than its whole volume (bulk). (This time 'super' is not the prefix meaning high.) Thus, the **superficial expansivity** of a substance is the change in area of unit area per unit temperature change; this is much more significant than other types of expansivity in a number of engineering contexts.

Superfluid liquid helium shows the property of **superfluidity** – apparently often fully friction-free flow (and also very high thermal conductivity). The transition to superfluid helium (helium II) occurs at 2.186 K (the lambda temperature); below this temperature, the fluid will climb out of a container, for instance. The explanation depends on quantum physics, as is the case with superconductivity (to which some people also apply the term). See also *Bose–Einstein condensate*, which is a type of superfluid. **Supergravity** is a theory – or, rather, a set of theories – which try to relate supersymmetry with all four basic interactions. It requires the graviton and the gravitino, neither of which have yet been found, and predicts even more fundamental particles than we have now. **Superheating** is, in effect, the reverse of *supercooling* (above): it involves taking a liquid above its boiling temperature without change of phase. The **superheavy elements** are the elements past uranium (proton number $A = 92$), the highest known by mid-2006 having $A = 116$. See *trans-uranic elements* and *ununxium*.

In a **superheterodyne** *receiver*, the input radio signal mixes with a signal generated inside the set to give an output that contains all the original information, though is still of a high frequency. The system amplifies and rectifies this intermediate signal in the usual way, but the end result is higher amplification and selectivity. Thus the superhet set, as it is called, is the most common type of radio receiver. See also *beat receiver*.

A **superhigh frequency** (shf) electromagnetic wave or signal is in the range 300–3000 MHz. A **superlattice** is a nano-structure of thin semiconductor layers, each a few atoms thick, used as the core of a saser, for instance. A **superlens** is a lens able to resolve below the wavelength of the light used; there are at least two designs, one using a material of negative refractive constant and the other using a very thin film of silver to amplify the input waves. It may also be possible to use such lenses for cloaking. There are various forms of **supermagnetism**, the magnetism of a substance to a higher degree than normal. A superparamagnetic substance is one that consists of such small ferromagnetic particles that each particle is a single domain. Superantiferromagnetism is much the same, except the tiny particles are antiferromagnetic. In both cases, the susceptibility is unusually high, especially at low temperatures. The name for a supermagnetic star – a neutron star with an extremely strong magfield, is magnetar.

A **supernova** is the explosive end of a star (a nova, on the other hand, doesn't kill the star, so it can explode again). During the event, the rate of light radiation can be billions of times greater than that of our Sun. This rate is so high that the event can be spotted even a huge distance away; it is also fixed, so it is common to use supernovas to measure distance. On the other hand, supernovas are very rare – there have been only three or four in our galaxy in the last thousand years, only one since the invention of the telescope in 1608:

- 1054, reported mainly in China (as bright as the Full Moon, and visible in daylight for a month); the Crab Nebula is all that remains, with a neutron star at the centre
- 1572, reported by Brahe in particular (as bright as Jupiter, so maybe just visible in daylight); there is no visible remnant, but there is an x-ray nebula in the right place
- 1604, studied mainly by Kepler (brighter than any other star); only an x-ray nebula remains for us to see
- 1987 (not strictly in our galaxy as in one of the star clouds that orbit it), studied in great detail by only two major telescopes (as it's visible only in the southern hemisphere) and the Hubble telescope. Thus, it is now clear that, after the explosion, its main source of energy is the decay of cobalt-56; this nuclide is the

product of the decay of nickel-56 produced in abundance during the explosion; in turn it decays to iron-56.

There are hopes of another supernova soon (in the next few tens of thousand years), in the case of the star RS Ophiuchi, which is only 5000 light years distant, therefore closer than the 1054 event (6000 light years). This white dwarf has shown nova outbursts six times since 1898, and is now thought to be close to the Chandrasekhar limit – it will then become a Type 1a supernova, which people should be able to see in daylight for up to a week.

For **superparamagnetism**, see *supermagnetism*, above. **Superplasticity** is a form of plastic behaviour in some fine-grained metals and ceramics under very high pressure and at high temperature: deformation is then fast acting, yet still takes place under low stress levels. The process is thought to follow slippage between the grains. See, for instance, *anti-crack*.

The principle of **superposition** is that when two (or more) waves of the same type pass through the same region, the amplitude at any point is the (algebraic) sum of the individual amplitudes at that point. However, the individual waves leave the region unchanged. See also *beats* and *interference*. A **supersaturated vapour** is a vapour whose pressure is greater than the saturated vapour pressure at the temperature concerned. As is the case with a supercooled or superheated liquid (above), the system is not stable. Both designs of cloud chamber depend on supersaturation. The **supersonic flow** of a fluid is flow at a speed greater than the speed of sound in the fluid. An object moving at **supersonic speed** through a fluid (in other words, at a speed greater than that of sound in the fluid) produces one or more shock waves, and therefore suffers high drag.

A **superstring** is supposed by some recent theories to be the fundamental component of matter a set of string-like particles that are extremely small and make up the particles in an atom. It is a ten-dimensional subatomic particle that lost the extra six dimensions around 10^{-43} s after the big bang (and thus became potentially visible to us 4D creatures). Based on highly complex maths – but maths of great beauty – superstrings can model all the known subatomic particles, the superstrings' 'dualities'. There are various superstring theories, however, and (as yet) it is not clear which – if any – will become supreme. **Superstring theory** is a major attempt to combine the theories of quantum physics and relativity, perhaps the biggest task of modern physics. The M- theory is a 'master' theory that would combine the five superstring theories by way of their dualities. For **supersymmetry**, a form of modern unifying theory, see *neutralino* (the massive superpartner of the Z-boson) and *susy particle*. Last, a **superwimp** is the decay product of a wimp (a weakly interacting massive particle and the prime candidate for dark matter); there are hopes that the new large hadron collider will be able to detect superwimps.

A **super-solid** is a proposed new state of matter observed by some – but not by others – in solid helium-4 cooled below about 250 mK. If it exists, it would act like a super-fluid as well as having an ordered solid structure.

supplementary unit a dimensionless unit with the features of a base unit in a unit system. In SI, the supplementary units are the radian and steradian, dimensionless units for angle and solid angle.

surface also called interface, the outside layer of a sample of matter, the layer with vacuum or some other substance on one side. Because of this, the properties of a surface often differ from those of a similar layer inside the bulk of the sample. In particular, as a result of the unfilled bonds at the surface, there is an energy barrier, and maybe energy levels inside the normal forbidden bands too. Such energy levels may also appear as the result of a layer of adsorbed matter on the surface. For all such reasons, surface physics is now a subject of great interest.

The **surface area** A (or, sometimes, S) of a surface is its area; if the surface is smooth, this is much the same as the area of the sample in contact with the outside: but the actual contact area of a rough surface may be far larger. A **surface barrier transistor** uses metal-semiconductor junctions (Schottky barriers) rather than p-n junctions. The **surface charge density** σ of a surface is the charge available per unit area. A **surface charge transistor** is an old name for charge coupled device.

The **surface energy** of a surface is the total energy per unit area (while the surface tension, see further below, is the free energy per unit area). A **surface force** between two objects is one that follows from their contact only. The deformations that always result may lead to forces (stresses) throughout each object, but contact forces differ in many ways from body forces such as gravity. The **surface mounting** of electronic components is

coming to replace the use of printed circuit boards. Rather than drilling holes through the board to take the pins (legs) of chips and other elements, surface mounted elements rest directly on the surface, attached by solder. The approach means that circuits need no longer sit on flat boards of a certain size or shape: they can go on any suitable surface (such as the curved inner panels of a car).

The **surface tension** γ (or σ) of a substance is an old, but still widely used, term for the free surface energy. The effect results from the fact that the cohesive forces between the particles of the sample are not in balance at the surface. There is therefore a tendency to reduce the surface area, and, on the other hand, a need for energy to create or extend a surface. Thus, we define the free surface energy, or surface tension, as the energy involved in increasing the surface by unit area; the unit is the joule per square metre, $J\,m^2$. γ depends on temperature: it tends to fall with temperature rise, and becomes zero at the critical temperature of the substance in question. That is the temperature at which one cannot distinguish between liquid and vapour forms.

If one attempts to increase the area of a surface, there is a tendency for the temperature to fall (as is easy to detect when stretching a rubber band). Extending a surface isothermally (without change of temperature) therefore involves the input of energy above that given by the free surface energy. Thus, the total energy of a surface is greater than the free surface energy γ.

The particles at the surface of the sample tend to be further apart than those inside as a result of the unbalanced cohesion; this causes an attractive force between them (in the plane of the surface). This force is a tension, hence the name: and hence too the impression of an elastic skin on a liquid surface

A **surface wave** is

a) any wave that transfers energy through the surface of an elastic substance, or

b) a radio wave that follows the curvature of the Earth's surface to some extent, so is of value in communications.

surface charge transistor a variant of the charge coupled device

surface plasmon a plasmon in the surface of a metal

surge sudden increase in power supply voltage lasting for some seconds. The cause may be nearby lightning, circuit switching, or a fault. A **surge generator** is a pulse generator, while **surge suppression** involves the use of an uninterruptible power supply system. Much the same as all this applies to a spike, a short surge.

susceptance B the reciprocal of reactance, the unit being the siemens. See also *admittance*.

susceptibility a) for a substance, χ, the ratio of the degree of magnetisation of a sample of the given substance (M unit: A m^{-1}) to the magnetic field applied (B – also in A m^{-1}). There is no unit. The value is 1 less than the relative permeability of the substance, and is approximately constant in many cases, as stated by the Curie law:

$$\chi = \mu_r - 1$$

This table below shows how the value of susceptibility relates to the magnetic properties of materials.

The **mass susceptibility** (or **specific susceptibility**) of a substance is the susceptibility of unit mass.

b) (rare) by analogy with the above, the value $\varepsilon_r - 1$. ε_r is the relative permittivity of the substance concerned.

susy particle a 'supersymmetric' counterpart

to a basic sub-atomic particle, as proposed by some modern theories that try to integrate bosons and fermions

svp standard vapour pressure

sweep a) rare name for the time base circuit of (for instance) an oscilloscope

b) the slow steady change of voltage of a time base signal (saw tooth) as opposed to the rapid flyback section of the cycle

switch ⏤ electric circuit control element with two discrete states (on and off, closed and open) or, in some cases, more. There are many forms of mechanical design, and relays, transistors and thyristors act as switches too. A switch should have zero resistance when closed and infinite resistance when open.

symmetry a) of a regular shape, feature of being identical with its image formed by reflection through a line, point or plane (the **axis of symmetry**, **centre of symmetry** or **plane of symmetry**). The **order of symmetry** is the number of ways one can place the object to give identical views; for a cube, for instance, the value is 6.

b) relationship between nuclei, atoms or molecules that depends on one or more of a given set of operations. Each such symmetry relates to a conservation law. For instance, **charge symmetry** relates to the law of constant charge, and **energy symmetry** to the law of constant energy. **Unitary symmetry** is a theory of elementary particles.

synchrocyclotron type of cyclotron in which the frequency of the electric field falls during particle acceleration to allow for relativistic mass increase. See *nuclear physics: accelerators*.

synchronous working at the same frequency (and often phase) as some other system. A **synchronous alternator** is an ac generator with a number of separate field coils to which there is dc applied in turn. The shaft carries one (or more) coils linked by a pair of slip rings with carbon brushes to the output; a

material	χ
diamagnetic	negative, small (around -10^{-5})
paramagnetic	positive, small (around 10^{-3})
antiferromagnetic	positive
ferromagnetic	positive, large

turbine or some other mechanical linkage spins the shaft at a given frequency. The coils turn in the magnetic field of one or more pairs of poles (often of electromagnets supplied by a separate dc source); induction causes a voltage to appear at the output. A **synchronous capacitor** is a synchronous motor (below) used in an ac circuit purely to improve the power factor of the circuit: it acts as a capacitor in that it takes a leading current. A **synchronous clock** uses a synchronous motor (below) to drive the hands, so depends only on the frequency of the power supply for its time keeping.

A **synchronous motor** runs at a frequency which depends only on its design and the frequency of its ac supply. In other words, its rate of turning does not depend on the load. The stator (field coil system) produces a rotating magfield as a result of the phase relationships between the currents in the coils. The rotor carries dc and turns with the rotating field. A **synchronous orbit** is the same as a geosynchronous one; it is an orbit of a satellite such that the craft stays at the same point in the sky as far as a viewer is concerned.

synchrotron cyclic accelerator in which the shape of the magfield changes and the frequency of the electric field falls during particle acceleration to allow for relativistic mass increase. **Synchrotron radiation** is the design output of a few high energy electron accelerators. It is very intense very parallel electromagnetic radiation in the infrared to x-ray range produced from the acceleration of circular motion of charged particles moving at relativistic speeds. For both, see also *nuclear physics: accelerators*.

syphon device using air pressure to transfer a liquid over a barrier to a lower level. The usual design involves an inverted U-tube with arms that differ in length; the tube must be full of the liquid before the siphon action can start.

system a) object or related set of objects we can treat as free from outside influence **b)** any combination of actions, procedures, hardware, software and links that helps someone to carry out a given task. In particular, in information technology, a system is such a combination concerned with the efficient handling of information. In practical science, a **systematic error** follows a fault in a meter or procedure; it results in all the readings of a set being shifted by a certain amount from the correct values.

Système International d'Unités the set of rational units and their relationships now used, SI for short – see *Appendix 1*.

Szilard, Leo (1898–1964) physicist who, with Fermi, led the development of the early US nuclear weapons programme. He then led the peace movement in 1945, before turning to molecular biology. The **Szilard–Chalmers reaction** is a chemical approach to separating a radioisotope from non-active but otherwise identical material. The **Szilard engine** is a theoretical device using a Maxwell demon to extract energy from a single particle in a cylinder.

t symbol for tonne (metric ton), 1000 kg

T a) unit prefix symbol for tera-, 10^{12}
b) symbol for tesla, unit for magnetic inductance (or flux density), i.e. magfield strength
c) symbol for the top quark
d) symbol for tritium, the hydrogen-3 isotope, and its nucleus, the triton
e) symbol for true (a logical state)
f) symbol for true (of a bearing in navigation)

t a) symbol for temperature (T preferred)
b) symbol for time
c) symbol for transport number

T a) symbol for kinetic energy (W or W_k preferred)
b) symbol for (thermodynamic) temperature, unit: kelvin, K, or degree Celsius, °C
c) symbol for the force of thrust (F preferred unless specific to aerodynamics etc)
d) symbol for time period, unit: second, s
e) symbol for torque, unit: newton metre, N m^{-1}
f) symbol for transmittance

τ symbol for the tau lepton

τ a) symbol for half-life (t$_½$ preferred)
b) symbol for proper time
c) symbol for time constant
d) symbol for torque (T preferred)

tachometer device able to measure the angular speed of a turning shaft. There is a large range of designs.

tachyon hypothetical particle that travels throughout its lifetime at a speed greater than the speed of light. This feature does not violate the statement of relativity that it is not possible to accelerate any object to or past the speed of light, as tachyons appear at their birth with a higher speed. A tachyon has imaginary mass amd negative proper time, and would move faster as its energy falls. Tachyons appear in some theories that result

from attempts to unify modern physics (which includes some string theories). No evidence for them has yet appeared in practice, however.

Talbot, William Fox (1800–1877) inventor of the calotype method of photography, the first process to involve the concept of a negative from which one can make any number of positive prints

Tamm, Igor (1895–1971) nuclear physicist who shared the 1958 Nobel Prize for his work with Ilya Frank on Cherenkov radiation

tandem generator an approach to the use of the van der Graaf generator which doubles the energy of the output beam. The input is of negative ions which accelerate to the energy available; the system then 'strips away' (removes) electrons from the ions to produce positive ions, which accelerate once again through the whole energy range.

tangential acceleration a_t unit: metre per second per second, m s^{-2} the component of an object's acceleration along the tangent to the curve in which it is moving. See also *angular acceleration.*

tantalum Ta transitional metal in period 6, with proton number 73, found in 1802 by Anders Ekeberg and named after Tantalus, king in a Greek myth whom the gods 'tantalised'; finding and proving this element took decades, so were tantalising tasks. Relative density 17, melting and boiling temperatures 3000 °C and 5400 °C; only stable isotope ^{181}Ta. Makes very compact capacitors and a range of alloys, including for use in surgical appliances.

tap or **tapping** fixed or variable contact to some point between the ends of a coil. The purpose is to provide some inductance lower than that of the whole coil, or smaller output voltage in the case of a voltage transformer.

457

Taylor, Joseph (1941–) professor of research student Russell Hulse, with whom he discovered the first binary pulsar – of great value for exploring general relativity. The two won the 1993 Nobel Prize for this work.

Taylor, Richard (1929–) physicist, joint winner (with Jerome Friedman and Henry Kendall) of the 1990 Nobel Prize for work on the structure of nucleons leading to the quark model. Carried out in the 1960s, this involved the deep inelastic scattering of electrons produced by the Stanford Linear Accelerator.

tdm time division multiplexing, a method of merging two or more separate signals into a single stream for transfer between two points

technetium Tc transitional metal, period 5, with proton number 43, found in 1877 by Serge Kern and confirmed sixty years later when Emilio Segrè found it again in a radioactive sample from a new cyclotron. Relative density 11.5, melting and boiling temperatures 2200 °C and 4500 °C; most stable isotope ^{98}Tc, beta emitter with half-life 4×10^6 years. Source of a large number of radiopharmaceuticals and used as a tracer and in some nuclear power cells.

technology the 'appliance of science' or applied science – the use of the knowledge gained from the study of science to solve problems and improve the quality of life

tele-

prefix that relates to working at a distance. **Telecommunications** is the whole field of passing information and data (as signals, streams, or packets) over a distance. There is no clear definition of the least distance involved, but all methods involve electricity at some stage, so are part of 'new' information technology.

Electric **telegraphy** first appeared on a commercial basis early in the nineteenth century (though the first invention was in 1727). This makes it much older than telephony; it was a well tried system by the time the threat of the voice telephone became real, and still remains in use in some contexts. Unlike telephony, which is in essence an analog signal transfer system, telegraphy works with digital (coded) signals. These are sets of pulses of current in a link between sender and receiver, with the earth providing the return path.

Note that although fax developed as a version of telegraphy, almost all of the millions of fax systems now in place communicate through a telephone network. This is because those networks are far more extensive, and therefore far cheaper, than the public access telegraph network (telex); this latter has only around 0.2% the number of lines world-wide.

Telemetry concerns making measurements at a distance. It ranges from the transfer of signals from sensors and such systems in survey satellites, through the use of remote sensing to cover all the needs of a large factory, down to meters (for instance, of gas and water usage) that a service company can interrogate for billing purposes. Early systems were photographic rather than electrical; one of the most important was that of Francis Ronalds, who set up

the automatic photographing of weather instruments at Kew
Gardens in 1845. A significant application of telemetry at the level of
the home is the growing use by energy supply firms (and others) of
systems that allow automatic remote reading of meters.

A **telephone** is a two-way voice communications system; it is
cheap, but because it links to a world-wide network ('the world's
biggest single machine') and therefore offers high-speed
communication almost anywhere, it is complex. A **telephone
handset** needs to have

a) a microphone to convert speech into an electric (analog) signal;
b) a speaker for the reverse process;
c) a switch to access the line or wireless channel to a local phone
   exchange (or node); and
d) a method of calling the number of the other handset wanted.

Many modern handsets, however, also offer a wealth of other
features as microelectronics becomes so cheap and effective. Many
of these require access to a digital phone exchange rather than to
the more traditional analog system.

The **telephone network** involves hundreds of thousands of
exchanges to link the nearly 10^{10} fixed, mobile and radio handsets
in the world (this much less than a century and a half after the first
effective telephone of Alexander Graham Bell, 1876). Each handset
has its own unique call number and must be able to connect to any
other one in the world for point to point communication. Most
countries now have a hierarchy of telephone exchanges between
local and international level, with automatic switching on the basis
of each line number called.

A **telephoto lens** is a camera lens that can produce a large
image of a distant object. It is in effect a form of Galilean telescope
fitted to the front of the camera. All **telescopes** have, in fact, the aim
of giving an enlarged (and/or intensified) image of a dim and
distant object. There are two main types for working with visible
light – the reflector and the refractor – these using respectively a
converging mirror and a lens to collect the light from the object in
the first place. A Galilean telescope is a type of refractor; see also
Schmidt lens (an effective add-on to a reflector) – but there are many
other designs.

The first telescopes, invented near the end of the sixteenth
century, were for use on Earth and worked with glass lenses on
visible light; terrestrial telescopes, binoculars and opera glasses are
modern versions. Astronomical telescopes – designed to study light
from objects outside the Earth – soon followed. There are two types
of these, each having a number of designs; refractors have an object

lens to capture the light, while reflectors use a mirror. (Isaac Newton developed, in 1704, the first of these latter designs.) In practice, all large astronomical telescopes are reflectors, as it is nearly impossible to build a large stable lens. The most powerful are those able to gather the most light as this gives greatest resolution; the most powerful of these is not the 2.4 m Hubble Space Telescope (though it suffers no aberration from air and its mirror is very accurately ground). (The size of a telescope is the diameter of its object lens or mirror.) The largest systems are ground-based – the 'Gran [= great] Telescopio Canarias' or GTC (a 10.4 m instrument which started work in 2006) and the Large Binocular Telescope or LBT (two 8.4 m mirrors, equivalent to 11.8 m, built in the Arizona desert and due to start work in 2007).

Astronomers use telescopes able to work with most other regions of the electromagnetic spectrum, as well as with beams of particles that reach the Earth from outside. See those regions and particle types for further information. Indeed, from the nineteenth century, people developed telescopes to work with other electromagnetic radiations and using reflecting or refracting systems. Telescopes working with wavelengths that can't pass through the Earth's atmosphere must travel on artificial satellites. Perhaps the most powerful is NASA's Swift system, a gamma-ray burst detector that has picked up events as much as 13 thousand million light-years away – that is distant enough to observe the explosive deaths of the first stars, forming the first black holes.

Television (tv) is a technique for the broadcast of moving video and sound signals to aerials linked to special receivers. See also *video* – sending such signals around some local system. Baird (in 1926) devised the first practical system; however, television as a concept went back to the early days of telegraphy (above), and Baird's system used techniques devised up to half a century before. The first commercial television broadcasts started in Britain in 1936. The principles are as follows.

a) By a process of modulation, the combined video/audio signal from the source adds to a high frequency radio wave (the carrier). This passes to a cable for transfer or to an aerial for broadcasting.

b) As each modulated carrier wave passes each receive aerial, it induces a current in this. The amplified current passes to the set; this selects the signal required (by a tuning process), amplifies it again, and then separates the carrier, video and audio parts.

c) Amplified further as required, the audio signal passes to the system's speaker(s).

d) Also amplified further as required, the video signal passes to the display (cathode ray tube for instance). In the case of the tube, during each fraction of a second, a beam of electrons scans the inner surface of the screen, dot by dot, and line by line, to produce a flying spot. The video signal modulates the brightness of the spot to build up each frame of the picture in that colour.

Those notes describe how a monochrome (black and white) set handles the input signal. In a colour set, there are three electron beams, and each dot on the screen is in fact a triplet to output light of the three primary colours. Work proceeds on a variety of developments to make television an even better channel of communication.

telex the major type of telegraphy as regards public usage. The name is short for telegraph exchange, this concept (much like that of the telephone network) dating back to the 1930s. It is still in quite wide use in less developed countries, as it is very reliable, although very slow.

teleporting in science fiction, some technique for the transfer of all the information that stands for an object (even a human body) so that one can re-assemble the object exactly at the other end. In science, there are various techniques that are much the same – such as teleporting atoms (with anything larger, there is loss of information about energy levels), quantum teleportation (which depends on quantum entanglement and does not allow the transfer of energy or matter, only information), and teleporting light (the record distance, late 2006, being half a metre).

Telford, Thomas (1757–1834) civil engineer renowned for many important bridges, harbours and canals of the period when the industrial revolution started to gather speed. His name is given to a large new town which includes the world's first iron bridge, in Shropshire (England).

Teller, Edward (1908–2003) nuclear physicist who made many contributions to the theory of nuclear fusion and to the practice of the hydrogen bomb

telluric of or like the Earth (Latin 'tellus'). Thus, **telluric currents** are natural, very long wavelength, currents in the Earth's crust and/or ocean; they result from changes in the Earth's magfield (e.g. in the solar wind), and from thunder storms on a more local scale. **Telluric lines** are Fraunhofer lines in the spectrum of the Sun that arise from absorption in the Earth's atmosphere. Most of these lines are in the red and infrared regions. A **telluric planet** is Earth-like (rather than being a gassy planet like Jupiter).

tellurium Te metalloid in group VI, period 5, number 53, found in 1782 by Franz Müller and confirmed by Martin Klaproth, who gave it its name, meaning the Earth, as it was 'taken from old Mother Earth'. Relative density 6, melting and boiling temperatures 450 °C and 990 °C; most common stable isotope ^{126}Te. Used in a number of alloys and has potential in solar panels.

temperature T (or θ or t) unit: kelvin, K, or degree Celsius, °C the property of a matter sample that determines whether there is a net energy flow to or from its surface. The second law of the thermodynamics covers this. There are various scales, the most widely used in physics being the scale of absolute temperature, whose zero is at the absolute zero, 0 K (kelvin). See also, for instance, the Celsius scale.

Many properties of a sample of matter depend on its temperature, and therefore there are many techniques of measurement. See, for instance, *thermometer*, the *kinetic theory*, and *states of matter*. The **temperature coefficient of resistance** α appears in the expression that relates the electrical resistance R of a sample to its temperature T:

$$R_T = R_0(1 + \alpha\, T + \beta\, T^2)$$

R_0 is the resistance of the sample at 0 °C, and R_T is that at T °C. There are two constants of proportionality, α and β; in most cases, β has negligible effect.

If different parts of a sample of matter differ in temperature, there is a **temperature gradient** between them; the usual symbol is dT/dl, the unit being the kelvin per metre, K/m. A **temperature inversion** is a state of the lower atmosphere in which temperature increases with height rather than falling (as is the norm). For **temperature scales**, see the various examples. See also *negative temperature*.

tensile of a sample of matter, associated with stretching deformation – deformation in which the particles become further apart than normal. The **tensile strength** of a sample is the tensile deformation it can withstand before fracture (in other words, before it breaks). See also *elasticity* and the *Young constant*.

tension a) old name, once in very common use but now rare, for voltage
b) the force or stress within an object that tends to stretch it, in other words to move its particles further apart than normal. It is the opposite of compression.
c) The deformation (strain) that results from such a force or stress

tensor a set of functions of the coordinates of a point in space that shows a linear transformation with a change of coordinate system. It is a higher order vector, a scalar being a tensor of rank 0 (one you can describe with a single number) and a vector having rank 1 (to describe it, you need two numbers). **Tensor analysis** is the branch of mathematics that deals with the properties of tensors and of **tensor fields**: regions of space in which we describe each point by a tensor, a 2D table of numbers.

Tensors are of great value to describe concepts in the science and engineering of 3D objects. For instance, we can give the stress at a point in an object as as a 3 × 3 array (table) – this is a tensor of rank 2; the same is true of the strain at the point; however, the relation between the stresses and strains throughout the objects is its elasticity, and this is a 4-tensor. **Diffusion tensor imaging** is a medical scanning technique that depends on tensors to give the organ's permeability to water in different directions in tensor form.

tera- T unit prefix that denotes 10^{12}, as in tera-watt, 10^{12} W

terbium Tb transitional rare earth in the lanthanide series, period 6, with proton number 65, isolated in 1843 by Carl Mosander and (like several other elements) named after Ytterby, site of the quarry in Sweden where its ore was found. Relative density 8.3, melting and boiling temperatures 1360 °C and 3100 °C; only stable isotope ^{159}Tb. The element has some uses in semiconductors and in phosphors.

terminal a) point in a circuit at which current enters or leaves. See also *termination*.
b) final. Thus, its **terminal speed** is the highest speed a freely falling object can reach in a fluid. At this speed, the friction becomes equal to the object's weight.

The terminal speed of a sphere of radius r and density ρ_s, falling through a fluid of friction (viscosity) ξ and density ρ_f in a gravitational field of strength g, is

$$c_t = 2\,(\rho_s - \rho_f)\, g\, r^2/9\,\xi$$

termination placing a load at the end of a transmission line in such a way as to provide impedance matching and thus prevent unwanted reflection

tesla T the name for the unit of magnetic flux density (field strength), the weber per square metre (= kg s^{-1} C^{-1}, i.e. the newton per ampere-metre). It is 10 000 gauss (the gauss being the former unit of magfield strength). The tesla is a large unit – so the field between the poles of a large strong horseshow magnet may be no more than 1 mT.

Tesla, Nikola (1856–1943) physicist, electrical engineer and prolific inventor who developed the first induction motor in 1888 and explored x-rays ten years before their 'official' discovery by Röntgen in 1895. A **Tesla coil** is any one of a large range of types of induction coil with a high frequency output that can produce large sparks (even half-metre sparks in air with a small device).

tetragonal system one of the crystal systems

tevatron one of the world's most powerful accelerators, first running in 1983, able to accelerate particles to energies as high as 1 TeV (10^{12} eV, almost a microjoule)

Thales of Miletus (c.625–c.545 bce) philosopher believed to be the first to attempt a scientific view of the Universe – one calling on nature, rather than the gods, as the cause of everything. Therefore, he has come to be called by some 'the father of science'. He was also the first to develop geometry as a theoretical study, to devise a coherent cosmology and to predict an eclipse of the sun. Thales had much influence on Aristotle.

thallium Tl metal, in group 3, period 6, with proton number 81, found from the bright green line in its spectrum by both William

Crookes and Claude Lamy in 1861; the name comes from the Greek 'thallos' = green shoot. Relative density 12, melting and boiling temperatures 304 °C and 1460 °C; most common isotope ^{207}Tl. Used in rectifiers, radiation sensors and nuclear medicine.

theatre glass(es) or **opera glasses** low magnification telescope (glass) or binocular telescope (glasses) based on the Galilean design

theorem statement (often in mathematics) which logically depends on a set of postulates, called axioms. In physics, the **theorem of parallel axes** relates the moments of inertia of an object about two parallel axes. If the object has a mass m and a moment of inertia I about an axis through the centre of mass, the moment about a parallel axis distant r from the first is $I^2 + m\ r^2$.

theory hypothesis or set of hypotheses supported by philosophical, mathematical, and/or practical arguments to explain some phenomenon or set of phenomena. The **theory of everything** (toe) is the physicist's dream: a universal single theory that people hope will one day cover all interactions and particles in all contexts. To achieve this may take a long time – the toe must bring together quantum physics and general relativity, but these two views describe the Universe in very different ways.

therm old unit of energy, used in particular in the gas supply industry, about 10^8 joules

---

thermal physics

the study of how 'heat' energy and matter relate to each other, 'heat' energy being energy associated with the temperature of a sample of matter. The field includes the kinetic model of matter; based on that is statistical mechanics on the one hand and thermodynamics on the other.

The word **thermal** itself means concerned with energy, in particular the energy associated with temperature and temperature change. For instance, **thermal agitation** is the name given to the random motion of particles that rises with temperature, or to such effects of its Brownian motion. The **thermal capacity** C of an object is the energy ΔW involved in changing its temperature by one kelvin. It is the product of the object's mass m and its specific thermal capacity c. The unit is the joule per kelvin, J K^{-1}. If the object's temperature change is ΔT, we have the relations in the box between energy transfer to/from a sample and the sample's change of temperature. The area of physics in practice that deals with thermal capacities, and their measurement, is calorimetry.

$$\Delta W = C \, \Delta T$$
$$= m \, c \, \Delta T$$

For the process of **thermal conduction**, the process of transfer of energy through matter, see *conduction*. The **thermal conductance** α of a sample of matter is the rate of energy transfer (power p) through unit area per unit temperature difference between the ends. The unit is the watt per square metre per kelvin, W m^{-2} K^{-1}.

Thermal conductance –

$$\alpha = p/A \, \Delta T$$
$$= W/A \, t \, \Delta T$$

Thermal conductivity k (or λ), on the other hand, is a property of a substance:

Thermal conductivity k (or λ) is the rate of energy transfer through unit area of a sample per unit temperature gradient dT/dl:

$$k = p/(A dT/dl)$$

In both cases, the energy transfer through the sides of the sample must be negligible.

It is also sometimes useful to use the **thermal diffusivity** (also α) of a substance; this is a measure of how quickly a sample reaches thermal equilibrium. The unit of thermal diffusivity is the metre squared per second, m^2 s^{-1}. However, the name implies an ancient view of energy as a fluid:

Thermal diffusivity:

$$\alpha = k/\rho \, c$$

Here ρ is the density of the substance, and c is its specific thermal capacity (above).

Thermal effusion is the appearance of a pressure gradient in a low pressure gas along a temperature gradient; the pressure must be low enough for the gas particles' mean free path to be not much less than the size of the container. Two (or more) nearby objects, or parts of a single system, are in **thermal equilibrium** if there is no net energy transfer between them; they are therefore at the same temperature. **Thermal expansion** is the change of size (length, area, volume) of an object with change of temperature; see *expansivity*. A **thermal neutron**, in particular inside a reactor, is one in thermal equilibrium with its surroundings. This means that its energy is of the order of kT; here k is the Boltzmann constant and T the temperature of the surroundings. In an electric circuit, or in its output signal, **thermal noise** (or Johnson noise) is a form of white noise that results from the random thermal interactions of the particles of the system.

Thermal pollution is the harmful release to the environment – the air or the water – of waste energy from industrial and other processes (for instance, at the output of factories). This can kill living species, often by affecting their reproduction, and reduce the oxygen content of surface water. A **thermal power station** is often a major source of such pollution. It is an electricity generating station whose input energy comes from burning a fuel in air; in most cases, the fuel is coal, gas or oil, but sometimes it is dung, peat or a crop waste such as straw). The high temperature in the system produces high pressure steam to turn a turbine linked to the electricity generator itself. **Thermal radiation** is a name for the infrared region of the electromagnetic spectrum; its waves are very effective at raising temperature (and we sense them as hotness through certain nerve endings in the skin). A **thermal radiator** is any device which is an efficient source of such radiation; note, however, that the space 'heating' systems called radiators raise the temperature of their environment mainly by convection.

A **thermal reactor** is any standard design of nuclear fission reactor, one whose energy output comes from the high temperature produced in the core. The **thermal resistance** of a sample of matter is the reciprocal of its *thermal conductance* (above); similarly, the **thermal resistivity** of a substance is the reciprocal of its *thermal conductivity*. Any solid sample suffers a **thermal shock**, the result of stress, upon a sudden temperature change; this may be enough to cause fracture (as is sometimes the case with glass beakers). The stress concerned, **thermal stress**, is the result of a high temperature gradient within the substance.

An object's **thermal transmittance** is the ratio of the power transfer through it to the power received at the surface. **Thermal transpiration** is an old name for *thermal effusion* (above). The **thermal vibrations** of matter are the harmonic vibrations of particles in a solid about their mean position, or the various forms of vibration of the particles of a fluid; both relate to temperature.

thermion general term for any charged particle that escapes from a surface at a high temperature

thermionic concerned with the release by a surface at high temperature of charged particles, in particular electrons. A **thermionic cathode** is such a source of electrons. As well as being raised to a high temperature (often several hundred or even thousand degrees Celsius), it is normally coated with a layer of substance (such as an oxide of barium or strontium). Such a layer has a low barrier ('work function') to the escape of electrons.

A **thermionic diode** is the simplest electron tube, used solely for its rectifying nature (though it is now rare, except for when dealing with very large currents). It consists of the heated cathode (electron source) and an unheated anode (electron sink) in a vacuum. Conduction occurs only when the cathode is negative with respect to the anode. Only then will the electrons released by the hot surface (during the evaporation-like process of **thermionic emission**) pass across to the other electrode.

This diode is a typical **thermionic valve**; the word valve means something that passes some fluid in one direction only, just like a diode. However, there is a whole family of valves – each being an electron tube that consists of a number of electrodes in a vacuum, one of which is the heated cathode. See also *triode*. Of all the thermionic valves, only the cathode ray tube is now in common use.

thermionics the science and applied science of thermions, thermionic emission, and the design and use of thermionic valves

thermistor semiconductor device used as a sensor – as its resistance changes significantly as its temperature rises: the graph of resistance against temperature is steep. (The design of a standard resistor aims to give it as flat a graph as possible.) The expression for the effect is

$$\Delta R = k \, \Delta T$$

Here ΔR is the resistance change associated with temperature change ΔT, while k is the first order temperature constant ('coefficient') of resistance for the substance concerned. (In most cases, we can ignore higher order constants.)

The first thermistors had a negative temperature coefficient of resistance; hence the name ntc sensors. There are now many – ptc sensors – with a positive slope to the graph. (Some people call these posistors.) Thermistors are very sensitive temperature sensors, and can cover a wide range; they have many uses in thermometry and in electrical control systems.

thermochemistry the study of the energy transfers involved in chemical change. These transfers tend to lead to temperature change, so that the techniques of calorimetry are common. Thermochemistry is very close to thermodynamics.

thermocouple a junction of two metals and/or semiconductors between the ends of which a voltage appears if it is heated or cooled, or a closed loop of two such junctions which behaves in the same way. This is the *thermoelectric effect* (below), made stronger by linking a number of such junctions to form a thermopile. The voltage, perhaps of a few millivolts, that appears between the ends of a thermocouple loop leads to a sensitive thermometric technique of value within the range (approximately) of 50–2000 K: one junction is kept at a known constant temperature, the other is at the unknown temperature, and the user measures the output voltage. Thermopiles are often used in local electricity generation (e.g. in systems based on isotope power).

The output voltage of a thermocouple does not relate in a linear fashion to temperature. In fact the curve is always close to a parabola. Using the device as a thermometer therefore means that one must be sure on which side of the neutral temperature (that at the apex of the parabola) the system is working.

thermodynamic associated with some aspect of thermodynamics. A **thermodynamic diagram** is a graph of two or three of the thermodynamic properties of a system that helps analyse and understand the system. Examples include phase diagrams (see *vapour*, for instance) and the pressure/volume graphs that show how a gas sample behaves.

The **thermodynamic efficiency** of an engine is the ratio of the useful energy output to the total energy input; it often appears as a percentage. **Thermodynamic equilibrium** is the state of a system whose free energy is at a minimum. This includes thermal equilibrium (when the parts of the system are at the same temperature) and the dynamic equilibrium between a saturated vapour and its liquid (for instance). The **thermodynamic potentials** of a system are a set of potential functions which describe aspects of the energy of the system; examples are enthalpy, entropy, the Gibbs and Helmholtz free energies, and internal energy. For **thermodynamic properties**, see *thermodynamics* (below).

We generally take the **thermodynamic temperature** T of a system to be its temperature on the absolute (kelvin) scale. That is because the **thermodynamic temperature scale** is based on the theoretical behaviour of a perfect engine, so cannot be used directly. The absolute scale, given by the behaviour of real gases when these are nearly ideal, is very close in practice: indeed, we define it to be identical at the various fixed temperatures. In fact, there are only two true fixed temperatures – absolute zero and the temperature of the triple point of water.

thermodynamics at the core of thermal physics, the study of energy transfer and its effect on systems that consist of large numbers of small particles. By 'system', we here mean any linked matter; a closed system is one which cannot exchange mass-energy with the outside. The thermodynamic properties that define the state of a system are statistical measures, as the system consists of many small particles; they include, in particular, its pressure p, volume V, and (thermodynamic, i.e. absolute) temperature T. The system's equation of state relates those properties in a useful way.

The modern study of thermodynamics does not now, as once it did, restrict itself to situations involving 'heat' energy only (as its name implies). Thus, the **laws of thermodynamics** are now of much wider relevance and value then they were. These laws are as follows (though note that there are many forms of their wording, some dating back over 150 years).

- The 'zeroth' law: if two systems are each in equilibrium with a third system, they are in equilibrium with each other. This is the basis of the concept of temperature and its measurement (thermometry).
- The first law: the total mass-energy of a closed system is constant. We can express this as

$$W_i = \Delta W + W_o$$

This states that the total energy input to a system is the sum of the change in the system's internal energy (in other words, the energy absorbed or 'released') and the total energy output; here energy, as elsewhere in such contexts, includes mass.

- The second law: there can be no free net energy transfer from a system to one at a higher temperature. This comes from Carnot's rule (1824) – his work in this area founded thermo-dynamics. In this context, we can define the entropy (disorder) S of the system, such that its change ΔS is $\Delta W/T$. Now, ΔS is zero for any reversible process, and positive for any other process. Therefore, entropy tends to increase with time.
- The third law: at the absolute zero of temperature, there is no entropy difference between the states of a system in equilibrium. This, the Nernst theorem, implies that it is not possible to reach absolute zero.

See also *perpetual motion*, the possibility of which these laws deny, and the *Maxwell demon* – long a basis for discussion as to how to break the second law.

All the above describes **classical thermodynamics**. The modern version, **quantum thermodynamics**, does not differ much, except that it takes into account the discrete quantum states of the particles of a system.

thermoelastic effect the effect of temperature change on the elastic stresses in a sample of matter

thermoelectric concerned with the interaction between temperature difference and electricity, in other words with **thermoelectricity**. There are various **thermoelectric effects** as follows (and see also *thermomagnetic effects*); note that the names differ somewhat between different authors.

- the Seebeck effect: there is a voltage in a circuit which consists of two junctions between two different metals and/or semi-conductors, if the junctions differ in temperature. This is the basis of the thermocouple and thermopile, which use this effect as thermometers – while there is now much research into using the pile to extract 10%, or even 20% of 'waste heat' from exhausts and chimneys. The main problem is that, to be most effective, a junction should be a poor conductor of thermal energy and a good conductor of electricity; a few semi-conductors approach an adequate 'figure of merit'.
- the reverse of the Seebeck effect, the Peltier effect: passing a current through such a circuit makes the junctions differ in temperature.
- the Kelvin effects: a current in a conductor in which there is a temperature gradient causes a flow of thermal energy. Also, a voltage appears between the faces of a metal sample if there is a temperature difference between them.

The **thermoelectric series** of metals and semiconductors lists these substances in such an order that, in a thermocouple made from any pair, there is a current at the high temperature junction from the substance higher in the series to the one lower. A **thermoelectric thermometer** is a thermocouple or thermopile used as a thermometer.

thermograph a thermometer with a hard copy output (in other words, one on paper) of temperature with time

thermojunction an old name for thermocouple

thermoluminescence form of luminescence from a surface at high temperature. The energy of the light output allows the technique of **thermoluminescent dating** of pottery samples. Alpha radiation absorbed since the pot was fired raises electrons to certain higher energy levels than normal; here they can remain for a long time. Raising the temperature of the sample during testing causes the electrons to revert to the normal state, with the emission of photons. The luminescent light output therefore depends on the age of the sample (and on the type of material and the degree of its exposure to alpha radiation since firing).

thermomagnetic effects thermoelectric effects which take place only in the presence of a magnetic field. See, for instance, the *Ettingshausen* and *Leduc* effects.

thermometer device whose design allows its use to measure the temperature of whatever it contacts. Each such system depends on a thermometric property: this is some property of some substance that depends in a known way on temperature. In daily life, the liquid-in-glass type of thermometer is perhaps still the most common (the liquid being coloured ethanol or, in a few cases in a few countries, mercury). The volume of the liquid sample depends on temperature (the liquid's expansivity is the thermometric property in this case). The liquid is in a bulb leading to a graduated capillary tube. As it expands and contracts, the level of the liquid in the tube rises and falls.

Variations on this standard design include:

- the traditional *clinical thermometer*: whose tube's graduations cover only a short range around human body temperature and which contains a kink to keep the recorded level fixed; and
- the *maximum and minimum thermometer*: with two ends to the liquid column, one to show high temperature and one to show low temperature, each bearing a spring-loaded pointer to record where it has reached during a period.

Electronic meters have largely replaced liquid-in-glass types – they cost more, but are much more robust (and mercury thermometers are now very rare because of the vapour hazard if they break). Thermochromic (band) thermometers are quite widespread in the home.

The other main designs of thermometer include the following, with their thermometric properties. Note, though, that there are very many types of thermometer not listed here, all of value in their own specific contexts.

- the constant volume gas thermometer, working with the pressure of a fixed mass of near perfect gas: this is the most important over a very wide range of temperatures, and has a temperature scale that relates closely to that of thermodynamic temperature.

- the resistance thermometer, working with the resistance of a coil of pure metal (often platinum): this is very easy to use and also offers a wide range.
- the thermocouple, working with the Seebeck thermoelectric effect: easy to use, though not very sensitive, but able to cover a good range of temperature, despite having a non-linear scale
- the various pyrometers for work with very high temperatures, by responding to the radiation output of the source; see, for instance, the *optical pyrometer*.

thermometric property a property of a substance that relates to temperature, so thus makes it of potential value in a thermometer. A good thermometric property varies in a linear fashion with temperature, has a large temperature gradient, is easy and quick to measure, and applies over a large temperature range.

thermometry the science and applied science of temperature measurement. If the temperatures concerned are very high (say, over 2000 K), the name often given to the field is pyrometry.

thermonuclear reaction old but still quite common name for a nuclear fusion reaction, one that involves extremely high temperatures, perhaps millions of kelvin

thermophone type of speaker whose sound wave (pressure wave) output follows the variation of temperature of the tiny central metal strip.

thermopile a number of thermocouples linked in series to give a larger output voltage. Thermopiles are widely used for radiation measurement and local power generation.

Thermos flask common term, from a trade name, no longer protected, for the vacuum flask (*Dewar flask*)

thermosphere old name: chemosphere, the layer of the atmosphere in which inbound uv rays ionise atoms. This is the outer region of the Earth's atmosphere, from about 80 km (to about 400 km). In this region, which contains the E and F layers of the ionosphere and the auroras, the temperature rises from about –100 °C to 25 °C (hence the name).

thermostat device the aim of whose design is to maintain an object or a space within a set narrow temperature range. There are many types, but all provide excellent early examples of automatic feed-back. In most cases, the sensor is some form of bi-metal strip, though electronic sensors are fast gaining ground.

theta pinch a common approach to forming a magnetic bottle in a fusion reactor: the plasma is in a toroidal (doughnut-shaped) tube, as in the zeta pinch system, but the applied magnetic field is along the axis of the plasma. This involves surrounding the walls of the torus with electromagnets.

Thévenin theorem It is possible to replace any electrical network, as viewed towards the input from any pair of terminals, by a single source in series with a single impedance. The techniques that follow in order to do this are of great value in the analysis and synthesis of complex circuits.

thick film circuit system part-way between the use of a printed circuit board and an integrated (thin film) circuit as on a chip. The circuit's links, resistors and inductors are laid down in the surface of a pot or glass sheet; the makers then solder the active components (e.g. sensors and transistors) and capacitors, often on chips, to this. This is a form of hybrid integrated circuit.

thick lens a lens the distance between whose faces is too large to ignore during analysis. Strictly, all lenses are thick in this sense, though in most cases no major problem will arise by treating them as infinitely thin. A **thick mirror** is a lens with one silvered surface.

thin film a film less than about a micrometre thick. To prepare a uniform thin film, one may use evaporation, electrolysis or sputtering. As many thin films have unique electrical, optical and/or magnetic properties, there are very many uses. **Thin film circuits**, for instance, laid down on a pot or glass surface, have features of integrated circuits. The **thin film transistors** often used in this context are a form of insulated gate field effect transistor.

thixotropy same as *anomalous viscosity*, the way the fluid friction of some kinds of liquid (so-called smart liquids) depends on electric field or shear stress

Thompson, Benjamin (Count Rumford, 1753–1814) scientist who carried out much early work on the nature of energy. He also helped to found the Royal Institution in London, where he set Humphry Davy on his career.

Thomson, George (1892–1975) physicist whose work on electron diffraction led him to the 1937 Nobel Prize, shared with Davisson for his independent discovery of the same process

Thomson, 'J J' (Joseph John, 1856–1940)

physicist deeply involved in the early days of modern physics, leader of the Cavendish Laboratory in Cambridge, making it into the greatest centre for atomic physics research in the world. He explored cathode radiation, and identified this in 1897 as a stream of negative subatomic particles, to be called electrons. This work gained Thomson the 1906 Nobel Prize. He then produced the first model of the atom to use subatomic particles. This was the 'plum pudding' model of 1904: he saw the atom as a sphere of positive matter in which electrons are embedded like fruit in a pudding. The model provides a good explanation of the **Thomson scattering** of electromagnetic radiation by free and loosely bound electrons; the electrons become radiators of a lower frequency radiation. The cross-section for Thomson scattering, σ, is

$$\sigma = (8\,\pi\,r_e^2)/3$$

Here r_e is the classical radius of the electron.

Thomson, William (later Baron Lord Kelvin, 1824–1907) natural philosopher (physicist in Scotland) whose main work was on electricity in submarine telegraph cables, but whose most significant research was in thermodynamics. His work in thermodynamics (partly with Joule) and electromagnetism are of major importance. For the **Thomson effects**, see *thermoelectric effects*, as seen with all metals except lead.

't Hooft, Gerardus (1946–) physicist who shared the 1999 Nobel Prize with Martinus Veltman for their work on quantum electroweak interactions

thorium Th transitional rare earth in the lanthanide series, period 7, with proton number 90, first found in 1829 by Jakob Berzelius and named after Thor, the Scandinavian god of thunder. Relative density 12, melting and boiling temperatures 1750 °C and 4700 °C; only natural isotope ^{232}Th, an alpha-emitter with half-life 10^{10} years. Main uses are as a reactor fuel, in electronic wiring, and in portable gas mantles and welding equipment.

thorium series one of the four radioactive series, which starts with thorium-232 and ends in stable lead-208. This is the $4n$ series.

thoron the isotope of radon gas, radon-220, which diffuses from matter during the decay of thorium-232. The former name was **thorium emanation**.

three-body problem using classical physics – Newton's laws of force and of gravitation – the analysis of the orbits of three objects which exist in each others' fields. This is a highly complex problem, not yet open to an exact result in all contexts, as the outcome can often be chaotic.

three-phase of an alternating electricity supply, one which consists of three equal components each 120° out of phase with the others. Such a supply is common in national power distribution networks and is brought to large sites, in particular factories. The user must try to balance usage of the three phases, the main methods of linking the three phases being delta and star.

threshold 'door step', the level of some measure at which some process starts to appear, and below which it does not occur. For instance, in photoelectricity, the **threshold frequency** is the lowest frequency reaching a given surface at which the effect occurs. The **threshold of hearing** is the lowest amplitude of sound waves the normal ear can detect at a given frequency.

thrust a) the force that drives a vehicle, in particular a plane or rocket, or the forward force from an engine
b) an old name for force in general, still used in 'upthrust' (the upward force on an object in a fluid) as well as the above.

thulium Tm transitional rare earth metal in the lanthanide series, period 6, proton number 69, first made by Per Cleve in 1879 and named after the Greek name for Scandinavia. Relative density 9.3, melting and boiling temperatures 1500 °C and 1900 °C; only stable isotope ^{169}Tm.

thunder the shock wave produced by the sudden expansion of a column of air during a lightning stroke. It is a very sharp crack; however, in practice, the sound observed becomes a rumble because of the length of the channel relative to the observer, and because of reflections from buildings, hills and the ground in general.

thyratron 'gas-filled relay', a type of hot-cathode electron tube ('valve') now rarely used, but once an important type of fast-acting switch. The only class of much significance now is the 'giant' thyratron, used in (for instance) radars, tv transmitters, high-energy gas laser circuits – they can handle thousands of amps at thousands of volts. In other contexts, the thyristor and the triac have replaced the thyratron.

thyristor common name for the silicon-controlled rectifier (scr). This is a p-n-p-n semiconductor device with (in the usual form) three electrodes: cathode, anode, and gate. A pulse at the gate will cause the device to conduct (one way only, as it is a rectifier); it will continue to pass current until the anode voltage falls below a certain value. The thyristor is therefore of great value as a switch or relay.

tidal power the use of the energy of the rising and falling sea level in the tides to produce electricity. The usual approach is to build a dam ('barrage') across an estuary whose tidal range (amplitude) is large. Turbines and generators in the barrage take energy from the water as it flows out of (and sometimes into) the estuary. The first, and still the largest (250 MW), such system started working in France in 1967. The prime candidate in Britain – as people have proposed for over a century – is the Severn (able to produce over 8 GW). Tides in the Severn estuary resonate because of the shape of the coastlines and their length being about a quarter of the wave length of the tides. (The Severn bore is an outcome of this responance.)

tide the periodic rise and fall of surface waters (and, to a much lesser extent, of the land) with change in relative position of the Moon (and, to a lesser extent, the Sun). In essence, a bulge of water moves round the Earth after the Moon, with a second, smaller, bulge on the far side. This motion of water round the Earth is making the day longer at the rate of about a milli-second per century – that's because there's a huge amount of energy involved (only a very tiny fraction being thought of for making tidal power). Tides have their effect on many other objects in the solar system and within double star systems.

Tidal flows are strongest when Sun, Moon and Earth are in a straight line (i.e. at Full and New Moon) – the tidal range is then the greatest (an average of 11m in Britain's Severn estuary). These are 'spring tides' (meaning 'jumping up'); between them are neap tides (meaning not known).

timbre the quality of a sound in music. It depends on the spectrum of the sound in terms of the amplitudes of the various harmonics.

time *t* unit: second, s one of the four coordinates of space–time, this being a dimension which we view as related to the sequence of events. If event A happens before a second event B, there is a time interval between the two events (but see *simultaneity*). We use some form of clock to measure time intervals; the SI unit is the second (s), with sixty seconds to the minute, and sixty minutes to the hour, etc. (Time measures are not yet decimal, though people have proposed decimal systems over the years.)

Philosophically, the concept of time still leaves us with many questions. For instance, the study of mechanics and of light rays shows that time is reversible, while that of thermodynamics shows that it is not. It is normal to take the forward direction of time as that in which the entropy of a system increases. However, even this does not always give a clear solution in practice. It is also possible that time is a quantised measure, the quantum being the chronon, of the order of 10^{-24} s. Note that quantum theory works fine in almost all contexts without quantised time – the chronon could help us devise a successful theory of quantum gravity. However, see *time reversal* (below).

The **time base** of an oscilloscope or tv set (for instance) is
a) the signal that moves the spot horizontally across the screen (in other words, in the x or t direction) at a steady speed, or
b) the time taken for one cycle of motion of the beam (or spot) of the set.

See *saw tooth* for the usual form. The **time constant** τ of an inductive or capacitive circuit is the time for the current (former case) or charge (latter case) to reach $(1 - 1/e)$ of its final value from switching on. The values are L/R and $R\,C$ respectively; here R is the circuit resistance, L the inductance, and C the capacitance. **Time delay integration** is a technique to improve an image (in any recording context) by using a computer to integrate all the views during a period of perhaps tens of seconds. This sharpens the image, cuts down grain, and removes blemishes and transient aspects such as noise.

From the special theory of relativity, we have a **time dilation** effect. This is the apparent slowing of a relativistic clock compared to a clock at rest in the frame of the observer. The slowing is $\sqrt{(1 - v^2/c^2)}$; here v is the speed of the moving clock and c is the speed of light in empty space. Amply verified

in practice, the effect leads to the twin paradox.

Time division multiplexing is the sharing of a single channel between a number of signals (that is multiplexing) by giving each signal in turn a time slot. The **time period** T of a harmonic motion is the time taken by one cycle. It is the reciprocal of the frequency. The crucial experiment of Cronin and Fitch in 1964 showed that the decay of neutral kaons is not symmetrical with respect to time reversal – the half-lives of the K^0 and its antiparticle are not quite the same; they should be the same if there is no preferred direction of time. For **time travel**, a concept dating 'back' to 1881 but forbidden by most philosophies and most of modern physics on the basis of Occam's razor, see *wormhole*.

tin Sn metal in group IV, period 5, with proton number 50, used in alloys since ancient times (at least 5500 years ago); its name and symbol come from the Latin 'stannum' and the Cornish 'sten' (though no one seems to know which of these came first – Cornwall/Devon were ancient Europe's main source of tin). Relative density 7.3, melting and boiling temperatures 232 °C and 2600 °C; most common stable isotope ^{120}Sn. Main uses – plating iron and steel to reduce corrosion, many alloys, float glass making and solders.

Ting, Samuel (1936–) physicist who shared the 1976 Nobel Prize with Richter for his discovery of the J/ψ (gipsy) particle.

tint an unsaturated colour – a pure hue mixed with some quantity of white

titanium Ti transitional metal of Period 4, proton number 22, found by William Gregor in 1791, confirmed by Martin Klaproth in 1795 (and named by him after the Titans, giant gods in early mythology), and isolated by Jakob Berzelius in 1825. Relative density 4.5, melting and boiling temperatures 1670 °C and 3300 °C; most common isotope ^{48}Ti. Titanium alloys have great tensile strength, even at high temperatures, yet are low density and resist corrosion very well, so have many uses in aerospace, shipping, pipework, and armour.

toe the (single) 'theory of everything' – the universal single theory that people hope will one day cover all interactions and particles in all contexts

tokamak a major design of fusion reactor, developed in USSR over recent decades (by, for instance, Sakharov), and now the most common used for research around the world.

(In 1968, a tokamak reached over 10 MK, far better than anything else in the world, so the design became the main one used.) The plasma is confined to a torus-shaped (doughnut-shaped) tube in some form of magnetic bottle (see theta pinch and zeta pinch); it is raised in temperature to the several million degrees needed for fusion by making the plasma the secondary of a transformer.

Tomonaga, Shin-Itiro (1906–1979) physicist who shared the 1965 Nobel Prize with Feynmann for his work in quantum electrodynamics

tone a) the sensation of the human ear-brain system of a pure (single-frequency) sound wave, or the wave itself
b) the quality or timbre of a sound, in other words its make up in terms of harmonics. A **tone control** is a device that allows the user to change the frequency response setting of an audio amplifier to provide the (subjectively) best mix of low and high harmonics.

tonne t large unit of mass, the mega-gram, 1000 kg

toric or **toroidal** of a surface, having the shape of a section of a torus. It is the locus of an arc of a circle turned about an axis through it which does not pass through the centre of the circle. Lenses with such a surface are of value in the treatment of astigmatism. A **toric winding** (or **toroidal winding**) is a coil wound on an iron ring (its core).

torque T (or N) the turning effect on an object of a force that does not act through the pivot (for instance, the centre of mass of the object); a push on the handle of a door is an example. The old name is moment. See also *couple*. A **torque-meter** is a device able to measure the torque of, for instance, a drive shaft.

torr old name for a unit of low pressure, around 133 pascals

Torricelli, Evangelista (1608–1647) Galileo's successor as professor of mathematics at Florence University, inventor of the first barometer. Still sometimes seen, this is a tube closed at the upper end that contains a column of mercury. The length of the column above the level in the tank at the open, lower end of the tube gives the pressure of the air. In the tube above the mercury is a vacuum called the **Torricellian vacuum** – the space in fact contains low pressure mercury vapour.

torsion a twisting force (stress), or the twist

deformation (strain) that results. There are various types of **torsion balance** in which one measures the twist in a fine suspension (for instance, of quartz fibre) to find the value of a small force.

torsional concerned with twisting (i.e. with torsion). It is possible to induce a metal bar to oscillate with **torsional vibrations**; here different parts of the bar twist cyclically relative to each other. This may produce (or be the effect of) a **torsional wave**, a wave that twists the matter through which it passes.

torus object in the shape of a ring or traditional doughnut

total complete, as opposed to partial. Thus, a **total eclipse** is one in which no light from the source reaches the observer; the observer is therefore in total shadow ('umbra'). The **total energy** (or **total 'heat'**) of a sample is its enthalpy, the sum of its internal energy and the product of its pressure and its volume. The process of **total internal reflection** (or **total reflection**) takes place when radiation tries to pass from one medium (e.g. glass) into a second in which its speed is greater (such as air), and does so at an angle greater than the critical angle. This would break the second law of refraction (Snell's law), so reflection takes place instead. This is the only case where all the input radiation is reflected; there are therefore many uses in optics and elsewhere.

tourmaline (or **schorl**) range of borosilicate minerals (whose name is Sri Lankan meaning 'stone catching dust'), once very important for their polarising effect as they are doubly refracting. They are also piezoelectric.

Townes, Charles (1915–) Nobel Prize winner, 1964, gaining this for his developments of the early masers and lasers (1951), part of his work in quantum electronics

tracer substance used in small amounts to allow one to follow the progress of a fluid through a system. A tracer should be easy to detect at the other end, even after heavy dilution; it is common, therefore, to use radioisotopes with quite short half-lives for the purpose. There are many examples of the technique in medicine and in research in geology and agriculture.

track the observable path of an ionising particle in a detector chamber (such as a bubble or cloud chamber) or within a photographic emulsion

trajectory the path of an object – sometimes called projectile – moving freely in a field. A satellite's orbit under gravity is therefore a trajectory, as long as there are no rocket burns and no air friction.

trans-actinide element element above 103, which is the end of the actinide series. Some people call trans-actinide elements superheavy elements.

transceiver any unit or terminal able to transmit and receive (output and input) signals or data

transconductance unit: milliamp per volt mA V^{-1} the ratio of the small change in the current output from an electronic device to the small change of voltage input. The measure is of great value when one wishes to compare two similar devices.

transducer a device whose input and output involve different forms of energy but have the same or related form. Sensors are a very important class of transducer; these all have an electrical output that relates closely to some other type of input.

transductor an inductor whose inductance varies with the current in its control coil, as in the case of a saturable reactor

transformation a) the change of a nuclide from one species to another, in particular as the result of the emission of an alpha or beta particle
b) change of phase (state)
c) change between coordinate systems

transformer, voltage electromagnetic machine used in alternating current contexts whose design output tends to be at a different voltage from the input. See *voltage transformer*.

transient a) temporary, lasting for only a short time, such as the signal that appears at the start of playing a note or saying a word; the water hammer (knock) in a pipe that follows a sudden change in liquid pressure; or the output of a supernova
b) a brief pulse or damped signal in a circuit as a result of a sudden change in voltage, current or load, including in (say) a radar or sonar circuit where it may mark a brief event
c) any brief change in the context of any system, in other words any brief disturbance, or its effect
For a non-ohmic dc circuit during switch-on and switch-off, there may be transient effects before the system reaches its new equilibrium. See *LC circuit* for simple cases.

transistor a semiconductor device whose output is under the control of signal(s)

applied to one or more input contacts. The main designs are for switching, amplification and oscillation.

The **bi-polar transistor** is the most common simple type. This has three layers of semiconductor, the central, very thin, layer being of opposite type ('polarity') to the other two. The figures show the two main kinds, pnp and npn (see *semiconductor*), with their symbols and the names for the three contacts.

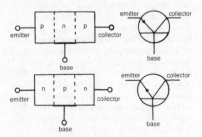

A transistor behaves, in essence, as two pn junctions (diodes) back to back. In normal use, the system forward biases the base-emitter junction (in other words, gives it low resistance), and reverse biases the collector-base junction (so this has a high resistance to current). Then a signal of small amplitude at the base (the input terminal) gives a large amplitude signal (output) on the emitter–collector current: the bi-polar transistor acts as a current amplifier. In a suitable circuit, it can also act as a very fast switch (i.e. a relay); its other main use is as an oscillator: with feedback between output and input it will produce an output wave of fixed frequency and amplitude.

A **field effect transistor** (fet) has a channel of semiconductor whose resistance depends on the voltage input(s) at one or more gate(s). Such transistors are therefore voltage-controlled. The sketch shows the switching action of an insulated gate fet: left: with no voltage on the gate, the device doesn't conduct; right: with a few volts on the gate, the device conducts. Note that this is a uni-polar transistor in that there is only one type of charge carrier. See also *field effect transistor* and *mos*.

An **organic transistor** is a design based on a nanotube broken in half with hydrocarbon molecules in the break. In one use, changing the tube's conductivity by exposing the hydrocarbon to other molecules makes the device a very sensitive chemical sensor.

Transistor–transistor logic (ttl) is a style for the design of logic circuits using a suitably linked pair of bi-polar transistors for each logic element.

transition change between significantly different states or energy levels, in most cases in a rather sudden way. The **transition elements**, or **transition metals**, include most of the important metals; they fill the short groups of the periodic table, and have properties that depend on their many unpaired d electrons. A **transition temperature** is a temperature at which a change of phase occurs, or the temperature at which superconductivity appears, for instance.

translation an object's displacement (movement along a straight line) with no turning taking place. In other words, all the parts of the object move the same way and for the same distance. **Translatory motion** is another term for this. The **translational energy** of a particle is that part of its energy associated with translation (as opposed, for instance, to that of its vibration). The value is $\frac{3}{2} k T$; here k is the Boltzmann constant and T is the absolute temperature of the particle.

translucent able to pass visible radiation, even if with a high degree of absorption and/or scattering – as opposed to transparent, where there is little absorption or scattering

transmission the transfer of energy from one place to a second, in particular through an engineered device or system (such as a gear train, other machine, or clutch), or through a given medium. The remaining definitions apply to this latter context only.

The **transmission constant** of a substance is the same as its transmittance. Its **transmission density** (or optical density) D is a measure of its opposition to the energy transfer concerned. It is the logarithm to base 10 of the reciprocal of its transmittance; there is no unit. For **transmission electron microscope**, see *electron microscope* and *scanning electron microscope*. The **transmission loss** along a channel between the sender and the receiver is the ratio of the power received to that sent; the decibel is the usual unit.

transmissivity measure of the ability of a substance to transmit a given radiation. It is the transmittance of a specified sample.

transmittance a measure of the ability of a sample of substance to pass a specified

radiation, sometimes called its transparency. It is the ratio of the power output by the sample to that input.

transmitter a device that can transmit (send out) a signal along a channel. In a radio system, the **transmitting aerial** is the aerial from which the signal leaves.

transmutation changing one chemical element into another. **Artificial transmutation** – e.g. changing 'base metals' (like lead) into 'noble metals' (like gold) was a dream of some people in the Middle Ages (see *alchemy*). As nuclear physics developed in the early decades of the twentieth century, artificial transmutation became real (though lead tends to be an end product rather than the reverse). Such a process in a nuclear pile or an accelerator mirrors the **natural transmutation** that occurs in stars and on Earth in radioactivity and as a result of cosmic rays.

A **transmuter** is a machine for transmuting some nuclear waste products to safer nuclides.

transparent able to pass a given radiation with little absorption, scattering or refraction. Note that a substance transparent to one form of radiation may be opaque to a second. Compare the word with *translucent*, though there is no clear divide between the two.

transponder device able both to receive and to transmit signals or data, where the output ('response') relates in some way to the input. It is the same in essence as a transceiver, except for that relation between input and output. Many types of vehicle, up to and including artificial satellites, have a transponder; this will automatically identify the vehicle on receipt of an interrogating signal.

In communications satellites, a transponder receives an input signal, amplifies it, and returns it at a different frequency. The difference of frequency is to prevent the strong output signal from interfering with the weak input. With compression, some such transponders can handle ten or more colour television channels, or 15 000 voice calls, or data at the rate of 500 million bits per second. A satellite may have up to sixteen transponders.

trans-uranic elements elements with greater proton numbers (Z) than uranium, number 92. Like technetium (43), promethium (61), astatine (85), and francium (87), none is found

93	neptunium	Np
94	plutonium	Pu
95	americium	Am
96	curium	Cm
97	berkelium	Bk
98	californium	Cf
99	einsteinium	Es
100	fermium	Fm
101	mendelevium	Md
102	nobelium	No
103	lawrencium	Lr
104	rutherfordium	Rf
105	dubnium	Db
106	seaborgium	Sg
107	bohrium	Bh
108	hassium	Hs
109	meitnerium	Mt
110	darmstadtium	Ds
111	roentgenium	Rg
112	ununbium	Uub
113	ununtrium	Uut
114	ununquadium	Uuq
115	ununpentium	Uup
116	ununhexium	Uuh

in nature as the only isotopes have half-lives that are short: i.e., all are radioactive, some so much so that we know very little about them. The reason for the low stability of these nuclides is that $Z = 105$ is the limit for the attractive strong forces between nucleons to be greater than Coulomb repulsion. Over recent decades, people have prepared the nuclides of these elements artificially, in most cases in accelerators, and in many cases in quantities of just a few atoms.

The table (left) shows all these elements known up to 2006.

The last few elements in the list, while having been discovered, as yet have no agreed names. See *ununxium*.

Those from 104, rutherfordium, onwards are the 'superheavy' elements, members of the trans-actinide series.

The table extends to the start of the proposed island of stability ($Z = 114$). However, that may only mean that some isotopes in that region are more stable than others nearby; in any event, none has yet been made (mid-2006).

transverse across the normal direction (as opposed to longitudinal, along that direction). Thus **transverse vibrations**, as found in the case of a **transverse wave** such as light, are at 90° to the direction of energy flow.

travelling moving from place to place. The design of a **travelling microscope** is so that the user can move it along a track at 90° to its axis. While it does not have high magnifying power, it can be very effective at measuring the distance it travels, perhaps within a micrometre, e.g. between the ends of a sample in the form of a rod.

A **travelling wave** is a progressive wave: one which can (potentially) transfer energy an infinite distance. On the other hand, a standing wave reflects to and fro between the ends of a confined space, as in the case of sound in a stretched string.

Trevithick, Richard (1771–1833) mining engineer well known for his significant improvements to the early steam engine and for the first steam locomotive (1804)

triac triode for alternating current, a particular type of silicon controlled rectifier. It consists of two thyristors joined at the gates, but back to back. There are many light current uses, as in dimmers and electric motor speed controls.

triangle rule When one draws to scale the three vectors concerned with a system in equilibrium, they form a closed triangle. If the vectors are forces, the triangle is a **triangle of forces**.

triatomic having molecules each of which consists of three atoms, as is the case with H_2O and ozone O_3

triboelectricity modern name for making (static) electricity by friction

tribology the study and application of concepts that relate to friction and overcoming it (lubrication)

triboluminescence form of luminescence that appears from a rubbed surface. It is not the same as friction-induced sparking.

trichromatic theory of vision the theory that human eyes can detect only three ('primary') colours, a full colour image arising from a mixing process in the brain

triclinic system a crystal system

trigger an input signal to an electronic circuit that causes a given output: a stimulus, in other words

trigonal system a crystal system

trimmer small variable capacitor, used to make fine adjustments

triode a three-electrode thermionic electron tube (vacuum tube) – now very rarely used, as the transistor can carry out all its functions much more cheaply and reliably. Between the high temperature cathode (electron source) and the anode (electron sink) of the vacuum diode is the third electrode; called the grid; this consists of a fine wire coil or mesh. The voltage on the grid controls the current through the device.

triple point the only pressure/temperature state at which a given substance can exist with the solid, liquid and gas (in fact, vapour) phases in equilibrium. In the case of H_2O, the triple point is at 273.16 K (0.01 °C) and 611.73 Pa (0.6% of an atmosphere); this is by definition, rather than by measuring – the triple point temperature in this case is one of the two fixed temperatures on the international temperature scale.

triplet group (multiplet) of three close related objects, such as a group of three close spectral lines or a set of three lenses used as one unit

tritiated with all or most of the hydrogen (^{1}H) atoms replaced with tritium (T = ^{3}H) atoms. **Tritiated water** is H_2O with a significant proportion of THO and/or T_2O molecules.

tritium T or ^{3}H the radioactive isotope of hydrogen with nucleon number 3 (and therefore with two neutrons in the nucleus).

The normal method of production is to bombard lithium-6 with neutrons. Tritium is a low energy beta-emitter with a half-life of 12.26 years. It is in wide use as a tracer, and is crucial to hydrogen bombs. Tritium has long been a widely used source of light in the form of radioluminescent sources (as well as in luminous paints); as the decay is low energy, there is no hazard from tritium unless you inhale it or take it in by mouth. The standard source has the gas at high pressure in a fluorescent tube. Brighter output is possible from a mixture of tritium and a phosphor.

triton T the tritium nucleus, a grouping of one proton and two neutrons

troposphere the layer of the atmosphere between the Earth's surface and the **tropopause**: that being the start, at about 20 km, of the stratosphere. All weather events and interactions take place within the troposphere.

Tsiolkovski, Konstantin (1857–1935) Russian physicist who was a pioneer in his country's development of space travel. He also built one of the first wind tunnels, in 1892.

Tsui, Daniel (1939–) one of the three winners of the 1998 Nobel Prize, for their discovery of a 'new type of quantum fluid with fractionally charged oscillations'

tube short name, mainly in the US, for an electron tube or valve

tuned circuit or **tuning circuit** circuit which contains capacitance and inductance, the values of either (or both) being set so that the circuit resonates to a given input ac frequency. **Tuning** is the process of setting the frequency, as when choosing a station with a dial radio set.

tungsten W transitional metal in period 6, proton number 74, identified in 1779 by Peter Woulfe, named in 1781 by Martin Klaproth (Swedish meaning is 'heavy stone'), and isolated in 1783 by Juan and Fausto de Elhuyar; the symbol comes from wolframite, the main ore (tungsten is 'wolfram' in most languages). Relative density 19, melting and boiling temperatures 3410 °C and 5600 °C; most common stable isotope ^{184}W. The metal is of wide use in filaments in lamps and discharge tubes, dense alloys (e.g. for armour and boat keels), super-alloys and composites, and glassware needing good seal between metal and glass.

tuning circuit better name for tuned circuit. See also *acceptor circuit*.

tuning fork two-pronged fork made of special steel; when gently struck, it resonates to produce a pure long-lived tone. There are forks like this that provide frequency standards over almost the whole audio frequency range.

tunnel for a particle of given energy, to pass across a barrier of higher energy in the process of **tunnelling**, as predicted by Josephson. While this process is not possible in classical physics, wave mechanics allows a small probability for it to happen: and there are many contexts where tunnelling occurs. A little later, a team observed electrons tunnel in real time for the first time: they used a very intense laser beam that causes orbital electrons of atoms to tunnel away in a few atto-seconds.

A **tunnel diode** is an extremely heavily doped p-n junction. When the junction has a small forward bias, electrons tunnel from the n-side across the junction barrier into empty levels on the other side. Over a small forward voltage range, therefore, the device is ohmic. Past that, it has a negative resistance. The tunnel diode has value in switching, amplifying, and oscillating circuits.

Tunnelling by photons was first found in 2006 – allowing thin but opaque films of silver metal to become transparent. The technique involves making the film lie between the zinc selenide coatings of two glass blocks; it should make organics leds more efficient.

turbine machine able to transfer the energy of a moving fluid (mainly air, water, steam or a hot gas) directly into rotary motion. The fluid pushes against the shaped blades of the turbine fan in the **impulse turbine**; this works in much the same way as the water mill and wind mill. The **gas turbine** is of this type; a version of it is the **turbo-jet**.

In the less common **reaction turbine**, the fluid passes at speed out of nozzles in the edge of the fan; this latter design is much like the first steam engine known, that of Hero of Alexandria in around 60 bce. The **steam turbine** is not like this, however – it is an impulse system.

turbulence the irregular motion, with a number of eddies, of a fluid moving too fast for regular laminar flow. In any given context, there is a critical fluid speed (or Reynolds number) at which turbulence starts; after that, the fluid motion at a point varies irregularly in both speed and direction.

Turbulent flow is the flow of a fluid at such a speed that turbulence appears – there are eddies and sometimes large changes in flow speed and direction from point to point.

turns ratio for a voltage transformer, the ratio of the number of turns in the output coil to the number in the input coil. If there are no losses (100% efficiency, the norm being around 98%), the voltage ratio equals the turns ratio.

tv short for television. In the recent decade or two, there has been much research into **digital tv** (the use of digital rather than analog systems, giving a much sharper image) and into **hdtv** (high definition television, with images with 1080 or more horizontal scan lines rather than a few hundred). Several 'standards' followed in each case. At the same time distribution of multi-channel signals by satellite and cable has become widespread; these systems give the users far more channels to watch. Research into flat screen systems has continued at the same time, and these are now coming to be quite cheap, even at large screen sizes. A fourth significant area of research has been into 3D tv, systems giving the appearance of depth, preferably without the viewer's having to wear special glasses or to sit in a single place.

tweeter speaker designed to reproduce high frequency signals with high fidelity

twinkling the variations in brightness of a star that result from turbulence high in the Earth's atmosphere

twin paradox confusion that results from a careless consideration of the effect of time dilation on a person who travels to a distant star and back – to find the twin left behind to be much more aged. The claim is that relativity means we can take either twin to be in the observer's frame of reference, therefore either twin will be the older at the end of the journey. However, relativity does not allow us to give artificial accelerations to objects just by changing the frame of reference: only the travelling twin experiences the accelerations, so only the other becomes more aged.

Tyndall effect the Rayleigh scattering of light by extremely small particles, including molecules in solutions, colloids and suspensions. The scattered light is polarised, and the scatterer seems cloudy. John Tyndall (physicist, 1820–1893) also explained the blues and reds of the sky using these concepts of scattering.

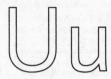

u a) old symbol for atomic mass unit (also out-dated) – now we use A for nucleon number instead

b) abbreviation for ultra-

c) symbol for the up quark

U a) in maths, symbol for unitary as in SU, for unitary symmetry, and for the unitary operator

b) symbol for uranium, element 92

u a) symbol for electric dipole moment (p preferred)

b) symbol for molecular velocity (v preferred)

c) symbol for relativistic mass (m preferred)

d) symbol for specific internal energy (w preferred), unit: joule per kilogram, $J\ kg^{-1}$

e) symbol for speed (c preferred)

f) symbol for velocity (v preferred) and for velocity in 4-space

U a) symbol for electric potential difference (V preferred)

b) symbol for internal energy (W preferred), unit: joule, J

c) symbol for potential energy (W or W_p preferred), unit: joule, J

uhf in the case of radio waves, ultra-high frequency – in the 30–300 MHz band, widely used for television

ultimate strength the same as the strength of a sample of substance: in other words, the stress at which it will fracture

ultra- prefix meaning extreme. For **ultra-high frequency** radio waves, see *uhf.* An **ultra-high vacuum** has a gas pressure below 10^{-7} Pa. An **ultra-microscope** is a light microscope (one working with visible wavelengths) whose resolving power is increased by some special technique or other. In most cases, the technique involves forming a Tyndall (scattering) cone using a very intense light source and a very dark background (e.g. a black body); very small objects (as in Brownian motion and the oil-drop experiment) then show as bright, small diffraction rings.

Ultra-sonic radiation consists of pressure waves (i.e. those with the same nature as sound waves) of frequency above the band the normal human ear can detect, that is above about 20 kHz (we define that as an **ultra-sonic frequency**). There are many uses. **Ultra-sonic scanning** (or **ultrasonography**) applies high-frequency pressure waves to scanning metal objects and people (for medical reasons in the latter case). The use of high frequency gives high resolution, but the range of the waves becomes low. Medical ultra-sonic scanning has a number of advantages over x-rays, however: there is much less damage to the patient and better contrast between soft tissues in particular. It can also safely provide moving images in real time (which is crucial for good ultra-sound scanning of the beating heart, for instance). There is also a use of higher power focused beams to cauterise capillaries and stop internal bleeding without hazardous surgery. **Ultra-sonics** is the study and use of ultra-sonic radiations, which we also call **ultra-sound**.

The **ultraviolet** region of the electromagnetic spectrum is that past the blue end of the visible band, and up to the start of the x-ray region. The wavelength range is 400 nm to 4 nm (both figures being approximate). The main artificial sources are certain discharges, such as arc and spark, while most detection methods involve fluorescence, photography or photoelectric effects. Transitions between the inner electron levels of atoms produce this radiation, which people sometimes split into several sub-regions for convenience. The **ultraviolet catastrophe** followed the derivation of several expressions giving the distribution of radiation from a *black body* at different temperatures. All these classical expressions showed that the radiation intensity becomes infinite at ultraviolet wavelengths; of course, this is not the case in practice, so the theories concerned led to a catastrophe in physics thinking. The quantum theory was able to provide a solution to the paradox; see, for instance, *Planck*.

umbra a region of full shadow, reached by no radiation at all from the source. This is as opposed to penumbra, a partial shadow region which varies from no radiation at the inside edge to full radiation at the outside edge.

Umklapp process or **Umklapp scattering** interaction between phonons, or between phonons and electrons, in a solid. The former is a major cause of thermal resistance, while the latter is a major cause of electrical resistance. In either case, a new phonon appears.

unbixium Ubx code name for a trans-uranic elements with *A* in the range 120–129, not yet found or made – see *ununxium*.

uncertainty principle an aspect of complementarity – see *Heisenberg's uncertainty principle*.

under-damping such a low level of damping that the vibrating system passes through a number of cycles before coming to rest

unified atomic mass unit u unit of mass on the atomic scale such that the mass of the nucleon is 1. We now use nucleon number *A* (no unit) instead.

unified field theory any modern theory that tries to bring together the physics of the different fields and interactions into a single set of equations. In particular, the aim is to unify the electromagnetic and nuclear field theories with relativity (and therefore with gravity). So far, no such theory has emerged, though unification has received a great deal of research effort since around 1915, and has made some progress.

uniform constant, or (in some cases) homogeneous. If an object moves with **uniform acceleration**, its acceleration is constant during the period concerned; we then describe how it moves as **uniform motion**: the equations of (uniform) motion apply. A **uniform temperature enclosure** is a closed space whose black body walls are at a constant temperature; a small hole in a wall

then gives a point source of black body radiation. An object's motion at **uniform speed** or **uniform velocity** involves constant speed or velocity, i.e. zero acceleration.

uni-junction transistor transistor that consists of a thin strip of n- or p-type semiconductor with a p- or n-junction near the centre. A common use is in pulse generators. A **programmable uni-junction transistor** is a type of thyristor, with a region of negative resistance in its characteristic curve.

uninterruptible power supply ups a device that sits between the mains power supply and an electronic device, with the aim of keeping the device running if a power supply fault occurs. An ups contains a battery and should filter out surges and spikes, top up power during a brown-out, and replace it during a cut for at least a few minutes.

uni-polar transistor a class of transistor whose action involves only one type of charge carrier, negative electrons or positive holes. The field effect transistors are the most important type in this class.

unit having a value of 1. The **unit cell** of a crystal is the basic structure which, repeated regularly in three dimensions, builds up the lattice. A **unit pole** is a monopole with unit magnetic pole strength; such a pole has not been found, but the concept is of value in definitions. A **unit vector** has unit length and lies along one of the three coordinate directions; the symbols are i, j, and k for the unit vectors in the x, y and z directions.

unitary symmetry the 'eightfold way', a theory of elementary particle groupings based on a (mathematical) special unitary (SU) group. The simplest such group is SU(2), described in mathematics by matrices of dimension 2 × 2. SU(3) group theory (which includes SU(2) as a subset) has received most attention in physics. Closely linked with qcd, the quark model of matter, this describes strongly interacting elementary particles as existing in multiplets (groups) with one, eight, ten or 27 members. The members of each group have the same baryon number, spin and parity; they differ in charge, iso-spin and hypercharge – but we could view them as different forms of the same basic particle. Thus unitary symmetry is a simplifying theory, there otherwise being hundreds of apparently separate particles.

However, it must be said that there is (as yet) no known physical reason why this approach works: it just does. This is so, even if one argues that unitary symmetry 'predicts' quarks.

The major triumph of unitary symmetry followed the discovery of the omega-minus particle in 1964: this particle, and its properties, had been predicted from SU(3). The name implies that people expected it would be the last particle to be found as it would complete the decuplet (set of ten particles). This set is the baryon decuplet, the groups of particles with baryon number 1, spin $\frac{3}{2}$ and parity +1. The particles fall symmetrically on axes of isospin I and hypercharge Y.

units a set of standard values for the set of standard measures on which we base all measures (in daily life as well as in physics). The standard measures are of mass, length, time (and a number more). Since ancient times there have been such sets; for instance, in Europe, a number of such systems arose from that used in Roman times. These include the Imperial system, still common in many countries: in this case, those three basic units are the pound, foot, and second. As time passed there was more and more standardisation from place to place, and more and more attempt to make the relationships rational; also, it has been crucial to make each standard unit reproducible. The world-wide standard in use now for science and engineering, and more and more for daily life, is SI, described in *Appendix 1*.

universal applicable most widely or generally. Thus, there are many **universal constants** in physics – so-called fundamental constants whose values do not depend on local conditions. Examples are the speed of light in empty space c, and the charge of the electron Q_e. The **universal gas constant** R applies to all gases we can take as behaving in an ideal way (in other words, any gas at low pressures and at temperatures above its critical values). R appears in the ideal gas equation that links the pressure p, volume V and temperature T of a sample:

$$p V = R T$$

The value of R is 8.314 41 J K^{-1} mol^{-1}.

A **universal motor** is able to run on either direct or alternating supply, while a **universal shunt** has a number of points at which one can tap (connect) it to vary its resistance.

Universe the closed system of all that now exists and happens, to our knowledge.

However, some theories hold that there are other quite separate Universes and/or that the current Universe is just one of a series in time.

unsaturated with a value less than what is possible. For instance, an **unsaturated vapour** has a pressure lower than that of the saturated vapour under the same conditions.

unstable equilibrium a state of equilibrium away from which the system will move if disturbed by a small amount

ununxium Uux temporary names of the radioactive trans-uranic metals found too recently for formal naming. These temporary names are from the Latin numbers, with un = 1, bi = 2, tri = 3 and so on.

112	Uub ununbium	1996
113	Uut ununtrium	2004
114	Uuq ununquadium	1999
115	Uup ununpentium	2004
116	Uuh ununhexium	2001
118	Uuo ununoctium	

On the same naming model, unbixium is the set of names given to elements from 120 onwards – such as unbihexium Ubh, whose isotope Ubh-310 people hope will be doubly stable, in the middle of the *island of stability*.

U-process short for the Umklapp process of scattering by phonons

upthrust common name for the upward force on an object fully or partly submerged in a fluid – as given by the principles of Archimedes and of flotation. See also *absolute weight*.

uranium U radioactive transitional rare earth metal in the actinide series, period 7, the proton number being 92 (the highest natural element), found by Martin Klaproth in 1789 and named after the planet Uranus, also newly found. Relative density 19, melting and boiling temperatures 1130 °C and 3800 °C; three natural isotopes, all alpha-emitters that also decay by spontaneous fission: 234U, half-life 250 000 years; ^{235}U, half-life 700 × 10^6 years; and ^{238}U, 4.5 × 10^9 years. ^{235}U is fissile, so is important for *nuclear power* and weapons, especially when enriched; on the other hand, depleted uranium (less than 0.25% ^{235}U) is of use for armour, penetrators, and inertial guidance.

The technique of **uranium–lead dating** depends on the decay of uranium-238 to lead-206 (see below); ^{238}U has a half-life of 4.5×10^9 years, so the technique is of value for dating even the oldest rocks. A variant depends on the decay of ^{235}U to ^{207}Pb. The **uranium series** is one of the four radioactive series of decays; called the $4n + 1$ series, it starts at uranium-238 and ends at lead-206.

Urey, Harold (1893–1981) chemist whose discovery of heavy water and then of deuterium in 1932 led to his Nobel Prize for chemistry in 1934. Amongst his other work is the exploration of dating, cosmology, entropy, isotopes and absorption spectra.

U-theory less common name for the M-theory, the master theory that would combine the five superstring theories by way of their dualities

Uuq symbol for ununquadium, temporary name for element 114; see *island of stability*.

uv ultraviolet, waves in the electromagnetic spectrum on the high frequency side of visible light. The wavelength range is 400 nm to 4 nm (approximately).

V a) symbol for any 'V particle', one that decays into a pair of particles that leave a shape like a V in a detector chamber; the term is no longer in use
b) symbol for vanadium, the element with proton number 23
c) symbol for volt, the unit of electrical potential and potential difference (i.e. 'voltage')

***v* a)** symbol for specific volume (i.e. volume per unit mass), unit: cubic metre per kilogram, $m^3\,kg^{-1}$
b) symbol for speed (c preferred)
c) symbol for velocity, unit metre per second, $m\,s^{-1}$
d) symbol for vibrational quantum number
e) symbol for volume (V preferred)

V a) symbol for the dispersive power (Abbe number or Abbe constringence) of a transparent substance to white light; no unit
b) symbol for electromotive force (E preferred)
c) symbol for electric potential and electric potential difference, unit: volt, V
d) symbol for potential energy (W preferred)
e) symbol for volume, unit: cubic metre, m^3

vacancy site in a solid sample where there are bonds for a particle, but the particle is missing. (See *defect* for vacancies in a lattice for atoms and ions (cores).) We call the vacancy for an electron a hole; this can migrate through the lattice when there is an electric field, though it does so at a slower net (drift) speed than a free electron.

vacuum space that contains no real matter. In practice, even intergalactic space contains a few particles per cubic metre, to say nothing of passing radiation (which has energy and therefore mass). In practice, we therefore mean by vacuum a volume with a low gas pressure (a **soft vacuum**) or a very low gas pressure (a **hard vacuum**). In addition, it is widely held that all space is at all times full of virtual particles, with an infinite total energy density (or at least an energy density like that inside a nucleus: see *zero point energy*). (A virtual particle is one that exists for a shorter time than would contravene the uncertainty principle.) This is a development of Dirac's early idea of vacuum as a sea of electrons of negative energy.

   Vacuum evaporation is a technique for coating a surface with a film; the surface is in a vacuum that contains a heated block of the coating substance. (In the same way, the inside of a lamp slowly becomes grey with a coat of tungsten.) A **vacuum flask** is a container designed to keep as low as possible the transfer of thermal energy between its contents and the outside. See *Dewar flask*, the name widely used in science. The purpose of a **vacuum pump** is to reduce the gas pressure within a space to as low a value as possible. Some such systems can achieve pressures as low as 10^{-12} pascal (10^{-17} atmospheres): but this still leaves a few million molecules per cubic metre.

   A **vacuum tube**, made of glass or metal, contains very low pressure gas and a set of electrodes that depend on the specific function; this may be for making a discharge of some type, or it may be to act as an electron tube ('valve').

valence (the older, and now less common, term being **valency**) a measure of how well an atom or ion can combine with others. Often this is the same as the number of unpaired (valence) electrons in the outer shell. In the band model, the band of energies that includes those unpaired electrons is the **valence band**; the electrons concerned are **valence electrons**.

validity measure of whether or not a value (or set of values) from some source is valid or not: in other words, is of the right type and within the right range. This is not the same as accuracy or precision.

valve a) mechanical device which controls the flow of fluid through a tube or channel. A **non-return valve** – perhaps the most common type – allows flow in one direction only. Valves are important in hydraulics and pneumatics.
b) electronic device which allows charge flow in one direction only, and may also control the rate of flow more precisely. The term mostly refers to electron tubes made of glass that contain a soft vacuum and a suitable set of electrodes. One of those electrodes is heated, hence the common name of **thermionic valve**.

Such valves are rare now, having been replaced in almost all cases by solid state systems; the cathode ray tube is the only design still in common use. The most important electronic valves of the past were the diode (a rectifier) and the triode (an amplifier or oscillator). A **valve voltmeter** was a voltmeter built from an amplifier designed round a valve, with a meter at the output. The modern solid state form is the dc amplifier.

vanadium V transitional metal in period 4, proton number 23, first found in 1801 by Andrés del Rìo (who felt he had to withdraw his claim in 1805 because of sceptics) and then in 1830 by Nils Sefström; he named it after the goddess of beauty because of the brilliant colours of its salts. Relative density 6.1, melting and boiling temperatures 1900 °C and 3400 °C; most common isotope ^{51}V. Its main use is in alloys; there are also uses in nuclear power, power supply, and superconductivity.

van Allen belts belts (regions) in the upper atmosphere high above the Earth (about 800–10 000 km); they contain high levels of radiation caused by the solar wind. James **van Allen** (physicist, 1914–2006) discovered them in 1958.

van de Graaff generator friction-based electrostatic generator of high voltages (in some cases as much as 15 MV). The system is still in wide use as a spark source, and to provide the electric field required in simple particle accelerators – **van de Graaff accelerators**. Robert van de Graaff (1901–1967) was a physicist and instrument maker.

van der Meer, Simon (1925–) accelerator physicist and Nobel Prize winner, sharing this in 1984 with Carlo Rubbia for the discovery of the W and Z particles the year before

van der Waals, Johannes (1837–1923) physicist who carried out much work on real gases (as compared with ideal gases), and thereby gained the 1910 Nobel Prize. The **van der Waals equation** is an important equation of state for a real gas rather than an ideal gas – it allows for the particles to attract each other (rather than interacting only during collisions), and to have finite size (rather than being points). These two aspects of the real gas as opposed to the ideal gas appear in the equation as the constants a and b respectively:

$$(p + a/V^2)(V - b) = R T$$

Here p, V and T are the pressure, volume and temperature of the gas sample, a and b are the constants mentioned in the text, and R is the gas constant.

The ideal gas equation – $p V = R T$ – follows from this by putting a and b equal to zero.

Van der Waals' forces (or **van der Waals' bonds**) weakly attract even neutral atoms and molecules to each other (as in the kinetic model of matter). They follow from the interactions between dipoles (and other multipoles), however these may be caused in practice. There are three main types of attractive process in the case of dipoles formed in particles A and B:
a) A and B are both permanent dipoles: they attract each other if they are nearby and suitably aligned, but the net attraction falls with temperature rie. This is the 'orientation force".
b) A is a permanent dipole, and induces an opposite dipole in B when this enters the field of A. This so-called 'induction force' is really therefore a variety of (a), but does not depend on temperature.
c) If neither A nor B is a permanent dipole, there will still be random motions in the electron cloud of each which will produce dipole effects and induce opposite attractive dipoles. In many gases, this 'dispersion effect' is stronger than the other two.
The **van der Waals radius** of an atom is the radius of a hard sphere that will behave in much the same way as the atom in contexts like those above. (There is also a **van der Waals volume** which derives from that.)

van't Hoff, Jacobus (1852–1911) the first
physical chemist who received the first Nobel
Prize in chemistry (that being in 1901) for
showing that the solute particles in very
dilute solutions behave much like the
particles of a gas. He also did major new
work in the physics of reactions, osmosis
and the study of crystals.

van Vleck, John (1899–1980) physicist who
shared the 1977 Nobel Prize for work on the
quantum theory of magnetism

vaporisation the change of phase from liquid
or solid to vapour. This includes the relevant
processes of evaporation, sublimation and
boiling.

vapour a gas below the critical temperature of
the substance, and thus able to be liquefied
by pressure. Humidity concerns the H_2O
vapour content of air.

Evaporation is the process by which
particles with energies above the mean leave
the free surface of a liquid (or, to a lesser
extent, of a solid); they then move around as a
vapour in the space above the surface. Some
will return into the surface in the process of
condensation. It is often useful to know the
vapour density and the **vapour pressure** for a
given sample of a given vapour in a given
context; both terms mean the same as for a
gas.

In most cases, the vapour in the space above
the liquid in a closed container reaches
equilibrium – then, as many particles enter
the liquid surface per second as leave it. In
other words, in this state, the evaporation
and condensation rates are the same. (This is
a good example of dynamic equilibrium.)
Now we describe the vapour as saturated: its
density and pressure are at the maximum

values for the temperature at the time. The
value of the vapour pressure is now the
saturated vapour pressure (svp). For a given
substance, this depends solely on the
temperature. The first graph below shows
how: at low temperatures, svp is low and
changes little with temperature rise; the curve
is, however, asymptotic to the critical
temperature value, that at T_c. That confirms
the definitions – it is not possible to have a
vapour above the critical temperature, in
other words, it is not then possible to
compress a gas into a liquid.

There are two other useful interpretations
of this graph. The first follows from the fact
that a liquid boils when its svp equals the
outside pressure. Thus, the same graph
shows how the boiling temperature of the
substance varies with pressure. It shows too
that it is impossible to have the substance in
liquid form above the critical temperature
(Andrews' experiment confirmed this). That
leads to the next interpretation of the graph:
it is a phase diagram. In other words, it
shows what phase (liquid, vapour or gas) the
substance is in for any given pressure/
temperature state. The line itself is the phase
boundary (called the steam line in the case
of H_2O).

The next graph extends this concept of
phase diagram to include the solid phase.
Now we have boundaries between solid and
liquid and between solid and vapour (which
reminds us that solids as well as liquids can
evaporate). The triple point P is the junction
between the three lines: P is the only
pressure/temperature state at which solid,
liquid and (saturated) vapour can exist
together in equilibrium. Note that

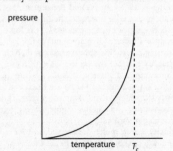

The saturated vapour pressure (svp) graph of a
substance becomes asymptotic at the critical
temperature.

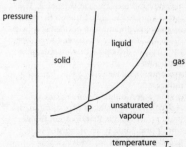

The complete phase diagram of a substance shows
the triple point as a unique pressure/temperature
value.

a) the vapour is saturated only at the vapour/solid and vapour/liquid phase boundaries, and

b) the solid/liquid phase boundary slopes the other way in the case of H_2O; this is part of the anomalous behaviour of water.

var rare name for the volt ampere (V A), the unit of power in reactive ac circuits; it stands for volt ampere reactive.

varactor sometimes called varicap, a variable capacitance diode: a reverse-biased junction diode whose capacitance depends on the bias voltage. In other words, it is a capacitor controlled by voltage (that also happens to be a diode). Such diodes are in wide use: for instance in parametric amplifiers, automatic frequency control systems, and **varactor tuning** (where turning the tuning knob changes the bias of the varactor).

variable a) able to be changed in value by the user or automatically, as part of some control system: as with the capacitance of a variable capacitor and the resistance of a variable resistor

b) able to change in some value, perhaps periodically: as with the brightness of a variable star

c) a measure whose value is not constant. Instead, the value of a **dependent variable** (common symbol y) is a function of the value of an **independent variable** (often x).

variance a) measure of the deviation of a value from the mean of the set

b) measure of the mean deviation of all values in the set from the mean, given by the square of the standard deviation

variation a) old term for the magnetic declination at a point on or near the Earth's surface – the angle between the magnetic and geographical meridians at the point. The meridians are the lines through the point to magnetic north and true north respectively.

b) in full, magnetic variation, the changes with time of the magnetic elements of the Earth's field

variometer confused name for a variable inductor

varistor a non-ohmic resistor made of semiconducting material (e.g. metal oxide), in other words, one that does not follow Ohm's law. Some types of varistor have negative resistance for either current direction.

The name is short for variable resistor, though a better one (quite often used) is voltage-dependent resistor – such a device has a very high resistance through a range of low voltage and a very low resistance past a certain higher voltage. This gives the varistor its main use – to protect sensitive parts of a circuit from voltage surges and spikes.

vector a measure to define which needs a direction as well as a size (magnitude); a scalar has size only. For instance, velocity is the vector form of the scalar speed. The sum of two vectors must also involve their directions as well as their sizes; we use the parallelogram rule for adding vectors to find their single resultant. In print, it is normal to show a vector in bold – we do this in these notes, for instance, using **F** rather than F for the vector force.

To define a **vector field**, we need an expression for the direction as well as the size of the relevant property at each point. A magnetic force field is an example, as is the field of wind strengths in a volume of the atmosphere.

The scalar product of two vectors **A** and **B** – written **A · B** (and, as a result, sometime called 'dot product') – is a scalar. For instance, when a force **F** moves an object a distance **s** (displacement) in the same direction, the product **F · s** is a scalar. See the energy transfer below.

The products of the vectors force **F** and displacement **s**

Energy transfer (unit: joule) is the scalar product $\qquad W = F \cdot s$

Torque (unit: newton metre) is the vector product $\qquad T = F \times s$

The **vector product** (or 'cross product') of two vectors **A** and **B** is itself a vector, its direction being at 90° to both **A** and **B**; we print this **A × B**. An example is torque, the turning effect of a force given by the product of the force and the distance (displacement) to the turning point: this appears last above. Note that **A × B** is not the same as **B × A**. In fact, the former is the vector $A\,B\,\sin x$, while the latter is $-A\,B\,\sin x$: the two are equal and opposite. (Here A and B are the sizes of the two vectors, and x is the angle between them.) See also *triangle rule*.

velocity v unit: metre per second, m s^{-1} the rate of change (with time) of the displacement of an object. This is a vector (as displacement is a vector). (Distance and speed are the scalar versions.)

In the process of **velocity modulation**, the

virtual

velocities of the charged particles in a beam are made to vary, using an applied radio frequency field, for instance. This leads to their bunching into groups. For the **velocity ratio** of a machine, which relates to its efficiency, see *distance ratio* (the modern term).

A **velocity/time graph** – as the name implies – plots an object's velocity (or speed) against time from some starting point. If the graph is horizontal, the speed is constant (no acceleration); if the graph is a straight line, the acceleration is constant – and, in all cases, the area between the graph and the time axis between two moments is the distance covered in that period.

Veltman, Martinus (1931–) physicist who shared the 1999 Nobel Prize with Gerardus 't Hooft for their work on quantum electroweak interactions

Venturi meter device used to measure the rate of flow of a fluid in a closed channel (pipe or tube). It depends on the **Venturi effect**, a special case of the Bernoulli effect. The meter consists of a narrow section in the channel (the **Venturi tube**) and a means to measure the loss of pressure in the fluid in that section of the channel. The name is that of the physicist, Giovanni Venturi (1746–1822).

vernier small side scale used alongside the main scale of a meter, by which one can much more accurately estimate fractions of a graduation. There are nine graduations on the vernier against ten on the main scale. The name comes from that of its inventor, mathematician Paul Vernier (1580–1637). A user's **vernier acuity** is a measure of the ability to decide whether a line on the vernier is or is not exactly lined up with one on the main scale.

vertex focal distance the distance between the surface of a thick lens or compound lens to the nearest focal point

very high frequency (vhf) (radio) frequency in the range 300 MHz to 30 MHz (wavelength 1 m to 10 m)

Very Large Telescope (VLT) the world's largest astronomical telescope, in Chile. It has four 8.2 m telescopes which can work together; this makes it equivalent to a single telescope with a mirror about 17 m across.

very low frequency (vlf) (radio) frequency in the range 30 kHz to 3 kHz (wavelength 10 km to 100 km)

vhf very high frequency radio waves

vibration (or oscillation) a cyclic to and fro motion of an object or part of an object. Some people use the term for any cyclic movement or even any cyclic change. See *simple harmonic motion*, which can describe such motions, and *sound sources*, all of which involve strict vibration as described above. For the **vibrational quantum number** v of a system, see *quantum number*.

vibrator a) electromagnetic device whose input is an alternating current and whose output is a matching physical vibration of some part of the device. A speaker is such a device.
b) electromagnetic system that uses a make-and-break device (as in an electric bell) to convert input direct current to pulsed dc (very crude ac). A buzzer is such a system.

video of an electronic system, able to work with signals that have enough band width to carry a full moving tv-type picture and sound. Most people take the **video bandwidth** to be several mega-hertz (though compression techniques may reduce this a great deal in practice).

virgin neutrons neutrons so fresh from a source that they have not yet been involved in any collisions

virial expression that appears in the calculation of the pressure of a gas. Such an expression may look like such a power series as $pV = a + bp + cp^2 + \ldots$ Here $a, b, c \ldots$ are **virial constants** and the expression as a whole – the virial – is a **virial equation** or **virial expansion**. The approach originated with Clausius, whose **virial law** states that the kinetic energy of a system depends only on its virial.

virtual not real, but with the same effect in at least some ways as the real equivalent. For instance, a **virtual cathode** may appear in the space charge of an electron tube; it has zero potential gradient and acts as a real cathode (i.e. as a source of electrons). A **virtual energy transfer** can take place if one disturbs a system by an infinitesimal amount; strictly, of course, the energy transfer is infinitesimal too (indeed, it is zero if the system is in equilibrium).

A **virtual image** in an optical system is one through which rays appear to pass, but really do not. The image of one's face behind a flat mirror is of this type. It is also possible in an optical system to have a **virtual object** (apparent source of radiation).

The flexibility allowed by the uncertainty principle makes it possible for a **virtual**

particle to appear for a very short time, apparently breaking the law of constant mass/energy. Such a particle (also called quasi-particle) has the properties of a real sub-atomic particle but has no real existence as it exists only for a shorter time than would go against the uncertainty principle. In fact, the product of the time of the existence of the particle and its energy must be less than $h/4 \pi$, h being the Planck constant. Modern theories propose that many real particles have a cloud of such virtual particles around them and that they fill even 'empty' space. If two such real particles are close enough, they may exchange virtual particles and thus experience a force (because of the change of momentum involved). For instance, **virtual photons** are the exchange particles of the electromagnetic force. This means that they provide the Coulomb force between charges, the Casimir effect between charged plates, and the van der Waals' bonds between atoms and molecules; they also cause real photons to appear when excited particles decay.

Virtual work is an old name for virtual energy transfer (above).

viscometer device used to measure the viscosity (friction) of a fluid. There are many types of design.

viscosity friction in the body and/or at the surface of a fluid that tends to reduce its flow – see *fluid friction*.

viscous with a high (or, at least, non-zero) value of viscosity (fluid friction). **Viscous damping** is the appearance of a friction force that varies with the relative speed of an object through a fluid (e.g. a boat through water).

visible able to be seen by the normal human eye. Thus **visible light** is the range of electromagnetic radiation the eye can detect, and the **visible spectrum** of such **visible radiation** is the colour spectrum.

vision the sense of sight, detailed sensitivity to visible light. In higher animals, the special organ involved is the eye. See also *colour vision*.

visual concerned with vision. Thus, a person's **visual acuity** is a measure of the resolving power of the eye.

vitreous humour classic name for the transparent jelly between the lens of the eye and the retina. The name means glassy bodily fluid.

vlf of radio waves, very low frequency – in the range 3–30 kHz

VLT 'Very Large Telescope' – the largest astronomical telescope, with four 8.2 m telescopes which can work together

voltamp, V A unit of a circuit's active power

voice frequency a) band of radio waves, 30–300 Hz
b) the same as audio frequency: in other words, in the range about 20 Hz to about 20 kHz
c) in the range to just over 2 kHz, this being the band width of voice telephone systems

Voigt effect the double refraction of light passing through a vapour across a strong magfield

volt V the unit of electrical potential difference (pd or voltage), which we now defined in terms of the base units metre, kilogram, second and ampere. The official definition is the volt is the pd between two points, a current of one ampere between which causes an energy transfer of one watt.

Volta, Alessandro (1745–1827) physicist who made many discoveries in the then new field of current electricity. Indeed, many argue that he founded this field by showing that one could produce the effects of 'animal electricity' (as relates to frogs' legs) by using only inorganic materials. This work involved the invention of the **voltaic pile**, the first electric battery. He also explored the reverse process, electrolysis, and carried out work in thermal physics and gases.

voltage V unit: volt, V common name for electrical potential difference, or pd. A **voltage amplifier** is an amplifier whose input signal is a varying voltage. A **voltage divider** is a circuit whose output voltage V_0 is a known, perhaps variable, fraction of the input voltage V_i. It is common to have R_b the resistance of a variable resistor – the output is then a variable voltage.

$$V_0 = V_i \left(R_b / (R_a + R_b) \right)$$

A **voltage doubler** is a rectifier circuit whose dc output has a voltage twice that of the ac input. The layout is as shown. In the form of **voltage multiplier**, the Cockcroft–Walton accelerator used this technique again and again in many stages to produce a dc voltage of around 10 MV in 1932 (and a Nobel Prize in 1951).

The **voltage drop** between two points is an old name for the potential difference between them: in other words, it is the same as the voltage between the points. A **voltage**

stabiliser is a circuit – perhaps based on a Zener diode or a *neon lamp*, sense b) – whose output voltage is constant (or at least lies within a small set range).

A **voltage transformer** is the full name of the standard transformer. It consists of a coil in the input circuit and a coil in the output, the corresponding numbers of turns being N_i and N_o. Both coils are linked by a magnetic circuit (in most cases a 'soft' iron core, laminated to reduce eddy current power losses). The supply is alternating, of value V_i. This causes an alternating current in the input coil (the 'primary'), which in turn causes an alternating magnetic field in the core. This in turn induces an alternating voltage V_i between the ends of the output coil (the 'secondary'). In a perfect machine, the magnetic field fully links the two coils, with no leakage in other words; then the output pd is $V_o = V_i (N_o / N_i)$.

Also, if there are no other losses of energy in the transfer between primary and secondary, the output power equals the input power. Therefore, the system acts as a current (*I*) transformer as well, as $I_o = I_i (N_i / N_o)$.

The efficiency of a large voltage transformer (e.g. one over about 50 MW) can be very close to 100%, perhaps 99.8%. However, on a domestic scale, such systems tend to be only around 85% efficient (which is why leaving tv sets, chargers and so on switched on when not in use causes great waste – they all have a small transformer at the power input). Apart from imperfections in the magnetic circuit, the transformer's main losses of input energy are due to

a) eddy currents in the core (reduced by lamination);

b) rise in temperature of the coils ('copper' loss, reduced by using wire of larger cross-section); and

c) the energy involved in cycling the magnetisation of the core, switching domains twice a cycle ('iron' loss, reduced by having 'softer' iron).

voltaic cell old name for any electric cell with an output voltage that depends on chemical action. The system used by Volta in his early work consisted of discs ('plates') of copper and zinc, with a sheet of paper soaked in salt solution between them. The **voltaic pile** is a battery of voltaic cells (in other words, a number of such cells in series to give a larger voltage output).

voltameter electrolytic cell used in former times to measure charge transfer and electric current. The term was more recently used for any electrolytic cell, but is now rare in any context.

voltamp or **volt ampere** V A the unit of active power – i.e., the unit of power in reactive circuits (ones where the current and voltage are not in phase). It is the product of the root mean square values of the current and voltage. The var was once proposed as an alternative name.

voltmeter ———Ⓥ——— meter used to measure the pd (voltage) between two points. In the traditional case, the usual design has an ammeter with a low value full scale deflection, in series with a high value resistance. The value of the series resistance fixes the voltage range over which one can use the meter.

volume a) V unit: cubic metre, m^3 – the space taken up by an object or offered by a container

b) the perceived loudness of a sound wave, or the intensity of an electrical signal (in particular one that feeds a speaker to produce a sound wave). A **volume control** is an electric circuit element (for instance a variable resistor) used to control the size of an audio current.

von Laue, Max (1879–1960) physicist whose work on x-rays led to his 1914 Nobel Prize. The main aspects of that work were predicting and demonstrating the diffraction of x-rays by crystals. The **Laue method** involves passing a beam of 'white' x-rays (or other diffracted radiation) through a thin crystal. A set of spots appears on the screen behind the crystal; the angles to and sizes of these spots gives much information about the crystal's nature and structure.

vortex a spiral motion of a fluid inside a certain volume. Large scale examples in nature are whirlpools in the surface of a body of water and tornadoes, whirlwinds, and dust devils in the atmosphere. While the vortices that trail behind an object moving through a fluid (produced by the process of **vortex shedding**) tend to be much smaller than these, they still involve significant energy transfer, and therefore drag. (Indeed, vortex shedding by a flying aircraft can cause turbulence for several kilometres behind it.) Streamlining cuts down vortex formation and shedding and thus can much reduce drag.

vorticity measure of the rate of rotation of a

vortex. The vorticity at a point in a vortex is twice the angular speed of a particle at that point.

V particle old name for any particle that decays into a pair of particles that leave a shape like a V in a detector chamber

w symbol for water (e.g. in chemical equations and, in physics, as a suffix)

W a) symbol for the W boson, the intermediate vector boson
b) symbol for watt (= joule per second), unit of power
c) symbol for wolfram (common name: tungsten), element 74

w a) symbol for range (R preferred)
b) symbol for velocity (v preferred)
c) symbol for width, unit: metre, m
d) symbol for work, i.e. energy transfer (W preferred)

W a) symbol for electric energy, energy in general, radiant energy, work (i.e. energy transfer), unit: joule, J
b) symbol for wavelength (λ preferred)
c) symbol for weight, unit: newton, N (as weight is a force)
d) symbol for work function, unit: joule, J (though the electron volt, eV, is quite common in some contexts)

W boson, W meson or **W particle** the intermediate vector boson, thought to be one of the two exchange particles of the weak interaction (hence the name). It is not a meson, and the name W meson is no longer in use; the other weak exchange particle is the Z boson. Discovered in 1983, the W has unit spin; a mass of 80 GeV/c^2 (almost a hundred times the proton mass and close to that of the molybdenum nucleus); unit charge (W^+ and the anti-W, W^-); and a mean life of around 10^{-25} s.

Waals, van der See *van der Waals.*

Wadsworth prism 60° prism with a plane mirror at 75° to the face from which the rays leave. Rays that enter the first face at the angle for minimum deviation then leave the system at about 90° to their original direction. This prism is in common use in infrared spectroscopy.

wafer thin disc cut from a long cylinder of very pure semiconductor material, used in chip making. Wafers are about 0.5 mm thick and between 25 mm and 300 mm across; each one carries several hundred identical chips during manufacture. Chip making involves adding ten or more separate layers of circuit elements and links.

wall boundary, for instance the surface of a container, or the interface between two regions. In the former case, a **wall effect** is any situation where interactions or reactions at the container wall affect what should happen inside. An example is the release of adsorbed gases. In the second case, **wall energy** is the free 'surface' energy per unit area of a **Bloch wall** – this boundary lies between two domains in a ferromagnetic sample.

Walton, Ernest (1903–1995) nuclear physicist who shared the 1951 Nobel Prize with Cockcroft for their development and use of the first particle accelerator

Wankel engine perhaps the most well known of the many rotary internal combustion engines, the first version being invented by engineer Fritz Wankel (1902–1988) in 1924 and at last designed and developed by him in 1954. Though the engine is more efficient, simpler, more smoothly running, and lighter than other petrol engines of the same power, the system has not been widely used. The main concerns are its lower fuel efficiency and higher pollution.

waste heat energy output from some process that is not used but released to the environment, through (for instance) an exhaust system or chimney. Combined heat and power systems attempt to make use of 40–80% of the otherwise waste energy in the exhaust from power generation; this can raise the overall efficiency of such stations from about 50% to as much as 90%. There are also moves to use thermocouples and thermopiles made from special semi-conductors to generate more electricity from cooling towers, chimneys, exhausts and braking systems on all scales.

water liquid hydrogen oxide, H_2O (in fact HOH), a crucially important substance for life on Earth. It seems fortunate that water has many unusual properties: it behaves unusually in a number of physical contexts in ways that greatly benefit life as we know it. For instance, it
a) is a very good solvent;
b) forms ions easily (H^+ and OH^-);
c) is amphoteric (it can act either as an acid or as a base);
d) contracts while melting and also as it warms through the temperature range 0 to 4 °C; and
e) has an unusual liquid structure.
People have proposed various models to explain such anomalous features; a big problem has always been how to test the models produced in practice. It seems that weak hydrogen bonding between neutral HOH molecules and hydronium ions, H_3O^+ (present to a certain extent in all water samples) is the cause of the behaviour. This bonding imposes some medium-range order (force) between water molecules, without which, for instance, the substance would be a gas at normal temperatures; instead, there appears to be a degree of clumping of liquid water molecules into small meta-stable groups. However, the forces are weak, so the bonds form and break thousands of millions of times a second: and they are thus not observable. (See also *poly-water*.)

Until a few years ago, the main attack on the observation problem was to freeze the water and test its structure using neutron and x-ray diffraction methods. The results have not been of much value, however. A more modern approach is to analyse structures trapped from the gas phase: but, again, there has not yet been much success.

http://en.wikipedia.org/wiki/Water and http://en.wikipedia.org/wiki/Water(molecule) have a lot of clear information about water and the science of water.

The **water equivalent** of an object is the mass of water with the same thermal capacity as the object. Numerically, the value is the product of the object's mass and its specific thermal capacity, but the unit is the kilogram. In a pipe of liquid, a **water hammer** is a sudden banging sound (or 'knock') that follows such a change as turning a tap off quickly. In the history of technology (up to the 'industrial revolution'), the **water mill** (a form of turbine driven by moving water) is as important as the wind mill. Indeed, one of the earliest known mass production factories, a fourth century Roman mill able to produce up to nine tonnes of milled flour a day, was powered by a sequence of water mills.

Water surface waves involve rather complex motions of the particles in the surface and for a distance below it. (Of course, this is true of any liquid: the strict name is gravity wave.) In the surface, there is a circular motion in a plane parallel to the direction of energy transfer. Deeper and deeper below the surface, the circles become ellipses, and then lines of shorter and shorter amplitude. Even this model is over simple, however, as it does not take account of surface tension (an important aspect of short wavelength, low amplitude waves, such as ripples).

Watson-Watt, Robert (1892–1973) physicist who played a major role in the development of radar. His first patent in the field was in 1919.

watt W the unit of power (the rate of transfer of energy), the name given to a rate of one joule (one newton metre) per second.

A **watt-hour** is a unit of energy (now very rarely used): it is the energy transfer of a system whose power is one watt and which is used for one hour; the **kilowatt-hour**, 1000 watt-hours, is a more practical unit (but still one that is in rapid decline). The **wattless** component of an alternating current is the same as the reactive current; the name refers to the zero power transfer theoretically possible when using purely reactive circuits. A **wattmeter** is a meter whose output shows the power transfer through a section of circuit, in watts. The most common type, as used in ac circuits, consists of two coils. The fixed coil carries the circuit current, so is in series with the circuit output (load); the coil that is able to move, fitted with a high series resistance, is in parallel with the output. The angle of deflection of the moving coil (and the attached needle in the analog type of meter) varies with the power.

Watt, James (1736–1819) engineer whose main work was with steam engines. Indeed, his first big invention (1765) was a much improved version of the Newcomen engine. From 1775, he worked in making these machines in partnership with Matthew Boulton. That date was the start of a highly productive quarter century for Watt, with many inventions and ideas. He also invented the rotary 'sun and planet' gear (1781) which allowed the use of static reciprocating engines (as for pumping water from mines) to drive transport. (The patent holders of the crank refused to let Watt use their simpler system.)

wave

method of energy transfer that involves some form of cyclic variation or vibration. Wave radiations show six types of behaviour; if a radiation shows enough of these, it is mainly wave-like. First, when any wave reaches the interface with a second medium, there may be
- reflection – it bounces back into the first medium;
- refraction – it enters and passes on through the second medium, in most cases with a change of direction;
- absorption – it loses its energy at or very near the interface, which the energy warms in most cases.

Second, when any wave passes near an interface, there may be
- diffraction – it bends towards the edge.

Third, when a wave passes through the same region as a second wave of the same type, there will be
- interference – the two waves add (reinforce each other) at some points (the antinodes) and subtract (tend to cancel each other) at points half-way between those (called nodes).

Last, a **transverse wave** (a 'shake' wave like light, one whose vibrations are across the direction of energy flow) may show
- polarisation effects – its vibration tends to stay in one plane.

A **longitudinal wave** (a 'push' wave such as sound) involves a vibration which is in the same direction as the direction of energy transfer. Some waves include both longitudinal and transverse components; **gravity waves**, such as waves at the surface of a body of water, are like that. Earthquake waves are of all three types.

Typical waves are electromagnetic radiation, sound waves (more generally pressure waves), and liquid surface waves (gravity waves). In each case, a graph of some suitable measure with time or position is a sine wave, or consists of the sum of a number of sine waves. Indeed, when input to a **wave analyser**, even a very complex wave can be broken into its component sine waves; see also *Fourier analysis*.

There are also several types of wave equation, as follows.
a) the relation between the speed c of a wave, its frequency (number of waves per second, v), and its wavelength, λ:

$$c = v \, \lambda$$

b) the relation giving the instantaneous displacement y in terms of the amplitude y_0, and the time t and distance x from some (arbitrary) start – (1) below. That is for a progressive (moving) wave; the relation it leads to for a standing wave appears as (2):

(1) $$y = y_0 \sin \left(2 \, \pi \, (v \, t - x/\lambda) \right)$$

(2) $$y = 2 \, y_0 \cos \left(2 \, \pi \, x/\lambda \right)$$

c) the Schrödinger wave function (below) of wave mechanics (below).

A **wave filter** is an electronic circuit which passes with ease waves in some band of frequencies, but blocks others – it's just the same as a light filter in effect, therefore. A **waveform** is the simple or complex sine wave graph of some suitable measure against time or distance mentioned above; there must be one (or more) for any wave. The **wave front** of a wave is a surface at all points of which the wave in question has the same phase at the same moment: a 'plane' wave has a plane wave front, as it is neither converging nor diverging; a 'spherical' wave has a spherical wave front, as it is diverging from, or converging to, a point.

A particle's **wave function** describes its motion through space and time according to the Schrödinger equation. The function deals with probability, and includes contributions from all possible paths between two points. Chaos theory predicts the possibility of discontinuities in wave functions (called scarring); these follow interference between the most likely path and the possible different but close others. Such an effect in practice was first observed in 1989, in the motion of a high energy orbital electron in hydrogen.

A **wave guide** is an artificially constrained path for electromagnetic radiation. This may be
- a metal tube (in particular for waves in the microwave region);
- a wire (as in the case of a wire that conducts an ac wave round an electric circuit); or
- a rod of dielectric material (insulator).

In any given case, the design of the guide fits it for use with only a few wave modes. A horn aerial is a type of wave guide, while the speaking tube (once common in large houses and ships) is an acoustic (sound) wave guide. A **wave guide lens** is a reflecting device designed to focus electromagnetic radiation; see *x-ray lens* for an example.

The **wave intensity** of a wave is the power transfer in the direction of the wave per unit area of wave front; it is the product of the wave's speed and its energy density.

The **wavelength** λ of a wave varies inversely with frequency as noted above, under wave equation a). (Energy content varies with frequency, so it varies inversely with wavelength.) Wavelength is the distance in space between any two successive points on the wave that have the same phase and amplitude (for instance, between two successive peaks or two successive troughs). **Wavelength constant** is the value $2\pi/\lambda$; it therefore relates closely to wave number (below) and often takes the name circular wave number because of the 2π.

Wave mechanics is now a branch of quantum mechanics. Originally, it was a new and separate area, explored first by Schrödinger in the 1920s. Its basis involves viewing any sample of matter (in particular, a particle) as a set of waves rather than a set of smaller particles. Each particle in a system has its own wave function; the square of this gives the probability of finding the particle at any given point in space–time. Thus, in the case of atomic electrons, the wave functions superpose to give a picture of the orbitals in the electron cloud. The link with quantum mechanics follows from the fact that the solutions to the wave functions (the so-called eigenvalues) relate to only certain energies – in other words, they are quantised.

The function of a **wave meter** is to give the frequency of an input wave; it is, therefore, a simple form of wave analyser (above). The **wave number** k of a wave is the reciprocal of its wavelength, $1/\lambda$, the unit being m^{-1}. The measure is of most use in spectroscopy. **Circular wave number** (also k) is a name for *wavelength constant* (above). A **wave packet** is a short pulse of wave radiation; it results from the superposition of a number of infinite waves of the same type but which differ slightly in wavelength. One may view this as a quantum of the radiation concerned, and assign to it a particle nature as well as a wave nature; that is what duality is about. For **wave plate**, see *retardation plate*.

Wave power involves using waves at sea as a low cost, but low concentration, source of energy; see also *tidal power*. It is the use of systems like Salter's ducks – special pumps and turbines – to generate electricity on a large scale by reducing wave heights by perhaps 10%. Sea-floor 'hubs' link the output from – say – twenty actual generators and feed the current to shore.

Wave surface is an old name for wave front (above).

wavicle name, now not as much used as earlier, for a quantum of radiation or wave packet; in other words, it is an object whose wave and particle features are both significant.

weak charge Q_w name for the strength of the weak interaction (force) between leptons and quarks; this charge between two such particles is much the same as the charge effect of – strength of the force between – their electric charges. See also the *CKM matrix*.

weak interaction former names **weak force** and **weak nuclear force**, the interaction between elementary particles (specifically quarks and left-handed leptons) that covers beta decay and the decay of some mesons and hyperons. Only the gravitational interaction is weaker, while the strong nuclear interaction is some 10^{12} times stronger. The W and Z bosons are the exchange particles for the weak interaction.

There are four important ways in which weak interactions differ from the other three types:
- they do not follow the law of constant parity (P-symmetry), or, therefore, CP-symmetry
- flavour can change during them
- the exchange particles are highly massive bosons (which means their range is very small, as their lifetimes must be small – no more than 10^{-18} m)
- they affect neutrinos (so, agreed, does gravity, but very, very weakly).

Weakly interacting massive particles, *wimps*, have been proposed for a number of reasons in astronomy and cosmology; none have yet been observed, however. A major astronomical problem is that of the 'missing mass' of the Universe, the so-called dark matter that could fill the gap between the Universe's calculated and observed masses. By many accounts, as much as 90–99% of the Universe's mass is missing in this sense. Dark matter, which could well include wimps, may also solve such problems as mismatch between the calculated age of some stars and the age of the Universe. People have also proposed wimps to explain why the Sun's output of neutrinos is only about a third of the level that theory predicts. While interacting only very weakly with matter, wimps may have masses as high as ten times that of the proton. If such massive particles exist in abundance in the Universe, it seems likely they would accumulate only in the cores of stars. Finding them is not an easy task, therefore. On the other hand, light dark matter particles – we could call them weakly interacting less massive particles – arise in some very new theories; it should be much less hard to find these (though they're less interactive than neutrinos), but no one has yet (2007).

weber Wb **1)** the unit of magnetic field strength. When a 1 Wb field links a circuit of one turn and falls steadily to zero in a second, 1 V is induced.
2) late twentieth century unit of pole strength, the strength of a pole able to produce 1 oersted at a distance of 1 cm (0.01 m)
c) late nineteenth century unit of current which came to be called the ampere

weight a) symbol W, unit: newton, N – the force on an object as a result of its being in a gravitational field; if there are no other forces on it, it will move in free fall (see also *weightlessness*). In everyday language, weight is the sensation caused by such another force (for instance, the upward force – equal to weight – from the floor on which the object stands). In physics, weight is the force of gravity on an object of mass m – $W = m\,g$. Here g is the gravitational field strength at the point concerned: given in newtons per kilogram or, as an acceleration (the acceleration of free fall), in metres per second per second. For **weights and measures**, see *SI* in particular.
b) or **weighting**, symbol w, the value put on a member of a set of data in recognition of its importance in the set (such as its frequency in a frequency distribution). The **weighted mean** of a set of values takes account not only of the frequencies of the elements (members of the set) but such a factor as their standard deviation. The weighted mean $\bar{x}_w$ of a set of values x_i, each having a frequency n_i and a weight w_i, is

$$\bar{x}_w = \Sigma\ w_i\ n_i\ x_i / \Sigma\ w_i\ n_i$$

weightlessness the sensation of having no weight in the case of any object in free fall in a gravitational field (such as a spacecraft in orbit, people inside it, and the tools they're trying to work with). In fact, unless the field strength is zero, such an object *does* have a weight (as always, it is equal to the product of the object's mass and the field strength); in free fall, however, there is nothing in the environment (such as the ground or a fixed floor) for the weight to react against. See also *microgravity* and the *general theory* of relativity.

Weinberg, Steven (1933–) physicist and Nobel Prize winner, 1979. He gained this with Salam and Glashow for coming up with the electroweak theory.

Weiss constant rare name for the Curie temperature T_C, the constant temperature that appears in the Curie(–Weiss) law for susceptibility. The **Weiss magneton** is the quantum of the magnetic moment of certain molecules, around 1.85×10^{-24} joule per tesla – which is about a fifth the value of the Bohr magneton.

well the name given to a region of space in which the potential energy is significantly below that outside. It is also normally the case that, at the edge of the region, the energy changes quickly (so that the walls of the well are steep). See *potential well* and also *barrier*.

well-behaved of a characteristic function, physically possible as a solution to the Schrödinger equation concerned

Wertheim effect one of the Wiedemann effects

Westinghouse, George (1846–1914) engineer who founded, in 1886, the electrical engineering company that bears his name. Prior to that, he was involved in developments in railway systems and the use of natural gas. The work of his company (which employed Tesla) led to the general acceptance of ac for mains power, rather than dc (for which Edison fought).

Weston cell also called cadmium cell, one of the most widely used standard cells, invented in 1893 by Edward Weston (1850–1936, chemist and competitor of Edison in the field of electricity supply). The electrolyte is cadmium sulphate solution, while one of the electrodes is a cadmium amalgam (alloy with mercury). The cell's output is very close to 1.019 V over a very wide temperature range; used with the standard resistor, people found it easy to set the standard ampere (i.e. obtain its value).

wet and dry bulb common design of hygrometer (device able to measure relative humidity). It contains two similar thermometers, the bulb of one of which is inside a wet cloth. The relative humidity relates to the difference in the readings of the thermometers.

wet cell electric cell whose electrolyte is a solution, rather than being a paste as in a dry cell

wetting or spreading, the degree to which a liquid drop spreads over a surface on which it is placed. This depends on the surface tension. The measure is the **wetting angle** (or

angle of contact): the angle between the tangent to the surface of the liquid and the solid surface at the edge of the drop.

Weyl, Hermann (1885–1955) mathematical physicist who made important contributions to the development of the theories of relativity and quantum mechanics

Wheatstone, Charles (1802–1875) physicist and engineer who invented the concertina (1829), electric telegraph (with Joseph Henry, 1837), and stereoscope (1838). The **Wheatstone bridge** is a circuit layout designed to measure resistance. In the usual diamond-shaped layout, the unknown resistance R_x is $R_1 R_2/R_3$ when the meter reading is zero (in other words, when the bridge is in balance). The circuit has many forms, such as the metre bridge (in which R_2 and R_3 are the two parts of a metre-long uniform wire divided by a sliding contact). In recent years, however, the ohm-meter and then cheap compact electronic systems have come to replace the Wheatstone bridge for measuring resistance and other types of impedance.

wheel and axle simple machine that is the basis of many types of winding gear, winch and transmission gear systems. The input force goes by a rope or chain to the large diameter wheel; the (smaller) axle of the wheel applies the output force.

whistler audio signal of falling frequency detected by a radio receiver. It is the effect of the radio output of a lightning stroke transferred along a line in the Earth's magnetic field.

white the colour obtained when all the wavelengths of the visible spectrum are present to much the same extent. **White light** is the visible electromagnetic radiation produced from a **'white hot'** surface: one at around 6000 K; it consists of near uniform output at all visible wavelengths. By extension to acoustic waves, **white noise** is the hissing sound of noise present uniformly over a wide range of audible wavelengths. In general, **white radiation** is radiation that is fairly uniform over a wide band of wavelengths. An example is the 'Bremsstrahlung' background of an x-ray spectrum.

white dwarf very small, dense star in the final stage of life: the hydrogen ('nuclear fuel') has gone, so the energy for radiation comes from gravitational collapse – until it is around the same size as the Earth. Perhaps most stars finish this way (certainly white dwarfs are very common): for only the most massive dying stars may become supernovas. Until recently, it was thought that a white dwarf absorbed all its system of planets, comets and so on, but study of one such star in 2005 showed that some such material may remain in at least some cases. It is also not necessarily the case that such a star fades away – 'recurrent novas' (stellar explosions that occur again and again in the same star) may be the outcome of a white dwarf absorbing matter at a steady rate from the atmosphere of a companion red giant.

white hole the time reversal of a black hole – a point mass that repels matter (whereas a black hole attracts mass). White holes appear in some solutions to the field equations for a kind of wormhole; in practice, they may not exist even there.

Wiedemann effects a set of so-called circular magnetic effects, interactions between a ferromagnetic rod and one or more magfields:

- Twisting a longitudinally magnetised rod produces a circular field round it.
- Twisting a circularly magnetised rod produces an axial field.
- Circular magnetostriction – applying axial and circular outside fields causes the rod to twist.

The **Wiedemann–Franz law** is that, at a given temperature, the ratio of the electrical to the thermal conductivity of a metal sample does not depend on the metal. Further, Lorenz found that the ratio varies with temperature. However, neither statement holds at very low temperatures.

Wieman, Carl (1951–) one of the three joint winners of the 2001 Nobel Prize, for being able to obtain Bose–Einstein condensates in 'dilute gases of alkali atoms', and for research into the properties of these condensates

Wien, Wilhelm (1864–1928) physicist whose work on black body radiation prepared the way for Planck's quantum theory – and gained him the 1911 Nobel Prize. The **Wien bridge** is a version of the Wheatstone bridge able to measure capacitance. The actual 'bridge' is the central part of the circuit, the branch with the meter; when the meter reads zero ('at balance'):

$$C_x = C_2 \left(R_4/R_3 - R_2/R_1 \right)$$

Other versions of the circuit allow the user to measure inductance and frequency. The name Wien bridge also applies to a full-wave

rectifier formed from four diodes, as shown under *rectifier*.

The **Wien effect** is the increased conductivity of an ionic solution at very high field strengths. The high field causes the ions to escape from their ionic atmospheres. The **Wien displacement law** is that for black body radiation, the product of the temperature and the wavelength at the point of maximum output is constant; the constant is the **Wien constant** σ. There is also a **Wien radiation law** giving the distribution of energy in the black body spectrum; the Planck radiation law later replaced this. See also *ultraviolet catastrophe*.

wiggler the way to arrange the magnets through which the electron beam passes in a free electron laser – the magnets alternate in their poles, so the electrons wiggle.

Wigner, Eugene (1902–1995) nuclear physicist who shared the 1963 Nobel Prize for his work on nuclear structure through the quantum theory, including reaching the law of constant parity (P-symmetry). The **Wigner effect** (or discomposition effect) is the change in the physical or chemical nature (composition) of a sample with high levels of radiation. It follows interactions which move particles from their proper positions. A major case of this is in the graphite used in a nuclear reactor – the graphite degrades due to the build-up of defects in the lattice. **Wigner energy** is the energy associated with these defects; heating can sometimes release the energy to restore the crystal lattice. Some people use either or both terms for the corresponding radiation damage in other types of substance. The **Wigner force** is a short range attractive force between nucleons that does not involve the exchange of a meson.

Wilczek, Frank (1951–) one of three joint winners of the 2004 Nobel Prize, for work on the theory of the strong interaction, in particular for discovering a new aspect

Wilkins, Maurice (1916–) biophysicist whose use of x-ray diffraction to study the structure of DNA led to his sharing the 1962 Nobel Prize for physiology with Crick and Watson

Wilson, Charles (1869–1959) physicist whose cloud chamber led to his 1927 Nobel Prize (shared with Compton). The design of the **Wilson cloud chamber** has not changed much since 1911; it remains a very important system for showing the tracks of ionising radiation quanta – see *cloud chamber*. The **Wilson effect** is the electric polarisation of a sample of insulator moving through a magnetic field; it is an example of the motor effect and relates to the Hall effect.

Wilson, Kenneth (1936–) theoretical physicist, whose 1982 Nobel Prize followed his use of group theory to model phase transformations

Wilson, Robert (1936–) Nobel Prize winner, 1978. He shared this for work in radio astronomy and radio physics, in particular the discovery of the cosmic background radiation.

wimps weakly interacting massive particles, these being devised to meet certain needs: but not yet observed. People do not expect to observe them yet, but their decay products – **superwimps** – may become accessible to the new accelerators of the late 2000s.

Wimshurst machine electrostatic generator, now very rarely used, able to produce pds of a megavolt or more with high currents, and thus strong sparking

wind body of air moving relative to the Earth's surface, an important feature of weather that results from air pressure differences. A **wind mill** is a machine, a form of turbine, used to capture wind power and convert it to the rotary motion of mill wheels for grinding corn (in most cases). Being complex and costly, wind mills may not have improved the lot of common people very much; all the same, their spread with water mills through Europe over the thousand years or so up to the early nineteenth century was a crucial aspect of the industrial revolution. This is early **wind power**, but in many parts of the world now groups of modern wind mills – **wind farms** – produce tens or hundreds of mega-watts; research goes on to make the modern mills and the farms even more efficient.

A **wind instrument** is a sound source whose output follows the vibration of a column of air in some kind of tube. In practice, the tubes are often very complex in shape; this much affects the quality (timbre) of the sounds they produce. A **wind tunnel** is a wide tube through which a smooth air stream passes in a controlled fashion. It allows the user to study the flow of fluid past objects (which may include models of buildings and aircraft, for instance). This is a major tool in aerodynamic research, some designs being able to work at several times the speed of sound; however, computer simulation is starting to take its place.

windage loss the loss of useful energy input

to an electric machine due to friction between the moving parts and the air

window a) the very thin sheet of mica (in most cases) at the end of a Geiger tube that allows passage of the radiation being studied **b)** term used to describe the fact that a normally opaque medium can pass radiations within a certain range of energies: we say that the medium has a window for the range passed. The term is most common in the case of the various windows through the Earth's atmosphere for electromagnetic radiation, in particular those used in astronomy.

wing one of the main lifting surfaces (foils) of an aircraft

wire metal sample drawn into a long uniform thread. In most cases, the radius lies between about 0.02 mm and 15 mm. For electrical work, copper and aluminium wires are the most common: they are highly ductile (easily able to be drawn into wire form), yet still good conductors and also fairly cheap.

wobbulator a test instrument – a signal generator whose output sweeps repeatedly over a set frequency range with near constant amplitude

Wollaston, William (1766–1828) chemist who developed the techniques of powder metallurgy, discovered several elements, was the first person to observe what we now call Fraunhofer lines (in 1802), and invented the camera lucida. A **Wollaston prism** separates the ordinary and extraordinary polarised rays produced by double refraction. It deviates both rays, by roughly the same angle in opposite directions, so is a source of polarised light. The Nicol prism had much the same purpose and was more widely used; both are now rare, since the use of Polaroid has become so common.

woofer large speaker able to reproduce low frequency sounds with high fidelity; a tweeter does the same with high frequency sounds.

work W unit: joule, J old name for energy transfer, in particular in mechanical contexts. The **work function** of a surface is the energy difference between the Fermi level in the substance and the vacuum energy level outside. It is, in other words, the energy involved in the transfer of an electron from inside the sample to outside the surface. See, in particular, *photoelectricity*.

working substance a) the substance whose temperature and pressure (and maybe state) change during a cycle of action of a 'heat' engine or refrigerator

b) in general, any substance whose changes of properties are crucial to the action of a machine or system

wormhole a bridge proposed to link two different points in space–time, thus giving a short cut that could, in theory, allow faster than light travel and time travel. No current theory rules out the wormhole; indeed, it is a valid solution to the equations of general relativity. On the other hand, it seems that a wormhole can form only at the core of a supernova or collapsed galaxy; also, the singularity produced at the end of the hole lies within the Cauchy horizon, so (according to most theories) could not be approached by normal matter objects.

A **quantum wormhole**, on the other hand, takes account of quantum effects and is viewed as the result of a small fluctuation in space–time; it would be far smaller than a sub-atomic particle. Perhaps all space (the 'multiverse', the set of all Universes in the bubble Universe theory) has a foamlike structure on that scale, acting as a barrier to the observation of various predicted particles (such as those that may form dark matter). Perhaps too the big bang was an event that caused our Universe to branch from the multiverse foam, still linked to it by a wormhole.

wow regular low frequency variations in the output of a system that results from some mechanical problem

wrench the resultant of the most general set of forces acting on an object: a couple plus a force in the direction of the axis of the couple

x type of radiation at first unknown (hence the name) near the high energy end of the electromagnetic spectrum, i.e. x-radiation

X a) name for the band of (radio) radiation between 8 GHz and 12 GHz
b) symbol for any (generic) element or nucleus, sometimes more specifically for a halogen (Cl, Br, etc.)
c) symbol for x-ray photon
d) the X boson, exchange particle for some interaction not yet proved

x a) symbol for distance (s preferred, unless the distance is in the x-direction)
b) symbol for mole fraction
c) symbol for rectangular coordinate, distance in the x-direction
d) symbol for an unknown value

X a) symbol for molar concentration
b) symbol for reactance, unit: ohm, A

Ξ the xi-particle, a cascade hyperon

X boson in some grand unified theories, an exchange particle for some interaction not yet shown to exist. Just like the X boson itself, in fact.

Xena the third dwarf planet of the solar system, announced in mid-2005 – three times more distant than Pluto and only slightly larger (2400 km across), but very bright: with the greatest albedo in the system, like that of fresh snow. Its energy source was a mystery for at least the first year after its discovery.

xenon Xe inert gas (group 0) in period 5, proton number 54, fifty parts per billion in air, found by William Ramsay and Morris Travers in 1898. Density 5.9 kg m^{-3}, melting and boiling temperatures – 112 °C and –108 °C; most common isotope ^{132}Xe. Used in a variety of light sources, bubble chambers and protein crystallography; the preferred fuel for ion spacecraft drives.

xerography the electrostatic process involved in the printing stage of most modern copiers. The word means 'dry writing': earlier copying processes involved working with messy chemicals. The first makers of these machines changed their name to Xerox because of the technique's great success.
Xeroradiography is much the same process used for x- and gamma-radiography.

xi particle Ξ a doublet of hyperons (baryons more massive than the nucleon), Ξ^0 and Ξ$^-$ and their antis. The masses are around 40% greater than that of the nucleon at about 1320 MeV/c^2, while there is a pair of **xi resonances** (**charmed xi particles**) at around 2470 MeV/c^2.

XOR gate logic circuit or 'gate' which gives an output if just one of the inputs is 1, as in this truth table for a two-input XOR gate

I_1	I_2	O
0	0	0
0	1	1
1	0	1
1	1	0

x-pinch method used to produce high-energy x-rays at low cost for research. The system involves vaporising one or mores pair of metal wires in the shape of an **x** by a very short very large current. The high temperature plasma (as much as 10 MK) produced can yield up to 100 J of soft x-rays in a small volume. The volume is a few micrometres, giving, in effect, a point source of x-rays.

still sometimes called Röntgen rays after the person who first explored them, high energy electromagnetic radiation produced by transitions between the inner electron shells of atoms. In practice, most artificial sources of these x-rays involve the absorption of high energy electrons ('cathode rays') by matter. The wavelength range of x-rays is from around 10^{-9} m to 10^{-13} m; the penetrating power that makes the radiation of such value in industry and medicine follows from the high energy density associated with the high frequency.

What also makes x-rays of value in industry and medicine is their high penetration into even very dense matter. The field of **x-ray absorption** is relevant here, as different substances (e.g. tissues in the body) absorb x-rays from a beam differently. Thus the shadow picture on the final screen or photographic film can give a lot of detail about the structure inside the body or object under test. See *absorption* sense c) and also *absorption edge*.

X-ray analysis allows the study of crystal structures by way of the diffraction or fluorescence they cause. See also x-ray crystallography and x-ray fluorescence, below. **X-ray astronomy** is the study of extraterrestrial x-rays; it uses satellites in the main (as the atmosphere absorbs x-rays). People working in the field soon found that there is an **x-ray background**; they found this hard to explain until better *x-ray telescopes* came into use.

X-ray crystallography is the use of **x-ray diffraction** to study crystal nature and behaviour and for analysis. The degree of diffraction in a given case depends on the wavelength of the radiation compared to the spacings of the planes of the crystal. There are various techniques, but in all cases the crystal sample diffracts the x-ray beam as it passes or reflects it on to a photographic film. Electronic and other forms of detector of ionising radiation are of great importance too, though the phosphor screens originally used (as by Röntgen) are now rather rare. **X-ray fluorescence** is also of great value as an analytical tool. The input radiation is a beam of gamma- or high energy x-radiation and the output is fluorescence in the x-ray region; the x-ray fluorescence spectrum typifies the substance concerned.

X-ray 'laser' action, first shown in 1984, with pulsed outputs of extremely high power of value in plasma research, is hard to produce. One recent technique involves scattering very high energy electrons from a grooved conducting surface, with the electron beam incident at almost 90°. Near the surface, each electron produces a positive image in the conductor, thus making a dipole. The grooves cause the dipole to oscillate; this forms electromagnetic radiation which the emerging electron beam amplifies perhaps a hundred times by interference or diffraction. For the radiation to be in the x-region, the electrons need at least 5 MeV energy, and the grooves should be around 10 nm apart. There are more and more uses of such systems, such as making images of very small objects, like the smallest cells and viruses.

Some form of **x-ray lens** is the basis of **x-ray microscopy**; this process is not easy to achieve either, but as the resolution is very high, a number of systems have appeared. An x-ray lens is in fact a mirror, used at glancing angle (as no substance has an adequate refractivity to allow normal lens action). The lens is converging and very finely polished to maximise efficiency in use. **X-ray microtomography** involves using a computer-assisted tomography scanning approach (as in a cat scan) with time delay integration of each section image to provide much greater and more accurate internal detail of an object.

Apart from the small *x-ray tube* (below), the synchrotron radiation from electron synchrotrons are the main **x-ray sources**. These are large and costly, however: even the most compact is over a metre across and costs many million pounds; various research projects aim to produce other types of machine. One such approach which may lead to success involves the x-pinch effect.

The **x-ray spectrum** produced from any source has two components. First, there is a line spectrum which is entirely characteristic of the source, as is the case with all line spectra. As well as this, however, there is a continuous background. The minimum wavelength (maximum frequency) of this continuous background relates directly to the greatest energy of the electrons in the beam that produces the x-rays. The other wavelengths in this spectrum result from bremsstrahlung (braking radiation).

Until the mid-1990s, we thought that diffuse hot gases throughout the galaxy cause the x-ray background in the sky; then results from more powerful **x-ray telescopes** show that in fact there are many discrete sources – possibly white dwarfs and other types of energetic star, more likely black holes sitting at the centres of galaxies and taking in vast amounts of matter.

Some form of **x-ray tube** is the normal source of x-rays in practice. It is a variation of the discharge (cathode ray) tube designed to maximise the efficiency of conversion of the high energy of the cathode rays to x-radiation. The target for the high energy electron beam is made of metal (often tungsten or copper), and is massive and water cooled (as stopping the electrons can raise it to a very high temperature). The target is shaped to produce a parallel beam of x-radiation; this passes through a suitable window in the side of the tube. The most well known x-ray tube design is the Coolidge type; still in wide use with little change after ninety years.

y unit prefix symbol for yocto-, 10^{-24}

Y a) the Y boson that links quarks and leptons

b) symbol for the luminance component of a video signal

c) unit prefix symbol for yotta-, 10^{24}

d) chemical symbol for yttrium, element 39

y symbol for a rectangular coordinate, distance in the y-direction

Y a) symbol for admittance (inverse of impedance), unit: siemens, S

b) symbol for the Planck function

c) symbol for the Young modulus (E preferred)

yagi aerial a highly directional aerial used in radio and television transmitting and for radio astronomy

Yang Chen (1922–) physicist who shared the 1957 Nobel Prize for Mrs Wu's experiment on the conservation of parity

yaw of a ship or other craft, to move back and forth around the vertical axis, as opposed to pitch and roll, movements about horizontal axes

Y boson in some grand unified theories, an exchange particle for some interaction not yet shown to exist

Y connection or Y link name for the star connection between the leads of a three-phase supply, as opposed to the delta link

yellow spot common name for the fovea, the sensitive central region of the eye. Here the cones of the retina are most dense, so that vision is keenest.

yield the onset of plastic flow (creep) after a sample has passed its elastic limit The yield point is the point on the elasticity (strain/stress) curve where this happens for the substance concerned. The yield strain and yield stress of the substance are the lowest values of the strain and stress at which a sample will start to yield; in other words, they are the coordinates of the yield point.

ylem term once used by Gamow and colleagues for the state of the Universe at the time of the big bang. Gamow thought that this primordial matter – or singularity or God – could have been a sea of high energy neutrons; or maybe it was a Bose–Einstein condensate with the mass of the present Universe. Or maybe it doesn't matter – in any event, the term is not much used now. The word comes from the Latin of the Middle Ages through Middle English; it means matter, or universal matter.

yoke piece of ferromagnetic material used to complete the magnetic circuit of an electromagnet, voltage transformer, or electromagnetic machine

Young, Thomas (1773–1829) physician and physicist whose most well known work is in the interference of light waves, though he also first used 'energy' in the modern sense and explained the accommodation of the eye. His major work was in:

- colour vision: he devised the Young–Helmholtz trichromatic, or three colour, theory, a theory which still stands in

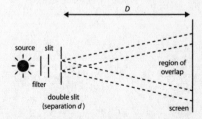

Young's experiment shows interference between the wavefronts produced at two close slits and provides crucial evidence of the wave nature of light.

essence (Helmholtz developed Young's ideas);

- elasticity: the **Young constant** or **Young modulus** E of a substance being the ratio of the tensile stress in an elastic sample to the tensile strain (deformation) produced; the unit is the newton per square metre;
- interference: the **Young fringes** observed with the **Young's double slit experiment** (1801) proving at last the wave nature of light; the sketch shows the layout of the experiment.

The wavelength of the light used, λ, is $\lambda = x\,d/D$. Here x is the fringe separation, the distance between one bright fringe and the next. Though the experiment is classic, it has not been much performed. It is not even clear when Young did it for the first time – the date of 1801 is common, but other dates up to about 1805 appear in the literature. In fact, it is widely seen to have been a thought experiment ('Gedankenexperiment') for 150 years. For instance, it does not seem to have been tried with radiations other than light until 1961.

ytterbium Yb transitional rare earth in the lanthanide series, period 6, with proton number 70, first claimed by Jean de Marignac in 1878 but properly found by Georges Urbain in 1907; the name comes from that of the village, Ytterby, where its ore was first found. Relative density 7.0, melting and boiling temperatures 820 °C and 1200 °C;

most common isotope ^{174}Yb. There are no major uses.

Ytterby small village near Stockholm, in one mineral (gadolinite) from a quarry in which seven new elements were found in the nineteenth century. Four are named after the village (erbium, terbium, ytterbium and yttrium) and the three others are gadolinium (from Johan Gadolin, the geologist who first studied the mineral concerned), holmium (for Stockholm) and thulium (for Thule, old name for Scandinavia).

yttrium Y transitional metal in period 5, proton number 39, identified by Johan Gadolin in 1794 and named for Ytterby, the village where its ore was found. Relative density 4.5, melting and boiling temperatures 1500 °C and 3300 °C; only natural isotope ^{89}Y. There are some uses for the metal in alloys, including a high temperature superconductor, while various garnets (complex silicate crystals) have value in the semiconductor, phosphor, and laser industries.

Yukawa, Hideki (1907–1981) physicist who proposed (in 1935) the exchange of a meson (the **Yukawa particle**) between a pair of nucleons as the cause of the strong nuclear force. The pion, discovered in 1947, fitted his predictions very closely, and Yukawa gained the 1949 Nobel Prize for the work. The **Yukawa potential** is a description of the variation of nuclear potential energy with distance between nucleons.

Zz

z unit prefix symbol for zepto-, 10^{-21}

Z a) in three-state (ternary digital) logic, the highest impedance level

b) the Z-boson, carrier of the weak interaction

c) unit prefix symbol for zetta-, 10^{21}

z a) symbol for angle of azimuth, the bearing on the Earth's surface in the direction of some object in the sky

b) symbol for height (h preferred)

c) symbol for ionic charge number

d) symbol for cylindrical coordinate

e) symbol for rectangular coordinate, distance in the z direction

f) symbol for the value of the red shift of an object in the sky

Z a) symbol for atomic number (now proton number)

b) symbol for impedance, unit: ohm, Ω

c) symbol for partition function (Q preferred)

d) symbol for proton number (was atomic number)

ζ symbol for the damping ratio of an oscillating system

Z boson with the pair of W bosons (also first found in 1983), the exchange particle involved with the weak interaction. The Z boson has no charge and is its own antiparticle; this means that all its quantum numbers are zero, so that in interactions it can exchange only momentum. All three bosons have very large masses (not far off 100 GeV/c^2); this is why the weak interaction has such a short range. The early electroweak theory was able to predict not just the W bosons (which have a crucial role in beta decay) but the Z (which has a crucial role in nothing); when they were found, it was time of triumph for that theory. However, it also predicts the Higgs boson – and no one has yet found that (mid-2007).

Zeeman, Pieter (1865–1943) physicist who shared the 1902 Nobel Prize with Lorentz for his work on the **Zeeman effect**. This is the splitting of spectral lines when the source is in a strong magfield. Lorentz predicted the effect on the basis of treating electrons as classical oscillators. The **anomalous Zeeman effect** is a similar splitting of spectral lines in a magnetic field, but not in the way described by the Lorentz theory. The cause is the effect of electron spin. In fact, this effect is the more common: the other is a special case.

Zeiss, Carl (1816–1888) industrialist who started an important optical workshop in Germany in 1846. This became very well known for its scientific research. See also *Abbe*, with whom Zeiss worked there in later years.

Zener breakdown the sudden increase in current passed by a reverse biased p-n junction when the reverse voltage goes higher than the threshold **Zener voltage**. The cause is an avalanche of current carriers (hence the common name of avalanche breakdown), as electrons jump from the valence band to the conduction band. The **Zener diode** uses the effect for regulating a system's supply voltage, in other words, for keeping it within a set range. While making such a diode, the level of doping controls the diode's Zener voltage; in use, the circuit design protects the device from damage. Clarence Zener (theoretical physicist, 1905–1993) was the first to describe and explore this process.

Zeno effect or **quantum Zeno effect** an almost paradoxical outcome of the behaviour of a particle's wave function: an unstable particle will not decay while observed. Each time the particle is observed (for instance, in a detector), the wave function collapses back into the form of the newly created particle. As a result, however short the particle's half-life period, continuous observation of it means it will never decay. Theory predicts this will be the case as long as, for a short period after creation (the **Zeno time**), the probability of decay increases with the square of the time. This is in fact the normal situation: and the effect was first observed experimentally in 1990. All the same, it is not easy to observe a given particle at the short intervals required to prevent decay. See also *quantum Zeno effect.*

The name comes from **Zeno of Elea** (*c.*490–*c.*430 bce), a philosopher some of whose paradoxes (which include those of the flying arrow and Achilles and the tortoise) had much influence, including on Aristotle's physics.

Zernicke, Frits (1888–1966) physicist whose 1953 Nobel Prize recognised his development of the technique of phase contrast microscopy. This is of great importance in biology.

zero point energy a) in quantum mechanics, the energy of the ground state of a system, that which remains even at a temperature of absolute zero (0 K). It is the lowest energy the system can have (so there can be no way to remove it). The value of the energy for each oscillator (e.g. vibrating atom) in the system is $\frac{1}{2} h \nu$. Here ν is the oscillation frequency. **b)** in quantum field theory, the (large) amount of energy even in a small volume of empty space even at absolute zero temperature (the '**zero point**', point being an old word for temperature). Theory and experiment (for instance, noise in microwave circuits, the Casimir effect and the Lamb shift) confirm its existence; its cause is less clear. It appears to be the result of the unpredictable but consistent random variations of the energy of the vacuum, as predicted by the uncertainty principle. The variations are enough to allow the creation of virtual particles (see further below).

Some people believe that the energy density of free space at 0 K is much the same as in an atomic nucleus, though it is convenient to take the value as zero (or, sometimes, infinite) in calculations. This very large value does not show up as the energy is highly uniform: we can observe only the effects of tiny variations; see also *Zitterbewegung*. It seems likely that the cause of the zero point energy is the radiation output as a result of random 'quantum fluctuations' throughout space:

virtual particles appear for short times (within that allowed by the Heisenberg uncertainty principle) and disappear again. Perhaps the **zero point fields** that arise from these virtual particles drive the motion of all matter in the Universe, and this motion in turn drives the zero point particles. Such a theory (as do others) readily explains Dirac's cosmological coincidence.

zero sound a high frequency pressure wave that can appear in liquid helium-3 (and in other quantum Fermi liquids, which exist only at temperatures close to absolute zero), when ordinary sound cannot transfer. The speed of these waves is somewhat greater than the speed of ordinary sound would be.

zero temperature a better name than zero point

ZETA one of the world's first fusion research reactors, working – though without success – in the 1950s. The **zeta pinch** (see *Z pinch*) was the method the staff tried to use to constrain the very hot plasma.

zeta potential the same as electrokinetic potential, the part of the potential barrier at the interface between a liquid and a solid that passes some distance into the liquid

zinc Zn transitional metal in period 4, with proton number of 30, used to make brass since the Bronze Age and identified in 1746 by Andreas Marggraf. Relative density 7.1, melting and boiling temperatures 420 °C and 910 °C; most common isotope ^{64}Zn. Main uses – in alloys, to reduce corrosion, in the larger organ pipes, and for engine parts.

zino the massive superpartner of the Z-boson – see *supersymmetry.*

zirconium Zr transitional metal in period 5 with proton number 40, found by Martin Klaproth in 1789 and named after zircon, one of the many semi-precious gemstones that contain it. Relative density 6.5, melting and boiling temperatures 1850 °C and 4400 °C; most common stable isotope ^{90}Zr. Main uses are in nuclear cladding, because of its low cross-section for thermal neutrons; also of value in alloys, to reduce corrosion, in artificial joints and limbs, and in incendiary devices.

Zitterbewegung The (German for) jiggling (jittery) motion of a particle in empty space (for instance, of an electron in an electron cloud); it is the result of tiny fluctuations in the zero point energy around it. It seems that the motion (first proposed by Schrödinger in

1930) causes particles' spin and magnetic moments. And it is possible that the coupling between the Zitterbewegung of close particles could explain gravity.

z-modulation semi-jargon name for the modulation of the brightness of the spot on the screen of a cathode ray tube. This compares with x and y for the position of the spot.

z-pinch or **zeta pinch** an approach to pinching the plasma in a fusion reactor: in other words, keeping it away from the walls of the toroidal (doughnut shaped) container. It involves passing a current along the axis of the plasma; the magnetic field that forms constrains the plasma. The names come from the ZETA machine, an early (1954–1958) fusion research reactor; this used the z-pinch enough to prove it could not work on such a large scale.

Physics is finished, young man. It's a dead-end street.
Max Planck's physics teacher (c. 1880)

Appendix 1: *Units of measurement*

The international system of units

The scientific community has agreed on SI for all units of measurement. There are other units and systems of units you may come across in your reading; for instance cgs and mks (which came before in the history of units), and fps (which came much before those but still hangs on in most parts of the world). There are also non-SI units that hang on in extreme cases (such as the electronvolt for particle energies and the light year and parsec for astronomical distances).

SI is *le système international d'unités*, the international system of units. This developed in the latter half of the twentieth century in an attempt to rationalise, metricate, and universalise units used in science and technology. The same applies to daily life worldwide – this system is the world's most widely used. There has been such progress that there is rarely any need to use the term 'SI units' now.

Note that, although the system should be universal – the same all round the world – there are some ways in which practice differs from country to country. Also, unit names have not been fixed for all time – for instance, the main dictionary entry defines weber as used now, but also notes two others (from the nineteenth and twentieth centuries) that you may come across in your reading.

Base units

A coherent system of units depends on a small number of well defined 'base' (fundamental) units; these must not relate to each other in any way. For instance, the metre (for length or distance) and the second (for time) are base units. On the other hand, the metre as unit for distance and the metre as unit for length could not both be base units as they relate to each other. The same applies to the second as unit for short times and the year as unit for long times – they relate to each other (there are 31 557 600 seconds in a year (well, roughly).

As the first table overleaf shows, SI has seven base units (the metre, kilogram, second, kelvin, ampere, mole and candela). There are also two 'supplementary' base units (the radian and steradian), which are not needed but are in the system for occasional convenience.

measure	unit	symbol	note
length/distance	metre	m	
mass	kilogram	kg	a problem here – see below[1]
time	second	s	
temperature (absolute)	kelvin	K	
electric current	ampere	A	
amount of substance	mole	mol	
luminous intensity	candela	cd	
angle	radian	rad	supplementary unit[2]
solid angle	steradian	sr	supplementary unit[2]

Notes

[1] The problem with the unit of mass? Its name is not that of a pure base unit as it has a prefix – this is the single main problem faced by the modern system. This is a historical accident (or a political one – it depends on your point of view): the name given the mass base unit kilogram when the French committee of weights and measures started in 1793 was grave, G. After the revolution, the new committee decided the base unit should be the gram g. In the 1950s and 1960s, when the mks and then SI unit systems started to take over from cgs, many people tried to have the kilogram called the giorgi, G.

[2] The rad and the sr are either supplementary units as they have no dimension and are not needed, as stated above the table; or they are derived units, being ratios of measures with the same units. It doesn't matter.

Multiples and submultiples

The 'kilo' of kilogram is a prefix (something that goes in front of some other words); kilo means 1000. The system uses such a prefix for each step up or down the scale from very large to very small. Unit systems should be 'rationalised'. That means you can logically form units larger or smaller than any of the above – by using a standard prefix to denote a multiplier which is a power of 10.

The next table sets out the unit prefixes we now use. It includes an attempt at giving the traditional names of the numbers in continental Europe (in brackets in the UK where they aren't the same), and in the US. However, it does seem that the US system is quite quickly taking over from the others.

value	prefix & symbol	Euro, etc. name (UK name)	US, etc. name	note
10^{24}	yotta Y	quadrillion	septillion	
10^{21}	zetta Z	trilliard (thousand trillion)	sextillion	
10^{18}	exa E	trillion	quintillion	
10^{15}	peta P	billiard (thousand billion)	quadrillion	
10^{12}	tera T	billion	trillion	
10^{9}	giga G	milliard (thousand million)	billion	
10^{6}	mega M	million	million	
10^{3}	kilo k	thousand	thousand	note[2]
10^{2}	hecto h	hundred	hundred	not strictly SI[1]
10^{1}	deca/deka da	ten	ten	not strictly SI[1]
10^{0}	*none*	*none*	*none*	
10^{-1}	deci d	tenth	tenth	not strictly SI[1]
10^{-2}	centi c	hundredth	hundredth	not strictly SI[1]
10^{-3}	milli m	thousandth	thousandth	
10^{-6}	micro μ (u)	millionth	millionth	
10^{-9}	nano n	milliardth (thousand millionth)	billionth	
10^{-12}	pico p	billionth	trillionth	
10^{-15}	femto f	billiardth (thousand billionth)	quadrillionth	
10^{-18}	atto a	trillionth	quintillionth	
10^{-21}	zepto z	trilliardth (thousand trillionth)	sextillionth	
10^{-24}	yocto y	quadrillionth	septillionth	

Notes

[1] The 'not strictly SI' is because, strictly, SI multiples and submultiples go up and down in steps of 10^{3}. This means that in science it is better to talk of 650 mm rather than 65 cm, to use kg m^{-3} rather than g cm^{-3}, and so on.

[2] One prefix per unit is enough. In particular, use megagram for 1000 kg, not kkg, and gram rather than mkg.

Derived unit

A unit that is neither a base unit nor a supplementary unit, but depends on some combination of those. For instance, the cubic metre, m^3, is a derived unit, as is the kilogram per cubic metre, $kg\ m^{-3}$. Unit systems should be coherent in that derived units follow simple multiplication, division and raising to the power of relevant base units; follow any defining formula to do that. For instance, the cubic metre is the unit of volume as $V = l^3$; density has the unit kilogram per cubic metre as $\varrho = m/V$. See also *dimensional analysis*.

In practice that can mean quite lengthy units sometimes. For instance, as $F = m\ a$, the unit of force must be the kilogram metre per second per second, $kg\ m\ s^{-2}$. Then we have pressure as force/area – unit: kilogram metre per second per second per square metre, which gives kilogram per metre per second per second. Such names are long – and they often have no clear meaning. Often, therefore, we give derived units a special name – the newton for force, and the pascal for pressure. (Indeed, we often have special names for base units – a litre for a thousandth of a cubic metre, a tonne for a thousand kilograms, etc.)

The next table lists those derived units with special names.

measure	expression	unit name	symbol	meaning in base units
absorbed dose	$= W/m$	gray	Gy	$m^2\ s^{-1}$
capacitance	$= Q/V$	farad	F	$m^{-2}\ kg^{-1}\ s^4\ A^2$
conductance	$= 1/R$	siemens	S	$m^{-2}\ kg^{-1}\ s^3\ A^2$
electric charge	$= I\ t$	coulomb	C	$A\ s$
energy	$= F\ s$	joule	J	$m^2\ kg\ s^{-2}$
equivalent dose	$= W/m$	sievert	Sv	$m^2\ s^{-1}$
force, weight, etc.	$= m\ a$	newton	N	$m\ kg\ s^{-2}$
frequency	$= n/t$	hertz	Hz	s^{-1}
illumination	$= I/A$	lux	lx	$cd\ m^{-2}$
inductance	$= V\ t/I$	henry	H	$m^2\ kg\ s^{-2}\ A^{-2}$
luminous intensity		lumen	lm	$cd\ sr$
magnetic field strength	$= W/I$	weber	Wb	$m^2\ kg\ s^{-2}\ A^{-1}$
magnetic induction	$= V\ t/A$	tesla	T	$kg\ s^{-2}\ A^{-2}$

power	= W/t	watt	W	m² kg s⁻³
pressure	= F/A	pascal	Pa	m⁻¹ kg s⁻²
resistance, reactance, etc.	= V/I	ohm	Ω	m² kg s⁻³ A⁻²
voltage	= W/Q	volt	V	m² kg s⁻³ A⁻¹

Here's the same table, but in order of unit name.

measure	expression	unit name	symbol	meaning in base units
electric charge	= I t	coulomb	C	A s
capacitance	= Q/V	farad	F	m⁻² kg⁻¹ s⁴ A²
absorbed dose	= W/m	gray	Gy	m² s⁻¹
inductance	= V t/I	henry	H	m² kg s⁻² A⁻²
frequency	= n/t	hertz	Hz	s⁻¹
energy	= F s	joule	J	m² kg s⁻²
luminous intensity		lumen	lm	cd sr
illumination	= I/A	lux	lx	cd m⁻²
force, weight, etc.	= m a	newton	N	m kg s⁻²
resistance, reactance, etc.	= V/I	ohm	Ω	m² kg s⁻³ A⁻²
pressure	= F/A	pascal	Pa	m⁻¹ kg s⁻²
conductance	= 1/R	siemens	S	m⁻² kg⁻¹ s³ A²
equivalent dose	= W/m	sievert	Sv	m² s⁻¹
magnetic induction	= V t/A	tesla	T	kg s⁻² A⁻²
voltage	= W/Q	volt	V	m² kg s⁻³ A⁻¹
power	= W/t	watt	W	m² kg s⁻³
magnetic field strength	= W/I	weber	Wb	m² kg s⁻² A⁻¹

Of course, many other derived units depend on the above derived units with special names as well as the base units. The unit of surface tension is the newton per metre, for instance, while that of radiance is the watt per square metre per steradian.

Relevant websites

International Bureau of Weights and Measures – the body looking after this area: http://www1.bipm.org

International Temperature Scale of 1990 – official reference documents: http://www.bipm.fr/en/publications/its-90.html

Strange units of measurement? – perhaps not entirely serious: http://en.wikipedia.org/wiki/listofstrangeunitsofmeasurement

Appendix 2: *The Greek alphabet*

lower case	upper case	name
α	A	alpha
β	B	beta
γ	Γ	gamma
δ	Δ	delta
ε	E	epsilon
ζ	Z	zeta
η	H	eta
θ	Θ	theta
ι	I	iota
κ	K	kappa
λ	Λ	lambda
μ	M	mu
ν	N	nu
ξ	Ξ	xi
ο	O	omicron
π	Π	pi
ϱ	P	rho
σ	Σ	sigma
τ	T	tau
υ	Y	upsilon
φ	Φ	phi
χ	X	chi
ψ	Ψ	psi
ω	Ω	omega

Appendix 3: *Circuit symbols*

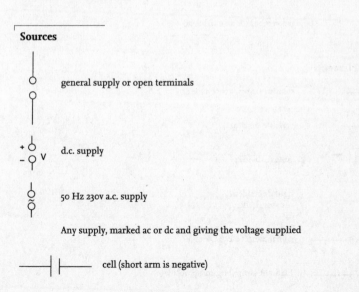

Connectors

—————— conductor with
current direction

——+—— crossing conductors,
no connection

——+—— conductor
junction

⌐ sliding conduct

⟋ —— simple switch*

•—— two-way switch*

* Note that there are many kinds of switch and therefore of switch symbol;
these two are the main ones

Sources

general supply or open terminals

+ ○
– ○ V   d.c. supply

50 Hz 230v a.c. supply

Any supply, marked ac or dc and giving the voltage supplied

——| |—— cell (short arm is negative)

Sources – *cont.*

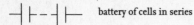 battery of cells in series

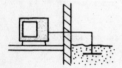 battery of cells in parallel

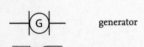 fuse

generator

transformer

object is earthed by connecting it to
a metal plate in the ground

standard symbol for an earth lead

Passive Elements

resistor or resistive element

variable resistor

voltage divider

light dependent
resistor (LDR)

signal lamp or light source

filament lamp (becoming obsolete)

Passive Elements – *cont.*

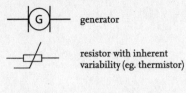 generator

resistor with inherent
variability (eg. thermistor)

inductor with magnetic core

inductor

diode (allows current to the right)

Zener diode

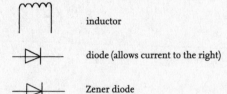

Meters

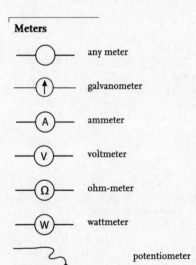

any meter

galvanometer

ammeter

voltmeter

ohm-meter

wattmeter

potentiometer

Transducers

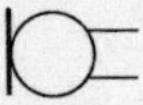

microphone

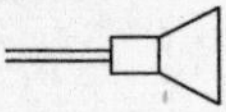

speaker

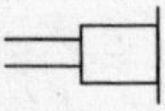

earphone

electric bell

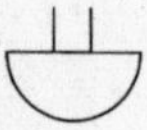

buzzer

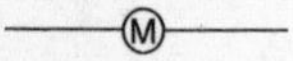

motor

alternative meter

thermocouple

light dependent
resistor (LDR)

photodiode

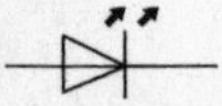

light-emitting diode

I know no general symbol for sensor or transducer. Sensors are, in essence,
variable resistors, however – if what you want isn't in this list, show it
as such.

Electron (Vacuum) Tubes

   triode

   oscilloscope

   Cathode ray tube

Gates

   gate (British standard but rarely used)

   AND gate

   NAND gate

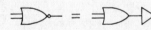   NOR gate

   NOT gate

   OR gate

   XOR gate

Semiconductor Elements and Networks

   light dependent resistor (LDR)

   resistor with inherent variability (eg. thermistor)

   Zener diode

   photodiode

Semiconductor Elements and Networks – *cont.*

 light-emitting diode

 pn junction (diode)

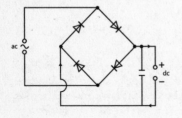 bridge rectifier

 amplifier

 pnp transistor

 npn transistor

 fet (n-channel)

 fet (p-channel)

Appendix 4: *Nobel Laureates in Physics*

The data here, and many of those used in the laureate entries in the body (e.g. citations and birth/death dates), are from http://nobelprize.org/physics/laureates/index.html. (Note that some mentions in the body of the book, however, are for prize winners in other areas than physics whose work relates to our subject.)

1901	x-rays	Wilhelm Conrad	Röntgen	1845–1923
1902	magnetism and radiation	Hendrik A	Lorentz	1853–1928
		Pieter	Zeeman	1865–1943
1903	natural radioactivity	Henri	Becquerel	1852–1908
		Pierre	Curie	1859–1906
		Marie	Curie	1867–1934
1904	gas physics/chemistry	John William Strutt	Lord Rayleigh	1842–1919
1905	cathode rays	Philipp	Lenard	1862–1947
1906	electricity in gases	J J (Joseph John)	Thomson	1856–1940
1907	optics	Albert A	Michelson	1852–1931
1908	colour photography	Gabriel	Lippmann	1845–1921
1909	radio communication	Karl Ferdinand	Braun	1850–1918
		Guglielmo	Marconi	1874–1937
1910	real gases	Johannes Diderik	van der Waals	1837–1923
1911	thermodynamics	Wilhelm	Wien	1864–1928
1912	automation of remote gas lighting	Gustaf	Dalén	1869–1937
1913	cryophysics	Heike Kamerlingh	Onnes	1853–1926
1914	x-ray diffraction	Max	von Laue	1879–1960
1915	x-ray crystallography	William	Bragg	1862–1942
		Lawrence	Bragg	1890–1971
1916	no award in physics this year			
1917	x-ray spectroscopy	Charles Glover	Barkla	1877–1944
1918	quantisation of energy	Max	Planck	1858–1947
1919	spectral line splitting	Johannes	Stark	1874–1957
1920	low expansivity alloys	Charles Edouard	Guillaume	1861–1938
1921	"services to theoretical physics"	Albert	Einstein	1879–1955
1922	atomic structure	Niels	Bohr	1885–1962
1923	the elementary charge of electricity	Robert A	Millikan	1868–1953

1924	x-ray spectroscopy	Manne	Siegbahn	1886–1978
1925	energy level quantisation	James	Franck	1882–1964
		Gustav	Hertz	1887–1975
1926	the quantisation of matter	Jean Baptiste	Perrin	1870–1942
1927	the quantisation of e-m waves	Arthur H	Compton	1892–1962
1927	invention of the cloud chamber	C T R (Charles)	Wilson	1869–1959
1928	thermionic emission	Owen Willans	Richardson	1879–1959
1929	wave nature of electrons	Louis	de Broglie	1892–1987
1930	scattering of light	Chandra	Raman	1888–1970
1931	no award in physics this year			
1932	quantum mechanics	Werner	Heisenberg	1901–1976
1933	atomic theory	Paul A M	Dirac	1902–1984
		Erwin	Schrödinger	1887–1961
1934	no award in physics this year			
1935	discovery of the neutron	James	Chadwick	1891–1974
1936	discovery of the positron	Carl D	Anderson	1905–1991
1936	discovery of cosmic radiation	Victor F	Hess	1883–1964
1937	electron diffraction	Clinton	Davisson	1881–1958
		George Paget	Thomson	1892–1975
1938	discovery of new radioactive elements	Enrico	Fermi	1901–1954
1939	invention of the cyclotron	Ernest	Lawrence	1901–1958
1940–2	no Nobel awards these years			
1943	magnetic moments	Otto	Stern	1888–1969
1944	atomic magnetic resonance	Isidor Isaac	Rabi	1898–1988
1945	the exclusion principle	Wolfgang	Pauli	1900–1958
1946	high pressure physics	Percy W	Bridgman	1882–1961
1947	physics of the upper atmosphere	Edward V	Appleton	1892–1965
1948	nuclear physics	Patrick M S	Blackett	1897–1974
1949	prediction of mesons	Hideki	Yukawa	1907–1981
1950	photographic emulsion as detector	Cecil	Powell	1903–1969
1951	the first accelerator	John	Cockcroft	1897–1967
		Ernest T S	Walton	1903–1995
1952	nuclear magnetic resonance	Felix	Bloch	1905–1983
		E M	Purcell	1912–1997
1953	phase contrast microscopy	Frits	Zernike	1888–1966
1954	quantum and wave mechanics	Max	Born	1882–1970
1954	particle detection	Walther	Bothe	1891–1957
1955	electron magnetic moment	Polykarp	Kusch	1911–1993
1955	hydrogen spectrum	Willis E	Lamb	1913–
1956	discovery of transistor action	John	Bardeen	1908–1991
		Walter H	Brattain	1902–1987
		William B	Shockley	1910–1989

1957	elementary particle parity	Tsung-Dao	Lee	1926–
		Chen Ning	Yang	1922–
1958	Cherenkov radiation	Pavel A	Cherenkov	1904–1990
		Ilya M	Frank	1908–1990
		Igor Y	Tamm	1895–1971
1959	discovery of the anti-proton	Owen	Chamberlain	1920–2006
		Emilio	Segrè	1905–1989
1960	invention of the bubble chamber	Donald A	Glaser	1926–
1961	nucleon structure	Robert	Hofstadter	1915–1990
1961	Mössbauer effect	Rudolf	Mössbauer	1929–
1962	liquid helium	Lev	Landau	1908–1968
1963	nuclear shell structure	Maria	Goeppert-Mayer	1906–1972
		J Hans D	Jensen	1907–1973
1963	particle symmetry	Eugene	Wigner	1902–1995
1964	quantum electronics	Nicolai G	Basov	1922–2001
		Aleksandr M	Prokhorov	1916–2002
		Charles H	Townes	1915–
1965	quantum electrodynamics	Richard P	Feynman	1918–1988
		Julian	Schwinger	1918–1994
		Shin-Itiro	Tomonaga	1906–1979
1966	laser action	Alfred	Kastler	1902–1984
1967	astro-physics	Hans	Bethe	1906–2005
1968	elementary particles	Luis	Alvarez	1911–1988
1969	elementary particles	Murray	Gell-Mann	1929–
1970	plasma physics	Hannes	Alfvén	1908–1995
1970	types of magnetic behaviour	Louis	Néel	1904–2000
1971	holography	Dennis	Gabor	1900–1979
1972	superconductivity	John	Bardeen	1908–1991
		Leon N	Cooper	1930–
		Robert	Schrieffer	1931–
1973	tunnelling in semiconductors	Leo	Esaki	1925–
	tunnelling in superconductors	Ivar	Giaever	1929–
1973	supercurrent tunnelling	Brian D	Josephson	1940–
1974	discovery of pulsars	Antony	Hewish	1924–
1974	radioastronomy	Martin	Ryle	1918–1984
1975	liquid drop model of nucleus	Aage N	Bohr	1922–
		Ben R	Mottelson	1926–
		James	Rainwater	1917–1986
1976	discovery of the J/psi particle	Burton	Richter	1931–
		Samuel C C	Ting	1936–
1977	structure of magnetic systems	Philip W	Anderson	1923–
		Nevill	Mott	1905–1996
		John H	van Vleck	1899–1980

1978	low-temperature physics	Pyotr	Kapitsa	1894–1984
1978	cosmic background radiation	Arno	Penzias	1933–
		Robert Woodrow	Wilson	1936–
1979	weak and electromagnetic interactions	Sheldon	Glashow	1932–
		Abdus	Salam	1926–1996
		Steven	Weinberg	1933–
1980	symmetry violations	James	Cronin	1931–
		Val	Fitch	1923–
1981	laser spectroscopy	Nicolaas	Bloembergen	1920–
		Arthur L	Schawlow	1921–1999
1981	electron spectroscopy	Kai M	Siegbahn	1918–
1982	phase transitions	Kenneth G	Wilson	1936–
1983	astro-physics	Subramanyan	Chandrasekhar	1910–1995
1983	astrophysical nuclear reactions	William A	Fowler	1911–1995
1984	discovery of the W and Z particles	Carlo	Rubbia	1934–
		Simon	van der Meer	1925–
1985	quantum Hall effect	Klaus von	Klitzing	1943–
1986	the scanning tunnelling microscope	Gerd	Binnig	1947–
		Heinrich	Rohrer	1933–
1986	the first electron microscope	Ernst	Ruska	1906–1988
1987	high temperature superconductivity	J Georg	Bednorz	1950–
		K Alex	Müller	1927–
1988	discovery of muon neutrino	Leon M	Lederman	1922–
		Melvin	Schwartz	1932–2006
		Jack	Steinberger	1921–
1989	ion traps in mass spectroscopy	Hans G	Dehmelt	1922–
		Wolfgang	Paul	1913–1993
1989	atomic clocks	Norman F	Ramsey	1915–
1990	inelastic scattering of electrons by nucleons	Jerome I	Friedman	1930–
		Henry W	Kendall	1926–1999
		Richard E	Taylor	1929–
1991	theory of liquid crystals	Pierre-Gilles	de Gennes	1932–
1992	multi-wire proportional detector	Georges	Charpak	1924–
1993	discovery of the first binary pulsar	Russell A	Hulse	1950–
		Joseph H Jr	Taylor	1941–
1994	development of neutron spectroscopy	Bertram N	Brockhouse	1918–2003
1994	development of neutron diffraction	Clifford G	Shull	1915–2001
1995	discovery of the tau lepton	Martin L	Perl	1927–
1995	detection of the neutrino	Frederick	Reines	1918–1998

1996	superfluidity in ^{3}He	David M	Lee	1931–
		Douglas D	Osheroff	1945–
		Robert C	Richardson	1937–
1997	using lasers to trap and cool atoms	Steven	Chu	1948–
		Claude	Cohen-Tannoudji	1933–
		William D	Phillips	1948–
1998	quantum fluid physics	Robert B	Laughlin	1950–
		Horst L	Störmer	1949–
		Daniel C	Tsui	1939–
1999	quantum electroweak interactions	Gerardus	't Hooft	1946–
		Martinus J G	Veltman	1931–
2000	semiconductor technology	Zhores I	Alferov	1930–
		Jack S	Kilby	1923–2005
		Herbert	Kroemer	1928–
2001	Bose–Einstein condensates	Eric A	Cornell	1961–
		Wolfgang	Ketterle	1957–
		Carl E	Wieman	1951–
2002	astrophysics	Raymond	Davis	1914–2006
		Riccardo	Giacconi	1931–
		Masatoshi	Koshiba	1926–
2003	superconductors and super-fluids	Alexei A	Abrikosov	1928–
		Vitaly L	Ginzburg	1916–
		Anthony J	Leggett	1938–
2004	strong interaction theory	David J	Gross	1941–
		H David	Politzer	1949–
		Frank	Wilczek	1951–
2005	quantum theory of optical coherence	Roy J	Glauber	1925–
2005	laser-based precision spectroscopy	John L	Hall	1934–
		Theodor W	Hänsch	1941–
2006	Cosmic background radiation	John C	Mather	1946–
		George F	Smoot	1945–

Appendix 5: *Some physics websites*

First, e-acknowledgements – very many thanks indeed to

- http://www.slcc.edu/schools/humsci/physics/whatis/quotations.html – where I found most of the quotations I thought of using in boxes in these pages, and http://www.gdargaud.net/Humor/QuotesScience.html#Physics where I found a few others and also had lots of laughs

- the amazing Wikipedia, the free on-line open-access encyclopedia with over a million entries, started in 2001 and tested by *Nature* in early 2006. That test showed its levels of accuracy not to differ greatly from those of the *Encyclopedia Britannica*: http://en.wikipedia.org/wiki/MainPage. (I used *Britannica* a lot too – have been a huge fan since school days – and agree.) In this wiki's address 'en' is English, but there are major versions for nine other languages as well as excellent and very fast translation facilities – and smaller versions for two hundred more languages (try surfing the old English wiki, for instance – wonderful!)

- for up-to-date – though not fully consistent – information on the chemical elements, to support Gerard Cheshire's Collins Gem *Chemical Elements* (2001), that wikipedia (in particular, http://en.wikipedia.org/wiki/Discoveryofthechemicalelements), and also http://www.vanderkrogt.net/elements/ and http://webelements.com

- Firefox – a great (and free of charge) browser that makes focused internet searches a joy . . . oh, maybe not a joy, as far too often I went down wonderful but very time consuming byways. No, I was wrong there – surfing in physics *is* a joy! http//:www.mozilla.com/firefox/

Second, my list of what I've found over the years to be the most entrancing physics websites. Physics *is* an entrancing subject, despite what some websites imply; *these* websites are all very much pro-physics, though.

American Institute of Physics – About physics

The area for school and post-school students physics.about.com/

Association for Science Education

UK's professional association for teachers of science www.ase.org.uk/

Astronomy and astronautics

Specialist popular science site, aimed at teenagers, but with some good hard physics too! http://www.space.com/

Astronomy etc. encyclopedia

One of several wiki-type encyclopedias, this one dealing (to quote in full) with astrobiology, astronomy and spaceflight http://daviddarling.info/encyclopedia

BBC science and technology news

(There's also a science/technology section on the BBC Ticker.) news.bbc.co.uk/1/hi/sci/tech/default.stm

Cool physics links

great range of sites a US college teacher recommends that his students browse http://www.cabrillo.edu/~jmccullough/links.html

Data sets

The standard British collection has long been 'Kaye and Laby'; free access to the on-line 1995 edition (which is unedited but kept up-to-date by the National Physical Laboratory): http://www.kayelaby.npl.co.uk/

The (US) Handbook of chemistry and physics, now in its 86th edition (2005), expects you to subscribe to the web version (though there is 30-day free trial): http://hbcpnetbase.com

Discovery

Here's the w-address of the family of digital tv channels:
http://www.discovery.com/

Exploratoriums

Sites with science activities and puzzles for younger people – why should us oldies
get all the fun? The first is US-based, the first on-line Exploratorium, in San
Francisco; the second is the on-line presence of what was Britain's first
Exploratorium, in Bristol: http://www.exploratorium.edu/
http://www.at-bristol.org.uk/

Institute of Physics (UK-based)

Main website:
www.iop.org
Area for school students:
learningphysics.iop.org
IOP's email lists archives (there are 26 such lists which have archives at the time
of writing):
http://networks.iop.org/archives/
'physics.org', a guide to physics on the internet, i.e. a kind of browser, but not very
precise:
www.physics.org
Primary school physics – There are very few physics specialists teaching in
primary schools (probably in any country), but children up to Britain's Year 6
already need some grasp of the subject. This site (new in 2006) aims to help.
teachingphysics.iop.org/primaryoutreach
physicsweb – the main Institute of Physics website – 'Physics news, jobs and
resources': free access and a free alerting service http://physicsweb.org
teaching advanced physics – for teachers of post-16 physics: www.tap.iop.org

The National Physical Laboratory (NPL)

A major role is metrology – measurement, units, and such like – but the site
ranges far www.npl.co.uk

Nature

Science and technology website – with news, hot news, e-alerts, podcasts, access to the journal, discussions, and an archive, hosted by UK's major general science journal; free access unless you want the full text of papers. etc. www.nature.com

New Scientist

Another science and technology website – with daily news, hot topics, feedback, letters, books, interviews, weblinks and an archive, hosted by UK's major science news weekly www.newscientist.com/

Nobel Prize winners in physics

A complete and fairly up-to-date list, with plenty of links (to much more than just these great physicists), 1901 (Röntgen) to date: www.slac.stanford.edu/library/nobel/

The 'official' website (mainly used for this book), which includes speeches and Nobel lectures as well as biographies: nobelprize.org/nobelprizes/physics/laureates

On-line teacher resource: physics

US-based teaching with technology site for years K-12 (which means school level) – no clear aims and style, but lots of useful links and ideas http://www.teach-nology.com/teachers/subjectmatter/science/physics/

physlink reference

many useful reference links, from symbols to Nobel prize winners, and even a glossary, within a generally interesting physics site you might also like to bookmark http://www.physlink.com/Reference/Index.cfm

phySpy

'Wiley-VCH's new international physics portal. Featuring THE search engine for physicists.' May be worth adding to your menu of browsers (though Google and Yahoo are far better and the IoP version is a lot better). www.physpy.com

Popular Science

Website of a US magazine, worth an occasional look
http://www.popsci.com/popsci/

Practical physics

the Nuffield Foundation's website of school experiments, mainly in worksheet
form www.practicalphysics.org

Royal Society

the independent scientific academy of the UK www.royalsoc.ac.uk/

S-cool

Revision for pre-university physics and equivalent in UK www.s-cool.co.uk

Science in School

New (2006) European magazine for science teachers, published mainly on the web
www.scienceinschool.org

Science World

Useful on-line encyclopedia (wiki, but not THE Wikipedia, nor – yet? – nearly as
comprehensive for physics) that specialises in astronomy, chemistry, physics and
biographies of scientists. It is supported by Wolfram Research, a software house,
as are many other sites of interest to people like you and me.
http://scienceworld.wolfram.com

Scientific American

US's main science magazine, once a gorgeous fat monthly, but now leaner and (no
doubt) fitter: http://www.sciam.com/

SETI – Search for Extra-Terrestrial Intelligence

Sharing the computer power of hundreds of thousands of idle computers to analyse radio signals from outer space setiathome.ssl.berkeley.edu

The SDV nuclear glossary

'The language of the nucleus', 'the world's largest nuclear glossary', though with generally very short definitions http://glossary.dataenabled.com/sdvglossaryP.html

Sky and telescope

Perhaps the best monthly for amateur astronomers http://skyandtelescope.com/

Stars and spacecraft

Watching artificial Earth satellites, including the International Space Station, from wherever you are www.heavens-above.com

Third, what about blogs?! The essence of these is in part their ephemeral nature – but people can make important, even authoritative, points in on-line discussions. Maybe you'll tell us a few words about blogs you come across that you feel other readers would find of particular value. Or is this an impossible task? Google came up with 18 000 000 hits when asked about 'physics blogs' in the spring of 2006, and nearly ninety million a few months later. Compare that with 'today there are still only a few dozen scientific bloggers', according to a very useful article in *Nature*, December 2005 – Declan Butler's 'Science in the web age: Joint efforts', http://www.nature.com/nature/journal/v438/n7068/full/438548a.html.